A SOURCE-BOOK of
# BIOLOGICAL NAMES and TERMS

# A SOURCE-BOOK of
# BIOLOGICAL NAMES and TERMS

### Third Edition
### Fifth Printing

## EDMUND C. JAEGER, D.Sc.

*Riverside Municipal Museum, Riverside, California*

*Illustrations by Merle Gish and the Author*

## CHARLES C THOMAS · PUBLISHER

SPRINGFIELD · ILLINOIS    U. S. A.

*Published and Distributed Throughout the World by*
CHARLES C THOMAS • PUBLISHER
Bannerstone House
301-327 East Lawrence Avenue, Springfield, Illinois, U.S.A.
Natchez Plantation House
735 North Atlantic Boulevard, Fort Lauderdale, Florida, U.S.A.

© 1944, 1950, 1955 *by* CHARLES C THOMAS • PUBLISHER

ISBN 0-398-00916-3
Library of Congress Catalog Card Number: 55-8867

First Edition, First Printing, July, 1944
First Edition, Second Printing, February, 1947
Second Edition, First Printing, January, 1950
Third Edition, First Printing, September, 1955
Third Edition, Revised Second Printing, June, 1959
Third Edition, Third Printing, September, 1962
Third Edition, Fourth Printing, August, 1966
Third Edition, Fifth Printing, February, 1972

*With* THOMAS BOOKS *careful attention is given to all details of
manufacturing and design. It is the Publisher's desire to present books
that are satisfactory as to their physical qualities and artistic possibilities
and appropriate for their particular use.* THOMAS BOOKS *will be true
to those laws of quality that assure a good name and good will.*

*Printed in the United States of America*
B-7

# TO DAVID STARR JORDAN

who had the good sense, when coining generic names, to explain their origin so that those who followed him could have no doubt concerning their exact connotation. He was seldom, if ever, given to the making of so-called nonsense names for he saw in every well-made scientific name a treasure house of meaning carrying valuable clues to identification, rich allusions to scientific history and discovery.

In many common and technical designations there is hidden a veritable romance of linguistic adventure where research leads across seas and sands to natural habitats and original appellations.

CARL SUMNER KNOPF
*Madroño*, Vol. VI, No. 7.

# Preface to the Third Edition

THE PRESENT BOOK is a careful revision of the Second Edition of 1950. It includes not only a number of corrections and emendations in the body of the text but also a supplement of more than a thousand entries which should increase very much its usefulness to those who wish to know the origin and significance of biological names and terms. In this supplement are included: (1) many new adjective and noun stems used in making trivial (specific) names of animals and plants; (2) more than one hundred additional suffixes and prefixes; (3) a limited number of much used geographical place-name stems employed in forming adjectival trivial and varietal names of plants and animals; and (4) a considerable number (more than 280) of short biographies of persons commemorated in botanical and zoological generic names. In selecting these commemorative names I have included only such as are of greatest importance and most often encountered by beginning students in the biological sciences. A heavy-type **S** placed after main entries in the front or original part of the Source-Book refers the student to additional helpful materials in the Supplement.

E. C. J.

Riverside, California

# Preface to the Second Edition

THE PRESENT work is a revised and enlarged edition of my *Source-Book of Biological Names and Terms*, first issued in 1944. To the Introduction have been added a number of explanatory paragraphs. A few necessary corrections and minor revisions of definitions of classical words also have been made. Of most importance is the embodiment of an addition of more than 1800 new word elements, increasing the total number of pages from 256 to 321.

In spite of the fact that I have taken much time in the revision, I cannot venture to hope that even now all imperfections have been removed. I will therefore deem it a favor if those who have found its pages useful will continue to make suggestions for further revision and send lists of terms, generic and specific names needing inclusion.

I am indeed grateful for the reception which has been accorded the first edition. I hope that the changes and additions to which I have alluded may increase many fold the usefulness of the book.

E. C. J.

Riverside College

# Preface to the First Edition

HERE ARE alphabetically listed fully 12,000 elements from which scientific biological names and terms are made. With them are given their Greek, Latin, or other origins and their concise meanings, together with numerous examples of their use in scientific nomenclature. The examples were not chosen at random but were selected to show as wide a variety of forms as possible and thus exhibit the many different "turns" a combining form might have. All the more commonly known genera and technical terms are included besides many that are only used by workers in special fields. There is also introduced a wide assemblage of prefixes and suffixes and this will be much appreciated by the young student who is constantly puzzled by the queer beginnings and endings of the words still so new to him.

This gives the student, who wishes to know the literal meaning of the words he uses, the largest assemblage of such formative elements, or combining forms ever brought together in a single volume. In scope it surpasses many times the most complete collections in unabridged dictionaries and scientific glossaries and gives a key which unlocks the treasury of meaning of more than a million technical names and terms. For the benefit of those unacquainted with the Greek characters, the Greek words which serve as bases for so many combining forms have been transliterated into English. That the student may understand and appreciate the methods governing the use of the classical stems and roots in the construction of scientific names, a comprehensive section on the philosophy of word building has been placed at the beginning of the volume.

All compound words serving as examples have been broken down into their simplest elements so as to make them easy to analyze. Each word-element and its meaning is found in its appropriate alphabetical place in the source book and with it are found numerous related words and illustrations of their use in scientific terminology. This segregation of word elements may serve as a guide to a more intelligent pronunciation in which the effort is to preserve the classical sources of the words rather than to conceal them by the rigid rules of euphony. It will also contribute to any easier remembrance of their correct spelling and a more facile understanding and appreciative use in spoken as well as written composition.

No attempt has been made to include geographical names, names based on modern personal names,* misspelled generic names, or a multitude of ill-coined terms of some of the recent ambitious yet careless insect-anatomists and ecologists who have proved themselves to be word-butchers of the mean-

---

* Generic names based on modern personal names are usually given the endings -a, -ai, -ea, -ia, (the -ia ending being the more frequent) or occasionally -ella, or -etta and thus are quite easily detected. Such names are legion both in botanical and zoological literature. Examples are: Westwoodia, Woodwardia, Boerhaavia, Blumea, Brandegea, Bradburya, Fendlera, Parishella, Helietta, etc.

est sort. In their effort to impress young students they have, after the manner of pedants, flooded recent literature with hundreds of new and useless compound terms. The origin of these terms they take no pains to explain and the definitions of them are often so involved that it is doubtful if the authors themselves remember them a week after they have been made. A cursory survey of CARPENTER's *An Ecological Glossary* and J. R. DE LA TORRE-BUENO's *A Glossary of Entomology* will reveal the type of degraded words to which I refer, words in which the beautiful classic roots have been chopped into halves, thirds, or quarters, and combined with other mutilated elements without following any rule of proper word-building. Such practices are inexcusable and should be condemned by all students who have any regard for the ethics and aesthetics of orthography.

Every effort has been made to make this a dependable source of information. For the form and definition of Greek words, reliance has been placed on the great lexicons of LIDDELL and SCOTT and of PICKERING. The comprehensive HARPERS' Latin Dictionary has served as the basis of definition for many of the Latin words.

That some errors and omissions may have inadvertently crept in is beyond doubt, and the author will appreciate it if his attention is called to any irregularities, so that in future editions of this work full correction can be made.

In some cases only an approximation toward accuracy was possible in making the application of meanings of the numerous generic examples given. Where the authors of generic names have concisely stated the origin, meanings, and application of their names, the task has been easy; but where no such aids are given, one intelligent guess is as good as another, and the chances for erroneous explanation are much increased. It is indeed unfortunate that the custom of explaining the derivation of scientific names is now so little observed. A few authors used to do it; practically none do it now. Whether it is due to indolence, carelessness or a sort of prosaic academic apathy I shall not say.

Books which have proved of most value as sources of information are: AGASSIZ' *Nomenclator Zoologicus*, NEAVES' *Nomenclator Zoologicus* (VOL. I–IV), JORDAN and EVERMANN's *Fishes of Middle and North America*, the *Challenger Reports* in which are ERNST HAECKEL's descriptions of Protozoa, DE DALLA TORRE's *Catalogus Hymenoptorum*, DON's *History of the Dichlamideous Plants*, FISCHER's *Manuel de Conchyliologie*, PALMER's *Index Generum Mammalium*, RICHMOND's *Lists of Generic Terms for Birds*, WATERHOUSE's *Index Generum Avium*, P. A. SACCARDO's monumental work, *Sylloge Fungorum*, HITCHCOCK's *Manual of the Grasses of the United States* and GRAY's *Synoptical Flora*. Liberal use has also been made of the *American Encyclopaedic* and the *Century* dictionaries as well as of numerous original descriptions in taxonomic works and journals.

The author wishes to express his deep gratitude to Dr. S. Stillman Berry of Redlands, California, and Julian K. Richards of Riverside College, for their critical reading of portions of the manuscript and for their many helpful suggestions. He wishes to acknowledge his indebtedness to the United States National Museum, Museum of Zoology of the University of Michigan,

Museum of Comparative Zoology at Harvard College, the American Museum of Natural History and other institutions for their cooperation in generously supplying technical information. Special thanks are also in order for the painstaking clerical work done by Mr. Lloyd M. Smith, Mr. Barton Barrier, Mr. Donald Worley, Mr. Chalmers MacIlvaine, Mr. Eugene Kozloff, Miss Carol McFarland, and Miss Doris Buttles.

Riverside College                                                   E. C. J.

# How Words Are Built

THAT PART of a derivative word which contains the principal idea is called the stem.* Thus, in the word pro-phase, *phase* is the stem meaning an appearance, aspect. The preceding syllable *pro* meaning before, is called the prefix. To the stem may be added both prefixes and endings called suffixes as in the words, ex-tract-ed and con-sist-ing.

Stems are of two kinds: (1) separable stems which may stand without prefix or suffix. Such a stem is *join* in conjoin: (2) inseparable stems, which though possessed of independent meaning, never stand alone but are always joined to prefixes or suffixes or joined to other stems to form compounds. Examples of such inseparable stems are *ject* (<L. *jacio*, to hurl) in eject, and *clude* (<L. *cludo*, to shut, close) in exclude.

If two stems are joined to form a compound, a vowel or combination of vowels (*a, ae, e, eo, i, io, o*) is generally added to the first stem as a joining agent to bring the two euphoniously together. Thus, in Sylvi-lagus, *i* is the connecting vowel; in Callo-mys, *o* is the joining agent.

In making generic and specific names the following rules outlined in the International Rules of Botanical Nomenclature have in general been applied:

When a new name for a genus is taken from the name of a person it is formed in the following manner: (a) When the name of the person ends in a vowel the letter *a* is added (thus *Bouteloua* after the brothers *Boutelou; Dalea* after Thomas *Dale*), except when the name already ends in *a*, then *ea* is added (e.g. *Jubaea*, after *Juba*, a king); (b) when the name ends in a consonant, the letters *ia* are added (e.g. *Wyethia* after Capt. N. J. *Wyeth, Encelia* after Christopher *Encel*) except when the name ends in -*er*, then *a* is added (e.g. *Frasera* after J. *Fraser, Viguiera* after Dr. A. *Viguier*). Exceptions are such as *Franseria* from Ant. Franser, etc. When a new specific name is taken from a personal name ending in a vowel, the letter *i* is added (e.g. *Glazioui* from *Glaziou*), except when the name ends in *a* then *e* is added (e.g. *Balansae* from *Balansa*). When the name ends in a consonant, the letters *ii* are added (*Parishii* from *Parish*), except when the name ends in -*er*, then *i* is added (thus *Kerneri* from *Kerner*). Exceptions to all of these rules may be found among the older names made by Linnaeus and other early taxonomists. The original spelling is generally retained except in a clear case of typographic error or of a clearly intentional error in spelling. Examples: Brodiaea* < *Brodie;* Jussiaea* < *Jussieu;* Tellaea* < *Tilli*.

In the formation of specific names composed of two or several roots taken from Latin or Greek, the vowel placed between the two roots becomes a connecting vowel, in Latin *i*, in Gr. *o* (thus *salviifolia* and *macrofolia*). When the second root begins with a vowel and euphony requires, the connecting vowel should be eliminated (e.g. *lepidantha*, not *lepidiantha*). The connecting vowels

---

* Many words contain a still more primitive form than the stem which is called the root. Thus the Latin stem *caed*-< from *caedo*, to cut, has as its root *cid*-.

*ae* should be retained only where this is required for etymological reasons (e.g. caricaeformis < *Carica*, in order to avoid confusion with cariciformis from *Carex*, genit. *Caricis*).

If an adjectival specific name ends in *us, a* or *um*, it may end in either of the other two endings to correspond with the gender of the generic name to which it is attached: as *Platystemon californicus* (Masculine), *Rhamnus californica* (Feminine), *Veratrum californicum* (Neuter). Or the specific name may end in *is* or *e*, the first agreeing with masculine or feminine generic names, the latter with neuter names. In this work all Latin adjectives are given in the masculine or feminine *-is* or masculine *-us* form.

No special case of the noun or adjective is used when combining words to make compounds, but only the stem. In some cases the stem has, through long use become blurred or obscured in the nominative but is still clear in the course of declension, particularly in the genitive, hence our frequent reference to the genitive case in certain instances, particularly in Greek.

# Types of Names Considered

THREE types of names are considered: (1) *specific names*,* generally derived from Latin; (2) *technical terms*, derived from both Greek and Latin; (3) *generic names*, usually derived from Greek stems.

(1) SPECIFIC NAMES are:

    (a) Adjectives, simple or compound, agreeing grammatically with the generic name. Example: *Felis marmorata;*

    (b) Nouns in the nominative in apposition with the generic name. Example: *Felis leo;*

    (c) Nouns in the genitive. Examples: *rosae, sturionis, antillarum, galliae, sancti-pauli, sanctae-helenae.*

In those specific names, called patronymics, honoring or commemorating persons, the genitive is always formed by adding, to the exact and complete name, an *i, ii* or *iana* if the person is a man, or an *ae* or *iae* if the person is a woman, even if the name has a Latin form. Such commemorative names are generally easily detected. The specific names most difficult to trace to their origin are those based on obscure geographical names, native names and names of little known deities. See foot-note to preface.

(2) TECHNICAL TERMS—These are usually combinations of Greek and Latin elements with appropriate Latin-derived English endings. Most physiological and many anatomical and medical terms fall in this group. A knowledge of their literal meanings aids much in remembering their orthography and leads to assurance and facility in their use.

(3) GENERIC NAMES—These consist of single words, simple or compound, written with an initial capital letter and used as a noun in the nominative singular. One of the most lucid and comprehensive statements ever made concerning word-formation of generic names was that of T. S. Palmer of the United States Biological Survey in his *Index Generum Mammalium* (North American Fauna No. 23, 1904). While it was written with special reference to generic names of mammals, its broad principles are equally applicable to other generic names. I have taken the liberty to further perpetuate Palmer's labors by quoting at length, with certain omissions, and added notes, the following paragraphs.

* I here use the specific term in the way it is commonly understood. Correctly speaking, to quote Dr. Charles H. O'Donoghue of the University of Reading: "The specific name of any animal or plant must consist of two words, i.e., it must conform with the Linnean binominal system.

"The specific name, i.e., nomen specificum, consists of a generic name (nomen genericum) followed by a trivial name (nomen triviale). What ordinarily is termed the specific name is actually the trivial name."

In zoology the trivial name should always commence with a small letter. The botanists do not always conform to this practice but often commence the trivial name with a capital letter when it is derived from a personal or place name, or when a generic name has been relegated to the position of a trivial name.

## KINDS OF GENERIC NAMES

### Mythological Names

"A considerable number of generic names are taken from mythology, both classical and Hindu, such as—Titanotherium, Vishnutherium.

### Geographical Names

"Geographical names have been used mainly in paleontology. In spite of the fact that they are mainly hybrid words, they have the advantage of convenience, as they are usually based on the type locality of one of the species. Such are: Atlantoxerus, Pampatherium, Uintatherium.

### Classical Names

"Apparently every name of an animal used in classical Greek has been made to do service in modern nomenclature, and these have been modified until they form a large number of the designations in common use. . . .

### Barbarous Names

"The recognition of generic names derived from barbarous words has given rise to much discussion. Several of the older systematists refused to recognize them, and regularly substituted new ones for those which they considered barbarous. . . .

"On the other hand, some authors not only frequently employed barbarous names, but also advocated their use. Lacepede apparently never missed an opportunity to use them, while Lesson and Gray are responsible for the introduction of many native names. . . .

"The objection to barbarous names has diminished of late, and many of those rejected by Illiger and others are now coming into general use for groups for which no earlier classical derivatives are available."

### Nonsense Names

"Reference should be made to names which have been 'coined' and which have no true derivation. These are merely arbitrary groups of letters* sometimes known as 'nonsense names'. They have been proposed by authors who, like Ameghino, Gray, and Lataste, in making many names have found the usual sources insufficient or unsatisfactory. These names may be divided into two groups:

(a) coined or nonsense names, like Azema, Blarina, Degonia, Kogia, and Tatera, and

(b) anagrams, such as—Senodon from Nesodon, Teonoma from Neotoma and Xotoprodon from Protoxodon.

### Derivative Names of Different Form

"A large number of names comprise derivatives and compounds of the same Greek or Latin words. These names may have different forms:

(a) According to *gender*, as Otostomus, Otostoma, Otostomum. Otostomis Menke, 1830, a mollusk; Otostomus Beck, 1837, a mollusk;

---

* It is quite possible for an arbitrary combination of letters to duplicate inadvertently a Greek or Latin root without carrying any of the classical meaning.

Otostoma Carter, 1856, a protozoan; Otostomum Ehrenberg, 1872, a protozoan, all evidently derived from the same Greek words, ους, ear, and στομα, mouth.

(b) According to the *particular Greek dialect* from which the words have been selected, as Lampronessa and Lampronetta. Both these names are compounds of λαμπρος, splendid, brilliant, and νηττα, duck; but in the former the Epic or Ionic form, νησσα, is used and in the latter its Attic equivalent.

(c) According to whether the original Greek form has been *preserved* or whether it has been *transliterated* into Latin form, as Hipposideros and Hipposiderus.

(d) According to whether the Greek *aspirate* has been *preserved or not,* as Abrothrix and Habrothrix; Reithrodon and Rhithrodon.

(e) According to whether *the connecting vowel i or o* has been used in compounding two classical roots, as Callorhinus and Callirhinus.

"It is therefore possible to make a number of compounds from the same words, all meaning the same thing, and differing from one another simply by a letter or two. This may be illustrated by compounds of καλος, beautiful, and μυς, mouse. The following list contains no less than sixteen variations compounded from these words in accordance with classical rules, two of which, Calomys and Callomys, have actually been proposed for different genera of mice.

| | | | |
|---|---|---|---|
| Calimys | Calimus | Kalimys | Kalimus |
| Calomys | Calomus | Kalomys | Kalomus |
| Callimys | Callimus | Kallimys | Kallimus |
| Callomys | Callomus | Kallomys | Kallomus |

"These sixteen variations of 'beautiful mouse' are all available as valid generic designations of mammals (if applied to different animals), according to those who 'regard all generic names as different unless originally spelled alike'. . . .

"It will be found that most generic names have been bestowed for the sake of drawing attention to some characteristic or resemblance of the animal, fancied or real. They may contain many facts of interest, descriptive, geographical, or historical, and the knowledge of such derivation may be, and often is, an aid in keeping in mind the relationship of the group. Unfortunately, very few authors have taken the trouble to give etymologies or explain the application of their generic names.

## Compound Derivatives

"A large proportion of modern generic names are compound words. Latin offers comparatively little opportunity for making compounds, and the number of such words is relatively small, although modifications by prefixes and suffixes are common. The Greek language lends itself almost as readily as the German to this kind of word-making, and nouns are coupled together or modified by adjectives and prepositions in almost endless variety. Formerly compounds seem to have been in disfavor, for Illiger, in 1811, following Linnaeus, rejected them, and quotes three Linnaean rules as authority for so doing. . . .

"It is difficult to understand this position, since compounds have the sanction of classical writers. Among numerous classical compound words which have been used as generic names of mammals may be mentioned Acanthonotus, Agricola, Camelopardalis, Cataphractus, Cynalopex, Hippopotamus, Hippotigris, Hydropotes, Nyctereutes, and Rhinoceros. At the present time compounds are considered not only unobjectionable, but highly desirable, for without them it would be almost impossible to coin designations for the ever-increasing multitude of genera and species without resorting to anagrams and arbitrary combinations of letters. They may have the advantage of indicating the relationship of a genus, and, what is even more important, of insuring it from being preoccupied in other groups. For example, compounds of *Mus* are usually restricted to rodents, and are not likely to be used in any class except mammals; the prefix *eu* is constantly used to distinguish the typical genus or sub-genus from groups which are aberrant, in contradistinction to such prefixes as *hemi-* and *para-* or the suffix *-oides*, which merely indicate resemblance; and the intensive *za* is used to call attention to some prominent or striking character.

"Nowhere have compounds been more constantly and more effectually employed than in paleontology. Indeed, we have here a certain approximation toward the standard which Coues has pictured as the ideal name when he says—

'Systematic zoology, or the practice of classification, has failed to keep pace with the principles of the science; we are greatly in need of some new and sharper "tools of thought", which shall do for zoology what the system of symbols and formulae have done for chemistry. We want some symbolic formulation of our knowledge. The invention of a practical scheme of classification and nomenclature, which should enable us to formulate what we mean by *Turdus migratorius* as a chemist symbolizes by $SO_4H_2$ what he understands hydrated sulphuric acid to be, would be an inestimable boon to working naturalists. (Key *N. Am. Birds*, 2d ed., 78, 1884.)'

"To a certain extent this is done in some paleontological names. Thus words compounded with *-therium*, or with the prefixes *amphi-*, *eo-*, *epi-*, *limno-**, *meso-*, *meta-*, *mio-*, *plesio-*, *plio-*, and *proto-*, are almost always used for extinct genera and should be reserved exclusively for them. Prefixes may be briefly and conveniently used to express relationship. Amphicyon, Epicyon, and Pseudocyon, all indicate groups more or less closely related to the dogs; Cimolestes, a marsupial from the Cretaceous; Eohippus, Miohippus, and Pliohippus were proposed for horses which existed in the Eocene, Miocene, or Pliocene; Protodichobune for a type of artiodactyl which preceded, and Metadichobune for one which followed, Dichobune.

"Pliohippus means an animal from the Pliocene related to the modern horse: Miosiren, an animal from the Miocene related to modern sirenians; and Limnofelis, an animal found in an old morass and related to living cats, etc. Here the names give (a) the designation of the genus, (b) its geological posi-

---

* Workers in groups other than mammals might not readily agree that *limno-* has its most apt connotation in the fossil field.

tion, and (c) its relationship; while their form indicates (d) that the genera are extinct."

## Double Generic Names

"A special class of compounds, which may be described as 'double generic names,' has been largely used in the case of mammals. Nearly 200 such names have been proposed, chiefly to denote resemblance or close relationship, and, when well chosen, serve the purpose admirably. They have the advantage of being self-explanatory, and are not apt to be preoccupied. They are ordinarily formed by combining two generic names into one, as Adapisorex and Cervalces, although some of them may be simply classical compounds of two names of animals, as Camelopardalis and Cynalopex. In either case the result is the same."

## APPLICATION OF NAMES

"Etymology in the widest sense of the term properly includes the *application of names*, but the latter subject is so broad as to merit special consideration. Although many generic names have been applied in such haphazard fashion or based on such apparently trivial or obscure characters that it is almost hopeless to attempt to explain their application unless the original author has furnished the key, still many others have been based on important characters or coined with a view of expressing relationships, indicating facts of distribution, or throwing light on their history, thus offering an interesting field for investigation.

"Apparently every conceivable character, external and internal, positive and negative, has been called into play in making generic names, and minute or imaginary resemblances have been utilized to such an extent that it is sometimes impossible to see the connection between the name and the animal even when the derivation is known. But the attempt to ascertain whether a certain term has originated in fact or fiction, or whether its application has been suggested merely by the fancy of the author, is at least interesting, and often successful. In classifying names, beginning with those which have an obvious application and passing to those which have none, at least ten subdivisions may be made: (1) classical names and their compounds; (2) native names; (3) geographical names; (4) personal names; (5) names indicative of age; (6) names indicating size, form, color, and resemblance; (7) names indicating habit and habitat; (8) names based on special characters; (9) names of fanciful or poetic application; (10) names founded on error. These numerous subdivisions may be arranged under two main headings: (1) names of obvious application, comprising the first seven groups, and (2) names of obscure application, comprising the last three groups.

## Names of Obvious Application

"(1) Classical names.—Words of classical derivation taken from names of animals, like Cebus, Gale, Mus, and Pithecus and compounded with such prefixes as *eu* (typical), *amphi* (on both sides), *para* (near), *pseudo* (false), *za* (intensive prefix), etc., need no special explanation. Similarly, words like

Alticola (high dweller), Terricola (ground dweller), Hydropotes (water drinker), etc., suggest their own application. But in some cases old names of animals have been transferred to groups entirely different from those to which they originally belonged. Thus Dasypus is now applied to the armadillos, which do not occur in the Old World, and Cebus (from κηβos, which meant any long-tailed monkey), is now restricted to neotropical monkeys, which name and its apparent compounds Arctocebus, Habrocebus, Microcebus, and Nycticebus, all applied to Old World lemurs, is not clear, except on the theory that the latter are not compounds of the modern generic name, but of the original κηβos. Similarly most of the compounds of πιθηκos, ape, are applied to Old World groups, while the root word in the form Pithecia is transferred to a South American monkey.

"(2) NATIVE NAMES.—Native names like Bandicota (pig rat) and Derivoula (plantain bat) constitute one of the most interesting groups (providing their derivation can be ascertained), and they are usually based on such prominent characters that little explanation is required.

"(3) GEOGRAPHICAL NAMES.—Geographical names are in most cases self-explanatory merely from the derivation, but, as already mentioned, they are usually hybrid words and are sometimes compounds of little used names of localities, so that their application is not evident at first sight. Typical examples are Ruscinomys from Ruscino, the Roman name of the modern town of Perpignan in France; Kasi, a subgenus of monkeys from India, named from Kasi, the ancient designation of Benares; and Argyrocetus, 'silver whale' of Argentina, used in the same sense of La Plata (silver) whale.

"(4) PERSONAL NAMES.—Personal names fall into two categories—those derived from the name of the collector of the species on which the genus was based, like Bruijinia, Carloameghinia, and Nelsonia, and those named in compliment to some distinguished person, as Garzonia, for Don Eleazar Garzon, governor of the province of Cordoba, Argentina; Capaccinius, for Monsignor Francesco Capaccini, under Secretary of State of Rome; and Romerolagus, for Don Matias Romero, formerly Mexican minister to the United States. In the first group, however, the application is by no means evident, and, without explanation, is often very obscure.

"(5) GEOLOGICAL NAMES.—Names indicative of age, or geological names, are frequently employed in paleontology to show the age of the deposits in which the animals were found. The most frequent are compounds of *eo-*, *mio-*, and *plio-;* thus Eopithecus, Miopithecus, and Pliopithecus represent apes from the Eocene, Miocene, and Pliocene. Similarly a few compounds have been made from *cimo-* and *caeno-;* thus Cimolestes, Cimolodon, and Comolomys indicate mammals from the Cretaceous (chalk); and Caenobasilius, Caenopithecus, and Caenotherium, mammals from recent or Quaternary beds. The prefixes *hyper-* (above), *infra-* (below), *proto-* (first), *pro-* (before), *meso-* (middle), and *post-* (after), are also employed to represent relative age, as Hypertragulus, Infrapithecus, Protohippus, Promeles, Mesohippus, and Postpithecus. Occasionally, names have been suggested by the character of the beds in which the fossils were discovered; thus Anthracotherium, is a genus from the anthracite or lignite beds of Tuscany; Argillotherium, one from the London clay; Chalicotherium, one from the iron ore deposits near Mosskirch, Baden.

"(6) DESCRIPTIVE NAMES.—Names indicative of *size, form, color, and resemblance* may be found in almost endless variety, and usually present few difficulties. *Size* is indicated in all gradations from the huge Megatherium to the pygmy Nannosciurus, but though the meaning of such names may be obvious, yet they convey no idea of absolute size to show that their selection is especially appropriate. Thus, while Megamys plainly refers to a large rodent, it does not show that the mammal thus named was supposed to have been as large as an ox; and while Microcebus is a small lemur, the fact that some species of the genus are only 5 inches in length (exclusive of the tail) is not shown. Similarly 'small shrew' does not suggest the fact that Microsorex is in reality one of the smallest mammals, with a total length of only $3\frac{1}{2}$ inches. *Form* is expressed in all gradations from fat, (Steatomys) to thin (Stenobalaena); from thick (Pachysoma) to slender (Leptomys); from short (Brachytherium) to long (Dolichotherium). *Color and markings* play a comparatively unimportant part in the formation of generic names, but are used as a basis in a few cases. References to color in general are found in Celacnomys, Chrotomys, and Chrotopterus; to red in Erythrocebus, Erythrosciurus, and Rousettus; to white in Beluga, Leucas, Leucocyon, Leucomitra, Leucopleura, and Leucorhamphus; to greenish yellow in Chloromys; and to yellow in Chryseus, Chrysochloris, Chrysocyon, Chrysomys, Chrysonycteris, Chrysospalax, Chrysothrix, Icterus, and Xantharpyia. *References to markings* may be either to spots (Balionycteris, Rhinostictus, Spilogale), to stripes (Lemniscomys, Strigocuscus), to bands (Histriophoca, Taeniogale), or to a combination of colors (parti-colored) or markings (Poecilogale, Poecilomys, Poecilophoca). *General resemblance* is indicated by compounds of *oides, ops,* and *opsis* (Petauroides, Dipodops, and Choeropsis), and by many double names of mammals, such as Antilocapra, Ovibos, and Taurotragus.

"(7) MISCELLANEOUS NAMES.—*Habits and habitat* form the basis of a great variety of names. Nearly every manner of progression is referred to directly or indirectly in the following examples: *Creeping* (Herpestes, Herpetomys), *walking* (Ocnobates), *digging* (Tachyoryctes), *groping about* (Pselaphon), *running* (Dromedarius and Dromicia), *flying* (Pteromys), and *swimming* (Nectomys); *living in the water* (Hydromys), *on land* (Terricola), and *underground* (Hypogeomys).

"*Habits and characteristics* of various kinds are illustrated by Chiropotes (hand drinker), Hydropotes (water drinker), Nyctereutes (night hunter), and Oxygous (shrill wailing).

"*Disagreeable odors* are suggested by such names as Bdeogale, Mephitis, Ozolictis, Osmotherium, and Putorius. *Character of habitat* is often indicated by a prefix or suffix. Thus we have names of animals of the water (Hydrochoerus, Hydrodamalis), sea (Halicore, Thalarctos), rivers (Hippopotamus, Potamochoerus), islands (Nesonycteris, Nesotragus), swamps (Helogale, Limnogale), rushes (Thryonomys), fields (Arvicola), gardens (Leimacomys), trees (Dendrolagus, Dryoryx), forests (Hylomys, Hylobates), plains (Pediotragus), pampas (Pampatherium), deserts (Xerospermophilus), sand (Ammospermophilus), mountains (Oreotragus and Oreamnos), of snow and ice (Chionobates, Pagophilus), and torrid heat (Helarctos, Heliophoca).

Finally, *character of the food* is sometimes expressed in the name, as when the animal feeds on seeds (Spermophilus), grain (Sitomys), rice (Oryzomys, Oryzoryctes), bark (Phloeomys), roots (Rhizomys), grass (Poebrotherium, Poephagomys, Poephagus), fruit (Carpomys, Carponycteris, Syconycteris, Trygenycteris), honey (Mellivora, Melursus), ants (Myrmecobius, Myrmecophaga), fish (Ichthyomys), or flesh (Sarcophilus and Sarcothraustes).

## NAMES OF OBSCURE APPLICATION

"(8) NAMES INDICATING RELATIONSHIP.—Many names denoting relationship or based on general characters or habits require further explanation than mere derivation to render them intelligible. Thus, *Mesomys* (middle mouse) and *Synaptomys* (connecting mouse) are evidently intermediate forms, but the names alone do not show that *Mesomys* is related both to the jumping rats and *Mus*, or that *Synaptomys* is a connectant form between the lemmings and field mice. *Aschizomys* (not splitting—in the sense of connecting mouse), which, in a somewhat different way, expressed the same idea of connection, does not show that it is based on a combination of the characters of *Microtus* and *Evotomys*. Likewise, *Orthriomys* (early mouse) and *Phenacomys* (deceptive mouse) are not self-explanatory. *Orthriomys* was so called from the fact that it suggests an ancient type intermediate between *Phenacomys* and the Microtine *Pedomys* and *Arvicola; Phenacomys* derives its name from the fact that externally it is almost indistinguishable from Microtus. *Hodomys* (road mouse), refers to the animal's habit of making trails, and not, as might be supposed, to its living along highways. *Monachus* (monk) is applied to the tropical seal, probably because of its more or less solitary habits, and *Semnopithecus* (sacred monkey) to a group of monkeys of India, because the type species is considered sacred by the Hindus. Less clear are such terms as *Cryptomys* (hidden mouse), *Dinomys* (terrible mouse), and *Xenomys* (strange mouse), which merely suggest the peculiarities of the groups to which they belong without explaining them. *Pectinator* (comber), and *Tamias*, the well-known genus of ground squirrels, have a habit of laying up stores of food.

"*Generic names based on special characters* are very numerous. They may refer to external characters, such as the skin, hair, head, nose, eyes, ears, tail, wings, or feet; to special characters of the teeth, skull, vertebrae, ribs, or limb bones; or to the soft anatomy. Those which owe their origin to characters of the teeth and skull are very common, while those based on the soft anatomy are comparatively rare. Not only do special parts of mammals suggest names, but a wide range of qualities of each part is represented in nomenclature.

"Thus, characteristics of the skin that give rise to names range from narrow (*Stenotherium*) to broad (*Megaderma*), probably in allusion to the size of the flying membranes; those of the hair from soft (*Abrothrix*) and wooly (*Lagothrix*—rabbit hair) to spiny (*Echiothrix*) and scaly (*Lepitherium*); those of the nose from allusions to a long nose (*Nasalis* and *Oxymycterus*) to noseless (*Arhinolemur*); those of the ears and tail from earless (*Aotus*) and tailless

(*Anoura*) to large eared (*Macrotis*) and feather-tailed (*Pteronura*); those of
limbs from short (*Brachytarsomys*) to long (*Megaptera, Macropus*).

"The examples just mentioned and some of the designations of teeth and
skulls require no special explanation and should perhaps not be classed with
names of obscure application. On the other hand, many of the terms derived
from special characters, both external and internal, are exceedingly obscure.
Among others may be mentioned *Heterodon* (different tooth), in allusion to
the inequality in size and form of the teeth; *Megantereon* (great chin), which
refers to the size of the lower jaw and not to that of the animal; *Ommato-
phoca* (eyed seal), which refers to the immense orbits; *Ommatostergus* (a
worker bereft of eyes), which antithetically refers to the apparent absence of
eyes; *Synotus*, based on the union of the inner margins of the ears on the forehead; and *Tomopeas* (stump awl), so called from the short, blunt tragus.
Names based on cranial or skeletal characters, and many of those based on
teeth, are of uncertain application unless explained. Such are *Caperea* (*capero*,
to wrinkle), from the rugulose character of the ear bones; and *Meganeuron*
(large nerve), from the large size of the neural passage in the atlas. *Deltathe-
rium, Lambdatherium*, and *Sigmodon* are suggested by the enamel patterns of
the molar teeth, which resemble the Greek letters $\Delta$, $\Lambda$, $\Sigma$, respectively; *Syc-
ium* refers to the bony walls of the pulp cavity, which close the lateral grooves
but do not close the pulp cavity below; *Nesodon* (island tooth) takes its name
from an island of enamel on the inner side of a molar; *Ochetodon* and *Reithrodon* (grooved tooth) refer to grooves on the upper incisors; *Plagiaulax* (oblique groove) refers to grooves on the lower premolars; while *Plagodontia*
(oblique tooth) refers to grooves on the molars.

"Perhaps *the most puzzling names are those of an indefinite character.* Such as
*Proteles* (complete in front), in reference to the full number of five toes on the
fore feet in contrast to four on the hind feet, and names suggesting characters which are rudimentary or absent, as *Ateles* (incomplete) and *Colobus*
(maimed), both referring to the rudimentary character of the thumb; *Perodic-
ticus*, referring to the rudimentary index finger; *Choloepus* (lame footed),
indicating a reduced number of toes, and *Olbodotes* (giver of bliss) on account
of the light which it throws on the development of the large incisors in the
primitive rodents.

"(9) FANCIFUL NAMES.—Names of fanciful or poetic signification comprise mythological designations and a few other names. In many instances
their application is exceedingly obscure, and although usually it is not difficult to ascertain the personage from whom the name is derived, the explanation of its application, unless given by the original describer, may easily
be erroneous.

"Mythological names are objectionable, not only because they have been
used so frequently in other classes that they are likely to be preoccupied,
but also because of this ambiguity in application. The explanations given
will be found unsatisfactory in many cases, but the difficulty in working
them out may be illustrated by a few examples. The genus of monkeys
called Diana is apparently so named from the white marking or line over the
forehead of the type species, which bears a fancied resemblance to the silver

bow of the goddess Diana. *Idomineus*, the name of a king of Crete, does not seem applicable to a genus of gerbilles, unless it is remembered that *Idomineus* and *Meriones* were companions in arms in the Trojan war, and Meriones having long been used for a genus of Gerbillinae, it was thought fitting that a subgenus of the same group should be named after his companion, *Idomineus*. *Adjidaumo*, as applied to an extinct genus of rodents from the western United States, requires for many the describer's explanation that the designation was the Indian name of a squirrel borrowed from Longfellow's poem *Hiawatha*.

"(10) NAMES FOUNDED ON ERROR.—Names founded on error or without application are comparatively few in number, but are still sufficiently numerous to warrant mention. Errors as to the relationships of animals are to be expected in the case of extinct forms described from fragmentary remains, and it is not surprising that additional specimens have sometimes shown that an animal belongs to a different family or order from the one to which it was originally referred. Such errors can be corrected when discovered, but the names in which they are sometimes embodied must stand as first published. Several generic names thus erroneously given are strikingly inapplicable. Thus *Aceratherium* was so named because it was supposed to be a hornless rhinoceros, but, according to Osborn, the animal probably did possess a rudimentary horn. *Ailuravus*, originally supposed to be an ancestral carnivore, is now regarded as a squirrel. *Aodon* (the toothless whale of Havre) really belongs to the toothed whales, but was described from an old specimen of *Mesoplodon bidens* which had evidently lost its teeth. The well-known genus *Zeuglodon*, originally described as *Basilosaurus* (king of the saurians) from its supposed reptilian characters, is now known to be a cetacean and not a reptile. *Condylura* (knotted tail) was applied to the star-nosed moles by Illiger, who based his description on a very imperfect figure, in which the tail was represented as having a series of nodes. *Hyperoodon*, and *Uranodon* (palate tooth) were applied to the same genus of ziphioid whales on account of the rough papillae on the palate, which were at first mistaken for teeth. *Paradoxurus* (strange tail) owes its name to the circumstance that the tail, which the animal has power to coil to some extent, was originally supposed to be prehensile, a character which would certainly be anamalous in the civet cats. *Protorhea*, based on an imperfect femur, was at first supposed to be an extinct struthious bird, but was afterwards regarded as a mammal related to the llamas. *Stemmatopus* (wreathed foot) was given to the hooded seal by Cuvier, though it is probable that the name intended was *Stemmatops* (wreathed face), in allusion to the hood, and that the insertion of a *u* by mistake transferred the allusion to the other extremity of the animal and destroyed the application of the name."

# Transliteration

In making transliterations from the Greek to the Latin and English, the following should be considered:

Most of the Greek consonants are represented by the Latin equivalents as shown in the following alphabet.

| CHARACTERS | | NAMES | EQUIVALENTS | |
|---|---|---|---|---|
| A | α | alpha | a | as in father |
| B | β | beta | b | as in bog |
| Γ | γ | gamma | g | as in gum |
| Δ | δ | delta | d | as in doll |
| E | ε | epsilon | e | as in get |
| Z | ζ | zeta | z | as in zone |
| Η | η | eta | e | as in fête |
| Θ | θ | theta | th | as in thing |
| I | ι | iota | i | as in machine |
| K | κ | kappa | k | as in keel |
| Λ | λ | lambda | l | as in lone |
| M | μ | mu | m | as in man |
| N | ν | nu | n | as in note |
| Ξ | ξ | xi | x | as in fox |
| O | o | omicron | o | as in obey |
| Π | π | pi | p | as in pill |
| P | ρ | rho | r | as in run |
| Σ | σ, s | sigma | s | as in suck |
| T | τ | tau | t | as in top |
| Υ | υ | upsilon | u | as in French u |
| Φ | φ | phi | ph | as in physical |
| X | χ | chi | ch | as in machen (German) |
| Ψ | ψ | psi | ps | as in whips |
| Ω | ω | omega | o | as in so |

| | | |
|---|---|---|
| κ | is changed to c | |
| ξ | is changed to x | |
| γ | occurring before γ, ξ, χ, and κ becomes n | |
| υ | is changed to y; occasionally u (when it is part of a diphthong) | |
| αι | becomes ae | |
| αυ | becomes au | |
| ει | becomes i or occasionally ē | |
| οι | becomes oe | |
| ου | becomes ū | |

The final -os in Greek nouns and adjectives of the second (o) declension is changed in Latin words to -us; and the neuter ending -on of the same declension becomes -um in Latin words.

Both short e (ε) and long ē (η) are transliterated e

Both short o (o) and long ō (ω) are transliterated o

# Concerning Greek Prefixes

A GREEK prefix ending in a consonant when joined to a stem beginning with a consonant often has its final letter changed for the sake of euphony.

The final vowel of a prefix is omitted before a stem beginning with a vowel. Thus ana-agōgē becomes anagōgē.

When a prefix ending in **n** is joined to a stem beginning in **b, ph,** or **t** it is changed to **m.** Thus *enbolē* becomes embolē, leading to embolism, in English.

When **n** occurs before a stem beginning with **m,** it is changed to **m.** Thus *syn-metria* becomes symmetry in English.

When **n** comes before a stem beginning with **l** it is changed into an **l.** Thus *syn-labē* becomes syllable in English.

When **n** occurs before a stem beginning with **s,** the **n** is dropped. Thus *syn-stēma* becomes system in English.

When a word beginning with an aspirated r (rh) is preceded by a prefix or element ending in a short vowel the initial letter **r** is repeated. Thus we have cata-rrh (Gr. *kata,* down+*rheō,* to flow), and Platy-rrhina (Gr. *platys,* flat+*rhinos,* nose).

# Concerning the Form of Latin Nouns

THE NAMES of genera and subgenera are usually nouns in the nominative singular. Some specific and subspecific names are nouns in the nominative singular; others may be nouns in the genitive singular or plural. The common forms are given in the table below.

There are five modes of declining Latin nouns, each distinguished from the other by the final letter of the stem and by the termination of the genitive singular. In the first mode or declension this genitive termination is *ae*, in the second *ī*, in the third *ĭs*, in the fourth *ūs*, and in the fifth *ēī, eī*.

Comparative view of nominative and genitive case endings of the five declensions.

## TERMINATIONS

### SINGULAR

|       | I    | II* M. | II* N. | III** M., F. | III** N. | IV M. | IV N. | V    |
|-------|------|--------|--------|--------------|----------|-------|-------|------|
| Nom.  | ă,   | ŭs, ĕr, | ŭm    | ŏr, etc.     | ĕ, etc.  | ŭs, – | ū,    | ēs,  |
| Gen.  | ae,  | ī,     |        | ĭs           |          | ūs,   |       | ēī,  |

### PLURAL

|       | I     | II* | II* | III** | III** | IV | IV | V    |
|-------|-------|-----|-----|-------|-------|----|----|------|
| Nom.  | ae,   | ī,  | ă,  | ēs,   | ă(iă) | ūs, | uă, | ēs,  |
| Gen.  | ārum, | ōrŭm, |   | ŭm,   | (iŭm) | uŭm, |   | ērŭm, |

\* Nouns in the second declension end in *er, ir, os, um* or *on*. Those ending in *um* and *on* are neuter, the others are masculine.

\*\* The endings of the nominative in the third declension are very many: *-a, -e, -i, -o, -y, -c, -l, -n, r, -s, -t,* and *-x.*

# Concerning the Form of Latin Adjectives

SINCE MANY specific names are Latin adjectives in the nominative or genitive, the following examples of adjective forms are given.

Latin adjectives take the forms of (1) First and Second Declension Nouns or of (2) Third Declension Nouns.

## ADJECTIVES OF THE FIRST AND SECOND DECLENSION

These form their masculine and neuter like nouns of the Second Declension, but form their feminine like nouns of the First Declension.

### SINGULAR

|             | *Masculine* | *Feminine* | *Neuter* |
|-------------|-------------|------------|----------|
| Nominative  | bonus, *good.* | bona    | bonum    |
| Genitive    | bonī        | bonae      | bonī     |

### PLURAL

|             | | | |
|-------------|-------------|------------|----------|
| Nominative  | bonī        | bonae      | bona     |
| Genitive    | bonorum     | bonārum    | bonŏrum  |

SINGULAR

|  | Masculine | Feminine | Neuter |
|---|---|---|---|
| Nominative | līber, *free.* | lībera | līberum |
| Genitive | līberī | līberae | līberī |

PLURAL

|  | | | |
|---|---|---|---|
| Nominative | līberī | līberae | lībera |
| Genitive | līberōrum | līberārum | līberōrum |

ADJECTIVES OF THE THIRD DECLENSION, STEMS IN -I

*a.* With three endings, one for each gender in the nominative singular.

SINGULAR

|  | Masculine | Feminine | Neuter |
|---|---|---|---|
| Nominative | ācer, *sharp.* | ācris | ācre |
| Genitive | ācris | ācris | ācris |

PLURAL

|  | | | |
|---|---|---|---|
| Nominative | ācrēs | ācrēs | ācria |
| Genitive | ācrium | ācrium | ācrium |

*b.* With two endings, masculine and feminine the same.

SINGULAR

|  | Masculine and Feminine | Neuter |
|---|---|---|
| Nominative | brevis, *short.* | breve |
| Genitive | brevis | brevis |

PLURAL

|  | | |
|---|---|---|
| Nominative | brevēs | brevia |
| Genitive | brevium | brevium |

SINGULAR

|  | Masculine and Feminine | Neuter |
|---|---|---|
| Nominative | regēns, *ruling.* | regēns |
| Genitive | regentis | regentis |

PLURAL

|  | | |
|---|---|---|
| Nominative | regentēs | regentia |
| Genitive | regentium | regentium |

*c.* With one ending, the same for all genders.

SINGULAR

| | *Masculine and Feminine* | *Neuter* |
|---|---|---|
| Nominative | audāx, *bold.* | audāx |
| Genitive | audācis | audācis |

PLURAL

| | | |
|---|---|---|
| Nominative | audācēs | audācia |
| Genitive | audācium | audācium |

SINGULAR

| | *Masculine and Feminine* | *Neuter* |
|---|---|---|
| Nominative | fēlīx, *happy.* | fēlīx |
| Genitive | fēlīcis | fēlīcis |

PLURAL

| | | |
|---|---|---|
| Nominative | fēlīcēs | fēlīcia |
| Genitive | fēlīcium | felicium |

## DECLENSION OF COMPARATIVES

Comparatives are declined as follows:

SINGULAR

| | *Masculine and Feminine* | *Neuter* |
|---|---|---|
| Nominative | superior | superius |
| Genitive | superioris | superioris |

PLURAL

| | | |
|---|---|---|
| Nominative | superiores | superiora |
| Genitive | superiorum | superiorum |

All comparatives except plus, *more,* are declined like superior.

# Concerning Latin Names of Anatomical Structures

ANATOMICAL NAMES are either simple, consisting of one word, such as *auris* (ear), *musculus* (muscle), or compounds, consisting of a noun in the nominative singular or plural, followed by other words (nouns, adjectives or participles used as adjectives) in the nominative or the genitive case, according to the sense to be conveyed.

The following are illustrations:

1. *Arteria dorsalis hallucis. Arteria* is a feminine noun followed by the genitive adjective, *dorsalis*, modifying the genitive form of the noun, *hallux*, which is *hallucis*. Translated literally the name is: artery of the back of the great toe.

2. *Arteria genus superior medialis.* Here the feminine noun *arteria* is followed by the neuter singular genitive form of the noun, *genus*, and two adjectives, both in the nominative singular feminine form and agreeing in number and gender with *arteria*. This is the superior median artery of the knee.

3. *Arteria auditiva interna. Arteria* is followed by two adjectives, both feminine, in agreement in number, case, and gender with the feminine noun, *arteria*. Translated the name is: internal auditory artery.

4. *Ligamentum arcuatum externum. Ligamentum* is a neuter noun followed by two adjectives with nominative singular neuter endings, in agreement with *ligamentum*. This is the external arcuate ligament.

5. *Ligamenta intercostalia. Ligamenta* is a neuter plural noun followed by *intercostalia*, a neuter plural adjective. Translated, it reads: intercostal ligaments.

6. *Nervus communicans tibialis. Nervus*, a masculine noun, is modified by the present participle, *communicans*, used as an adjective. *Tibialis* is a noun in the genitive case. This is the communicating nerve of the tibia. It is so-named because it joins (communicates) with the anastomotic branch of the common peroneal to form the sural nerve.

7. *Os frontale. Os* is a neuter noun modified by *frontale*, the neuter form of *frontalis*. This is the frontal bone of the cranium.

8. *Venae interlobares renis. Venae* is the plural form of the feminine noun *vena*. It is followed by the plural adjective, *interlobares*, which agrees in case and number with *venae*, which it modifies. *Renis* is a genitive singular noun. Translated, it reads: interlobular veins of the kidney.

The following are adjective endings most often appearing in compound anatomical names.

| SINGULAR | | PLURAL | |
|---|---|---|---|
| *M., F.* | *N.* | *M., F.* | *N.* |
| -aris | -are | -ares | -aria |
| -alis | -ale | -ales | -alia |
| -is | -e | -es | -ia |

| *M.* | *F.* | *N.* | *M.* | *F.* | *N.* |
|---|---|---|---|---|---|
| -icus | -ica | -icum | -ici | -icae | -ica |

# Abbreviations

AS.......Anglo Saxon
Abor.....Aboriginal
Acanth...Acanthocephala
Adel......Adelochorda
adj.......adjective
Amph....Amphibia
Ammon...Ammonite
Ann......Annelida
Anthro....Anthropology
app.......apparently
Ar........Arabic
Arach.....Arachnida
Arth......Arthropoda
As........Anglo-Saxon
Asc.......Ascidian
Att.......Attic
Av.......Aves

Blast.....Blastoidea

Chin......Chinese
Chord....Chordata
Coel......Coelenterata
comp.....comparative
contr.....contraction
Crust.....Crustacea
Cten......Ctenophora
Cycl......Cyclostomata

Dan......Danish
Dor......Doric

Ecol......Ecology
Elasm....Elasmobranchia
Eng......English
err.......error, erroneously
Ethn.....Ethnology
Ex.......Example

fem.......feminine
Fr........French

Ger......German
genit.....genitive
Geph.....Gephyra
Gr........Greek

Icel......Icelandic
impf......imperfect
Ins......Insecta
It........Italian

Jap.......Japanese

L.........Latin
L. Ger....Low German
LL.......Late Latin
lit........literally
Low L....Low Latin

Mam.....Mammalia
masc......masculine
ME.....Middle English
Med......Medicine
Mes......Mesozoa
Mex......Mexican
ML.......Modern Latin
Moll......Mollusca
Myr......Myriopoda

Nemat....Nemathelminthes
Nemert...Nemertes
NL.......New Latin
Neut.....Neuter

OFr......Old French
OGer.....Old German

part......participle
Path......Pathology
Pers......Persian
Phor......Phoronida
Pisc......Pisces

Platy.....Platyhelminthes
plu......plural
Por......Porifera
Port.....Portuguese
pp.......past-participle
ppr......present participle
pref.....prefix
prob.....probably
prop.....properly
Prot.....Protozoa
Protoch...Protochordata

Rept......Reptilia

Sanskr....Sanskrit
Serv......Servian
sing.....singular

Sp.......Spanish
suff......suffix
sup......supine
super.....superlative
Sw.......Swedish

Tril......Trilobata
Troch.....Trochelminthes
Tun.....Tunicata

Verm.....Vermes

Xiphos....Xiphosura
<.......derived from
>.......giving rise to
*.........Plant names, mostly
          generic

# Explanation of Language Abbreviations

**Gr. Greek,** divided chronologically into (1) **Classical Greek,** to the year A.D. 200. At first mainly Attic Greek but finally the standard literary tongue of the country and embodying the best of the Greek Literature. It is the Greek of the standard Greek Lexicons such as Liddell and Scott's,
(2) **Late Greek,** from A.D. 200 to about A.D. 600.
(3) **Middle Greek,** from about A.D. 600 to A.D. 1500.
(4) **New Greek,** from A.D. 1500 to the present.

**L. Classical Latin,** the standard Latin of the grammars and lexicons. It includes the Latin used by the classical writers of the Golden Age of Latin literature and taken as the standard of purity; the period beginning about 75 B.C. and ending about A.D. 200.

**LL. Late Latin.** The Latin of the period just after the Classical, i.e., about A.D. 200 to 600; the Latin of early church fathers.

**Low L. Middle Latin.** The Latin of the Middle Ages, A.D. 600 to 1500; called **LOW LATIN** because it incorporated so many foreign word elements.

**NL. New Latin** or **Modern Latin.** The Latin of modern times, from A.D. 1500 to the present time. It includes a host of scientific names, especially in the field of biology. These names are made by casting many Greek words and modern words of various origins into Latin form.

**O.Ger. Old German** or **Old High German.** The German of the 8th to 12th Centuries.

**Ger.** The **New High German** or **Modern German,** from A.D. 1500 to the present.

**Low Ger. Low German.** The dialectic German of North Germany and The Low Countries.

**Middle German.** The **High German** of the middle or late 12th Century to the year A.D. 1500.

**AS. Anglo-Saxon,** from A.D. 450 to about A.D. 1200.

**ME. Middle English (Old English)** from A.D. 1200 to A.D. 1500.

**Eng. Modern English,** from A.D. 1500 to the present time.

**Fr. French,** "the language of the Franks"; one of the most highly developed of the Romance Languages; descended from the Latin. From A.D. 1500 to the present time.

**OFr. Old French,** from A.D. 800 to A.D. 1300.

**MFr. Middle French,** from A.D. 1300 to A.D. 1500.

A SOURCE-BOOK of
# BIOLOGICAL NAMES and TERMS

# To the Student

HERE ARE found many of the thousands of word elements or combining forms from which scientific names are made. After them are found the Greek, Latin or other names in which they have their origin; then follow the examples (generic names, specific names and technical terms, each dissected into its proper parts) to illustrate their use in the building of words. By combining these word elements in the numerous ways possible more than a million words have been formed and are now used in biological treatises.

In attempting to learn the literal meanings of words you may have in mind, first attempt to *find the actual word in heavy-faced type* in its alphabetical place. If you do not find it, it is quite possible that it is a compound and you will need to break it up into the elements of which it is composed. The next step is to seek for each of these elements in its proper place in the Sourcebook.

As you continue to study words in this manner, you will gradually become "word conscious" and find yourself taking new and intelligent delight in finding their meanings and the stems from which they came. It is one of the best ways to increase your vocabulary and in turn will lead to more rapid reading of technical literature.

Some of the most puzzling words are those which have at their beginning some of the simple prefixes such as *a-, an-, ba-, di-, dia-, e-, em-, en-, per-, pro-,* and which one might easily suspect as being inseparable parts of a longer element.

Other perplexing words are those in which a certain letter or letters is shared by the two elements of the compound. Examples of such words are: (1) Symplocarpus* (derived from Gr. *symplokos,* entwined and *karpos,* a fruit) in which the shared letter is *c,* (2) Ambonychia (Moll.) (from Gr. *ambōn,* a ridge and *onychos* of a claw, nail) in which the shared letters are *on,* (3) Aphidecta (Ins.) (from NL. *aphis,* genit. *aphidis,* an aphid and Gr. *dēktēs,* an eater) in which the shared letter is *d.*

Equally vexatious are those words in which some of the letters of one or both elements are omitted as in Arisaema* (from Gr. *aris,* arum and *haema,* blood) in which *h* is omitted from the beginning of the second element, or in Aglamyla* (from Gr. *aglama,* delight and *hylē,* wood, forest) in which *h* is wanting.

It is exceedingly important that you read the introductory section in which are stated the elementary principles of word-building.

Be most careful to follow up the cross references.

Classificatory names of plants serving as examples are indicated by an asterisk (*), animal names, by appropriate abbreviations, (Arth.) for Arthropoda, (Moll.) for Mollusca, etc.

# A

a- —1. Gr. *a-* (when placed before words of Greek origin beginning with a consonant), or *an-* (when placed before words of Greek origin beginning with a vowel) may (a) signify a negation or absence of something. *Ex:* a-biogenesis; A-cephala (Moll.); A-pogon*; An-ura (Amph.). (b) act as an intensive. *Ex:* A-blas (Av.); a-xylos, very woody; (c) have the force of *ama*, with, together with; or (d) have no significance, in the case of roots beginning with two consonants, or be merely euphonic as is A-splenium*. **2.** L. *a-* (when placed before roots beginning with a consonant), or *ab- = abs-* (when placed before roots beginning with a vowel) is used as a prefix to Latin words to mean from, without, away. *Ex:* A-diantum*; A-di-dactis (Ins.); a-phyll-ous; ab-errant; ab-oral; ab-ortive.

-a— -*a*, fem. ending for Latin first declension sing. nouns and fem. ending for 1st and 2nd declension adjectives: *Ex:* Cyparae-a*; Daedale-a*; Diomede-a*; also an ending for a considerable class of Greek neuters (ending in -*ma*); also some other Greek nouns (ending in -*a*).

**aalge**—Dan. *aalge*, the murre.

**aapt**—Gr. *aaptos*, unapproachable. *Ex:* Aaptus (Av.).

**abac**—Gr. *abax*, genit. *abakos*, an abacus, a board for arithmetical calculation, a trencher. *Ex:* Abaco-crinus (Echin.).

**abact**—L. *abactus*, driven away.

**abama**—Gr. *abama*, ancient name for the bog asphodel, perh. < *a*, not + *bama*, Doric for *bēma*, step, it being thought to produce lameness in cattle. *Ex:* Abama*.

**abastor**—NL. *abastor*, a coined name. *Ex:* Abastor (Rept.).

**abat**—LL. *abatus*, beaten down, lowly. *Ex:* Abatus (Ins.).

**abbreviat**—L. *abbreviatus*, shortened.

**abder**—Gr. *Abdēra*, a town of Thrace, its inhabitants accounted stupid, hence Abderite was a term of reproach. *Ex:* Abderites (Mam.); Abdero-spira (Moll.).

**abdit**—L. *abditus*, hidden, secret, removed < *abdo*, to put away.

**abdomen**—See abdomin.

**abdomin**—L. *abdomen*, genit. *abdominis*, the abdomen, paunch, perh. < *abdo*, to conceal. *Ex:* abdomen; Abdomin-ales (Pisc.).

**abelmosch**—NL. *abelmoschus* < Ar. *abu-al-misk*, father of musk. *Ex:* Abelmoschus*.

**abelter**—Gr. *abelteros*, silly, stupid. *Ex:* Abelterus (Av.).

**aberran**—L. *aberrans*, genit. *aberrantis*, wandering from the way, ppr. of *aberro*, to go astray; *aberratio*, genit. *aberrationis*, a diversion. *Ex:* aberration; aberrant.

**abies**—See abiet.

**abiet**—L. *abies*, genit. *abietis*, name of the fir-tree; *abietinus*, living on the fir. *Ex:* Abies*; abieti-col-ous; Abietin-aria (Coel.).

**abisp**—NL. *abispa* < Sp. *abispa*, wasp < L. *vespa*, a wasp, hornet. *Ex:* Abispa (Ins.).

**abit**—L. *abitus*, vanished, departed, pp. of *abeo*, to go away.

**abject**—See abjicien.

**abjicien**—L. *abjicio*, to throw away, to throw down; ppr. *abjiciens*, throwing down; pp. *abjectus*, cast down, rejected, low, mean, worthless.

**ablact**—L. *ablactus*, weaned; pp. of *ablacto*, to wean.

**ablat**—L. *ablatus*, removed, withdrawn < *aufero*, to bear away.

**ablaut**—Gr. *ablautos*, unslippered. *Ex:* Ablautatus (Ins.).

**-able**—L. -*abilis* > Eng. -*able*, suffix meaning tendency toward, able to be, that may be, worthy of, fit to be. *Ex:* cur-able; malle-able; reli-able.

**ablus**—L. *ablusus*, unlike < *abludo*, to disagree.

**ablut**—L. *ablutus*, cleansed < *abluo*, to wash.

**abobr**—Braz. *abobra*, native name of a gourd. *Ex:* Abobra*.

**abol**—Gr. *abolos*, a young horse that has not shed the foal teeth. *Ex:* Abolo-ceras (Mam.).

**aboma**—Guianan *aboma = boma*, name applied to large tree-dwelling snakes of tropical America.

**abnod**—L. *abnodo*, to clear of knots (of trees). *Ex:* abnod-ation.

**abnorm**—L. *abnormis*, irregular, abnormal.

**abort**—L. *abortivus*, born prematurely; *abortio* to miscarry. *Ex:* abort-ion.

**abr**—1. Gr. *habros*, soft, delicate, splendid. *Ex:* Abr-inus (Ins.): not Abra (Mam.) which comes from a native Tibetan name; Abro-gnathus (Ins.); Abro-stomus (Pisc.); Abronia*; Abrus*; An-abrus (Ins.); Habro-cytus (Ins.); Habro-thamnus*: **2.** Gr. *abrōs*, genit. *abrotos = nēstis*, fasting, causing hunger.

**abrad**—L. *abrado*, to shave; pp. *abrasus*, scraped away, shaved, peeled off; *abrasio*, genit. *abrasionis*, the act of abrading, wearing-off. *Ex:* abrasion.

abram—Gr. *abramis*, genit. *abramidos*, name of a sea-fish. *Ex:* Abramis (Pisc.).

abras—See abrad.

abrax—*abraxas*, mystical word formed from the Greek letters α, β, ρ, α, ξ, α, s. Gems with this word inscribed became charms. *Ex:* Abraxi-morpha (Ins.); Abraxas (Ins.).

abronia—NL. *abronia* < Gr. *habros*, graceful. *Ex:* Abronia*.

abrot—1. Gr. *abrōtos*, not fit to be eaten. *Ex.* Abroto-crinus (Echin.): 2. Gr. *abrotos*, immortal, divine: 3. Gr. *abrotēs*, delicacy, splendor.

abrupt—L. *abruptus*, abrupt, separated, disconnected.

absciss—L. *abscissio*, genit. *abscissionis*, the act of cutting-off, a breaking-off < *abscindo*, to tear off; *abscissus*, clipped. *Ex:* abscission.

abscondit—L. *absconditus*, secret, concealed.

absidat—L. *absidatus*, arched.

absinthi—L. *absinthium;* also *absinthius* < Gr. *apsinthion*, wormwood. *Ex:* absinthium.

absit—L. *absitus*, distant.

absterg—See absters.

absters—L. *abstersus*, wiped away, expelled, pp. of *abstergeo.* to wipe off, banish, expel.

abund—L. *abundo*, to overflow, to abound with plants, grow up with luxuriance. *Ex:* floribundus (the *a* omitted for the sake of euphony).

abuta—Tupi *abuta*, name of a plant from which is derived the violent poison, curare. *Ex:* Abuta*.

abutilon—Ar. *abūtīlūn*, one of the names of the mulberry which has leaves resembling those of malvaceous plants belonging to the modern genus Abutilon. *Ex:* Abutilon*.

abyss—Gr. *abyssos*, bottomless. *Ex:* abyss-al; Abyss-ascidia (Tun.); Abyssi-cola (Pisc.).

ac—1. Gr. *akē* also *akis*, genit. *akidos*, a point, needle, splinter > L. *acus*, dim. *acicula*, a point > *aculeus*, dim. *aculeolus*, a needle, point, prickle, sting; *acutus*, sharp, pointed; *aculeatus*, furnished with spines or prickles. *Ex:* Ac-odont-aster (Echin.); Aci-carpha*; aci-form; acicul-ar; acicul-ate; aciculum; Aco-diplosis (Ins.); Aco-mys (Mam.); aco-spore; Aculeata (Ins.), (Mam.); aculeate; aculeol-ate; acute; Acuto-mentum (Pisc.). See also ad: 2. Gr. *akos*, genit. *akeos*, cure. *Ex:* aceo-logy (Med.); aut-ac-oid; Pan-ax*.

aca—See -acus.

acac—Gr. *akakia*, the acacia < *ukis*, *akē*, a point, thorn. *Ex:* Acacia*.

acaen—Gr. *akaina*, a thorn, goad < *akē*, a thorn. *Ex:* Acaena*; An-acaena (Ins.); Acin-onyx (Mam.).

acajub—NL. *acajuba* < Portuguese *acaju*, the cashew. *Ex:* Acajuba*.

acaleph—Gr. *akalēphē*, a nettle. *Ex:* Acalepha (Coel.); Acalepha*.

acalypha—See acaleph.

acamant—Gr. *akamas*, genit. *akamantos*, untiring, unwearied.

acan—Gr. *akan*, genit. *akanos*, a prickly shrub, kind of thistle, briar, thorn < *akē*, a point and akin to *akis*, a thorn, prickle. *Ex:* acan-aceous; Acan-archus (Pisc.); Akania*; Acanus (Pisc.)

acanth—Gr. *akantha*, a thorn, prickle; *akanthikos*, of thorns; *akanthion*, a kind of thistle; also the porcupine, a hedgehog; *akanthinos*, thorny,

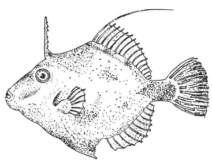

Hispid Single-spine Fish, *Monocanthus hispidus*, a small-scaled herbivorous Trigger fish. Redrawn from Fishes of North and Middle America—Jordan.

made of thorns. *Ex:* Acanth-id-ops (Av.); Acanthico-lepis (Ann.); Acanthion (Mam.); Acantho-phrynus (Arach.); Acanthus*; Gaster-acanthus (Arach.); Pyr-acantha*.

acanthis—Gr. *akanthis*, a bird fond of thistles, the linnet < *akantha*, a thorn. *Ex:* Acanthis (Av.).

acar—1. Gr. *akari* > L. *acarus*, a kind of mite. *Ex:* Acar-ina (Arth.); acar-in-osis; Acaro-talpa (Arach.); Acarus (Arach.): 2. Gr. *akarēs*, small, short, tiny. *Ex:* Acare-mys (Mam.); Acara (Pisc.).

acat—Gr. *akatos*, a ship. *Ex:* Acat-odes (Ins.).

acatalect—Gr. *akatalēktos*, without end, incessant; also not defective at the end. *Ex:* Acatalectus (Ins.).

acceden—L. *accedens*, genit. *accedentis*, resembling, ppr. of *accedo*, to resemble, approach.

accent—L. *accentor*, one who sings with another. *Ex:* Accentor (Av.).

accessor—ML. *accessorius* < L. *accessus*, moved toward. *Ex:* accessor-y.

accipit—L. *accipiter*, a bird of prey, a hawk; *accipitrinus*, hawk-like. *Ex:* Accipiter (Av.); accipitrine.

acclin—L. *acclino*, to lean against, to incline. *Ex:* acclini-dens.

accliv—L. *acclivis*, ascending, steep. *Ex:* acclivous.

accol—NL. *accolus*, dwelling near<L. *accola* a neighbor.

accult—L. *accultus*, approximated, made near; pp. of *accolo*, to dwell near.

accumb—L. *accumbo*, to lay one's self down, to recline at a table; ppr. *accumbens*, genit. *accumbentis*, reclining. *Ex:* accumbent.

-ace—Gr. *akē*, a point. *Ex:* Prion-ace (Ins.).

-aceae—L. *-aceae*, fem. pl. adj. suffix added to botanical generic names to indicate a family. *Ex:* Cact-aceae*; Orobanch-aceae*.

-aceous—See -aceus.

acer—1. L. *acer*, genit. *aceris*, the maple tree, wood of the maple tree. *Ex:* Acer*; aceri-folia, Rub-acer*: 2. L. *acer*, genit. *acris*, sharp, active, violent, ardent. See also cer.

acerb—L. *acerbus*, rough, tart.

acerv—L. *acervus*, dim. *acervulus*, a heap. *Ex:* acervul-ine; acervulus cerebri.

acestr—Gr. *akestra*, a darning needle. *Ex:* Acestra (Ins.).

acet—L. *acetum*, sour wine, wine-vinegar; also sagacity, keeness of mind. *Ex:* aceto-acetate; Aceto-bacter*.

acetabul—L. *acetabulum*, a vinegar cup. *Ex:* Acetabuli-fera (Moll.); acetabulum.

acetos—NL. *acetosa*, acid, full or sourness<L. *acetum*, vinegar. *Ex:* Acetos-ella.*

aceum—See -aceus.

-aceus—L. *-aceus*, adj. suffix meaning of or pertaining to. *Ex:* herb-aceus; junc-aceus; ros-aceus.

achaen—See achen.

achat—1. Gr. *achatēs*, agate. *Ex:* Achat-ina (Moll.); Achatia (Ins.); Achatin-ella (Moll.): 2. L. *Achates*, faithful friend of Aeneas.

acheloi—Gr. *Achelōis*, a mythological name. *Ex:* Achelois (Moll.).

achen—1. Gr. *achanēs*, not gaping<*a*, not+*chainō*, to gape. *Ex:* Achen-opsis (Ins.); achene; acheno-carp: 2. Gr. *achēn*, poor, needy.

acherd—Gr. *acherdos*, a wild prickly hedge-shrub. *Ex:* Acherdo-cerus (Ins.).

acheron—Gr. *Acherōn*, genit. *Acherontos*, a river of the underworld; *Acheronteios*, pertaining to the river Acheron<*achos*, a pain+*rhoos*, a stream. *Ex:* Acheront-emys (Rept.); Acherontia (Ins.).

achet—L. *acheta*=Gr. *ēchetēs*,=Dor. *achetēs*, the male cicada. *Ex:* Acheta (Ins.).

acheus—Gr. *Achaios*=L. *Achaeus*, a proper name of Greek fable. *Ex:* Acheus (Mam.).

achill—Gr. *Achilleus*, hero of Homer's "Iliad" who is said to have discovered the virtues of a certain plant. *Ex:* Achill-ides (Ins.); Achillea*.

achly—Gr. *achlys*, mist, dimness, darkness; *Achlys*, goddess of obscurity. *Ex:* Achlyo-geton*; Achlys*; Achlys-ictis (Mam.).

achlys—See achly.

achn—Gr. *achnē*, anything shaved off, froth, chaff. *Ex:* Hydr-achna (Ins.).

achor—Gr. *achōr*, genit. *achoros*, scurf, dandruff. *Ex:* Achoro-cephalus (Ins.).

achoreut—NL. *achoreutes*<Gr. *a*, priv.+*choreutēs*, a dancer. *Ex:* Achoreutes (Ins.).

achrad—Gr. *achras*, genit. *achrados*, a kind of wild pear. *Ex:* Achrado-crinus (Echin.); Achras*.

achras—See achrad.

achrest—Gr. *achrēstos*, useless, unprofitable. *Ex:* Achresto-coris (Ins.).

achrochord—Gr. *achrochordōn*, a wart. *Ex:* Achrochordo-pus (Av.).

achroo—Gr. *achroos*, without color. *Ex:* achroö-dextrin.

achrost—Gr. *achrōstos*, untouched, colorless. *Ex:* Achrostus (Ins.).

achth—Gr. *achthos*, a weight, burden; *achtheinos*, annoying; *achthērēs*, burdensome, troublesome. *Ex:* Achtheinus (Crust.); Achtheres (Arth.); Hom-achthes (Ins.).

achyr—Gr. *achyron*, chaff, husks. *Ex:* Achyr-a-chaena*; Achyr-odon (Mam.), "in the sense of pointed"; achryo-phytum; Amphi-achyris*; Phil-achyra (Ins.).

aci—See acies.

acicul—See ac.

acid—L. *acidus*, sour, tart.

acidal—L. *Acidalia*, a proper name. *Ex:* Acidalia (Ins.).

acidot—Gr. *akidōtos*, pointed<*akis*, genit. *akidos*, a point.

acies—L. *acies*, a sharp edge or point, mental acuity, sharpness of vision.

acin—L. *acinus*, a grape seed, a berry; *acinosus*, like grapes. *Ex:* acino-dendrus; Acino-pterus (Ins.); acinose; not Acin-onyx (Mam.) which is partly derived from Gr. *akaina*, thorn.

acinac—L. *acinaces*, a kind of short sword or scimitar<Gr. *akinakēs*. *Ex:* acinaci-folia.

acinet—Gr. *akinētos*, fixed, motionless. *Ex:* Acinet-actis (Moll.); Acinet-aria (Prot.).

-acious—Eng. *-acious*<L. *-ax*, genit. *-acis*+Eng. *-ous*, a suffix meaning tending to, abounding in. *Ex:* rap-acious.

acipenser—L. *acipenser*, the sturgeon, perhaps< *aci-* (<Gr. *ōkys*), swift+*penna*=*pinna*=OL. *pesna*, a wing, fin *Ex:* Acipenser (Pisc.).

acis—Gr. *akis*, a point. *Ex:* Lasi-acis*; Heter-akis (Rot.).

acm—Gr. *akmē*, a point, edge. *Ex:* Acmea (Moll.); Acme-pteron (Ins.); Acmo-pupa (Moll.).

acmae—Gr. *akmaios*, in full bloom, flourishing; also entire, mature, perfect. *Ex:* Acmae-idae (Moll.); Acmaea (Moll.); Acmaeo-dera (Ins.).

acmen—L. *Acmena*, one of the names of Venus. *Ex:* Acmena*.

**acmon**—1. Gr. *akmōn*, anvil. *Ex:* Acmono-rhynchus (Av.):    2. L. *Acmon*, companion of Aeneas.

**acoc**—Gr. *akokē*, a point, edge. *Ex:* Acoc-anthera\*; Sperma-acoce\*.

**acoet**—Gr. *akoitēs*, a bed-fellow, husband. *Ex:* Acoetes (Ann.); Acoetus (Crust.).

**acol**—Gr. *akolos*, a bit, morsel. *Ex:* Acol-oides (Ins.); Acolus (Ins.).

**acolast**—Gr. *akolastos*, intemperate, given to sensual pleasures. *Ex:* Acolastus (Ins.).

**acon**—1. Gr. *akonē*, whetstone. *Ex:* Akone-mys (Mam.):    2. Gr. *akōn*, dart. *Ex:* Adr-acon (Mam.).

**aconit**—Gr. *akoniton*, a kind of poisonous plant, monk's hood. *Ex:* Aconitum\*.

**acont**—Gr. *akontion*, a dart, javelin; *akontistēs*, a darter, javelin thrower; *acontias*, a quick-striking serpent. *Ex:* acontia; Acontias (Rept.); Acontio-stoma (Crust.); Aconto-dactylus (Ins.); Acontistes (Ins.).

**acor**—1. Gr. *akoros*, the sweet-flag. *Ex:* Acorus\*: 2. Gr. *akoros*, without ceasing. *Ex:* Acoro-tricha (Ins.).

**acoust**—Gr. *akoustos*, heard, audible; *akoustikos*, pertaining to hearing<*akouō*, to hear. *Ex:* acousticus.

**acqu**—See **aqu**.

**acquir**—L. *aquiro*, to acquire. *Ex:* acquir-ed.

**acr**—Gr. *akron*, extremity, summit, highest point; *akros*, at the end or edge. *Ex:* acr-andry; acro-gyn-ous; Acro-notus (Pisc.); Ep-acris\*, living upon hills; Mon-acrum (Mam.). See also acer 2.

**acrag**—L. *Acragas*, genit. *Acragantis*, a proper name. *Ex:* Acrag-idae (Ins.); Acraga (Ins.).

**acrant**—Gr. *akrantos*, fruitless, mutilated, idle. *Ex:* Acrant-ophis (Rept.); Acrantus (Ins.).

**acrasi**—Gr. *akrasia*, a bad mixture<*akratos* un-mixed. *Ex:* Acrasia\*.

**acredul**—L. *acredula*, name of some bird.

**acricul**—L. *acriculus*, somewhat sharp, pointed.

**acrid**—1. Gr. *akris*, genit. *akridos*, dim. *akridion*, a locust. *Ex:* Acris (Amph.); Acridi-idae (Ins.); Acridium (Ins.); Acrido-carpus\*; Pseud-acris (Amph.):    2. Gr. *akris*, genit. *akrios*, a hill-top, peak.

**acris**—See **acrid**.

**acrit**—Gr. *akritos* unarranged, undistinguish-able. *Ex:* Acrit-illus (Av.); Acrita; Acrito-cera (Ins.).

**acrobat**—Gr. *akrobatos*, going to the top. *Ex:* Acrobates (Mam.).

**acrochord**—Gr. *akrochordōn*, a wart. *Ex:* Achro chordus (Rept.).

**acrom**—Gr. *akrōmion* = *akrōmia*, the point of the shoulder blade<*akros*, at the top+*ōmos*, the shoulder with the upper arm. *Ex:* acromio-clavicular; acromion.

**acron**—Gr. *akron*, peak, extremity, highest pitch. *Ex:* Acroni-ops (Arach.); Acrono-lepia (Ins.); Meg-acronus (Ins.). See acr.

**acrot**—Gr. *akrotos*, unapplauded. *Ex:* Acrotus (Pisc.).

**acroton**—Gr. *akrotonos*, strained to the utmost, muscular. *Ex:* Acrotona (Ins.).

**acrul**—Gr. *akroulos*, curled at the end. *Ex:* Acrulia (Ins.): Acrulo-cercus (Av.).

**act**—1. Gr. *aktē*, a rocky promontory, sea shore; *aktios*, of the sea-shore. *Ex:* Acte-dium (Ins.), see idio; Acti-ornis (Av.); Acto-phila (Av.); Phil-acte (Av.):    2. Gr. *aktē*, corn, meal.

**actaea**—1. Gr. *aktaia*, erroneous form of *aktea*, contr. *aktē*, the elder tree. *Ex.* Actaea\*: 2. Gr. *aktaia*, a Persian robe.

**actaeon**—Gr. *Aktaiōn*, a huntsman, grandson of Cadmus, *Ex:* Actaeon (Moll.): Actaeon-ella (Moll.).

**actin**—Gr. *aktis*, genit. *aktinos*, a ray, beam. *Ex:* Act-idion (Ins.); Act-issa (Pisc.)\*; Actin-istia (Pisc.); Actin-ophrys (Prot.); Actini-aria (Coel.); Actino-zoa (Echin.); Actino-chir (Pisc.); Actino-glena (Prot.); Elo-actis (Coel.); Haren-actis (Coel.).

**actis**—See **actin**.

**actit**—Gr. *aktitēs*, a dweller of the sea coast< *aktē*, the sea shore. *Ex:* Actitis (Av.).

**actuos**—L. *actuosus*, full of activity, agile.

**acu**—L. *acus*, a needle. *Ex:* acu-puncture; Acu-palpus (Ins.).

**aculeat**—See **ac**.

**aculo**—See **ac**.

**acumen**—See **acumin**.

**acumin**—L. *acumen*, genit. *acuminis*, a point to sting with; *acuminatus*, taper-pointed, pp. of *acumino*, to sharpen. *Ex:* acumen; acumini-folius; acuminate.

**-acus**—L. -*acus*, suffix added to noun stems to form adjectives with the sense of belonging. *Ex:* auranti-acus.

**acut**—L. *acutus*, sharp, pointed. *Ex:* acuti-palpus. See ac.

**acyr**—Gr. *akyros*, without authority, incorrectly. *Ex:* Acyro-gonia (Moll.).

**ad-**—L. *ad*, towards> *ad-*, prefix changed to *ac-*, *af-*, *ag-*, etc., following the Latin rule that the final consonant of a prefix may be changed to the same letter which follows it. *Ex:* ac-cli-mate; ag-gregation; as-similation; an-nectens.

**-ad**—1. -*ad*<Gr. -*as*, genit. -*ador*, a suffix mean-ing daughter of. *Ex:* gae-ad, daughter of earth: 2. -*ad*, suffix of Greek origin used in the forma-tion of collective nouns. *Ex:* mon-ad; tri-ad; 3. NL. -*ad*, adverbial suffix<L. *ad*, to, to-ward. *Ex:* caud-ad; cephal-ad; dors-ad.

**adal**—Gr. *adēlētos* = Dor. *adalēs*, unhurt. *Ex:* Adalia (Ins.), or perhaps from Sp. *adalia*, chief of a troop.

**adamant**—L. *adamanteus*, hard as steel, iron; *adamantinus*, steel-blue.

**adap**—ML. *adapis*, name applied by Gesner to the common European rabbit, perh.<Gr. *a*, intensive+*dapis*, a rug. *Ex:* Adapi-sorex (Mam.); Adapi-soric-ulus (Mam.); Adapis (Mam.).

**adaptat**—L. *adaptatus*, fitted, adapted; Fr. *adaptation*, fitness. *Ex:* adaptation.

**adarc**—Gr. *adarkēs*, a matter gathering on plants in salt marshes, pickle, brine. *Ex:* Adarces (Coel.).

**addend**—L. *addendum*, added; gerund of *addo*, to add. *Ex:* addendum.

**ade**—Gr. *adeō*, to be sated; *adēn*, enough, abundantly. *Ex:* Ade-orbis (Moll.); Ade-phagus (Ins.). See aden.

**adel**—Gr. *adēlos*, concealed, not seen or known. *Ex:* Adel-ura (Ins.); Adela (Ins.); Adelia*; Adelo-chorda; adelo-morphic. Opl-adelus (Pisc.).

**adelg**—NL. *adelges*<Gr. *adelos*, hidden+*gē*, earth. *Ex:* Adelges (Ins.).

**adelph**—Gr. *adelphos*, a brother. *Ex:* Adelphe (Ins.); Adelpho-myia (Ins.).

**ademon**—Gr. *adēmōn*, sore, troubled. *Ex:* Ademon (Ins.); Ademonia (Ins.).

**aden**—Gr. *adēn* genit. *adenos*, a gland. *Ex:* Aden-anthera*; Aden-area*; aden-oid; adeniform; Adeno-stoma*; Cer-adia*; Zyg-adenus*. See ade.

**adeps**—See adip.

**aderc**—Gr. *aderkēs*, invisible, unexpected. *Ex:* Aderces (Ins.).

**-ades**—L. *-ades*, masculine patronymic ending. *Ex:* Aene-ades, son of Aeneas.

**adet**—Gr. *adetos*, free, unbound. *Ex:* adeto-pneustic.

**adhatod**—Singhalese or Tamil *adhatoda*, native name for a plant used to expel the dead fetus in abortion. *Ex:* Adhatoda*.

**adhes**—L. *adhesus*=*adhaesus*, adhesion.

**adia**—See aden.

**adiant**—Gr. *adiantos*, the maidenhair fern< *adiantos*, unwetted<*a*, priv.+*diantos*, capable of being wet. *Ex:* Adiantum*.

**adiaphor**—Gr. *adiaphoros*, pure, uncorrupted, not different. *Ex:* Di-adiaphorus (Mam.).

**adiastalt**—Gr. *adiastaltos*, ambiguous. *Ex:* Adia staltus (Mam.).

**adiastol**—Gr. *adiastolos*, not separated, mixed, joined. *Ex:* Adiastola (Ins.).

**adin**—Gr. *adinos*, crowded. *Ex:* Adina*; Adino-therium (Mam.).

**adip**—L. *adeps*, genit. *adipis*, the soft fat of animals, suet. *Ex:* adip-ose; adipo-cere.

**adipat**—L. *adipatus*, fat, greasy.

**adit**—L. *aditus*, an approach, entrance. *Ex:* adit-al; aditus.

**adject**—L. *adjectus*, placed near, increased <*adjicio*, to place near, to add to.

**adlumin**—L. *adlumino*, to fringe with purple; pp. *adluminus*, fringed with purple.

**admet**—Gr. *admētē*, fem. of *admētos*, not tamed, unwedded. *Ex:* Admete (Moll.); Admetus (Ins.).

**adminicul**—L. *adminiculum*, a prop, support. *Ex:* adminicula.

**adnasc**—L. *adnascor*(=*agnascor*=*adgnascor*); pp. *adnatus*, ppr. *adnascens*, genit. *adnascentis*, to be born in addition to, grow to<*ad*, to+ *nascor*=*gnascor*, to be born. *Ex:* adnascent; adnate.

**adnat**—See adnasc.

**adnex**—L. *adnexus*, bound or tied to, neut. pl. *adnexa*, pp. of *adnecto*, to tie, bind. *Ex:* adnexa.

**adocet**—Gr. *adokētos*, unexpected. *Ex:* Adoceta (Ins.).

**adol**—Gr. *adolos*, without fraud, genuine. *Ex:* Adolo-pus (Ins.); Adolus (Ins.).

**adolescen**—L. *adolesco*, to grow up; ppr. *adolescens*, genit. *adolescentis*, growing up; *adolescentia*, adolescence.

**adon**—Gr. *Adōnis*, beloved of Venus. He was slain by a wild boar and after death changed into a flower. *Ex:* Adonis*.

**adornat**—L. *adornatus*, equipped, prepared< *adorno*, to provide, furnish.

**adox**—Gr. *adoxos*, obscure, insignificant. *Ex:* Adox-ellus (Ins.); Adoxia (Ins.); Adoxo-phyes (Ins.).

**adr**—Gr. *hadros*, thick, swollen, bulky, strong, great. *Ex:* Adr-ac-odon (Mam.); Adri-cara (Ins.); Adro-therium (Mam.); see hadros.

**adran**—Gr. *adranēs*, inactive, feeble. *Ex:* Adranes (Ins.); Adrano-chelia (Arach.).

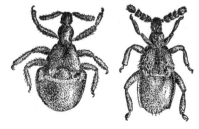

Leconte's Feeble Guest Beetle, *Adranes lecontei* (left). Testaceus Club-bearing Guest Beetle, *Claviger testaceus* (right). Redrawn from Wheeler, Ants, by permission of Columbia University Press.

**adrast**—1. Gr. *adrastos*, not running away, not inclined to run away. *Ex:* Adrastis (Arach.); Adrasto-therium (Mam.); Adrastus (Ins.): 2. L. *Adraste*, daughter of Jupiter. *Ex:* Adrast-aea*.

**adscenden**—L. *adscendens*, genit. *adscendentis*,

ascending, ppr. of *adscendo* (*ascendo*), to ascend.

adscit—L. *adscitus*, approved.

adspers—L. *adspersus*=*aspersus*, a sprinkling upon<*aspergo*, to scatter, sprinkle.

adsurg—L. *adsurgens*, ascending, standing up; ppr. of *adsurgo*, to rise up.

adult—L. *adultus*, grown up.

adumbrat—L. *adumbratus*, counterfeited, false.

adunat—L. *adunatus*, united.

adunc—L. *aduncus*, hooked>ML. *aduncatus*, hooked, see -atus.

adust—L. *adustio*, a burning<L. *aduro*, to scorch, pp. *adustus*, scorched.

adven—L. *advenio*, to come, to arrive; *advena*, a stranger, one who has reached a place; NL. *advenus*, newly arrived, adventive.

adyt—Gr. *adytos*, a place not to be entered, the innermost shrine. *Ex:* Adyt-aster (Echin.).

-aea—See -ea.

aechm—Gr. *aichmē*, the point of anything, a spear. *Ex:* Aechm-odus (Pisc.); Aechm-aea*; Aechmo-rhynchus (Av.).

aeci—Gr. *aikia*, injury. *Ex:* aecia, aecio-spores.

aedeag—NL. *aedeagus*<Gr. *aidoia*, the genitals +*agos*, leader. *Ex:* aedeagus.

aedes—Gr. *aēdēs*, disagreeable. *Ex:* Aedes (Ins.).

aedil—L. *aedilis*, a Roman magistrate of public works. *Ex:* Aedilis (Ins.).

aedoe—Gr. *aidoia*, the genitals. *Ex:* aedoe-tomy: aedoe-logy.

aedon—Gr. *aēdōn*, genit. *aēdonos*, a songstress, applied by Hesiod to the nightingale; *Aēdōn*, daughter of Pandareus, who was changed into a nightingale. *Ex:* Aedon (Av.); Aedon-opsis (Av.).

aeg—Gr. *aix*, genit. *aigos*, a goat. *Ex:* Aeg-idae (Ins.); Aega (Ins.); Aego-cerus (Mam.); Aex-toxicon*; Aigo-cerus (Mam.); Ego-cerus (Mam.).

aegagr—Gr. *aigagros*, the wild goat<*aix*, genit. *aigos*, a goat+*agrios*, wild. *Ex:* aegagro-pilae.

aegeir—See aegir.

aeger—1. L. *Aegeria*=*Egeria*, a nymph, said to have given to Numa Pompilius his laws. *Ex:* Aegeria (Ins.): 2. L. *aeger*, fem. *aegra*, weak, troubled.

aegial—Gr. *aigialos*, the seashore. *Ex:* Aegial-ornis (Av.); Aegialia (Ins.); Aegial-ites (Ins.), (Av.); Aigialo-saurus (Rept.).

aegialit—See aegial.

aegilips—Gr. *aigilips*, destitute even of goats, hence steep, sheer. *Ex:* Aegilips (Ins.); Acanth-aegilips (Ins.).

aegilop—Gr. *aigilops*, a disease of the eye frequently found in goats<*aix*, genit. *aigos*, a goat+*ops*, an eye; also a kind of wild oat; a kind of oak with edible acorns. *Ex:* aegilops (Med.); Aegilops*, (Moll.).

aegin—Gr. *Aigina*, nymph beloved of Zeus; also the name of an island. *Ex:* Aegin-opsis (Coel.); Aegin-idae (Coel.); Aegina (Coel.); Aegin-uva (Coel.).

aeginth—NL. *aeginthus*, erroneously formed< Gr. *aegithos*, a sparrow *Ex:* Zon-aeginthus (Av.); Spor-aeginthus (Av.).

aeginura—Gr. *Aiginura*, a proper name. *Ex:* Aeginura (Coel.).

aegipan—Gr. *Aigipan*, the goat-footed Pan. *Ex:* Aegipan (Ins.).

aegir—1. *Aigeiros*, a proper name. *Ex:* Aegires (Moll.): 2. Gr. *aigeiros*, the black poplar tree.

aegith—Gr. *aigithos*, a hedge sparrow. *Ex:* Aegitho-gnath-idae (Av.).

aegithal—Gr. *aigithalos*, a kind of bird, the tit. *Ex:* Aegithalus (Av.).

aegl—Gr. *aiglē*, radiance, brightness. *Ex:* Aegle (Crust.).

aegr—See aeger 2.

aegrot—L. *aegrotus*, weak, diseased.

aeidi—1. Gr. *aeidia*, deformity. *Ex:* Pro-edium (Mam.): 2. Gr. *aeidios*, everlasting.

aeletes—NL. *aeletes*, an insect name, etym. unknown. *Ex:* Aeletes (Ins.).

aell—Gr. *Aellō*, Storm-swift, one of the Harpies <*aella*, a stormy wind. *Ex:* Aello (Mam.).

aelur—Gr. *ailouros*, a cat. *Ex:* Aeluro-idae (Mam.); Aeluro-saurus (Rept.); Ailur-avis (Mam.); Aelurus (Mam.); Ict-aelurus (Pisc.).

aelus—See aeol.

aem—NL. *aem*<Gr. *haima*, blood. *Ex:* Andros-aemum*.

aemul—L. *aemulus*, emulating, rivalling.

aen—1. L. *aeneus*=*aenus*, of bronze or copper *Ex:* aenei-ceps; aeneo-micans; Chlor-aenos (Av.): 2. Gr. *ainos*, grim, terrible. *Ex:* Aeno-cyon (Mam.); ?Aeno-plex (Ins.): 3 Gr. *ainos*, a tale, proverb.

-aena—Gr. fem. suffix -*aina*. *Ex:* Hy-aena (Mam.); Oxy-aena (Mam.), for Amphisbaena, see baen.

aene—See aen.

aenea—L. *Aenea*=*Aneas*, son of Venus.

aenict—Gr. *ainiktos*, enigmatical, alluded to. *Ex:* Aenicto-soma (Ins.); Aenictus (Ins.).

aenigm—Gr. *ainigma*, genit. *ainigmatos*, an obscure question, a riddle. *Ex:* Aenigmato-cystis (Echin.).

aeno—See aen 2.

aeol—Gr. *aiolos*, quick moving, flexible, changeable, variegated; *aiollō*, to vary>*Aiolos*, god of the winds. *Ex:* aeoli-an; aelo-tropism; Aeol-agrion (Ins.); Aeolis (Moll.); Aeol-id-idae (Moll.); Aeolo-metris (Ins.); Aeolo-morphus (Ins.); Aeolo-soma (Ann.); Aelus (Ins.); Coryth-aeolus (Rept.).

aeore—Gr. *aiōreō=airō*, to hover, flit about. *Ex:* Aeores-tes (Mam.).

aep—See aepy.

aept—Gr. *aeptos*, not able to fly, unfledged, crippled, feeble. *Ex:* aepti-pod-ous.

aepy—Gr. *aipys*, high, difficult, steep; *aipos*, genit. *aipeos*, height. *Ex:* Aepy-ceros (Mam.); Aepy-ornith-idae (Av.); Aepeo-mys (Mam.); Aepi-gnatha (Arach.); Aepi-saurus (Rept.); Aepy-ornis (Av.).

aequabilis—L. *aequabilis*, equal, similar.

aequor—L. *aequoreus*, of the sea. *Ex:* Aequor-ella (Coel.); Aequore-idae (Coel.); Aequorea (Coel.).

aer—1. Gr. *aēr*, genit. *aeros*, the air, atmosphere; *aerios*, pertaining to the air, in the air, lofty. *Ex:* Aer-ides*; aero-bic, see bi; Aero-nautes (Av.): 2. Gr. *airō*, to lift up. *Ex:* Pyg-aera (Ins.).

aere—L. *aĕreus* also *airius*, pertaining to the air, airy, aerial<Gr. *aēr*, air; Gr. *aerios*, high in air, also in the air.

aereum—L. *aereum*, copper color.

aereus—L. *aereus*, made of copper, furnished or covered with copper or bronze.

aeri—See aere.

aerope—Gr. *Aeropē*, wife of Atreus. *Ex:* Aerope (Arach.).

aert—Gr. *aertazō*, to lift up. *Ex:* Cerc-aertus (Mam.).

aerug—L. *aerugo*, genit. *aeruginis*, copper rust, verdigris. *Ex:* Aeruga (Ins.).

aesal—Gr. *aisalōn*, a small kind of hawk. *Ex:* Aesalon (Av.); Aesalus (Ins.).

aeschn—See aeschr.

aeschr—Gr. *aischros*, ugly, deformed>NL. *aeshna* and *aeschna*, perh. errors for *aeschra*. *Ex:* Aeschna (Ins.); Aeshna-soma (Ins.); Aeschro-cnemis (Ins.); Aeschro-domus (Moll.).

aeschyn—Gr. *aischynē*, shame. *Ex:* Aeschyn-anthus*.

aeschynomen—Gr. *aischynomenē*, a sensitive plant, the mimosa<*aischynē*, shame. *Ex:* Aeschynomene*.

aescul—L. *aesculus*, a kind of oak. *Ex:* Aesculus*.

aeshn—See aeschr.

aesi—Gr. *aisios*, lucky, correct. *Ex:* Aesio-copa (Ins.).

aesio—Gr. *aisiōs*, auspicious, fortunate. *Ex:* Aesio-crinus (Echin.).

aest—L. *aestas*, summer, the hot season; *aestivus*, pertaining to summer; *aestivo*, to spend the summer; related to *aestus*, fire and to *aestuosus*, burning with heat, glowing<*aestuo*, to burn, glow, be inflamed. *Ex:* aestiv-ation.

aesthem—Gr. *aisthēma*, genit. *aesthēmatos*, sensation. *Ex:* aesthemato-logy.

aesthesi—Gr. *aisthēsis*, sensation. *Ex:* aesthesio-graphy.

aesthet—Gr. *aisthēsis*, sense, perception; *aesthētikos*, sagacious, sensitive, perceptive. *Ex:* aesthesis; an-aesthetic.

aestiv—See aest.

aestu—L. *aestuans*, genit. *aestuantis*, burning, tossing with rage, ppr. of *aestuo*, to rage, toss; *aestuosus*, full of fury.

aestuar—L. *aestuarium*, a tidal estuary.

aesur—Gr. *aēsuros*, light as air, nimble. *Ex:* Aesurus (Mam.).

aesylacr—Gr. *Aesylacris*, a proper name. *Ex:* Aesylacris (Ins.).

aet—1. Gr. *aetos*=Ionic *aeitos*, an eagle; also a kind of ray>NL. *aitotis*, of the eagles. *Ex:* Aet-opsis (Av.); Aeto-batis (Elasm.); Aeto-phorus (Ins.); Gyp-aetus (Av.): 2. Gr. *aētēs*, a blast, breeze; *aētos*, stormy, furious.

aeterus—NL. *aeterus*<Gr. *oietēs*, equality. *Ex:* Phil-aeterus (Av.).

aeth—1. Gr. *aēthēs*, irregular, curious, unusual. *Ex:* Aeth-urus (Mam.); Aethe-cerus (Ins.); atheo-gamous; Aethio-merus (Ins.); Aetho-lepis (Pisc.): 2. Gr. *aithos*, burnt; also of a red-brown color, dusky; *aithō*, to scorch; *aithalos*, smoke, soot; *aithinos*, burning. *Ex:* Aethal-orus (Ins.); Aethalion (Pisc.); Aeth-alium*; Aethalo-ptera; Aethin-odes (Ins.); Aethio-nema*; Aetho-chroi (Ethn.); Aetho-prora (Pisc.); Aethusa*: 3. Gr. *aithya*, a sea bird of some kind. *Ex:* Aethia (Av.).

aethal—See aeth 2.

aether—Gr. *aithēr*, the upper air, sky, heaven; *aitherios*, on high, ethereal pertaining to the upper air, heavenly. *Ex:* aethereal=ethereal; Aetheria (Moll.).

aethin—See aeth 2.

aethio—Gr. *aithiōps*, an Ethiopian, negro. *Ex:* Aethiops (Mam.).

aethr—Gr. *aithrios*, clear, calm, in the open air<*aithrē*, the open sky<*aithēr*, air, the sky. *Ex:* Aithria-manta (Ins.).

aeti—Gr. *aitia*, a cause; *aitios*, causing, originating. *Ex:* aetio-logy=etio-logy;

aetr—NL. *aetro*<Gr. *ētron*, the abdomen. *Ex:* Aetro-cantha (Arach.), Gr. *akantha* something sharp.

aex—See aeg.

af—See ad.

afer—L. *Afer*, African. *Ex:* Afer-ulus (Moll.); Afr-odonta (Moll.); Afro-mysis (Crust.).

affin—L. *affinis*, related, adjacent<*ad*, to+*finis* border, end; *affinitas*, alliance, relationship by marriage.

afr—See afer.

ag-—See ad-, also aga-.

ag—1. *agō*, to bring on, to carry, to induce. *Ex:* Agu-stylus (Mam.); Gal-ago*: 2. Gr. *agē*, Dor. *aga*, jealousy, hate; also wonder, awe. *Ex:* Aga-thaumas (Rept.): 3. Gr. *agos*, chief.

*Ex:* Ago-seris*: **4.** Gr. *agos,* veneration, respect: **5.** Gr. *agē,* the beach, seashore. *Ex:* ag-ad (Ecol.): **6.** Gr. *agē,* a fragment, splinter: **7.** Gr. *agē,* a curving, bending.

**aga-** —*aga-,* an intensive prefix < Gr. *agan,* very much, too much. *Ex:* Aga-belus (Mam.); Agorophius (Mam.); Aga-phelus (Mam.); Agapet-idae (Ins.); Aga-sphaer-ops (Ins.). See also agaon.

**agae**—Gr. *agaios,* elegant, neat, admirable. *Ex:* Agaeo-cera (Ins.); Agaeus (Ins.); Pan-agaeus (Ins.).

**agall**—Gr. *agallō,* to adorn, make glorious. *Ex:* Agalli-opsis (Ins.); Agallia (Ins.); An-agallis*.

**agalm**—Gr. *agalma,* genit. *agalmatos,* a delight, an ornament, a pleasing gift. *Ex:* Agalm-yla*, see hyl; Agalmato-saurus (Rept.).

**agam**—**1.** Dutch Guianan *agama,* name of a lizard. *Ex:* Agam-idae (Rept.); Agama (Rept.); not A-gama (Moll.), nor A-gamae*, nor a-gamous, see gam: **2.** Galibi (Guianan) *agamy,* native name of a bird. *Ex:* Agami (Av.).

**agan**—**1.** Gr. *aganos,* gentle, agreeable. *Ex:* Aganites (Moll.); Agan-osma*; Agano-crossus (Ins.). See also aga: **2.** Gr. *agan,* very much.

**agaon**—Gr. *agaō,* to admire, ppr. *agaon,* beloved *Ex:* Agaon (Ins.); Agaon-ella (Ins.).

**agap**—Gr. *agapē,* love; *agapētos,* beloved. *Ex:* Agap-anthus*; Agape (Ins.); Agapet-idae (Ins.); Agapet-ornis (Av.); Agapetes*; Agapetus (Ins.); Agapo-stemon (Ins.).

**agapet**—See agap.

**agaric**—Gr. *agarikon,* a mushroom. *Ex:* agaric; Agarico-chara (Ins.); Agaricus.*

**agarist**—L. *Agarista,* beautiful daughter of Clisthenes. *Ex:* Agarista*.

**agast**—Gr. *agastos,* admirable, wonderful. *Ex:* Agastachys* < *agastos,* admirable + *stachys,* an herb like horehound; or perhaps < *aga,* many + *stachys;* Agasto-cerus (Ins.).

**agasthen**—Gr. *agasthenēs,* powerful. *Ex:* Agasthenes (Ins.); Agastheno-poda (Ins.).

**agath**—**1.** Gr. *agathos,* excellent, well, good; *agathotēs,* goodness. *Ex:* Agath-osma*; Agathea*; Agathi-stega (Prot.); Agatho-phyllum*; Agathotes*; Phil-agathes (Ins.): **2.** Gr. *agathis,* a knot, a ball of thread, a round head. *Ex:* Agath-engis (Ins.); Agath-idium (Ins.); Agathis*, (Ins.); Agathis-anthes*.

**agav**—Gr. *Agauē,* a proper name, meaning the noble, illustrious one. *Ex:* Agave*.

**agchist**—Gr. *agchistos,* properly *anchistos,* nearest; *agchisteus* properly *anchisteus,* next of kin. *Ex:* Agchisteus (Verm.). See anchist.

**agdest**—L. *Agdestis,* an hermaphrodite, descendant from Jove and the Agde rock. *Ex:* Agdestis*.

**agel**—Gr. *agelē,* a herd; *agelaios,* gregarious, feeding at large; *agelastikos,* living in herds, disposed to live together. *Ex:* Agel-opsis (Ins.); Agela-discus (Echin.); Agelaius (Av.); Agelastes (Av.); Agelastica (Ins.); Agele-cyathus (Coel.).

**agelae**—Gr. *agelaios,* gregarious. *Ex:* Agelaeus (Av.).

**agen**—Gr. *ageneios,* beardless, young. *Ex:* Ageneo-tettix (Ins.); Agenia (Ins.).

**agenor**—**1.** Gr. *agenor,* noble, manly, splendid. *Ex:* Agenor (Ins.), (Pisc.): **2.** L. *Agenor,* father of Cadmus and Europa.

**ageratum**—See ger 2.

**ageroch**—Gr. *ageröchos,* noble, haughty.

**agged**—Gr. *angeidion,* a small vessel, erroneously transliterated *aggeidion. Ex:* agged-ula.

**aggest**—L. *aggestus,* an accumulation; also a small elevation, a mound.

**agglutin**—L. *agglutino,* to glue, to fasten with aid of glue. *Ex:* agglutin-ation.

**aggregat**—L. *aggregatus,* assembled, brought together in flocks < *ad,* toward + *grego,* to herd together.

**agi**—See hagi.

**agil**—L. *agilis,* easy to move, agile, busy. *Ex:* agili-ty.

**agitat**—L. *agitatus,* stirred, up, moved, disturbed.

**agkistr**—Gr. *agkistron (ankistron),* a fish hook, a bait, an enticement. *Ex:* Agkistr-odon = Ancistr-odon (Rept.). See ancistr.

**agl**—Gr. *aglaos,* splendid, bright, majestic. *Ex:* Agl-antha (Coel.); Agl-aspis (Arth.); Aglao-stigma (Ins.); Aglao-zonia*; Aglo-phema (Mam.).

**aglai**—Gr. *aglaia,* splendor, beauty > *Aglaia,* one of the three graces. *Ex:* Aglaia*; Aglaio-cerus (Av.).

**aglao**—See agl.

**aglaophem**—Gr. *aglaophēmos,* of splendid fame; *Aglaophēmē,* one of the sirens. *Ex:* Aglaophemia (Coel.) = Aglaophenia (Coel.).

**aglaopheni**—NL. *Aglaophenia,* erroneously made from Gr. *Aglaopheme,* a goddess, one of the sirens, *Ex:* Aglaophenia (Coel.).

**aglaur**—Gr. *Aglauros,* a mythological name < *aglauros,* splendid. *Ex:* Algaura (Ann.), (Ins.), (Coel.).

**agn**—**1.** Gr. *agnos,* pure, innocent; *agnisma,* genit. *agnismatos,* a cleansing; *agnotēs,* purity, chastity; *agnitēs,* a purifier, also one who needs cleansing. *Ex:* Agnito-gaster (Ins.); Agno-pterus (Av.); Agnoto-myia (Ins.); Elae-agnus*: **2.** Gr. *agnōs,* unknown, uncertain. *Ex:* Agno-cyon (Mam.).

**agnat**—L. *agnatus,* connected by birth, blood-related.

**agnin**—L. *agninus,* pertaining to a lamb, fleecy.

**agnism**—See agn.

agnit—See agn 1.

agnost—Gr. *agnōstos*, unknown, unheard of. *Ex:* Agnostus (Brach.).

agnot—See agn.

-ago—NL. *-ago*, botanical suffix, usually signifying resemblance with the word that precedes it. *Ex:* Gith-ago*. See also ag 1.

agog—Gr. *agōgē*, a carrying or leading away; *agōgos*, leading, guiding < *agō*, to carry. *Ex:* Agogo-blatt-ina (Ins.); chlor-agogen; chol-agogue (Med.).

agon—1. Gr. *agōn*, genit *agōnos*, a gathering, assembly (not to be confused with *agon*, from *a*, priv.+*gōnia*, an angle). *Ex:* Agonio-neurus (Ins.); Agono-phorus (Ins.); not Siphagonus (Pisc.), see gon: 2. Gr. *agōn*, a struggle, contest.

agoseris—See ag 3.

agost—Gr. *agostos*, the hollow of the hand; also the elbow, hence, an angle, a recess. *Ex:* Agosto-pus (Mam.).

agr—1. L. *ager*, genit. *agri*, a field; *agrestis*, pertaining to land, rural; *agrarius*, pertaining to a field < Gr. *agros*, a field. *Ex:* agri-culture; agrest-ic; Agrion (Ins.); Agro-pyron*; Agrotis (Ins.): 2. Gr. *agrios*, wild, savage, living in the fields; *agrotēs*, a rustic; *agriotēs*, wildness. *Ex:* agria (Med.); Agri-opis (Ins.), *agrios*, wild+NL. *opis*, face, countenance < Gr. *ōps*; Agriotes (Ins.).

agra—1. Gr. *agra*, a catching, a hunting for prey or food; also the prey itself. *Ex:* Agra (Ins.); Agra-philydrus (Ins.); Ther-agra (Pisc.): 2. It. *agra*, rough. *Ex:* pell-agra (Med.).

agraul—Gr. *agraulos*, living in the fields. *Ex:* Agraulis (Ins.); Agraulos (Tri.).

agrost—See agr 1.

agreu—Gr. *agreuō*, to catch, ensnare > *agreutēr* and *agreutēs*, a hunter. *Ex:* Agreuo-coris (Ins.); Agreuter (Ins.); Agreutes (Av.).

agril—NL. *agrilus* < Gr. *agrios*, living in the country. *Ex:* Agrilus (Ins.).

agriot—See agr 2.

agroec—Gr. *agroikos*, living on the land, dwelling in the country, uncouth. *Ex:* Agroeca (Arach.); Agroeco-tettix (Ins.); Mes-agroecus (Ins.).

agron—NL. *agronus* < Gr. *a-*, without+*grōnē*, a cavern, cavity or *grōnos*, cavernous.

agrost—1. Gr. *agrōstis*, a kind of grass < *agros*, a field. *Ex:* Agrostis*; Agrosto-bia (Ins.); agrosto-logy; Calam-agrostis*: 2. Gr. *agrōstēs*, a hunter, a rustic. *Ex:* Agrost-ichthys (Pisc.).

agrot—See agr 2.

agrypn—Gr. *agrypnos*, watchful, incessant. *Ex:* Agrypn-ella (Ins.); Agrypnus (Ins.).

agyrt—Gr. *agyrtēs*, a begger, an imposter. *Ex:* Agyrt-idia (Ins.); Agyrtes (Ins.); Agyrto-myia (Ins.).

ai—1. Gr. *aei*, ever, always, forever. *Ex:* ai-anthous; Ai-zoon*: 2. Gr. *aiōn*, a period of existence, an age, generation. *Ex:* aeon; ai-phyllium.

aig—See aeg.

aigial—Gr. *aigialos*, the seashore; also a cliff. *Ex:* aigialo-phyta; Aigialo-saurus (Rept.). See also aegial.

ailant—Moluccan *ai lanit* < Malay *kayu langit*, native name for the "tree of heaven." *Ex:* Ailantus = Ailanthus*.

ailin—Gr. *ailinos*, a plaintive dirge.

ailur—See aelur.

aim—Gr. *aima*, genit. *aimatos*, blood; also courage. *Ex:* Aimo-phila (Av.).

ain—See aen.

-aina—Gr. *-aina*, a fem. suffix, see lept, also -aena.

aiol—See aeol.

aip—See aepy.

aipt—Gr. *aeptos*, one not able to fly, a fledgling. *Ex:* Aipt-asia (Coel.), the last element of unknown meaning.

aipys—Gr. *aipys*, hanging straight down, sheer, mighty. *Ex:* Aipys-urus (Rept.).

aira—Gr. *aira*, a hammer (*airō*, to destroy) also an injurious plant in wheat. *Ex:* Aira; Aira-philus (Ins.).

aire—Gr. *aireō*, to grasp, overpower. *Ex:* Aireo-graptus (Coel.).

aist—Gr. *aistos*, unseen. *Ex:* Aisto-poda (Amph.).

aith—Gr. *aithō*, to light up, to glow; *aithousa*, burning, glowing, fem. of *aithōn*, ppr of *aithō*. *Ex:* Aeth-urus (Av.); Aeth-usa (Av.); Aethio-nema*; Aith-urus (Av.). See also aeth.

aithal—Gr. *aeithalos*, ever green. *Ex:* aithal-ium (Ecol.).

aithr—See aethr.

aix—Gr. *aix*, a water bird. *Ex:* Aix (Av.). See also aeg.

aizoon—L. *aizoon* < Gr. *aeizoon*, a plant called house-leek.

ajaj—S. Amer. *ajaja*, native name of a bird, the roseate spoonbill. *Ex:* Ajaja = Ajaia (Av.).

akan—See acan.

akeb—Jap. *akebi*, name of a climbing plant. *Ex:* Akebia*.

akis—See acis.

-al—Eng. *-al*, suffix meaning, when compounded with a noun, relating to or belonging to, as in juven-al, person-al; when compounded with a verb it means the act of, as in renew-al.

al—L. *ala*, pl. *alae*, dim. *alula*, a wing; *alatus*, winged. *Ex:* al-ar; Al-aria*; alaeo-blast; alate; Ali-gera*; ali-sphen-oid; alula; Eury-alae (Echin.).

alacer—See alacr.

alacr—L. *alacer*, genit. *alacris*, active, eager, joyful.

alactag—Mongolian Tartar *alactaga*, a spotted colt. *Ex:* Alactaga (Mam.).

alaeo—See al.

alalung—Sardinian *alalunga*, meaning long wing and applied to the long-finned albacore.

alao—Gr. *alaos*, blind, short-sighted, obscure *Ex:* Alao-morphus (Ins.); Alao-tanais (Crust.).

alastor—Gr. *Alastōr*, Zeus, the unforgetting avenging one < *alastōr*, a punisher, an avenger. *Ex:* Alastor (Mam.), (Ins.).

alat—See al.

alathet—Gr. *alathētos*, conspicuous. *Ex:* Alathetus (Ins.).

alaud—L. *alauda*, the lark; NL. *alaudinus*, larklike. *Ex:* Alauda (Av.); Saxi-lauda (Av.).

alaus—NL. *alaus*, name applied to a genus of click beetles < Gr. *a-*, priv.+*laō*, to look at eagerly (with a view to seizing) or perhaps < *alaomai*, to wander. *Ex:* Alaus (Ins.). See also alos.

alax—NL. *alaxa*, Alaska. *Ex:* alax-ensis.

alb—L. *albus*, dim. *albulus*, white; *albineus*, white; *albico*, to make white; ppr. *albicans*, genit. *albicantis*, making white; *albesco*, to become white; ppr. *albescens*, genit. *albescentis*, becoming white. *Ex:* albi-gula; Albulina (Pisc.); Albula (Pisc.).

albat—L. *albatus*, whitened, clothed in white.

albatros—Pg. *alcatraz*, a sea-fowl, cormorant, albatross, pelican; Fr. *albatros*, Ger. *albatross*, Eng. *albatross*, etc. *Ex:* Albatros (Av.); Albatross-aster (Echin.); Albatrus (Av.).

albid—L. *albidus*, white.

albug—L. *albugo*, genit. *albuginis*, a white spot, also a disease of the eye characterized by whiteness of the cornea; *albuginosus*, whitespotted. *Ex:* albugo (Med.); Albugo*; tunica albuginea.

albul—L. *albulus*, whitish.

albumen—See albumin.

albumin—L. *albumen*, genit. *albuminis*, white of egg. *Ex:* albumen; albumin-ous.

alburn—1. L. *alburnum*, sap-wood. *Ex:* alburn-ous: 2. L. *alburnus*, a white fish: 3. L. *Alburnus*, a mountain of Lucania.

alc—Icel. *alka*, the auk. *Ex:* Alc-idae (Av.); Alca (Av.). See also alce.

alcae—Gr. *alkaios*, strong, noble, generous. *Ex:* Alcaeo-rrhynchus (Ins.).

alcali—Fr. *alcali* < Ar. *al-qaliy* (*al*, the+*qaliy*, ashes of the salt wort). *Ex:* alcalio-trop-ism.

alce—1. Gr. *alkē*, the elk. *Ex:* Alc-elaphus (Mam.); Alces (Mam.); Alci-cephalus (Mam.); Cere-alces (Ins.): 2. Gr. *alkē*, prowess,

boldness: 3. Gr. *alkea*, the mallow. *Ex:* Alce-coris (Ins.); alcei-folia; Sphaer-alcea*.

alced—L. *alcedo*=*halcedo*, the kingfisher. *Ex:* Alcedo (Av.). See alcyon.

alchem—MGr. *chēmeia*=*chēmia*, alchemy < Ar. *alkīmīa*, alchemy > *alkemelyeh*, name of a kind of plant. *Ex:* Alchemilla*.

alci—See alce 1.

alcim—Gr. *alkimos*, strong, brave. *Ex:* Alcim-ochthes (Arach.); Alcima (Ins.); Alcimo-sphenus (Arach.); Alcimus (Ins.).

alcipp—Gr. *Alkippē*, daughter of Ares. *Ex:* Alcipp-ornis (Av.); Alcippe (Av.); Alcippus (Ins.).

alcyon—1. Gr. *alkyōn*, the kingfisher > L. *alcyon* = *halycon*, the kingfisher. *Ex:* Alcyon-otus (Ins.); Jacamar-alcyon (Av.): 2. Gr. *alkyonion*, name of a kind of sponge, so called from its resemblance to the nest of the kingfisher < *alkyōn*, the kingfisher. *Ex:* Alcyon-aria (Coel.); Alcyon-ella (Moll.); Alcyon-idium (Bry.); Alcyoni-cola (Crust.); Alcyonio-lithes (Por.); Alcyonium (Coel.).

aleator—L. *aleator*, a gamester; *aleatorius*, pertaining to a gamester. *Ex:* Aleator (Av.).

alec—See halec.

alect—1. Gr. *alektryōn*, poet. *alektōr*, a cock. *Ex:* Alector-urus*; Alectrion (Moll.); Alectoro-podes (Arth.); Alectro-morpha (Av.); Alectr-urus (Av.); Alectryo-pelia (Av.); Alectryon*: 2. Gr. *Alēktō*, one of the three Furies. *Ex:* Alect-ops (Mam.); Alecto (Echin.): 3. Gr. *alēktos*, unceasing.

alectr—See alect.

alegein—See alegin.

alegin—Gr. *alegeinos*, mournful, annoying. *Ex:* Alegina (Ins.); Alegeino-saurus (Amph.).

aleo—1. Gr. *aleos*, hot, warm; *alea*, temperate, lying open to the sun, warm, hot. *Ex:* Aleo-chara (Ins.); Aleo-drilus (Ann.): 2. Gr. *aleos*, foolish.

-ales—L. *-alis*, pl. *-ales*; fem. suffix used in forming plant order names. *Ex:* Pin-ales; Ros-ales, etc. See also -alis.

alesc—L. *alesco*, to grow up; also to increase; ppr. *alescens*, genit. *alescentis*, growing up. *Ex:* co-alesce; co-alesc-ence.

alet—1. Gr. *aletos*, a grinding; *aletēs*, a grinder. *Ex:* Aletes*, (Moll.), or perh. it is from the following: 2. Gr. *alētēs*, a wanderer, vagabond. *Ex:* Alet-ornis (Av.); aleto-phytes.

aleth—Gr. *alēthēs*, real, authentic. *Ex:* Alethe (Av.); Alethe-saurus (Rept.); Aletho-pteris*.

aletris—Gr. *aletris*, a female slave who grinds corn. *Ex:* Aletris*, because of the mealy pubescence.

aleur—Gr. *aleuron*, wheaten flour; *aleurōdēs*, like flour. *Ex:* Aleur-acanthus (Ins.); Aleur-ites*; Aleuro-daphnis (Ins.); Aleurodi-phagus (Ins.); Aleyrodes (Ins.); aleurone.

**alex**—1. Gr. *alexō*, to ward off, protect; *alexētēr*, one who keeps off or defends; *alexētērion*, a protection. *Ex:* Alex-urus (Pisc.); Alexeter (Ins.); Alexeterion (Ins.); alexin: 2. NL. *alexia* < Gr. a- not +*lexis*, a speaking. *Ex:* alexia; Alexia (Ins.).

**aleyr**—See **aleur.**

**alg**—1. L. *alga*, pl. *algae*, a sea-weed. *Ex:* algology; algo-phagus: 2. L. *algeo*, to be cold; *algidus*, cold. *Ex:* Algidus (Arach.): 3. Gr. *algos*, pain; *algēsis*, sense of pain. *Ex:* algesi-receptors; an-algesic.

**alges**—See **alg 3.**

**algid**—See **alg 2.**

**alhag**—Mauretanian *alhag*, the camel's thorn. *Ex:* Alhagi*.

**ali**—NL. *ali* < Gr. *halys*, the sea. *Ex:* Ali-gena (Moll.).

**alia**—1. L. *alius*, fem. *alia*, another, other. *Ex:* Alia (Moll.); Alia-typus (Arach.): 2. Gr. *alia*, a gathering, collection *Ex:* Alia-nta (Ins.), the last element without meaning; Bass-alia (Zoo-geo.).

**alibil**—L. *alibilis*, nutritive.

**alien**—L. *alienus*, belonging to another person, strange, unsuitable, hostile. *Ex:* alien; alienate; alieni-cola.

**alieus**—See **alieut.**

**alieut**—Gr. *alieutēs* also *alieus*, a fisherman, a seaman. *Ex:* Di-loph-alieus (Av.); Zal-ieute (Pisc.).

**alima**—Gr. *halimos*, pertaining to the sea. *Ex:* alima; Alima (Crust.).

**aliment**—L. *alimentum*, food, nourishment < *alo*, to nourish. *Ex:* aliment-ary.

**alimon**—L. *alimonia* also *alimonium*, nourishment.

**alind**—Gr. *alindō*, to roll, to roam far and wide. *Ex:* Alind(r)ia (Ins.).

**aliquot**—L. *aliquot*, some, in numbers.

**alis**—1. Gr. *alis*, in heaps, abundant. *Ex:* Alisodon (Pisc.): 2. Gr. *aleison*, cup, goblet. *Ex:* Aliso-crinus (Echin.).

**-alis**—L. *-alis*, suffix added to noun stems to form adjectives meaning pertaining to. *Ex:* radic-alis; farin-alis; fontin-alis.

**alism**—Gr. *halisma*, salt loving > Gr. *halisma*, a kind of water plant. *Ex:* alism-ellus; Alisma*; alismae-folius.

**alkann**—Sp. *alcana* < Ar. *al kennā*, henna. *Ex:* Alkanna*.

**all**—Gr. *allos*, other, another, different, strange. *Ex:* All-ac-odon (Mann.); All-odon (Mann.); Allo-carya*; Allo-lobo-phora (Ann.); Allo-plexus*; not Allo-saurus (Rept.), see hall; Allo-teuthis (Moll.).

**allach**—Gr. *allachē*; elsewhere.

**allactag**—Mongol *alak-daagha*, a variegated colt. *Ex:* Allactaga (Mam.).

**allag**—Gr. *allagē*, a change, an exchange, alteration. *Ex:* Allage-crinus (Echin.); allagostemon.

**allagm**—Gr. *allagma*, genit. *allagmatos*, an exchange. *Ex:* En-allagma*.

**allant**—Gr. *allas*, genit. *allantos*, a sausage. *Ex:* allanto-id; allanto-is; Allanto-nema (Nem.); Allanto-soma (Prot.); Allasia*.

**allas**—See **allant.**

**allass**—Gr. *allassō*, to change. *Ex:* Allasso-stomata (Platy.); allasso-tonic.

**allax**—Gr. *allax*, crosswise. *Ex:* Allaxi-theca (Ins.).

**allelo**—Gr. *allēlōn*, of one another, mutually, each other, in turn. *Ex:* allelo-morph; allelo-tropic.

**allex**—See **hallux.**

**alli**—See **allium.**

**alligat**—L. *alligatus*, bound; pp. of *alligo*, to bind, attach; NL. *alligator*, one who binds, a slave-maker. See also next entry.

**alligator**—Sp. *el lagarte*, "the lizard" < L. *ille*, that +*lacertus*, a lizard. *Ex:* Alligator (Rept.); Alligator-ellus (Rept.).

**allium**—L. *allium* = *alium*, garlic. *Ex:* alli-aceous; Alli-aria*; Allium*; not Allionia*, named after Charles Allioni, Italian botanist.

**allo**—See **all**, also **hall.**

**allocot**—Gr. *allokotos*, unusual, irregular. *Ex:* Allocot-ops (Av.); Allocota (Ins.); Allocoto-cerus (Ins.).

**allodap**—Gr. *allodapos*, belonging to another people, a stranger. *Ex:* Allodape (Ins.); Allodapo-suchus (Rept.).

**alloe**—See **alloi.**

**alloi**—Gr. *alloios*, of another sort, different; *alloioō*, to alter. *Ex:* Alloca (Ins.); Alloeo-carpa (Tun.); Alloeo-cnemis (Ins.); alloi-genesis; Alloio-rhabdus (Crust.).

**allom**—Gr. *allomai*, to leap, to bound upwards. *Ex:* An-allomes (Ins.).

**alloprosall**—Gr. *alloprosallos*, fickle. *Ex:* Allo porsallo-crinus (Echin.).

**allotri**—Gr. *allotrios*, belonging to another, foreign, unsuitable. *Ex:* allotrio-morphic; Allo trius (Ins.).

**alluv**—L. *alluvius*, neut. *alluvium* > NL. *alluvium*, a sedimentary deposit resulting from the action of a stream or river < *alluvio*, genit. *alluvionis*, an overflowing < *alluo*, to flow upon. *Ex:* alluvi-al; alluvius.

**alm**—L. *almus*, nourishing, bountiful.

**aln**—L. *alnus*, the alder. *Ex:* alni-folia; Alni-phagus (Ins.); Alnus*.

**aloc**—See **aulac.**

**alocasia**—NL. *alocasia,* an alteration of *Colocasia* a genus of aroid plants. *Ex:* Alocasia*.

**aloch**—Gr. *alochos,* a wife. *Ex:* Heter-alocha (Av.).

**aloe**—Gr. *aloē,* name of a plant, the aloe. *Ex:* Aloe*; Aloe-xylon*; Hesper-aloe*.

**alog**—L. *alogus,* irrational.

**alop**—See **alopec.**

**alopec**—**1.** Gr. *alōpēx,* genit. *alōpekos,* a fox; *alōpos,* a fox; also, fox-like. *Ex:* Al-opsis (Mam.); Allopec-urus*; Alopo-notus (Rept.): **2.** Gr. *alōpekia,* a disease like the mange of foxes, a falling of the hair < *alōpēx,* a fox. *Ex:* alopecia (Med.): **3.** Gr. *alōpēx,* a kind of shark, the fox-shark < *alōpēx,* a fox; *alōpekias,* the thresher shark. *Ex:* Alopias (Elasm.), shortened from Alopecias (Elasm.); Alopi-idae (Elasm.); Alopecias (Elasm.).

**alos**—L. *alosa = alausa,* name of a fish, the shad. *Ex:* Alausa (Pisc.); Alosa (Pisc.); Alos-ina (Pisc.); alos-oides.

**alot**—Gr. *alōtos,* easy to be taken, accessible, subdued. *Ex:* Alota (Ins.). Dicty-alotus (Ins.).

**alouatt**—Guianan *alouata,* name for the red howling monkey. *Ex:* Alouatta (Mam.).

**alpae**—L. *alpaeus,* living in or of the Alps. *Ex:* Alpaei-xena (Ins.).

**alpestr**—ML. *alpestris,* pertaining to the Alps: *Ex:* alpestre, neut. of alpestris.

**alph**—Gr. *alphos,* farinose, white as with meal. *Ex:* Alphus (Ins.).

**alphe**—NL. *Alpheus* < Gr. *Alpheios,* a river in Greece. *Ex:* Alphe-idae (Crust.); Alpheus (Crust.).

**alphest**—Gr. *alphēstēs,* a finder, inventor, a leader; also a kind of fish that went in pairs. *Ex:* Alphestes (Pisc.).

**alphit**—Gr. *alphiton,* barley-meal, also any other meal or groats. *Ex:* Alphito-bius (Ins.); Alphito-phagus (Ins.).

**alpigen**—NL. *alpigenus,* alpine.

**alpin**—L. *Alpinus,* alpine.

**als**—**1.** Gr. *alsos,* a grove; *alsōdēs,* like a woodland. *Ex:* Als-eu-osmia*; also-cola; Also-phila (Ins.): **2.** L. *alsus = alsius,* cold, chilly.

**alsin**—Gr. *alsinē,* a kind of plant, perh. the chickweed. *Ex:* Alsine.*

**alt**—L. *altus,* high. *Ex:* Alti-camelus (Mam.); Alti-cola (Mam.).

**altern**—**1.** L. *alterno,* to change, vary, alternate. *Ex:* Altern-anthera*; Altern-aria*; Altern-echinus (Echin.); Alterno-cidaris (Echin.): **2.** L. *alternus,* every other. *Ex:* alterni-petalous: alterni-pinnate.

**alternan**—L. *alternans,* genit. *alternantis,* interchanging, alternating < *alterno,* to change.

**althae**—Gr. *althainō,* to heal. *Ex:* Althaea*.

**altilis**—L. *altilis,* fat, full, fattened.

**altric**—L. *altrix,* genit. *altricis,* pl. *altrices,* a nour-

isher. *Ex:* atrici-al; artrices.

**aluc**—L. *alucus = ulucus,* a screech owl. *Ex:* Aluco (Av.).

**alucit**—L. *alucita,* a gnat. *Ex:* Alucita (Ins.).

**alul**—See **al.**

**alumn**—L. *alumnus,* strong, brought-up, wellnourished, flourishing < *alo* to rear, nourish.

**alut**—L. *aluta,* leather; LL. *alutacius,* pertaining to soft leather. *Ex:* Alut-era (Pisc.); Aluta (Crust.).

**alutari**—L. *alutarius,* made of leather.

**alv**—L. *alvus,* the womb, the belly. *Ex:* alv-ary; alv-arius; alvus. See also alveol.

**alvear**—L. *alvearium,* a beehive or something resembling one < *alveus,* a hollow. *Ex:* alvear-y; alvearium.

**alveat**—L. *alveatus,* hollowed out like a trough.

**alveol**—L. *alveolus,* a pit, small hollow; dim. of *alveus,* a hollow. *Ex:* alveol-ar; alveolo-labial; alveus.

**alyct**—Gr. *alyktos,* to be shunned. *Ex:* Alyctus (Ins.).

**alypi**—Gr. *alypia,* freedom from grief or pain. *Ex:* Alypia (Ins.).

**alys**—**1.** Gr. *alysis,* a chain, fetter. *Ex:* Alys-elminthes (Platy.); Alysia (Pisc.); not paralysis, see lys: **2.** Gr. *alysis,* distress, anguish.

**alysc**—Gr. *alyskō,* to escape, to be uneasy, to wander. *Ex:* Alyscum (Prot.).

**alyss**—Gr. *alysson,* name of a kind of plant, perh. < *a,* not + *lysson,* madness, or < *a,* not + *lyzō,* to hiccup. *Ex:* alyss-oides; Alyssum*.

**alyt**—**1.** Gr. *alytēs,* a police officer, one who attends a magistrate. *Ex:* not Alytes (Amph.), see 2: **2.** *alytos,* continuous, firm. *Ex:* Aly to-pistis (Ins.); Alytes (Amph.), "in allusion to the connected egg mass the animal carries about."

**alyx**—Gr. *alyxis,* an escape. *Ex:* Alyxia*.

**ama**—Gr. *ama,* together. *Ex:* ama-sthenic; amalthea (althea).

**amabil**—L. *amabilis,* neut. *amabile,* lovely.

**amal**—Gr. *amalos,* soft, weak, tender. *Ex:* Amalo-rrynchus (Ins.); Amalus (Ins.).

**amalth**—Gr. *Amaltheia,* the goat that suckled Zeus. *Ex:* Amalthea (Moll.).

**aman**—L. *amans,* genit. *amantis,* loving, ppr. of *amo,* to love. *Ex:* salic-amans.

**amanit**—Gr. *amanitai,* a kind of fungus. *Ex:* Amanit-opsis*; Amanita*.

**amar**—**1.** Gr. *amara,* a trench, conduit; also the hollow of the ear. *Ex:* Amara-carpus*; Amaro-dytes (Ins.); Amaro-rhynchus (Mam.). **2.** NL. *amara,* name applied to a genus of hemipterous insects, perh. < Gr. *a-,* priv. + *mairō,* to shine. *Ex:* Amara (Ins.). **3.** L. *amarus,* bitter.

**amarac**—Gr. *amarakos,* name of dittany of Crete. *Ex:* Amaracus*.

amaranth—Gr. *amarantos*, amaranth, the unfading. *Ex:* Amaranthus*.

amart—Gr. *amartanō*, to fail; *amartia*, a failure. *Ex:* Amartus (Ins.).

amarygm—Gr. *amarygma*, genit. *amarygmatos*, a sparkle, twinke. *Ex:* Amarygmus (Ins.).

amaryll—Gr. *Amaryllis*, genit. *Amaryllidos*, name of a shepherdess < *amaryssō*, to sparkle, twinkle. *Ex:* Amaryllid-aceae*; Amaryllis*.

amath—Gr. *amathos*, sandy soil; *amathitis*, dwelling in the sand. *Ex:* Amathitis (Ins.); amatho-phyta.

amathi—Gr. *amathia*, a state of ignorance, stupidity; *amathēs*, stupid, unlettered. *Ex:* Amathia (Coel.).

amathus—L. *Amathusia*, a proper name. *Ex:* Amathusia (Ins.).

amaur—Gr. *amauros*, dark, obscure, mean. *Ex:* Amaur-onia (Ins.); Amaur-ornis (Av.); Am aura (Moll.); Amauro-bius (Arach.); Amauro limnas (Av.).

amb—1. L. *ambi-* (other variants are *ambe-, amb-, am-, an-*), inseparable prefix meaning around, round about, on both sides. *Ex:* ambi-dextrous: 2. L. *ambo*, both. *Ex:* ambo-ceptor.

ambe—Gr. *ambōn* = Ionic *ambē*, a ridge, rim. *Ex:* Gaur-ambe (Ins.).

ambes—L. *ambesus*, wasted, gnawed.

ambien—L. *ambiens*, genit. *ambientis*, going round about, ppr. of *ambio*, to surround, go round about. *Ex:* ambient.

ambigen—L. *ambigens*, genit. *ambigentis*, wandering, hesitating, ppr. of *ambigo*, to go around.

ambigu—L. *ambiguus*, doubtful. *Ex:* ambiguiflorus.

ambit—L. *ambitus*, a going round < *ambio*, to encircle. *Ex:* ambitus.

ambl—Gr. *amblys*, blunt, stupid. *Ex:* Ambliramphus (Av.); Amblo-blattus (Ins.); amblyopia; Ambly-opsis (Pisc.); Ambly-teles (Ins.).

ambleo—NL *ambleo-* < Gr. *ambloō*, to be abortive. *Ex:* ambleo-carpus.

amblot—Gr. *amblōsis*, abortion; *amblōtikos*, fit to produce abortion. *Ex:* Amblotis (Mam.).

ambly—See ambl.

ambo—See ambon; also amb.

ambon—Gr. *ambōn*, the ridge or crest of a hill, a raised edge, rim of a cup. *Ex:* ambo; ambon; Ambon-ychia (Moll.), see onych; Ambonostola (Ins.).

ambros—Gr. *ambrosia*, fabled food of the gods conferring immortality; *ambrotos*, immortal, divine. *Ex:* Ambrosia*; Ambrot-odes (Ins.).

ambrot—See ambros.

ambul—L. *ambulo*, to walk; *ambulacrum*, a covered way, an alley, a walk planted with trees; *ambulatorius*, movable, suitable for walking. *Ex:* ambulacr-al; ambulatory.

ambulacr—See ambul.

amby—Gr. *ambyx*, genit. *ambykos*, a cup, the rounded top of a cup. *Ex:* Amby-stoma (Amph.), sometimes erroneously Amblystoma; Cer-ambyx (Ins.).

ambyx—See amby.

ameiv—Abor. *ameiva*, name of a kind of lizard. *Ex:* Ameiv-idae (Rept.); Ameiva (Rept.).

amel—1. Gr. *amelēs* neglected; *amelētēs*, not worthy of attention. *Ex:* Ameles (Ins.); Amelo-ctnonus (Ins.); Amelet-ia*; Ameletus (Ins.): 2. OFr. *amel*, enamel. *Ex:* ameloblast.

amelanchier—Savoy *amelanchier*, name of the medlar-tree. *Ex:* Amelanchier*.

amelet—See amel.

amell—NL. *amelloides*, amellus-like < L.*amellus*, the Italian starwort. *Ex:* amelloides; Amellus*.

ament—L. *amentum*, a strap. *Ex:* ament; amentaceous.

ametr—Gr. *ametros* and *ametrētos*, without measure, immense. *Ex:* Ametro-somus (Av.).

ametrid—Gr. *amētris*, genit. *amētridos*, a reaper, destroyer. *Ex:* Ametrida (Mam.); Ametris (Ins.).

ametris—See ametrid.

ami—Gr. *amia*, name of a kind of fish, the perch, the scomber. *Ex:* Ami-idae (Pisc.); Amia (Pisc.). See also ham.

amiant—L. *amiantus*, asbestos < Gr. *amiantos*, pure and without spot.

amic—L. *amicus*, friendly, kind.

amict—L. *amictus*, wrapped up.

amicul—L. *amiculus*, a dear friend.

amin—1. Gr. *ameinōn*, better, superior. *Ex:* Amino-ornis (Av.): 2. NL. *amino* < L. *Ammon*. *Ex:* amino-acid, see ammon.

-amine—NL. *-amine* (< *ammonia+ine*), name of a group of chemical compounds derived from ammonia, see ammon. *Ex:* vit-amine, later vitamin. See vit.

amiss—1. L. *amissus*, omitted, dismissed < *amitto*, to let go, dismiss: 2. L. *amissus*, a loss.

amm—1. Gr. *amma*, genit. *ammatos*, a knot. *Ex:* Ammato-campa (Ins.); Sten-amma (Ins.): 2. Gr. *ammos*, sand. *Ex:* Ammo-spermo-philus (Mam.); Ammo-selinum*: 3. Gr. *ammi*, an African plant. *Ex:* Ammi*.

ammat—See amm.

ammon—Gr. *Ammōn*, the Egyptian ram-like deity < Egypt. *Amen* = *Amun*, the sun god > NL. *ammonites*, name applied to a genus of extinct cephalopods, so called from their resemblance to a ram's horn or horn of Ammon; *ammōniakon*, sal ammoniac (ammonia salts) first prepared from camel's dung near the Temple of Ammon; also gum ammonias, taken from an umbelliferous plant. *Ex:* ammon-oid; Ammonea (Moll.); Ammoni-crinus (Echin.); ammonia; Ammono-cerina (Moll.); Ammonit-ella (Moll.); Ammonites (Moll.).

**amn**—1. Gr. *amnos*, a lamb>dim. *amnion*, the membrane around the fetus; also the bowl for catching the blood of sacrificial victims>NL. *Amniota*, vertebrates developing an amnion. *Ex:* amnion; Amniota; Ore-amnos (Mam.): 2. L. *amnis*, a stream. *Ex:* Amni-genia (Moll.).

**amnic**—L. *amnicus*, pertaining to a river.

**amnicol**—L. *amnicola*, that dwells by a river. *Ex:* Amnicola (Moll.).

**amoeb**—Gr. *amoibē*, a change, alternation. *Ex:* amoeb-oid; Amoeba (Prot.); End-amoeba (Prot.).

**amoen**—L. *amoenus*, pleasing, lovely. *Ex:* Amoena (Ins.); in-amoenus.

**amom**—1. Gr. *amōmon*, an aromatic shrub from which the Romans made a fragrant balsam; *amōmis*, a plant like the amomum. *Ex:* Amomis\*; Amomum\*: 2. Gr. *amōmos*, blameless.

**amomph**—NL. *amomphus* from a supposed Gr. *amomphos*, blameless, erroneously derived< Gr. *amōmos*, blameless. *Ex:* Amomphus (Ins.).

**amor**—Gr. *amoros*=*amoiros*, incomplete, unlucky, wretched. *Ex:* Amor-oecium (Tun.); Amoro-myza (Av.).

**amorph**—Gr. *amorphos*, without form. *Ex:* Amorpha\*; Amorpho-chilus (Mam.).

**amot**—L. *amotus*, withdrawn, banished<*amoveo*, to banish.

**amped**—Gr. *ampedaō*=*anapedaō*, to spring up. *Ex:* Ampedus (Ins.).

**ampel**—1. Gr. *ampelos*, a vine, the grape vine. *Ex:* Ampel-oeca (Ins.); Ampel-opsis\*; Ampelo-glypter (Ins.): 2. Gr. *ampelis*=*ampelion*, a kind of singing bird which frequents vines <*ampelos*, a vine. *Ex:* Ampel-idae (Av.); Ampeli-ceps (Av.); Ampelio (Av.); Ampelis (Av.).

**ampelin**—L. *ampelinus*, of the vine.

**amph**—Gr. *amphi*=*amphis*, both sides of, double; also apart, asunder, round about; when used to denote relationship it may mean doubtful, ambiguous. as in Amph-ictis (Mam.). *Ex:* amph-anthium; Amphi-oxus (Cephal.); Amphi-spiza (Av.); Amphio-cnida (Echin.); Amphis-baena (Rept.); for Amphiuma, see amphium.

**amphac**— Gr. *amphakēs*, two edged, ambiguous. *Ex:* Amphaces (Ins.).

**ampheke**—Gr. *amphēkēs*, double-edged, ambiguous. *Ex:* Ampheke-pubis (Rept.).

**amphelikt**—Gr. *ampheliktos*, coiled around. *Ex:* Amphelicto-gon (Myr.); Amphelictus (Ins.).

**ampherist**—Gr. *amphēristos*, doubtful, disputed. *Ex:* Ampheristo-crinus (Echin.).

**amphibi**—Gr. *amphibios*, leading a double life. *Ex:* Amphibia; Amphibio-philus (Nemat.).

**amphibol**—Gr. *amphibolos*, doubtful, ambiguous. *Ex:* Amphibola (Moll.); Amphibolo-thrips (Ins.).

**amphicneph**—Gr. *amphiknephēs*, dark all around. *Ex:* Amphicnephes (Ins.).

**amphidoz**—NL. *amphidoz*< Gr. *amphidoxos*, doubtful. *Ex:* Amphidozo-therium (Mam.).

**amphigy**—Gr. *amphigyos*, double-pointed, pointed at each end. *Ex:* Amphigyus (Ins.).

**amphis**—See **amph**.

**amphisbaen**—Gr. *amphisbaina*, a kind of serpent. *Ex:* Amphisbaena (Rept.).

**amphisil**—NL. *amphisile*, name for a fish genus, perh.<Gr. *amphis*, around+an ending of uncertain meaning. *Ex:* Amphisil-idae (Pisc.); Amphisile (Pisc).

**amphitrit**—Gr. *Amphitritē*, a sea nymph, wife of Poseidon. *Ex:* Amphitrite (Ann.).

**amphium**—NL. *amphiuma*, name applied to a genus of amphibia, probably <Gr. *amphi*, on both sides+*pneuma*, breath. *Ex:* Amphium-idae (Amph.); Amphiuma (Amph.).

**ampho**—Gr. *amphō*, both. *Ex:* ampho-genic.

**amphor**—Gr. *amphoreus*=L. *amphora*, a two-eared pitcher, flask. *Ex:* Amphor-ella (Prot.); Amphora-cystis (Echin.); Amphoro-phora (Ins.).

**amphoter**—Gr. *amphoteros*, each, both. *Ex:* Amphotero-cotyle (Platy.); Amphoterus (Ins.).

**ampl.**—L. *amplio*, to make large or ample; *amplus*, large; *ampliatus*, made larger, wider. *Ex:* Ampli-cephalus (Ins.); Amplo-rhinus (Rept.).

**amplect**—L. *amplectens*, genit. *amplectentis*, encircling, ppr. of *amplecto*, to wind around, encircle.

**amplex**—L. *amplexus*, an embracing, encircling. *Ex:* amplexi-caulis; Amplexis (Moll.).

**ampliat**—See **ampl**.

**amplificat**—L. *amplificatio*, an enlarging. *Ex:* amplificat-ion.

**ampull**—L. *ampulla*, a jug or flask; LL. *ampullosus*, turgid; *ampullaceus*, big bellied, flask-shaped. *Ex:* ampullace-al; Ampull-aria (Moll.); ampull-ous; ampulli-form.

**ampyx**—Gr. *ampyx*, a head band; anything circular. *Ex:* Ampyx (Arth.); ampyx.

**amudr**—See **amydr**.

**amus**—1. Gr. *amousos*, unpolished, without elegance. *Ex:* Amusus (Ins.). 2. Gr. *amousia*, lack of harmony. *Ex:* amusia.

**amyc**—L. *Amycus*, son of Neptune. *Ex:* Amycus (Ins.).

**amych**—Gr. *amychē*, a scratch.

**amyd**—L. *amyda*, a turtle. *Ex:* Amyda (Rept.).

**amydr**—Gr. *amydros*, indistinct, indistinctly marked, dusky. *Ex:* Amudr-ogmus (Ins.); Amydro-cerus (Ins.); Amydro-neura (Ins.); Amydrus (Av.).

**amygdal**—Gr. *amygdalē*, the almond; *amygdalos*, the almond tree. *Ex:* amygdal-oid; amygdali-form; Amygdalo-ptera (Ins.); Amygdalus*.

**amyl**—Gr. *amylos*, neut. *amylon*, starch, any fine meal. *Ex:* amyl-ase; amylo-clastic.

**amymon**—Gr. *amymōn*, blameless, noble; *Amym-ōnē*, a mythological name. *Ex:* Amymona (Ann.); Amymone (Crust.).

**amyn**—Gr. *amynō*, to aid, defend; *amyna*, defense. *Ex:* Amyn-odon (Mam.); Not-amynus (Mam.).

**amyris**—NL. *amyris* < Gr. *a*, not + *myron*, balsam. *Ex:* Amyris*.

**amyst**—1. Gr. *amystis*, a deep drinking; also a large cup. *Ex:* Amystes (Rept.): 2. Gr. *amystos*, profane.

**amyt**—Gr. *Amytis*, daughter of Astyages. *Ex:* Amyt-ornis (Av.).

**amythestin**—NL. *amythestinus*, violet-colored.

**amythet**—Gr. *amythētos*, very many or great, wonderful. *Ex:* Amythetes (Ins.).

**amyx**—Gr. *amyxis*, a tearing, scratching, cut, wound. *Ex:* Amyx-odon (Mam.).

**an-** —Gr. *an-* (*a-*, before a consonant), inseparable negative prefix meaning not, without; like Eng. *-un* or L. *-in*. See also a- and anus. *Ex:* an-acro-gyn-ous; an-aero-bios-is; an-aerob-ic; An-ota (Rept.); a-pais.

**-an** —Eng. *-an*, suffix ( < L. *-anus*) meaning one who, when it forms a noun, and belonging to, when it forms an adjective or an adjective used as a substantive. *Ex:* proboscide-an, crustace-an.

**an**—L. *anus*, an old woman. *Ex:* an-ile. See also anus.

**ana-** —Gr. *ana-*, prefix meaning up, upon, throughout, back, again, similar to. *Ex:* ana-bol-ism; Ana-camptis*; Ana-cardium*; Ana-cyrtus (Av.); ana-phase; ana-tomy. See also -anus.

**anabant**—See **anabas**.

**anabas**—Gr. *anabas*, genit. *anabantos*, gone up, 2nd. aorist part. of *anabainō*, to go up. *Ex:* Anabant-idae (Pisc.); Anabas (Pisc.).

**anabi**—Gr. *anabiōsis*, recovery of life. *Ex:* anabiosis.

**anacampser**—Gr. *anakampserōs*, an herb, a kind of sedum the touch of which was said to bring back love < *anakamptō*, to return + *erōs*, love. *Ex:* Anacampseros*.

**anacampt**—See **anacampser**.

**anache**—Gr. *anacheō*, to spread abroad, to pour forth. *Ex:* Anaches (Ins.).

**anachores**—Gr. *anachōrēsis*, a going back.

**anacol**—Gr. *anakōlos*, small, short, defective. *Ex:* Anacolo-blatta (Ins.).

**anact**—Gr. *anax*, genit. *anaktos*, a chief, leader; *anaktoria*, rule, sway; *anaktorios*, regal. *Ex:* Anactoria (Av.); Anax-onchium (Nemat.).

**anaere**—Gr. *anaireō*, to take away, to abolish, to refute. *Ex:* Anaerea (Ins.).

**anaeret**—NL. *anaeretes* < Gr. *anairetikos*, destructive < *anaireō*, to abolish. *Ex:* Anaeretes (Ins.).

**anaesthet**—Gr. *anaisthētos*, stupid, without sense; *anaisthēsia*, insensibility. *Ex:* anaesthet-ic.

**anagall**—Gr. *anagallis*, a kind of plant, sea-purslane, also pimpernel. *Ex:* anagall-oides; Anagallis*.

**analc**—Gr. *analkēs* = *analkis*, genit. *analkidos*, feeble, slothful. *Ex:* Analci-morphus (Mam.); Analcis (Rept.).

**analeps**—Gr. *analēpsis*, a recovery; *analēptikos*, restorative. *Ex:* analepsis.

**analys**—Gr. *analysis*, a releasing, a dissolution.

**anamps**—NL. *anampsis* = *anampses*, name applied to a genus of fishes < Gr. *anakampsis*, a bending back. *Ex:* Anampsis = Anampses (Pisc.).

**anan**—Tupi *ananas* = *anass* = *nanas*, native name of the pineapple. *Ex:* Ananas*.

**ananch**—NL. *ananchytes*, etym. unknown, name applied to a genus of echinoderms. *Ex:* Anancho-thuria (Echin.); Ananchytes (Echin.).

**anapaus**—Gr. *anapausa* also *anapausis*, rest, repose. *Ex:* Anapausa (Ins.).

**anaped**—See **amped**.

**anapet**—Gr. *anapetēs*, expanded, with parts wide open. *Ex:* Anapetes (Ins.).

**anaph**—Gr. *anaphēs*, not to be touched, insipid. *Ex:* Anaphes (Ins.); Anapho-thrips (Ins.).

**anaphalis**—NL. *anaphalis*, name of some Composite plant, perh. an anagram of Gnaphalium. *Ex:* Anaphalis*.

**anapt**—Gr. *anaptō*, to fasten, hang. *Ex:* Anapto-mecus (Arach.); Anapto-pora (Bry.).

**anarmost**—Gr. *anarmostos*, unsuitable, disproportionate. *Ex:* Anarmosto-dera (Ins.); Anarmostus (Ins.).

**anarrhich**—Gr. *anarrhichaomai*, to clamber up, scramble up. *Ex:* Anarrhichas (Pisc.).

**anarsi**—Gr. *anarsios*, hostile, unfit. *Ex:* Anarsia (Ins.).

**anas**—See **anat**.

**anass**—Gr. *anassa*, a lady, mistress; also a queen, fem. of *anax*, master, ruler. *Ex:* Hydr-anassa (Av.); Nyct-anassa (Av.).

**anastat**—1. Gr. *anastatos*, uprooted, overturned < *anastasis*, erection, uprising. *Ex:* Anastatica; Anastatus (Ins.): 2. Gr. *anastatēr*, a destroyer.

anastomos—Gr. *anastomōsis*, an opening; also a bringing to a point, a coming together, a union of one sea with another <*ana*, again +*stoma*, mouth; *anastomoō*, to furnish with a mouth. *Ex:* anastomosis; Anastomus (Av.), (Pisc.).

anat—L. *anas*, genit. *anatis*, a duck; *anatinus*, pertaining to a duck. *Ex:* Anas (Av.); Anatidae (Av.); Anatin-acea (Moll.); Anatina (Moll.).

anath—Gr. *anatheō*, to run up. *Ex:* Anath-itus (Mam.), the last element strangely derived from Gr. *this*, sand.

anathrot—NL. *anathrotus* < Gr. *anathrōskō*, to spring up, rebound, to awake. *Ex:* Anathrotus (Ins.).

anatol—Gr. *anatolē*, a rising, a growing, the dawn, the east > L. *Anatola*, a mythological name. *Ex:* Anatola (Moll.); Anatolo-mys (Mam.).

anaud—Gr. *anaudos* = *anaudēs*, dumb, unheard of. *Ex:* Anaudus (Ins.).

anax—See anact.

anc—1. Gr. *ankōn*, also *ankos*, a bend or hollow, an angle. *Ex:* Anc-odon (Mam.); Anco-coelus (Arach.); Ancon; ancon-eus; An-anconia (Ins.); An-ancus (Mam.): 2. Gr. *ankos*, a valley; also a crag.

anceps—L. *anceps*, two-headed, twofold; also doubtful, dangerous.

anch—1. Gr. *anchi*, near. *Ex:* Anchi-saurus (Rept.); Anchi-therium = Angchi-therium (Mam.); Ancho-gnatha (Arach.): 2. Gr. *anchō*, to choke, strangle, lace up. *Ex:* Anchisomus (Pisc.); Ancho-desmus (Ins.); Cynanchum*; Hex-anchus (Elasm.).

anchial—Gr. *anchialos*, maritime, bordering on the sea. *Ex:* Anchialus (Ins.).

anchist—Gr. *anchistos*, next, nearest; *anchisteus* next of kin. *Ex:* Anchista (Ins.); Anchistea*; Anchisto-cephalus (Platy.).

anchon—Gr. *anchonē*, a choking, a cord for hanging. *Ex:* Anchon-idium (Ins.); Anchonium*.

anchor—L. *anchor* < Gr. *ankyra*, an anchor < Sansk, *anc*, to bend. *Ex:* Anchor-ella (Arth.); Anchora-carp-acea (Arth.); Anchori-fera(Ins.); Ancor-ella (Por.); Ancor-ina (Por.).

anchus—Gr. *anchousa*, paint for the skin; also the alkanet, a plant from which is derived a red coloring matter. *Ex:* Anchusa*.

ancill—L. *ancilla*, a maid-servant; *ancillaris*, relating to maid-servants. *Ex:* Ancillaria (Moll.).

ancis—L. *ancisus*, cut away, cut around.

ancistr—Gr. *ankistron* (*agkistron*), a fish-hook. *Ex:* Ancistr-ella (Prot.); Ancistr-omma (Ins.); Ancistria (Ann.); Ancistro-cladus*; Ancistro-teuthis (Moll.); Ancistrona (Ins.).

ancon—See anc 1.

ancor—See anchor.

ancul—See ancyl.

ancyl—Gr. *ankylos*, bent, crooked, curved, *ankylis*, a hook. *Ex:* Anculo-pus (Ins.); Ancylis (Ins.); Ancylo-ceras (Moll.); Ankylo-stoma (Nemat.); Ancylus (Moll.).

ancyr—Gr. *ankyra*, an anchor. *Ex:* Ancyr-onis cus (Crust.); Ancyro-crinus (Echin.); Ancyro-monas (Prot.).

andin—NL. *andinus*, Andean, of the Andes.

andr—Gr. *anēr*, genit. *andros*, a man, male. *Ex:* andr-oecium; andro-gen-ous; Andro-pogon*; Andro-sace*; Andros-aemum*; gyn-andro-morph; Heter-andria (Pisc.).

andren—NL. *andrena* < Gr. *anthrēnē*, a hornet, wasp. *Ex:* Andren-idae (Ins.); Andrena (Ins.).

androsac—Gr. *androsakes*, said to be "an unknown sea-plant," but prob. a madrepore < *anēr*, *andros*, man +*sakos*, a shield. *Ex:* Androsace*.

aneb—Gr. *anēbos*, beardless, impotent; *anēbotēs*, childhood. *Ex:* Anebo-caris (Crust.).

anec—Gr. *anēkō*, to reach up. *Ex:* Aneco-rhamphus (Av.).

anec- —Gr. *anēk-*, prefix meaning not. *Ex:* Anec-physis (Ins.); Anec-toma (Ins.).

anech—Gr. *anechō*, to hold up something, to rise up, to emerge. *Ex:* Anech-ura (Ins.).

anect—Gr. *anektos*, bearable, able to be tolerated. *Ex:* Anectus (Ins.).

aneid—Gr. *aneidēs*, shapeless. *Ex:* Aneides (Amph).

anem—Gr. *anemos*, the wind. *Ex:* Anem-opsis*; Anemedo-philus* (a genus of plants living in hollows made by wind; the name is badly formed); anemo-tropism. See also eim and haem.

anemoc—Gr. *anemōkēs*, swift as the wind. *Ex:* Anemoces (Pisc.).

anemone—Gr. *anemōnē*, the wind flower, perh. < *anemos*, the wind. *Ex:* Anemon-ella*; Anemone*.

anepsiot—Gr. *anepsiotēs*, the relationship of cousins. *Ex:* Anepsiota (Ins.).

aner—See andr.

anerist—Gr. *aneristos*, undisputed. *Ex:* Aneristus (Ins.).

anet—Gr. *anetos*, relaxed, set free. *Ex:* Aneto-gnatha (Arach.).

aneth—Gr. *anēthon*, anise, dill. *Ex:* Anethum*.

aneu—Gr. *aneu*, without. *Ex:* Aneu-rrhinus (Ins.).

-aneus—L. -*aneus*, suffix added to noun stems to form adjectives denoting made of or belonging to. *Ex:* sub-terr-aneus.

anfract—L. *anfractus*, a curving, a bending; as an adj., winding, bending, crooked; *anfractuosus*, twisted, roundabout, full of bends. See -osus.

ang—Gr. *angeion*, a vessel, reservoir, dim. of *angos*, a jar. *Ex:* Angio-spermae*; Astr-angia (Coel); Hydr-angea*; pyl-angium.

angch—See anch.

ange—See ang.

angelic—Gr. *angelikos*, angelic, heavenly, divine. *Ex:* Angelica*.

angl—L. *angulus*, angle, corner. *Ex:* angle.

angui—L. *anguis*, a snake > *anguilla*, an eel. *Ex:* Angu-idae (Rept.); angui-cida; Anguill-idae (Pisc.); Anguill-ula (Nem.); not Anguillaria* named after Luigi Anguillara, It. botanist; Anguis (Rept.).

anguill—See angui.

angul—L. *angulus*, angle, corner > *angularis*, having corners. *Ex:* Anguli-scala (Moll.); Angulo-crinus (Echin.); Angulus (Moll.).

angur—Gr. *angyria*, a cucumber < *angos*, a vessel. *Ex:* Anguria*.

angust—L. *angustus*, narrow, small; *anguste*, within narrow bounds. *Ex:* Angust-ella (Moll.); angusti-folia.

anhelat—L. *anhelatus*, puffing, breathing with difficulty, pp. of *anhelo*, to pant, puff. *Ex:* Anhelata (Ins.).

anhinga—NL. *anhinga*, a South American native (Tupi) name for the water-turkey. *Ex:* Anhinga (Av.).

aniar—Gr. *aniaros* wretched, grievous, annoying. *Ex:* Aniar-ophron (Ins.); Aniaros (Ins.).

aniba—Port. *aniba* < Tupi Indian *anhoaiba*, a plant name. *Ex:* Aniba*.

anicet—Gr. *aniketos*, unconquerable. *Ex:* Anicetus (Ins.).

anidryt—Gr. *anidrytos*, inconstant, unsettled. *Ex:* Anidrytus (Ins.).

anilast—Gr. *anilastos*, not appeased, merciless. *Ex:* Anilastus (Ins.).

anili—Gr. *anileōs*, cruel, pitiless. *Ex:* Anilius (Rept.).

anima—L. *animus*, soul, spirit. *Ex:* not Animasaurus (Rept.) which refers to Las Animas, Colorado, from which specimens were shipped to the describer, E. C. Case.

animos—L. *animosus*, full of spirits, bold.

anir—Gr. *-aneira*, man < *aner*, man. *Ex:* Calli-anira (Cten.).

anis—1. Gr. *anison*, anise, dill. *Ex:* anis-atus; Anison*:    2. Gr. *anisos*, unequal. *Ex:* Anis-odon (Mam.); Anis-ol-ornis (Av.); Aniseia*, the sepals being unequal; Aniso-lambda (Mam.); Aniso-meles*, see mel 6.

ankyl—See ancyl.

ankylos—Gr. *ankylōsis*, a stiffening of the joints. *Ex:* ankylosis = anchilosis.

anlage—Ger. *Anlage*, foundation < *anlagen*, to found. *Ex:* anlage.

annecten—L. *annectens*, genit. *annectentis*, joining, connecting, ppr. of *annecto*, to link, join. *Ex:* annectent.

annel—See annelid.

annelid—Fr. *annelide* < *anneler*, to arrange in rings; *annela*, ringed < L. *anulus* = *annulus*, dim. *anellus* = *annellus*, a ring. *Ex:* annel-ism; annel-oid; Annelida.

annon—L. *annona*, victuals, a year's harvest, the annual income of the field < *annus*, a year. *Ex:* Annona (Ins.); not Annona = Anona*, nor Annon-aceae*, which are said by some to be derived from some native name although Don suggests that Linnaeus derived it from the L. *annona*.

annos—L. *annosus*, full of years, old, aged.

annotin—L. *annotinus*, a year old, of the previous year.

annu—L. *annus*, annual.

annul—L. *anulus* = *annulus*, a ring; *annulatus*, furnished or ornamented with a ring; *annularis*, relating to a ring. *Ex:* annular ligament; Annularia (Moll.); Annulata.

ano-—1. Gr. *ano-*, prefix meaning up, upward, on high, aloft < *ana-*, up, upward. *Ex:* Ano-bium.

The Depressed Anostoma, *Anostoma depressum*, showing the aperture opening upward. In crawling, however, the animal carries the shell with "mouth" and spire downwards.

(Ins.); Ano-cysti (Echin.); Ano-glypta (Moll.); Ano-stoma (Moll.):    2. Gr. *aneu*, without. *Ex:* Ano-tylus (Ins.).

anoda—Ceylonese *anoda*, native name of a plant, the abutilon. *Ex:* Anoda*.

anoect—Gr. *anoixis*, an opening, doorway; *anoiktos*, opened. *Ex:* Anoecto-chilus*; Anoicto-stoma (Platy.).

anogra—Anagram of *Onagra*, a plant genus. *Ex:* Anogra*.

anoict—See anoect.

anoig—Gr. *anoigō*, to expand, lay open, unlock. *Ex:* Anoig-anthus*.

anolis—West Indian *anoli*, a lizard. *Ex:* Anolis (Rept.).

anom—Gr *anomos*, without law, irregular, unequal < *a*, priv. + *nomos*, law; *anomia*, lawless conduct. *Ex:* Anom-odontia (Rept.); Anom-ura (Ins.); Anoma-theca*, Anomia (Moll.); Anomo-neura (Arth.).

anomadus—See mad.

anomal—Gr. *anōmalos*, uneven, irregular < *an*, priv. + *homalos*, even. *Ex:* Anomal-anthus (Echin.); Anomal-urus (Mam.); anomali-ped; Anomalo-cardia (Moll.): Anomalo-ceras (Moll.).

**anomoe**—Gr. *anomoios*, unlike, dissimilar. *Ex:* Anomoeo-cera (Ins.); Anomoeus (Ins.).

**anona**—See **annon.**

**anonym**—Gr. *anōnymos*, nameless, unknown, inglorious. *Ex:* Anonymus (Platy.).

**anopai**—Gr. *anopaia*, unnoticed. *Ex:* Anopaia (Av.).

**anophel**—Gr. *anōphelēs*, useless, troublesome. *Ex:* Anopheles (Ins.).

**anopl**—Gr. *anoplos*, unarmed < *an*, priv. + *hoplon*, a weapon. *Ex:* Anopl-ura (Ins.); Anopleta (Ins.); Anoplo-therium (Mam.); Mer-anoplus (Ins.).

**anost**—Gr. *anostos*, not to be taken back. *Ex:* Anosto-stoma (Ins.).

**anoter**—Gr. *anōterō*, superior, upper. *Ex:* Anoteropsis (Arach.), see ops. 2; Anotero-pora (Bry.).

**anour**—See **an** and **ur.**

**anous**—Gr. *anous*, stupid, without regard. *Ex:* Anous (Av.).

**ans**—L. *ansa*, dim. *ansula*, a handle; *ansatus*, having a handle. *Ex:* Ans-perus (Rot.); ansulate; Ansulus (Moll.).

**ansat**—See **ans.**

**anser**—L. *anser*, genit. *anseris*, a goose; *anserinus*, of or pertaining to geese. *Ex:* Anser (Av.); Anseri-formes (Av.); Merg-anser (Av.).

**ansul**—See **ans.**

**ant-** —Gr. *ant-* (used before roots beginning with a vowel; *anti-*, used before roots beginning with

The Antirrhinum-like Bush Penstemon, *Penstemon antirrhinoides.*

a consonant), prefix meaning against, set against, opposite, opposed to, instead, in return; also signifies in composition resemblance to the word that follows it as in Anti-chorus*. *Ex:* Ant-echinus (Mam.); anti-mere; Antipatharia (Coel.). See also anta.

**-ant**—Eng. *-ant*, adj. and noun suffix, in adjectives meaning, being and in nouns, one who; sometimes equivalent to *-ent* as in pendent.

**anta**—Gr. *antaō*, to meet face to face, to partake of. *Ex:* Phorb-antus (Mam.).

**antagonism**—Gr. *anatagōnisma*, a struggle.

**antar**—Gr. *Antarēs*, name of a bright star < *ant-*, rivalling, similar to + *Arēs*, Ares, Mars. *Ex:* Antares (Por.), (Crust.).

**ante-** —L. *ante-*, prefix meaning before (either in place or in time). *Ex:* ante-brachium; antedorsal; Ante-pithecus (Mam.).

**antedon**—Gr. *anthēdōn*, a nymph "the flowery one," hence, also, a bee; a kind of medlar tree, cf. *antheō*, to blossom. *Ex:* Antedon (Echin.).

**anteli**—Gr. *antolē*, poetic for *anatolē*, sunrise, dawn, the east; *antēlios*, eastern. *Ex:* Anteliomys (Mam.); Antole-therium (Mam.).

**antenn**—L. *antenna*, a sail yard > NL. *antenna*, a feeler. *Ex:* Antenn-aria*, (Ins.); Antenn-ellopsis (Coel.); antenna.

**antero-** —NL. *antero-* (as if derived from a L. *anterus*), prefix meaning fore, prior, preceding, anterior; L. *anterior*, fore, going before. *Ex:* antero-lateral; antero-parietal.

**anth**—1. Gr. *anthos*, a flower; *anthēros*, flowery *antheō*, to blossom. *Ex:* Anth-otium*; Antherochalina (Por.); Anthi-pes (Av.); Antho-nomus (Ins.); Crypt-antha*; Cun-oct-antha (Coel.); hyp-anthium; Schiz-anthus*: 2. Gr. *anthos*, brightness, brilliancy, excellence: 3. Gr. *anthos*, a kind of bird like the bunting. *Ex:* Anthus (Av.).

**anthel**—Gr. *anthēlē*, dim. *anthēlion*, the downy plume of the reed. *Ex:* Diss-anthelium*.

**anthem**—1. Gr. *anthemis*, genit. *anthemidos*, a flower; also an herb similar to chamomile; *anthemous*, flowery. *Ex:* Anthemis*; Anthemocrinus (Echin.); Anthemus (Ins.).

**anther**—See **anth.**

**antherix**—Gr. *antherix*, an awn. *Ex:* An-antherix*.

**anthes**—Gr. *anthēsis*, the flower or bloom of a plant, the full bloom of a plant. *Ex:* anthesis.

**anthia**—Gr. *anthias*, name of a certain sea-fish. *Ex:* Anthia (Ins.); Anthias (Pisc.).

**anthic**—Gr. *anthikos*, pertaining to flowers. *Ex:* Anthicus (Ins.).

**anthist**—Gr. *anthistēmi*, to stand against, resist, obstruct. *Ex:* Anthisteria*.

**anthod**—Gr. *anthōdēs*, flower-like, flowery. *Ex:* anthod-ium.

**anthonom**—Gr. *anthonomos*, having its flowers fed on by bees. *Ex:* Anthonomus*.

**anthrac**—Gr. *anthrax*, genit. *anthrakos*, coal, carbon. *Ex:* Anthrac-idae (Ins.); Anthraco-saurus (Amph.); Anthrax (Ins.); anthrax (Med.).

**anthrax**—See **anthrac.**

**anthren**—Gr. *anthrēnē*, a wasp, hornet. *Ex:* Anthrena (Ins.); Anthreno-soma (Ins.).

**anthrib**—NL. *anthribus*, a coined name < Gr. *anthos*, a flower + an element of unknown origin and meaning. *Ex:* Anthribus (Ins.); Anthribo-scyla (Ins.).

**anthrisc**—Gr. *anthriskos*, name of some parsley-like plant. *Ex:* Anthriscus*.

**anthrop**—Gr. *anthrōpos*, a man. *Ex:* Anthropoides (Av.); anthropo-logy; anthropo-morph-ic.

**anthus**—See **anth.**

**anthyll**—Gr. *anthyllis*, name of some plant. *Ex:* Anthyllis*.

**anti-** —See **ant-.**

**antiad**—Gr. *antias*, genit. *antiados*, a tonsil, one of the glands of the throat. *Ex:* antiad-itis (Med.).

**antiaris**—Javenese *antjar*, name of the gum-resin from the upas tree. *Ex:* Antiarus*.

**antic**—L. *anticus*, frontal, foremost. *Ex:* tibialis anticus.

**antigon**—Gr. *Antigonē*, Greek ideal of noble womanhood, heroine of many plays. *Ex:* Antig one (Av.); not Antigonon*, see ant and gon.

**antil**—Gr. *antholops*, genit. *antholopos*, a kind of horned animal, probably the antelope > O Fr. *antelop* > Eng. *antelope* = *antilope*, name of a kind of antelope. *Ex:* Antilope (Mam.); Antilocapra (Mam.) = Antilope-capra (Mam.).

**antilect**—Gr. *antilektos*, questionable, subject to doubt.

**antillar**—NL. *antillarus*, of the Antilles.

**antio**—Gr. *antios*, neut. *antion*, set against, opposite. *Ex:* Antio-bactrum (Ann.); An-antios-odon (Mam.).

**antiopa**: —Gr. *Antiopē*, wife of Lycus, King of Thebes.

**antiqu**—L. *antiquus*, old, of old time; *antiquorum*, of old times.

**antlia**—1. L. *antlia*, a machine for drawing water, a pump. *Ex:* antli-ata; antlia: 2. Gr. *antlia*, filth; also the hold of a ship. *Ex:* Antlia-rhinus (Ins.).

**antole**—See **anteli.**

**antr**—Gr. *antron*, a cave > L. *antrosus*, full of caves or cavities. *Ex:* Antro-demus (Rept.); Antro-zous (Mam.); Hyph-antro-phaga (Ins.).

**antrors**—NL. *antrorsus*, turned backwards < L. *ante* + *versum*, turned. See retro.

**antus**—See **anta.**

**anub**—L. *Anubis*, Egyptian god of the hunt.

**-anum**—See **-anus.**

**-anus**—L. *-anus*, fem. *-ana*, neut. *-anum*, suffix added to noun stems to form adjectives meaning belonging to. *Ex:* californi-anus; mexic-ana, etc.

**anus**—L. *anus*, the fundament, anus. *Ex:* an-al; ano-coccygeal; anus.

**anusi**—1. NL. *anusia* < a supposed Gr. *anousios*, treacherous, cruel. *Ex:* Anusia (Ins.): 2. Gr. *anosos* = Ion. *anousos*, healthy, without defect. *Ex:* Anusio-ptera (Ins.).

**anxi**—L. *anxius*, distressed, uneasy.

**aochlet**—Gr. *aochlētos*, undisturbed, calm. *Ex:* Aochleta (Ins.).

**aocn**—Gr. *aoknos*, tireless. *Ex:* Aocnus (Ins.).

**aonid**—NL. *Aonidia*, a proper name. *Ex:* Aonidia (Ins.).

**aor**—1. Gr. *aōros*, without youthful freshness, deformed, ugly. *Ex:* Aoro-crinus (Ech.): 2. Gr. *aor*, genit. *aoros*, a sword. *Ex:* Aor-urus (Ann.).

**aorat**—Gr. *aoratos*, invisible, blind. *Ex:* Aorato-thrips (Ins.).

**aort**—Gr. *aortē*, the aorta < *aeirō*, to lift. *Ex:* aort-ic; aort-itis; aorta.

**aosseter**—Gr. *aossētēr*, one who aids, a helper. *Ex:* Aosseterus (Ins.).

**ap**—L. *apis*, a bee; *apiarius*, relating to bees. *Ex:* Ap-idae (Ins.); api-culture; Apiaria (Ins.); apiary; Apis (Ins.).

**ap-** —Gr. *apo-*, prefix meaning from, away from, separate. *Ex:* ap-helio-tropism; Apo-glossum*; apo-physis; apo-pyle.

**apaeletic**—NL. *apaeleticus* < Gr. *apaiolaō*, to perplex, to confuse. *Ex:* Apaeleticus (Ins.).

**apal**—Gr. *apalos*, soft, tender, gentle, weak. *Ex:* Apalo-trius (Ins.), the second element < Gr. *thrix*, a hair; Apalo-pteron (Av.).

**apantes**—Gr. *apantēsis*, an encounter, reply. *Ex:* Apantesis (Ins.).

**aparg**—Gr. *apargia*, a kind of dandelion < *apo*, up + *argia*, idleness (of the cultivator). *Ex:* Apargia*.

**aparin**—Gr. *aparinē*, bedstraw, cleavers < *apairō*, to take or lay hold of.

**apat**—Gr. *apatē*, illusion, error; *apatēlos*, erroneous, producing illusion; *apatēlios*, wily. *Ex:* Apat-ornis (Av.); Apat-ura (Ins.); Apate (Ins.); Apate-mys (Mam.); Apateo-lepis (Ins.); Apato-carabus (Ins.); apatel-ic; Apateles (Ins.).

**apatel**—See **apat.**

**apatem**—Gr. *apatēma*, deceit, cunning. *Ex:* Apatema (Av.).

**apech**—Gr. *apechō*, to withhold, be far from. *Ex:* Apecho-neura (Ins.).

**apeches**—Gr. *apēchēs*, quarrelsome. *Ex:* Apeches (Pisc.).

**apechth**—Gr. *apechthēs*, hateful, hated. *Ex:* Apechthis (Ins.).

**apenes**—Gr. *apēnēs*, tough, hard-hearted, cruel. *Ex:* Apenesia (Ins.).

**aper**—L. *aper*, wild boar. *Ex:* Aper (Mam.).

**aperien**—L. *aperiens*, genit. *aperientis*, opening < *aperio*, to uncover.

**apert**—L. *apertus*, opened, uncovered. *Ex:* Aperti-rostra (Av.); apertum.

**apertur**—L. *apertura*, an opening. *Ex:* aperture.

**apex**—See **apic.**

**aph**—Gr. *haphē*, touch. *Ex:* ambly-aphia. See **haph**.

**aphaeret**—Gr. *aphairetos*, separable<*aphaireō*, to take from, prevent. *Ex:* Aphaereta (Ins.).

**aphan**—Gr. *aphanēs*, invisible, secret, unknown; *aphanistikos*, destroying, putting out of sight. *Ex:* Aphan-odon (Ins.); Aphanes*; Aphanisti cus (Ins.); Aphano-stoma (Platy.).

**aphare**—See **phar**.

**aphat**—Gr. *aphatos*, not named, nameless. *Ex:* Aphatum (Ins.).

**apheil**—Gr. *apheilon*, aor. 2 act. of *aphaireō*, to deprive, cut-off, separate. *Ex:* Apheilo-cheira (Ins.).

**aphel**—Gr. *aphelēs*, smooth, sleek. *Ex:* Aphel-enchus (Nem.); Aphel-inus (Ins.); Aphel-ops (Ins.); Aphelae-ceras (Moll.); Apheli-desmus (Myr.); Aphelo-coma (Av.); Aga-phelus (Mam.).

**aphelus**—See **aphel**.

**aphet**—Gr. *aphetos*, free, roaming at large. *Ex:* Apheto-cera (Moll.).

**aphid**—ML. *aphis*, genit. *aphidis*, a plant louse. *Ex:* Aph-idae (Ins.); Aphid-opsis (Ins.); Aphi-decta (Ins.); Aphidius (Ins.); Aphis (Ins.).

**aphis**—See **aphid**.

**aphobet**—Gr. *aphobētōs*, without fear.

**aphod**—Gr. *aphodos*, a going away; also excrement. *Ex:* Aphodites (Ins.); Aphodius (Ins.); Aphodo-derus (Pisc.).

**aphos**—Gr. *aphosioō*, to dedicate, to purify. *Ex:* Ther-aphosa (Arach.).

**aphr**—Gr. *aphros*, froth, sea foam. *Ex:* Aphri-za (Av.) (<*aphros*+*zaō*, to live); Aphro-phora (Ins.); aphro-stase. See aphrit; also aphrodit.

**aphrast**—Gr. *aphrastos*, wonderful. *Ex:* Aphrast-ura (Av.).

**aphrit**—Gr. *aphritis*, the foam-fish<*aphros*, foam. *Ex:* A-canth-aphrites (Pisc.).

**aphrodis**—*Gr. Aphrodisios*, pertaining to Aphroditē; *aphrodisiakos*, exciting sexual pleasure; *aphrodisia*, sexual pleasure. *Ex:* aphrodisiac.

**aphrodit**—Gr. *Aphroditē*, name for Venus, goddess of love who sprang from the foam of the sea<*aphros*, foam. *Ex:* Aphrodite (Ann.); herm-aphrodit-ism.

**aphron**—Gr. *aphrōn*, genit. *aphronos*, silly, senseless. *Ex:* aphronia; Aphron-astes (Ins.); Aer-aphron (Ins.).

**aphth**—Gr. *aphtha*, pl. *aphthai*, the "thrush," an eruption in the mouth>NL. *aphthosus*, full of eruptions. *Ex:* Aphtho-monas (Prot.).

**aphthart**—Gr. *aphthartos*, uncorruptible, undecaying. *Ex:* Aphthartus (Crust.).

**aphthit**—Gr. *aphthitos*, undecaying, imperishable.

**aphthon**—Gr. *aphthonos*, plentiful. *Ex:* Aphthon-ella (Ins.); Apthona (Ins.).

**aphy**—Gr. *aphyō*, to suck. *Ex:* Aphyo-rrhamphus (Ins.).

**api**—L. *apium*, parsley. *Ex:* Api-aceae*; Apium*. See also ap.

**apiar**—See **ap**.

**apiastr**—L. *apiastrum*, false celery. *Ex:* Apias trum*.

**apic**—L. *apex*, genit. *apicis*, dim. *apiculus*, a tip or point. *Ex:* Apex (Moll.); apic-al; apiculus.

**apicat**—L. *apicatus*, wearing a priest's cap.

**apiculat**—NL. *apiculatus*, small or abruptly pointed<L. *apex*, genit. *apicis*, apex.

**apin**—Gr. *apinēs*, neat, clean. *Ex:* Apines (Ins.).

**apio**—1. Gr. *apion*, a pear. *Ex:* Apio-crinites (Echin.); Apio-merus (Ins.); Apion (Ins.); Apios*; Apio-porthe*: 2. L. *apios*, distant. *Ex:* Apio-cera (Ins.).

**apis**—See **ap**.

**apist**—Gr. *apistos*, faithless, false. *Ex:* Apisto-calamus (Rept.); Apistus (Ins.).

**apium**—L. *apium*, celery<Celtic *apon*, water, because of the plant's habitation. *Ex:* Apium.*

**aplat**—Gr. *aplatos*, terrible. *Ex:* Aplat-acris (Ins.).

**aplerot**—Gr. *aplērōtos*, deficient, greedy, insatiable. *Ex:* Aplerotus (Ins.).

**aplo**—See **hapl**.

**aplud**—L. *apluda*, chaff. *Ex:* Apluda*.

**aplustr**—L. *aplustre*, the curved stern of a ship together with its ornaments. *Ex:* Aplustrum (Moll.).

**aplys**—Gr. *aplysia*, filthiness>*aplysias*, a kind of sponge of dirty color. *Ex:* Aplys-in-opsis (Por.); Aplysi-opsis (Moll.); Aplysia (Moll.).

**apo-**—See **ap-**.

**apocopt**—Gr. *apokoptō*, to cut off. *Ex:* Apocopt-oma (Ins.). See om 2.

**apodem**—Gr. *apodēmos*, away from home. *Ex:* Apodemus (Mam.).

**apolaustic**—Gr. *apolaustikos*, agreeable, tasting well.

**apolem**—NL. *apolemia*, a generic name of uncertain origin. *Ex:* Apolemia (Coel.).

**apoll**—Gr. *Apollōn*, god of manly youth and beauty, of poetry and music, also of wisdom and oracles. *Ex:* Apollo-phanes (Arach.).

**aponeuros**—Gr. *aponeurōsis*, the end of the muscle where it passes into a tendon. *Ex:* aponeurosis.

**aponogeton**—NL. *aponogeton*, etym, uncertain, cf. Potamogeton*. *Ex:* Aponogeton*.

**aporrhai**—Gr. *aporrhaiō*, to devastate. *Ex:* Aporrhais (Moll.).

**aporrhys**—Gr. *aporrhysis*, a flowing off. *Ex:* aporrhysa.

**apostasia**—Gr. *apostasia*, a revolt, departure, distance, interval. *Ex:* Apostasia*.

apothec—Gr. *apothēkē*, a storehouse. *Ex:* apothec-ium.

append—L. *appendo*, to hang something; *appendix*, that which hangs to anything. *Ex:* append-ent; append-ec-tomy; Appendic-ularia (Tun.); appendix.

appens—L. *appensus*, weighed, hung upon something, pp. of *appendo*, to hang.

applanat—NL. *applanatus*, flattened. *Ex:* applanat-ic; applanate.

applicat—L. *applicatus*, fastened, joined to, attached, close to.

appos—L. *appositus*, united, placed near, applied, pp. of *appono*, to unite, to add to. *Ex:* apposi-folli-ar.

appress—NL. *appress*<L. *ad*, toward +*pressus*, kept down. *Ex:* appress-orium, see -sorius.

approximat—L. *approximatus*, approached. *Ex:* approximate.

apr—L. *aper*, genit. *apri*, a wild boar; *apra*, genit. *aprae*, the wild sow.

apric—L. *apricus*, sunny, exposed to the sun; *apricatus*, sunniness, sunshine. *Ex:* apric-arium; Apric-(c)ardia (Moll.).

aprican—L. *apricans*; ppr. of *aprico*, to expose to the sun's warmth, to warm in the sun.

aproid—Gr. *aproidēs*, impoverished, unforseen. *Ex:* Aproida (Ins.)

aprosict—Gr. *aprosiktos*, not to be attained. *Ex:* Aprosictus (Ins.).

aprosmict—Gr. *aprosmiktos*, isolated, disassociated. *Ex:* Aprosmictus (Av.).

apsaust—Gr. *apsaustos*, not touching, sacred.

apsid—Gr. *apsis*, genit. *apsidos*, a loop, mesh. *Ex:* Apsi-nota (Ins.), Apsido-ceras (Moll.); Par-apsida (Rept.); Ther-apsida (Rept.).

apsis—See apsid.

apt—See hapt.

apten—Gr. *aptēn*, genit. *aptēnos*, unable to fly <*a*, priv.+*ptēnos*, feathered, winged. *Ex:* Apteno-dytes (Av.); Aptinus (Ins.).

apteryg—Gr. *apterygos*, without wings. *Ex:* Apteryg-ida (Ins.).

aptin—See apten.

aqu—L. *aqua*, water; *aquaticus*, found in the water. *Ex:* aqu-arium; aquatic.

aquat—See aqu.

aquatil—L. *aquatilis*, living in or near water.

aque—NL. *aqueus*, watery<L. *aqua*, water.

aquil—L. *aquila*, an eagle, possibly <*ac*, sharp, swift. *Ex:* Aquil-aria*; Aquil-astur (Av.); Aquila (Av.); Aquilo-fusus (Moll.).

aquileg—NL. *aquilegia*, name applied to a genus of plants, prob. <*aquilegus*, a water-drawer, but perhaps <*aquila*, an eagle, whose claws the spurs of the petals are supposed to resemble. *Ex:* aquilegi-folia; Aquilegia*.

aquilin—L. *aquilinus*, eagle-like, of or pertaining to the eagle.

aquilon—L. *aquilo*, genit. *aquilonis*, the north-wind; *aquilonarius*, northern. *Ex:* Aquilonaria (Moll.).

ar—Gr. *aron*, the plant arum. *Ex:* ar-oid; Araceae*; Ari-saema*; Arum*; Hedys-arum*.

-ar—Eng. -*ar*, adjectival suffix of Latin origin, meaning like, pertaining to, of the nature of. *Ex:* lamin-ar; vol-ar.

ara—1. Gr. *Ara*, goddess of destruction and revenge: 2. Gr. *ara*, strong. *Ex:* Ara-migus (Ins.): 3. Tupi, *ara*, a bird. *Ex:* Ara (Av.).

arab—L. *Arabs*, genit. *Arabis*, from Arabia, Arabian.

arabesc—Sp., Pg. *arabescos*, Arabian or resembling the Arabian in style<*Arabo*, Arab. *Ex:* Arabesc-ula (Por.).

arabis—Gr. *Arabis*, Arabian <*Arabia*, Arabia *Ex:* Arabis*.

aracang—Tupi *aracanga*, name of a kind of macaw. *Ex:* Aracanga (Av.).

arach—Gr. *arachos*=*arakos*, dim. *arakis*, name of a leguminous plant. *Ex:* Arachis*.

arachn—Gr. *arachnē*=*arachnēs*, a spider; *Arachnē*, a Lydian maiden changed by Minerva into a spider. *Ex:* Arachn-ida (Arach.); Arachne (Arach.); Arachnio-phyllum (Coel.).

arad—Gr. *arados*, a rattling. *Ex:* Aradus (Ins.).

arae—Gr. *araios*, rare, thin, narrow. *Ex:* Araeo-gnatha (Ins.); Din-araea (Ins.); Sten-araeus (Ins.).

arali—NL. *aralia* a plant name, etym. unknown. *Ex:* Arali-aceae*; Arali-ae-phyllum*; Aralia.*

aram—NL. *aramus*, an avian generic name of unknown origin, perhaps< *aramo*, a name which the Persians gave to the palace of their kings. *Ex:* Aram-idae (Av.): Aramus (Av.).

arane—L. *aranea*, a spider. *Ex:* Arane-inae (Arach.); arane-ose; Aranea (Arach.); araneiform.

arat—L. *aratus*, ploughed, pp. of *aro*, to plough. *Ex:* ex-arate.

araucan—Chilean *Araucanos*, the name of a tribe of Indians inhabiting the southern parts of Chili, see araucar.

araucar—NL. *araucaria*, a name applied to a genus of coniferous trees found in S. America, Australia, and certain islands in the Pacific Ocean<Chilean *Araucanos*, name of an Indian tribe. *Ex:* Araucaria*; Auracario-xylon*.

arbac—Gr. *Arbakēs*, first king of Media. *Ex:* Arbacia (Echin.).

arbel—Gr. *arbēlos*, a rounded knife. *Ex:* Arbel-odes (Ins.); Arbelo-rhina (Av.).

arbor—L. *arbor*, genit. *arboris*, a tree; dim. *arbuscula*; *arbustivus*, planted with trees. *Ex:* Arbor-cornus (Ins.); arbore-al; arbor-escent; arbor-etum; arbuscle.

**arbuscul**—See arbor.

**arbust**—See arbor.

**arbut**—L. *arbutus*, name of the wild strawberry tree < Celtic *ar boise*, rough bush, because of the granular berry. *Ex:* Arbutus*; arbuti-folia.

**arc**—1. L. *arca*, dim. *arcella*, a box. *Ex:* Arcopsis (Moll.); Arca (Moll.); Arcella (Prot.); Scaph-arca (Moll.): 2. L. *arcus*, a bow. *Ex:* Arci-dens (Moll.); arci-form; arco-centrum; Arco-ptera (Moll.). See also arcy.

**arcan**—L. *arcanus*, shut up, hidden; *arcanum*, a secret. *Ex:* Arcano-pora (Bry.).

**arce**—Gr. *arkeō*, to satisfy, support. *Ex:* Zo-arces (Pisc.).

**arcest**—NL. *arcestes*, name applied to a genus of ammonites (etym. unknown, perh. < Gr. *archeō*, to stand back). *Ex:* Arcest-idae (Moll.); Arcestes (Moll.).

**arceuth**—Gr. *arkeuthos*, a juniper-bush; *arkeuthinos*, of the juniper-tree. *Ex:* Arceuthobium*.

**arch-**—1. Gr. *arch-* (used before roots beginning with a vowel) = *arche- archi-* (used before roots beginning with a consonant), prefix meaning first (in time), primitive. *Ex:* Arch-aelurus (Mam.); arch-enteron; arche-bios-is; Archi-annelida (Ann.); archi-zoo-ic: 2. Gr. *arch-* (used before roots beginning with a vowel), *archi-*, *archo-*, (used before roots beginning with a consonant), prefix meaning chief, first in importance < *archos*, a chief, ruler; *archikos*, pertaining to rule. *Ex:* Archigonus (Arth.); Archo-blatt-ina (Ins.).

**arch**—1. Gr. *archē*, beginning. *Ex:* mes-arch; end-arch; ex-arch; xer-arch (Ecol.). See also arch-: 2. Gr. *archos*, the rectum. *Ex:* Archo-term-opsis (Ins.): Acan-archus (Pisc.); Not-archus (Moll.)

**archae**—Gr. *archaios*, ancient, primeval. *Ex:* Archaeo-cetus (Mam.); archaeo-cyte = archeocyte; Archaeo-teuthis (Moll.).

**archeg**—Gr. *archēgos*, originating. *Ex:* Archegosaurus (Rept.).

**archeget**—Gr. *archēgetēs*, a first leader, first cause. *Ex:* Archegetes (Ins.), (Platy.).

**archegon**—Gr. *archegonos*, first of a race, original. *Ex:* archegoni-um. archegoni-ate.

**archelon**—NL *archelon* < Gr. *archōn*, ruler + *chelōnē*, tortoise. *Ex:* Archelon (Rept.).

**archemor**—L. *Archemorus*, son of Lycurgus, killed by an adder. *Ex:* Archemora*.

**archeo**—See archae.

**archi-**—See arch-.

**archidi**—Gr. *archidion*, a petty office or position. *Ex:* Archidium*; meg-archidium.

**archiget**—See archeget.

**archit**—L. *Archyta*, ancient philosopher of Tarento. *Ex:* Archita*.

**archo-**—See arch-.

**archon**—See archont.

**archont**—Gr. *archōn*, genit. *archontos*, a ruler. *Ex:* Archon-desa (Ins.); Archonta (Moll.); Archonto-phoenix*.

**archyl**—NL. *archylo-*. said to be from Gr. *agkylos*, a hook. *Ex:* Archylo-onyx (Crust.).

**arct**—1. L. *arctus*, more correctly *artus*, narrow, straight > *co-arctatus*, confined drawn close together. *Ex:* Coarcto-termes (Ins.): 2. Gr. *arktos*, a bear. *Ex:* Arct-idea (Mam.); Arctium*; Arcto-cebus (Mam.); Arcto-mecon*; Arcto-mys (Mam.); Arcto-staphylos*: 3. Gr. *arktos*, a bear, also a name applied to northern constellations known as the "Greater and Lesser Bear" > *arktikos* and *arktōos*, of the bear; northern; "land of the bear." *Ex:* Arct-alia (Zoo-geo.); Arct-er-anthis*; Arctica (Moll.); Arcto-gaea (Zoo-geo.).

**arcu**—1. NL. *arcuo* < Gr. *arkys*, a net. *Ex:* Arcuo-thrix (Prot.): 2. L. *arcuo*, to bend.

**arcuali**—NL. *arcualia* < L. *arcus*, a bow. *Ex:* arcualia.

**arcuat**—L. *arcuatio*, a curvature, an arch. *Ex:* arcuat-ion.

**arcuo**—See arcuat.

**arcy**—Gr. *arkys*, a net. *Ex:* Arcy-ptera (Ins.); Poly-mit-arc-idae (Ins.), -mit- < *mitos*, thread.

**ard**—Gr. *ardō*, to water; in ecological terms -*ard* signifies water-content. *Ex:* ard-ella; chres-ard (Ecol.); ech-ard (Ecol.); ho-lard (Ecol.). See also ardis.

**ardal**—Gr. *ardalos*, dirty, foul. *Ex:* Ardalus (Ins.).

**arde**—L. *ardea*, a heron. *Ex:* Arde-idea (Av.); Arde-omega (Av.); Ardea (Av.); Ardei-cola (Arth.).

**arden**—L. *ardens*, genit. *ardentis*, growing warm, glowing, hot, ardent, ppr. of *ardeo*, to be on fire.

Native Bear or Koala, *Phascaloarctus cinerus*, of Australia.

**ardis**—Gr. *ardis*, the point or head of a thing, a sting. *Ex:* Ardis (Ins.); Ardisia*; Din-arda (Ins.).

ardosiac—ML. *ardosiacus*, slate-gray.

-are—NL. *-are*, a suffix used by Clements to indicate a community. *Ex:* gili-are (Ecol.).

area—L. *area*, a space.

areca—East Indian vernacular name, *areca*, the betel-nut. *Ex:* Areca*; Arec-astrum*.

arefact—L. *arefactus*, withered, dried up.

aren—L. *harena=arena*, sand, a sandy place; *arenarius*, pertaining to sand. *Ex:* aren-ose; Areni-cola (Ann.); Areno-chalina (Por.).

arens—L. *arens*, genit. *arentis*, drying, ppr. of *areo*, to be dry.

areol—L. *areola*, a small open space>NL. *areolatus*, with small spaces or areoles. *Ex:* areol-ar; areol-et; areolate; Areolo-pristomerus (Ins.).

arest—Gr. *arestos*, pleasing, acceptable.

arethus—Gr. *Arethousa*, one of Diana's nymphs who was transformed into a fountain. *Ex:* Arethus-ina (Moll.); Arethusa*, (Amph.).

areus—Gr. *areios*, war-like, brave. *Ex:* Areus (Ins.).

arg—1. Gr. *argēs*, also *argos*, bright, white; also swift-footed. *Ex:* Arg-idae (Pisc.); Arges (Pisc.); Argo-cebus (Mam.); Argo-pus (Ins.): 2. Gr. *argos*, NL. dim. *argulus*, an idler, a lazy one. *Ex:* Argulus (Crust.). See also argia.

argaleo—Gr. *argaleos*, difficult. *Ex:* Argaleo-cichla (Av.).

argas—NL. *argas*, prob. <Gr. *argos*, a contraction of *aergos*, idle, or perh. <Doric *argas*, shining, glittering, bright. *Ex:* Argas (Arth.); Argas-idae (Arth.).

argemon—Gr. *argemon*, a small white speck or ulcer on the cornea>*argemōnē*, name of a kind of poppy, the medicinal properties of which were supposed to act as a cure for white specks on the eye. *Ex:* Argemone*; Argimonia*; Agrimonia* (a corruption of Argimonia).

argente—L. *argenteus*, of silver, silvery.

argeus—Gr. *argeus*, a hunter. *Ex:* Nyct-argeus (Av.).

argia—Gr. *argia=aergia*, laziness. *Ex:* Argia (Ins.); Ap-argia*.

argil—Gr. *argilos=argillos*, white clay, potter's earth. *Ex:* Argill-ornis (Av.); Argillo-chelys (Rept.).

argill—See argil.

argimon—See argemon.

argiop—Gr. *Argiopē*, name of a nymph. *Ex:* Argiop-idae (Arach.); Argiope (Arach.).

argonaut—Gr. *Argonautēs*, a sailor in the ship Argo. *Ex:* Argonaut-ites (Moll.); Argonauta (Moll.).

argul—See arg 2.

argus—Gr. *argos*, shining, bright>*Argos*, a giant of vast strength who had a hundred shining eyes: upon his death, these eyes were placed by Juno on the tail of a peacock. *Ex:* Argus (Av.); Argusi-ana (Av.); Pod-argus (Av.). See arg 1.

argut—L. *argutus*, bright, clear, quick.

argutul—L. *argutulus*, rather noisy or talkative, somewhat subtle.

argynn—NL. *argynnis*, a name applied to a genus of Lepidoptera, perhaps<Gr. *Argynnis*, a name for Aphrodite, or <*argyros*, silver, in allusion to the silvery spots on the underside of the wings. *Ex:* Argynn-idae (Ins.); Argynnis (Ins.).

argyr—Gr. *argyros*, silver; *argyreios*, silvery. *Ex:* Argyr-odes (Arach.); Argyreia*; Argyro-pelecus (Pisc.); Argyro-ploce (Ins.).

argyrae—NL. *argyraeus*, silvery.

ari-—Gr. *ari-*, prefix strengthening the notion conveyed by its compound; chiefly denoting excellence, goodness. *Ex:* Ari-manus (Av.); Ari-otus (Ins.); Ari-phrades (Ins.); Ari-zostus (Mam.).

ari—Gr. *areios*, warlike, pugnacious. *Ex:* Ari-idae (Pisc.); Arius (Pisc.).

-aria—L. *-aria*, suffix added to noun stems to denote a thing like or connected with something. *Ex:* argent-aria; Ulm-arla*; Utricul-aria*.

ariadne—Gr. *Ariadnē*, daughter of Minos who, when falling in love with Theseus, gave him a ball of thread to guide him out of the labyrinth in case he slew the Minataur. *Ex:* Ariadne (Arach.).

-arian—Eng. *-arian*, compound suffix of Latin derivation denoting occupation, office, or belief. *Ex:* veget-arian; agr-arian.

arid—L. *aridus*, dry, withered, unadorned.

aridit—L. *ariditas*, genit. *ariditatis*, dryness.

aries—See ariet.

ariet—L. *aries*, genit, *arietis*, a ram. *Ex:* Aries (Mam.); arieti-form; ariet-inus, like a ram's head, of a ram.

aril—NL. *arillus*, a wrapper of a seed, an aril> LL. *arilli* (pl.), dry grapes<*aridus*, dry. *Ex:* aril; arill-ate; arill-ode; arilli-form.

arill—See aril.

ario—1. NL. *arion*, name applied to a genus of slugs<Gr. *Ariōn*, musician of Lesbos, rescued from drowning by a dolphin. *Ex:* Ario-limax (Moll.); Arion (Moll.); Arion-idae (Moll.); Hesper-arion (Moll.); Micr-arionta (Moll.): 2. Gr. *Arionios*, belonging to *Ariōn*, see 1. above. *Ex:* Arionus (Mam.).

arion—See ario.

ariophant—NL. *ariophanta*<Gr. *areios*, brave+ *phant-*, visible<*phantazō*, to be visible. *Ex:* Ario-phanta (Ins.).

-aris—L. *-aris*, adjectival suffix meaning pertaining to. *Ex:* Pedicul-aris*.

aris—L. *aris*, a kind of arum. *Ex:* Aris-aema* (*aris*+Gr. *haima*, blood).

arist—L. *arista*, dim. *aristula*, the awn or beard of grain, *aristatus*, awned, bearded. *Ex:* Arist-ida*; aristi-form; aristate.

aristo—Gr. *aristos*, best, noblest. *Ex:* Aristolochia*; Aristo-netta (Av.).

-arium—L. *-arium*, suffix added to noun stems to form adjectives denoting place of a thing. *Ex:* herb-arium; serpent-arium.

-arius—1. L. *-arius, -a, -um*, suffix added to noun stems to form adjectives denoting belonging to. *Ex:* ordin-arius; set-arius: 2. L. *-arius*, suffix added to noun stems to denote a person employed about anything. *Ex:* argent-arius; aucup-arius.

arizel—Gr. *arizēlos*, distinct. *Ex:* Arizelo-myia (Av.).

arm—1. Gr. *armos*, a joint; *armonia*, a fastening; also order, harmony. *Ex:* Armi-ger (Moll.); Di-armus (Ins.): 2. Gr. *arma*, genit. *armatos*, food: 3. Gr. *arma*, genit. *armatos*, a chariot: 4. L. *armatus*, armed. *Ex:* Armata (Geph.); Armato-balanus (Crust.).

armal—Gr. *armalia*, food<*arma*, food. *Ex:* Armalia (Ins.).

armen—Gr. *armenos*, adapted, proper, suitable; also, pleasing, agreeable. *Ex:* Armeno-soma (Ins.).

armeniac—NL. *armeniacus*, apricot-colored, dull orange<*Prunus armeniaca*, the apricot.

armer—NL. *armeria*<Fr. *armoires*, a name for various species of Dianthus. *Ex:* Armeria*.

armill—L. *armilla*, a ring, bracelet; *armillatus*, consisting of rings. *Ex:* Armill-aria*.

armillar—NL. *armillaris*, with a bracelet or collar<L. *armilla*, an arm band.

armoraci—Gr. *armorakia*, the horseradish. Ex: armoracia.

armost—Gr. *armostos*, well-fitted, joined, suitable. *Ex:* Armostus (Ins.).

arn—Gr. *arnos*, a lamb. *Ex:* Arno-gnathus (Rept.); Arno-seris*.

arnica—NL. *arnica*, etym. unknown, perh.< *Ptarmica*<Gr. *ptarmikē*, the yarrow. *Ex:* Arnica*.

arnio—Gr. *arnion*, dim. of *arnos*, a lamb. *Ex:* Arnio-ceras (Moll.); Arnion (Pisc.).

arnis—Gr. *arneios*, a young ram. *Ex:* Arnis-ceras (Moll.).

aro—Gr. *aroō*, to plow, to cultivate, bear seed. *Ex:* Aro-aethrus (Mam.); Aroa (Moll.); Aroaphila (Ins.); Aroa-pyrgus (Moll.).

arog—Gr. *arōgē*, help, protection; *arōgos*, aiding, serviceable.

arolium—NL. *arolium* (etym. unknown), a pad between the claws of certain insects. *Ex:* arolium, not<Gr. *arōlē* (a mistake for *arōgē*, protection) as given in Henderson's Dict. of Scientific Terms.

arom—Gr. *arōma*, a spice; *arōmatikos*, fragrant. *Ex:* Aromo-chelys (Rept.); aromatic.

aromatic—See arom.

arot—Gr. *arotēs*=*arotēr*, a ploughman; *arotron*, a plough. *Ex:* Arotes (Ins.); Arotr-ura (Ins.); Arotro-coris (Ins.); Arotro-phora (Ins.).

arotr—See arot.

arpact—Gr. *harpaktēr*=poet. *harpaktēs*, a robber>*harpaktikos*, pillaging, rapacious. *Ex:* Arpacto-philus (Ins.); Arpactus (Ins.).

arpag—Gr. *harpagē*, seizure, plunder. *Ex:* Arpagodus (Pisc.).

arpe—Gr. *harpē*, a sickle. *Ex:* Arpe-phorus (Rept.).

arped—Gr. *arpedēs*, level, flat. *Ex:* Arpedi-um (Ins.).

arquat—NL. *arquata*, a curlew<L. *arcuatus*, bowed, curved. *Ex:* Arquat-ella (Av.); Arquata (Av.).

arracacia—Sp. *arracacha*<native name for an umbelliferous plant of Mexico. *Ex:* Arracacia*.

arraph—Gr. *arraphos*, without seam, of one piece. *Ex:* Arrapho-gaster (Ins.).

arrect—L. *arrectus*, set upright; animated; pp. of *arrigo*, to erect.

arrector—NL. *arrector*, pl. *arrectores*<L. *arrectus* steep, straight up, upright, raised excited; ppr. of *arrigo*, to raise up. *Ex:* arrectores pilorum.

arrem—Gr. *arrhēmōn*, silent. *Ex:* Arremon (Av.); Arremon-ops (Av.).

arren—See arrhen.

arret—Gr. *harrētos*, inexpressible, mysterious *Ex:* Arreto-cera (Ins.); Arreto-therium (Mam.).

arrhen—1. Gr. *arrhēn*, genit. *arrhenos*, male; *arrhenōtos*, manly. *Ex:* Arren-urus (Arth.); Arrhen-atherum*; Arrheno-thrix (Ins.); ar rheno-toky: 2. Gr. *arrhēnēs*, masculine, strong, fierce. *Ex:* Arrhenes (Ins.); Arrhenophagus (Ins.); Lept-arrhena*.

arrhost—Gr. *arrhōstos*, invalid, *Ex:* Arrhostos (Ins.).

arrig—L. *arrigens* genit. *arrigentis*, erecting, ppr. of *arrigo*, to arouse.

ars—Gr. *arsis*, an elevation. *Ex:* Arsis*.

arsen—Gr. *arsēn*, genit. *arsenos*, masculine, strong. *Ex:* Arseno-xenus (Ins.); Hemipt-arsenus (Ins.).

arsin—Gr. *Arsinoē*, one of the ancient Egyptian queens whose supposed palace was in Fayum. *Ex:* Arsinoe (Ins.); Arsinoi-therium (Mam.).

arsis—See ars.

art—1. Gr. *artos*, dim. *artiskos*, a cake, loaf of wheat bread. *Ex:* Artiscus (Prot.); Arto-bius (Mam.); Arto-carpus*; Arto-phantia (Prot.); Cann-artus (Prot.); Desm-artus (Prot.: 2. Gr. *artaō*, to suspend, hang up, hang on. *Ex:* Arta-botrys*; Arta-nema*; Arto-colax (Arth.): 3. L. *artus*, straight, narrow: 4. Gr. *arti*, straight, exactly fitted. *Ex:* Arti-beus (Mam.), see bain: 5. L. *artus*, a joint. *Ex:* artiphyllus: 6. *L. ars*, genit. *artis*, art. *Ex:* arti-fact; arti-ficial. See also arti-.

**arta**—See art 2.

**artam**—Gr. *artamos*, a butcher, cook, butler < *artos*, bread + *temnō*, to cut. *Ex:* Artam-ides (Av.); Artama (Arach.).

**artem**—1. Gr. *Artemis*, a goddess usually identified with the Roman Diana. *Ex:* Artemi-dora (Arth.); Artemia (Arth.); Artemis-ina (Por.): 2. Gr. *artēma*, an earing, something suspended.

**artemis**—Gr. *artemisia*, name of a herb similar to wormwood < *Artemis*, the Greek goddess corresponding to the Roman Diana. *Ex:* Arte misia*; not Artemis-ina (Por.), see artem.

**artemon**—Gr. *Artemōn*, a personal name. *Ex:* Artemon (Moll.); Artemon-opsis (Moll.); Odont-artemon (Moll.).

**arter**—L. *arteria*, an artery. *Ex:* arteri-al.

**arthmi**—Gr. *arthmios*, united. *Ex:* Arthmius (Ins.).

**arthr**—Gr. *arthron*, a joint; *arthrōdēs*, well-jointed; *arthritikos*, of or in the joints. *Ex:* arthri-um; Arthro-poda; arthrodi-al; Arthrod-osis (Ins.); Condyl-arthra (Mam.); Di-arthrono-myia (Ins.).

**arti**—Gr. *artios*, entire, even-numbered; also sound, healthy. *Ex:* Artia-zontes (Ins.); Artio-cotylus (Platy.); Artlo-dactyla (Mam.).

**arti-** —Gr. *arti-*, prefix meaning lately, newly. *Ex:* arti-gam-ous. See also art 4.

**articul**—L. *artus*, dim. *articulus*, a joint; *articulatus*, divided into joints. *Ex:* Articulata; articulat-ed.

**artificial**—L. *artificialis*, according to the rules of art.

**artisc**—See art.

**arum**—See ar.

**arunc**—L. *aruncus* < Gr. *ēryngos*, the plant called goat's beard. *Ex:* Aruncus.*

**arundin**—L. *harundo* = *arundo*, genit. *arundinis*, a reed. *Ex:* Arundin-aria*; Arundini-cola (Av.); Arundo*.

**arv**—L. *arvum*, a field, ploughed land; *arvalis*, pertaining to a cultivated field; NL. *arvensis*, of or belonging to a field. *Ex:* Arvi-cola (Mam.); amb-arvalis.

**arval**—See arv.

**arvens**—See arv.

**-ary**—Eng. *-ary*, suffix meaning, in nouns, one who or that which, as in the noun vision-ary; in adjectives meaning relating to. *Ex:* evolu-tion-ary; avi-ary. Also Eng. *-ary* < L. *-arium*, place where, place of. *Ex:* gran-ary, place where grain is stored.

**aryst**—Gr. *arystis*, genit. *arystidos* = *aryster*, genit. *aryteros*, a ladle, cup. *Ex:* Arysti-dictya (Por.).

**aryten**—Gr. *arytainu*, a pitcher, fem. of *arytēr*, a ladle or cup. *Ex:* aryten-oid.

**aryter**—See aryst.

**-as**—L. *-as*, suffix added to noun stems to form adjectives meaning belonging to, as in *infimas*, of lowest rank; it may also be used as a feminine patronymic indicating descent or relationship. See also -ad.

**asaph**—Gr. *asaphēs*, indistinct, dubious. *Ex:* Asaph-idion (Ins.); Asaph-oid-ichnus (Tril.) Asapho-ceras (Moll.); Asaphus (Tril.).

**asar**—Gr. *asaron*, the asarabacca, a low stemless shrub. *Ex:* asari-folia; Asarum*.

**asarc**—Gr. *asarkos*, lean, without flesh. *Ex:* Asarc-ornis (Av.).

**asbest**—Gr. *asbestos*, unquenchable, inextinguishable. *Ex:* Asbesto-pluma (Por.).

**asbol**—Gr. *asbolos*, soot. *Ex:* Asbolus (Ins.).

**asc**—Gr. *askos*, dim. *askidion*, a leathern bottle, bag or bladder. *Ex:* ascidi-form; Ascidia (Tun.); Ascidio-clava (Coel.); asco-carp; Asco-glossa (Moll.); ascon; ascus.

**ascalab**—Gr. *askalabos* = *askalabōtēs*, the spotted lizard. *Ex:* Ascalabos (Rept.); Ascalabotes (Rept.).

**ascalaph**—Gr. *askalaphos*, a word used by Aristotle, apparently meaning a kind of owl. *Ex:* Ascalapha (Ins.); Ascalaphus (Av.).

**ascar**—1. Gr. *askaris*, an intestinal worm, the maw-worm. *Ex:* Ascaris (Nem.); Scaris (Ins.): 2. Gr. *askaros*, a slipper (Fr. *babouche*). *Ex:* Ascaro-sepion (Moll.).

**ascel**—Gr. *askelēs*, imperfect, immature, soft, tender, also dried up, withered, without legs. *Ex:* Ascel-ichthys (Pisc.).

**ascet**—Gr. *askētos*, curiously wrought; complete. *Ex:* Asceta (Por.); Ascetta (Por.).

**aschemon**—Gr. *aschēmōn*, misshapen, base, unseemly. *Ex:* Aschemon-ella (Prot.).

**aschet**—Gr. *aschetos*, intolerable, violent.

**asci**—1. L. *ascia*, a hatchet. *Ex:* ascii-formis. 2. ML. *ascus*, a wine-skin. *Ex:* asci-fer-ous; asci-ger-ous. See also ascia.

**ascia**—Gr. *askia*, without shadow < *a*, not + *skia*, a shadow. For ascia, pl. of *ascus*, a botanical term, see asc.

**ascid**—See asc.

**ascio**—Gr. *askios*, shady, dusky. *Ex:* Asci-odes (Ins.); Ascio-dermal (Ins.).

**asclep**—Gr. *asklēpias*, the swallow wort, one of the milk-weeds, named for *Asklēpios*, god of medicine and healing. *Ex:* Asclepi-ad-aceae*; Asclepias*; Asclepio-dora*.

**ascyr**—Gr. *askyron* = *skyron*, a name used by Dioscorides for a kind of St. John's wort. *Ex:* Ascyrum*.

**-ase**—NL. *-ase*, suffix used in forming names of enzymes. *Ex:* lip-ase; prote-ase.

**asell**—L. *asellus*, a little ass. *Ex:* Aselli-cola (Prot.); Asella (Mam.); Asellus (Crust.).

**asem**—Gr. *asēmos*, shapeless, obscure, indistinct. *Ex:* Asemus (Ins.).

**asemant**—Gr. *asēmantos*, without a leader, insignificant. *Ex:* Asemantus (Ins.).

**asil**—L. *asilus*, a gad-fly>NL. *asilid*, like a gad-fly. *Ex:* Asil-idae (Ins.); Asilido-dexia (Ins.); Asilus (Ins.).

**asilid**—See **asil**.

**asimin**—Abor. (Algonkin) *asimina*, corruption of Abor. (Illinois) *rassimina*, native name of the papaw. *Ex:* Asimina*.

**asin**—L. *asinus*, an ass. *Ex:* Asino-cleonus (Ins.); Asinus (Mam.).

**asio**—**1.** *Asia*, a continent. *Ex:* Asio-cricetus (Mam.): **2.** L. *asio*, a kind of horned owl. *Ex:* Asio (Av.). See **si**: **3.** Gr. *asis*, genit. *aseōs*, slime, mud. *Ex:* Asio-bates (Ins.).

**-asis**—See **-iasis**.

**asmod**—Heb. *Asmodoi*, a demon known as the Destroyer. *Ex:* Asmodeus (Mam.).

**asop**—Gr. *Asōpos*, name of the river-god < *Asōpos*, a name applied to several rivers. *Ex:* Asop-ella (Arach.); Asopus (Ins.).

**asot**—Gr. *asōtos*, lost, reprobate, destructive. *Ex:* Asota (Ins.); Asoto-cerus (Ins.).

**asp**—See **aspid**.

**aspalath**—Gr. *aspalathos*, a kind of shrub, yielding a fragrant oil. *Ex:* Aspalathus*.

**aspalax**—See **spalax**.

**aspalo**—See **spalax**.

**asparag**—Gr. *asparagos*=*aspharagos*, ancient name for the asparagus<the Persian. *Ex:* Asparago-bius (Ins.); Asparagus*.

**aspasi**—Gr. *aspasios*, pleasing, acceptable; *Aspasia*, companion of Pericles. *Ex:* Aspasia*, (Av.).

**aspasm**—Gr. *aspasmos*, a greeting, an embrace. *Ex:* Aspasmo-gaster (Echin.).

**aspect**—L. *aspectus*, sight, view.

**asper**—L. *asper*, rough, thorny; *asperum*, an uneven or rough place. *Ex:* Asper-ula*; Asperi-corv-ina (Pisc.); Aspro-gramme (Ins.).

**aspergill**—ML. *aspergillum*, a holy-water brush <L. *aspergo*, to scatter, sprinkle. *Ex:* aspergilli-form; Aspergillus*.

**asperug**—L. *asperugo*, genit. *asperuginis*, a plant with prickly leaves<L. *asper*, rough. *Ex:* Asperugo*.

**asphodel**—Gr. *asphodelos*, asphodel. *Ex:* Asphodelus*.

**aspid**—**1.** Gr. *aspis*, genit. *aspidos*, a shield; *aspidiōtēs*, one armed with a shield. *Ex:* Aspi-carpa*; Aspid-apion (Ins.); Aspidi-stra*, see **astr**; Aspidium*; Aspido-cotylea (Platy.); Aspidotus (Ins.); Cephal-aspis (Ins.); Cyath-aspis (Pisc.): **2.** Gr. *aspis*, a viper, asp. *Ex:* Aspi-carpi*.

**aspides**—Gr. *aspidēs* broad, also rugged. *Ex:* An-aspides (Artho.).

**aspidisc**—Gr. *aspidiskos*=*aspidiskē*, a boss, a small shield, dim. of *aspis*, shield. *Ex:* Aspidisca (Prot.).

**aspir**—L. *aspiro*, to breath toward or upon, pp. *aspiratus*. *Ex:* aspirate.

**aspis**—See **aspid**.

**aspist**—Gr. *aspistēs*, one armed with a shield, shield-bearing. *Ex:* Aspistis (Rept.); Agath-aspistes (Rept.).

**asplen**—L. *asplenum*, spleenwort<Gr. euphonic *a-*+*splen*, the spleen. *Ex:* Asplenium*: asplenoid.

**aspr**—See **asper**.

**aspratil**—L. *aspratilis*, rough.

**aspred**—L. *aspredo*, roughness. *Ex:* Aspredo (Pisc.).

**aspris**—Gr. *aspris*, a kind of oak. *Ex:* Aspris*.

**assess**—L. *assessor*, an aid<L. *assideo*, to aid, assist.

**-assic**—Eng. *-assic*<Fr. *-assique*, adj. ending meaning of or pertaining to. *Ex:* Jur-assic; Tri-assic, from Ger. *Trias*, a geological system.

**assimil**—L. *assimulo*, to make one thing like another, compare; *assimulatio*, genit. *assimilationis*, likeness. *Ex:* assimilation.

**asso**—See **att**.

**assument**—L. *assumentum*, a patch.

**astac**—Gr. *astakos*, a kind of lobster or crayfish. *Ex:* Astacus (Crust.); Astaco-morphr-ops (Crust.); Astacus (Crust.); Par-astacus (Crust.).

**astag**—Gr. *astagēs*, hard-frozen, now sometimes taken to mean dry. *Ex:* Astago-bius (Ins.).

**astart**—Gr. *Astartē*, Phoenician goddess of fertility and sexual love. *Ex:* Astart-idae (Moll.); Astart-opsis (Moll.); Astarte (Moll.).

**astas**—Gr. *astasia*, unstableness; *astatos*, unstable, unsteady. *Ex:* Astasia (Prot.); Astata (Ins.).

**astat**—See **astas**.

**astath**—Gr. *astathēs*, unstable. *Ex:* Astathes (Pisc.); Astatho-mima (Ins.).

**-aster**—L. *-aster*, suffix sometimes added to noun stems to form diminutives. *Ex:* parasit-aster.

**aster**—**1.** Gr. *astēr*, genit. *asteros*, a star; also a star-fish; *astron*, a star; *asteroeidēs*, like a star; *asterias*, and *astraios*, neut. *asterion*, starred, starry. *Ex:* Aster-ina (Echin.); Aster-iscus (Echin.); Asterias (Echin.); Asterion-ella*; Astero-idae (Echin.); Astero-stemma (Mam.); Astraea (Moll.); Astr-angia (Coel.); Astron-esthes (Pisc.); Astro-caryum*; Astronia*; not Aster (Av.) nor Asterias (Av.), see **astur**; not Asternata (Echin.), see **stern**;Pis-aster (Echin.); Zoro-aster (Echin.), see **zor** 3: **2.** Gr. *astēr*, a kind of plant, the aster<*astēr*, a star. *Ex:* Aster*; Aster-aceae*.

**asterict**—Gr. *astēriktos*, unstable.

**astes**—Gr. *astēs*, a singer. *Ex:* Aphron-astes (Ins.).

asthen—Gr. *asthenēs*, weak. *Ex:* Asthen-actis (Echin.); asthen-odont; Astheno-ceras (Moll.).

asti—Gr. *asteios*, beautiful. *Ex:* Asti-anthus*.

astic—Gr. *astikos*, native to a city, hence polite, sensible. *Ex:* Astico-stena(Ins.).

astr—See aster.

astrab—1. Gr. *astrabē*, a saddle, pair of panniers. *Ex:* Astrabe (Pisc.): 2. Gr. *astrabēs*, even, straight, well made, regular. *Ex:* Astrab-odus (Pisc.).

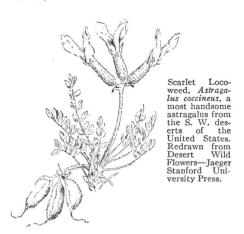

Scarlet Loco-weed, *Astragalus coccineus*, a most handsome astragalus from the S. W. deserts of the United States. Redrawn from Desert Wild Flowers—Jaeger Stanford University Press.

astrae—Gr. *Astraia*, goddess of justice changed into the constellation Virgo < *astraios*, starry. *Ex:* Astraea (Coel.); astraei-form.

astragal—1. Gr. *astragalos*, the ankle bone. *Ex:* astragalus: 2. Gr. *astragalos*, name of a kind of leguminous plant. *Ex:* Astragalus*.

astragalin—Gr. *astragalinos*, name of a kind of bird, a gold-finch. *Ex:* Astragalinus (Av.).

astrap—Gr. *astrapē*, lightning; also the brightness of flowers; *astrapaios*, of lightning. *Ex:* Astrap-odon (Mam.); Astrap-aea*; Astrape-phora (Ins.); Astrapo-therium (Mam.).

astrean—L. *astreans*, genit. *astreantis*, gleaming, shining like a star.

astron—See aster.

astrop—Gr. *astropos*, star-like. *Ex:* Astropo-lithon (Prot.), Gr. *lithos*, a stone.

astur—L. *astur*, a hawk. *Ex:* Astur (Av.) = Aster (Av.) hence also Asterias (Av.); Astur-aetos (Av.); Astur-ina (Av.).

astut—L. *astutus*, cunning < *astus*, craft, dexterity. *Ex:* Astutus (Av.).

astyc—Gr. *astykos*, inhabiting a city, a citizen; also one refined, sensible, shrewd. *Ex:* Astyco-phobus (Ins.); Astycus (Ins.).

asyndet—Gr. *asyndetos*, loose, unconnected. *Ex:* Asyndetus (Ins.).

-ata—NL. *-ata*, suffix used in zoological divi-sional names which are to be understood as adjectives modifying L. *animalia*, animals. *Ex:* Annul-ata; Branchi-ata; Pallio-branchi-ata (Brach.).

atal—Gr. *atalos*, delicate. *Ex:* Atalo-triccus (Av.).

atav—L. *atavus*, ancestor. *Ex:* atav-ism; atav-istic; Atava (Ins.).

-ate—Eng. *-ate*, suffix of Latin origin used in forming verbs, 1. especially those taken from Latin, as: motiv-ate, castr-ate, 2. but also some taken from other sources, as: vaccin-ate; aere-ate. See also -atus and -tus. S.

atel—Gr. *atelēs*, imperfect, unproductive. *Ex:* atel-ism; Atele-crinus (Echin.); Ateleo-pterus; Ateles (Mam.); atelo-cardia.

atelest—Gr. *atelestos*, to no purpose, imperfect, unfinished. *Ex:* Atelesto-crinus (Echin.); Atelestus (Ins.).

atelios—Gr. *ateleiōsis*, uncompleted. *Ex:* ateliosis.

atemeles—Gr. *atēmelēs*, neglected. *Ex:* Atemeles (Ins.).

ater—L. *ater*, fem. *atra*, neut. *atrum*, black; *atratus*, blackened; *aterrimus*, the deepest black. *Ex:* atri-cillus; Atri-mitra (Moll.); atro-punctatus.

aterrim—See ater.

ateuch—Gr. *ateuchēs*, unarmed. *Ex:* Ateuches (Ins.).

athal—Gr. *athalēs*, not green, withered. *Ex:* Athalia (Ins.), named in allusion to the devastation produced by its larvae.

athen—Gr. *Athēnē*, Olympic deity. *Ex:* Micr-athene (Av.).

ather—1. Gr. *athērē*, a form of *atharē*, groats, a porridge of meal > *athērōma*, a tumor full of gruel-like matter. *Ex:* atheroma (Path.): 2. Gr. *athēr*, an awn or beard of an ear of wheat. *Ex:* Ather-mantus (Ins.); Ather-urus (Mam.); Atheri-cera (Ins.); Athero-sperma*; Arrhen-atherum*.

atherin—Gr. *atherinē*, a kind of smelt. *Ex:* Atherina (Pisc.); Atherino-morus (Pisc.).

atherom—See ather 1.

athet—Gr. *athetos*, annulled, rejected, invalid, unfit; not in its place. *Ex:* Atheta (Ins.); Atheto-cephus (Ins.).

athlet—Gr. *athlētēs*, a combatant. *Ex:* Athleta (Moll.); athlet-ic.

athlo—Gr. *athlos*, toil, a contest. *Ex:* Athlo-pecten (Moll.); Athlo-phorus (Ins.).

athol—Gr. *atholos*, clear, not turbid. *Ex:* Atholus (Ins.); Athol-ister (Ins.), see hister.

athous—Gr. *athōos*, unpunished, harmless. *Ex:* Athous (Ins.).

athr—Gr. *athroos*, crowded together, in heaps. *Ex:* Athr-odon (Mam.); Athro-stictus (Ins.); Athro-taxis*.

**athyr**—Gr. *athyros*, without door, open<*a*, without+*thyreos*, door, shield. See thyr 2. *Ex:* Athryr-ium*.

**-atic**—L. *-aticus, -a, -um* (<Gr. *atikos*), an adjectival ending of words mostly formed from Gr. nouns and meaning pertaining to. See -ic. *Ex:* Toreum-atica (Echin.).

**-atilis**—L. *-atilis*, adj. termination denoting belonging to pertaining to, found in, as: *fluviatilis* belonging to a river<*fluvius*, a river; *aquatilis*, relating to water<*aqua*, water.

**atim**—Gr. *atimos*, unhonored. *Ex:* Atimo-blatta (Ins.); Atimus (Ins.).

**atimast**—Gr. *atimastos*, neglected, dishonored. *Ex:* Atimast-illas (Av.).

**atimet**—Gr. *atimētos*, unhonored, despised. *Ex:* Atimeta (Ins.).

**-ation**—Eng. *-ation* (<Fr. *-ation*<L. *-ationem*), suffix used in forming nouns from verbs ending in *-ate, -ize*, etc. It may denote action, state or condition, or result. *Ex:* discolor-ation, sublimation; zon-ation.

**atis**—See batis.

**-ative**—Eng. *-ative*<NL. *-ativus*, suffix (<L. pp. stem *-at-*+*-ivus*) meaning, has a tendency to; as in carmin-ative, purg-ative, etc.

**atlant**—See atlas.

**atlantic**—See atlas.

**atlas**—1. Gr. *Atlas*, genit. *Atlantos*, the gigantic god who bore up the pillars of heaven (or, according to some legends, bore up the earth). *Ex:* Atlanto-saurus (Rept.): 2. Gr. *atlas*, genit. *atlantos*, one of the cervical vertebrae which supports the head. *Ex:* atlas: 3. Gr. *Atlas*, genit. *Atlantos*, a mountain in North Africa, regarded as the pillar of heaven<Atlas (see 1.): 4. Gr. *Atlantikon*, the Atlantic (Ocean)<*Atlas*, name of a mountain. *Ex:* Atlanta (Moll.); atlanticus.

**atm**—1. Gr. *atmis*, genit. *atmidos*, steam, vapor, smoke; *atmos*, steam, vapor. *Ex:* Atmo-ceras (Ins.); atmo-genic; atmo-sphere.

**atom**—See atomar.

**atomar**—NL. *atomarius*, covered with atoms or spots<Gr. *atomos*, indivisible, uncut. *Ex:* atom; atom-ic; Atomaria (Ins.).

**atop**—Gr. *atopos*, strange, anomalous, out of place. *Ex:* Atop-ornis (Av.); Atopo-gnathus (Ins.).

**-ator**—L. *-ator*, suffix added to verb, adjective and noun stems to denote an agent or doer. *Ex:* putator<*puto*, to trim; nitidulator<*nitidulus*, rather neat trim; flavator<*flavis*, yellow.

**atr**—See ater.

**atract**—Gr. *atraktos*, a spindle, a shaft, arrow.

*Ex:* Atract-odes (Ins.); Atracto-glymma (Ins.); Ptych-atractus (Moll.); Atractus*.

**atractyl**—Gr. *atraktylis*, a thistle-like plant, the wooly carthamus. *Ex:* Atractylis*; atractyl-oides.

**atragen**—Gr. *atragenē*, name of a tree from which tinder is made. *Ex:* Atragene*.

**atrament**—L. *atramentum*, ink, anything black; *atramentarium*, an inkstand. *Ex:* atramentous.

**atrat**—L. *atratus*, clothed in black as for mourning.

**atrec**—Gr. *atrekēs*, real, true. *Ex:* Atrecus (Ins.).

**atrest**—Gr. *atrestos*, fearless.

**atret**—Gr. *atrētos*, imperforate, without opening. *Ex:* Atreta (Moll.).

**atri**—L. *atrium*, a hall, entrance-room. *Ex:* atrio-pore; atrium.

**atriplex**—L. *atriplex*=*atriplexum*, an orach, a saltbush<Gr. *atraphaxys*, an orach plant, saltbush. *Ex:* Atriplex*.

**atriplic**—L. *atriplex*, genit. *atriplicis*, a plant genus. *Ex:* atriplici-folium.

**-atrix**—L. *-atrix*, genit. *-atricis*, fem. ending. *Ex:* Buccul-atrix (Ins.).

**atroc**—L. *atrox*, genit. *atrocis*, dark, hideous, savage.

**atrom**—Gr. *atromos*=*atromētos*, fearless, calm. *Ex:* Atrom-opsis (Ann.); Atrometus (Ins.).

**atromet**—See atrom.

**atrop**—Gr. *Atropos*, one of the Fates, the unbending one<*a*, not +*tropos*, turning. *Ex:* Atropa*.

**atroph**—Gr. *atrophos*, not flourishing>L. *atrophus*, in a state of decline. *Ex:* atroph-ic.

**atrox**—See atroc.

**atryton**—Gr. *Atrytōnē*, the tameless one. *Ex:* Atrytone (Ins.).

**att**—L. *Atta*, a surname applied to persons who walk on their shoetips, perhaps<Gr. *attō*=*assō*, to spring, hop. *Ex:* Atta (Ins.); Att-idae (Arach.); Att-opsis (Arach.); Att-ulus (Arach.); Attus (Arach.).

**attac**—Gr. *attakos*=*attakēs*, a kind of locust. *Ex:* Attaco-bius (Ins.); Attacus (Ins.).

**attact**—L. *attactus*, touched; pp. of *attingo* to touch.

**attag**—Gr. *attagēn*=*attagēs*, a kind of bird, perhaps the francolin. *Ex:* Attagis (Av.).

**attagas**—Gr. *attagas*, name of a partridge-like bird of reddish color and spotted on the back. *Ex:* Attagas (Av.).

**attagen**—Gr. *attagēn*, genit. *attagēnos*, name of a kind of grouse, the francolin, similar to the partridge. *Ex:* Attagen-inae (Av.); Attagenus (Ins.).

**attalea**—L. *Attalus*, king of Pergamum. *Ex:* Attalea*.

attelab—Gr. *attelabos*, a kind of wingless locust. *Ex:* Attelabus (Ins.).

attenuat—L. *attenuatus*, weakened, reduced. *Ex:* attenuate.

attingens—L. *attingens*, touching; ppr. of *attingo*, to touch.

attolen—L. *attolens*, genit. *attolentis*, rising up, ppr. of *attolo=attollo*, to rise up, elevate.

attonit—L. *attonitus*, astonished.

attons—L. *attonsus*, sheared, clipped; pp. of *attondeo*, to shear.

attrahen—L. *attrahens*, genit. *attrahentis*, drawing to, dragged with force, ppr. of *attraho*, to drag, draw. *Ex:* attrahent.

attrit—L. *attritio*, genit. *attritionis*, friction, a rubbing against; *attritus*, rubbed, worn away, weakened < *attrero*. *Ex:* attrition.

-atus—L. *-atu.* = Eng. *-ate*, suffix added to noun stems to form adjectives meaning provided with. *Ex:* barb-atus; capit-atus; cune-ate; ligul-ate.

au—Gr. *au-*, prefix signifying besides, again, back. *Ex:* au-gnathus.

auc—See aux.

auchen—Gr. *auchēn*, genit. *auchenos*, the neck. *Ex:* Auchenia (Mam.); Aucheno-ceros (Pisc.); Macr-auchenia (Mam.).

auchm—Gr. *auchmos*, dryness, drought; *auchmēros*, dry, without rain, hence squalid, dull, dusky; *auchmeō*, to be unwashed, squalid. *Ex:* Auchmer-esthes (Ins.); Auchmo-phoba (Ins.).

auchmer—See auchm.

auct—L. *auctus*, increased, augmented.

aucup—L. *aucupor*, to catch birds, chase. *Ex:* aucup-arious.

audac—L. *audax*, genit. *audacis*, bold, rash. *Ex:* Audax-lyto-ceras (Moll.).

audax—See audac.

auden—L. *audens*, genit. *audentis*, daring, ppr. of *audeo*, to venture, dare.

aug—Gr. *augē*, sunlight, any bright light; *augeō*, to shine, glitter. *Ex:* Aug-astes (Av.); Auga (Av.); Augo-mono-ctenus (Ins.); Ur-auges (Av.).

augasm—Gr. *augasmos*, splendor. *Ex:* Augasma (Ins.).

augesc—L. *augescens*, genit. *augescentis*, increasing < *augesco*, to grow.

augur—L. *augur*, a soothsayer.

august—L. *augustus*, notable, majestic, august, worthy of honor.

aul—1. Gr. *aulē*, a court, open court; *aulikos*, pertaining to a court. *Ex:* aula; aulic: 2. Gr. *aulos*, a pipe, tube; any wind instrument, like a flute. *Ex:* Aul-actinia (Coel.); Aul-acantha (Coel.); Aulo-stoma (Pisc.): 3. Gr. *aulis*, a tent or place to spend the night in.

aulac—Gr. *aulax*, genit. *aulakos=alox*, genit. *alokos*, a furrow, a mark, wound, the womb. *Ex:* Aulac-aspis (Ins.); Aulaco-sternum (Ins.); Aulacus (Ins.); Aulax-odon (Mam.); Aloconota (Ins.); Periss-aulax (Moll.).

aulax—See aulac.

aulet—Gr. *aulētēs*, a flutist, musician, *aulētris*, a flute-girl. *Ex:* Auletes (Amph.); Auletris (Amph.).

auletris—See aulet.

auleum—L. *auleum=aulaeum*, a curtain.

aulic—1. L. *aulicus*, noble: 2. L. *aulix*, genit. *aulicis*, a furrow. Same as *aulax*. *Ex:* not-aulices.

aulon—Gr. *aulōn*, a pipe, a channel; also a level plain, meadow. *Ex:* Aulonium (Ins.); Aulono-gyrus (Ins.).

aur—1. L. *aura*, air, breath. *Ex:* Auro-physa (Coel.): 2. L. *auris*, an ear; *auritus*, eared; *auricula*, the ear, external ear; ML. *auricularis*, pertaining to the ear or auricle of the ear. *Ex:* auri-puncture; auricul-ate; Plex-aur-ella (Coel.); Auricul-ina (Moll.); Aurio (Moll.): 3. L. *aurum*, gold, the color of gold; *aureus*, dim. *aureolus*, golden, splendid; *aureatus*, adorned with gold. *Ex:* Auro-cores (Ins.).

aurant—NL. *aurantium*, an orange < *Citrus aurantium*, the orange tree > *aurantiacus*, of the color of the orange, see -acus. *Ex:* Auranteae*; auranti-aceous.

aurat—L. *auratus*, rich in gold, ornamented with gold.

aureat—See aur 3.

aurel—L. *aurelia*, a gold-colored pupa < *aurum*, gold. *Ex:* Aurel-issa (Coel.); Aurelia (Coel.).

aurelian—L. *Aurelianus*, Roman Emperor. *Ex:* Aurelian-aster (Echin.).

aureol—See aur 3.

auric—See aur 2.

aurigine—L. *aurigineus*, yellowish.

auror—L. *aurora*, dawn, morning; *Aurora*, goddess of the dawn. *Ex:* auror-eus.

auros—L. *aurosus*, of the color of gold, golden.

aurulent—L. *aurulentus*, of the color of gold.

auscult—L. *ausculto*, to listen to with attention, care; *auscultatio*, a listening. *Ex:* auscult; auscultat-ion.

auspicat—L. *auspicatus*, favorable, lucky.

auster—L. *austerus*, harsh, severe.

austr—L. *auster*, genit. *austri*, the south wind; *australis*, of the south wind, southern > NL. *Australia*, the southern continent. *Ex:* Australanthus*; australi-oid (Ethn.); Australo-mantis (Ins.); Australo-pithecus (Mam.); Austritragus (Mam.); Austro-cyclus (Moll.).

austral—See austr.

aut- —Gr. *aut-* (used before roots beginning with a consonant), *auto-* (used before roots beginning with a vowel), prefix meaning self < *autos,* self. *Ex:* aut-ac-oid; aut-oeci-ous; Auto-grapha (Ins.); auto-scop-y (Med.); auto-tom-y.

**autacoid**—NL. *autacoid* < Gr. *autos,* self + *akos,* remedy. *Ex:* autacoid.

**authad**—Gr. *authadēs,* stubborn.

**autochthon**—Gr. *autochthōn,* sprung from his native land, a primitive inhabitant < *autos,* self + *chthon,* land. *Ex:* Autochthon (Ins.); autochthon-ous; Autochthonus (Ins.).

**autumnal**—L. *autumnalis,* belonging to autumn.

**aux**—Gr. *auxē,* a growth, an increase; *auxēsis,* a growth, an increase; *auximos,* promoting growth; *auxanō,* to increase. *Ex:* Aux-ornis (Av.); aux-in; Auxi-ceros (Ins.); Auximo-basis (Ins.); auxo-spore.

**auxan**—See **aux.**

**auxim**—See **aux.**

**av**—L. *avis,* dim. *avicula,* pl. *aves,* a bird. *Ex:* Aves; avi-an; avi-fauna; Avicul-aria (Bry.); Avicula (Moll.); Aviculi-pecten (Moll.); Avi culo-mon-otis (Moll.).

**avar**—L. *avarus,* greedy, avaricious.

**avellan**—L. *avellana,* a filbert; *avellaneus,* drab, the color of a fresh hazel-nut shell. *Ex:* Avel lana (Moll.).

**aven**—L. *avena,* an old name for oats. *Ex:* Avena*.

**averrunc**—L. *Averruncus,* a deity which wards off. *Ex:* Averruncus (Pisc.).

**avers**—L. *aversus,* turned back, turned away from, removed.

**avia**—L. *avia,* grandmother; cf. *avus,* a grandfather.

**avicul**—See **av.**

**avid**—L. *avidus,* hungry, greedy.

**avium**—L. *avium,* a desert, a place of wildness. See avius.

**avius**—L. *avius,* deserted, solitary. See avia.

**avocett**—It. *avocetta* = *avosetta;* name of a bird, the avocet. *Ex:* Avocetta (Av.); Avocett-ula (Av.).

**avuls**—L. *avulsus,* pp. of *avello,* to pluck, to tear out or away; *avulsis,* a tearing apart. *Ex:* avul-sion.

**avus**—L. *avus,* a grandfather. *Ex:* Lemur-avus (Mam.); Nimr-avus (Mam.); Lutr-avus (Mam.).

Gill's Avocet-headed eel, *Avocettina gilli,* one of the deep-sea snipe-eels. Redrawn from Fishes of North and Middle America—Jordan.

**-ax**—L. *-ax,* suffix, denoting a faulty or aggressive tendency. *Ex:* pugn-ax, aud-ax.

**ax**—1. Gr. *axōn,* an axle, kindred to L. *axis,* an axis, axle. *Ex:* axi-al; Axio-kersos (Ins.); axo-posium; Axo-stylus (Prot.); Axona (Ins.), (Arach.); Axono-pus*; Plan-axis (Moll.): 2. Gr. *axiōs,* worthily, of like value. *Ex:* Anth-axia (Ins.). See also ac 2.

**axill**—L. *axilla,* the armpit. *Ex:* axill-ar; axil-lary; axilla.

**axin**—Gr. *axinē,* an axe-head, a battle-axe. *Ex:* Axin-aea*; Axine(Nemat); Axino-palpus(Ins.).

**axio**—See **ax 2.**

**axiogast**—Gr. *axiogastos,* remarkable, worthy of notice. *Ex:* Axiogastus (Ins.).

**axon**—See **ax 1.**

**axyr**—1. Gr. *axyrias,* one unshorn; *axyrēs* and *axyros,* uncut, unshorn. *Ex:* Axyrias (Pisc.): 2. NL. *axyrum,* which according to Don is derived from Gr. "*axyra,* an anchor"; compare with Gr. *ankyra,* an anchor. *Ex:* Sperm-axyrum*.

**az**—Gr. *azō,* to dry. *Ex:* Azolla*, *azō* + *ollyō,* to kill, since the plants are killed by drought.

**azale**—Gr. *azaleos,* dry, parched, severe. *Ex:* Azalea*, which once was thought to grow best in comparatively dry situations; Azale-odes (Ins.).

**azoll**—See **az.**

**azor**—Sp. *Azores,* islands of the North Atlantic < Sp. *azor,* a hawk. *Ex:* Azor (Moll.); Azor-anchus (Ins.); Azor-ica (Por.); Azori-pupa (Moll.).

**azot**—NL. *azotum,* nitrogen < Gr. *a-,* priv. + *zōtikos,* fit for maintaining life (hence, not fit for breathing). *Ex:* azot-emia (Med.); Azoto-bacter*.

**aztec**—Nahuatl *Azteca,* original home of the Aztecs. *Ex:* Azteca (Ins.); Aztec-ula (Pisc.); Aztecus (Pisc.); Azteko-philus (Myr.).

**azur**—LL. *azureus,* sky blue, see lazul. *Ex:* Azur-ina (Pisc.); Aruzia (Av.).

# B

ba—Gr. *ba*, an intensive particle. *Ex:* Ba-en-odon (Mam.), see en.

babyruss—Port. *babirosa = babirussa* (<Malay *babi*, hog+*rusa*, deer), name of the horned hog. *Ex:* Babyrussa (Mam.) = Babirussa (Mam.).

bacc—L. *bacca*, more correctly *baca*, a small round fruit such as a berry; also that which is berry-like in shape such as a pearl, dung of sheep and goats, etc. *Ex:* bacc-ate; Bacc-aurea*; Bacca-coccus (Ins.); Bacci-spongia (Por.).

bacch—Gr. *Bakchē*, a mythological name. *Ex:* Baccha (Ins.); Bacchi-opsis (Ins.).

baccharis—Gr. *bakkaris*, name of an unknown plant with an aromatic root yielding oil. *Ex:* Baccharis*.

bacidi—NL. *bacidium*, a little rod <L. *baculum*, a staff, stick, rod+the dim., -*idium*. *Ex:* Bacidia*.

bacill—See bacul.

bacop—Abor. *bacopa*, name of some plant. *Ex:* Bacopa*.

bacter—See bactr.

bactr—Gr. *baktron = baktēria*, dim. *baktērion*, a staff, stick, club. *Ex:* Bactr-idium (Bry.); Bactr-ites (Moll.); bacteria; bacterio-logy; Bactri-ola (Ins.); Bactris*; Bactro-mantis (Ins.); A-bactrus (Ins.).

bacul—L. *baculum = baculus*, dim. *bacillus*, a staff, support. *Ex:* Bacill-aria (Prot.); Bacilli-desmus (Myr.); Bacillo-gaster (Ins.); Bacillus (Ins.); Bacul-aria*; Bacul-ites (Moll.); Baculi-palpus (Ins.); Baculo-cystis (Echin.); baculum = os penis; Saccino-baculus (Prot.).

bacuncul—NL. *bacunculus* <L. *baculum*, a rod +*unculus*, a small hook. *Ex:* Bacunculus (Ins.); Bacuncul-idae (Ins.).

bad—Gr. *bados*, a step, path; also a collection. *Ex:* Bad-ura (Ins.); Bado-glio-pora (Bry.).

badio—Gr. *badioumai*, fut. ind. of *badizō*, to walk, to advance slowly step by step > *badisma*, a step, gait. *Ex:* Badio-felis (Mam.); Badismus (Ins.); Badizo-blax (Ins.).

badism—See badio.

badist—Gr. *badistēs*, a goer, walker, quick runner; *badistikos*, good at walking. *Ex:* Badistes (Ins.); Badistica (Ins.); Anti-badistes (Ins.).

badius—L. *badius*, chestnut-colored, brown.

badiz—See badio.

bae—See bai.

baen—See bain, also ba.

baet—L. *Baetis*, name of a river in Spain. *Ex:* Baet-idae (Ins.); Baetis (Ins.).

bago—Gr. *bagōas*, a eunuch, a guard of women; without seed or fruit. *Ex:* Bagous (Ins.).

bagr—NL. *bagrus* <Sp. *bagre*, a fish. *Ex:* Bagrus (Pisc.).

bagrad—NL. *bagrada*, prob. <a native name for some insect. *Ex:* Bagrada (Ins.).

bai—Gr. *baios*, slim, small, feeble; also dry, concise. *Ex:* Baeo-myces*; Baeo-semus (Ins.); Baeus (Ins.); Baio-sci-urus (Mam.); Boeo-cerca (Av.); Boeo-scelis (Ins.); Cerato-baeus (Ins.).

baical—See baikal.

baikal—Russ. *Baikal*, name of a lake in southern Siberia. *Ex:* Baicali-ella (Moll.); Baicalia (Moll.); Baicalo-cottus (Pisc.); Baikalo-dendron (Prot.).

bain—Gr. *bainō = baō*, to go, walk, step. *Ex:* baeno-mere; baeno-pod; Halo-baena (Av.).

bajul—L. *bajulus*, a burden bearer, a carrier; *bajulator*, a porter; *bajulo*, to carry heavy things. *Ex:* Bajulus (Crust.).

bal—See ball.

balaen—L. *balaena*, a whale. *Ex:* Balaen-odon (Mam.); Balaena (Mam.); Balaeni-ceps (Av.); Balaeno-ptera (Mam.).

balan—Gr. *balanos*, an acorn; also the male glans; *balaninos*, made of acorns; *balanitēs*, acorn-shaped. *Ex:* Balanites (Por.); Balane-phagus (Ins.); Balani-ceps (Av.); Balaninus (Ins.); Balanites*; Balanito-zoon (Prot.); Balano-bius (Ins.); Balanus (Mam.).

balanat—L. *balanatus*, perfumed with balsam.

balanit—See balan.

balant—Gr. *balantion = ballantion*, a bag, purse. *Ex:* Balant-idium (Prot.); Balantia (Mam.); Balantio-stoma (Bry.).

balaust—Gr. *balaustion*, the wild pomegranate flower >NL. *balausta*, a many-celled fruit like the pomegranate. *Ex:* Balaustion*; Balaustium (Arach.).

balearic—L. *Balearicus*, Balearic <Gr. *Baliareis*, the Balearic Islands in the Mediterranean. *Ex:* Balearica (Av.).

bali—Gr. *balios*, spotted, various; also swift, nimble. *Ex:* Balio-nycteris (Mam.); Balio-pygus (Rept.); Balius (Ins.). See also ball.

balist—L. *balista = ballista = ballistra*, a military engine, catapult <Gr. *ballō*, to throw. *Ex:* Ballist-ura (Ins.); Balistes (Pisc.); Balisto-pus (Pisc.).

ball—Gr. *ballō*, to throw, emit. *Ex:* Ballo-gnatha (Arach.); Ballo-philus (Myr.); Ec-balium*. See also bol.

ballism—Gr. *ballismos*, a dance, leaping.

ballist—See balist.

ballot—Gr. *ballōtē*, a name for black horehound < *ballō*, to emit. *Ex:* Ballota*.

balsad—NL. combining form of Provincial American Sp. *balsas*, a place full of swamps, noxious weeds>Rio Balsas, a river in Mexico. *Ex:* Balsad-ichthys (Pisc.).

**balsam**—L. *balsamum.*<Gr. *balsamon,* balsam, the balsam tree; *balsameus,* of balsam, *balsaminus* of balsam. *Ex:* Balsamo-citrus\*; Balsamo-rrhiza\*.

**balteat**—L. *balteatus,* girdled.

**bambus**—E. Ind. *bambu,* bamboo. *Ex:* Bambus-acris (Ins.); Bambusa\*; Bambusi-cola (Av.).

**bandicot**—Telegu *pandi-kokku,* the pig-rat. *Ex:* Bandicota (Mam.).

**baph**—Gr. *baphē,* a dyeing, the art of enameling. *Ex:* Baphia\*; Bapho-nulina (Prot.); Em-baph-ion (Ins.); Eu-baphis (Amph.); phlo-baph-ene, see -ene and -enus.

**bapt**—Gr. *baptos,* dyed, tinged; *baptisis,* a dipping; *baptizō,* to dip in or under water, dye. *Ex:* Bapt-an-odon (Rept.); Bapt-ornis (Av.); Baptisia\*; Bapto-doris (Moll.).

**bar**—1. Gr. *baris,* genit. *baridos,* dim. *baridion,* an Egyptian flat-bottomed boat, also a castle or other similar structure. *Ex:* Baris (Ins.); Bari-scelis (Ins.); Baridi-aspis (Ins.); Baridius (Ins.): 2. Gr. *baros,* weight; *barytēs,* weight, heaviness; *barys,* heavy, burdensome; *bary-peithēs,* slow to believe. *Ex:* Bar-ornis (Av.); Bar-osma\*; Barita (Av.); baro-taxis; Bary-cnemis (Ins.); bary-morphosis; Barypeithes (Ins.).

**barathr**—Gr. *barathroō,* to cast into a pit; *barathron,* a gulf or deep abyss; *barathrōdēs,* like a pit, abysmal. *Ex:* Barathro-bius (Coel.); Barathro-demus (Pisc.); Barathronus (Pisc.).

**barb**—L. *barba,* a beard; *barbatus,* bearded; Fr. *barbastelle=barbetelle,* a little beard. *Ex:* Barbastell-us (Mam.); Barbat-ula (Av.); Barb-odon (Pisc.); barbel; barbell-ate; Barbi-cornus (Ins.); Barbi-lanius (Av.); Barbo-nema (Nem.).

**barbar**—Gr. *barbaros,* strange, foreign. *Ex:* Barbar-orthis (Brach.); Barbaro-pus (Ins.).

**barbat**—See **barb.**

**barbiton**—Gr. *barbitos=barbiton,* a stringed musical instrument. *Ex:* Barbiton-ia (Moll.).

**bard**—L. *bardus,* stupid.

**bardist**—Gr. *bardistos,* very slow, stupid. *Ex:* Bardistus (Ins.).

**barid**—See **bar.**

**bary**—See **bar.**

**barydotir**—Gr. *barydoteira,* a giver of ill gifts. *Ex:* Barydotira (Ins.).

**baryntic**—Gr. *baryntikos,* weighing down. *Ex:* Baryntica (Ins.).

**bas**—Gr. *basis,* base, bottom, foundation; also rarely a stepping or a going, as in Cata-basis (Pisc.); NL. *basilaris,* pertaining to or situated at the base, especially of the skull. *Ex:* Bas-ommato-phora (Moll.); base-ost; Baseo-neura

(Ins.); Basi-ceros (Ins.); basi-occipital; basi-plast; basi-pod-ite; basilar; basilaris; Cata-basis (Pisc.); Gonio-basis (Moll.); Stylo-basium\*.

**basanistes**—Gr. *basanistēs,* an examiner, torturer. *Ex:* Basanistes (Av.).

**bascan**—1. Gr. *baskanion,* a charm, amulet. *Ex:* Bascanion (Rept.): 2. Gr. *baskanos,* bewitching, malignant, slanderous; *baskania,* slander, envy. *Ex:* Bascan-ichthys (Pisc.).

**baseo**—See **bas.**

**basid**—L. *basidium,* a small pedestal. *Ex:* Basidio-mycetes\*; basidio-phore.

**basil**—Gr. *basileus=basileutōr,* a king; *basileō=basileuō,* to be king, to rule; *basilikos,* royal, kingly; *basileutos,* kingly>NL. *basilaris,* regal. *Ex:* Basil-emys (Rept.); Basileo-stylus (Moll.); Basilicus (Rept.); basilic vein; Basileuterus (Av.); Basilio-chiton (Moll.); Basilo-saurus (Rept.). See also bas.

**basilar**—NL. *basilaris,* basilar, see bas; also NL. *basilaris,* regal, pertaining to royalty<Gr. *basileus,* king.

**basilisc**—Gr. *basiliskos,* a petty king; also a kind of serpent with a spot on its head like a crown. *Ex:* Basiliscus (Rept.).

**basiliss**—Gr. *basilissa=basilis,* genit. *basilidos,* a queen. *Ex:* Basilissa (Moll.).

**basim**—Gr. *basimos,* approached, exposed; also passable, acceptable. *Ex:* Basimus (Crust.).

**basm**—Gr. *basmos,* a step, threshold.

**bass**—1. LL. *bassus,* deep, perh.<Gr. *bassōn,* deep. See bathy. *Ex:* Bass-alia (Zoo-geo.), Basso-cryptus (Ins.); Basso-gigas (Pisc.)`; not Bassia\*, named after Ferdinand Bassi, Italian botanist: 2. L. *Bassus,* a proper name. *Ex:* Bassus (Ins.).

**bassar**—Gr. *bassara=bassaris,* a fox. *Ex:* Bas saris (Mam.); Bassar-iscus (Mam.); Bassarisc-ops (Mam.).

**bast**—1. AS. *baest,* ME. *bast,* the inner bark of various trees, especially the linden, from which rope was made. *Ex:* bast cells: 2. NL. *-bastes,* a carrier<Gr. *bastazō,* to carry, to handle. *Ex:* Melo-bastes (Ins.).

**bastes**—See bast 2.

**bat**—1. Gr. *batēs,* one that treads or haunts; also a climber. *Ex:* Brachy-batus (Ins.); Hydro-bates (Ins.); Hydro-bat-idae (Ins.); Pelo-bates (Ins.): 2. Fr. *baton,* a staff. *Ex:* Bato-lites (Moll.): 3. Gr. *batos,* a bramble bush, thorn. *Ex:* Bato-crinus (Echin.); Bato-mys (Mam.); Bato-phila (Ins.); Chamae-batia\*; Citrio-batus\*; Sarco-batus\*: 4. Gr. *batos,* pervious, passable. *Ex:* Bato-pora (Bry.). See also batis.

**batat**—Haytian *batata,* the sweet potato. *Ex:* Batatas\*.

**bath**—Gr. *bathos,* genit. *batheos,* depth or height, greatness. *Ex:* Bath-ornis (Av.); Batho-theca (Por.).

**bathm**—Gr. *bathmos*, a step or degree, rank; *bathmēdon*, by steps. *Ex:* bathm-ism; Bathm-ochtha (Ins.); Bathmedonia (Av.); Bathmi-syrma (Av.); Bathmo-cercus (Av.); A-bathm-odon (Mam.).

**bathr**—Gr. *bathron*, a base, bench, pedestal. *Ex:* Bathr-odon (Mam.); Bathro-pyramis (Prot.); Ana-bathra (Ins.); Brachy-bathron (Moll.); Hypo-bathrum*.

**bathy**—Gr. *bathys* = Doric *bassōn*, deep, low, broad; *bathysma*, a depth; *bathyergeō*, to plow deep. *Ex:* Bathy-actis (Coel.); bathy-pelagic; bathy-al; Bathyergus (Mam.); bathysm-al; Basso-zetus (Pisc.). See also bass.

**bathyergus**—See **bathy.**

**batis**—1. Gr. *batis*, a ray or skate. *Ex:* Bat-oidea (Elasm.); Dasy-atis (Elasm.); Dasy-batis (Elasm.); Mylio-batis (Elasm.): 2. Gr. *batis*, a plant name. *Ex:* Batis*.

**batrach**—Gr. *batrachos*, a frog; *batracheios*, belonging to a frog. *Ex:* Batrachyla (Amph.) < *batrachos* + *Hyla*, a frog genus; Batrach-ichthys (Pisc.); Batrachium*; Batrachio-saurus (Rept.); Batracho-stomus (Av.).

**bave**—Fr. *bave*, drivel. *Ex:* bave.

**baz**—Ar. *baz*, a hawk. *Ex:* Baza (Av.).

**bdals**—Gr. *bdalsis*, genit. *bdalseōs*, a milking. *Ex:* Bdalsi-podo-batrachi (Amph.).

**bde**—Gr. *bdeō*, to stink, to break wind. *Ex:* Bdeo-gale (Mam.).

**bdell**—Gr. *bdella*, a leech, sucker. *Ex:* Bdell-idium (Arach.); Bdell-our-idae (Platy.); Bdello-stoma (Cycl.); Amphi-bdella (Platy.).

**bdelygm**—Gr. *bdelygma*, genit. *bdelygmatos*, an idol, abomination. *Ex:* Bdelygma (Mam.).

**bdesm**—Gr. *bdesma*, a foul odor, a stench.

**bdol**—Gr. *bdolos*, a fetid scent. *Ex:* Caleo-bdolon*.

**beat**—L. *beatus*, fortunate, blessed.

**bebae**—Gr. *bebaios*, firm, established, durable > NL. *bebaeus*. *Ex:* a-bebaeus.

**bebel**—Gr. *bebēlos*, profane. *Ex:* Bebelo-thrips (Ins.); Bebelus (Ins.).

**beber**—L. *beber*, the beaver; *bebrimus*, of or pertaining to a beaver.

**bebr**—Gr. *bebros*, stupid. *Ex:* Bebr-ornis (Av.); Bebro-ptera (Ins.). See also **beber.**

**behavior**—Ger. *gehaben*, to have; *sich gehaben*, to carry oneself, to behave; AS. *behabban*, to hold, restrain; ME. *behaven*; Eng. *behave*, to conduct oneself in proper manner > Eng. *behavior*.

**bel**—Gr. *belos*, genit. *beleos*, a dart, sting. *Ex:* Bel-ideus (Mam.); Bel-odon (Rept.); Beleo-phorus (Ins.); Belo-perone*; Belo-teuthis (Moll.).

**belemn**—Gr. *belemnon*, a dart, javelin. *Ex:* Belemn-acanthus (Pisc.); Belemn-ites (Moll.); Belemno-cyst-ites (Echin.).

**bell**—L. *bellus*, neat, charming, handsome. *Ex:* Bella-spina (Moll.).

**bellac**—L. *bellax*, genit. *bellacis*, warlike, bellicose.

**bellator**—L. *bellator*, a warrior.

**bellatul**—L. *bellatulus*, neat, pretty.

**bellax**—See **bellac.**

**bellid**—See **bellis.**

**bellis**—L. *bellis*, genit. *bellidis*, the white daisy. *Ex:* bellidi-forme; Bellid-ium*; Bellis*.

**bellon**—L. *Bellona*, goddess of war. *Ex:* Bellona (Av.).

**bellul**—L. *bellulus*, beautiful, lovely.

**belon**—Gr. *belonē*, a needle, any sharp point; also a kind of fish. *Ex:* belon-oid; Belono-ptera (Ins.); Belono-staurus (Prot.).

**beltist**—Gr. *beltistos*, the absolute best. *Ex:* Beltista (Ins.).

**belug**—Russ. *bieluga*, a species of fish, the great sturgeon < *bieluĭ*, white; also the white whale. *Ex:* Beluga (Mam.), (Pisc.).

**belui**—L. *beluilis*, brutal; *beluinus*, bestial.

**bema**—Gr. *bēma* = *bama*, genit. *bēmatos*, a step, pace; also a raised place, pedestal; *bēmatistēs*, one who measures by paces. *Ex:* Bema (Ins.); Bematicus (Mam.); Bematistes (Ins.); Calo-bemon (Av.); Lepi-bema (Pisc.).

**bembex**—See **bembic.**

**bembic**—Gr. *bembēx*, genit. *bembēkos* = *bembix*, genit. *bembikos*, a buzzing insect; also a whirlpool, top > NL. *bembex*. *Ex:* Bembex (Ins.); Bembic-idium (Ins.); Bembico-soma (Arth.); Bembix (Ins.); Bembyxium (Moll.); Bathy-bembix (Moll.).

**bembyx**—See **bembic.**

**bene**—L. *bene*, well, agreeable, good.

**beneolens**—L. *beneolens*, smelling agreeably.

**benign**—L. *benignus*, kind, favorable, pleasing.

**benth**—Gr. *benthos*, depth of the sea. *Ex:* Benth-euphausia (Crust.); Bentheo-caris (Crust.); Bentho-desmus (Pisc.); archi-benth-al.

**berber**—ML. *berberis* = *barbaris*, the barberry > Eng. *berberid*. *Ex:* Berberid-aceae*; Berberid-opsis*; Berberis*.

**bergami**—L. *bergamium*, bergamot; appar. < *Bergamo*, an Italian town. *Ex:* oleum bergamii; Citrus aurantia bergamia.

**bernicl**—NL. *bernicla*, name applied to a genus of geese < ML. *bernicla*, a barnacle. *Ex:* Bernicla (Av.)

**bero**—Gr. *Beroē*, one of the nymphs, daughter of Oceanus. *Ex:* Beroe (Coel.); Bero-ides (Coel.); Bero-soma (Coel.).

**beros**—Gr. *Berōssos*, celebrated priest of Balus in Babylon. *Ex:* Berosus (Ins.).

**berul**—L. *berula*, water-cress. *Ex:* Berula*.

**berus**—ML. *berus*, a name applied to the water snake, prob. Natrix natrix, and said to have first been used by Albertus Magnus Vincent de Beauvais.

**beryc**—NL. *beryx*, genit. *berycis*, name of a fish genus. *Ex:* Beryc-idae (Pisc.); Beryx (Pisc.).

**beryllin**—L. *beryllinus*, green-colored.

**beryx**—See **beryc**.

**bess**—Gr. *bēssa*, a mountain glen, thicket, wood; also a drinking cup. *Ex:* Bess-ornis (Av.); Besso-bia (Av.); Besso-phora (Ins.).

**beta**—1. L. *beta*, the beet. *Ex:* Beta*: 2. Gr. *β*, second letter of the Greek alphabet. *Ex:* Beta-suchus (Rept.).

**bethyl**—NL. *bethylus*, evidently an invented name made by Latreille. *Ex:* Bethylus (Ins.).

**betonic**—L. *betonica* (<*vettonica* < *Vettones*, a people in Spain), name of a labiate plant, the wood betony. *Ex:* Betonica*; betony*.

**betul**—L. *betula*, the birch. *Ex:* Betula*; Betul-aphis (Ins.); Betul-ites*; betul-oides; betuli-folia.

**bi-** —L. *bi-*, combining form of *bis* (Gr. *di-*, *dis-*), prefix meaning two, two-, twice, twofold, double, etc. *Ex:* Bi-cuculla*; Bi-dens*; bi-fid; bi-pinn-aria; bi-ram-ous; bi-vium. See also bin.

*Nyetes bidens*, the Two toothed Puncturer, a curculionid beetle from Australia. After Tillyard.

**bi**—Gr. *bios* and *biotē*, life; *biotos*, life, manner of living, poet. for *bios*; *biosis*, the act of living > NL. *biota*, the faunal and floral features of a particular region or period, or a treatise on the same. *Ex:* bio-gen; bio-logy; biot-ic; biota; Bius (Ins.); A-bia (Ins.); anaero-bic; Clu-biona (Arach.); Nycti-bius (Av.); sym-biosis.

**biai**—Gr. *biaios*, forced, violent; *biastikos*, violent. *Ex:* biaio-meta-morphosis; Biasticus (Ins.).

**biast**—Gr. *biastēs* = *biatas*, strong. *Ex:* Biastes (Ins.), (Av.). See also biai.

**biator**—NL. *biatora*, a name of unknown meaning for a genus of mosses. *Ex:* Biatora*.

**bibi**—LL. *bibio*, a small insect generated in wine. *Ex:* Bibio (Ins.); Bibi-opsis (Ins.); Bibio-dites (Ins.), see dyt.

**bibl**—Gr. *biblos*, bark; also a book made of the bark of papyrus; *biblion*, a paper, scroll, book, *Ex:* Biblo-plectus (Ins.); litho-biblion; Stylo-biblium (Prot.).

**bicos**—Gr. *bikos*, a drinking cup or bowl. *Ex:* Bicos-oeca (Prot.).

**bidu**—L. *biduus*, two days long.

**bifor**—L. *biforus* = *biforis*, having two doors or openings. *Ex:* bifor-ate.

**bil**—L. *bilis*, bile, gall. *Ex:* bili-rubin.

**-bilis**—See **-ilis**.

**bim**—L. *bimus*, lasting two years.

**bin**—L. *bini*, two at a time, two, two by two, double. *Ex:* bin-ary; bin-ate; bin-ocular; Bin-voluta (Moll.); bini-folius.

**bio**—See **bi**.

**biona**—See **bi**.

**bis**—L. *bis*, twice, double. *Ex:* Bis-torta*.

**biscot**—Fr. *biscotin*, a small biscuit easily broken. *Ex:* biscoti-form.

**bison**—Gr. *bisōn*, the original wild ox, the aurochs. *Ex:* Bison (Mam.).

**bithynia**—Gr. *Bithynia*, a former country of Asia Minor; *Bithynis*, pertaining to or of *Bithynia*. *Ex:* Bithinis (Crust.); Bithynia = Bythinia (Moll.); Bythin-ella (Moll.). Some would derive the molluscan genera cited here from Gr. *bythios*, of the deep.

**bittac**—Gr. *bittakos*, a variant of *psittakos*, a parrot. *Ex:* Bittacus (Ins.); Bittaco-morpha (Ins.); Aptero-bittacus (Ins.).

**bitumin**—L. *bitumen*, genit. *bituminis*, bitumen; *bitumineus*, of bitumen. *Ex:* bitumin-osus.

**bivium**—See **bivius**.

**bivius**—L. *bivius*, having two ways < *bi*, two + *via*, a way. *Ex:* bivium (neut. of *bivius*).

**bix**—Sp. *bixa* < an aboriginal name for some plant. *Ex:* Bixa*; Bix-aceae*.

**blab**—Gr. *blabē*, harm, damage. *Ex:* Blabe-phorus (Ins.); Blabi-rhinus (Ins.); Blabo-phanes (Ins.); A-blab-ophis (Rept.); A-blabus (Ins.).

**blaber**—Gr. *blaberos*, harmful, pernicious. *Ex:* Blaberus (Ins.).

**blac**—Gr. *blax*, genit. *blakos*, worthless, without significance, stupid; *blakikos*, lazy. *Ex:* Blac-ops (Av.); Blacicus (Av.); Blacus (Ins.); Blax (Av.); not Blaki-aster (Echin.), which was based on the name of the ship "Blake" whose extensive collection of sea-stars Perrier studied in the early 1880's; Blax (Ins.), (Av.).

**blacic**—See **blac**.

**bladap**—Gr. *bladapos*, flaccid. *Ex:* bladap-id-ous.

**blaes**—Gr. *blaisos*, crooked, *Ex:* Blaeso-spira. (Moll.). See spir. 1.; Blaeso-xipha (Ins.).

**blaidot**—Gr. *blaidotēs*, crookedness.

**blan**—Gr. *blanos*, blind. *Ex:* Blanus (Rept.).

bland—L. *blandus*, smooth, of a smooth tongue.

blaps—Gr. *blapsis*, genit. *blapseōs*, a hurting, injuring. *Ex:* Blap-isa (Ins.), see isos; Blaps (Ins.); Blapsi-dotes (Ins.); A-blapsis (Ins.).

blapt—Gr. *blaptō*, to disable, hinder. *Ex:* Blapto-campus (Ins.).

blaptic—Gr. *blaptikos*, hurtful. *Ex:* Blaptico-xenus (Ins.); Blapticus (Ins.).

blarin—NL. *blarina*, a coined name for certain shrews. *Ex:* Blarina (Mam.); Blarino-mys (Mam.).

blas—See blaz.

blast—Gr. *blastos*, a germ, bud, shoot, branch, blossom; *blastēma*, an outgrowth, shoot. *Ex:* Blast-oidea (Echin.); blast-ula; blastema; blasto-derm; Blasto-phaga (Ins.); blasto-pore; cyto-blastema, epi-blast; meso-blastic.

blastic—Gr. *blastikos*, inclined to shoot upward. *Ex:* Blastico-toma (Ins.).

blastis—NL. *blastis*<A.S. *blāest*, *a* blowing> Eng. *blast*, to injure, to destroy by blowing. *Ex:* Cacto-blastis (Ins.).

blatt—L. *blatta*, some insect that shuns light, a cockchafer or some other beetle. *Ex:* Blatt-elytron (Ins.); Blatta (Ins.); Blatti-cola (Nem.); Blatto-morpha (Ins.).

blax—See blac.

blaz—Gr. *blazō*, to be silly. *Ex:* A-blas (Av.), a- in the sense of very.

blechn—Gr. *blēchnon*, a kind of fern. *Ex:* Blech num*.

blechr—Gr. *blēchros*, gentle, sluggish. *Ex:* Blechro-pus (Av.); Blechrus (Ins.).

blem—Gr. *blēma*, a coverlet. *Ex:*Blemus (Ins.); Epi-blema (Ins.); peri-blem.

blemm—Gr. *blemma*, genit. *blemmatos*, a look, glance, appearance. *Ex:* Blemmatia (Ins.).

blenn—Gr. *blennos*, slime, mucous; also a fish, the blenny which has mucous on its scales. *Ex:* Blenni-ophidium (Pisc.); Blennius (Pisc.); Blenno-generis (Ins.); Blenno-sperma*.

blep—Gr. *blepō*, to look, to understand; also to resemble. *Ex:* Ana-blepas (Mam.); Cata-blepas (Mam.); Cato-bleps (Pisc.).

blephar—Gr. *blepharis*, genit. *blepharidos*, an eyelash; *blepharon*, an eyelid. *Ex:* Blephar-idia (Ins.); Blephari-pappus*; Blepharido-pterus (Ins.); Blepharo-cer-idae (Ins.); blepharo-plast; Blepharo-stoma (Prot.).

bleps—Gr. *blepsis*, sight, view. *Ex:* Blepsis (Pisc.). See also plep.

blepsias—Gr. *blepsias*, name of an unknown fish. *Ex:* Blepsias (Pisc.).

blept—Gr. *bleptos*, worth seeing, to be seen. *Ex:* Blepti-phora (Ins.); Blepto-nema (Pisc.); Bleptus (Ins.).

blet—Gr. *blētos*, stricken, wounded; also as subst.

a beast that strikes. *Ex:* Bleta (Ins.); Bleto-gona (Ins.); Para-bleta (Ins.).

blit—Gr. *bliton*, name of a plant of southern Europe, the strawberry-blite. *Ex:* Blito-phaga (Ins.); Blito-pertha (Ins.); Blitum*.

blom—Gr. *blōmos*, a morsel.

blos—1. Gr. *blōsis*, approach, arrival:   2. Gr. *blōsio*, genit. *blosiōs*, meat; also corrosion, rust.

blosyr—Gr. *blosyros*, grim, frightful. *Ex:* Blosyro pus (Ins.); Blosyrus (Ins.).

blothr—Gr. *blōthros*, tall, stately. *Ex:* Blothro-phyllum (Coel.); Blothrus (Arrch.); Ideo-blothrus (Arach.).

blysm—Gr. *blysma*=*blysis*, a bubbling-up. *Ex:* Blysmia (Ins.); Cata-blysmia (Ins.); Blysmus*.

bo—See boo, also boa.

boa—L. *boa*, a kind of water snake. *Ex:* Boa (Rept ); Aspido-boa (Rept.); Bo-avis (Rept.).

boanerg—Gr. *Boanergēs*, sons of thunder. *Ex:* Boanerges (Av.).

bodo—NL. *bodo*, a name evidently without meaning invented by Stein (1878) for a genus of protozoans. *Ex:* Bodo (Prot.); Bodo (n)-idae (Prot.).

bodon—See bodo.

boea—See bai.

boeo—See bai.

boeth—Gr. *boētheō*, to come to the rescue. *Ex:* Boetho portia (Arach.); Boethus (Ins.).

bol—1. Gr. *bolē*, a throw; dart, stroke; also a thunderbolt, a wound<*boleō*, to throw. *Ex:* Boleo-soma (Pisc.); cata-bol-ism; meta-bol-ism; Trema-bol-ites (Por.):   2. Gr. *bolis*, genit. *bolidos*, a dart. *Ex:* Bol-idium (Por.):   3. Gr. *bōlos*, a lump. *Ex:* Bol-odon (Mam.).

bolax—Gr. *bolax*, synonym of *bōlos*, a lump. *Ex:* Bolax*.

bolb—Gr. *bolbos*, a bulb; also the name of a certain plant with bulbous root that grew wild in Greece. *Ex:* Bolbo-ceras (Moll.); Bolbo-coelon*.

bolelia—NL. *bolelia*, anagram of *Lobelia*. *Ex:* Bolelia*.

bolet—Gr. *bōlitēs*, a superior kind of mushroom >L. *boletus*. *Ex:* Boleti-cola (Ins.); Boletus*.

bolid—See bol 2.

bolimnius—Anagram of *Limnobius*. *Ex:* Bolim nius (Ins.).

bolin—Gr. 1. *Bolina*, name of a nymph. *Ex:* Bolin-opsis (Coel.); Bolina (Cten.):   2. Gr. *bōlinos*, full of lumps of earth. *Ex:* Bolinia*.

bolit—Gr. *boliton*=*bolitos*, cow-dung. *Ex:* Bolito-bius (Ins.); Bolito-glyphus (Ins.).

bolitaen—Gr. *bolitaina*=*bolbidion*, a small kind of cuttle-fish. *Ex:* Bolitaen-ella (Moll.); Boli taena (Moll.).

**bom**—Gr. *bōmos*, a raised place, stand, foundation. *Ex:* A-boma (Pisc.); for Aboma (Rept.), see aboma.

**bomb**—Gr. *bombos*, a heavy deep tone, humming > ML. *bombino*, to buzz; *bombinator*, a buzzer; Gr. *bombylios*, a buzzing insect. *Ex:* Bombornis (Av.); Bombilio-mya (Ins.); Bombinator (Amph.); Bombylia (Ins.); Bombylius (Ins.); Bombylo-myia (Ins.); Bombus (Ins.). **S.**

**bombac**—LL. *bombax*, genit. *bombacis*, cotton. *Ex:* Bombac-idae*; Bombax*.

**bombax**—See **bombac.**

**bomby**—See **bombyc.**

**bombyc**—1. L. *bombyx*, genit. *bombycis*, the silkworm; *bombycinus*, silky. *Ex:* Bomby-cilla (Av.); Bombyx (Ins.): 2. Gr. *bombyx*, a name for cotton. *Ex:* Bombax*, see bombac.

**bombyl**—See **bomb.**

**bombyx**—See **bombyc.**

**bomoloch**—Gr. *bōmolochos*, one lying in wait about an altar to beg or steal food offered to the gods < *bōmos*, an alter + *lochaō*, to lie in wait. *Ex:* Bomolochus (Crust.).

**bonas**—1. L. *bonasum*, a kind of buffalo, the aurochs < Gr. *bonasos*, a wild ox. *Ex:* Bonasus (Mam.): 2. NL. *bonasa*, a name applied to a genus of birds, perhaps < L. *bonus*, good + *assum* a roast. *Ex:* Bonasa (Av.).

**boo**—Gr. *bous*, genit. *boos*, an ox, bull > L. *bos*, genit. *bovis*, an ox. *Ex:* Bo-ops (Mam.) (Pisc.); Boo-philus (Arach.); Bos (Mam.); bov-ine; Bovi-cerus (Ins.); Bu-ceros (Av.); Bu-pleurum*; Bu-tomus*; Bus-elaphus (Mam.); Ictio-bus (Pisc.); Ovi-bos (Mam.).

**boop**—Gr. *boōpis*, ox-eyed. *Ex:* Boop-inus (Ins.); Boopis (Moll.); Boops (Mam.), (Pisc.).

**bopyr**—ML. *Bopyrus*, a proper name. *Ex:* Bopyr-idae (Crust.); Bopyrus (Crust.).

**bor**—Gr. *boros*, greedy, a glutton; *bora*, food, flesh. *Ex:* Boro-mys (Mam.); Boro-phagus (Mam.); Sym-bor-odon (Mam.); Yucca-borus (Ins.)

**borag**—LL. *borago*, genit. *boraginis*, the borage, related to Fr. *bourre*, the hair of beasts; Don says: "altered from *cor*, the heart, and *ago*, to affect". *Ex:* Boragin-aceae*; Borago*.

**borass**—Gr. *borassos*, the fruit of the palm. *Ex:* Borassus*.

**borbor**—Gr. *borboros*, mud, filth. *Ex:* Borboro-coete (Amph.); Borboro-pora (Ins.); Borborus (Ins.).

**borborygm**—Gr. *borborygmos*, intestinal rumblings.

**bore**—Gr. *boreas*, the north wind; *boreios*, northern > L. *borealis*, northern. *Ex:* boreal; Bore-aspis (Pisc.); Boreo-nymphon (Arth.); Bori-oikon (Mam.); Borio-gale (Mam.); Boreus (Ins.).

**bori**—See **bore.**

**bos**—See **boo.**

**bosc**—1. Gr. *boskas*, a kind of duck. perhaps the teal. *Ex:* bosci-anus: 2. Gr. *boskō*, to feed, nourish. *Ex:* pro-boscis

**bostrich**—See **bostrych.**

**bostrych**—Gr. *bostrychos* = *botrychos*, a small curl; anything twisted; also a kind of insect (supposed by some to be the male glow-worm). *Ex:* Bostrychus (Ins.); bostrych-oid; Bostrycho-ceras (Moll.); bostryx.

**botan**—Gr. *botanē*, an herb, grass, fodder; also later the study of herbs; related to *boskō*, to graze and *bous*, ox; *botanikos*, belonging to herbs, of herbs. *Ex:* botanic; Botano-bius (Ins.); Botano-phila (Ins.); botany.

**botaur**—NL. *botaurus*, name applied to a genus of birds, etym. uncertain, but probably allied to words arising from L. *butio*, genit. *butionis*, a bittern. *Ex:* Botaurus (Av.).

**both**—NL. *bothus*, a fish name, meaning and origin unknown. *Ex:* Both-idae (Pisc.); Bothus (Pisc.).

**bothin**—See **bothr.**

**bothr**—Gr. *bothros* (= *bothynos*), dim. *bothrion*, a hole, trench. *Ex:* Bothin-odontes (Ins.); Bothri-deres (Ins.); Bothri-odon (Mam.); bothria; Bothrio-pupa (Moll.); Bothyno-stethus (Ins.); Laemo-bothrion (Ins.); Phyllo-bothrium (Platy.).

**bothyn**—See **bothr.**

**botrio**—See **botry.**

**botry**—Gr. *botrys*, a cluster, bunch of grapes > NL. dim. *botryllus*. *Ex:* Botrio-cyrtis (Prot.); botry-oid-al; Botryllus (Tun.); Botrytis*; Poly-botrya*; Botrychium*.

**botrych**—See **bostrych.**

**botul**—L. *botulus*, a sausage. *Ex:* Botul-ina (Moll.); botul-ism (Med.); botuli-form.

**botum**—Gr. *boutomos*, name of some water plant. *Ex:* Botumus*.

**bov**—See **boo.**

**bracat**—L. *bracatus*, wearing trousers.

**brachi**—Gr. *brachiōn*, genit. *brachionos*, the upper part of the arm > L. *brachiatus*, having arms; *brachiolatus*, with arms. *Ex:* Brachi-aria*; Brach-ina (Echin.); brachi-plex; brachio-cephalic; Brachio-poda; Brachionus (Rot.); Cirro-brachium (Moll.).

**brachin**—See **brachi, brachy.**

**brachiol**—See **brach.**

**brachist**—See **brachy.**

**brachy**—Gr. *brachys*, short; *brachistos*, shortest; *brachynō*, to shorten; *brachytēs*, shortness. *Ex:* Brachinus (Ins.); Brachista (Ins.); brachisto-cephalic; brachy-cephalic; Brachy-deuterus (Pisc.); Brachy-teles (Mam.), see atel; Brachynus (Ins.); Brachyta (Ins.); Brachytes (Ins.).

brachyn—See **brachy.**

brachyt—See **brachy.**

bracon—NL. *bracon,* name applied to a genus of ichneumon-flies (etym. uncertain). *Ex:* Bracon (Ins.); Bracon-idae (Ins.); Habro-bracon (Ins.).

bract—L. *bractea,* a thin plate of metal, gold-leaf, veneer; *bracteatus,* covered with gold-plate, shining, shining only on the surface. *Ex:* bract; bracteate; bractei-form; Bracteon (Ins.).

brad—Gr. *brados = bradytēs,* slowness. *Ex:* Brado-ponera (Ins.).

bradi—See **brady.**

bradin—Gr. *bradinos,* Aeolic for *rhadinos,* pliant, slender; also nimble. *Ex:* Bradina (Ins.); Bradino-pyga (Ins.).

brady—Gr. *bradys,* slow. *Ex:* Bradi-cebus (Mam.); Brady-cinetus (Ins.); Brady-pus (Mam.).

brahma—1. NL. *brahma,* name applied to a genus of sea-breams. *Ex:* Brahma (Pisc.): **2.** Hind. *brahma,* the impersonal and absolute divinity, the divinity conceived as a god, the creator. *Ex:* Brahma-therium (Mam.).

branchell—NL. *branchellion,* a genus of leeches < Gr. *branchia,* gills+*ell*<*bdella,* a leech+*-ion.* *Ex:* Branchelli-idae (Ann.); Branchellion (Ann.).

branchi—Gr. *branchion,* a fin; pl. *branchia,* the gills of fishes. *Ex:* Branchi-obdella (Ann.); Branchi-pus (Crust.); Branchio-ellion (Ann.) <*branchio*+*bdella,* a leech; Branchio-pneusta (Moll.); Nudi-branchia (Moll.).

brant—NL. *branta*<Eng. *brant,* the brant goose <AS. *bernan, brennan,* to burn; the goose being so named because of the reddish brown coloration. *Ex:* Branta (Av.).

brassic L. *brassica, cabbage*<Celtic *bresic,* a cabbage. *Ex:* Brassica*.

braul—NL. *braula,* a generic name of unknown origin. *Ex:* Braul-idae (Ins.); Braula (Ins.).

brech—Gr. *brechō,* to wet, to water, to shower rain. *Ex:* Brech-ites (Moll.).

brechm—Gr. *brechmos = bregma,* the top part of the head, the head. *Ex:* Brechmo-tri-plax (Ins.).

brect—Gr. *brekieon,* one must soak. *Ex:* Halo-brecta (Ins.).

bregm—Gr. *bregma,* genit. *bregmatos,* the front of the head. *Ex:* bregma; Bregmato-thrips (Ins.); Bregmo-cer-ella (Crust.); Scali-bregma (Ann.).

brem—NL. *bremus,* prob.<Gr. *bremō,* to rage in anger, roar. *Ex:* Bremus (Ins.).

brent—See **brenth.**

brenth—Gr. *brenthos,* an unknown water bird of stately bearing; also haughtiness, pride; *brenthuomai,* to be proud, walk with stately grace. *Ex:* Brenthis (Ins.); Brenthus = Brentus (Ins.); Brenthus (Av.).

breph—Gr. *brephos,* a foetus, a new-born child; also a cub, whelp; *brephikos,* childish. *Ex:* brephic; Brepho-cotosia (Ins.); Brephos (Ins.).

brev—L. *brevis,* dim. *breviculus,* short. *Ex:* Brevi-pecten (Ins.); Brevi-pennes (Av.).

Short-headed Frog, *Breviceps,* an African ant- and termite-eating species.

brex—Gr. *brexis = bronchē,* a shower, rain, a wetting. *Ex:* Brexia*; Brexius (Ins.).

bri—Gr. *briaō,* to make or be strong. *Ex:* iso-bri-ous.

briar—Gr. *briaros,* strong, robust, hardy> *Briareos,* the mighty hundred-handed giant of Grecian mythology. *Ex:* Briar-axis (Ins.); Briareum (Coel.); Briareus (Coel.); Briaro-mys (Mam.).

brim—Gr. *brimē,* strength, bulk. *Ex:* Brimo-saurus (Rept.).

briseis—Gr. *Brisēis,* Briseus or Hippodamia, wife of Pelops

brising—NL. *Brisinga*<Icel. *Brisinga men,* the necklace of the mythological Brisingas of Scandinavia. *Ex:* Brising-opsis (Echin.); Brisinga (Echin.).

briss—Gr. *bryssos,* a kind of sea urchin. *Ex:* Briss-opsis (Echin.); Brisso spatangus (Echin.); Brissus (Echin.).

briz—Gr. *brizō,* to nod, be sleepy>NL. *briza,* a name applied to a genus of grasses (Briza) which includes the so-called "quaking grass"; this name may have been suggested by Gr. *briza,* a kind of rye-like grain growing in Macedonia. *Ex:* Briza*.

bro—Gr. *broō,* to eat. *Ex:* Poe-bro-therium (Mam.).

broch—1. Gr. *brochos,* a cord, mesh, loop. *Ex:* broch-ido-dromus, see -id 3; brochi-dromus; Brocho-sphaera (Por.); ?Brocho-peplus (Ins.); ?Campto-brochus (Ins.): **2.** L. *brochus,* (pertaining to animals) with projecting teeth; *brochatus,* having projecting teeth. *Ex:* Bro chata (Mam.); Brochus (Mam.): **3.** Gr. *brochē,* a shower.

brom—1. Gr. *brōma,* genit. *brōmatos,* food. *Ex:* bromatia; Ammo-broma*; Theo-broma*; theo-brom-ine: **2.** Gr. *bromos,* ancient Greek name for the oat. *Ex:* Bromus*: **3.** Gr. *Bromios,* a name for Bacchus. *Ex:* Bromius

(Ins.): **4.** Gr. *brōmos*, a rank smell, a poison. *Ex:* brom-idrosus (Med.); brom-ine; Hippo-bromia\*, poisonous to horses.

**bronch**—Gr. *bronchos*, wind-pipe; *bronchion*, pl. *bronchia*, the bronchial tube. *Ex:* bronchi-al; bronchio-stenosis (Med.); broncho-tomy (Med.). See also brex.

**bront**—Gr. *brontē*, thunder > *Brontēs*, the Thunderer, one of the three cyclopes. *Ex:* Bront-ornis (Av.); Bronto-saurus (Rept.), so named because of its giant size; Bronto-therium (Rept.).

**bros**—Gr. *brōsis*, genit. *brōseos*, meat, food; also a canker. *Ex:* Broseo-cnemis (Por.).

**brosim**—Gr. *brōsimos*, edible, esculent. *Ex:* Brosimum\*.

**brosm**—Norw. *brosme*, vernacular name of a certain fish. *Ex:* Brosme (Pisc.); Brosmo-phycis (Pisc.).

**brot**—**1.** Gr. *brōteos*, esculent, eatable; *brōtikos*, inclined to eat, voracious. *Ex:* Broteo-chactas (Arach.); Broto-mys (Mam.); Dia-brotica (Ins.): **2.** Gr. *brotos*, mortal, frail, infirm, i.e., human. *Ex:* brot-ium (Ecol.); broto-chore (Ecol.): **3.** Gr. *brotos*, blood, gore.

**broteas**—L. *Broteas*, one of the Lapithae slain by the centaur, Gryneus, at the marriage of Perseus. *Ex:* Broteas (Crust.).

**brotul**—Sp. *brotula*, name of a Cuban fish. *Ex:* Brotul-idae (Pisc.); Brotulus (Pisc.).

**bruc**—**1.** NL. *brucus* < Gr. *broukos*, a wingless locust. See bruch: **2.** NL. *brucus* < Gr. *brux*, the depth of the sea or, *bruchios*, of the depths of the sea.

**bruch**—L. *bruchus* < Gr. *brouchos* = *broukos*, name of a wingless locust. *Ex:* Bruchi-gavia (Av.); Brucho-ptinus (Ins.); Bruchus (Ins.).

**brum**—L. *bruma*, the winter solstice, hence winter; *brumalis*, pertaining to winter. *Ex:* brum-ous; Brumus (Ins.).

**brunne**—ML. *brunneus* = *bruneus*, dark brown < *brunus*, brown. *Ex:* brunn-escens; brunnei-capillus; brunneus.

**brut**—L. *brutus*, stupid, unwieldy, heavy.

**bry**—**1.** Gr. *bryō*, to be full of, to swell, sprout up, burst forth > Gr. *bryon*, a lichen, tree-moss, sea-weed. *Ex:* Bry-anthus\*; Bryo-nympha (Ins.); Bryo-phyllum\*; Bryo-phyta\*; Bryo-zoa; Bryum\*; em-bryum; Entomo-brya (Ins.): **2.** Gr. *bryōnia* = *bryōnē*, a kind of cucurbitaceous plant < *bryō*, to sprout up. *Ex:* Bryonia\*.

**bryc**—See brych.

**brych**—**1.** Gr. *brychō* = *brykō*, ppr. *brykon*, to gnaw, to eat with much noise, tear in pieces; also to roar or bellow. *Ex:* Brycon (Pisc.); Brycon-odon (Pisc.); Megalo-brycon (Pisc.); Ixo-brychus (Av.); Ono-brychis\*: **2.** Gr. *brychios*, deep in water, from the depths. *Ex:* Brychius (Ins.).

**brycon**—See brych.

A blennid fish, *Bryostemma polyactocephalum*, from marine waters of North America.

**bryx**—Gr. *bryx*, depth of the sea. *Ex:* Em-bryx (Pisc.).

**bu**—See boo.

**bubal**—Gr. *boubalos*, a kind of gazelle, buffalo. *Ex:* Bubalo-cephalus (Ins.); Bubalus (Mam.).

**bubo**—L. *bubo*, the great horned-owl. *Ex:* Bubo (Av.). See also bubon.

**bubon**—Gr. *boubōn*, the groin, a swelling in the groin > ML. *bubo*, genit. *bubonis*, a tumor. *Ex:* bubo; bubon-algia (Med.); bubon-ic; bubono-cele (Med.).

**bubul**—L. *bubulus*, of or concerning cattle.

**bucc**—L. *bucca*, the cheek, mouth cavity. *Ex:* bucc-al; bucc-ate; bucco-labi-al. See also bucco.

**buccin**—**1.** L. *bucinum* = *buccinum*, a shell-fish used in dyeing purple. *Ex:* Buccinum (Moll.); Buccino-triton (Moll.); Argo-buccinum (Moll.): **2.** L. *bucina* = *buccina*, a trumpet. *Ex:* buccini-form.

**buccinator**—L. *bucinator* = *buccinator*, a trumpeter < *bucina* = *buccina*, a trumpet. *Ex:* buccinator muscle.

**bucco**—L. *bucco*, genit. *bucconis*, a babbler < *bucca*, the cheek. *Ex:* Bucco (Av.); Bucconidae (Av.).

**bucculent**—L. *bucculentus*, with full cheeks or a large mouth.

**bucul**—L. *buculus*, a steer.

**bud**—Late ME. *budde*, a bud; Danish *bot*, a bud. *Ex:* bud.

**budyt**—Gr. *boudytēs*, name of some small bird, perh. the wagtail. *Ex:* Budyt-anthus (Av.); Budytes (Av.).

**buf**—L. *bufo*, genit. *bufonis*, a toad. *Ex:* Bufo (Amph.); Bufo-cephalus (Brach.); Bufon-idae (Amph.); Bufon-acris (Ins.); Bufon-ita (Echin.).

**bufon**—See buf.

**bugul**—L. *bugulus*, a female ornament such as a bunch of flowers. *Ex:* Bugula (Bry.); possibly Oken, who described the genus, had in mind a bunch of flowers in miniature. The Spanish word *bugula* refers to the labiate plant *Ajuga reptans*, which it might have been thought to resemble

**bulb**—L. *bulbus*, a bulb; *bulbosus*, full of bulbs, having bulbs. *Ex:* Bulbi-pora (Bry.); Bulbo-gaster (Ins.).

**bulg**—L. *bulga*, a leather sac or bag. *Ex:* Bulgaria\*; Bulga (Amph.).

**bulim**—1. Gr. *boulimos*, dearth, extreme hunger. *Ex:* bulimia (Med.): 2. NL. *bulimus*, name applied to a genus of mollusks, apparently a typographical error for *bulinus*=*bullinus*<L. *bulla*, a bubble, blister+dim. suffix *-inus*. *Ex:* Bulim-opsis (Moll.); Bulim-ulus (Moll.); bulimi-form; Bulim-inus (Moll.); Bulimus (Moll.).

**bull**—L. *bulla*, genit. *bullae*, pl. *bullae*, a large bubble, blister; *bullatus*, blistered. *Ex:* Bullaria (Moll.); Bull-oideus (Moll.); bulla ossea; Bullae-pus (Arach.); Bullata (Moll.); Bulli-opsis (Moll.); Bullo-pora (Prot.); Bullus (Moll.).

**bumel**—Gr. *boumelia*, a kind of ash. *Ex:* Bumelia*.

**bun**—Gr. *bounos*, a hill, mound. *Ex:* bun-odont; Buni-stygnus (Arach.); Bunium*; Buno-dactis (Coel.); Lepto-bunus (Arach.).

Lower bunodont molar of *Elotherium* (left). Lower lophodont molar of *Hyrachyus* (right). Redrawn from A Text-Book of Palaeontology—Zittel. The Macmillan Co.

**bund**—See **abund**.

**bupal**—Gr. *boupalis*, hard-struggling, obstinate, stubborn. *Ex:* Bupalo-mima (Ins.); Bupalus (Ins.).

**buprest**—Gr. *bouprēstis*, name of a poisonous beetle which, when eaten by cattle, caused them to swell up and die<*bous*, cow+*prēthō*, to swell. *Ex:* Buprestis (Ins.).

**burr**—L. *burrus*, red.

**burrescen**—NL. *burrescens*, genit. *burrescentis*, becoming red<L. *burrus*, red.

**burs**—Gr. *bursa*, a hide, skin>ML. *bursa*, a pouch, purse made of skin>ML. *bursarius*, a treasurer. *Ex:* bursa; Bursa (Moll.); Bursaria*, (Prot.); not Bursera*, named after Joachim Burser, 17th century German botanist; bursiform; Bursi-spongia (Por.).

**bus**—See **boo**.

**but**—L. *buteo*, a kind of falcon or hawk. *Ex:* But-aetos (Av.); Bute-aetos (Av.); Buteo (Av.); Buteo-astur (Av.); Archi-buteo (Av.).

**buteo**—See **but**.

**butio**—See **butorid**.

**butorid**—NL. *butorides*, name applied to a genus of birds, etym. uncertain, but probably allied to words arising from L. *butio*, genit. *butionis*, a bittern. *Ex:* Butorides (Av.).

**bux**—L. *buxus*, the box-tree. *Ex:* Bux-aceae*; Buxus*.

**byas**—Gr. *byas*, an owl. *Ex:* Byas (Ins.), (Av.).

**bybl**—Gr. *Byblis*, daughter of Miletus, who was changed into a fountain. *Ex:* Byblis*.

**byct**—Gr. *byktēs*, swelling, blustering; also whistling. *Ex:* Byct-iscus (Ins.).

**byrrh**—L. *byrrhos*, flame colored, yellowish-red <Gr. *pyrrhos*, red. *Ex:* Byrrho-morphus (Ins.), i.e., shaped like the insect, Byrrhus; Byrrhus (Ins.).

**byrs**—Gr. *byrsa*, a skin, hide. *Ex:* Byrs-ops (Ins.); Byrsa-lepsis (Ins.); Byrso-crypta (Ins.); Byrso-nima*, see nimi.

**byrsonim**—Gr. *byrseuō*, to tan, dress hides+L. *nimius*, beyond measure, excessive, but here taken to mean much used. *Ex:* Byrsonima*.

**byss**—Gr. *byssos*, a fine yellowish flax, fine thread. *Ex:* byss-aceus; Byssi-fera (Moll.); bysso-gen-ous, byssus.

**byth**—Gr. *bythos*, depth; *bythios*, of the deep, sunken; *bythitis*, a deep-sea animal. *Ex:* Bythites (Pisc.); Bythinus (Ins.); Bytho-cypris (Arth.). For Bythinia (Moll.) see bithynia.

# C

**caball**—L. *caballus*, horse. *Ex:* Caballus (Mam.).

**cabomb**—Guianan, *cabomba*, native name of the water-shield. *Ex:* Cabomba*.

**cac**—1. Gr. *kakos*, bad, mean, ignoble; *kakia*, badness, cowardice; *kakotēs*, badness. *Ex:* cacogenesis; Cacus (Ins.); A-caco-crinus (Echin.):

**2.** L. *caco*, pp. *cacatus*, to go to stool. *Ex:* cacat-ory (Med.).

**cacain**—ML. *cacainus*, chocolate brown < the name *Theobroma cacao*\*.

**cacali**—Gr. *kakalia*, a plant called the colt's foot. *Ex:* Cacali-opsis\*; Cacalia\*.

**cacat**—Malay. *kakatua*, the cockatoo parrot. *Ex:* Cacatoes (Av.); Cacatua (Av.). Not cacatory (Med.), see cac 2.

**cacc**—**1.** Gr. *kakkē*, ordure, excrement. *Ex:* Cacco-philus (Ins.): **2.** NL. *cacc*, sometimes used for *cac* < *kakos*, bad. *Ex:* Cacco-pitta (Av.).

**caccab**—**1.** Gr. *kakkabis*, a collateral form of *kakkabē*, < Skt. *kukkubha*, a partridge. *Ex:* Caccabis (Av.): **2.** Gr. *kakkabē;* also *kakkabos*, a pot, earthen vessel.

**cachinn**—L. *cachinno*, to laugh loudly; ppr. *cachinnans*, genit. *cachinnantis*, laughing, imitative. *Ex:* Cachinna (Av.).

**cachr**—Gr. *kachrys*, parched barley; also a catkin, a cone. *Ex:* cachri-form; Cachry-phora (Ins.); Cachrys\*, also said to be derived < Gr. *kaiō*, to burn, because of the carminative qualities of the plant.

**cacic**—Haytian *cacique*, chief. *Ex:* Cacicus (Av.).

**cact**—Gr. *kaktos*, a prickly plant. *Ex:* Cact-acae\*; Cact-ales\*; Cact-ornis (Av.); Cacto-phagus (Ins.); Echino-cactus\*.

**cacumen**—See cacumin.

**cacumin**—L. *cacumen*, genit. *cacuminis*, the very end, the limit.

**cad**—**1.** L. *cado*, to fall, pp. *cadens*, genit. *cadentis*, falling; *caducus*, falling early, deciduous. *Ex:* Caduca (Ins.); caduci-branchi-ate: **2.** Gr. *kados*, an urn. *Ex:* cado-phore.

**cadaver**—L. *cadaver*, a corpse < *cado*, to fall. *Ex:* cadaver; cadaver-ous.

**cadent**—See cad.

**cadmic**—NL. *cadmicus*, metallic, like tin < L *cadmia* < Gr. *kadmia*, an ore of zinc.

**caduc**—See cad.

**cae**—See cai.

**caec**—L. *caecum*, blind gut; also obscurity, uncertainty; *caecus*, devoid of light < *caecus*, blind. *Ex:* Caec-idotea (Crust.); Caeco-spaeroma (Crust.); caeco-stomy (Med.); Caecum (Moll.).

**caecator**—L. *caecator*, one who obstructs a fountain, one who makes blind.

**caecil**—L. *caecilia*, a kind of lizard, probably the "blind worm" < *caecus*, blind. *Ex:* Caecil-ophis = Coecil-ophis (Pisc.); Caecilia (Amph.); Caecilia = Coecilia (Pisc.); Caecilian-ella (Moll.).

**caecutien**—L. *caecutio*, to be blind; ppr. *caecutiens*, genit. *caecutientis*.

**cael**—**1.** L. *caelo*, to engrave, to carve; *caelum*, a graving tool; *caelatus*, carved in relief; *caelator*, a carver, engraver: **2.** NL. *cael*-,

sometimes used for *coel* < Gr. *koilos*, hollow *Ex:* Bi-cael-otus (Ins.): **3.** NL. *cael*-, also used for *coel* < L. *coelum* = *caelum*, sky.

**caeleb**—L. *caelebs*, genit. *caelibis* = *coelebs*; not married.

**caen**—Gr. *kainos*, new, recent. *Ex:* Caeno-crinus (Echin.); Caeno-lesthes (Mam.); Caeno-zo-ic; ceno-genesis. See also cen.

**caenis**—L. *Caenis*, genit. *Caenidis* a proper name. *Ex:* Caenis (Ins.).

**caeom**—NL. *caeoma*, a rust fungus < Gr. *kaiō*, to burn. *Ex:* Caeoma\*.

**caer**—See cair.

**caereb**—NL. *caereba*, an avian generic name < Braz. *quira* = *coereba*, name of some bird. *Ex:* Caereb-idae (Av.); Caereba (Av.).

**caerul**—L. *caeruleus* = *caerulus* = *coeruleus*, dark-colored, dark-blue.

**caesi**—L. *caesius*, the gray of the eye, light gray, bluish gray. *Ex:* Caesio-soma (Pisc.); not Caesia\* nor Caesio (Pisc.) which are derived from personal names.

**caesp**—L. *caespes* = *cespes*, sod, turf, somewhat tufted like sod-grass; *caespitosus*, tufted like the grass of a sod; *caespiticius*, made of turf.

**cafer**—L. *cafer*, of Caffraria, country of the Kafirs.

**caffr**—Ar. *kafir*, an infidel, pagan; also the name of a S. African tribe. See cafer. Caffr-anthus (Av.); Caffro-blatta (Ins.).

**cai**—Gr. *kaiō*, to burn. *Ex:* Cae-oma\*; cae-omo-spore; Caio-phora\*.

**caia**—L. *caia*, a cudgel, a club.

**caiman**—Guiana *cayman*, Sp. *caimen*, name for one of the American alligators. *Ex:* Caimen (Rept.).

**cain**—See caen.

**cair**—**1.** Gr. *kairos*, important, opportune, convenient. *Ex:* Epi-caerus (Ins.): **2.** Gr. *kairos*, a warp, a web in the loom.

**cakile**—Ar. *kakile*, a kind of sea-rocket. *Ex:* Cakile\*.

**cal**—Gr. *kalos*, beautiful. *Ex:* Cal-aster (Echin.); Cali-gorgia (Coel.); Calo-chortus\*; Calo-soma (Ins.) Micro-cale\*; See cali, also kal and call.

**caladi**—Malay *kalādi*, a plant name. *Ex:* Caladium\*.

**calam**—L. *calamus*, a reed; Gr. *kalamitēs*, reed-like < Gr. *kalamos*, a stalk, reed; L. *calamistrum*, an iron rod for curling the hair. *Ex:* Calam-agrostis\*; Calam-icthys (Pisc.); calami-folia; calamistrum; calamit-oid; Calamites\*; Calamo-spiza (Av.); calamus scriptorius.

**calamistr**—See calam.

**calan**—NL. *calanus*, a name for a genus of copepods (etym. unknown). *Ex:* Calan-idae

(Crust.);   Calanus   (Crust.);   Acro-calanus (Crust.).

**calandr**—**1.** Gr. *kalandros* = *kalandra*, a kind of lark. *Ex:* Calandria (Av.):    **2.** NL. *Calandra*, a genus of weevils < Fr. *calandre*, weevil. *Ex:* Calandr-idae (Ins.).

**calapp**—NL. *calappa*, an invented name for a crab genus. *Ex:* Calappa (Crust.); Calapp-idae (Crust.).

**calar**—L. *calo*, inf. *calare*, to call, to proclaim > *intercalo*, to proclaim that something has been inserted; *calatus*, called. *Ex:* intercalar-y.

**calat**—See **calar.**

**calath**—Gr. *kalathos*, a vase-shaped basket. *Ex:* Calath-odes*; Calathea*; Calathi-iscus (Por.); Calatho-spongia (Por.); Calathus (Ins.).

**calc**—L. *calx*, genit. *calcis*, lime, chalk; also the heel-bone; *calceus*, chalk-white; *calculus*, a pebble. *Ex:* calci-fic-ation; calcar-eous; calci-fuge; calculi-form; os calcis.

**calcane**—L. *calcaneum*, the heel < *calx*, chalk. *Ex:* calcaneo-cuboid; calcaneum.

**calcar**—**1.** L. *calcar*, genit. *calcaris*, a spur; *calcaratus*, furnished with a spur. *Ex:* Calcar-ina (Prot.); calcarate; calcari-formis; Cal-carius (Av.):    **2.** L. *calcarius*, pertaining to or of lime. *Ex:* Calcarea (Por.).

**calcarat**—See **calcar.**

**calce**—L. *calceus*, dim. *calceolus*, a shoe, slipper; *calceolarius*, one who makes a shoe. *Ex:* Calce-spongia (Por.); calcei-form; Calceo-crinus (Echin.); Calceolaria*.

**calceol**—See **calce.**

**calchaen**—Gr. *kalchainō*, to make purple; also to consider, to ponder. *Ex:* Calchaen-esthes (Ins.).

**calcin**—Low L. *calcino*, to reduce lime to powder with aid of heat (< L. *calx*, genit *calcis*); pp. *calcinatus*, calcined > Fr. *calcination*. *Ex:* calcine, calcin-osis; calcination.

**calcitr**—L. *calcitrans*, genit. *calcitrantis*, kicking, ppr. of *calcitro*, to kick.

**calcul**—See **calc.**

**calculi**—L. *calculus*, pl. *calculi*, a small stone, a pebble used in counting. *Ex:* renal calculi.

**calend**—L. *Kalendae*, the first day of the month. *Ex:* Calend-ula*, "monthly blooming."

**calendari**—L. *calendarium*. an account-book. *Ex:* calendarium florae.

**cali**—Gr. *kalia*, a dwelling, a bird's nest; *kalias*, genit. *kaliados*, a hut. *Ex:* calio-logy; Collo-calia (Av.); Eu-calia (Pisc.). See also cal.

**calic**—L. *calix*, genit. *calicis*, a cup (< Gr. *kalyx*, a cup), dim. *caliculus; caliculatus*, cupped. *Ex:* Calici-um*; caliculate.

**calid**—L. *calidum*, a hot drink; in general, hot, rash, spirited; *calide*, quickly, with promptness.

**calidr**—Gr. *kalidris*, name of some beach bird. *Ex:* Calidris (Av.).

**calig**—**1.** L. *caliga*, a boot; *caligatus*, booted. *Ex:* Caligus (Crust.):    **2.** L. *caligo*, darkness; *caliginosus*, full of darkness. *Ex:* caligin-ous; Caligo (Ins.).

**caliulus**—NL. *caliulus* a genus of millipedes, < *Cal.*, California + *iulus*, a milipede. *Ex:* Cal-iulus (Arthr.).

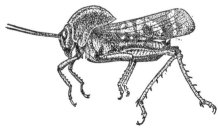

Crest-bearing Handsome-backed Locust, *Callonotacris lophophora*. Redrawn from Volume 35, Proceedings of United States National Museum.

**call**—Gr. *kalos*, beautiful; *kallos*, a beauty; *kallistos*, most beautiful. *Ex:* Call-eida (Ins.), see -eid; Call-othrus (Av.), see molothr; Calliope (Ave.), see op. 1.; Calli-tris*, the last element without meaning; Callistus (Av.).

**callae**—Gr. *kallaion*, a cock's comb. *Ex:* Callaeas (Av.); Callae-ops (Av.).

**callar**—Gr. *kallarias*, a kind of codfish. *Ex:* Callarias (Pisc.).

**callid**—L. *callidus*, experienced, skillful, cunning.

**callim**—Gr. *kallimos*, beautiful, *Ex:* Callimo(m)-idae (Ins.).

**callinic**—Gr. *kallynikos*, triumphant, with glory. *Ex:* Callinicus (Ins.).

**callio**—Gr. *kallion*, more beautiful, comp. of *kalos*. *Ex:* Callio-stoma (Moll.); Callio-tectum (Moll.).

**calliope**—Gr. *Kalliopē*, "the beautiful voiced," Muse of eloquence.

**callipare**—Gr. *kallipareios*, with beautiful cheeks. *Ex:* Callipareus (Ins.).

**callipheng**—Gr. *kalliphengēs*, beautiful, shining. *Ex:* Calliphenges (Ins.).

**callirrhoe**—Gr. *Kallirrhoē*, wife of Alcmaeon < *kalli*, beautiful + *rhoē*, to flow. *Ex:* Callirrhoe*.

**callist**—Gr. *kallistos*, most beautiful > *Kallisto*, daughter of Lycaon, changed by Juno into a she-bear. See call.

**callisthen**—Gr. *Kallisthenēs*, philosopher and youthful friend of Alexander the Great. *Ex:* Callisthenes*.

**callitris**—NL. *callitris*, generic name of the cypress pines < Gr. *kallos*, a beauty. *Ex:* Callitris*.

**callos**—L. *calleo*, to be callous, thick-skinned; *callum* = *callus*, hard skin; *callosus*, full of hard spots, hard; *callositas*, hard-skinned. *Ex:* cal-losity; callose; callus; corpus callosum.

**callun**—Gr. *kallynō*, to beautify, to look becomingly, to make clean. *Ex:* Calluna.*

**calor**—L. *calor*, genit. *caloris*, heat>Fr. *calorique*. *Ex:* calor-escence; calori-tropic; caloric.

**calp**—Gr. *kalpis*, genit. *kalpidos*, dim. *kalpion*, an urn. *Ex:* Calpe (Ins.); Calpi-carpum*; Calpia (Por.); Calpido-pora (Bry.); Acro-calpis (Prot.).

**calth**—L. *caltha*, the marsh marigold<Gr. *kalathos*, a vase-shaped basket, a goblet. *Ex:* Caltha*; calthi-folia.

**caltrop**—AS. *coltraeppe*, a thistle. *Ex:* caltrop =calthrop.

**calumni**—L. *calumniosus*, full of tricks<*calumnia*, craft, trickery.

**calv**—L. *calvus*, hairless, smooth; *calvatus*, made bare, bald; *calvaria*, the skull; *calvescens*, genit. *calvescentis*, becoming bald, ppr. of *calvesco*, to become bald. *Ex:* calvarium; calvus.

**calyc**—Gr. *kalyx*, genit. *kalykos*, a calyx, cup. *Ex:* Calyc-adenia*; Calyc-anthus*; Calyc-opsis (Coel.); calyci-form; calycle; Calyco-nectae (Coel.).

**calycin**—NL. *calycinus*, cup-like<L. *calyx*, genit. *calycis*+-*inus* an adj. ending.

**calycul**—L.*calyculus*, a small flower bud; also the hide of an animal.

**calymm**—Gr. *kalymma*, veil, hood; also the skull. *Ex:* Calymma (Moll.); Calymna (Moll.) and several other generic names of similar spelling are probably all errors for *calymma*.

**calymmene**—NL. *calymmene*<Gr. *kekalymmenos*, fem. *kekalymmenē*, covered, pp. pass. of *kalyptō*, to hide, cover. *Ex:* Calymmene (Tri.) =Calymene (Tri.).

**calymn**—See **calymm**.

**calyps**—Gr. *Kalypsō*, goddess of silence. *Ex* Calypso*.

**calypt**—1. Gr. *kalyptos*, covered, hidden; *kalyptēr*, a sheath, a covering; also a small chest. *Ex:* Calypto-rhynchia (Platy.); Eu-calyptus*: 2. Gr. *Kalyptē*, a proper name. *Ex:* Calypte (Av.).

**calyptr**—Gr. *kalyptra*, a covering for the head, veil. *Ex:* Calyptr-ata (Ins.); Calyptr-idium*; Calyptr-ura (Crust.); Calyptr-ur-aea (Moll.); calyptro-gen; Calyptro-phorus (Av.).

**calyx**—L. *calyx*<Gr. *kalyx*, the bud cup or calyx of a flower. See calyc. *Ex:* calyx.

**camar**—Gr. *kamara*, a vaulted chamber; *kamarōtos*, vaulted; L. *cameratio*, vaulting; *cameratus*, vaulted. *Ex:* Cama-rhynchus (Av.); Camara-saurus (Rept.); Camarata (Echin.); camero-stome; Camarota (Ins.); Camer-ina (Prot.); camer-ation.

**camarot**—See **camar**.

**camass**—Abor. *quamash*=*camass*, name of a food plant. *Ex:* Camassia*; Quamasia*.

**camax**—Gr. *kamax*, a stake. *Ex:* Camax (Ins.); Ophio-camax (Echin.).

**camb**—L. *cambio*, to exchange>LL. *cambium*, exchange. *Ex:* cambi-form; cambio-genetic; cambium.

**cambar**—NL. *cambarus*<Gr. *kammaros*, a sea crab. *Ex:* Cambarus (Crust.).

**cambr**—Welsh *Cambria*, Wales. *Ex:* Cambri-an.

**came**—Gr. *chamai*, on the ground, dwarf. *Ex:* Came-lina*, see lin 2. See also chamae.

**camel**—Gr. *kamēlos*, camel. *Ex:* Camel-ops (Mam.); Camelo-strongylus (Nem.); Camelus (Mam.); not Camelina* which is said to be derived from Gr. *chamai*, on the ground +*linon*, flax.

**camelin**—NL. *camelinus*, camel-colored<*camelus*, a camel. See also camel.

**camer**—See **camar**.

**camin**—Gr. *kaminos*, oven, kiln, chimney; also fire, warmth. *Ex:* Camin-ella (Por.); Camino-sphaera (Prot.).

**camp**—1. Gr. *kampē*, caterpillar, also a bending. *Ex:* Campe-philus (Av.); Campo-dea (Ins.), see eido; Campo-stoma (Pisc.); campodei-form; Clisio-campa (Ins.): 2. Gr. *kampos*, a sea animal. *Ex:* Proto-campus (Pisc.): 3. L. *campus*, a plain; *campestris*, relating to a plain, growing in a field. *Ex:* campestr-al; Epi-campes*; Lasio-camp-idae (Ins.).

**campal**—Gr. *kampaleos*, bending, winding, flexible. *Ex:* Campal-ita (Ins.).

**campan**—L. *campana*, dim. *campanula*, a bell. *Ex:* Campan-opsis (Coel.); campani-form; Campanula (Coel.),*; campanul-ate.

**campestr**—See **camp 3**.

**campil**—See **campyl**.

**campilo**—See **campyl**.

**campode**—See **camp 1**.

**camps**—Gr. *kampsis*, a curve, a bending. *Ex:* Camps-urus (Ins.); Campsis*; Campso-scolia (Ins.).

**campt**—Gr. *kamptō*, to bend; *kamptos*, flexible, bent; *kamptikos*, flexible. *Ex:* Campt-elasmus (Ins.); campto-drome; Campto-laemus (Av.); Campto-sorus (Pisc.); A-campto-pappus*; Trachelo-camptus (Arach.).

**campul**—See **campyl**.

**campyl**—Gr. *kampylos*, bent, curved. *Ex:* Campilo-rhynchus (Av.); Campuli-pus (Ins.); campylo-dromus; Campylo-neuron*; Campy lus (Ins.).

**camur**—L. *camur*, crooked, turned inwards.

**can**—1. L. *canis*, dog; *caninus*, pertaining to or belonging to a dog. *Ex:* Can-idae (Mam.); Canis (Mam.): 2. L. *cano*, to sing; *canor*, song, melody; *canorus*, tuneful: 3. L. *canus*, white, hoary, foamy, gray. *Ex:* cani-ceps: 4. Gr. *kanon*, a straight rod, bar, a rule. *Ex:* Cano-pholis*; canon bone.

The silent hunting, wolf-like, wild Dingo Dog, *Canis dingo*, of Australia. Dingo is a native word.

**canabin**—Gr. *kanabinos*, slender, thin as a rod. *Ex:* Canabina (Av.).

**canace**—L. *Canace*, a fem. proper name. *Ex:* Canace (Ins.), (Av.); Canace-idae (Ins.).

**canach**—Gr. *kanacheō*, to make a noise, to crow; *kanachos*, noisy. *Ex:* Canach-ites (Av.).

**canad**—NL. *Canada*, in North America. *Ex:* Canado-ceras (Moll.); Canadi-pterus (Pisc.).

**canal**—L. *canalis*, dim. *canaliculus*, a channel, conduit; *canaliculatus*, channeled. *Ex:* Canali-pora (Bry.); Canali-spira (Moll.); canalicul-ar.

**canari**—Malay *kanari*, the Java almond. *Ex:* Canarium*.

**canaster**—L. *canastor*, semi-gray, grizzled.

**cancell**—L. *cancelli*, pl. of *cancellus*, lattice-work, an enclosure of gratings or bars. *Ex:* Cancell-aria (Moll.); Cancello-carabrus (Ins.); Cancella (Moll.).

**cancer**—L. *cancer*, genit. *cancris*, also *canceris*, a crab; also an ulcer. *Ex:* Cancer (Crust.); cancer (Path.); Cancri-cepon (Crust.); cancri-form; Cancro-phagus (Av.).

**cancr**—See cancer.

**candescen**—L. *candescens*, genit. *candescentis*, shining, ppr. of *candesco*, to shine, dazzle. *Ex:* in-candescent.

**candican**—L. *candicans*, genit. *candicantis*, whitening, ppr. of *candico*, to be whitish or white.

**candid**—L. *candidus*, white, clear, shining < *candeo*, to be brilliant, shine.

**candidat**—L. *candidatus*, one dressed in white robes, a candidate for office < *candidus*, white.

**cane**—Gr. *kaneon*, a wicker basket; *kanēs*, a basket, mat of reeds. *Ex:* Cane-phora*; Cane-phorus (Ann.).

**canell**—Fr. *canelle*, cinnamon. *Ex:* Canella*.

**canen**—L. *canens*, genit. *canentis*, graying, ppr. of *caneo*, to be gray or hoary.

**canescen**—L. *canescens*, genit. *canescentis*, becoming gray, ppr. of *canesco*, to become white or hoary.

**canin**—See can 1.

**canistr**—Gr. *kanistron*, a wicker basket; also an earthen vessel. *Ex:* Canistro-crinus (Echin.); Canistrum (Moll.).

**canit**—L. *canities* = *canitia*, a gray or grayish-white color.

**canitud**—L. *canitudo*, genit. *canitudinis*, a gray color.

**cann**—L. *canna*, dim. *cannula*, a reed, tube, tunnel < Gr. *kanna*, reed, a reed-mat. *Ex:* Cann-ella*; Canna*; Canno-capsa (Prot.); Canno-rhapis (Prot.); Cannul-aria (Moll.).

**cannab**—Gr. *kannabis*, a kind of plant, hemp > L. *cannabis*, hemp; *cannabinus*, pertaining to hemp. *Ex:* Cannabin-aceae*; Cannabis*.

**canon**—L. *canon*, genit. *canonis*, a rule, model, a measuring line; also a channel in a hollow reed; Gr. *kanonikos*, regular. *Ex:* Canon-anthus*; Canon-ura (Ins.).

**canop**—NL. *canopium*, a tent: *Ex:* canop-y.

**canopus**—Gr. *Kanōpos*, brightest star in the constellation Argo; also, a town in lower Egypt. *Ex:* Canopus (Prot.).

**canor**—See can 2.

**canotia**—Mex. *canotia*, a native plant name. *Ex:* Canotia*.

**cansu**—Chinese *Kansu*, name of a province of southern China. *Ex:* Cansu-mys (Mam.).

**cantan**—L. *cantans*, genit. *cantantis*, singing, ppr. of *canto*, to sing.

**canth**—Gr. *kanthos*, corner of the eye. *Ex:* Canth-orbis (Moll.); Cantho-stomus (Ins.); canthus; not Acanthus*. In a number of fish genera described by Craig (Canthi-leptes, Canthi-rynchus, etc.), also in the molluscan subgenus Canthidomus (Swainson), the first element of the generic name is derived < Gr. *akantha*, a spine.

**canthar**—1. L. *cantharis*, the Spanish fly < Gr. *kantharis*, a "beetle-fly." *Ex:* Canthar-idae (Ins.): 2. Gr. *kantharos*, a scarab beetle; also a drinking cup, a boat, canoe. *Ex:* Canthar-ellus*; Canthari-ella (Moll.); Cantharo-spyris (Prot.); Cantharus (Prot.), (Moll.).

**canthium**—Malabar, *canti*, a plant name. *Ex:* Canthium*.

**canthon**—Gr. *kanthōn*, a pack-ass, a name applied in jest to a beetle. *Ex:* Canthon (Ins.); Canthono-soma (Ins.).

**cantor**—L. *cantor*, a singer, a poet.

**canut**—L. *canutus*, white, frosty, gray-haired. S.

**cap**—Gr. *kapē*, a manger, crib. *Ex:* Formici-capa (Av.).

**capac**—L. *capax*, genit. *capacis*, wide, large, roomy. *Ex:* capaci-ous.

**capax**—See capac.

**capell**—L. *capella*, a she-goat; also the name of a star in the constellation Auriga; *capellianus*, of or pertaining to kids, goats. *Ex:* Capell-aria (Pisc.); Capell-inia (Brach.); Capella (Mam.).

**caper**—L. *caper*, goat; *caperatus*, provided with goats. *Ex:* Caper (Mam.); Caperea (Mam.).

**caperat**—L. *caperatus*, drawn into folds, wrinkled <*capero*, to wrinkle. See **caper.**

**capill**—L. *capillus*, hair; *capillaris*, of or pertaining to the hair, considered collectively. *Ex:* Capill-aria (Nem.); capillary; Capilli-rhynchia (Brach.); Capillo-spir-ura (Nem.).

**capillit**—L. *capillitium*, the hair (in a collective sense.): *Ex:* capillitium, pl., -tia.

**capistr**—L. *capistrum*, a halter, band, muzzle, mask; *capistratus*, provided with a mask. *Ex:* Capistro-cardia (Moll.); capistrum.

**capit**—L. *caput*, genit. *capitis*, dim. *capitulum* = *capitellum*, head; *capitatus*, having a head; *capito*, one that has a large head. *Ex:* Capit-ell-idae (Ann.); Capito (Av.); Capito(n)-idae (Av.); capitulum; caput.

**capitan**—L. *capitaneus*, of large size.

**capn**—Gr. *kapnos*, smoke, vapor; *kapnōdēs*, smoky; also a plant called fumitory and in this sense often used in plant names. *Ex:* Capnoides*; Capnia (Ins.); Capno-botes (Ins.); Capnodes (Ins.); Capnodium*; a-capnia; Dactylo-capnos*.

**cappar**—Gr. *kapparis*, a kind of plant, the caper <Pers. *kabar*, the caper. *Ex:* Cappar-idaceae*; Capparis*.

**capr**—1. L. *caper*, a goat, dim. *caprella*; *capra*, a she-goat; *caprinus*, relating to goats. *Ex:* Capr-aria*; Capri-mulg-idae (Av.); capri-ped; Caprina (Ins.); Antilo-capra (Mam.): 2. Gr. *kapros*, dim. *kapriskos*, a wild boar; *kaprios*, like a wild boar; also a name applied to a seafish. *Ex:* Capr-osma*; Caprios (Mam.); Capro-idae (Pisc.); Capro-mys (Mam.); Cap rus (Pisc.).

**capreol**—1. L. *capreolus*, a tendril; *capreolatus* tendrilled, in twisting manner: 2. L. *capreus*, dim. *capreolus*, a wild goat. *Ex:* Capreolus (Mam.).

**capron**—L. *capronae*, the hair hanging over the forehead, forelock.

**caprotin**—L. *Caprotina*, a name for Juno: *Ex:* Caprotina (Moll.).

**caps**—1. Gr. *kaptō*, to gulp down, eat quickly. *Ex:* Capsus (Ins.); Carpo-capsa (Ins.): 2. L. *capsa*, dim. *capsula*, a box, case < Gr. *kapsa*, a box; L. *capsulatus*, boxed. *Ex:* Caps-icum*, the last element evidently without meaning; Capsa*; Capsul-aria (Brach.); capsule.

**capsul**—See **caps.**

**capt**—L. *captus*, taken, captured <*capio*, to take, seize. *Ex:* capt-ula; capta-culum.

**capul**—L. *capulus*, a handle; also a tomb: *Ex:* Capulus (Moll.).

**caput**—See **capit.**

**capy**—Gr. *kapyō*, to breath, gasp. *Ex:* Capya (Av.).

**car**—Gr. *kara*, head, the top of anything. *Ex:* Car-acanthus (Pisc.); Car-ilias (Ins.); Care-mitrus (Pisc.); Aulo-cara (Ins.); Cono-cara (Pisc.); zesto-carus. See also carid.

**carab**—Gr. *karabos* = *karabis*, genit. *karabidos*, a horned beetle; also a spiny lobster. *Ex:* Carab-idae (Ins.); carab-oid; Carab-ops (Ins.); Carabi-phagus (Ins.); Carabo-morphus (Ins.); Carabus (Ins.).

**caracall**—L. *caracalla*, a hooded cloak or tunic. *Ex:* Caracalla (Moll.).

**caracoll**—Sp. *caracol*, a snail. *Ex:* Caracolla (Moll.); Caracoll-ina (Moll.).

**caragan**—Tartar *caragan*, name of a tree. *Ex:* Caragana*.

**carang**—Sp. *caranga*, a flatfish of the West Indies >NL. *caranx*. *Ex:* Carango-morus (Pisc.); Caranx (Pisc.); Usa-caranx (Pisc.).

**caranx**—See **carang.**

**carap**-1. Guianan *carapa*, name of the crab-wood tree. *Ex:* Carapa*: 2. Tupi *carapo*, name of a fish. *Ex:* Carapus (Pisc.).

**carapac**—Fr. *carapace*, a shell, shield. *Ex:* carapace; carapaci-al.

**carausi**—L. *Carausius*, masc. proper name. *Ex:* Carausius (Ins.).

**carb**—L. *carbo* >Fr. *carbone*, coal. *Ex:* Carb-oxydo-monas*; carb-uncle.

**carcer**—L. *carcer*, NL. dim. *carcerulus*, a prison; *carceralis*, pertaining to a prison. *Ex:* carcerule.

**carchar**—Gr. *karcharos*, sharp, jagged >*karcharias*, a kind of dog-fish having jagged teeth. *Ex:* Carcha-rhinus (Elasm.); Carchar-odon (Elasm.); Carcharias (Elasm.).

**carches**—Gr. *karchēsion*, bowl, drinking-cup; also a ship. *Ex:* Carchesium (Prot.).

**carcin**—Gr. *karkinos*, a crab; also an eating sore, a snare. *Ex:* Carcin-aspis (Crust.); carcin-oma (Path.); Carcino-coris (Ins.); carcino-logy; Carcinus (Crust.).

**carcith**—NL. *carcithium* < Gr. *karkinousthai*, to become entangled, as roots. *Ex:* carcithium.

**card**—L. *cardo*, genit. *cardinis*, a hinge; related to *cardinalis*, pertaining to a hinge, that on which something turns, depends. *Ex:* Cardio-derma (Mam.); cardo; E-cardines (Brach.).

**cardamin**—Gr. *kardaminē*, name cited by Dioscorides for a species of cress. *Ex:* Cardamine*.

**cardi**—Gr. *kardia*, heart; *kardiakos*, belonging to the heart. *Ex:* Card-issa (Moll.); card-itis (Med.); Cardi-soma (Crust.); cardiac; cardio-lith; Cardit-opsis (Moll.); Cardita (Moll.); Cardium (Moll.); Mexi-cardia (Moll.).

**cardinal**—L. *cardinalis*, chief, principal; also later meaning red. *Ex:* cardinal vein; Cardi nalis (Av.), so named because the plumage of the birds belonging to this genus assumes a red color much like that with which the garb of cardinals is dyed.

**cardit**—NL. *cardita*<Gr. *kardia*, heart+Gr. *-itēs*, of the nature of. *Ex:* Cardit-idae (Moll.); Cardita (Moll.).

**cardu**—L. *carduus*<Gr. *kardos*, a thistle. *Ex:* Cardu-ella (Av.); Cardui-folia (Prot.); Cardu us*.

**carebar**—Gr. *karēbarēs*, top-heavy, heavy in the head. *Ex:* Carebar-ella (Ins.); Carebara (Ins.).

**caren**—Gr. *karēnon*, the head. *Ex:* Caren-archus (Ins.); Careno-scaphus (Ins.); Ambylo-care num (Arach.).

**carett**—Sp. *carey*, name of a turtle>NL. *caretta*. *Ex:* Caretta (Rept.); Caretto-chelys (Rept.).

**carex**—See caric.

**cariac**—SA. native name *cariacou*, a kind of deer. *Ex:* Cariacus (Mam.).

**cariam**—Tupi *cariama* = *ceriema* = *sariama*, name of a bird, the crested screamer. *Ex:* Cariama (Av.).

**caric**—1. L. *carex*, genit. *caricis*, sedge. *Ex:* Carex*; caric-etum; carico-log-ist. See also cary. 2. L. *carica*, a kind of dry fig. *Ex:* caric-ous; Carica*; Carici-cola (Av.).

**carid**—L. *caris*, genit. *caridis*, a shrimp. *Ex:* Caris (Crust.); Car-idae (Crust.); Cari-cyphius (Crust.); Carid-erus (Ins.), see der; Carid-ops (Ins.); Carid-opthalmus (Ins.); Carides (Crust.); Carido-pus (Ins.); Hymeno-caris (Crust.).

**caries**—See carios.

**carin**—L. *carina*, a keel; *carinatus*, keel-formed. *Ex:* carin al; Carin-aria (Moll.); Carin-atae (Av.); Carin-oma (Nemert.); Carini-ana*.

**carios**—L. *caries*, decay>*cariosus*, much decayed. *Ex:* caries (Path.); carioso-cancellate.

**caris**—See carid.

**cariss**—NL. *carissa*, perh.<Sanskr. name for some plant. *Ex:* Carissa*.

**carminat**—1. L. *carminatus*, carded, cleansed; pp. of *carmino*, to card, comb, cleanse. *Ex:* carminat-ive. 2. Fr. *carmin*, carmine. *Ex:* carmin-atus.

**carn**—L. *caro*, genit. *carnis*, flesh; *carneus*, fleshy, flesh-colored; *carnosus*, fleshy, pulpy; *carnatio*, genit. *carnationis*, fleshiness; *carnifex*, an executioner; Fr. *carnassier*, flesh-eating. *Ex:* car nassi-al; carneus; Carni-vora (Mam.); Carni fex (Av.); A-carnus (Por.).

**carnat**—See carn.

**carnosit**—NL. *carnositas*, fleshiness.

**carot**—L. *carota*, carrot<Gr. *karōton*, a carrot. *Ex:* carot-ine.

**carotid**—Gr. *karōtides*, the great arteries of the neck<*karos*, heavy sleep, stupor. *Ex:* carotid artery.

**carp**—1. L. *carpus*, the wrist-joint. *Ex:* carp-al; carpo-pod-ite; carpus: 2. Gr. *karpos*, fruit. *Ex:* carpel; Carpo-capsa (Ins.); Carpo-dacus (Av.); Carpo-mys (Mam.); carpo-phore; Adeno-carpus*.

**carpalim**—Gr. *karpalimos*, quick. *Ex:* Carpali mus (Ins.).

**carpel**—See carp 2.

**carpesi**—Gr. *karpēsion*, an aromatic medicinal wood from Asia. *Ex:* Carpesium*.

**carph**—Gr. *karphos*, a splinter, twig, chaff, straw. *Ex:* Carph-ophis (Rept.); Carphe-phorus*; Carpho-soma (Ins.); Hemi-carphos*.

**carphal**—Gr. *karphaleos*, dry. *Ex:* Carphalea*.

**carpi**—Low L. *carpio*, the carp. *Ex:* Carpi-odes (Pisc.).

**carpin**—L. *carpinus*, a kind of plant, the horn-beam. *Ex:* Carpinus*.

**carpt**—L. *carptus*, divided, separated; also plucked.

**cars**—Gr. *karsis*, a shearing, clipping. *Ex:* Carso-saurus (Rept.); Dino-carsis*.

**carsio**—Gr. *karsios*, crooked, sideways, oblique.

**cart**—Gr. *kartos*, strong. *Ex:* Carto-dere (Ins.)

**cartall**—Gr. *kartallos*, a basket with narrow bottom. *Ex:* Cartallum (Ins.).

**carter**—Gr. *karteros*, strong. *Ex:* Carter-odon (Mam.); Cartero-phonus (Ins.).

**carteric**—Gr. *karterikos*, patient, enduring. *Ex:* Carterica (Ins.).

**cartham**—Ar. *quartam*, the safflower. *Ex:* Car-thamus*.

**cartilag**—L. *cartilago*, genit. *cartilaginis*, carti-lage, gristle; *cartilagineus*, gristly. *Ex:* carti lage; cartilagin-ous.

**carum**—Gr. *karon*, name for the caraway. *Ex:* Carum*.

**caruncl**—See caruncul.

**caruncul**—L. *caro*, dim. *caruncula*, a piece of flesh. *Ex:* caruncle; Caruncul-ina (Moll.).

**cary**—Gr. *karyon*, a nut, a walnut. *Ex:* Cary-amoeba (Prot.); cary-opsis; Carya*; caryo-car; caryo-kinesis; caryo-lite; Caryo-phyllus*; caryo-some; Karyo-spora (Prot.); mono-caric; Ptero-carya*.

**caryedon**—Gr. *karyēdon*, like a split or splintered nut. *Ex:* Caryedon (Ins.).

**caryophyll**—Gr. *karyophyllon*, a kind of plant, the clove-tree, altered from Ar. *Qarumfel*. *Ex:* Caryophyllum*.

**caryota**—Gr. *karyōtos*, a kind of palm with nut-like fruit. *Ex:* Caryota*.

**casarc**—South Russian *kasarka*, a goose. *Ex:* Casarca (Av.).

**casc**—L. *cascus*, old.

**cascara**—Sp. *cascara*, bark. *Ex:* Cascara*.

**case**—L. *caseus*, cheese. *Ex:* casein.

**casignet**—Gr. *kasignētos*, a brother. *Ex:* Casig neta (Ins.); Casignetus (Ins.).

**casm**—See chasm.

**cass**—1. L. *cassis*, genit. *cassidis*, a helmet; *cassideus*, helmet-shaped. *Ex:* Cassi-gobius

(Pisc.); Cassida (Ins.); Cassidix (Av.); Cassis (Moll.); Cassis-ul-ina (Prot.): **2**. L. *cassus*, empty, hollow > *incassum*, in vain.

cassia—Gr. *kassia*, ancient name for some leguminous plant. *Ex:* Cassia*.

cassid—See cass 1.

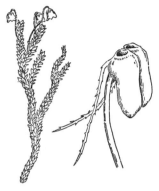

Mertens's Cassandra, *Cassiope mertensiana*, with details of anther.

cassiop—Gr. *Kassiopeia* = *Kassiopē*, wife of Cepheus and mother of Andromeda. *Ex:* Cassiope*; Cassiopea (Coel.).

cassiter—L. *cassiterum* < Gr. *kassiteros*, tin; L. *cassiterinus*, made of tin.

cast—L. *castus*, pure, spotless, without blemish. *Ex:* caste.

castal—Gr. *Kastalia*, famed fountain of Mt. Parnassus. *Ex:* Castali-ella (Moll.); Castalia (Moll.).

castan—Gr. *kastanos*, the chestnut tree > NL. *castaneus*, of chestnut color. *Ex:* Castan-idium (Prot.); Castan-opsis*; Castanea*; Castanocrinus (Echin.).

caste—L. *castus*, pure. *Ex:* caste.

castell—L. *castellum*, a castle; ML. *castellanum*, keeper of a castle. *Ex:* Castellania (Prot.); Castello-trachia (Moll.).

castni—NL. *castnia*, a Fabrician generic name of unknown meaning. *Ex:* Castni-idae (Ins.); Castnia (Ins.).

castor—Gr. *kastōr*, the beaver. *Ex:* Castor (Mam.); Castor-oïdes (Mam.); Castoro-mys (Mam.).

castr—L. *castro*, to deprive of generative power, emasculate > *castratus*, gelded. *Ex:* Castr-ado (Rot.); castr-ation.

casual—L. *casualis*, occasional, casual.

casuari—ML. *casuarius*, the cassowary < Malay *kasuari*. *Ex:* Casuar-ina*; Casuarius (Av.).

cat—**1**. Gr. *kata*, downward, inferior; *katai* poet. form of *kata*, motion from above, downwards. *Ex:* Cat-herpes (Av.); cata-bol-ism; cata-crotic; Cata-rrhina (Mam.); Cata-tropis (Platy.); Cato-stomus (Pisc.); Kata-mysis (Crust.); Katae-tyx (Pisc.); Kato-tropa (Tun.);

Phago-cata (Platy.): **2**. L. *catus*, dim. *catulus*, cat. *Ex:* Cato-lynx (Mam.); Cato-puma (Mam.); Catulus (Pisc.); Catus (Mam.): **3**. L. *catus*, as adj., means sharp-eyed, keen.

catabros—Gr. *katabrōsis*, a devouring. *Ex:* Cata brosa*.

catact—Gr. *kataktēs*, a lodger, one who turns into and lodges in a house < *katagō*, to lead into a place, to lodge; *kataktēs* may also be derived from *katagnumi*, to break, to shatter. *Ex:* caryo-catactes, a nut-cracker.

catalepsis—Gr. *katalēpsis*, a seizure. *Ex:* cata lepsis.

catalpa—Am. Indian *catalpa*, the catalpa tree. *Ex:* Catalpa*.

catant—Gr. *katantēs*, downward. *Ex:* Catanto-stoma (Moll.).

cataphan—Gr. *kataphaneia*, clearness, transparency. *Ex:* Cataphania (Av.).

cataphoric—Gr. *kataphorikos*, violent. *Ex:* Cata phoricus (Ins.).

cataphract—Gr. *kataphractos*, clad in armor, decked. *Ex:* Cataphractus (Mam.).

catapon—Gr. *kataponeō*, to tire out. *Ex:* Cata ponera (Av.).

catari—NL. *catarius*, of cats < L. *cata*, a cat.

cataulac—NL. *cataulacus*, furrowed < Gr. *kataulakizō*, to plough, till. *Ex:* Cataulaces (Ins.); Cataulax (Ins.).

cataulax—See cataulac.

catech—NL. *catechu* < Malay *kachu*, a plant extract.

caten—L. *catena*, dim. *catenula*, a chain; *catenatus*, chained, connected by chains; *catenatio*, a band, clamp. *Ex:* Catenula (Platy.); catenation; Cateni-pora (Coel.); Cateno-taenia (Platy.).

cateor—Gr. *kateōros*, hanging down. *Ex:* Cateo rus (Mam.).

cateres—Gr. *katērēs*, fitted out, furnished. *Ex:* Lopho-cateres (Ins.).

cathamm—Gr. *kathamma*, anything tied. *Ex:* cathamm-al.

cathar—Gr. *katharos*, pure; *kathartēs*, a purifier, cleanser; *kathartikos*, cleansing. *Ex:* Catharista (Av.); Catharo-peza (Av.); Cathartes (Av.); cathartic (Med.); Catharto-carpus*; Catharus (Av.).

catharm—Gr. *katharma*, refuse, the residuum < *kathairō*, to cleanse; *katharmos*, a cleansing. *Ex:* catharma (Med.).

cathart—See cathar.

cathea—See cathem.

cathedr—Gr. *kathedra*, a seat, chair. *Ex:* Cathe dra (Ins.); cathedrus.

cathem—Gr. *kathēmai*, to sit. *Ex:* La-cathea*.

cathestec—Gr. *kathestēkotōs*, stationary. *Ex:* Cathestecum*.

cathet—Gr. *kathetos*, hanging down, perpendicular; *kathiēmi*, to send down, thrust in; *kathetēr*, a thing put in. *Ex:* Cathet-urus (Av.); cathe ter; Catheter-onchyia (Ins.); Catheto-rhinus (Rept.).

cathor—Gr. *kathoraō*, to look down. *Ex:* Cathor-ops (Pisc.).

cathormi—Gr. *kathormion*, a chain, necklace. *Ex:* Cathormio-ceras (Ins.).

catill—Gr. *katillō*=*kateileō*, to force into a narrow space, to enclose, roll up, fold up. *Ex:* Catillo-crinus (Echin.); Catillus (Moll.), or perh. from L. *catillo*, a glutton. See also catin.

catin—L. *catinus*, dim. *catinulus*=*catillus*, a bowl, a hollow in a rock; *catillo*, a glutton< *catillo*, to lick a plate. *Ex:* Catin-ella (Moll.); Catin-ulus (Moll.), (Prot.); Catinus (Moll.).

catkin—Eng. *catkin* (<*cat*, a cat+dim. -*kin*), an ament, so called from its resemblance to the tail of a kitten.

catom—L. *catomus*, the shoulders.

catopt—Gr. *katoptos*, conspicuous. *Ex:* Catopto-metra (Echin.).

catoptr—Gr. *katoptron*, a mirror. *Ex:* Catoptria (Ins.); Catoptro-pteryx (Ins.).

caucalis—Gr. *kaukalis*, name of the hedge-parsley. *Ex:* Caucalis*.

caucas—Gr. *Kaukasos*, the Caucasus; *Kaukasianoi*, the inhabitants of the Caucasus. *Ex:* Caucas-ella (Brach.); Caucasi-gena (Moll.); Caucaso-cressa (Moll.).

caud—L. *cauda*, tail; *caudatus*, having a tail. *Ex:* caud-ad; caud-al; Caud-orchis (Platy.); cauda equina; Caudata (Amph.); caudate; Caudolanius (Av.).

caudescens—NL. *caudescens*, becoming stem-like<*caudex*, a stem.

caudex—See caudic.

caudic—L. *caudex*, genit. *caudicis*, dim. *caudiculus*, the stem of a tree. *Ex:* caudex cerebri; caudicle.

caul—1. *caulis*, dim. *cauliculus*<Gr. *kaulos*, the the stem of a plant, a cabbage stalk, cabbage. *Ex:* Caul-anthus*; cauli-flower; cauli-fer-ous; caulis; caulo-mer; hydro-caulus: 2. ME. *calle*, covering. *Ex:* caul: 3. Gr. *kaulos*, the penis. *Ex:* macro-caulia.

caulesc—Fr. *caulescent*, provided with an evident stem < L. *caulis*, stem. *Ex:* a-caulescent.

caum—Gr. *kauma*, genit. *kaumatos*, burning heat.

caur—L. *Caurus*, the northwest wind; *caurinus*, of the northwest wind, northwestern.

caus—Gr. *kausos*, heat<*kaiō*, fut. *kausō*, to burn; *kaustikos*, burning; *kautēriazō*, to burn with a branding-iron. *Ex:* caustic; Caustis*; Causto-loma (Ins.); cauterize; Helio-causus (Ins.). See also caut.

caust—See caus.

caut—1. L. *cautus*, guarded, avoided: 2. Gr.

*kautos*=*kaustos*, burned. *Ex:* Epi-cauta (Ins.).

cav—1. L. *cavea*, a cave, an excavated place; *cavus*, hollow, excavated; *cavatus*, hollowed; *cavator*, an excavator. *Ex:* cavi-colous; cavi-corn; Cavi-lucina (Moll.); cavit-y; Cavo-tettix (Ins.); Cavus-gnathus (Ann.); vena cava: 2. Braz. *cavia*, name for a rodent. *Ex:* Cavi-odon (Mam.); Cavia (Mam.).

cavat—See cav 1.

cavatic—L. *cavaticus*, living in hollows, caves.

caveat—L. *caveatus*, encaged, cooped up.

cavern—L. *caverna*, a cavern, hollow; *cavenrosus*, full of hollows. *Ex:* Cavern-acris (Ins.); Cavernae-cola (Ann.); Cavern-ul-aria (Coel.).

cavill—L. *cavillator*, a joker, jester.

-ce—Eng. -*ce*, adj. suffix denoting quality or state of being. *Ex:* distan(t)-ce; sibilan(t)-ce; silen(t)-ce.

ce—See ceo.

ceanoth—Gr. *keanōthos*, name used by Dioscorides for some spiny plant. *Ex:* Ceanothus*.

ceb—Gr. *kēbos*, a long-tailed monkey. *Ex:* Ceb-id-ichthys (Pisc.); Ceb-idae (Mam.); cebo-cephalia (Med.); Cebu-gale (Mam.); Cebus (Mam.).

ceble—Gr. *keblē*, a contraction of *kephalē*, head. *Ex:* Ceble-pyrus (Av.); Neo-ccble (Ins.).

cebrio—NL. *cebrio*, perh.<Gr. *kebrionēs*, an unknown bird of ancient Greece. *Ex:* Cebrio (Ins.); Cebrio-rhiphis (Ins.); Cebrion-idae (Ins.).

cebrion—See cebrio.

cebu—See ceb.

cec—Gr. *kēx*, genit. *kēkos*, a kind of sea-bird. *Ex:* Ceco-morphae (Av.). See also caec.

cechen—Gr. *kechēnōs*, yawning, gaping, silly; *kechēnōdes*, forming a hiatus; *kechēnotōs*, open-mouthed. *Ex:* Cecheno-sternum (Ins.); Cechenus (Ins.).

cecid—Gr. *kēkis*, genit. *kēkidos*, an oak apple, gall nut; also smell, rarely a smoke, juice. *Ex:* cecid-ium; cecido-gen-ous; Cecido-myi-idae (Ins.); Cecido-nomus (Ins.).

cecrop—Gr. *Kekrops*, fabulous king who founded Athens. *Ex:* Cecropia*; Cecropia (Ins.).

ced—Gr. *kēdos*, near by marriage; also care of, concern; *kēdō*, to be concerned for. *Ex:* Cedio-psylla (Ins.).

cedr—Gr. *kedros*, a cedar, resinous tree; *kedrinos*, pertaining to or of the cedar; *kedron*, the fruit of the cedar. *Ex:* Cedr-ela*; Cedro-crypta (Ins.); Cedron-ella*; Cedrus*; Libo-cedrus*.

ceir—Gr. *keiras*, genit. *keirados*, shorn. *Ex:* Ceira (Ins.).

cel—1. Gr. *kēlē*, a tumor, swelling. *Ex:* celo-tomy; Concho-cele (Moll.): 2. Gr. *kēlis*, genit. *kēlidos*, a spot, stain; *kēlidoō*, to spot, stain. *Ex:* Celid-ota (Ins.); Celido-phylla

(Ins.); Haemato-celis (Ins.): **3.** Gr. *kēlos,* dry, parched; *kēleos,* burning. *Ex:* Celosia*.

celaen—Gr. *kelainos,* black, dark. *Ex:* Celaen-ura (Ins.); Celaeno-mys (Mam.).

celastr—Gr. *kēlastros* = *kēlastron,* an evergreen tree, privet. *Ex:* Celastr-aceae*; Celastr-ina (Ins.); Celastrus*.

celat—L. *celo,* to conceal > *celatus,* concealed and *celator,* a hider. *Ex:* Celato-conus (Moll.); Celatoria (Ins.).

cele—Gr. *kēleō,* to charm, bewitch. See also celer.

celeb—Gr. *kelebē,* a jar, cup, pan. *Ex:* Celebo-mastax (Ins.).

celeo—Gr. *keleos,* a kind of bird, the green wood-pecker. *Ex:* Celeo-morphae (Av.); Celeo-psyche (Ins.).

celer—L. *celer,* swift, fleet; akin to Gr. *kelēs,* a racer. *Ex:* Cela (Av.); Cele-saurus (Rept.); celer-ity; Celeri-pes (Ins.); Celerio (Ins.).

celest—L. *celestis* = *caelestis,* heavenly, blue. *Ex:* ?Celestus (Rept.).

celet—L. *celetus,* hidden, covered < *celo,* to hide. *Ex:* Celeto-thrips (Ins.); Celet-or (Ins.).

celid—See cel 2.

celis—See cel 2.

cell—**1.** L. *cella,* dim. *cellula,* storeroom, chamber. *Ex:* cell; Celle-pora (Bry.); cellul-ar; cellul-ose; celluli-fer-ous; Cellulo-monas*: **2.** Gr. *kellō,* to drive, urge > *dikella,* a two-pronged hoe. *Ex:* Dicello-ceras (Ins.).

cellar—L. *cellarius,* pertaining to a store-room > *cellarium,* a pantry > *cellararius,* a butler. *Ex:* Cellar-ina (Bry.); Cellari-ana (Moll.).

cellul—L. *cellula,* a small store-room; a cell. *Ex:* cellul-arius. See cellar.

celos—See cel 3.

cels—L. *celsus,* high, elevated.

celt—L. *celtis,* an African species of lotus. *Ex:* Celti-phaga (Ins.); Celtis*.

celyph—Gr. *kelyphos,* a husk, pod, shell of a fruit. *Ex:* Celyphus (Ins.).

cem—Gr. *kēmos,* a muzzle. *Ex:* Cemo-phora (Rept.). See cemad.

cemad—Gr. *kemas,* genit. *kemados,* a young deer. *Ex:* Cemas (Mam.); Stephano-cemas (Mam.).

cemas—See cemad.

cement—L. *cementum,* chips of stone used in making mortar. *Ex:* cement; cementum; cemento-blast.

cen—Gr. *kenos,* empty. *Ex:* Cen-angium*; cen-anthous; Cen-ellipsis (Prot.); Cenia*; Ceno-crinus (Echin.). See also caen.

cenchr—**1.** Gr. *kenchros,* millet, anything in small grains, seeds. *Ex:* Cenchrus*: **2.** Gr. *kenchris* = *kenchros,* a kind of serpent with bead-like protuberances < *kenchros,* millet. *Ex:* Cenchr-ina (Rept.); Cenchris (Rept.).

-cene—Eng. *-cene* < Gr. *kainos,* new, recent; a

word element meaning a special period of time. *Ex:* Mio-cene.

cent—**1.** Gr. *kenteō,* to prick; *kentron,* a point, spine; also the center of a circle (because it is the stationary point of a pair of compasses); *kentēsis,* a puncture; *kentētēs,* one who pierces. *Ex:* Cent-ella*; Centetes (Mam.); Centet-odon (Mam.); Centr-archus (Pisc.); Centri-cnemus (Ins.); Centro-madia*; Centro-pristes (Pisc.); centro-some; A-kentetus (Ins.); Ken tro-chona (Prot.); Tri-centes (Mam.): **2.** L. *cento,* patchwork; *centonarius,* pertaining to a patchwork. *Ex:* Cent-unculus*; Cento-ptera (Ins.); centon-ate.

centaur—Gr. *kentauros,* mythical creature, half man and half horse, a centaur > *kentaureion,* a kind of plant, centaury, the medicinal proper-ties of which were said to be discovered by the centaur, Chiron. *Ex:* Centaurea*; Centaurium*.

centet—See cent.

centr—See cent.

ceo—Gr. *keiō,* to split, cleave. *Ex:* Ce-onyx (Mam.); Ceo-phloeus (Av.).

cep—**1.** L. *cepa* = *caepa,* dim. *cepula* = ML. *ce-pola,* an onion. *Ex:* cep-aceous; Cepola (Pisc.): **2.** Gr. *kēpos,* a garden, plantation. *Ex:* Cancri-cepon (Crust.). See also ceps.

cephal—Gr. *kephalē,* a head; *kephalōtos,* with a head. *Ex:* Cepha-lophus (Mam.); Cephaelis*, in ref. to the disposition of the flowers in heads; cephal-ad; Cephal-inus; Cephalo-poda (Moll.); Cephalotes (Mam.); en-cephalon; Globi-cephala (Mam.).

cephalus—L. *Cephalus* < Gr. *Kephalos,* young huntsman beloved by Aurora; Gr. *kephalos,* name of a kind of fish. *Ex:* Cephalus (Pisc.).

cephe—Gr. *Kēpheus,* mythological king of Ethi-opia. *Ex:* Cephea (Coel.); Cepheus (Arach.).

cephen—Gr. *kēphēn,* genit. *kēphenos,* a drone. *Ex:* Cepheno-myia (Ins.); Cepheno-plosus (Pisc.); Cephenus (Ins.); Cephus (Ins.).

cepol—See cep.

cepor—Gr. *kēpōros,* keeper of a garden. *Ex:* Cepora (Ins.).

cepph—Gr. *kepphos,* a kind of sea bird, a booby. *Ex:* Cepphus (Av.).

ceps—NL. *ceps,* genit. *cipitis,* head < L. *caput,* genit. *capitis,* head; cf. *anceps,* two-headed. *Ex:* Bothri-ceps (Amph.); Cordy-ceps*; Flori-cep-idae (Rot.); Rani-ceps (Pisc.); Rani-cipit-idae (Pisc.); for Podiceps (Av.), see podic. See particeps.

cer—**1.** Gr. *keras,* genit. *keratos,* dim. *keration,* a horn, a bow; *keraos,* horned; *kerastēs,* horned; *keratophyēs,* having horns; *keratinos,* made of horn. *Ex:* Cer-ascaris (Nemat.); Cerasti-psocus (Ins.); Cerastium*; Ceratino-stoma (Ins.); Ceratium (Pisc.); Cerato-blepharum (Av.); Ceratosa (Por.); Ceri-anthus (Coel.); Ceri-ops*; Cerio-pora (Bry.); A-cerates*; A-cero-therium (Mam.); di-cerus; Calo-cerinus

(Ins.): **2.** Gr. *kēros*, beeswax; *kērion*, honeycomb>L. *cera*, wax and *cereus*, a wax candle or torch and NL. *cerumen*, earwax. *Ex:* Cer-adia\*, see aden; cer-ine; Cer-oxylon\*; Cereus\*; ceri-fer-ous; Cero-coccus (Ins.); Ceroplastes (Ins.); cerumen.

**cerae**—Gr. *keraia*, a horn, a projection, an antenna. *Ex:* Ceraeo-cercus (Ins.); Ceraeo-tricha (Ins.); Rhodino-ceraea (Ins.).

**ceram**—1. Gr. *keramos*, dim. *keramion*, a vessel, earthen pot, pitcher. *Ex:* ceram-idium; Cerami-opsis (Ins.); Ceramium\*; Ceramophylla (Bry.); Ino ceramus (Moll.): **2.** Gr. *keramis*, a roof tile, tiling, something imbricated; *kerameia*, pottery, the potter's art. *Ex:* Ceramo-pora (Bry.).

**ceras**—Gr. *kerasos*, the cherry tree; L. *cerasinus*, cherry-red. *Ex:* Cerasina (Moll.); Ceraso-phila (Av.); Cerasus\*.

**cerast**—See cer 1.

**cerat**—See cer 1.

**ceraton**—Gr. *keratōnia* = *keratea*, the carob-tree. *Ex:* Ceratonia\*.

**ceraun**—Gr. *keraunos*, a thunder-bolt. *Ex:* Cerauno-cochlis (Moll.).

**cerauni**—L. *ceraunius*, pertaining to thunder.

**cerc**—1. Gr. *kerkis*, genit. *kerkidos*, a shuttle, a peg, pin. *Ex:* Cerco-carpus\*, or perhaps it is from 2. See also cercis: **2.** Gr. *kerkos*, tail, a handle, membrum virile. *Ex:* cerc-aria; Cerco-cebus (Mam.); Cerco-monas (Prot.); cercus; Schisto-cerca (Ins.).

**cercer**—NL. *cerceris* < Gr. *kerchō* = *krechō*, to clatter, rustle, to be rough, harsh. *Ex:* Cer ceris (Ins.).

**cerchale**—Gr. *kerchaleos*, dry, rough, harsh.

**cerchn**—Gr. *kerchnos*, roughness, hoarseness; as adj. harsh, rough; *kerchnōtos*, roughened. *Ex:* Cerchnotus (Ins.); cerchnus.

**cerchne**—Gr. *kerchnē* = *kerchnēs*, genit. *kerchnēdos*, a kind of hawk, perhaps the kestrel. *Ex:* Cerchne (Av.); Cerchnei-picus (Av.); Cerchneis (Av.).

**cercid**—Gr. *kerkidion* (dim. of *kerkis*, a comb), a small comb, weaver's shuttle. *Ex:* Cercidia (Arach.); Cercidium\*, (Prot.); Cercido-cerus (Ins.).

**cercis**—Gr. *kerkis*, a kind of poplar; also, perhaps the Judas-tree. *Ex:* Cercis\*.

**cercop**—Gr. *kerkōps*, a long-tailed monkey. *Ex:* Cercopis (Ins.).

**cercyon**—Gr. *Kerkyōn*, the robber Theseus killed. *Ex:* Cercyon (Ins.).

**cerd**—1. Gr. *kerdō*, a fox; *kerdaleos*, of the fox, crafty. *Ex:* Cerd-odon (Rept.); Cerdale (Pisc.); Cerdo-cyon (Mam.); Galeo-cerdo (Elasm.): **2.** Gr. *kerdos*, gain, advantage, tricks.

**cere**—L. *cereus*, waxen, of wax; also a wax candle < Gr. *kēros*, wax. *Ex:* Cereus\*. See cer. 2.

**cereal**—NL. *cereal*, grain < L. *Cerealis*, pertaining to *Ceres*, goddess of vegetation; NL. *cerealis*, pertaining to grain or the arts of agriculture.

**cerebell**—See cerebr.

**cerebr**—L. *cerebrum*, dim. *cerebellum*, the brain. *Ex:* cerebell-ar; Cerebr-at-ulus (Nemer.); Cerebr-ina (Platy.); cerebri-form; cerebropedal.

**cereus**—See cere.

**cerin**—L. *cerinus*, yellowish, the color of yellow wax.

**cerinth**—1. Gr. *kērinthē*, a kind of plant, the borage-wort. *Ex:* Cerinthe\*: **2.** Gr. *kerinthos*, bee-bread.

**cerion**—Gr. *kērion*, a honeycomb. *Ex:* Cerion (Moll.).

**cerith**—NL. *cerithium* < Gr. *kērykion*, a shellfish. *Ex:* Cerithi-idae (Moll.); Cerithi-opsis (Moll.); Cerithium (Moll.).

**cermat**—Gr. *kerma*, genit. *kermatos*, a mite, a small coin. *Ex:* Cermatia (Myr.); Cermatobius (Myr.).

**cernu**—L. *cernuus*, turned towards the earth. *Ex:* cernu-ous.

**cerom**—Gr. *kērōma*, ointment; anything made of wax. *Ex:* ceroma.

**certh**—Gr. *kerthios*, some small bird, the tree creeper. *Ex:* Certhi-lauda (Av.); Certhia (Av.); Certhio-parus (Av.).

**certomi**—Gr. *kertomios*, jocular, taunting, delusive; *kertomikos*, jeering.

**ceruch**—Gr. *kerouchos*, horned. *Ex:* Ceruchus (Ins.).

**cerul**—L. *ceruleus* = *cerulus*, sky blue. *Ex:* cerulei-gul-aris.

**cerumen**—NL. *cerumen*, earwax < L. *cera*, wax. *Ex:* cerumen; cerumen-ous = cerumin-ous; cerumeni-fer-ous.

**cerumin**—See cerumen.

**ceruss**—L. *cerussa*, white lead; *cerussatus*, white, as though painted with white lead.

**cerv**—L. *cervus*, ML. dim. *cervulus* = *cervillus*, deer; *cervinus*, tawny, like a deer. *Ex:* Cerv-equus (Mam.); cervi-corn; cervini-ventris; Cervo-cerus (Mam.); Cervulus (Mam.); Cervus (Mam.).

**cervic**—L. *cervix*, genit. *cervicis*, the neck. *Ex:* cervic-al; cervico-thoracic.

**cervin**—See cerv.

**cervix**—See cervic.

**ceryl**—Gr. *kērylos*, a sea bird, the kingfisher. *Ex:* Ceryle (Av.).

**ceryx**—Gr. *kēryx*, a herald; also a kind of shellfish used by criers. *Ex:* Ceryx (Ins.); An-ceryx (Ins.); Hiero-ceryx (Ins.).

**cesp**—See caesp.

**cest**—Gr. *kestos*, a girdle; as adj. embroidered, variegated. *Ex:* Cest-oda (Platy.), Poly cesta (Ins.).

cestr—1. Gr. *kestra*, dim. *kestrakion*, a kind of weapon, a hammer; also a kind of fish. *Ex:* Cestra-phora (Elasm.); Cestracion (Elasm.); Cestreus (Pisc.); Cestro-rhinus (Pisc.): **2.** Gr. *kestron*, a kind of plant, betony. *Ex:* Cestrum*.

cet—Gr. *kētos*, a whale, sea monster>*kēteios*, monstrous. *Ex:* Cetus (Mam.); Cetio-therium (Mam.); Cetio-saurus (Rept.); ceto-lith; Ceto-therium (Mam.); Mysti-ceti (Mam.).

ceter—L. *ceterus*, the other; *ceterum*, the remainder.

ceterach—Fr. *ceterach*, a plant name. *Ex:* Ceterach*.

ceton—NL. *cetonia*, name for a certain metallic flower beetle. *Ex:* Cetoni-schema (Ins.); Cetonia (Ins.).

cetr—L. *cetra=caetra*, a sort of leather shield. *Ex:* Cetra (Arach.); Cetr-aria*.

cetrat—L. *cetratus*, bearing armor.

ceuth—Gr. *keuthō*, to hide; *keuthos*, depths. *Ex:* Ceutho-philus (Ins.); Ceutho-rhynchus (Ins.); Acro-ceuthes (Ins.).

ceyx—Gr. *kēyx*, a kind of bird, perhaps the male kingfisher; *Kēyx*, king of Trachis, husband of Halcyone who was changed into a bird. *Ex:* Ceyx (Av.); Ceyxia (Ins.); Clyto-ceyx (Av.).

chaem—See chamae.

chaen—Gr. *chainō=chaskō*, to gape. *Ex:* Chaen-actis*; Chaen-ophrys (Mam.); Chaeno-cetus (Mam.); Chaeno-delphinus (Mam.); Chaenon (Ins.); Chaino-dictyon (Bry.); a-chene; a-chaeno-carp.

chaer—Gr. *choiros*, a young pig, porker; also a fish of the Nile. *Ex:* Chaero-potamus. (Mam.); Chaero-phyllum*; Chaero-pus (Mam.). See also char.

chaet—NL. *chaeta*, a bristle<Gr. *chaitē*, long flowing hair, a mane. *Ex:* Chaet-etes (Coel.); Chaet-ura (Av.); Chaito-iulus (Myr.); Chato-gaster=Chaeto-gaster (Ann.); Oligo-chaeta (Ann.); poly-chaete.

chai—Gr. *chaios*, genuine, good.

chain—See chaen.

chait—See chaet.

chalar—Gr. *chalaros*, slack, supple<*chalaō*, ppr. *chalōn*, to slacken, loosen. *Ex:* Chalar-aspis (Crust.); Chalaro-pegma (Por.); chalone.

chalasm—Gr. *chalasma*, genit. *chalasmatos*, a relaxation, a gap, dislocation.

chalast—Gr. *chalaō*, to relax; *chalastos*, loose; *chalastikos*, relaxing, making supple. *Ex:* Chalast-inus (Ins.); chalastic; Chalasto-gastra (Ins.).

chalaz—Gr. *chalaza*, hail, a hail-stone, a tubercle. *Ex:* chalaza; Chalaz-acanthus (Pisc.); Chalaz-odes (Prot.); chalazo-gamy.

chalc—1. Gr. *chalkis*, genit. *chalkidos*, a fish; also a kind of lizard. *Ex:* Chalcides (Rept.); Chalcido-lepis (Rept.); Chalcis (Rept.): **2.**

Gr. *chalkos*, copper>NL. *chalcis*, genit. *chalcidis*, a name applied by Fabricius to a group of metallic colored parasitic Hymenoptera. *Ex:* Chalc-angium (Ins.); Chalcid-idae (Ins.); Chalcis (Ins.); Chalco-phaps (Av.).

chalcid—See chalc.

chalcis—Gr. *chalkis*, a bird of the night. *Ex:* Seto-chalcis (Av.), Gr. *sēs*, genit. *sētos*, a moth. See also chalc 2.

chalic—1. Gr. *chalix*, genit. *chalikos*, pebbles, gravel. *Ex:* Chalico-doma (Ins.); Chalico-mys (Mam.): 2. ME. *chalic*, a cup, chalice. *Ex:* chalic-ed.

chalin—Gr. *chalinos*, a bit, strap, bridle. *Ex:* Chalina-ura (Pisc.); Chalina (Por.); Chalino-rhaphis (Por.); chalines; Pachy-chalina (Por.).

chalon—See chalar.

chalyb—Gr. *chalyps*, genit. *chalybos*, steel>L. *chalybeius*, of steel>NL. *chalybeatus*, qualified by the presence of iron; *chalybaeus*, steel-colored. *Ex:* Chalyb-ura (Av.); Chalybeo-themis (Ins.); chalybeus; Chalybion (Ins.).

cham—1. Gr. *chamos*, a bit, rein, bridle. *Ex:* Mono-chamus (Ins.): **2.** L. *chama*, a cockle<Gr. *chainō*, to gape. *Ex:* Chama (Moll.). See chamae.

chamae—Gr. *chamai*, on the ground, dwarf; in botany *chamae*- sometimes signifies false. *Ex:* Chamae-batia*; Chamae-cyparis*; Chamae-leon (Rept.); Chamae-pelia (Av.); Chamira*.

chamaele—L. *chamaeleon*<Gr. *chamaileōn*, "ground-lion." *Ex:* Chamaeleon (Rept.).

chamomill—L. *chamomilla*, chamomile<Gr. *chamaimēlon*, a kind of plant, "earth apple," chamomile. *Ex:* Chamomilla*.

champs—Gr. *champsai*, a crocodile<an Egyptian word for the crocodile. *Ex:* Champs-odon (Pisc.); Champse (Rept.); Champso-cephalus; Champso-myrmex (Ins.).

chan—Gr. *chanos*, the open mouth. *Ex:* Chan-omphalus (Moll.); Chano-stoma (Prot.); Chanos (Pisc.); Myio-chanes (Av.).

chann—Gr. *channē=channos*, a kind of wide-mouthed fish of the sea. *Ex:* Neo-channa (Pisc.).

chao—1. Gr. *chaos*, genit. *chaeos*, empty space, an abyss. *Ex:* Chao-laimus (Nemat.); Chaos (Prot.): **2.** Gr. *chaos*, good, illustrious: **3.** Gr. *chaoō*, to entirely destroy, ruin. *Ex:* Chao-borus (Ins.).

chaparr—Basque *chabarra*, name for a scrub oak of the Pyrenees Mts.>Sp. *chapparo*, a dwarf evergreen oak>Mod. Span. *chuppar-al*, lit. place of evergreen scrub oaks. The Span. suffix *-al*, meaning place of, is seen in such words as *pin-al*, place of pines; *encin-al*, place of oaks; *alis-al*, place of sycamores; *chamis-al*, place of small shrubs, lit. place of small sticks for firewood.

char—Gr. *chairō*, to rejoice, *charis*, delight, grace, beauty, favor; *charitos*, to make graceful: *charieis*, graceful; *Charitēs*, the three Graces

*Ex:* Chaero-drys (Ins.); Chaero-phyllum*; Char-ales*; Chara*; Chari-anthus*; Chari-theca (Ins.); Charieis*; Charis-asterea (Coel.); Charites (Moll.); Charito-metra (Echin.); Eleocharis*; Helo-chara (Ins.).

**charac**—Gr. *charax*, genit. *charakos*, a sea-fish; the word is perhaps akin to *charax*, a pointed stake. *Ex:* Charac-odus (Pisc.); Characopygus (Ins.); Charaxi-cephalus (Platy.); Manno-charax (Pisc.).

**characin**—NL. *characin* < Gr. *charax*, a sea-fish. *Ex:* Characin (Pisc.); characin-oid.

**character**—Gr. *charakter*, something engraved > L. *character*, an instrument for marking, also a characteristic mark. *Ex:* character.

**charadr**—Gr. *charadra*, a cleft, gully > *charadrios*, a cleft-dwelling bird. *Ex:* Charadrius (Av.); Charadro-bia (Moll.).

**charag**—Gr. *charagē*, the figure or impress on a coin. *Ex:* Charago-tettix (Ins.); Charagus (Ins.).

**charagm**—Gr. *charagma*, genit. *charagmatos*, any mark graven or imprinted; *charagmos*, an incision, notch. *Ex:* Charagmo-phorus (Ins.); Charagmus (Ins.).

**charass**—Gr. *charassō*, to make sharp or pointed, to engrave. *Ex:* Charasso-carcinus (Crust.).

**charax**—See **charac.**

**charidot**—Gr. *charidōtēs*, a giver of joy. *Ex:* Charidot-ella (Ins.); Charidotis (Ins.).

**charism**—Gr. *charisma*, genit. *charismatos*, a gift. *Ex:* Charisma (Moll.).

**charit**—See **char.**

**charm**—Gr. *charma*, genit. *charmatos*, joy, delight; *charmonē*, gladness; *charmosynos*, causing joy, agreeable. *Ex:* Charmato-metra (Ins.); Charmon (Ins.); Charmosyna (Av.).

**charmat**—See **charm.**

**charmosyn**—See **charm.**

**charon**—Gr. *Charōn*, ferryman of the Styx. *Ex:* Charon (Prot.); Charoni-ella (Moll.).

**charop**—Gr. *charopos*, glad-eyed, bright-eyed; also serene, tawny. *Ex:* Charopa (Moll.).

**chartac**—L. *chartaceus*, made of paper, papery.

**charte**—L. *charteus*, pertaining to paper.

**charybd**—Gr. *Charybdis*, a dangerous whirl-pool on the Sicilian coast. *Ex:* Charybd-ella (Coel.); Charybdi-teuthis (Moll.).

**chas**—Gr. *chasis*, a separation, a chasm. *Ex:* Chasi-empis (Av.); pleio-chasium.

**chasc**—Gr. *chaskō*, to open. *Ex:* Chasco-thyris (Brach.); a-chasco-phytum.

**chascac**—Gr. *chaskax*, genit. *chaskakos*, a gaper. *Ex:* Chascax (Amph.).

**chascax**—See **chascac.**

**chasm**—Gr. *chasma*, genit. *chasmatos*, an opening, expanse, chasm, the open mouth. *Ex:* Casm-erodius (Av.); Chasm-odes (Pisc.); Chasma-rhynchus (Av.); chasmato-plasm;

Chasme (Ins.); Chasmias (Ins.); chasmophyte; Echino-chasmus (Platy.); Macrochasma (Moll.).

**chasmod**—Gr. *chasmōdēs*, yawning. *Ex:* Chasmodes (Pisc.).

**chat**—See **chaet.**

**chaul**—Gr. *chauliodous* and *chauliōdōn*, genit. *chauliōdōntos*, with projecting teeth > a supposed *chaulios*, protuberant. *Ex:* Chaulelasmus (Av.); Chauliodonta (Rept.); Chauliodus (Av.).

**chaun**—Gr. *chaunos*, gaping; also soft, empty, loose, bloated. *Ex:* Chaun-angium (Por.); Chauna (Av.); Chauno-graptus (Coel.); Chaunus (Rept.).

**chaunac**—Gr. *chaunax*, genit. *chaunakos*, liar, cheat. *Ex:* Chaunax (Pisc.).

**chaunax**—See **chaunac.**

**chavica**—South Sea Island *chavica*, name of the long-pepper. *Ex:* Chavica*.

**che**—Gr. *cheia*, a cave, a serpent's den. *Ex:* Cheporus (Ins.).

**cheil**—See **chil.**

**cheim**—Gr. *cheimōn*, genit. *cheimōnos*, winter; *cheimarros*, winter-flowing; also a torrent caused by winter thaws; *cheimatikos*, of winter, wintry. *Ex:* Cheimarr-ichthys (Pisc.); Cheimarr-ornis (Av.); Cheimato-bius (Ins.); Cheimo-phila (Ins.); Cheimono-phila (Ins.); Chima-phila*; chimo-pelagic.

**cheimarr**—See **cheim.**

**cheimat**—See **cheim.**

**cheimon**—See **cheim.**

**cheir**—See **chir.**

**chel**—Gr. *chēlē*, a hoof, claw; also as an adj., cloven, parted; *chēlargos*, with fleet hoofs. *Ex:* Chel-ura (Crust.); Chela; Chelarga (Av.); Chele-tropis (Ins.); cheli-cera; Cheli-notus (Moll.); cheli-ped; Chelo-meles (Rept.); Chelophore; Chelo-teuthis (Moll.); Chelo-trop-ella (Por.); not Chelo-dina (Rept.), see chely. See also **chelon.**

**chelid**—Gr. *chelidōn*, genit. *chelidonos*, a swallow; *chelidonios*, of or like the swallow; colored like the swallow's throat, russet. *Ex:* Chelidomyia (Ins.); Chelidon-ichthys (Pisc.); Chelidonium*; Hydro-chelidon (Av.).

**chelio**—Gr. *chēlion*, dim of *chēlē*, hoof. *Ex:* Chelio-myrmex (Ins.); Cheliones (Mam.).

**chelon**—Gr. *chelōnē*, a tortoise. *Ex:* Chel-ina (Rept.); Chelo-dina (Rept.); Chelon-echinus (Echin.); Chelon-iscus (Mam.); Chelone*, (Rept.); Cheloni-coccus (Ins.); Chelonia (Rept.); Chelono-bia (Crust.); Chelonus (Rept.).

**chely**—Gr. *chelys*, a tortoise. *Ex:* Cheli-notus (Moll.); Chelo-dinus (Rept.); Chely-therium (Rept.); Chelyo-soma (Tun.); Dermo-chelys (Rept.).

**chelydr**—Gr. *chelydros*, a water serpent. *Ex:* Chelydr-opsis (Moll.); Chelydra (Rept.).

**chem**—Gr. *chēmeia*, an infusion; *chemikos*, of or concerning juices, infusions>Eng. *chemistry*, "the infusory art." *Ex:* chemic-al; chemo-nastic; chemo-taxis; bio-chemistry: 2. Gr. *chēmē*, yawning, gaping. *Ex:* Chema (Av.)

**chen**—Gr. *chēn*, a goose>*chēniskos*, the part of a ship's prow which is turned up like a goose's neck. *Ex:* Chen-endro-scyphia (Por.); Cheniscus (Av.); not a-chene, see chaen; Cheno-podi-um*.

**chenisc**—See chen.

**chenni**—Gr. *chennion*, a kind of quail. *Ex:* Chennium (Ins.).

**cher**—See chaer, also chir and choer.

**cherad**—Gr. *cherados*, genit. *cheradotos*, silt, mud, gravel, rubbish. *Ex:* cherado-philus.

**cheram**—Gr. *chēramos*, a hole, a cleft, hiding place. *Ex:* Cheram-oeca (Av.).

**therm**—1. Pers. *kermes*, crimson>NL. *chermisinus*, dyed with crimson. *Ex:* Chermes (Ins.); Kermes (Ins.); Kermo-coccus (Ins.): 2. Gr. *chermas*, genit. *chermados*, a stone for slinging, a large pebble.

**chermast**—Gr. *chermastēr*, a slinger.

**cherne**—Gr. *chernē*, genit. *chernētos*, a needy man, day-laborer; *cherna*, need, poverty. *Ex:* Chernes (Arach.).

**chernet**—See cherne.

**chernozem**—Russ. *chiernoziem*, black earth. *Ex:* chernozem.

**cherr**—See chers.

**chers**—Gr. *chersos*=*cherros*, dry land; *chersinos*, pertaining to land tortoises. *Ex:* Cherrus (Ins.); chers-ad; Chers-ad-aula (Ins.); Chersina (Rept.); Cherso-bius (Rept.); cherso-phyte (Ecol.).

**chesis**—Gr. *cheseiō*, to desire to ease one's self. *Ex:* Xenos-chesis (Ins.).

**chevr**—Fr. *chevre*, dim. *chevrot*, a goat; *chevron*, a rafter, a chevron. *Ex:* Chevro-latia (Ins.); chevrot-ian.

**chia**—Gr. *chia*, a hiding place, a hole. *Ex:* Chia-mela (Rept.).

**chiasm**—Gr. *chiasma*, the mark of χ (chi) or a cross<*chiazō*, to mark as false, to mark with a cross. *Ex:* Chiasm-odus (Pisc.); Chiasmo-neura (Ins.); chiasmo-typy; Chiasmus (Ins.).

**chiast**—Gr. *chiastos*, arranged diagonally. *Ex:* Chiasto-lepis (Pisc.); chiasto-neur-al; Chiastos-ella (Bry.).

**chil**—1. Gr. *cheilos*, a margin, lip, brim. *Ex:* Cheil-anthes*; Chil-opsis*; Chilo-branchus (Pisc.); Chilo-stoma (Moll.); a-cheil-ary; Craspedo-chilus (Moll.); Eu-cheil-ota (Coel.); Mega-chile (Ins.): 2. Gr. *chilos*, fodder. *Ex:* Ceto-chilus (Arth.).

**chilari**—Gr. *cheilarion*, a small lip. *Ex:* chilarium.

**chili**—Gr. *chilias*, also *chilios*, a thousand. *Ex:* Chilio-stigma (Moll.).

**chilitic**—NL. *chiliticus*, lip-like, said to be from Gr. *cheilos*, lip.

**chim**—See cheim.

**chimaer**—See chimair.

**chimair**—Gr. *chimaira*, a she-goat; a fire spouting monster of fantastic form; *chimaros*, a goat. *Ex:* Chimaera (Pisc.); Chimaero-psylla (Ins.); Chimaira (Pisc.); Chimaro-cephala (Ins.).

**chimar**—See chimair.

**chimarr**—Gr. *cheimarros*, a torrent. *Ex:* Chimarro-gale (Mam.).

**chimen**—Gr. *cheimōn*, winter; *cheimōnikos*, stormy. *Ex:* A-chimenes*.

The Desert Willow, *Chilopsis linearis*, with lipped flowers and linear leaves. Redrawn from Desert Wild Flowers—Jaeger. Stanford University Press.

**chio**—Gr. *chiōn*, snow; *chioneos*, white as snow. *Ex:* Chio-cocca*; Chio-genes*; Chion-anthus*; Chione*; Chionea (Ins.); chiono-philus; Hedychium*.

**chion**—See chio; also cion.

**chir**—Gr. *cheir*, hand; *cheironomos*, one who moves the hands with regularity as in a pantomine. *Ex:* Cheir-inia*; Chiro-nectes (Mam.); Chir-acanthium (Arach.); Chir-otes (Rept.); Chiri-dota (Echin.); Chiro-petes (Mam.); Chiro-ptera (Mam.); Acro-chira (Coel.); Cheironomus (Ins.).

**chiragric**—Gr. *cheiragrikos*, afflicted with gout in the hand.

**chirodrop**—Gr. *cheirodropos*, plucking by hand (cf. *chedropa*, legumes, beans); <*cheir*, the hand+*drepō*, to pluck. *Ex:* Chirodropus (Coel.).

**chiron**—1. Gr. *Cheirōn*, a centaur famous for his knowledge of plants. *Ex:* Chironia*; Hespero-chiron*: 2. Gr. *cheirōn*, inferior, in want.

**chirotonet**—Gr. *cheirotonētēs*, a voter; *cheirotonētos*, a magistrate elected by show of hands. *Ex:* Chirotonetes (Ins.).

**chirox**—NL. *chirox*<Gr. χ (*chi*), the letter χ, hence a cross+*rhōe*, a cleft or fissure. *Ex:* Chirox-com-idae (Ins.).

**chit**—Gr. *chitōn*, an outer covering, a kind of garment, coat of mail>NL. *chitin*. *Ex:* chitin; Chitini-poma (Ann.); chiton-ize; Chitino-lepis (Platy.); Chito-balanus (Crust.); Chiton (Moll.); Chitona (Ins.); Chitonia*; Crypto-chiton (Moll.); Diplo-chita*; Eu-chitonia (Prot.); Lysi-chitum*; Sarco-chitum (Prot.).

**chitin**—See chit.

**chiton**—See chit.

**chlaen**—See chlain.

**chlain**—Gr. *chlaina*, cloak. *Ex:* Chlaenio-ctenus (Ins.); Eu-chlaena*; Eu-chlan-idae (Rot.); Exo-chlaenus (Ins.).

**chlamyd**—Gr. *chlamys*, genit. *chlamydos*, cloak, mantle. *Ex:* Chlamydo-saurus (Rept.); Chlamydo-selachus (Pisc.); chlamydo-spores; Chlamys (Moll.).

**chlamys**—See chlamyd.

**chlan**—See chlain.

**chlanid**—Gr. *chlanidōtōs*, clad in a woolen garment. *Ex:* Chlanidota (Ins.); Chlanido-phora (Ins.).

**chled**—Gr. *chlēdos*, slime, rubbish. *Ex:* chledo-philus; chledo-phyta.

**chlid**—Gr. *chlidē*, softness, delicacy, pride. *Ex:* Chlid-anthus*. See also chlidon.

**chlidon**—Gr. *chlidōn*, an ornament such as a bracelet, anklet. *Ex:* Chlidoni-brya (Bry.); Chlidonia (Ins.); Chlidonias (Av.).

**chlo**—Gr. *chloē*=Dor. *chloa*, the young shoots of grass, grass, young herbage. *Ex:* Chl-opsis (Pisc.); Chloe (Ins.); Chloe-phaga (Av.); Chloo-phaga (Av.); Antho-chloa*.

**chloan**—Gr. *chloanos*, greenish.

**chloanth**—Gr. *chloanthēs*, budding. *Ex:* Chloan thes*.

**chloasm**—Gr. *chloazō*, to be or become pale green. *Ex:* chloasma (Med.).

**chlor**—1. Gr. *chlōros*, green, greenish-yellow>L. *chloroticus*, pale yellowish green; Gr. *chlōreus*, a kind of greenish or yellowish bird. *Ex:* chlor-agogen; Chlora*; Chloreus (Av.); Chloro-galum*; Chloro-stoma (Moll.); Zoo-chlor-ella*: 2. Gr. *Chloris*, goddess of flowers. *Ex:* Chloris*.

**chlorion**—Gr. *chlōriōn*, some yellowish bird, perhaps the golden oriole. *Ex:* Chlorion (Ins.); Chlorion-inae (Ins.).

**chlorotic**—See chlor.

**chnoo**—Gr. *chnoos*, wool. *Ex:* Chnoo-triba (Ins.); Sporo-chnus*.

**chnus**—See chnoo.

**choan**—Gr. *chōnē*=*choanē*, funnel, tube. *Ex:* Choan-omphalus (Moll.); choana; Choane-phora*; choano-cyte; Chona*; Chone-morpha*; Chone-rhinus (Pisc.); Chonio-stoma (Crust.); Chono-phorus (Pisc.); Spiro-chona (Prot.).

**choenic**—Gr. *choinikē*, a shackle. *Ex:* Choenico-sphaera (Prot.).

**choer**—Gr. *choiros*, pig. *Ex:* Choer-opsis (Mam.); Choere-laphus (Mam.); Choiro-potamus= Koiro-potamus (Mam.); Myxo-cherus (Mam.); Poly-choerus (Platy.).

**choic**—L. *choicus*, of earth or clay.

**choir**—See choer.

**chol**—Gr. *cholē*, bile, and in this sense most often used in anatomical terms; also anger, loathing; *cholēdochos*, containing bile; *cholos*, bitter anger, wrath. *Ex:* chol-agoge (Med.); chole-sterin; choledo-graphy; ductus choledochus. See choler.

**choler**—Gr. *cholera*, the cholera, a bilious disease prob.<*cholē*, bile, see chol. *Ex:* Cholera-stoma (Ins.); choleri-genous (Med.); cholero-mania (Med.); Cholero-phytum (Nem.).

**cholo**—Gr. *chōlos*, lame, maimed; *chōlōma*, a lameness. *Ex:* Cholo-gaster (Pisc.); Choloe-pus (Mam.).

**cholom**—See cholo.

**chom**—Gr. *chōma*, genit. *chōmatos*, a mound, heap of rubbish. *Ex:* Chomato-aster (Echin.); Chomato-bius (Myr.); chomo-phyte; chasmo-chomo-phyte; Exo-chomus (Ins.).

**chomat**—See chom.

**chon**—See choan.

**chonax**—NL. *chonax*, an error for Gr. *phonax*, eager for blood<*phonaō*, to be athirst for blood. *Ex:* Chonaxis (Coel.).

**chondr**—Gr. *chondros*, dim. *chondrion*, a grain, something granular, grit, a lump of salt; also, gristle, cartilage. *Ex:* Chondr-illa*; chondrio-some; chondro-cranium; Chondro-phora (Moll.); Di-chondra*; mito-chondria.

**chonet**—NL. *chonetes*<Gr. *chōnō*=*choanē*, a funnel. *Ex:* Chonetes (Brach.); Choneti-pustula (Brach.).

**chor**—1. Gr. *choreia*, a dance with music<*choros*, a dance. *Ex:* Choreia (Ins.); Choreo-dromia (Ins.): 2. Gr. *chōris*, asunder, apart; *chōrismos*, a place apart; *chōristos*, separated; *chōrizō*, to separate, to spread. *Ex:* Choreo-colax*; chori-petalous; Chori-zema*; not Chorisia*, named after J. L. Choris, eminent artist who went around the world with Kotzebue; Choriso-neura (Ins.); Chorism-agrion (Ins.); Chorisma*; Chorismo-dactylus (Pisc.); Chorist-idae (Ins.); Choristo-poda (Crust.); Choriz-anthe*; Chorizo-agrotis (Ins.): 3. Gr. *chōros*, place, region; *chōrētēs*, a rustic, a country-man. *Ex:* Choretis*; Choretrum*; choro-logy; bio-chore (Ecol.); Meso-chorus (Ins.): 4. Gr. *chōros*, the north-west wind.

**chorag**—Gr. *choragos*, leader of a chorus. *Ex:* Choragus (Ins.).

**chord**—Gr. *chordē*, guts, a string, musical instrument>NL. *chordata*, having a notochord or "back-string." *Ex:* Chord-aria*; Chord-eiles

(Av.); Chordata; noto-chord; Uro-chorda (Protoch.).

**chordon**—NL. *chordonia* < Gr. *chordē*, a string, chord. *Ex:* Chordonia (Chord.).

**-chore**—Gr. *chōreō*, to spread. *Ex:* zoo-chore. See also chor 3.

**chores**—Gr. *chōrēsis*, taking, receiving; also capacity.

**choret**—See chor 3.

**choreut**—Gr. *choreutēs*, a ballet dancer, a choral dancer. *Ex:* Choreutes (Ins.).

**chorio**—Gr. *chorion*, skin, the membrane that encloses the fetus. *Ex:* chorio-retin-al; Choriotherium (Mam.); chorion; chorion-ic.

**chorion**—See chorio.

**choris**—See chor 2.

**chorism**—See chor 2.

**chorist**—See chor 2.

**choriz**—See chor 2.

**choroid**—Gr. *choroeidēs*, the grape-colored coating of the eye. *Ex:* choroid plexus; choroid-itis (Med.).

**chort**—Gr. *chortos*, an enclosure for plants; grass, green herbage; cf. L. *hortus*, a garden. *Ex:* Chorto-phila (Ins.); Calo-chortus*; Rhodo-chorton*.

**chos**—Gr. *chōsis*, a heaping up of earth. *Ex:* Chos-ornis (Av.).

**chrei**—See chri 2.

**chrem**—Gr. *chrēma*, genit. *chrēmatos*, property, wealth. *Ex:* Chremon (Ins.).

**chreo**—1. Gr. *chreō*, to want, need; *chreios*, useful, needful. *Ex:* Chreo-noma (Ins.). See chri 2: 2. Gr. *chreos*, a debt, need; *chreōstēs*, a debtor. *Ex:* Chreostes (Ins.).

**chreost**—See chreo 2.

**chres**—Gr. *chrēsis*, use, utility, profit < *chraō*, to use. *Ex:* chres-ard.

**chrest**—Gr. *chrēstos*, good, useful, lucky; *chrēstotēs*, kindness, good nature. *Ex:* Chrestotes (Ins.); Chresto-sema (Ins.), Iso-chresta (Ins.).

**hri**—1. Gr. *chriō*, to bedaub, to besmear; to touch lightly, wound on the surface, to puncture. *Ex:* Chri-acus (Mam.); Chri-odes (Ins.). See also chreo 2: 2. Gr. *chreia*, want, use, advantage. *Ex:* Chrio-lepis (Pisc.).

**chro**—Gr. *chroa* or *chroia*, genit. *chrōtos*, = *chrōs*, genit. *chroos*, the superficial appearance or color of a thing; also the skin or surface of the body; *chroiakos*, colored. *Ex:* Chloro-chroa (Ins.); Chroeco-cephalus (Av.); Chroico-ptera (Ins.); Chroo-cephalus (Av.); Chroto-mys (Mam.); Chroto-pterus (Mam.); allo-chrous; Di-chroa (Ins.).

**chroec**—See chro.

**chroic**—See chro.

**chrom**—Gr. *chrōma*, genit. *chrōmatos*, the surface of the body, the color of the skin surface, color; *chrōmatikos*, relating to colors, soft, harmoni-

ous; *chrōmatinos*, colored. *Ex:* chromatic; chromatin; chromato-phore; Chromo-doris (Moll.); chromo-some.

**chron**—Gr. *chronos*, time > *chronikos*, of or concerning time. *Ex:* chron-oste-al; chrono-graph; chrono-logy; Chrono-zoon (Mam.).

**chroni**—Gr. *chronios*, late. *Ex:* chroni-zoo-spore.

**chroo**—See chro.

**chros**—Gr. *chrōsis*, a coloring. *Ex:* hema-chrosis. See also chro.

**chrot**—See chro.

**chrys**—Gr. *chrysos*, gold; *chryseos*, golden, rich, superb; *chrysion*, a piece of gold; *chrysitēs*, gold-colored; *chrysallis*, genit. *chrysallidos*, the golden colored pupa of certain butterflies, a moth. *Ex:* Chrys-anthemum*; Chrys-emys (Rept.); chrysalid = chrysalis; Chrysalido-car pus*; Chrysallo-teuthis (Moll.); Chryso-chloris (Mam.); Chryso-thamnus*.

**chrysall**—See chrys.

**chrysid**—Gr. *chrysidion*, a bit of gold. *Ex:* Chrysidi-astrum (Prot.); Chrysidium (Ins.).

**chrysomitr**—Gr. *chrysomētris*, the goldfinch. *Ex:* Chrysomitr-id-ops (Av.).

**chthamal**—Gr. *chthamalos*, on or near the ground. *Ex:* Chthamal-idae (Arth.); Chthamalo-pteryx (Pisc.); Chthamalus (Arth.).

**chthe**—Gr. *chthes*, yesterday. *Ex:* Calo-chthebius (Ins.).

**chthon**—Gr. *chthōn*, genit. *chthonos*, the earth. *Ex:* Chthon-ergus (Mam.); Chthon-erpeton (Amph.); Chthoni-cola (Av.); Chthonius (Arach.); Chthono-bdella (Ann.); auto-chthon-ous; Hypo-chthon (Amph.).

**chunga**—Argentine *chunga*, native name for a crane-like bird. *Ex:* Chunga (Av.).

**chyl**—Gr. *chylos*, juice, the chyle. *Ex:* chyli-ferous; Chylo-cladia*; chylo-phylly.

**chym**—1. Gr. *cheō*, to pour > Gr. *enchyma*, an infusion, something poured in. *Ex:* mes-en-chyme; par-en-chyma: 2. Gr. *chymos*, juice. *Ex:* chyme; Chymo-carpus*.

**chys**—Gr. *chysis*, a diffusion, melting, pouring forth. *Ex:* Chysis*.

**chyt**—1. Gr. *chytos*, liquid, fluid, flowing. *Ex:* Chyt-onix (Ins.); Chyto-desmus (Myr.): 2. Gr. *chytos*, a heap < *cheō*, to heap up. *Ex:* Asco-chyta*.

**chytr**—Gr. *chytros*, also *chytra*, dim. *chytrion*, a pitcher, flower-pot. *Ex:* Chytr-idium*; Chytra (Moll.); Chytrio-dinium (Prot.); Chytro-crinus (Echin.); Chloro-chytrium*; Enchytraeus (Ann.).

**cian**—See cyan.

**cib**—L. *cibus*, food; *cibarius*, suitable for food, pertaining to food; *cibo*, to feed, pp. *cibatus*, fed; Fr. *cibation*, taking of food. *Ex:* cibation; Cibi-corbis (Prot.); cibr mania (Med.).

**cibar**—See cib.

cibdel—Gr. *kibdēlos*, disguised; also filthy, spurious, base. *Ex:* Cibdela (Ins.); Cibdelis (Ins.).

cibicid—L. *cibicida*, a bread eater. *Ex:* Cibicidella (Prot.); Cibicides (Prot.).

cibis—Gr. *kibisis*, a pouch, wallet.

cibor—Gr. *kibōrion*, a drinking cup. *Ex:* ciborium.

cibot—Gr. *kibōtos*, a wooden chest; dim. *kibōtion*, a little vessel, a cup, chest. *Ex:* Cibotion (Pisc.); Cibotium*; Ciboto-gaster (Ins.).

cicad—L. *cicada*, a tree-cricket. *Ex:* Cicad-ella (Ins.); Cicada (Ins.); Cicado-morpha (Ins.).

cicatric—L. *cicatrix*, genit. *cicatricis*, a scar. *Ex:* cicatrix; cicatrici-al.

cicatrix—See cicatric.

ciccab—Gr. *kikkabē*, the screech owl. *Ex:* Ciccaba (Av.).

cicer—L. *cicer*, the chick-pea. *Ex:* Cicer*; Cicerocrinus (Echin.).

cichl—Gr. *kichlē*, a bird like a thrush; also kind of sea fish. *Ex:* Cichla (Pisc.); Cichla-dusa (Av.); cichlo-morhpic; Eu-cichla (Av.).

cichor—Gr. *kichorion = kichorē = kichora = kichoreia*, chicory. *Ex:* cichori-aceous; Cichorium*.

cicindel—L. *cicindela*, a glow worm. *Ex:* Cicindelopsis (Ins.); Cicindela (Ins.).

cicinn—Gr. *kikinnos*, a ringlet. *Ex:* cicinn-al; Cicinno-cnemis (Ins.).

cicon—L. *ciconia*, a stork. *Ex:* Ciconi-opsis (Av.); Ciconia (Av.).

cicut—L. *cicuta*, name of a poison (hemlock) derived from certain plants of the family Umbelliferae. *Ex:* cicut-arius; Cicuta*.

cicy—Gr. *kikys*, force, strength.

cid—L. *cid*, the root of *caedo*, to cut. *Ex:* Bicid-ium (Coel.); loculi-cid-al; Stipuli-cida*.

cidar—Gr. *kidaris*, the jewelled turban of a Persian king. *Ex:* Cidario-phanes (Ins.); Cidarotropius (Echin.); Bothrio-cidaris (Echin.).

cigcl—Gr. *kigklis = kinklis*, a latticed gate. *Ex:* Cigcli-rhina (Moll.); Cigclis-ula (Bry.).

cil—NL. *cilium*, dim. *ciliolum*, pl. *cilia*, dim. pl. *ciliola*, a hair or hair-like process; *ciliatus*, furnished with cilia; *ciliosus*, fringed, full of hairs <L. *cilium*, an eyelid. *Ex:* cilia; Ciliata (Prot.); cilii-form; Cilio-spina (Prot.); ciliola. See also cilion.

cilic—Gr. *kilikion*, a cloth of goat's hair. *Ex:* Cilicio-sporum*.

cilion—L. *cilio*, genit. *cilionis*, a chisel, graver. See also cil.

cilla—NL. *cilla*, tail <L. *cillo*, to move. *Ex:* Mota-cilla (Av.); Rubi-cilla (Av.); Ruti-cilla (Av.).

cim—See cym.

cimbex—See cimbic.

cimbic—Gr. *kimbēx*, genit. *kimbēkos*, and *kimbix*, genit. *kimbikos*, a miser; also a kind of bee-like or wasp-like insect. *Ex:* Cimbex (Ins.); Cimbic-idae (Ins.). See also cymb.

cimel—Gr. *keimēlion*, goods, chattels, treasure. *Ex:* Cimel-idium (Prot.); Cimelia (Ins.).

cimex—See cimic.

cimic—L. *cimex*, genit. *cimicis*, a bug; *cimicinus*, smelling like or of bugs. *Ex:* Cimex (Ins.); Cimic-idae (Ins.); Cimici-fuga*.

cimol—Gr. *Kimōlia*, Cimolian earth (clay) from an island of the Cyclades, *Kimolē*, known for its chalky soil; the combining form *cimol-* now stands for clay. *Ex:* Cimol-odon (Mam.); Cimoli-ornis (Av.); Cimolio-chelys (Rept.).

cin—Gr. *kineō*, to move; *kinēsis*, movement; *kinētēs* and *kinētēr*, a mover, disturber; *kinētikos*, pertaining to motion. *Ex:* Cin-ura (Ins.); Cino-sternon = Kino-sternon (Rept.); A-cinet-ae (Prot.); caryo-kinesis; Eu-cinostomus (Pisc.).

cinaed—L. *cinaedus*, without shame, lewd.

cinar—L. *Cinara*, fem. proper name. *Ex:* Cinara (Ins.).

cincinn—L. *cincinnus*, a curl of hair; *cincinnatus*, with curled hair; *cincinnalis*, curled.

cincinnat—L. *cincinnatus*, with curls.

cincl—1. Gr. *kinklis*, genit. *kinklidos*, a lattice or opening. *Ex:* cincl-ides; Kinklido-blatta (Ins.): 2. Gr. *kinklos*, name of some bird, the water ouzel or wagtail. *Ex:* Cinclo-soma (Av.); Cinclus (Av.).

cinct—L. *cinctus*, banded, girdled <*cingo*, to bind about. *Ex:* Cinct-odonta (Av.); Cincti-pora (Bry.); Cincto-nema (Nemat.); cincto-planular.

ciner—L. *cinis*, genit. *cineris*, ashes; *cinereus*, ash-colored; *cinerarius*, pertaining to ashes; *cinerascō*, to turn to ash. *Ex:* Ciner-ascens (Moll.); Cineraria*.

cinet—See cin.

cing—L. *cingens*, genit. *cingentis*, surrounding < *cingo*, to encircle, go around.

cingul—L. *cingulum*, a collar, girdle <*cingo*, to surround, to gird. *Ex:* cingul-ate; cingulum.

ciniflon—L. *ciniflo*, genit. *ciniflonis*, a hair-curler. *Ex:* Cliniflo (Arach.); Ciniflon-idae (Arach.).

-cinium—L. *-cinium*, suffix added to noun stems to form abstract nouns. *Ex:* latro-cinium.

cinn—Gr. *kinna*, name for a kind of grass. *Ex:* Cinna*.

cinnabar—Gr. *kinnabari*, a red pigment, cinnabar. *Ex:* cinnabar-inus.

cinnamo—Gr. *kinnamōmon*, the cinnamon-tree. *Ex:* Cinnamo-dendron*; Cinnamomum*.

cinnamome—NL. *cinnamomeus*, cinnamon color.

cinnyr—Gr. *kinnyris*, a small bird. *Ex:* Cinnyra (Av.); Cinnyris (Av.); Cinnyro-rhyncha (Av.).

cinygm—Gr. *kinygma*, genit. *kinygmatos*, a moving object, anything moved about. *Ex:* Cinygm-ula (Ins.); Cinygma (Ins.).

cion—1. Gr. *Chionē*, a demi-goddess. *Ex:* Chione (Moll.); Ciona (Tun.); Cionea (Tun.): 2. Gr. *kiōn*, genit. *kionos*, a pillar, column; also the uvula. *Ex:* Cion-odon (Rept.); Ciono-bryssus (Echin.); ciono-cranial; Cionus (Ins.); Sy-cium (Mam.), see syn.

cipit—See ceps.

cir—L. *ciris*, a greedy sea-fowl. *Ex:* Ciro-rrhinchus (Ins.).

circ—1. Gr. *kirkos*=*kirkinos* a circle. *Ex:* Circ-omphalus (Moll.); Circo-peltis (Echin.): 2. Gr. *kirkos*, a falcon that moves in a circle. *Ex:* Circ-aetus (Av.); Circus (Av.): 3. L. *circum*, about. *Ex:* circ-ovarian. See also circa.

circa—L. *circa*, about, near to, almost, approximately; represented in botanical writings by the abbreviation *ca*. See also circ.

circaea—See circe.

circe—L. *Circaeus*=*Circe*<Gr. *Kirkē*, fabled enchantress. *Ex:* Circaea*; Circe (Moll.); Circe-aster (Echin.).

circell—L. *circellus*, a small ring.

circin—L. *circino*, to make round, pp. *circinatus*, made round<Gr. *kirkinos*, a circle. *Ex:* circin-al; Cicinal-ium (Tun.); circinate.

circinan—L. *circinans*. genit. *circinantis*, making round, ppr. of *circino*, to make round.

circuit—L. *circuitus*, going around, revolving.

circulan—L. *circulans*, genit. *circulantis*, making round, ppr. of *circulo*, to make round.

circulat—L. *circulatio*, a revolution, a going around. *Ex:* circulat-ion.

circum—L. *circum*, around. *Ex:* circum; circum-flex; circum-vallate.

circumdat—L. *circumdatus*, put around, pp. of *circumdo;* to put around.

ciris—Gr. *keiris*, a kind of bird into which Scylla was supposed to have been transformed. *Ex:* Ciris (Ins.).

cirr—1. L. *cirrus*, a curl>*cirrh-*, "a wrong but very common form of *cirr-*, from the mistaken notion that the L. *cirrus*, a curl of hair was represented by a Gr. *kirrhos*, which is not found."; *cirratus*, curled. *Ex:* cirrat-ulus (Ann.); cirrate; Cirrh-aea*; Cirrh-ites (Pisc.); Cirrho-branchiata (Moll.); Cirrho-petalum*; Cirri-pathes (Coel.); Cirro-drilus (Ann.); Cirro-teuthis (Moll.); cirrus: 2. Gr. *kirrhos*, yellow. *Ex:* cirrh osis (Path.): 3. Gr. *kirris*, genit. *kirrithos*, a sea-fish.

cirrat—See cirr.

cirrh—See cirr.

cirs—1. Gr. *kirsion*, a kind of thistle. *Ex:* Cirsium*: 2. Gr. *kirsos*, a swollen vein, a welt, varicocele. *Ex:* Cirs-omphalus (Moll.); Cirso-chilus (Moll.); cirso-tome (Med.); Cirso-trema (Moll.).

cis—1. L. *cis*, on this side. *Ex:* cis-montane (Ecol.); cis-oceanic: 2. Gr. *kis*, a wood-worm. *Ex:* Cis (Ins.); Cis-arthron (Ins.); Cis-urgus (Ins.).

ciss—1. Gr. *kissos*=*kittos*, ivy. *Ex:* Ciss-ampelos*; Ciss-ites (Ins.); Cisso-phyllus (Nemat.); Cissus*; Citto-bium (Ins.); Partheno-cissus*: 2. Gr. *kissa*, a magpie. *Ex:* Cissa (Av.).

cist—1. Gr. *kistē*, a box, chest>L. *cista*, dim. *cistella*=*cistula*, a box, chest. *Ex:* Cist-udo (Rept.), *cista*+*testudo*, a tortoise; Cista (Rept.) Ciste-cephalus (Rept.); Cistella (Brach.); Cisti-cola (Av.); Cistello-morpha (Ins.); Cisto-pteris*; cistome, see stom. 2. Gr. *kistos*=*kisthos*, a flowering shrub, rock-rose<*kistē*, a box, capsule. *Ex:* Cist-aceae*; Cisti-cola (Av.); Cisto-thurus (Av.); Cistus*.

cit—L. *citus*, swift. *Ex:* Citi-gradae (Arach.).

citell—L. *citellus*, a ground squirrel, ziesel. *Ex:* Citell-nema (Nem.); Citello-philus (Ins.); Citellus (Mam.).

cithar—1. Gr. *kithara*, a lyre. *Ex:* Cithar-acanthus (Arach.); Cithar-opsis (Moll.); Cithara (Moll.); Cithare-xylium*; Citharo-ceps (Arach.); cithar-arius: 2. Gr. *kitharos*, a kind of fish<*kithara*, a lyre. *Ex:* Cithar-ichthys (Pisc.); Anti-citharus (Pisc.).

citharod—Gr. *kitharōdos*, a harp player.

citr—L. *citrus*, the citron-tree; *citratus*, steeped in citrus oil; ML. *citrinus*, lemon-colored>NL. dim. *citrinella;* Gr. *kitrion*, citron-fruit; *kitria*, the citron-tree. *Ex:* Citr-opsis*; Citri-phaga (Ins.); Citri-vir; Citrinella (Av.); Citrino-phaga (Ins.); Citro-myses*; Citrus*.

citra—L. *citra*, on this side. *Ex:* citra-montane. See also cis.

citrat—L. *citratus*, adj. form of *citra*, of this side. See also citr.

citrull—NL. *citrull*<OF. *citrulle*, a cucurbit> NL. *citrullus*. *Ex:* Citrullus*.

citt—Gr. *kitta*=*kissa*, a kind of chattering bird. *Ex:* Citt-ura (Av.); Citto-cincla (Av.); Cyano-citta (Av.). See also ciss 2.

cium—See cion 2.

civet—Fr. *civette*<It. *cibetto*=*zibetto*<Ar. *zab-bad*. *Ex:* Civettea (Mam.); Civett-istis (Mam.).

cixi—Gr. *kixios*, a cicada. *Ex:* Cixi-idae (Ins.); Cixio-soma (Ins.); Cixius (Ins.).

clad—Gr. *klados*, a branch, sprout; dim. *kladion*, a club, baton; *kladōdēs*, with many branches. *Ex:* Clad-rastes*, see thraust; Clade-monas (Prot.); Cladio-drilus (Ins.); Cladio-phleps (Ins.); Cladium*; Cladius (Ins.); Clado-cera (Coel.); Cladodi-um*; Poly-clada (Platy.); Tri-clad-ida (Platy.).

cladar—Gr. *kladaros*, frail, easily broken. *Ex:* Cladar-odes (Ins.); Cladara (Ins.).

clam—L. *clamo*, to complain, ppr. *clamitans*, genit. *clamitantis*, loud-calling; *clamator*, a noisy, shrill declaimer. *Ex:* Clamatores (Av.).

clamator—See clam.

**clamb**—Gr. *klambos*, mutilated, deficient. *Ex:* Clambus (Ins.).

**clamitans**—See **clam**.

**clancul**—L. *clanculum*, clandestinely, secretly. *Ex:* Clanculus (Moll.).

**clandestin**—L. *clandestinus*, secret, hidden. *Ex:* Clandestinum (Moll.).

**clang**—Gr. *klangē*, a noise, creak, sound>L. *clangor*, NL. dim. *clangula*, clang, sound. *Ex:* Clango-cycnus (Av.); Clangula (Av.).

**clao**—Gr. *klaō*, to break. *Ex:* Clao-rhynchus (Rept.); Clao-saurus (Rept.).

**clar**—L. *clarus*, clear, brilliant, shining, renowned. *Ex:* Clar-avis (Av.).

**clas**—Gr. *klasis*, a fracture, break. *Ex:* clasi-leucite.

**clasm**—Gr. *klasma*, genit. *klasmatos*, a fragment, remnant. *Ex:* Clasma (Ins.); clasmato-cyte.

**class**—L. *classis*, a class, a division; also an assembly of people, fleet>NL. *classifico*, to classify; *classificatio*, genit. *classificationis*, the act of classifying or grouping together. *Ex:* classification.

**clast**—Gr. *klastos*, broken in pieces<*klaō*, to break. *Ex:* Clasto-ptera (Ins.); osteo-clast.

**claster**—Gr. *klastērion*, a knife. *Ex:* Clastero-sporum*.

**clathr**—L. *clathri*, lattice work; *clathratus*, latticed. *Ex:* Clathr-aria*; Clathr-ina (Por.); Clathr-ul-ina (Prot.); Clathria (Por.); Clathrio-dendron (Por.); Clathro-neuria (Ins.); Clathr-ella (Moll.).

**claud**—1. L. *claudo*, to shut, ppr. *claudens*, genit. *claudentis*, closing. *Ex:* claudent; Claudi-concha (Moll.): 2. L. *claudus*, lame< *claudico*, to limp, ppr. *claudicans*, genit. *claudicantis*, limping. *Ex:* claudicant.

**claus**—L. *clausum*, a closed space>NL. *claucilium*, a subspiral appendage fitting into a groove of the columella of certain mollusks. *Ex:* Claus-asterea (Coel.); Claus-idium (Crust.); Clausili-opsis (Moll.); Clausilia (Moll.).

**claustr**—L. *claustrum*, a bar, that which closes. *Ex:* Claustra (Coel.).

**clav**—1. L. *clava*, NL. dim. *clavula*=*clavella*, club; NL. *clavatus*, dim. *clavellatus*, club-shaped. *Ex:* Clav-aria*; Clava (Por.); clavate; Clavelia (Ins.); Clavi-ceps*; Clavi-cornia (Ins.); Clavul-inae (Prot.): 2. L. *clavis*, dim. *clavicula*, a key>NL. *clavicle*, the collar bone. *Ex:* clavicle; clavicul-arium: 3. L. *clavus*, a nail, tumor. *Ex:* clavus; Clavus (Moll.).

**-cle**—Eng. *-cle*, dim. ending (<L. *-culus*, *-a*, *um*, a noun suffix used in forming diminutives.). *Ex:* denti-cle<*dens*, genit. *dentis;* particle< *pars*, genit. *partis;* see also *-culus*.

**cle**—Gr. *kleos*, a rumor, good report; also, fame. *Ex:* Cle-onymis (Ins.); Cli-anthus*; Eu-clea*; not Hymeno-clea; see clea.

**clea**—Gr. *kleiō*, to enclose. *Ex:* Nau-clea*, with reference to the hull-shaped capsule. See also clei; Hymeno-clea*.

**cleav**—Eng. *cleave*, to divide<AS. *cleofan*, to cut. *Ex:* cleav-age.

**clei**—Gr. *kleis*, genit. *kleidos*, a key; also the clavicle; *kleidoō*, to lock, to shut up. *Ex:* Clei-gaster (Ins.); cleido-mastoid; Cleido-theca (Moll.); Cleio-crinus (Echin.); Clido-thaerus (Moll.); Eu-clidium*; Hydro-cleis*; Ptero-cles (Av.).

**cleid**—See **clei**.

**cleist**—See **clist**.

**cleit**—Gr. *kleitos*, renowned. *Ex:* Cleito-diplosis (Ins.).

**cleithr**—Gr. *kleithron* and *klēthron*, a bar, a key for closing. *Ex:* Cleithro-lepis (Ins.); cleithrum.

**clem**—Gr. *klēma*, dim. *klēmatis*, a branch, twig; *klēmatis*, genit. *klēmatidos*, a young shoot, tendril. *Ex:* Clematis*, (Coel.); Clemato-crinus (Echin.).

**clemat**—See **clem**.

**clemen**—L. *clemens*, placid, calm; *clementia*, mildness.

**clemm**—See **clemmyd**.

**clemmyd**—See **clemmys**.

**clemmys**—Gr. *klemmys*, a tortoise. *Ex:* Clemmus (Ins.); Clemmyd-opsis (Rept.); Clemmys (Rept.); Cy-clemmys (Rept.).

**cleob**—Gr. *Kleobis*, a man's name. *Ex:* Cleobis (Arach.).

**cleodora**—Gr. *Kleodōra*, name of a Danaid and also of a nymph. *Ex:* Cleodora (Moll.).

**cleom**—L. *cleome*, ancient name of the mustard plant<Gr. *kleiō*, to shut up, enclose. *Ex:* Cleom-ella*; Cleome*.

**cleon**—Gr. *Kleōn*=L. *Cleonus*, Athenian demagogue. *Ex:* Cleon-urus (Ins.); Cleono-lithus (Ins.); Cleonus (Ins.).

**cleps**—Gr. *klepsia*, theft. *Ex:* Clepsine (Ann.). See also clepsydr.

**clepsydr**—Gr. *klepsydra*, a device for measuring time by the amount of water discharged from a vessel through a small aperture<*klepsia*, theft+*hydor*, water. *Ex:* Clepsydra (Moll.).

**clept**—Gr. *kleptēs*, a thief; *kleptikos*, thievish. *Ex:* Clept-idea (Ins.); Clept-ornis (Av.); Cleptes (Ins.); Clepticus (Pisc.); clepto-brosus; Cleptodromia (Ins.).

**cler**—Gr. *klēros*, a lot, chance; also a kind of predaceous insect in bee-hives. *Ex:* Cler-oides (Ins.); Clero-dendron*; Clerus (Ins.); syn-clero-bi-osis.

**cles**—Gr. *klēsis*, a closing, a shutting up. *Ex:* di-cles-ium. See also clei.

**clest**—Gr. *klēstos*, old Attic for *kleistos*, shut, closed. *Ex:* clest-ine; Clesto-bothrium (Platy.).

**clet**—1. Gr. *Klēta*, who in Spartan mythology

was one of the Charities, personification of grace and beauty. *Ex:* Cleta (Crust. ); Cleto-pontius (Crust.): **2.** Gr. *klētos*, named, called, invited. *Ex:* Cleto-campus (Ins.).

**clethr**—**1.** Gr. *klēthra*, the alder tree. *Ex:* Clethra*: **2.** Gr. *klēthron=kleithrion*, a bar for closing. *Ex:* Clethriono-mys (Mam.); Clethro-gyna (Ins.).

**cli**—Gr. *kleiō*, to shut; *kleistos*, shut, closed. *Ex:* Di-cli-ptera*; not Cli-anthus*, see cle.

**cliban**—Gr. *klibanos*, pot, earthenware vessel. *Ex:* Clibanus (Av.).

**clidem**—Gr. *Klidemi*, ancient Greek botanist. *Ex:* Clidemia*.

**cliens**—See client.

**client**—L. *cliens*, genit. *clientis*, a companion, client.

**climac**—Gr. *klimax*, genit. *klimakos*, a ladder, staircase, climax; *klimactēr*, the step of a ladder, also the critical period of life. *Ex:* Climaci-ella (Ins.); Climaco-stomum (Prot.); climac ter-ic; Climacteris (Av.); Climax-odus (Pisc.).

**climacod**—Gr. *klimacōdēs*, like stairs, terraced.

**climax**—See climac.

**clin**—Gr. *klinē*, a bed<*klinō* to slope, recline. *Ex:* clin-andrium; clino-logy; clinic; Clino-pleura (Ins.); patro-clin-ous; Stylo-cline*; syn-cline.

**cline**—NL. *cline*, an ecological term<L. *clino*, to lean. *Ex:* cline.

**clint**—Gr. *klintēr*, genit. *klintēros*, sofa. *Ex:* Clintero-cera (Ins.).

**clinter**—See clint.

**clio**—**1.** L. *Clio*<Gr. *Kleiō*, a sea nymph, sister of Beroe>NL. *Clion*. *Ex:* Clio (Moll.); Clio-dora (Moll.); Clion-ites (Por.); Cliona (Por.); Clione (Moll.); Cliono-lithes (Por.); Archae(o)-cliona (Por.): **2.** L. *Clio*<Gr. *Kleiō*, the Muse of History.

**clion**—See clio.

**clis**—**1.** Gr. *klisia*, a place for lying down, a hutr *klision*, a small chamber. *Ex.* Clisio-camp: (Ins.); Clisio-phyllum (Coel.); Tri-clis (Mam. a **2.** Gr. *klisis*, a bending, an inclination<G.) *klinō*, to incline. *Ex:* cliseo-meter.

**clism**—Gr. *klismos*, a couch, an inclination.

**clist**—Gr. *kleistos*, that can be closed, enclosed <*kleiō*, to close. *Ex:* Cleisto-crinus (Echin.); Cleisto-yucca*; clisto-gam-ous; Clisto-olynth-etta (Por.); physo-clist-ic.

**clit**—Gr. *klitos=klitys*, a slope, hill-side; *klitos* the lower part of a place. *Ex:* Clit-onyx (Av.) Clito-cybe*; Clito-pygus (Echin.); Gnatho clita (Ins.); hetero-clitus.

**clitell**—L. *clitellae*, a pack saddle. *Ex:* Clitell-aria (Ins.); Clitello-xenia (Ins.); clitellum.

**clitor**—Gr. *kleitoris*, genit. *kleitoridos*, the cli-toris<*kleiō*, to close. *Ex:* Clitoria*; clitorid-ec-tomy; clitoris.

**cliv**—L. *clivus*, a hill, slope. *Ex:* cliv-al; Clivi-cola (Av.); Cliv-ina (Ins.); clivus.

**cloac**—L. *cloaca*, a sewer; *cloacalis*, pertaining to a sewer. *Ex:* Cloac-ina (Nemat.); cloaca; Cloaci-trema (Nemat.).

**cloe**—Gr. *kloios*, a collar. *Ex:* Cloe-otis (Mam.); Cloeo-siphon (Geph.): Cloio-ceras (Moll.); cloio-choan-itic.

**cloi**—See cloe.

**clon**—**1.** Gr. *klōn*, dim. *klōnion*, a branch, twig. *Ex:* Clon-orchis (Platy.); Clono-thrix*; A-clono-phlebia (Ins.); Poly-clonus (Ins.); Tropido-clonion (Rept.): **2.** Gr. *klonis*, the os sacrum: **3.** Gr. *klonos*, a turmoil, tumult <*kloneō*, to put to flight, drive into confusion. *Ex:* clon-ic; clonus.

**clone**—NL. *clone*<Gr. *klōn*, a twig or slip used in propagation. *Ex:* clone.

**clopim**—NL. *clopimus*<Gr. *klopimos*, thievish.

**clor**—See chlor.

**clost**—**1.** Gr. *klōstos*, spun, coiled. *Ex:* Clost-ophis (Moll.): **2.** Gr. *klostēs*, a spinner. *Ex:* Calamo-clostes (Ins.); Clostes (Ins.).

**closter**—Gr. *klōstēr*, genit. *klōstēros*, a spindle; *klōstron*, a clue>NL. dim. *clostridium*, a clue of thread. *Ex:* Closterium*; Clostero-cerus (Ins.); Clostridium*.

**clostr**—See closter.

**cloth**—Gr. *klōthō*, to spin, twist> *Klōthō*, one of the three Fates or Destinies who spun the thread of life. *Ex:* Cloth-elaps (Rept.); Clotho (Rept.); Clothonia (Rept.); Amphi-clotho (Crust.).

**clu**—Gr. *kleos*, glory. *Ex:* Clu-biona (Arach.).

**clup**—L. *clupea*, some small river fish. *Ex:* Clupe-ops (Pisc.); Clupea (Pisc.); Clupeo-labrus (Pisc.); Clupi-soma (Pisc.).

**clus**—L. *clusus*, closed, pp. of *claudo*, to close. *Ex:* oc-clus-ion; retro-clus-ion; sub-in-clus; not Clusia*, named after C. de Lecluse, French botanist.

**clymen**—**1.** Gr. *klymenos*, famous, celebrated; sometimes meaning scandalous, infamous: **2.** Gr. *Klymenē*, name of several mythological personages, perhaps<*klymenos*, famous (infamous). *Ex:* Clymene (Mam.); Clymenia (Moll.).

**clype**—L. *clypeus*, NL. dim. *clypeolus*, a shield. *Ex:* Clype-aster (Echin.); clypei-form; Clypeo-brissus (Echin.); clypeol-ate; Clypeolum (Prot.); clypeus.

**clysi**—NL. *clysium*<Gr. *klyzō*, to dash against, to wash. *Ex:* clysium (Ecol.).

**clyst**—Gr. *klystēr*, a syringe, pipe. *Ex:* Clysto-gaster (Ins.); Halo-clystus (Por.).

**clyt**—Gr. *klytos*, heard of, glorious. *Ex:* Clytia (Coel.); Clyto-stoma*; Clytus (Ins.); Neo-clytus (Ins.).

**clythr**—NL. *clythra*, a word with meaning unknown, perh. < Gr. *kleithron*, bar, bolt. a beam. *Ex:* Clythr-opsis (Ins.); Clythra (Ins.); Clythr cerus (Crust.).

**cnec**—See **cnecos.**

**cnecos**—Gr. *knēkos*, pale yellow, tawny; also a plant of the thistle kind. *Ex:* Cnecoso-phagus (Ins.); Cnecus (Ins.); Cnicus*.

**cnem**—Gr. *knēmis*, genit. *knēmidos*, a legging; *knēmē*, the knee, lower part of the leg; *knēmidophoros*, equipped with leggings. *Ex:* cnem-idium; Cnem-idium (Coel.), (Ins.); Cnemidophorus (Rept.); Macro-cnemum*; Phrixocnemis (Ins.): 2. *knēma*, genit. *knēmatos*, a fragment, chip, shavings; 3. Gr. *knēmos*, shoulder of a mountain. *Ex:* Cnem-archus (Av.); Cnemo-philus (Av.).

**cnemid**—See **cnem.**

**cnemidot**—Gr. *knēmidōtos*, with leggings on. *Ex:* Cnemidotus (Ins.).

**cneo**—Gr. *knaō*, to scratch (one's self), scrape, tickle. *Ex:* Cneo-glossa (Ins.); Cneo-rrhinus (Ins.).

**cneor**—Gr. *kneōron*, a plant like the nettle. *Ex:* Cneor-idium*; Cneorum*.

**cnepha**—Gr. *knephas*, genit. *knephatos*, darkness; *knephaios*, dark. *Ex:* Cnephaeus (Mam.); Cnephaio-philus (Mam.); Cnephata (Ins.).

**cnephal**—Gr. *knephallon* = *knaphallon*, stuffing for cushions; *knaphalōdēs*, soft as wool. *Ex:* Cnephalo-gonia (Ins.); Cnepahlodes (Ins.).

**cnest**—Gr. *knēstis*, a grater, scratcher, knife for scraping; *knēstēr*, a rake, scraper; also a killer; *knēstos*, shaved, cut, mangled, rasped. *Ex:* Cnestis*; Cnesto-cera (Ins.); a-cnestis.

**cnestr**—Gr. *knēstron* = *knēstis*, a scraping knife. *Ex:* Cnestro-stoma (Pisc.); Cnestrum (Ins.).

**cneth**—Gr. *knēthō*, to scratch. *Ex:* Cnetho campa (Ins.).

**cnic**—See **cnecos.**

**cnid**—Gr. *knidē*, nettle. *Ex:* cnido-blast; cnido-cil; Cnido-pus (Coel.); Hespero-cnide*.

**cnip**—Gr. *knips*, genit. *knipos*, nom. plu. *knipes*, an insect living under bark; *knipolegos*, a gatherer of wood insects < *knips* + *legō* to choose, select, gather. *Ex:* Cnipo-dectes (Av.); Cnipolegus (Av.).

**cnodac**—Gr. *knōdax*, genit. *knōdakos*, pivots upon which a body turns as on an axis, a linchpin. *Ex:* Cnodaco-phora (Ins.); Arthro-cnodax (Ins.).

**cnodax**—See **cnodac.**

**co-** —See **con-.**

**coacervat**—L. *coacervatus*, heaped up, collected in a heap. *Ex:* coacervate.

**coagul**—L. *coagulo*, to cause a fluid to curdle < *cogo*, to bring together to one point; *coagulatio*, a curdling. *Ex:* coagulat-ion.

**coalesc**—L. *coalesco*, to grow together, to unite. *Ex:* coalesc-ed.

**coalit**—L. *coalitus*, united < *coalesco*, to join.

**coarct**—See **arct.**

**coati**—Native Tropical Amer. *coati*, coati. *Ex:* Coati (Mam.).

**cobio**—Gr. *kobios*, some fish of the gudgeon kind. *Ex:* Cobio-morus (Pisc.).

**cobit**—Gr. *kōbitēs*, fem. *kōbitis*, a fish like the gudgeon. *Ex:* Cobit-idae (Pisc.); Cobitis (Pisc.).

**cobra**—Pg. *cobra*, a serpent. *Ex:* Cobra (Rept.); Cobra-cephalus (Moll.).

**coc**—L. *coquere* to cook > *praecox*, genit. *praecocis*, ripe beforehand, premature. *Ex:* precoci-ous.

**cocc**—1. Gr. *kokkos*, a kernel, grain > L. *coccus*, NL. dim. *cocculus* = *coccidium*, a kernel, grain, berry. *Ex:* cocci-genic; Coccidi-phaga (Ins.); Coccido-myia (Ins.); Cocco-discus (Prot.); Cocco-myces*; Cocculus*; Tetra-coccus*: 2. Gr. *kokkos*, name of the female cochineal or kermes insect, which yields a red coloring principle and which was formerly thought to be a berry or grain of a plant > *kokkinos*, scarlet. *Ex:* Coccin-ella (Ins.); Cocco-chloris (Ins.); Coccus (Ins.).

**coccid**—See **cocc 1.**

**coccin**—L. *coccineus*, scarlet, red like a berry < Gr. *kokkinos*, scarlet; L. *coccinatus*, clothed in scarlet.

**coccul**—See **cocc 1.**

**coccy**—Gr. *kokkyx*, genit. *kokkygos*, a cuckoo; *kokkyzō*, to cry "cuckoo". *Ex:* coccyge-al; Coccygo-morphae (Av.); coccyx; Coccyzus (Av.); Geo-coccyx (Av.).

**cochl**—Gr. *kochlos*, a mollusk with a spiral shell < *kochlō*, to wind, turn > L. *cochlea*, a snail shell. *Ex:* Cochl-ic-ella (Moll.); Cochl-ops (Mam.); Cochlea (Moll.); cochle-ar; Cochleo-phorus (Arth.); Cochli-carina (Moll.); Cochlio-podium (Prot.); Cochlo-spira (Moll.).

**cochlear**—L. *cochlear*, a spoon < *cochlea*, a snail's shell. *Ex:* cochleari-form; Cochlearia*; Cochlearius (Av.); if used in connection with names of organs of hearing, as cochlear duct, cochlear canal, etc., see cochl.

**cocles**—L. *Cocles*, genit. *Coclitis*, masc. proper name. *Ex:* Cocles (Ins.).

**cocoon**—Fr. *cocon*, shell. *Ex:* cocoon.

**cocos**—NL. *cocos*, cocoanut, prob. < Gr. *kouki*, a cocoa-tree, cocoanut, but perhaps < Pg. *coco*, cocoanut < *macoco* = *macaco*, a kind of monkey, to the face of which the cocoanut, with the three scars upon one end of it, was thought to bear a resemblance. *Ex:* Cocos*.

**cocyt**—L. *Cocytus*, masc. proper name *Ex:*. Cocyt-inus (Amph.).

**cod**—Gr. *kōdeia*, a head, a little ball, a poppy head. *Ex:* Cod-atractus (Ins.); Codia-crinus (Echin.); Codio-soma (Ins.); not Cod-aster (Echin.); Codium*. See also codo.

**codi**—Gr. *kōas*, dim. *kōdion*, a sheepskin, fleece. *Ex:* Bulbo-codium\*. For Codium\*, see cod.

**codo**—Gr. *kōdōn*, dim, *kōdōnion*, a bell. *Ex:* Cod-aster = Codon-aster (Echin.); Codo-cera (Ins.); Codo-siga (Prot.); Codon-opsis\*; Codonia\*; Lepto-codon\*; Platy-codon (Moll.).

**coecil**—See caecil.

**coel**—**1.** L. *coelum* = *caelum*, the sky, heavens. *Ex:* Coeli-gena (Av.): **2.** Gr. *koilos*, hollow; *koilia*, the belly; *koiliakos*, suffering in the belly > L. *coeliacus*, pertaining to the belly. *Ex:* Coel-ambus (Ins.); coel-enteron; Coelia\*; coeliac; Coelo-genys (Mam.); Coelo-gyne\*; Coil-antha\*; hydro-coele; spongo-coel.

**coelebs**—L. *coelebs* = *caelebs*, single, without a partner.

**coelestin**—L. *coelestinus*, heavenly (blue).

**coelom**—Gr. *koilōma*, genit. *koilōmatos*, a hollow. *Ex:* coelom, pl. coelomata.

**coen**—**1.** L. *coenum*, dirt; *coenosus*, filthy. *Ex:* obs-cen-ity: **2.** Gr. *koinos*, common, shared in common. *Ex:* coen-enchym; Coeno-bit-idae (Crust.); caeno-cyte = cenocyte; bio-coen-oses. See also caen.

**coept**—L. *coeptus*, begun.

**coereb**—Braz. *coereba*, name of some small bird. *Ex:* Coereb-idae (Av.); Coereba (Av.).

**coerul**—See caerul.

**coet**—Gr. *koitē*, a bed; *koitos*, a bed; also sleep; *koitōn*, a bed-chamber. *Ex:* Coeto-mys (Mam.); Ammo-coetes (Pisc.); Exo-coetus (Pisc.); not coitus, see coit.

**coetan**—L. *coaetaneus*, of the same age.

**coffea**—NL. *coffea* < Ar. *kahwah*, coffee. *Ex:* Coffea\*.

**coffeat**—NL. *coffeatus*, of the color of the coffee berry.

**cognat**—L. *cognatus*, related < *co* = *con*, with + *nascor*, pp. *natus*, to be born.

**cohort**—L. *cohors*, genit. *cohortis*, an enclosure; also a crowd, throng. *Ex:* cohort.

**coil**—See coel.

**coit**—L. *coitus*, a coming together, sexual intercourse = *coitio*, genit. *coitionis*, a coming together, sexual congress, both < *coeo*, to go or come together, to copulate. *Ex:* coition; coito-phobia; coitus.

**coix**—Gr. *koix*, a kind of palm. *Ex:* Coix\*.

**col**—**1.** L. *colo*, to inhabit > *incola*, an inhabitant. *Ex:* deserti-cola, Limi-cola (Av.); Petri-cola (Moll.): **2.** Gr. *kōlon*, a limb, member of a body. *Ex:* Mcta-colus (Ins.): **3.** Gr. *kolon*, the colon. *Ex:* col-ec-tomy (Med.); Col-ydium (Ins.), see ideo: **4.** Gr. *kolos*, defective, maimed, hornless. *Ex:* Col-aspis (Ins.); Col-ur-ella (Troch.); Colo-cephali (Pisc.).

**cola**—African native name *cola*, the cola plant. *Ex:* Cola\*.

**colab**—Gr. *kolabos*, a morsel, piece, a mouthful of bread. *Ex:* Colabo-telus (Ins.); Colabus (Ins.).

**colac**—Gr. *kolax*, genit. *kolakos*, a flatterer; often used in sense of an imitator or a parasite. *Ex:* Colaco-nema\*; Theo-colax (Ins.).

**colaph**—Gr. *kolaphos*, a buffet; *kolaphizō*, to buffet. *Ex:* Colapho-ptera (Ins.); Colaphus (Ins.).

**colapt**—Gr. *kolaptēr*, a hammer, a chisel; *kolaptō*, to peck with the bill, chisel. *Ex:* Colaptes (Av.); Geo-colaptes (Av.), see -tes.

**colast**—Gr. *kolastēs*, a corrector, one who reproves. *Ex:* Colastes (Ins.).

**colax**—See colac.

**colchic**—Gr. *kolchikon*, a plant with a poisonous bulbous root < *Kolchis*, a country in Asia, east of the Black Sea, of which Medea, sorceress and poisoner of ancient legend, was said to have been a native. *Ex:* Colchicum\*.

**cole**—Gr. *koleos*, a sheath. *Ex:* Cole-anthus\*; Cole-ura (Mam.); Coleo-gyne\*; Coleo-ptera (Ins.); Coleos-anthus\*; Coleus\*; Bolbo-coleon\*.

**colen**—Gr. *kōlēn*, genit. *kōlēnos*, thigh, leg. *Ex:* Colenis (Ins.).

**colens**—See colent.

**colent**—L. *colens*, genit. *colentis*, honoring, respecting, ppr. of *colo*, to cultivate, to care for.

**colep**—Gr. *kōlēps*, the hollow bend of the knees. Coleps (Prot.).

**coli**—Gr. *kolios*, a kind of woodpecker. *Ex:* Colio-morphae (Av.); Colius (Av.).

**colias**—Gr. *Kōlias*, Aphrodite, a name for Venus. *Ex:* Colias (Ins.).

**colic**—Gr. *kōlikos*, of the colon, affecting the bowels, suffering in the colon. *Ex:* Colico-dendron\*.

**colin**—Sp. *colin* < Nahuatl *zolin*, a partridge. *Ex:* Colina (Av.) Colinus (Av.).

**colius**—NL. *colius* prob. < Gr. *kolios*, a woodpecker. *Ex:* Colius (Av.); Hypo-colius (Av.).

**coll**—**1.** L. *collum*, neck. *Ex:* colli-form; sub-rufi-collis: **2.** Gr. *kolla*, glue; *kollētos*, glued together; *kollēma*, that which is glued; *kollōdēs*, like glue; *kollētēs*, one who glues or fastens. *Ex:* Coll-embola (Ins.); Collema\*; Colletes (Ins.); not *Colletia*\*, named in honor of Philibert Collet, French botanist; Colleto-ptera (Av.); collo-blast; Collo-theca (Troch.); Collod-aria (Por.); Collomia\*: **3.** L. *collis*, a hill.

**collaben**—L. *collabens*, genit. *collabentis*, crumpling up < *collabor*, to fall together. *Ex:* collabent.

**collaps**—**1.** L. *collapsus*, flattened side-wise: **2.** L. *collapsio*, a falling together. *Ex:* collaps-ion.

**collar**—L. *collare*, a collar; *collaris*, pertaining to the neck, with a collar. *Ex:* Collaria (Ins.); Collaris (Av.).

collect—L. *collectio*, a gathering together. *Ex:* collect-ors; collect-ion.

collem—See coll 2.

colleno—NL. *colleno* < *coll-en* (chyme) < Gr. *kolla*, glue+*engchyma*, something poured in. *Ex:* colleno-cyte, a poorly formed word.

collet—See coll 2.

collicul—See collin

colligat—L. *colligatus*, fastened together.

collin—1. L. *collino*, to smear over, defile, cover over. *Ex:* Collina (Moll.); Collino-myia (Ins.); not Collinsia*, named after Zaccheus Collins of Philadelphia:      2. L. *collis*, dim. *colliculus*, a hill; *collinus*, hill-loving; *Collina*, goddess of the hills. *Ex:* collicul-ate; Colliculus (Moll.).

collinit—L. *collinitus*, smudged, covered with slime.

collod—See coll 2.

collom—See coll 2.

colludens—L. *colludens*, keeping up a false appearance; ppr. of *colludo*, to play or sport with, to counterfeit.

collur—Gr. *kollyriōn*, some bird of the thrush kind. *Ex:* Colluri-soma (Av.); Collurio (Av.); Collyrio-cincla (Av.).

collyb—Gr. *kollybos*, a small coin. *Ex:* Collybia*; Collybus (Pisc.).

collyr—Gr. *kollyra*, a small cake, loaf of coarse bread. *Ex:* Collyr-opsis (Echin.); Collyria (Ins.). See also collur.

collyrit—NL. *collyrit* < Gr. *collyris*, a loaf of bread+*ilēs*, of nature of, like. *Ex:* Collyrit-idae (Echin); Collyrites (Echin.).

colob—Gr. *kolobos*, mutilated, stunted. *Ex:* Colob-antha*; Colob-otis (Mam.); Colobi-cones (Ins.); Colobo-cephalus (Moll.); Colobus (Mam.).

colocasia—Gr. *kolokasia*, an Egyptian water-plant. *Ex:* Colocasia*.

colon—1. Gr. *kolonō*, to curtail, shorten > NL. *colono*. *Ex:* Colono-mys (Mamm.); Colono-cera (Inst.):      2. L. *colonia*, a colony: *colonus*, a husbandman, colonist, *Ex:* Colonia (Moll.); Colonus (Arach.):      3. Gr. *kolon*, the colon. *Ex:* colon-ic:      4. Gr. *kōlon*, a limb, a member of the body:      5. Gr. *kolōnos*, a hill. *Ex:* Colono-saurus (Rept.).

colonic—L. *colonicus*, pertaining to farming. See also colon 3.

color—L. *color*, tint, hue; *coloratus*, colored, having color.

coloran—L. *colorans*, genit. *colorantis*, stained, dyed, tinged, ppr. of *coloro*, to color.

coloss—Gr. *kolossos*, a colossus, a gigantic statue once at Rhodes. *Ex:* Colloso-chelys (Rept.).

colostr—L. *colostrum*, the first milk secreted by the breasts after childbirth. *Ex:* colostr-ation; colostrum.

colp—Gr. *kolpos*, bosom, breast; also a bay, the womb > *kolpōdēs*, winding, sinuous, embosomed. *Ex:* colp-enchyma; Colpo-cephalum (Ins.); Colpoda (Prot.); A-colpus (Ins.); Peri-colpa (Coel.).

coluber—See colubr.

colubr—L. *coluber*, fem. *colubra*, a serpent; *colubrinus*, like a serpent. *Ex:* Coluber (Rept.); Colubr-ina (Rept.).

colum—L. *columna*, dim. *columella*, a pillar. *Ex:* Column-astrea (Coel.); Column-ites (Por.); Columno-pora (Coel.).

columb—L. *columba*, a dove, pigeon. *Ex:* Columba (Av.); Columbi-formes (Av.); Columbi-perdix (Av.).

columbarius—NL. *columbarius*, dove-like, of or pertaining to a dove < L. *columba*, a dove +*arius*, pertaining to.

colutea—Gr. *koloutea*, some pod-bearing plant. *Ex:* Colutea*.

colym—Gr. *kōlyma*, genit. *kōlymatos*, a hindrance. *Ex:* Colyma (Moll.).

colymb—Gr. *kolymbos*, a diving bird; *kolymbētēs*, a diver. *Ex:* Colymbetes (Ins.); Colymbi-formes (Av.); Podi-lymbus (Av.); see podilymbus.

com—1. L. *com*, together, with. *Ex:* com-mensal:      2. Gr. *komē*, hair > L. *coma*, hair; *comosus*, hairy; *comatulus*, having hair neatly curled. *Ex:* Com-andra*; Com-aster (Echin.); Coma-phorus (Pisc.); Comatula (Echin.); Come-phorus (Mam.); Aphelo-coma (Av.):      3. Gr. *kōma*, deep sleep: *Ex:* coma.

coman—L. *comans*, genit. *comantis*, with long hair, hairy.

comar—Gr. *komaros*, the arbutus. *Ex:* Comarus*.

comastes—Gr. *kōmastēs*, a reveler. *Ex:* Comastes (Mam.).

comat—L. *comatus*, long haired, furnished with long hairs < *como* to provide with long hair. See also com 2.

comaz—Gr. *kōmazō*, to go to a holiday-feast. *Ex:* Comazus (Ins.).

comb—Gr. *kombos*, a roll, knot, bag, purse, band, girth. *Ex:* Combo-ceras (Ins.).

combin—L. *combinatus*, joined, combined < *combino*, to unit, join.

combret—L. *combretum*, a kind of climbing plant. *Ex:* Combretum*.

comeden—L. *comedens*, genit. *comedentis*, eating up, destroying < *comedo*, to consume.

comes—L. *comes*, companion. *Ex:* comes.

comet—1. L. *cometa*, a comet < Gr. *komētēs*, long-haired, hairy; a comet. *Ex:* Comet-ura (Ins.); Cometo-dendron (Prot.):      2. Gr. *kōmētēs*, a dweller, a rustic. *Ex:* Hypsi-cometes (Pisc.).

comis—L. *comis*, kind, pleasing, friendly.

comit—L. *comito* to accompany; *comes*, genit. *comitis*, a companion. *Ex:* comit-alia; comites, pl. of *comes*.

comitial—L. *comitialis*, one who has epilepsy, fits.

comium—NL. *comium* < Gr. *komē* hair+-*ium*. *Ex:* Hylo-comium*.

comm. 1. Gr. *kommos*, ornamentation; *kommōsis*, embellishment; *kommōtēs*, a beautifier. *Ex:* Commo-ptera (Ins.); Commosia (Ins.); Tylocommus (Ins.): 2. Gr. *kommi*, gum. *Ex:* Commi-phora*.

comma—Gr. *komma*, dim. *kommation*, a fragment, that which is cut off. *Ex:* myo-coma = myo-commata.

commatic—Gr. *kommatikos*, concise. *Ex:* Commatica (Ins.).

comminat—L. *comminator*, one who threatens; *comminativus*, threatening.

comminct—L. *comminctus*, defiled < *commingo*, to pollute.

comminut—L. *comminutus*, broken in bits, crumbled < *comminuo*, to break in pieces. *Ex:* comminut-ion.

commissur—L. *committo*, to join, put together > *commissura*, a joint, seam. *Ex:* commissure.

commixt—L. *commixtus*, mingled, mixed < *commisceo*, to mix.

commos—See comm.

communis—L. *communis*, growing in a society, common, general.

commut—L. *commuto*, to completely alter; *commutatus*, changed.

comp—Gr. *kompeō*, to make a noise, to clatter; *kompos*, a noise; *kompastēs*, one who boasts. *Ex:* Compastes (Ins.); Compo-suchus (Ins.); Hypero-compa (Ins.).

compactil—L. *compactilis*, compact, joined together, snugly fitted.

companion—Eng. *companion* < L. *com*, with+ *pan*(< L. *panis* bread)+-*ion*. *Ex:* companion.

compar—L. *comparo*, to match, join, set in good order; *comparis*, well matched, set in good order; *compar*, genit. *comparis*, an equal, companion, one related.

compensat—L. *compensatio*, a weighing together. *Ex:* compensat-ion.

comper—Gr. *kompēros*, boastful.

compital—L. *compitalis*, pertaining to cross roads. *Ex:* compital.

complan—L. *complano*, to make level; *complanatus*, flattened. *Ex:* Complan-aria (Moll.).

composit—L. *compositus*, put together, joined, pp. of *compono*, to put or place together. *Ex:* Composit-ae*.

comps—Gr. *kompsos*, becoming, elegant, ornate. *Ex:* Comps-aspis (Por.); Comps-idia (Ins.); Compso-gnathus (Rept.); Compso-thamnion*; Compsus (Ins.).

compt—L. *comptus*, a band, an ornament for the head; *comptus*, decked; *comptulus*, luxuriously decked. *Ex:* Compt-ella (Ins.); Compto-lampra (Ins.).

con-—L. *con-* prefix meaning with; when combined with words beginning with r, changed to *cor*, as in *Cor-rodentia* (Ins.). *Ex:* co-arctatus; con-jug-ation; con-volv-ent; Con-volv-ulus*.

con—1. L. *conus*, NL. dim. *conulus*, a cone < Gr. *kōnos*, a pine-cone, a cone, dim. *kōnarion*, the pineal gland. *Ex:* Con-alia (Ins.); Con-anthus*; Con-idae (Moll.); Con-ur-opsis (Av.); coni-fer; Coni-rostres (Av.); Cono-chilus (Troch.); Cono-medusae (Coel.); not Cono-morpha*, see chon.; Conul-aria (Coel.); Conus (Moll.); Crypto-conus (Moll.): 2. Gr. *konis*, dim. *konidion*, dust, ashes; *konios*, dusty. *Ex:* Coni-ornis (Av.); conidii-fer-ous; conidium; Conio-mycetes*; Conio-pteryg-idae (Ins.); conio-sperm-ous.

concaterv—L. *concatervatus*, heaped together < *caterva*, a crowd; NL. *concatervans*, in heaps, crowding together.

concept—L. *concepto*, to become pregnant > *conceptaculum*, a receiver, a vessel. *Ex:* concep tacle; conception.

conch—Gr. *konchē*, dim. *konchion*, a shell, shellfish. *Ex:* Conch-astraca (Crust.); concho-logy; conchi-form; conchi-ol-in; Concho-chelys (Rept.); Chlamydo-concha (Moll.).

concinn—L. *concinnus*, neat, skillfully joined.

concolor—L. *concolor*, of the same color, one-colored (as opposed to L. *discolor*, of different colors, parti-colored).

cond—L. *condo*, to hide. *Ex:* testi-cond.

conduplic—L. *conduplicans*, doubling; ppr. of *conduplico*, to double; *conduplicatus*, folded together lengthwise, doubled. *Ex:* conduplication.

condyl—Gr. *kondylos*, the knob of a joint, knuckle, bump. *Ex:* Condyl-arthra (Mam.); condyl-oid; Condyl-ura (Mam.); condyle; Condylo-crinus (Echin.); Mono-condyla.

conepat—NL. *conepatus* < Nahuatl. *conepatl*, the white-backed skunk. *Ex:* Conepatus (Mam.).

conex—L. *conexus*, joined, connected, coherent.

confert—L. *confertus*, crowded, dense.

conferv—L. *conferva*, some healing water-plant. *Ex:* Conferv-ales*; Conferva*.

configurat—L. *configuratus*, formed, fashioned.

confinis—L. *confinis*, related.

conflect—NL. *conflectus*, crowded, thickly clustered < L. *con-*, together+*flecto*, to turn, bend.

confult—L. *confultus*, pressed together.

confus—L. *confusus*, perplexed, confused.

congener—L. *congener*, genit. *congeneris*, of the same sort or kind. *Ex:* congener-ic.

congenit—L. *congenitus*, born together. *Ex:* congenit-al.

conger—L. *conger*, ancient name of the eel. *Ex:* Conger-muraena (Pisc.).

conglob—L. *conglobatus*, made like a ball.

congru—L. *congruus*, similar, agreeing, in harmony.

coniat—Gr. *koniatos*, plastered, covered with pitch; *koniatēs*, a plasterer. *Ex:* Coniatopenia (Por.); Coniatus (Ins.).

conic—Gr. *kōnikos*, cone-shaped. *Ex:* Conic-odon (Mam.); Conicus (Moll.).

conid—NL. *conidium*, a propagative body of fungi < Gr. *konis*, dim. *konidion*, dust. *Ex:* conidii-ferous; conidio-phore; not Conidae (Moll.), see con 1.

conifer—L. *conifer*, cone-bearing. *Ex:* Coniferae*.

conil—*konilos*, an error for Gr. *koniklos* = *kyniklos*, a rabbit, cony > L. *cuniculus*, a rabbit, cony. *Ex:* Conil-urus (Mam.); Cuniculus (Mam.).

conio—See con 2.

conistr—Gr. *konistra*, a dusty rolling place < *konis*, dust. *Ex:* Conistra (Ins.).

conium—1. L. *conium*, hemlock < Gr. *kōneion*, hemlock. *Ex:* Conium*: 2. Gr. *konia*, dust, sand. *Ex:* oto-conium.

conjug—L. *conjugatus*, united; pp. of *conjugo*, to join, couple. *Ex:* conjugat-ion.

conn—Gr. *konnos*, beard; *konnophrōn*, silly, foolish, one who is thought to be wise because he has a beard. *Ex:* Conn-odontus (Ins.); Conno-chaetes (Mam.); Connophron (Ins.).

connar—Gr. *konnaros*, an evergreen thorny tree like Celastrus; not the modern Connarus*. *Ex:* Connar-aceae*; Connarus*.

connat—L. *connatus*, born at the same time, having blood connection.

connex—L. *connexus*, joined, connected.

conniv—L. *conniveo*, to wink at, ppr. *connivens*, genit. *conniventis*, winking at, overlooking, shutting one's eyes. *Ex:* connivent.

connophron—See conn.

connub—L. *connubium*, wedlock. *Ex:* connubial.

connudat—L. *connudatus*, wholly naked, nude.

conop—Gr. *kōnōps*, genit. *kōnōpos*, a gnat, mosquito. *Ex:* Conop-idae (Ins.); Conopi-soma (Ins.); Conopo-deras (Av.); Conops (Ins.).

conquisit—L. *conquisitor*, a recruiting officer.

consci—L. *conscius*, a partaker, an accomplice.

consert—L. *consertus*, connected, joined < *consero*, to unite.

consimil—L. *consimilis*, wholly similar, exactly like. *Ex:* consimil-ar.

consit—L. *consitus*, planted < *consero*, to sow, plant.

consobrin—L. *consobrinus*, related.

consoc—L. *consociatus*, united. *Ex:* consociat-ion.

consocies—NL. *consocies*, poorly made ecological term < L. *con-* + Eng. *society*.

consonans—L. *consonans*, agreeing, harmonizing.

consor—L. *consors*, genit. *consortis*, a brother, sister, consort.

conspecif—ML. *conspecific*, belonging to the same species.

conspect—L. *conspectus*, a mental view, survey, pp. of *conspicio*, to look at.

conspers—L. *conspersus*, spotted, speckled, pp. of *conspergo*, to sprinkle.

conspic—L. *conspicuus*, conspicuous, distinguished, pp. of *conspicio*, to look at, see, observe.

consput—L. *consputus*, held in contempt, pp. of *conspuo*, to spit upon in derision.

constipat—L. *constipatio*, a binding together. *Ex:* constipat-ion.

consuet—L. *consuetus*, usual, customary, pp. of *consuesco*, to accustom.

consut—L. *consutus*, sewn together, pp. of *consuo*, to stitch together.

cont—1. Gr. *kontos*, short. *Ex:* Conto-pus (Av.): 2. Gr. *kontos*, a pole, spear, arrow.

contabesc—L. *contabesco*, to waste away. *Ex:* contabesc-ence.

contact—L. *contactus*, touching.

context—L. *contextus*, interwoven, united, connected.

contigu—L. *contiguus*, near together, neighboring, adjoining.

contort—L. *contortus*, full of turns, twisted; also *contortus*, hurled, full of motion < *contorqeo*, to twist, turn. *Ex:* contort-ed.

contour—Fr. *contour*, the outline of a figure or body < ML. *contorno*, to go round, turn round. *Ex:* contour feather.

contra—L. *contra*, against, opposite. *Ex:* Contra-cavia (Mam.); contra-ception (Med.).

contract—L. *contractus*, drawn together, made short, narrow.

contrectat—L. *contrectatio*, genit. *contrectationis*, a touching, handling. *Ex:* contrectation.

controvers—L. *controversus*, controversial, in opposite directions.

conul—See con 1.

conviva—L. *conviva*, a table guest, one who feeds with another.

convolvul—L. *convolvo*, to roll around > *convolvulus*, a kind of plant called bindweed. *Ex:* Convolvulus*.

conyz—Gr. *konyza*, a strong smelling plant, fleabane, a thorn; perhaps < *kōnōps*, a gnat. *Ex:* Conyza*.

cop—1. Gr. *kōpē*, handle, oar. *Ex:* Cope-poda (Arth.); Eu-cop-ella (Coel.); Eu-copia (Coel.); Kopio(n)-ella (Moll.): 2. Gr. *kopis*, genit. *kopidos*, a cleaver, a dagger. *Ex:* Copid-ita (Ins.); Copido-soma (Ins.): 3. Gr. *kopos*, fatigue, uneasiness, toil. *Ex:* copi-opsia (Med.);

copo-dys-kinesia (Med.): **4.** Gr. *kopē*, a division, cut, cutting, collision. *Ex:* Meli-cope*. See also copt.

copai—Tupi Indian *copa-iba*, the copabia tree yielding a medicinal resin. *Ex:* Copai-fera*.

copan—Gr. *kopanon*, a pestle; also a sword, hatchet. *Ex:* Copano-pachys (Ins.); Copanon (Por.); Copanum (Por.).

cope—See cop 1.

copelat—Gr. *kōpēlatēs*, a rower. *Ex:* Copelata (Tun.).

copeo—1. Gr. *kōpēeis*, provided with a handle< *kōpē*, an oar, haft. *Ex:* Copeo-gnatha (Arth.): 2. Gr. *kōpeōn*, a fragment.

coph—Gr. *kōphos*, dumb, deaf also obtuse, blunt, dull. *Ex:* Coph-osus (Ins.); Coph-yla (Amph.), see hyl; Cophi-xalus (Amph.); Copho-saurus (Rept.).

cophias—Gr. *kōphias*, the deaf-adder. *Ex:* Cophias (Rept.).

cophin—Gr. *kophinos*, a basket. *Ex:* Cophino-ceras (Moll.); Cophinus (Echin.).

copid—See cop 2.

copios—L. *copiosus*, plentiful.

copr—Gr. *kopros*, dung. *Ex:* Copr-inae (Ins.); Copr-inus*; Copris (Ins.); Copr-osma*; copro-lite; copro-phagous; Copro-theres (Av.).

copt—Gr. *koptō*, to smite, pierce, cut up; *koptos*, cut small>*koptē*, a biscuit. *Ex:* Coptis*; Copto-chilus (Ins.); Copto-thyris (Brach.); Coptosia (Ins.); Antho-coptes (Ins.); Chori-optes (Arth.); Dendro-copus (Av.); Gastro-copta (Moll.); Pro-copt-odon (Mam.); Psor-optes (Arth.); Xylo-copa (Ins.).

copul—L. *copula*, a band or link; *copulo*, to join, pp. *copulatus*, joined, united>*copulatio*, genit. *copulationis*. *Ex:* copul-arium; copula; copulat-ory; copulation.

cor- —See con-.

cor—1. Gr. *korē*, pupil of the eye. *Ex:* core-dia-stasis (Med.); Core-gonum (Pisc.); coreo-plasty (Med.): 2. Gr. *korē*, a maiden, girl; also a doll or puppet> *Korē*, Cora, the name under which Proserpine was worshipped in Attica. *Ex:* Hali-core (Mam.); Spongo-core (Prot.). See cori 2. 3. Gr. *koris*, a bed-bug. Core-opsis*; Cori-arachne (Arach.); Cori-melaena (Ins.); Corisa (Ins.); Corixo-gryllus (Ins.), NL. *corixa*<*corisa*<Gr. *koris*, a bug; Hydro-cores (Ins.): 4. Gr. *koris*, a kind of St. John's wort. *Ex:* Coris*: 5. Gr. *koris*, a kind of fish. *Ex:* Coris* (Pisc.). See cori: 6. Gr. *koreō*, to purge.

corac— See corax.

coral—See corall.

corall—Gr. *korallion*, coral, esp. red. coral>LL. *corallinus*, coral-red. *Ex:* Coralli-ophis (Rept.); Corallina (Ann.)*; Corallio-phila (Moll.); Corallo-rhiza*; Corallo-spondylus (Moll.); corallum; Corallus (Rept.).

corax—Gr. *korax*, genit. *korakos*, a raven, crow; *korakias*, a kind of raven; *korakinos*, like a raven, raven-black; also a young raven. *Ex:* corac-oid; Coraci-ura (Av.); Coracias (Av.); Coracii-formes (Av.); Coracinus (Pisc.); coraco-brachialis; Phalacro-corax (Av.).

corb—L. *corbis*, a basket, dim. *corbula*=LL. *corbicula*=NL. *corbicella*. *Ex:* Corbi-sema (Prot.); Corbicell-opsis (Moll.); Corbicul-ina (Moll.); corbicula; Corbis (Moll.); Corbuli-pora (Bry.); Corbulo-mya (Moll.).

corbic—See corb.

corbit—L. *corbita*, a slow sailing ship carrying goods. *Ex:* Corbit-ella (Por.).

corbul—See corb.

corchor—Gr. *korchoros*, a plant of bitter taste, chickweed, pimpernel. *Ex:* Corchorus*.

corcul—L. *corculus*, dim. of *cor*, the heart.

cord—L. *cor*, genit. *cordis*, the heart. *Ex:* cord-ate; cordi-form; not Cordi-ceps*, see cordy; not Cordia*, named after Euricius Cordius, German botanist.

cordat—1. L. *cordatus*, prudent, wise: 2. NL. *cordatus*<*cord* (stem of *cor*, genit. *cordis*, the heart+*atus*.).

cordul—See cordyl.

cordy—See cordyl.

cordyl—Gr. *kordylē*, a cudgel, bump, swelling, tumor; also a headdress. *Ex:* Cordule-gaster (Ins.); Cordulia (Ins.); Cordy-ceps=Cordi-ceps*; Cordyl-ine*; Cordylo-phora (Por.)· Cordylo-porus (Myr.).

core—See cor 3.

corem—Gr. *korēma*, genit. *korēmatos*, filth, refuse; also a broom and in this sense most often used. *Ex:* corem-oid; Corema*; Coremat-ura (Ins.); coremium; Coremo-thrips (Ins.); Korem-aster (Echin.).

corethr—Gr. *korēthron*, a bunch of twigs, broom. *Ex:* Corethr-ura (Av.); Corethro-gyne*; Corethro-machilis (Ins.).

cori—1. L. *corium*, leather, skin>*excoriatus*, peeled, stripped of its skin. *Ex:* cori-aceous; Cori-aria; coria; anta-coria: 2. Gr. *korē*, a maiden. *Ex:* Cori-sternum (Mam.). See also cor. 3.

coriandr—L. *coriandrum*, ancient name for coriander, an umbelliferous plant<Gr. *korian-non*=*korion*, coriander<*koris*, a bug. *Ex:* Coriandrum*.

corin—NL. *corinus*<L. *cor*, the heart+*-inus*, adj. ending meaning like. *Ex:* corinus.

coris—See cor 3, 4, 5.

corix—See cor 3.

corm—Gr. *kormos*, a stump, log. *Ex:* corm; Corm-ias (Av.); Corm-ura (Mam.); Cormo-dactylus (Rept.); cormo-phyte; Hypso-cormus (Pisc.).

corn—L. *cornu*, a horn; *cornutus*, horned; *corneus*, horny. *Ex:* cornea; Corni-gramphus (Ins.); Cornu-aspis (Ins.); Cornu-copia (Moll.); Cornu-lites (Ann.); Cornus*; Lamelli-cornia (Ins.).

cornic—L. *cornix*, genit. *cornicis*, a crow.

cornicin—L. *cornicen*, genit. *cornicinis*, blower of a trumpet.

cornix—See cornic.

The Cornucopia Crater Fungus, *Craterellus cornucopioides*. Redrawn from The Romance of The Fungus World —Rolfe. J. B. Lippincott Co., Chapman & Hall, Ltd.

cornubic—NL. *cornubicus*, of or pertaining to *Cornubia*, a Latinized name of Cornwall.

cornul—LL. dim *cornulum*<L. *cornu*, a horn. *Ex:* Cornul-aria (Coel.).

coroll—See coron.

coron—1. L. *corona*, dim. *coronula*=*corolla*,= NL. *coronilla*, a crown, wreath, garland; *coronatus*, crowned; *coronarius*, pertaining to a crown, wreath, or garland. *Ex:* coroll-atus; corolla; Corolli-florae*; Coron-illa*; coron-oid; Coronula (Crust.); Acro-coronis (Prot.):    2. Gr. *korōnē*, a sea-crow, crow, raven. *Ex:* Corone (Av.).

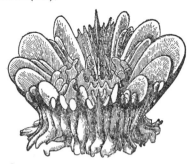

Crowned Wheel-cup Coral, *Trochocyathus coronatus*. Redrawn from Voyage of The Challenger—Thomson. Courtesy of Joseph McDonough Co.

coronop—L. *coronopus*, a plant called crowfoot. *Ex:* coromopi-folia. See coron 2.

coroph—Fr. *corophie*, the name of some crustacean. *Ex:* Corophium (Crust.). See also coryph.

corp—L. *corpus*, genit. *corporis*, dim. *corpusculum*, a body; *corporalis*, bodily. *Ex:* corpora quadrigemina; corpore-al; corpus luteum; cor puscle.

corporal—See corp.

correct—L. *correctus*, improved, set in order.

corrh—Gr. *korsē*=new Att. *korrhē*, the temple, temporal bone. *Ex:* Lypro-corrhe (Ins.).

corrigi—L. *corrigia*, a shoestring, a thong. *Ex:* Corrigi-ola*; Corrigiol-aceae*.

corrod—L. *corrodo*, to gnaw, ppr. *corrodens*, genit. *corrodentis*, gnawing. *Ex:* Corrodentia (Ins.).

corros—L. *corrosus*, gnawed<*corrodo*, to gnaw, eat away.

corrug—L. *corrugis*, having folds or wrinkles.

corryss—Gr. *korryssō*, to equip, provide. *Ex:* Corrysso-merus (Ins.).

cors—Gr. *korsē*=new Att. *korrhē*, head, temple, side of the forehead. *Ex:* Caeno-corse (Ins.).

cortader—Argentine *cortaderia*, native name for the Pampas grass<Sp. *cotadera*, cutting. *Ex:* Cortaderia*.

cortex—See cortic.

corthyl—Gr. *korthylos*, a crested bird. *Ex:* Cor thylio (Av.).

cortic—L. *cortex*, genit. *corticis*, bark, cork; *corticinus*, of bark-like texture. *Ex:* cortex; Cortic-aria (Ins.); Cortici-thoa (Coel.); cortico-spinal; Cortico-tomus (Ins.).

corticat—L. *corticatus*, covered with bark, having a bark<*cortex*, genit. *corticis*, bark.

corticium—See cortic.

cortin—1. L. *cortina*, a round vessel or kettle: 2. L. *cortina*, a curtain. *Ex:* Cortin-arius*; cortin-ate; cortina.

corus—L. *Corus*=*Caurus*, the Northwest wind.

corusc—L. *coruscus*, waving; *coruscans*, genit. *coruscantis*, flashing, vibrating, ppr. of *corusco*, to flash suddenly, to sparkle.

corv—L. *corvus*, a raven, crow. *Ex:* Corv-aspis (Pisc.); corv-ine; Corvi-vulture (Av.); Corvospongilla (Por.); Corvus (Av.).

cory—Gr. *korys*, genit. *korythos*, a helmet; *korystēs*, a helmeted man, warrior; *korythaix*, helmet-shaking; i.e., with waving plumes. *Ex:* Cory-lophus (Ins.); Corystes (Ins.), (Crust.); Coryth-opis (Av.); Oto-corys (Av.); Tricorythus (Ins.); Tricho-coryes (Mam.). See also coryn.

coryb—Gr. *Korybas*, genit. *Korybantos*, the priests of Corybantes known for their mad frenzy. *Ex:* Coryb-issa (Ins.); Corybantes (Ins.); Corybas (Por.).

coryc—Gr. *kōrykos*, a sack of leather; also the scrotum; *kōrykōdēs*, like a sack. *Ex:* Coryc-ia (Ins.); Coryc-ella (Arth.); Corycodus (Crust.).

corycae—Gr. *kŏrykaios*, a spy, an inhabitant of Corycus in Lydia. *Ex:* Corycaeus (Arth.).

coryd—1. Gr. *korydŏn* = *korydos*, the crested lark < *korys*, a crest. *Ex:* Coryd-onyx (Av.); Corydo-spiza (Av.):    2. Gr. *Korydŏn*, a shepherd's name. *Ex:* Corydon (Av.).

corydal—Gr. *korydallis* = *korydalos*, a plant with flower spur resembling the spur of the lark, fumitory; also a lark; an extended form of *korydos*, the crested lark. *Ex:* Corydalis*; Corydalla (Av.); Corydalus (Ins.).

coryl—L. *corylus* which is said to be derived from a supposed Gr. *korylos*, name of the hazel (which has a helmet-shaped involucre) < *korys*, a helmet. *Ex:* Coryl-opsis*; Corylus*.

corymb—L. *corymbus*, a flower cluster < Gr. *korymbos*, the head, highest point; also a cluster of fruits or flowers < Gr. *korys*, a helmet. *Ex:* corymb; Corymb-ites (Ins.); Corymb-osa (Bry.); corymbi-fer-ous; Corymbo-crinus (Echin.).

coryn—Gr. *korynē*, a club, club-shaped bud or shoot; *korynētēs*, a club-bearer. *Ex:* Cory-morpha (Coel.); Coryn-idae (Coel.); Coryne (Coel.); Coryne-soma (Platy.); Corynet-ops (Ins.); Corynetes (Ins.); Coryno-philus (Ins.); Leuco-coryne*; Syn-coryne (Coel.).

coryneum—See coryn and -um.

coryph—Gr. *koryphē*, the head, summit, vertex, principal point. *Ex:* Coryph-ella (Moll.); Coryph-idae (Av.); Coryph-odon (Mam.); Corypha*; Di-coryphe*; Tri-coryphus (Ins.).

coryphister—Gr. *koryphistēr*, the upper ring of a net used by hunters. *Ex:* Coryphister-a (Av.).

coryst—See cory.

coryth—See cory.

corythaeol—Gr. *korythaiolos*, with helmet moving quickly. *Ex:* Corythaeola (Av.).

cos—1. Gr. *kŏs*, Ionic for *pŏs*, an interrogative. *Ex:* Cos-oryx (Mam.):    2. Gr. *kŏs*, a covering. *Ex:* Diple-cosia*.

-cosa—NL. -*cosa* < *Lycosa*, a genus of spiders *Ex:* Schizo-cosa (Arach.); Allo-cosa (Arach.).

coscin—Gr. *koskinon*, dim. *koskinion*, a sieve. *Ex:* Coscinio-pterus (Ins.); Coscinium*; Coscino-derm (Por.); Coscino-discus*.

coscorob—1. Trinidad *coscorob*, native name for a fish of the genus Cichlasoma.    2. NL. coscoroba < Tupi *cosaroba* = *saroba*, native name of a swan-like diving bird. *Ex:* Coscoroba (Av.).

cosil—NL. *cosila*, anagram of *Scolia*. *Ex:* Cosila (Ins.).

cosm—Gr. *kosmos*, order, form, ornament, a well-ordered system (such as the world or universe); *kosmios*, well-ordered, well-behaved; *kosmētos*, trim, adorned; *kosmēsis*, trimming, adornment; *kosmētēs*, one who adorns; *kosmētikos*, skilled in decorating. *Ex:* cosm-ine; Cosmet-idae (Arach.); Cosmia (Ins.); cosmic;

Cosmio-ceramus (Moll.); Cosmo-coma (Ins.); cosmo-politan; Cosmus*; A-cosmia*; micro-cosm.

cosmet—See cosm.

coss—1. L. *cossus*, a kind of larva found under the bark of trees > Fr. *cosson*, worm. *Ex:* Coss-idae (Ins.); Cossus (Ins.):    2. Gr. *kossos*, a box on the ear.

cossyph—Gr. *kossyphos*, a kind of singing bird; also, a kind of sea-fish. *Ex:* Cossyph-icula, (Av.); Cossyph-odes (Pisc.); Cossyphus (Ins.), (Av.), (Pisc.).

cost—L. *costa*, a rib; *costatus*, having lines or rib-like ridges > NL. *costalis*, pertaining to ribs or sides of the body. *Ex:* Costato-crinus (Echin.); Costata (Amph.); Costato-scala (Moll.); Costi-fer (Por.); costo-xiphoid.

cosymbot—Gr. *kosymbŏtos*, fringed, fimbriated. *Ex:* Cosymbotus (Rept.).

cot—1. Gr. *kotis*, genit. *kotidos*, the top and back of the head, cerebellum. *Ex:* Cot-aster (Ins.). 2. Gr. *kotos*, anger, rancor; *koteŏ*, to be angry at. *Ex:* Xylo-cota (Av.). See cotos.

coth—1. Gr. *kŏthŏn*, cup, bowl; also a quay, dock (later applied to the inner harbor at Carthage). *Ex:* Cotho-crinus (Echin.); Cothon-aspis (Ins.); Cothono-laimus (Nemat.):    2. Gr. *kothŏ*, to damage, hurt. *Ex:* Coth-urus (Mam.).

cothon—See coth.

cothurn—Gr. *kothornos*, a boot or buskin, shaped to fit either foot, and a part of the garb of tragic actors; also a changeable-minded person, this meaning derived from the fact that the buskin could be worn on either foot. *Ex:* Cothurnia (Prot.); Cothurno-cystis (Echin.).

cotid—See cot.

cotin—L. *cotinus*, name of some shrub furnishing a purple dye. *Ex:* Cotinus*.

coting—Braz. *cotinga*, < a stem meaning to wash white, native name for several birds. *Ex:* Cotinga (Av.).

cotone—ML. *cotonea* = L. *cydonia*, a quince. *Ex:* Cotone-aster*; cotoneus; Cydonia*.

cotos—Gr. *kotos*, grudge, envy, ill-will. *Ex:* Brepho-cotosia (Ins.).

cott—Gr. *kottos*, a cock; a horse; also a river fish, perhaps the bull-head or miller's thumb. *Ex:* Cott-idae (Pisc.); Cotto-gaster (Pisc.); Argyro-cottus (Pisc.); Cottus (Pisc.).

cotul—See cotyl.

coturnic—L. *coturnix*, genit. *coturnicis*, a quail. *Ex:* Coturnic-ops (Av.); Coturnix (Av.); Turnix (Av.).

coturnix—See coturnic.

cotyl—Gr. *kotylē*, cup-shaped; *kotylēdŏn*, a cavity, any cup-shaped hollow; also a plant, probably the navel-wort. *Ex:* Cotula*; cotyla; Cotylea (Platy.); cotyledon; Cotyledon*; Cotylo-genes (Platy.); Cotylo-phora (Mam.); hecto-cotylus.

couend—L. *couendi*, coming together, meeting; gerund of *coeo*, inf. *coire*, to go, come together.

counter-—Eng. *counter-*, against<L. *contra-*, against. *Ex:* counter-irritant.

covert—Eng. *covert*, a cover>Fr. *couvrir*, to cover, pp. *couvert*, covered. *Ex:* covert.

cox—L. *coxa*, hip, the hipbone. *Ex:* cox-al; coxopodite; coxo-sternal. See also coc.

coyp—*coypus*, S.A. native Indian name for some rodent. *Myocastor coypus*, is the aquatic coypu.

crabro—L. *crabro*, genit. *crabronis*, a hornet. *Ex:* Crabro (Ins.); Crabron-idae (Ins.).

crac—Gr. *krazō*, to caw like a crow or raven> NL. *crax*, genit. *cracis*, a screamer. *Ex:* Crac-idae (Av.); Crax (Av.).

cracc—L. *cracca*, name of a vetch. *Ex:* Cracca*

cracen—L. *cracens*, genit. *cracentis*, neat, slender·

cract—Gr. *kraktēs*, a screamer. *Ex:* Cractes (Av.).

crad—Gr. *krados*, the black blight of the wild fig tree; also a twig. *Ex:* crad-ina.

craer—Gr. *kraira*, a point, top, head. *Ex:* Pro-craerus (Ins.).

cram—Gr. *krama*, genit. *kramatos*, a mixture. *Ex:* Cram-auchenia (Mam.).

cramb—1. Gr. *krambē*, cabbage, kale. *Ex:* Cramb-aphis (Ins.); Crambe*: 2. Gr. *krambos*= *chēros*, parched, hot, shrivelled; also clean, neat, delicate. *Ex:* Crambo-morphus (Ins.); Crambus (Ins.): 3. Gr. *krambos*, loud, ringing.

crampon—O.H.Ger. *chramph*, crooked. *Ex:* crampon.

cran—1. Gr. *kranos*, a helmet. *Ex:* Crano-ceras (Moll.); Crano-cephalus (Crust.); Cran-odus (Pisc.), Gr. *odous*, tooth: 2. Gr. *kranon*, the dog wood tree=*kraneia*: 3. Gr. *kranion*, the skull>NL. *cranium*, the skull. *Ex:* Craneo-phora (Ins.); crani-al; Crani-aspis (Prot.); Crani-ata; Crania (Brach.); Cranio-lithes (Brach.); Crano-pelagrus (Av.).

crang—Gr. *krangōn*, a shrimp. *Ex:* Crang-opsis (Ins.); Crangon (Crust.); Crangono-bdella (Ann.); Glypho-crangon (Crust.).

crantor—L. *Crantor*, fabled armor-bearer of Peleus. *Ex:* Ipo-crantor (Av.).

cras—Gr. *krasis*, genit. *kraseōs*, a mixing, blending. *Ex:* Craseo-mys (Mam.).

crasped—Gr. *kraspedon*, pl. *kraspeda*, an edge, border; NL. *craspedotus*, as if from a Gr. *kraspedotos*. *Ex:* Crasped-acusta (Coel.); Craspedo-opsis (Ins.); Craspedo-cephalus (Rept.); craspedote; Craspid-aster (Echin.).

craspid—See crasped.

crass—L. *crassus*, thick, heavy. *Ex:* Crass-ula*; Crassi-pedia (Moll.); Crasso-femuria (Ins.); in-crass-ate.

Fat-tailed Pouched Mouse, *Sminthopsis crassicaudata*. Redrawn from The Wild Animals of Australasia—Le Souef and Byrrell.

crat—1. L. *cratis*, a wicker work; also a joint rib. *Ex:* Cratis (Moll.). 2. Gr. *kratos*, might, power; *krateros*, poet. *krataios*, strong, mighty; *kratistos*, strongest. *Ex:* Cratae-pus (Ins.); Cratero-pus (Av.); Cratero-mys (Mam.); Crato-xylon*; En-crates (Ins.); Pan-cratium (Ins.): 3. Gr. *krata*, head.

crataeg—Gr. *krataigos*, a kind of thorny flowering shrub. *Ex:* Crataegus*.

crataepus—Gr. *krataipous*, stout-footed. *Ex:* Crataepus (Ins.).

crater—Gr. *kratēr*, a cup, the mouth of a volcano. *Ex:* crateri-form; crateria; Cratero-lampas (Echin.); see crat 2.

craticul—L. *craticulus*, composed of reeds, lattice work<*cratis*, wicker work. *Ex:* Craticula (Ins.); craticul-ar.

cratist—See crat 2.

craur—Gr. *krauros*, hard, fragile, brittle. *Ex:* Crauro-thrix (Mam.).

crax—See crac.

cre—Gr. *kreas*, genit. *kreatos*, flesh. *Ex:* Cre-odonta (Mam.); creat-ine; Creo-philus (Ins.).

cread—Gr. *kreadion*, a bit of flesh, meat; dim. of *kreas*, flesh. *Ex:* Allo-creadium (Platy.).

creagr—Gr. *kreagra*, a flesh hook<*agreuō*, to seize. *Ex:* Creagrius (Av.); Creagro-ceros (Mam.).

creat—See cre.

crebr—L. *creber*, *-bra*, *-brum*, thick, crowded, frequent. *Ex:* crebri-spinus.

creber—See crebr.

crec—See crex.

credul—L. *credulus*, believing, confiding.

crem—Gr. *kremaō*, to hang; *kremastos*, hung, hung up; *kremastēr*, a suspender, hanger>*oi kremasterēs*, the muscles by which the testicles are suspended. *Ex:* Cremanium*; cremaster; Cremasto-saurus (Rept.); cremo-carp; Cremo-lobus*.

cremast—See crem.

cremn—Gr. *krēmnos*, an overhanging rock, a cliff, declivity<*kremaō*, to hang. *Ex:* cremn-ad; Cremno-bates (Pisc.); cremno-phyte; Holo-cremnus (Ins.).

cremnod—Gr. *krĕmnŏdēs*, steep. *Ex:* Cremnodes (Ins.).

cremor—L. *cremor*, genit. *cremoris*, soup, broth.

cren—1. NL. *crena*, dim. *crenula*, a notch; *crenatus*, notched. *Ex:* crenate; crenati-flora; Creni-labrus (Pisc.); Creno-lepis (Pisc.): 2. Gr. *krēnē*, a spring. *Ex:* Creno-thrix*.

crenat—See cren 1.

crep—Gr. *krēpis*, genit. *krēpidos*, a shoe, a half-boot, a slipper worn by men > L. *crepida*, dim. *crepidula*, a slipper. *Ex:* Crepi-limne-bius (Ins.); Crepi-pora (Bry.); Crepid-aspis (Ins.); Crepido-bothrium (Platy.); Crepidula (Moll.).

creper—L. *creper*, dark, dusky, uncertain.

crepid—See crep.

crepis—L. *crepis*, a name for some unknown plant. *Ex:* Crepis*; Hippo-crepis*.

crepit—L. *crepito*, to rattle, to clatter; *crepitans*, clattering < *crepo*, to rattle. *Ex:* crepit-ation; Crepiti-termes (Ins.).

crepuscul—L. *crepusculum*, twilight. *Ex:* crepuscul-ar.

cresc—L. *cresco*, to increase, grow, be born of; *cretus*, arisen, born of, increased. *Ex:* ac-crescent; ac-cretion.

cresson—Gr. *kressōn*, powerful. *Ex:* Cressona (Ins.).

crest—OF. *creste* < L. *crista*, the crest of a helmet, the plume; the tuft of hair on the head of animals.

cret—1. L. *cerno*, to separate, to sift; pp. *cretus*, separated. *Ex:* ex-cret-ion; se-cret-ion. See also cresc; 2. L. *creta*, chalk; *cretaceus* chalky. *Ex:* Cret-aspis (Crust.); Creta-gryphaea (Moll.); Cretac-echinus (Echin.); Cretaceous: 3. Gr. *Krētē*, Crete > L. *creticus*, of the island of Crete. *Ex:* cret-an cret-ensis; Creticus (Moll.). See also cresc 4. L. *cretus*, born of, sprung from.

cretac—See cret 2.

cretic—See cret 3.

creurg—Gr. *kreourgos*, a butcher. *Ex:* Creurgus (Av.).

crex—Gr. *krex*, genit. *krekos*, a kind of long-legged bird. *Ex:* Crex (Av.); Crec-opsis (Av.); Mega-crex (Av.).

cribell—See cribr.

cribr—L. *cribro*, to sift; pp. *cribratus*, sifted; *cribrum*, dim. *cribellum*, sieve. *Ex:* cribellum; Cribr-aria*; Cribr-ina (Coel.); Cribrat-ina (Prot.); Cribrat-ores (Av.).

cric—Gr. *krikos*, a ring; *krikōma*, a ring, circle; *krikōtos*, made of rings. *Ex:* cric-oid; Crico-cephalus (Platy.); Cricotus (Amph.); Kriko-gonia (Ins.); Kriko-nema (Ins.).

cricet—ML. *cricetus*, the hamster < It. *criceto* or < Polish *krecek*. *Ex:* Cricet-omys (Mam.); Criceto-dipus (Mam.); Uro-cricetus (Mam.).

cricot—See cric.

crin—1. L. *crinis*, hair; *crinalis*, pertaining to hair; *crinatus*, hairy, long-haired. *Ex:* Crinala (Ins.); Crini-ger (Av.): 2. Gr. *krinon*, a lily. *Ex:* Crin-oidea (Echin.); Crino-cidaris (Echin.); Crinum*; Leuco-crinum*: 3. Gr. *krinō*, to separate, pick out. *Ex:* endo-crine.

crio—Gr. *krios*, a ram, goat. *Ex:* Crio-ceris (Ins.); Crio-therium (Mam.).

cris—1. Gr. *Krisiē*, a mythological feminine name. *Ex:* Cris-idia (Bry.); Crisi-idae (Bry.); Crisia (Bry.): 2. Gr. *krisis*, a separating, alteration, a means of distinguishing.

crisim—Gr. *krisimos*, scrupulous, judicial, decisive, fatal. *Ex:* Crisimus (Ins.).

crisp—L. *crispo*, to curl; *crispus*, curled, uneven, wrinkled, quivering. *Ex:* Crisp-ella (Brach.); Crispi-spongia (Por.).

criss—1. L. *crisso*, to move the haunches > NL. *crissum*, that part of a bird surrounding the cloaca, the under-tail coverts; *crissalis*, pertaining to the crissum or under-tail coverts. *Ex:* crissal; crissum: 2. Gr. *krissos = kirsos*, a swelling of a blood vessel due to excess blood.

crist—L. *crista*, a crest; *cristatus*, dim. *cristulatus*, crested. *Ex:* crista galli; Crista-saura (Rept.); Cristat-ella (Bry.); Cristato-gobius (Pisc.); Cristi-vomer (Pisc.).

cristat—See crist.

crit—Gr. *kritēs*, a judge; *kritikos*, a critic; *kritos*, chosen, select, separated, notable > *akritos*, unarranged, undetermined. *Ex:* crit-en-chyma; critic-al; Crito-merus (Ins.); Ana-crites (Av.); Acrita (Ins.).

crith—Gr. *krithē*, barley-corns, barley. *Ex:* Crith-idia (Prot.); Crithe-phaga (Ins.).

crithm—Gr. *krēthmos = krithmos = krithmon*, the name of an umbelliferous plant, the samphire. *Ex:* Crithmum*.

crius—NL. *crius* < Gr. *krios*, a goat. *Ex:* Crius (Pisc.).

crobil—See crobyl.

crobyl—Gr. *krōbylos*, a braid, a curl; also a knot of hair on the head. *Ex:* Crobilo-cerus (Ins.); Crobyl-ura (Prot.); Crobylus (Arachn.).

croc—1. Gr. *krokos*, the saffron plant, saffron; *krokōtos = L. croceus*, saffron-colored. *Ex:* Crocinus (Ins.); Croc-osmia*; Croco-zona (Ins.); Crocota (Ins.); Crocus*; Peri-crocotus (Av.): 2. Gr. *krokē*, a pebble; *krokalē*, the beach, a pebble of the beach. *Ex:* Croce-thia (Av.); Crocalia (Ins.): 3. Gr. *krokis* genit. *krokidos*, dim. *krokidion*, the woof or weft, loose threads, lint. *Ex:* Crocid-ura (Mam.); Crocidium*; Crocido-laemus (Rept.).

crocal—See croc 2.

crocat—L. *crocatus*, saffron-yellow.

croce—See croc 1 and 2.

crocid—See croc 3.

crocodil—L. *crocodilus*, a crocodile. *Ex:* Croco dili-cola (Platy.); Crocodilia (Rept.); Crocodilus (Rept.).

crocosmi—NL. *crocosmia* (<Gr. *krokos* saffron+ *osmē*, smell), a genus of iridaceous plants. *Ex:* crocosmi-folia; Crocosmia*.

crocot—See croc 1.

crocus—Gr. *krokos*, the saffron plant, saffron. *Ex:* Crocus*. See croc 1.

crocut—L. *crocuta*, an unknown animal, prob. the hyaena.

croes—Gr. *Kroisos*, Croesus, rich king of Lydia. *Ex:* Croeso-myrmex (Ins.); Croesus (Ins.).

cromy—Gr. *kromyon=krommyon*, an onion. *Ex:* Cromy-echinus (Echin.); Cromyo-crinus (Echin.); Stylo-cromyum (Prot.).

cronart—NL. *cronartium*, a fungus genus, etym, unknown. *Ex:* Cronartium.

crop—AS. *cropp*, *crop*, a top, also a bird's craw; Dutch *krop*, a craw. *Ex:* crop

cross—1. Gr. *krossoi*, a fringe, tassels; *krossōtos*, fringed. *Ex:* Cross-aster (Echin.); Crosso-pteryg-idae (Pisc.); Crosso-pus (Mam.); Crosso-soma*: 2. Gr. *krōssos*, a pail, pitcher, jar.

crot—1. Gr. *krotōn*, a tick, bug. *Ex:* Croto-phaga (Av.); Croton*: 2. Gr. *krotos*, a beat, rattling noise. *Ex:* di-crot-ic (Med.).

crotal—Gr. *krotalon*, a rattle, little bell<*kroteō*, to rattle. *Ex:* Crotal-aria*; Crotal-inae (Rept.); Crotal-cephalus (Arth.); Crotalus (Rept.).

crotaph—Gr. *krotaphos*, the side of the face, the temple of the head; also the head of a hammer; *krotaphitēs*, relating to the side of the head. *Ex:* Crotaphitis (Amph.); Crotaphytus (Rept.); Gymno-crotaphus (Av.).

croton—See crot.

cruc—L. *crux*, genit. *crucis*, a cross; *crucio*, to torment, pp. *cruciatus*, tormenting. *Ex:* crucial ligaments; Cruci-brissus (Echin.); Cruci-fer-ae*; Cruc(ian)-ella*.

cruciator—L. *cruciator*, a tormentor.

crucibul—ML. *crucibul*, an earthen pot. *Ex:* Crucibulum*.

cruciens—L. *cruciens*, torturing, afflicting; ppr. of *crucio*, to torment, torture.

crudel —L. *crudelis*, hard-hearted, cruel.

cruent—L. *cruento*, to make bloody; pp. *cruentatus*, stained with blood, cruel. *Ex:* Cruentata (Moll.).

-crum—L. *-crum*, suffix added to verb stems to form nouns denoting means or instrument. *Ex:* ful-crum.

crumen—L. *crumena*, pl. *crumenae*, a purse, small money-bag. *Ex:* crumen; Crumen-aria*; Crumen-opthalmus (Pisc.); Crumenae-crinus (Echin.); Crumeni-fera (Amph.).

crun—Gr. *krounos*, spring, well. *Ex:* Cruno-mys (Mam.).

cruor—L. *cruor*, blood. *Ex:* cruor; cruor-in.

crur—See crus.

crus—L. *crus*, pl. *crura*, the leg, thigh; *cruralis*, pertaining to the hind limb or leg. *Ex:* crural; Cruri-thyris (Brach.); Cruro-saurus (Rept.); crus; Brachy-crus (Mam.).

crust—L. *crusta*, dim. *crustula*, the tough, hard surface of a body, shell; *crustaceus*, having a shell or rind; *crustosus*, covered by a shell or crust. *Ex:* Crust-acea (Arth.); Crusti-pora (Bry.); crustose; Crustul-ina (Arach.).

cry—Gr. *kryos*, cold, chilly. *Ex:* Cry-aster (Echin.); Cryo-philus (Arach.); cryo-plankton; Haemato-crya.

crybel—Gr. *krybēlos*, hidden. *Ex:* Crybelo-cephalus (Crust.); Crybelus (Av.).

crybetes—Gr. *krybētēs*, one hidden in the earth.

crym—Gr. *krymos*, frost, cold, ice; *krymōdēs*, icy cold, frozen. *Ex:* Crymo-bia (Ins.); Crymo philus (Av.); Crymodes (Ins.); Crymus (Ins.).

cryph—Gr. *kryphaios* also *kryphios*, hidden; *kryphos*, a hiding place. *Ex:* Cryph-ops (Arth.); Cryphal-ops (Ins.); Cryphalus (Ins.); Cryphaes (Tri.); Cryphia*; Cryphi-phorus (Ins.); Cryphio-crinus (Echin.); Crypho-cricos (Ins.); Eu-cryphia*.

cryphal—See cryph.

cryps—Gr. *krypsi-*, when joined with other stems means secret, in secret, hidden, concealed; *krypsis*, a hiding; *kryptikos*, hidden. *Ex:* Crypsi-rhinus (Av.); cryptic; Crypsis*.

crypt—Gr. *kryptos*, secret, hidden; *kryptē*, a crypt; *kryptikos*, fit for concealing; *kryptadios*, secret. *Ex:* Crypt-otis (Mam.); Crypta (Ins.); Crypticus (Ins.); Crypto-branchus (Amph.); Crypto-glaux (Av.); Crypto-stegia*; Hypo-cryptadios (Av.). See also cryps.

cryptad—See crypt.

cryptic–See crypt.

crystall—Gr. *krystallos*, clear ice, glass. *Ex:* Crystall-aria (Pisc.); Crystallo-gobius (Pisc.); Crystallo-teuthis (Moll.).

ctamen—Gr. *ktamenos*, killed. *Ex:* Helio-ctamenus (Ins.).

cte—See cten.

ctedon—Gr. *ktēdōn*, a comb. *Ex:* Ctedonia (Ins.).

ctein—Gr. *kteinō*, to kill. *Ex:* cteino-phyte.

cten—Gr. *kteis*, genit. *ktenos*, comb; *ktenistēs*, one who uses a comb. *Ex:* Cten-ac-odon (Mam.); Cten-idae (Arach.); Cten-odus (Pisc.); cten-oid; Ctenistes (Ins.); ctenium; Cteno-phora; Cteno-plana (Cten.).

cteniz—Gr. *ktenizō*, to comb. *Ex:* Cteniza (Arach.).

ctist—Gr. *ktistēs*, a settler. *Ex:* Oreo-ctistes (Av.).

cton—Gr. *ktonos*, a murderer, slayer. *Ex:* Dendro-ctonia (Ins.); Dendro-ctonus (Ins.); Uro-ctonus (Arth.).

cub—1. L. *Cuba*, goddess who presides over sleeping children: 2. *Cuba*, an island in the Atlantic Ocean; NL. *cubensis*, of Cuba: 3. Gr. *kybos*, a cube. *Ex:* cub-oid; Cubi-ceps (Pisc.); Cubo-medusae (Coel.); Acro-cubus (Prot.).

cubeba—Ar. *kababah*; Fr. *cubebe*, the cubeb. *Ex:* Cubeba*.

cubit—1. L. *cubitus = cubitum*, the elbow, ulna; *cubitalis*, pertaining to the elbow, forearm or ulna. *Ex:* cubital; cubiti-digital; cubito-carpal; cubitus: 2. L. *cubito*, to lie down, *cubitor*, one who reclines.

cucl—See cycl.

cucub—L. *cucubo*, to hoot like the screech-owl.

cucuj—Braz. *cucujo*, name for some bupestrid beetle. *Ex:* Cucuj-idae (Ins.); Cucujo-calli-merus (Ins.); Cucujus (Ins.).

cucul—L. *cuculus*, the cuckoo. *Ex:* Cucul-anus (Nemat.); Cucul-idae (Av.); Cuculi-philus (Ins.); Cuculus (Av.).

cucull—L. *cucullus*, a hood; *cucullatus*, hooded. *Ex:* cucullate; cuculli-form; Cucullo-thorax (Ins.).

cucum—L. *cucumis*, genit. *cucumeris*, a cucumber; also the name of some sea plant with color and odor like the cucumber. *Ex:* Cucum-aria (Echin.), (Moll.); Cucumer-unio (Moll.); Cucumeria (Moll.); Cucumis*.

cucurbit—L. *cucurbita*, a gourd. *Ex:* Cucurbit-aceae*; Cucurbit-ina (Platy.); Curcubita*.

cuiller—Fr. *cuiller*, a spoon. *Ex:* cuiller.

-cule—See -culus.

culex—See culic.

culic—L. *culex*, genit. *culicis*, a gnat, small fly, midge. *Ex:* Culex (Ins.); Culic-idae (Ins.); culici-vor-ous.

culici—Gr. *kulikion*, a small cup. *Ex:* Culicia (Coel.). See also culic.

culm—L. *culmus*, a stem of grain, straw; *culmeus*, of straw.

culmen—L. *culmen = columen*, genit. *culminis = columinis*, a ridge, the top of anything. *Ex:* culmen.

culmin—See culmen.

culp—L. *culpa*, a defect, fault; *culpatio*, genit. *culpationis*, a reproach.

cult—L. *cultus*, cultivated < *colo*, to tend the vine, to cultivate. See also cultr.

cultorum—L. *cultorum*, of the gardeners, faulty; genit. pl. of *cultor*, a cultivator of land.

cultr—L. *culter*, genit. *cultri*, dim. *cultellus*, a knife, plow. *Ex:* Cultellus (Moll.); Cultri-rostres (Av.); cultri-formis.

-culum—L. *-culum*, suffix added to verb stems to form nouns denoting means or instrument. *Ex:* vehi-culum. See -culus.

-culus—L. *-culus, -a, -um*, suffix added to noun stems to form diminutives. *Ex:* auri-cula; diverti-culum; homun-culus; mole-cule; oper culum; Ringi-culo-spongia (Por.).

culus—L. *culus*, the buttocks, anus. *Ex:* culus.

cum—Gr. *kyma*, genit. *kymatos*, wave. *Ex:* Cum-acea (Crust.); Cum-idae (Crust.); Cuma (Crust.); Cumato-tom-icus (Ins.).

cumat—See cum.

cumb—L. *cumbo* (nasalized form of *cubo*), to recline, ppr. *cumbens*, genit. *cumbentis*, lying down. *Ex:* ac-cumbent; pro-cumbent.

cuminum—Gr. *kuminon*, the aromatic herb called cumin. *Ex:* Cuminum*.

cumul—L. *cumulo*, to pile-up; *cumulatus*, heaped-up; *cumulus*, a heap. *Ex:* cumulat-ive.

cun—1. L. *cunae*, a cradle. *Ex:* Cun-antha (Coel.); Cun-archa (Coel.); Cun-oct-antha (Coel.): 2. L. *cuneus*, a wedge; *cuneatus*, wedge-shaped. *Ex:* cunei-form; Cunei-signa (Ins.); Cuneo-corbula (Moll.).

cunabul—L. *cunabula*, a cradle, the lair or nesting place of young animals.

cunctan—L. *cunctans*, genit. *cunctantis*, lingering, slow; ppr. of *cunctor*.

-cundus—L. *-cundus*, suffix added to verb stems to denote a continuance of the act or quality expressed by the verb. *Ex:* fe-cundus.

cuneat—See cun 2.

cunicul—L. *cuniculus*, a rabbit; *cuniculus = cuniculum*, an underground passage, cavity; *cunicularius*, a miner, burrower; *cuniculatus*, having the form of a tube. *Ex:* Cuniculus (Mam.). See also conil.

cunil—L. *cunile*, a plant called origanum, said to be from Gr. *kōnos*, a cone. *Ex:* Cunilum*.

cup—L. *cupa*, a tub, vat.

cupes—L. *cupes*, fond of dainties. *Ex:* Cupes (Ins.); Cupes-idae (Ins.).

cuph—See cyph.

cupid—L. *cupidus*, eager, with desire.

cupidine—L. *cupidineus*, belonging to cupid.

cupr—L. *cuprum*, copper; *cupreus*, coppery; *cuprinus*, of copper. *Ex:* cupreous; cuprinus.

cupress—L. *cupressus*, collat. form *cyparissus*; Gr. *kyparissos*, the cypress. *Ex:* Cupresso-crinus (Echin.); Cupressus*.

cupul—L. *cupa*, dim. *cupula*, a cask, tub; also a small crooked handle; *cupulatus*, cask- or cup-shaped. *Ex:* Cupul-ita (Coel.); cupula; cupulate; Cupuli-fer-ae*; Cupulo-chonia (Por.).

cur—Gr. *kyroō*, to make valid, sure. *Ex:* Curo-treron (Av.).

curat—L. *curatus*, taken care of; *curator*, a caretaker; *curatorius*, of or pertaining to a manager, guardian.

curcas—Sp. *curcaso*, the physic nut from Curcas purgans.

curcul—L. *curculio*, genit. *curculionis*, a corn weevil. *Ex:* Curculigo*; Curculio (Ins.); Curculionidae (Ins.).

curcum—Ar. *kirkum*>Sp. *curcuma*, turmeric. *Ex:* Curcuma*.

curim—Gr. *kourimos*, shorn off, cut off. *Ex:* Curimus (Ins.).

curimat—NL. *curimatus*, generic name applied by Cuvier to a group of South American fishes, etym. unknown. *Ex:* Curimat-opsis (Pisc.); Curimatus (Pisc.).

currac—L. *currax*, genit. *curracis*, swift, quick.

currax—See **currac**.

curruc—Sp. *curruca*, the linnet. *Ex:* curruc-oides.

curs—L. *cursio*, a running; *cursor*, pl. *cursores*, a runner. *Ex:* Cursi-pes (Amph.); Cursor (Av.); Cursores (Av.); cursori-al; Cursorius (Av.).

cursitans—L. *cursitans*, running about; ppr. of *cursito*, to run hither and thither.

curt—1. L. *curtus*, short, mutilated, *curtatus*, shortened, clipped. *Ex:* Curti-cephalus (Ins.); Curto-notus (Moll.); Sole-curtus (Moll.): 2. Gr. *kyrtos*, curved. *Ex:* Curt-odon (Mam.); Curto-gyne*.

curtat—See **curt**.

curv—L. *curvo*, to curve, pp. *curvatus*, curved; *curvus*, curved, bent. *Ex:* curvat-ose; Curvitermes (Ins.).

-cus—L. *-cus*, suffix added to noun stems to denote possession. See -icus.

cusc—NL. *cuscus*=Fr. *couscous*, from native Moluccan name for a kind of phalanger. *Ex:* Cuscus (Av.).

cuscut—NL. *cuscuta*<Ar. *cochout*=*keshut*, dodder. *Ex:* Cuscuta*.

cusp—L. *cuspis*, a point; *cuspidatus*, made pointed. *Ex:* Cuspi-cora (Moll.); Cuspid-aria (Moll.); bi-cuspidate; not Cusparia*, which is from a vernacular name for a tropical American plant.

cust—L. *custos*, genit. *custodis*, a guard. *Ex:* Custa (Av.); Custi-phorus (Moll.); Craspedacusta (Coel.).

cut—L. *cutis*, dim. *cuticula*, the skin; NL. *cutaneus*, pertaining to the skin. *Ex:* Cutiterebra (Ins.); cutin; cutis; Cuto-coris (Ins.); cutaneous; sub-cutaneous.

cutan—See **cut**.

cutic—See **cut**.

-cy—Eng. *-cy* (<L. *-cia*, *-tia*; Gr. *-kia*, *-keia*, *-tia* and *-teia*), noun suffix, especially of abstract nouns, denoting a state, rank, condition. *Ex:* malignan(t)-cy.

cy—Gr. *kyos*=*kyēma*, a foetus. *Ex:* Cyo-bius (Ins.); cyo-phoris; Acro-cyum (Ins.).

cyam—Gr. *kyamos*, a bean, a pebble; *kyamōn*, a bean field. *Ex:* Cyam-ops (Ins.); Cyamo-bolus (Ins.); Cyamon (Por.); Hyos-cyamos*.

cyan—Gr. *kyaneos*, dark blue<*kyanos*, a dark blue substance, lapis-lazuli; also the blue cornflower>L. *cyaneus*, NL. dim. *cyaneculus*, dark blue. *Ex:* Cyanea (Coel.); Cyanecula (Av.); Cyano-citta (Av.).

cyanecul—See **cyan**.

cyasm—Gr. *kyasma*, a fragment.

cyath—Gr. *kyathos*, a cup. *Ex:* Cyath-ina (Coel.); Cyath-aspis (Pisc.); Cyath-ea*; Cyatho-cephalus (Platy.).

cyb—1. Gr. *kybē*, the head. *Ex:* Ino-cybe*: 2. Gr. *kybos*, a cube; *kybikos*, cubical. *Ex:* Cybi-anthus*; Cybo-cephalus (Ins.); Iso-cybus (Ins.): 3. Gr. *kybion*, the flesh of the tunny salted and cut in square pieces. *Ex:* Cybium (Pisc.).

cybel—Gr. *Kybelē*, a Phrygian goddess. *Ex:* Cybele (Arach.).

cybernet—Gr. *kybernētēs*, a helmsman, one who steers a vessel. *Ex:* Cybernetes (Av.).

cybister—Gr. *kybistētēr*, one who plunges head foremost, a diver, tumbler. *Ex:* Cybister (Ins.).

cycad—Gr. *kykas*, genit. *kykados*, a form of Gr. *kiakas*, acc. pl. of *koix*, name for a kind of palm tree growing in Egypt. *Ex:* Cycado-filic-ales*; Cycas*.

cycas—See **cycad**.

cychr—Gr. *Kychreus*, son of Poseidon and Salamis. *Ex:* Cychro-cephalus (Ins.); Cychrus (Ins.).

cychram—Gr. *kychramos*, a kind of bird, the corn-crake. *Ex:* Cychramus (Ins.).

cycl.—Gr. *kyklos*, a circle; *kyklas*, genit. *kyklados*, circular; *kyklōtos*, rounded; *kyklōsis*, a shutting in, enclosing. *Ex:* Cycl-adenia*; Cycl-ura (Rept.); Cyclas (Moll.); Cyclo-stomata; cyclosis; Cyclot-urus (Mam.); Physo-cyclus (Arach.).

cyclamen—Gr. *kyklaminos*=*kyklamis*, the cyclamen, a bulbous plant, appar.<*kyklos*, a circle. *Ex:* Cyclamen*.

cyclop—Gr. *Kyklōps*, a one-eyed giant<*kyklos*, a circle+*ops*, eye. *Ex:* Cyclop-idius (Mam.) Cyclop-ia*; Cyclops (Crust.).

cyclot—See **cycl**.

cycn—Gr. *kyknos*=L. *cycnus*=*cygnus*, a swan. *Ex:* cygn-et; Cygnus (Av.); Poly-cycnis*.

cydim—Gr. *kydimos*, glorious, renowned. *Ex:* Cydimon (Ins.).

cydipp—Gr. *Kydippē*, mythological name of an ancient priestess, a Nereid, etc. *Ex:* Cydippe (Ins.), (Arach.), (Cten.), etc.

cydist—See **cydn**.

cydn—Gr. *kydnos* and *kydros*, renowned, famous; *kydistos*, most glorious, most renowned. *Ex:* Cydisto*; Cydno-coris (Ins.).

cydon—Gr. *kydōnia*, a kind of tree, the quince< *Kydōnia*, a town in Crete. *Ex:* Cydonia*; Cydonium (Por.); Cydono-crinus (Echin.).

**cydr**—See **cydn.**

**cyem**—Gr. *kyēma*, genit. *kyēmatos*, embryo. *Ex:* Cyema-genia (Ins.); Di-cyema (Mes.); Di-cyemat-idae (Mes.); Di-cyem-ella (Coel.).

**cyes**—Gr. *kyēsis*, gestation, pregnancy. *Ex:* cyesis; cyesio-logy (Med.); meta-cyesis.

**cygn**—See **cycn.**

**cylic**—Gr. *kylix*, genit. *kylikos*, a cup. *Ex:* Cylico-mastiges (Prot.).

**cylichn**—Gr. *kylichnē*, a cup, a little box for ointment. *Ex:* Cylichna (Moll.); Cylichn-ella (Moll.); Cylichno-stomum (Nem.).

**cylindr**—Gr. *kylindros*, a cylinder, roller. *Ex:* Cylindr-ella (Moll.); Cylindr-opuntia*; Cylindro-copt-urus (Ins.).

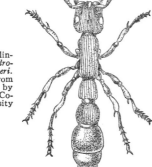

Whymper's Cylinder Ant, *Cylindromyrmex whymperi.* Redrawn from Wheeler, Ants, by permission of Columbia University Press.

**cylist**—Gr. *kylistos*, rolled, tumbled, fit for rolling, twined in a circle. *Ex:* Cylisto-soma (Ins.); Cylistus (Ins.).

**cyll**—Gr. *kyllos*, crippled, maimed, crooked. *Ex:* Cyll-aster (Echin.); Cyllo-metra (Echin.); Trema-cyllus (Mam.).

**cyllar**—Gr. *kyllaros*, the hermit-crab. *Ex:* Cyllarus (Ins.).

**cyllene**—L. *Cyllene*, the mountain where Mercury was born. *Ex:* Cyllene (Ins.).

**cym**—Gr. *kyma*, genit. *kymatos*, a wave, swelling; also a sprout or bud; a fetus; dim. *kymation*, a waved molding, cyma. *Ex:* Cimo-tomus (Av.); cyme; Cymateo-phyllum (Coel.); Cymatium (Moll.); Cymato-gaster (Pisc.); Cymato-nautilus (Moll.); Cymato-syrinx (Moll.); Cymo-pterus*.

**cymaen**—Gr. *kymainō*, to swell, to rise in waves. *Ex:* Cymaenes (Ins.).

**cymat**—See **cym.**

**cymb**—Gr. *kymbos* = *kymbē*, dim. *kymbion*, a cup, boat. *Ex:* Cymb-aria*; cymbae-form; Cymbi-rynchus (Av.); Cymbio-dyta (Ins.); Cimbo-metopia (Ins.): **2.** Gr. *kymbe*, the head.

**cymbal**—Gr. *kymbalon*, a cymbal < *kymbos*, a hollow vessel. *Ex:* Cymbal-aria*.

**cymbax**—NL. *cymbax* < Gr. *kymbachos*, head foremost, landing on the head.

**cymin**—Gr. *kyminon*, cumin-seed. *Ex:* Cymin-osma*.

**cymind**—Gr. *kymindis*, a kind of bird of prey. *Ex:* Cymindis (Ins.); Cymindus (Av.).

**cymodoc**—L. *Cymodoce* or *Cymodocea*, a nereid. *Ex:* Cymodocea*.

**cyn**—Gr. *kyōn* genit. *kynos*, a dog; *kynēdon*, ravenously, like a dog. *Ex:* Cyn-anchum*; Cyn-od-ictis (Mam.); Cyno-glossum*; Cyno-mys (Mam.); cyno-pod-ous; Cyno-therium (Mam.); Pro-cyon (Mam.); Uro-cyon (Mam.).

**cynar**—Gr. *kinara*, an artichoke. *Ex:* Cynara; Cynaro-cephalae*.

**cyneget**—Gr. *kynēgetēs*, fem. *kynēgetis*, a hunter; *kynēgetikos*, fond of hunting. *Ex:* Cynegetis (Ins.).

**cynic**—Gr. *kynikos*, like a dog, snarling; a cynic.

**cynip**—Gr. *sknips*, genit. *sknipos*, an insect living under tree bark. *Ex:* Cynip-idae (Ins.); Cynips (Ins.).

**cynth**—Gr. *Kynthos*, a mountain of Delos, birth-place of Diana. *Ex:* Cynthi-opsis (Tun.); Cynthia (Tun.).

**cyon**—See **cyn.**

**cypad**—See **cypas.**

**cyparis**—Gr. *kyparissos*, cypress. *Ex:* Cyperus*; Chamae-cyparis*. See also cupress.

**cypas**—Gr. *kypassis* = *kypas*, genit. *kypados*, a tunic. *Ex:* Cypassis (Prot.); Pelio-cypas (Ins.).

**ypass**—See **cypas.**

**cyper**—Gr. *kypeiros*, a rush, sedge. *Ex:* Cyperus*, Cypero-bia (Ins.).

**cyph**—Gr. *kyphos*, humped, sloped, curved. *Ex:* Cuphea*; Cyphia*; Cypho-myrmex (Ins.); Noto-cyphus (Ins.); Rhino-cypha (Ins.).

**cyphant**—Gr. *Kyphanta*, sea-port in Laconia. *Ex:* Cyphanta (Prot.).

**cyphell**—Gr. *kyphella*, the hollow of the ears. *Ex:* Cyphella*; Di-cyphell-ion*.

**cyphin**—Gr. *kyphinos*, a roundish vessel < *kyphos*, curved. *Ex:* Cyphinus (Prot.).

**cyphom**—Gr. *kyphoma*, genit. *kyphomatos*, a hunch back. *Ex:* Cyphoma (Moll.).

**cyphon**—Gr. *kyphōn*, a crooked piece of wood < *kyphos*, bent. *Ex:* Cyphon (Ins.); Cyphon-ium (Prot.); Cyphono-cephalus (Ins.).

**cypr**—Gr. *Kypris*, a name for Venus or Aphrodite; *Kypridios*, belonging to Aphrodite. *Ex:* Cypr-ina (Moll.); Cypr-idae (Crust.); Cypraea (Moll.); Cyprae-ovula (Moll.); Cypri-pedium*; Cyprid-ina (Crust.); Cypris (Crust.).

**cyprid**—See **cypr.**

**cyprin**—Gr. *kyprinos*, a kind of carp. *Ex:* Cyprin-odon (Pisc.); not Cyprina (Moll.); see cypr; Cyprino-cirrh-ites (Pisc.); Cyprinus (Pisc.).

**cypsel**—**1.** Gr. *kypselē*, a beehive; a hollow vessel; the hollow of the ear. *Ex:* cypsela; Cypselea*; Cocco-cypsellum*: **2.** Gr. *kyp-*

*selos*, name of the sand martin or European swift. *Ex:* Cypsel-idae (Av.); Cypsel-urus (Pisc.) = Cypsil-urus (Pisc.); Cypselo-metra (Echin.); Cypselus (Av.).

**cypsil**—See **cypsel.**

**cypt**—Gr. *kyptō*, to stoop forward. *Ex:* Cypt-onychia (Ins.); Cypto-bunus (Arach.); Halo-cyptena (Av.).

**cyren**—Gr. *Kyrēnē*, a name applied to several nymphs. *Ex:* Cyren-odonta (Moll.); Cyrena (Moll.); Cyreno-donax (Moll.).

**cyrio**—Gr. *kyrios*, chief, lord; also as adj., authentic, regular. *Ex:* Cyrio-crates (Ins.); Cyrio-gonus (Arach.).

**cyrt**—1. Gr. *kyrtē*, a fish basket, cage. *Ex:* Podo-cyrtis (Prot.): 2. Gr. *kyrtos*, arched, curved; *kyrtōma*, curvature. *Ex:* Curto-podium*; Cyrt-idae (Ins.); Cyrtomium*; Cyrt-onyx (Av.); Cyrti-capsus (Ins.); Cyrto-calpis (Prot.); Tri-cyrtis*.

**cyrtom**—Gr. *kyrtōma*, genit. *kyrtōmatos*, a curve, a swelling. *Ex:* Cyrtoma (Ins.).

**cyrton**—Gr. *kyrtōn*, a hunch-back. *Ex:* Cyrtonus (Ins.).

**cyst**—Gr. *kystis* genit. *kysteōs*, a bladder, pouch. *Ex:* Cyst-idi-cola (Nemat.); Cysteo-demus (Ins.); cysti-cercus; Cysti-phyllum (Coel.); Cystia (Moll.); Cysto-pteris*; cyston; Echino-cystis*.

**cyt**—1. Gr. *kytos*, a hollow vessel, now often taken to mean a cell. *Ex:* Cyt-aster (Echin.); cyt-ula; cytes; cyto-logy; cyto-phagous; cyton; oo-cyte; spermato-cyte: 2. Gr. *kytis*, genit. *kytidos*, a small chest, trunk.

**cyther**—Gr. *Kythereia*, a name for Venus or Aphrodite < *Kythēra*, an island where Aphrodite was worshipped. *Ex:* Cythere (Crust.); Cytherea (Moll.).

**cytid**—See **cyt** 2.

**cytin**—Gr. *kytinos*, the calyx of the pomegranate < *kytos*, a hollow. *Ex:* Cytinus*.

**cytis**—Gr. *kytisos*, name for a kind of shrubby clover. *Ex:* Cytisus*.

**cyttar**—Gr. *kyttaros*, a partition, a division or comb in bees' or wasps' nests; also any cavity such as the cell in which the larvae of bees or wasps are deposited. *Ex:* Cyttaro-myia (Ins.); A-cyttaria (Prot.); Mischo-cyttarus (Ins.).

# D

**da-**—Gr. *da-*, an intensive prefix. *Ex:* Da-centrus (Pisc.).

**daboec**—NL. *daboecia*, name applied to a genus of plants < *St. Daboec. Ex:* Daboecia*.

**daboia**—Hindu *daboya*, that lies hidden. *Ex:* Daboia (Rept.).

**dac**—See **dak.**

**dacel**—NL. *dacelo*, anagram of L. *alcedo*, name of the kingfisher. *Ex:* Dacelo (Av.).

**dacet**—See **dak.**

**dacn**—*daknō*, to bite, sting; *daknēros*, biting; *daknistēr*, a biter. *Ex:* Dacn-usa (Ins.); Dacne (Av.); Dacno-mys (Mam.); Daknister (Ins.); Tri-dacna (Moll.).

**dacry**—Gr. *dakry*, also *dakyron*, a tear; *dakry-dion*, the resinous plant called scammony; in anatomical terms the combining form refers to the tear glands or ducts. *Ex:* Dacry-therium (Mam.), Dacrydium*; dacryo-cele; Dacryo-phorum (Av.).

**dacryd**—See **dacry.**

**dactyl**—Gr. *daktylos*, a finger, toe; *daktylōtos*; having fingers, finger-shaped; *dactylēthra*, a

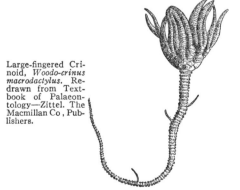

Large-fingered Crinoid, *Woodo-crinus macrodactylus.* Redrawn from Textbook of Palaeontology—Zittel. The Macmillan Co, Publishers.

finger-sheath. *Ex:* Dactylethra (Amph.); Dactylis*; Dactylo-metra (Coel.), Dactylotus (Por.); Artio-dactyla (Mam.); ptero-dactyl.

**dad**—Gr. *das*, genit. *dados*, a contracted form of *dais*, a torch. *Ex:* Dado-crinus (Echin.); Dado-xylon*; Das-ornis (Av.). See also daed.

**dae**—Gr. *daios*, hostile, destructive. *Ex:* Daeodon (Mam.); Daeo-chaeta (Ins.). See also daed and daet.

**daed**—Gr. *dais*, genit. *daidos*, a fire-brand, pinetorch. *Ex:* Dais*.

**daedal**—See daidal.

**daemon**—See daimon.

**daet**—Gr. *dais*, genit. *daitos*, a feast. *Ex:* Daetora (Ins.).

**daethm**—See daithm.

**daeum**—See deum.

**dafila**—NL. *dafila*, a coined name. *Ex:* Dafila. (Av.).

**daict**—Gr. *daiktēr* = *daiktēs*, a murderer, a butcher, one that cuts in pieces. *Ex:* Daictes (Ins.).

**daidal**—Gr. *daidaleos* = *daidalon*, ingenious, beautifully wrought; *Daidalos*, builder of the labyrinth in Crete. *Ex:* Daedalea*; Daedalo-pelta (Por.); Daidalo-metra (Echin.); Daedalus (Ins.).

**daimon**—Gr. *daimōn*, a deity, ghost; *daimonios*, belonging to a god, divine, grand, marvelous, strange. *Ex:* Daemon-arthra (Ins.); Daemon-or-ops*; Daemono-crinus (Echin.); Daimono-helix (Moll.).

**dais**—See dad, also daed.

**daithm**—Gr. *daithmos*, a division, boundary.

**dak**—Gr. *dakos* = *daketon*, a noxious animal whose bite is venemous to man; also a bite, a biting. *Ex:* Daceton (Ins.); Dako-saurus (Rept.); Capro-dacus (Av.).

**dakrum**—NL. *dakruma*, a word of unknown meaning, perh. a nonsense word. *Ex:* Dakruma (Ins.).

**dal**—Gr. *dalos*, a fire-brand; also a shining, brightness. *Ex:* Dal-ophis (Rept.).

**dama**—L. *dama*, the fallow deer.

**damal**—Gr. *damalis*, a calf. *Ex:* Damal-ichthys (Pisc.); Damal-iscus (Mam.); Damalis (Mam.).

**dammar**—Javan *damar*, a name for a tree, the agathis. *Ex:* Dammar = Dammara*; Dammarobius (Ins.).

**damn**—Gr. *damnaō*, to conquer. *Ex:* Damn-acantha*.

**damnos**—L. *damnosus*, full of injury, hurtful.

**damul**—L. *damulus*, a little fallow deer.

**dan**—Gr. *danos*, burnt; also as a substantive, a gift, present, torch. *Ex:* Danis (Ins.); Dano-soma (Ins.); Peuce-danum.*

**dana**—Gr. *Danaus*, mythical king of Arabia. *Ex:* Dana-idae (Ins.); Danaus (Ins.).

**dapan**—Gr. *dapanos* = *dapanēros*, prodigal, having the power of consuming. *Ex:* Dapanera (Ins.); Dapano-ptera (Ins.).

**daped**—Gr. *dapedon*, any level surface such as a pavement, also a meadow. *Ex:* Dapeda (Ins.); Dapedo-glossus (Pisc.); Hypero-dapedon (Rept.).

**daphn**—Gr. *daphnē*, the laurel; *Daphnē*, the nymph Diana, said to have been turned into a laurel tree. *Ex:* Daphne*; Daphneo-derma (Moll.); Daphni-opsis (Crust.); Daphnia (Crust.); Daphno-derma (Moll.).

**daphoen**—Gr. *daphoinos*, red, red-streaked, gory. *Ex:* Daphoeno-citta (Av.); Daphoenus (Mam.).

**dapsil**—Gr. *dapsilēs*, plentiful, ample. *Ex:* Dapsilo-toma (Ins.); Dapsilus (Ins.).

**dapt**—Gr. *daptō*, to devour; *daptēs*, a blood sucker, an eater. *Ex:* Daptes (Av.); Daption (Av.); Dapto-cephalus (Rept.); Daptus (Ins.).

**daptr**—Gr. *daptria*, fem. of *daptēs*, a biter, bloodsucker; *daptrios*, devouring, tearing. *Ex:* Daptrius (Av.).

**darm**—Ger. *Darm*, gut, intestine. *Ex:* blasen-darm (Ger. *blasen*, to blow.).

**dart**—Gr. *dartos*, skinned, excoriated. *Ex:* Dartus*.

**das**—Gr. *dasos*, forest, thicket. See also dad.

**dasci**—Gr. *daskios*, much-shaded, bushy. *Ex:* Dascia (Ins.); Dascio-pteryx (Ins.).

**dascill**—Gr. *daskillos*, the name of a mud-dwelling fish. *Ex:* Dascillus (Ins.), (Pisc.).

**dasi**—See dasy.

**dasplet**—Gr. *dasplētis*, horrid, frightful. *Ex:* Daspletis (Ins.).

**dasy**—Gr. *dasys*, hairy, shaggy. *Ex:* Dasi-ornis (Av.); Dasy-lirion*; Dasy-procta (Mam.); Dasy-ure (Mam.).

**dasyt**—Gr. *dasytēs*, roughness, hairiness. *Ex:* Dasytes (Ins.).

**dat**—Gr. *dateomai*, to divide, cut in pieces; *datētēs*, a distributor. *Ex:* Dato-micra (Ins.).

**datisc**—NL. *datisca*, a plant name, eytm. unknown. *Ex:* Datisca*.

**datur**—NL. *datura*, a genus of solanaceous plants <Hindustani *dhatūrā*; Skr. *dhattūra*; Arabic *tatorah*. Many Hindustani words are of Arabic or Persian origin.

**dauc**—Gr. *daukos*, name of an umbelliferous plant of the carrot kind. *Ex:* Daucus*.

**daul**—Gr. *daulos*, thick, shaggy; also dark, impervious. *Ex:* Daulo-pogon (Ins.); Eri-daulus (Ins.).

**daulias**—Gr. *Daulias*, an epithet of Philomela, who was changed into a nightingale. *Ex:* Daulias (Av.).

**de-** —L. *de-*, prefix meaning down, down from, away, off; also indicating an undoing of an action. *Ex:* de-cidua; de-compose; de-sensitize; de-tectus.

**dealbat**—L. *dealbatus*, made white, whitewashed.

debil—L. *debilis*, disabled, crippled, weak.

deca—Gr. *deka*, ten. *Ex:* Deca-conus (Mam.); Deca-dactylo-crinus (Echin.); deca-gyn-ous; Deca-poda (Crust.).

decat—Gr. *dekatos*, the tenth. *Ex:* Decato-cerus (Ins.).

decen—L. *decens*, genit. *decentis*, decent, proper <*decere*, to be seemly, becoming.

decept—L. *deceptus*, deceiving.

decidu—L. *deciduus*, deciduous, falling off, ready to fall.

decipien—L. *decipiens*, genit. *decipientis*, deceiving, ppr. of *decipio*, to deceive.

declin—L. *declinatus*, turned aside.

decliv—L. *declivis*, sloping, inclining.

decolor—L. *decolor*, faded <*de*-, without+*color*.

decor—1. L. *decorus*, elegant, decorative, suitable: 2. L. *decoris* or *decor*, adorned with ornaments.

decrepitan—L. *decrepitans*, genit. *decrepitantis*, creaking, rattling, ppr. of *decrepo*, to creak.

decrescen—L. *decrenscens*, genit. *decrescentis*, growing shorter, waning<*decresco*, to make smaller.

decret—1. L. *decretus*, decreed, decided <*decerno*. 2. L. *decretus*, decreased <*decresco; decrescens*, genit. *decrescentis*, decreasing.

dect—1. Gr. *dektēs*, a beggar, receiver; *dēktikos*, able to bite, capacious, wide, capable of receiving; *dektos*, acceptable. *Ex:* Dectico-gaster (Ins.); Decticus (Av.); Disso-dectes (Av.): 2. Gr. *dēktēs*, a biter. *Ex:* Latro-dectus (Arach.); Smilo-dectes (Mam.); Stereo-dectes (Mam.); Tricho-dectes (Ins.).

dectic—See dect.

decumb—L. *decumbo*, to recline, to lie down, ppr. *decumbens*, genit. *decumbentis*, lying down, recling. *Ex:* decumbent.

decurren—L. *decurro*, to run down; prr. *decurrens*, genit. *decurrentis*, running down.

decurt—L. *decurto*, to cut short>*decurtatus*, mutilated.

decus—L. *decus*, ornament, splendor. *Ex:* en-decus.

decuss—L. *decusso*, to cross, divide crosswise; pp. *decussatus*, crossed, divided crosswise; *decussatio*, genit. *decussationis*, the act of crossing or intersecting. *Ex:* decussation; Decussi-scala (Moll.).

deducans—L. *deducans*, leading forth.

deduct—L. *deductus*, led apart, split, separated.

defect—L. *defectus*, loosened, worn-out, imperfect, weak; pp. of *deficio*, to set free; *defectus*, a failure, weakness.

definit—L. *definitus*, clear, precise. *Ex:* definite.

defix—L. *difixus*, fastened; pp. of *defigo*, to fix.

deflet—L. *deflētus*, deplored, wept over<*defleo*, to bewail.

deflorat—L. *defloratus*, without flowers, plucked.

deform—L. *deformis*, misshapen. *Ex:* deform-ity.

defoss—L. *defossus*, digging deeply; NL. *defossor*, a deep digger.

degener—L. *degener*, not genuine.

degm—Gr. *dēgma*, genit. *dēgmatos*, a bite or sting; *dēgmos*, the act of biting. *Ex:* Degma-ptera (Ins.); Acro-degmia (Ins.); Poly-degmon (Ins.).

dehisc—L. *dehisco*, to divide, ppr. *dehiscens*, genit. *dehiscentis*, parting, dividing, yawning. *Ex.* dehiscent.

deil—1. Gr. *deilē*, the cool of the day, evening. *Ex:* Deile-mys (Mam.); Deile-phila (Ins.); Chor-deiles (Av.): 2. Gr. *deilos*, weak, cowardly. *Ex:* Deilo-therium (Mam.); Dilus (Ins.); Helio-dilus (Av.).

deilias—Gr. *deiliasis*, fear, dread<*deilos*, wretched, viscious.

deim—Gr. *deima*, genit. *deimatos*, an object of fear; *deimos*, fear, terror. *Ex:* Dima-don (Mam.); Dima (Ins.).

dein—See din 1.

deipn—Gr. *deipnon*, food, a feast. *Ex:* Deipno-psocus (Ins.).

deir—Gr. *deiras*, genit. *deirados*, a hill, summit, hump. *Ex:* Deirado-ceras (Moll.); Deiro-chelys (Rept.); chryso-deirus.

deirad—See deir.

deiroped—Gr. *deiropedē*, a collar, necklace. *Ex:* Deiropeda (Rept.).

deject—L. *dejectus*, thrown down, debased.

del—Gr. *dēlos*, visible>*adēlos*, not manifest. *Ex:* Delo-glyptus (Ins.); Delo-rrhips (Arach.); Adelo-demus (Ins.); Spiro-dela*; Opl-adelus (Pisc.); uro-dele.

delea—Gr. *deleazō*, to bait, ensnare; *deleasma*, a bait; *deleastikos*, enticing. *Ex:* Deleaster (Ins.).

deleast—See delea.

delet—L. *deleo*, to destroy, delete; pp. *deletus*, destroyed; Gr. *delētēr*, a destroyer. *Ex:* Deleter (Ins.).

deletr—Gr. *deletron*, a lantern. *Ex:* Deletro-cephalus (Nemat.).

deletric—L. *deletrix*, genit. *deletricis*, she that destroys.

deletrix—See deletric.

delicat—L. *delicatus*, giving pleasure, delighting, providing dainties, alluring, tender, delicate.

delicios—L. *deliciosus*, delicate, delicious< *delicio*, to allure. *Ex:* delici-ous.

delim—NL. *delima*<*de*, from +*limo*, to file; L. *delimatus*, filed off; *delimator*, one who files. *Ex:* Stigilo-delima (Moll.)

deliphr—NL. *deliphrum* from a supposed Gr. *deliphron*, dumb. *Ex:* Deliphrum (Ins.).

deliquesc—L. *deliquescens*, genit. *deliquesentis*,

becoming fluid < *deliquesco* to melt, to run out. *Ex:* deliquescent.

**delir**—L. *delirus*, silly, crazy.

**delitesc**—L. *delitesco*, to hide, to conceal oneself; ppr. *delitescens*, genit. *delitescentis*, hiding.

**delius**—NL. *delius*, of the laurel.

**delph**—Gr. *delphys*, the womb. *Ex:* Delph-astus (Ins.); A-delpho-therium (Mam.); Mono-delphia (Mam.). See **adelph.**

**delphac**—Gr. *delphax*, genit. *delphakos*, a little pig. *Ex:* Delphaci-xenos (Ins.); Delphacio-gnathus (Rept.); Delphax (Mam.).

**delphax**—See **delphac.**

**delphin**—1. Gr. *delphis* = *delphin*, genit. *delphinos*, a dolphin. *Ex:* Delphin-idae (Mam.); Delphin-ula (Moll.); Delphinus (Mam.): **2.** Gr. *delphinion*, larkspur < *delphis*, a dolphin. *Ex:* Delphin-astrum*; delphini-folia; Delphinium*.

**delt**—Gr. *delta*, fourth letter of the Greek alphabet (Δ); also anything so shaped, especially a triangular island or any similarly shaped tract formed at the mouth of large rivers. *Ex:* delt-oid; Delta-therium (Mam.); Delto-cephalus (Ins.); pseudo-delt-idium.

**dem**—1. Gr. *demas*, body, frame. *Ex:* Dem-ochrus (Ins.); Demo-dex (Arach.); apo-dema; Rhyncho-demus (Platy.); Trachy-demus (Rot.): **2.** Gr. *dēmos*, people, multitude. *Ex:* Demo-spongiae (Por.); en-dem-ic: **3.** Gr. *demō*, to build. *Ex:* Demo-crinus (Echin.). See also **demat.**

**demat**—Gr. *dema*, genit. *dematos*, dim. *demation*, a band, a bundle. *Ex:* Dematium*; Demato-bactron (Platy.).

**demers**—L. *demersus*, submerged, plunged under.

**deminut**—L. *deminutus*, diminutive, small.

**demiss**—L. *demissus*, let down, fallen, pp. of *demitto*, to let down.

**demors**—L. *demorsus*, bitten off, pp. of *demordeo*, to bite off.

**dendr**—1. Gr. *dendron* and *dendros*, a tree, stick. *Ex:* Dendr-aster (Echin.); Dendro-coel-ida (Platy.); Dendr-oica (Av.); Dendro-hyrax (Mam.); Phoro-dendron*: **2.** Gr. *dendrōn*, a thicket.

**dendrit**—Gr. *dendritēs*, fem. *dendritis*, of a tree < *dendron*, a tree. *Ex:* dendrit-ic; dendrite.

**denigrat**—L. *denigratus*, blackened.

**dens**—L. *densus*, dense, compact. See also **dent.**

**dent**—L. *dens*, genit. *dentis*, a tooth; *dentatus*, toothed = LL. *dentarius*; *dentiens*, developing teeth < Gr. *odons*, a tooth. *Ex:* Dent-alium (Moll.); Dentaria*; Dentato-nema (Nemat.); Denti-fibula (Ins.); Dento-stoma (Moll.); Bi-dens*; Duplici-dentata (Mam.).

**dentane**—L. *dentaneus*, threatening.

**dentiens**—L. *dentiens*, cutting teeth.

**denud**—L. *denudatus*, stripped, made bare, pp. of *denudo*, to lay bare. *Ex:* denudate.

**deo**—1. Gr. *deō*, to link. *Ex:* Deo-mys (Mam.): **2.** Gr. *deō*, to want, need, require.

**deodar**—Skt. *devadaru*, the divine tree, the deo-dar (Cedrus deodara). *Ex:* Deodara*.

Tufted Mistletoe, *Phoradendron densum*, with branches in dense tufts. Redrawn from Desert Wild Flowers—Jaeger. Stanford University Press.

**deors**—L. *deorsum*, also *deorsus*, downwards, down, below.

**depas**—Gr. *depas*, = *depastron*, a cup, beaker. *Ex:* Depaso-phyllum (Coel.); Depastro-morpha (Coel.).

**depast**—L. *depastus*, eaten off, eaten down < *depasco*, to consume.

**depastr**—See **depas.**

**depauperat**—LL. *depauperatus*, impoverished, low, less than natural size.

**depend**—L. *dependens*, genit. *dependentis*, hanging down; ppr. of *dependeo* to hang down.

**deperdit**—L. *deperditus*, ruined, lost.

**deph**—Gr. *dephō*, to soften by working with the hand, knead, to masturbate. *Ex:* Depho-mys (Mam.).

**depil**—L. *depilo*, to despoil of feathers or hair; ppr. *depilans*, genit. *depilantis*. *Ex:* depilant.

**deplanat**—L. *deplanatus*, flattened; also expanded; *deplanans*, making level < *deplano*, to level off.

**deplet**—L. *depletus*, emptied.

**deposit**—L. *depositus*, laid aside.

**der**—1. Gr. *derē* = *deirē*, the neck, throat. *Ex:* Dere-taphrus (Ins.); Dero-cephalus (Ins.); Dero-tremata (Amph.); Crypto-dera (Rept.): 2. Gr. *deros* = *deras*, genit. *deratos*, poet. form for *derma*, hide, leather. *Ex:* Echino-der-idae (Ann.); Hetero-dera (Nem.): **3.** Gr. *dēros*, long, too long, long-lived, ancient: **4.** Gr. *derō*, to flay, cudgel. *Ex:* Conopo-deras (Av.).

**derc**—Gr. *derkiomai*, to look, to see clearly. *Ex:* Derco-thoe (Crust.); Oxy-derces (Pisc.); Pyro-derces (Ins.).

dercet—Gr. *Derkitis*, Syrian goddess represented as half woman, half fish. *Ex:* Dercetis (Pisc.), (Ins.); Dercetum (Myr.).

deres—See der 1.

derm—Gr. *derma*, genit. *dermatos*, skin, leather; *dermēstēs*, a worm which eats leather or skin; *dermatinos*, made of leather > NL. *dermatinus*, skinny. *Ex:* Derma-ptera (Ins.); dermatophyte; Dermestes (Ins.); Dermi-pus (Mam.); Dermo-chel-idae (Rept.); Dermo-ptera (Mam.).

Fungus Gall on Yellow Pine, caused by *Peridermium harknessii*. Redrawn from Forest Tree Diseases Common in California and Nevada. —Meinecke.

dermest—See derm.

derr—Gr. *derris*, fur, hair-cloth, a leather covering. *Ex:* Derris (Verm.); Sclero-derris*.

-des—Gr. *-des*, Gr. patronymic ending. Gr. stems (in *ā-*) of the first declension shorten *ā* and add *-des*, as: *Borea-des*, son of Boreas.

des- —1. NL. *des-*, prefix from L. *de*, indicating lack or privation in chemical terminology. *Ex:* des-oxalic: 2. L. *-des*, prefix meaning not. *Ex:* des-crepant < L. *descrepo*, noiseless, very old.

deses—See desid.

desid—L. *deses*, genit. *desidis*, inactive, quiet, indolent. *Ex:* Desidi-opsis (Arach.).

designat—L. *designatus*, described, specified, elected.

desinen—L. *desinens*, genit. *desinentis*, ceasing; ppr. of *desino*, to desist.

desit—L. *desitus*, ceasing < *desino*, to stop, end.

desm—Gr. *desma*, genit. *desmatos*, similar to *desmos*, a chain, bundle, tie, band, ligament; *desmios*, confined. *Ex:* Desm-id*; Desm-odont-idae (Mam.); Desmat-ippus (Mam.); Desmia*; Desmo-scolec-idae (Nem.); A-syn-desmus (Av.); Ento-desma (Moll.).

desman—NL. *desmana*, perh. < Sw. *desman*, musk. *Ex:* Desmana (Mam.).

desmat—See desm.

desmid—Gr. *desmis*, genit. *desmidos*, a bundle. *Ex:* Desmido-crinus (Echin.). In Desm-id*, the

-id is a diminutive ending and not part of the genitive stem.

desmot—Gr. *desmōtēs*, a prisoner; as adj., captive, confined. *Ex:* Desmot-aulius (Ins.).

despect—See despic.

despic—L. *despicio*, to despise, ppr. *despiciens*, genit. *despicientis*, despising, pp. *despectus*, despised; *despicus*, despised.

desultor—L. *desultor*, a leaper; also a fickle lover.

det—Gr. *dēta*, manifestly, to be sure. *Ex:* Detodesmus (Myr.). See also -detic. S.

detect—L. *detectus*, laid bare.

determin—L. *determino*, to limit. *Ex:* determiner.

deters—L. *detersus*, cleaned, removed.

-detic—NL. *-detic* as if from Gr. *detos*, bound, tied. *Ex:* amphi-detic; opistho-detic.

detons—L. *detonsus*, sheared, pp. of *detondeo*, to clip.

detrit—L. *detritus*, a rubbing or wearing away. In geology it means alluvial material; in zoology, waste parts of the integument, etc. *Ex:* detriti-vor-ous (Ecol.).

-deum—NL. *-deum*, suffix, perhaps < Gr. *daiō*, to divide, cut, or < Gr. *odaios*, on the way, by the way < *odos*, way. *Ex:* procto-deum = proctodaeum; stomo-deum = stomo-daeum.

deust—L. *deustus*, burned up, pp. of *deuro*, to burn up.

deut—See deuter.

deuter—1. Gr. *deuteros* (contracted form, *deut-*, *deuto-*), second, secondary. *Ex:* deut-ovum; deutero-coel; Deutero-spinolia (Ins.); deuto-scolex (Ins.): 2. Gr. *deutēr*, genit. *deutēros*, a vessel for cooking.

deva—Skr. *dēva*, name of a god.

deverr—L. *Deverra*, goddess of the broom, housewivery. *Ex:* Deverra*.

devex—L. *devexus*, sloping.

devi—L. *devius*, lying off the high-road, out of the way, lonely.

devon—Eng. *Devon* or *Devonshire*, a maritime county in England. *Ex:* Devoni-an.

dex—1. Gr. *dēx*, genit. *dēkos*, a worm found in wood < *daknō*, to bite. *Ex:* Demo-dex (Arach.). Demo-dic-idae (Arach.), in forming the syllable *dic*, the Gr. *dex* has been treated as a L. noun with genitive *dicis*: 2. Gr. *dexia*, the right hand, a covenant; *dexios*, on the right hand or side, favorable; *dexiteros*, pertaining to the right hand, right > L. *dexter*, genit. *dextris*, right, to the right; also, skillful; *dextratus*, lying to the right. *Ex:* Dexi-idae (Ins.); Dexia-dora (Ins.); Dexia (Ins.); Dexio-gyra (Ins.); dextralis; dextr-ose; Dextri-dens (Ins.); dextro-rotary.

dexamen—Gr. *dexamenē*, a receptacle for water, reservoir. *Ex:* Dexamene (Crust.)

dexio—See dex 2.

**dexius**—L. *Dexius*, a proper name.

**dextr**—See **dex** 2.

**di-** —Gr. *di-*<*dis*, prefix meaning two, double. *Ex:* di-ac-odon (Mam.); di-androus; ?Di-anthus\*; Di-chondra\*; di-delphic; di-morphic; Di-oden (Pisc.); di-oico-poly-gam-ous; Di-opsis (Ins.). See also dia-, dios. **S.**

**dia-** —Gr. *dia-*, much used prefix found in many curious compounds meaning through, through-out, during, over, across; Di-odia\* (*dia*+*odos*, way); Dia-brotica (Ins.); Dia-lophus (Mam.); Dia-pedium\*; Dia-pensia\*, see pent; dia-pedesis; dia-stase, see diastas.

**diabet**—Gr. *diabētēs*, diabetes; also a compass, a spiphon. *Ex:* diabet-ic; diabetes.

**diabol**—Gr. *diabolos*, a devil; *diabolikos*, mali-cious. *Ex:* Diabolus (Mam.).

**diabor**—Gr. *diaboros*, gnawed, eaten through. *Ex:* Diaborus (Ins.).

**diabrotic**—Gr. *diabrōtikos*, able to eat through. *Ex:* Diabrotica (Ins.); Diabroticus (Mam.).

**diad**—Gr. *diadeō*, to bind around. *Ex:* Diad-ophis (Rept.).

**diadet**—Gr. *diadetos*, bound securely, well fastened. *Ex:* Diadeto-gnathus (Rept.).

**diaene**—NL. *diaene*. See triaen.

**dial**—L. *dialis*, ethereal, aerial. *Ex:* Dialis (Av.).

**dialeg**—Gr. *dialegō*, to discern, to distinguish. *Ex:* Dialeges (Ins.).

**dialy**—Gr. *dialyō*, to part asunder; *dialysis*, a separating. *Ex:* Dialy-cera (Ins.); dialysis.

**dialysis**—See **dialy.**

**diamphid**—Gr. *diamphidios*, utterly different. *Ex:* Diamphidia (Ins.).

**dian**—L. *Diana*, ancient Greek goddess, goddess of the chase. *Ex:* Dian-ella\* (Moll.); Diana (Pisc.).

**diant**—Gr. *diantos*, capable of being wetted< *diainō*, to wet, moisten. *Ex:* A-diantum\*.

**diapens**—NL. *diapensia*, name applied to a genus of plants<Gr. *dia*+*pente*, by five, in allusion to the arrangement of the petals. *Ex:* Diapensia\*.

**diaper**—Gr. *diapeirō*, to perforate. *Ex:* Diaperis (Ins.).

**diaphor**—Gr. *diaphoros*, different. *Ex:* Diaphor-illus (Av.); Diaphoro-cetus (Mam.); Diaphorus (Mam.).

**diaphragm**—See **dia-** and **phragm.**

**diasc**—Gr. *diaskeō*, to adorn; also to practice. *Ex:* Diascia\*; Diasco-rhynchus (Platy.).

**diaspas**—Gr. *diaspasis*, a pulling asunder. *Ex:* Diaspasis\*.

**diastas**—Gr. *diastasis*, a separation<*dia*, apart +*histanai*, to set, place. *Ex:* diastase.

**diastema**—Gr. *diastēma*, an interval. *Ex:* diastema; A-diastemus (Mam.).

**diastol**—Gr. *diastolē*, a drawing asunder, an ex-panding, dilatation. *Ex:* diastole.

**diator**—Gr. *diatoros*, piercing; also pierced.

**diatrop**—Gr. *diatropos*, different. *Ex:* Diatrop-ura (Av.).

**dibam**—Gr. *dibamos*, two-footed, on two legs. *Ex:* Dibamus (Rept.).

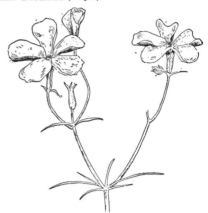

Evening Snow, *Gilia dichotoma*, with "twice-cut" i.e., forked branches. Redrawn from Des-ert Wild Flowers—Jaeger. Stanford University Press.

**dic**—See **dex.**

**dicae**—NL. *dicaeum*, a name for the flower wood-pecker. *Ex:* Dicae-idae (Av.); Dicaeum (Av.).

**dicell**—Gr. *dikella*, a two-pronged hoe, a spade. *Ex:* Dicello-ceras (Ins.). See **cell.**

**dicer**—Gr. *dikerōs*, genit. *dikerōtos*, two-horned; *dikeraios*, having two prongs. *Ex:* Diceraeus (Ins.).

**dich**—Gr. *dicha*=*dichē*, in two, asunder, at vari-ance; *dichērēs*, dividing in twain. *Ex:* Dich-odon (Mam.); Dicha-tomus (Ins.); Diche-cephala (Ins.); Dicheres (Ins.); dicho-gamy; Dicho-meris (Ins.); dicho-tom-ous.

Horn of Elegant Forked-horn Deer, *Dicroceras ele-gans*, of the Miocene. This is the earliest deer in which horns have been found. Re-drawn from Textbook of Palaeontology—Zittel. The Macmillan Co.

**dichel**—Gr. *dichēlos*, with split hoof; *dichēlon*, a forceps. *Ex:* Dichelo-tarsus (Ins.); Dichela-cera (Ins.).

**dicher**—See **dich.**

dichostat—Gr. *dichostatēs*, uncertain, doubtful, in disagreement. *Ex:* Dichostates (Ins.).

diclid—Gr. *diklis*, genit. *diklidos*, a double or folding door, two-valved. *Ex:* Diclid-urus (Mam.); Spiro-diclis*.

diclis—See diclid.

dicr—Gr. *dikroos*, forked. *Ex:* Dicr-urus (Av.); Dicro-cerus (Mam.).

dicrae—L. *dikraios*, forked. *Ex:* Dicraeus (Ins.).

dicran—Gr. *dikranon*, a pitchfork; *dikranos*, two-headed. *Ex:* Dicrano-carpus*; Dicrano-chirus (Arach.).

dict—See dyct.

dictamn—Gr. *diktamnos*, the plant called dittany < Mt. Dicte in Crete. *Ex:* Dictamnus*.

dictic—Gr. *deiktikos*, able to show, proving, serving to point out. *Ex:* Pero-dicticus (Mam.).

dicty—Gr. *diktyon*, dim. *diktydion*, a net. *Ex:* Dictio-clostus (Brach.); Dictydium (Prot.); Dictyna (Arach.); Dictyn-idae (Arach.); Dictyo-caulus (Nemat.); Dictyon-ina (Por.).

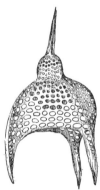

*Dictyopodium*, a Radiolarian with delicate minutely fenestrated shell of silica. Redrawn from Voyage of The Challenger—Thomson. Courtesy of Joseph McDonough Co.

dictyd—See dicty.

dictyn—See dicty.

dictynn—*Dictynna*, Cretan goddess. *Ex:* Dictynna (Ins.).

dictyot—Gr. *diktyōtos*, netted, reticulated. *Ex:* Dictyota (Ins.).

did—NL. *didus*, name applied to a genus of birds < Pg. *doudo*, a dodo. *Ex:* Did-idae (Av.); Did-unculus (Av.); Didus (Av.).

didact—Gr. *didaktos*, learned, teachable. *Ex:* didactic; A-didactus (Ins.).

dido—L. *Dido*, genit. *Didonis*, legendary queen of Carthage. *Ex:* Didonia (Ins.).

didon—See dido.

didym—Gr. *didymos*, double, twofold; as substantive, the testicles. *Ex:* Didym-aspis (Pisc.); Didymo-chelia (Crust.); Didymo-graptus (Coel.); epi-didymis.

diedam—L. *Diedamia*, mythical daughter of Lycomedes, king of Scyros. *Ex:* Diedemia*.

diedr—Gr. *diedros*, separated, sitting apart. *Ex:* Diedro-notus (Ins.); Diedrus (Ins.).

diel—Gr. *dieileō*, to unroll a book. *Ex:* Dielocerus (Ins.).

dier—1. Gr. *diērēs*, double. *Ex:* Lebe-dier-opsis*: 2. Gr. *dieros*, nimble, quick. *Ex:* Diero-bia (Ins.). 3. Gr. *dierō*, to say with distinctness.

dieram—Gr. *dierama*, a strainer, a funnel. *Ex:* Dierama*, (Moll.).

diet—Gr. *diaita*, a way of living, a diet; *diaitētikos*, pertaining to diet. *Ex:* diet; dietet-ics; Diet-opsis (Arach.); Dieta (Arach.); Phyto-dietus (Ins.).

difficil—L. *difficilis*, troublesome, difficult to control.

diffin—NL. *diffinis = disfinis*, the opposite of *adfinis*, i.e., not near.

difflu—L. *diffluo*, to flow apart, pp. *diffluxus*, parted, ppr. *diffluens*, genit. *diffluentis*, flowing apart. *Ex:* diffluence; Difflugia (Prot.), improperly formed from *diffluxus*.

digest—L. *digestus*, separated, dissolved > Fr. and Sp. *digestion*. *Ex:* digest; digestion.

digit—L. *digitus*, a finger, toe; *digitalis*, pertaining to a finger or digit; *digitatus*, having fingers or toes. *Ex:* Digit-aria*; digiti-grade; Digitalis*; Digital-ina (Prot.); digitate.

dign—L. *dignus*, worthy, fit.

dil—See deil 2.

dilat—L. *dilato*, to dilate, pp. *dilatatus*, dilated, spread out, extended < *differo*, pp. *dilatus*, to carry apart, scatter. *Ex:* dilat-ion; Dilati-labrum (Moll.).

dilect—1. L. *dilectus*, delightful, beloved, prized < *deligo*, to value: 2. L. *dilectus = delectus*, a selecting, choosing.

dilut—L. *dilutus*, diluted, weak.

dim—See deim.

dimens—L. *dimensus*, measured.

dimer—Gr. *dimerēs*, two-parted. *Ex:* Dimero-stephanos (Mam.).

dimidiat—L. *dimidiatus*, halved, divided, pp. of *dimidio*, to divide.

din—1. Gr. *deinos*, terrible, powerful, mighty. *Ex:* Dein-amoeba (Prot.); Dein-anara*; Deino-stoma (Ins.); Din-acrida (Ins.); Din-ictis (Mam.); Din-opis (Arach.) < *deinos+ops*, eye; Dino-ptera (Ins.); Dino-sauria (Rept.): 2. Gr. *dinos*, full of eddies, whirling; *dinētos*, whirled about, rolled around; *dinōdēs*, eddying < *dineō*, to whirl round. *Ex:* Dineto-morpha (Ins.); Dineta (Ins.); Dino-bryon*; Dino-flagellata (Prot.); Dino-philus (Ann.); Dinodes (Ins.); Melo-dinus*.

dinet—See din 2.

dinot—Gr. *denōtos*, turned, rounded, rotund. *Ex:* Dinoto-saurus (Rept.); Eury-dinote (Ins).

dioch—Gr. *diochē*, distance, (separated). *Ex:* Diocho-tichus (Mam.).

**diocl**—Gr. *Dioklēs*, a name of Karystios, an ancient Greek botanist. *Ex:* Dioclea\*.

**dioct**—Gr. *diōktēs, dioktēr*, one who pursues. *Ex:* Dioctes (Ins.).

**diod**—Gr. *diodos = diodeia*, a passage through, a thoroughfare. *Ex:* Diodia\* (so called because many species belonging to the genus frequent waysides); diodo-phyte.

**diogen**—Gr. *diogenēs*, divine, sprung from Zeus. *Ex:* Diogena (Ins.).

**diomed**—Gr. *Diomēdēs*, one of the warriors before Troy. *Ex:* Diomedea (Av.).

**dion**—Gr. *Diōnē*, a name for the mother of Venus. *Ex:* Dion-aea\*; Dione (Ins.).

**diond**—NL. *dionda*, a coined name. *Ex:* Dionda (Pisc.).

**dioptr**—Gr. *dioptra*, an optical instrument for measuring heights < *dia*, through + *optomai*, to see; *dioptēr*, a scout. *Ex:* Dioptr-ornis (Av.).

**dior**—Gr. *dioros*, a divider. *Ex:* Dioro-therium (Mam.).

**dioratic**—Gr. *dioratikos*, clear-sighted. *Ex:* Dioratica (Ins.).

**dioritic**—NL. *dioriticus*, greenish like diorite.

**dioryg**—Gr. *diōryx*, genit. *diōrygos*, a channel, canal. *Ex:* Dioryx (Moll.).

**dioryx**—See dioryg.

**dios**—1. Gr. *Zeus*, genit. *Dios*, Jupiter, Zeus Jove; *dios*, sprung from Jove, hence, divine excellent. *Ex:* Dios-pyros\*; ?Di-anthus\*, see di-:  2. Gr. *diōsis*, a pushing apart; also a delaying.

**dioscorea**—NL. *dioscorea*, a plant name < Dioscorides, ancient Greek naturalist. *Ex:* Dios corea\*.

**diospyr**—Gr. *diospyros*, a kind of plant < *Dios*, Zeus + *pyros*, wheat. *Ex:* Diospyros\*, see dios.

**diotreph**—Gr. *diotrēphēs*, beloved of Zeus. *Ex:* Diotrepha (Ins.).

**diph**—Gr. *diphao*, to search after. *Ex:* Diphaglossa\*.

**diphy**—Gr. *diphyēs*, double. *Ex:* diphy-cercal; diphy-odont; Diphyes (Coel.).

**dipl**—1. Gr. *diploos*, double. *Ex:* Dipl-ac-odon (Mam); Diplo-docus (Rept.); dipl-oid; Diplo-poda (Myr.); Di-diplis\*:  2. Gr. *diplon*, a fold > Fr. *diploé*, the spongeous substance separating the tables of the skull. *Ex:* dilpoe; diplo-ic.

**diplac**—Gr. *diplax*, genit. *diplakos*, two-fold, double. *Ex:* Diplac-odon (Mam.); Diplax (Ins.).

**diplasi**—Gr. *diplasios*, double, twice as much as.

**diplax**—See diplac.

**diplazi**—Gr. *diplazō*, to double or be doubled *Ex:* Diplazi-um\*.

**dipn**—Gr. *deipnon*, a meal, food. *Ex:* Dipno-lepis (Pisc.); Dendro-dipnis (Ins.).

**dips**—1. Gr. *dipsa*, thirst; *dipsakos*, a kind of diabetes attended with violent thirst; also a kind of teasel, the leaf axils of which hold water. *Ex:* Dipso-saurus (Rept.); Dipsacus\*; Haema-dipsa (Ann.); Haemo-dipsus (Ins.):  2. Gr. *dipsas*, genit. *dipsados*, a venemous serpent whose bite caused intense thirst; also a thorn. *Ex:* Dipsado-morphus (Rept.); Dipsas (Rept.).

**dipsac**—See dips 1.

**dipsad**—See dips 2.

**dipther**—Gr. *dipsthera*, membrane, hide. *Ex:* diptheria.

**dipyren**—NL. *dipyrenus*, two seeded or stoned < Gr. *di* two + *pyrēn*, a precious stone.

**dir**—L. *dirus*, dreadful, ill-omened. See deir.

**diranch**—Gr. *deiranchēs*, throttling. *Ex:* Diranchis (Ins.).

**dirc**—Gr. *Dirkē*, a fountain near Thebes < *dircē*, a fountain, spring. *Ex:* Dirca\*; Dirc-aea (Ins.).

**dirempt**—L. *diremptio*, a separation; *diremptus*, divided. *Ex:* dirempt-ion.

**dirept**—L. *direptus*, torn apart; pp. of *diripio*, to tear in pieces.

**dirin**—NL. *dirina* < Gr. *deirē*, ridge of a hill, neck + *-ina*. *Ex:* Dirina\*.

**dirot**—NL. *diroto-* < Gr. *dis + ressō*, to break into fragments. *Ex:* Diroto-gnathus (Ins.).

**dis-** —1. *dis-*, prefix corresponding to Eng. prefix *un-*; added to words to signify a negation. *Ex:* Dis-aster (Echin.); dis-bud. See dys:  2. Gr. *dis-*, twice, like L. *bis*; also, an intensive prefix meaning very, again and again < *dis*, twice, again. *Ex:* Dis-andra\*.

**dis**—Gr. *deisa*, moistness, filth. *Ex:* Diso-chara (Ins.).

**disc**—Gr. *diskos*, a disc. *Ex:* Disc-ina (Brach.); Disc-ino-caris (Crust.); Disci-spongia (Por.); Disco-gloss-idae (Amph.); Cephalo-discus (Adel.).

**dischid**—Gr. *dischides*, cloven, parted. *Ex:* Dischides (Moll.); Dischidia\*; Ana-dischidus (Ins.).

**discin**—See disc.

**discio**—Gr. *diskios* (< *dis*, very + *skia*, shade), very shady, sheltered.

**discolor**—L. *discolor*, variegated, of different colors, parti-colored (as opposed to *concolor*.).

**discor**—L. *discors*, genit. *discordis*, disagreeing, inharmonious, different; *discordans*, genit. *discordantis*, ppr. of *discordo*, to disagree.

**discret**—L. *discretus*, parted.

**discuss**—L. *discussus*, scattered, shattered, broken; pp. of *discutio*, to scatter in pieces, break.

**diserm**—L. *disermus*, wavy.

**dispar**—L. *dispar*, genit. *disparis*, unlike, different, unequal

disparal—L. *disparalis*, dissimilar, unlike.

disparat—L. *disparatus*, separated, divided.

dispers—L. *dispersus*, scattered, to disperse; Fr. *dispercion*. *Ex:* dispersion.

diss—Gr. *dissos* = Attic. *dittos*, double, in pairs, twins. *Ex:* Diss-acanthus (Ins.); Diss-ura (Av.); Disso-dectes (Av.); Disso-steira (Ins.); Ditto-pora (Bry.).

dissemur—NL. *dissemurus*, prob. <L. *disseminatus*, to spread abroad, to sow+*ura*, tail, because of the spreading forked tail. *Ex:* Dissemurus (Av.).

dissep—L. *dissepimentum*, a partition, a dividing <*dissepio* = *dissaepio*, to divide. *Ex:* dissepiment.

dissil—L. *dissilio*, to fly apart, ppr. *dissiliens*, genit. *dissilientis*, flying apart, flying asunder. *Ex:* dissilient; Dissil-aria*.

dissit—1. LL. *dissitus*, lying apart. 2. L. *dissitus*, sown, pp. of <*dissero*, to scatter seed, sow.

distal—Eng. *distal*, situated away from the center of a body, at the end (opposite of proximal) <*dist* (*ance*)+-*al*, an analogy of *central*.

distan—L. *distans*, genit. *distantis*, separated, apart, ppr. of *disto*, to separate. *Ex:* Distans-eschar-ella (Bry.).

distens—L. *distensus*, parted, stretched apart extended.

distent—L. *distentus*, filled full, ample <*distendo*, to fill.

distich—See distoech.

distinct—L. *distinctus*, separated, distinguished, marked apart.

distingu—L. *distinguo*, to separate; gerundive, *distinguendus*, separating, see -nduo.

distoech—Gr. *distoichos* = *distichos*, in two rows. *Ex:* Distoech-urus (Mam.)

distract—L. *distractus*, pulled apart, pulled two ways. *Ex:* distract-ile.

ditt—See diss.

diuretic—Gr. *diourētikos*, promoting the flow of urine. *Ex:* diuretic.

diurn—L. *diurnus*, daily, of the day; *diurnalis*, daily, pertaining to the daytime. *Ex:* Diurna (Ins.); Diurnae (Av.); Diurnis (Av.); diurnal.

dival—L. *divalis*, god-like, king-like. *Ex:* Divales (Ins.).

divar—See divari.

divari—L. *divarico*, to spread apart, pp. *divaricatus*, spreading apart. *Ex:* Divari-cardium (Moll.); divaricate; Divarico-lima (Moll.).

diverg—NL. *divergens*, genit. *divergentis*, widespreading, bending apart <*di-* = *dis-*+ppr. of *vergo*, to bend.

divers—L. *diverto*, to separate, pp. *diversus*, separated, turned. *Ex:* Diversi-color (Moll.); diversi-loba; Diverso-sexus (Ins.).

diverticul—L. *diverticulum* = *deverticulum*, a digression <*diverto*, to turn aside, turn away.

divid—L. *dividus*, separated, divided.

divis—L. *divisus*, divided, pp. of *divido*, to divide, separate.

divuls—L. *divulsus*, torn apart, torn; pp. of *divello*.

dix—Gr. *dixoos*, forked, in two, asunder. *Ex:* Di-dix-odon (Pisc.).

dixa—NL. *dixa* <Gr. *dicha* (dixa) into, apart. *Ex:* Dixa (Ins.), from which the family name *Dix-idae* is derived.

dnoph—Gr. *dnophos*, darkness, *dnopheros*, gloomy, dark, *Ex:* Dnophera (Av.).

doc—1. Gr. *dokos*, beam, spear, lance. *Ex:* Doc-odon (Mam.); Doco-glossa (Moll.); Doco-phoroides (Ins.); Diplo-docus (Rept.): 2. Gr. *dokos* = *dokēsis*, an opinion.

doch—Gr. *dochē*, a receptacle; *dochos*, containing. *Ex:* Doche-phora (Ins.); chole-doch; elaeo-dochon; haemato-docha; sporo-dochium.

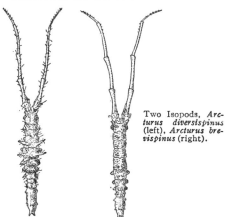

Two Isopods, *Arcturus diversispinus* (left), *Arcturus brevispinus* (right).

dochm—Gr. *dochmos*, slant-wise. *Ex:* Dochmo-nota (Ins.).

docidi—Gr. *dokidion* (dim. of *dokos*), a small beam, door-bar; *dokis*, genit. *dokidos*, a stick, rod. *Ex:* Docidium*.

docim—Gr. *dokimos*, proof, trial. *Ex:* Docimo-cephalus (Tri.).

docimast—Gr. *dokimastikos*, proving; *dokimastēs*, an examiner, one who scrutinizes. *Ex:* Docimastes (Av.).

dodec—Gr. *dōdeka*, twelve; *dōdekatheon*, a medicine of twelve-ingredients. The plant, dodocatheon is said to have been named in honor of twelve great gods; NL. *dodecem*, twelve. *Ex:* Dodeca-stichus (Ins.).

dodo—See did.

dodran—L. *dodrans*, genit. *dodrantis*, about 3/4 of a Roman foot, equal to 8.73 English inches. *Ex:* dodrant-alis.

**doedic**—Gr. *doidyx* genit. *doidykos*, a pestle, a spoon, ladle. *Ex:* Doedic-urus (Mam.); Doedyco-rrhinus (Ins.); Doidyx-odon (Pisc.).

**doedyc**—See **doedic**.

**doidyx**—See **doedic**.

**dol**—1. Gr. *dolos*, deceit, a trap. *Ex:* Dolo-mys (Mam.); Dolo-philus (Ins.): 2. Gr. *dolōn*, a dagger. *Ex:* Dolo-cerus (Ins.); Dol-ortho-ceras (Moll.); Dolonus (Arach.).

**dolab**—See **dolabr**.

**dolabr**—L. *dolabra*, dim. *dolabella*, a hatchet, pick-ax; *dolabratus*, hatchet-shaped. *Ex:* Dolabella (Moll.); Dolabri-fera (Moll.).

**dolen**—L. *dolens*, genit. *dolentis*, suffering, ppr. of *doleo*, to suffer, lament.

**doler**—Gr. *doleros*, deceptive. *Ex:* Doler-orthis (Brach.); Dolerus (Ins.).

**doli**—See **dolium**.

**dolich**—Gr. *dolichos*, long, lengthy. *Ex:* Dolich-onyx (Av.); dolicho-cephalic; Dolicho-cer-inae (Ins.); Dolycho-rhynch-ops (Rept.). 2. Gr. *dolichos*, a kind of bean. *Ex:* Dolichos*.

**dolio**—Gr. *dolios*, deceitful < *dolos*, deceit. *Ex:* Dolio-cherus (Mam.); Dolio-carpus; Dolio-malus (Arach.).

**doliol**—See **dolium**.

**dolium**—L. *dolium*, dim, *doliolum*, a large jar, cask. *Ex:* Doli-opsis (Moll.); Doliolum (Tun.); Dolium (Moll.).

**dolom**—Gr. *doloma*, genit. *dolomatos*, a trick. *Ex:* Doloma (Ins.).

**dolomed**—Gr. *dolomēdēs*, wily. *Ex:* Dolomeda (Arach.); Dolomedes (Arach.).

**dolomet**—Gr. *dolomētēs* and *dolomētis*, treacherous. *Ex:* Dolometis (Av.).

**dolon**—See **dol** 2.

**dolopes**—NL. *dolopes* < Gr. *dolopis*, crafty. *Ex:* Dolopes (Rept.).

**dolophon**—Gr. *dolophonos*, slaying by treachery. *Ex:* Dolophon-odus (Pisc.); Dolophona (Arach.).

**dolophr**—Gr. *dolophroneō*, to plot, scheme; *dolophradēs*, treacherous. *Ex:* Dolophrades (Ins.); Dolophron (Ins.); Dolophro-syne (Ins.).

**dolops**—Gr. *dolops*, one hiding in ambush, a spy. *Ex:* Dolops (Ins.); Poly-dolops (Mam.).

**dolor**—L. *dolor*, pain, *dolorosus*, full of pain. *Ex:* dolori-fer-ous.

**dolos**—1. L. *dolosus*, cunning, false < Gr. *dolos*, deceit: 2. Gr. *dolōsis*, a fraud. *Ex:* Dolosis (Ins.).

**dolych**—See **dolich**.

**dom**—1. Gr. *doma*, genit. *domatos*, a gift. *Ex:* Domato-ceras (Moll.): 2. Gr. *domos* and *domē*, a house, structure > L. *domus*, a house, household; *domesticus*, belonging to household. *Ex:* domesticus; Domo-myza (Ins.); Domo-spongia (Por.); Chalico-doma (Ins.); Litho-domus (Moll.).

**domabil**—L. *domabilis*, tamable.

**domat**—See **dom** 1.

**domestic**—See **dom** 2.

**domin**—See **domn**.

**domit**—L. *domitus*, tamed.

**domn**—L. *domnus* = *dominus*, ruler. *Ex.* Domnina (Mam.); Domnus (Ins.).

**don**—See **odon**.

**donac**—L. *donax*, genit. *donacis*, a sort of reed, a pen, pipe, an arrow made from a reed; also the male scallop or pecten. *Ex:* Donac-ias (Av.); Donaco-bius (Av.); Donaco-philus (Av.); Donax (Moll.).

**donax**—See **donac**.

**dor**—1. Gr. *dora*, a hide, skin; *doros*, a leather bottle, sack. *Ex:* Calli-dora (Ins.); Echino-dorus (Echin.); Mono-dora*; Lio-dora (Ins.): 2. Gr. *dōron*, a gift; *dōrēma*, a gift. *Ex:* Dorema*; Eu-dor-ina (Prot.). Meno-dora*. See doris.

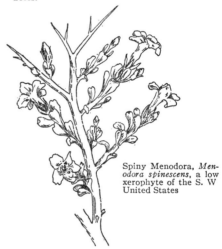

Spiny Menodora, *Menodora spinescens*, a low xerophyte of the S. W United States

**dorat**—Gr. *doration*, a small spear. See also **dory**.

**dorc**—Gr. *dorkas* (dim. *dorkadion*) and *dorx*, genit. *dorkos*, a gazelle, the steinbock. *Ex:* Dorc-elaphas (Mam.); Dorca-toma (Ins.); Dorca-therium (Mam.); Dorcadion (Ins.); Dorco-tragus (Mam.); Dorx (Ins.); Anti-dorcas (Mam.).

**dorcad**—See **dorc**.

**dorem**—See **dor** 2.

**dorid**—See **doris**.

**dorimarg**—Gr. *dorimargos*, furious in battle. *Ex:* Dorimargus (Ins.).

**-dorina**—NL. *-dorina*, < *Pandorina*, a genus of Protozoa. *Ex:* Pheo-dorina (Prot); Eu-dorina (Prot.). See **pandor**.

doripon—Gr. *doriponos*, warlike. *Ex:* Doriponus (Av.).

doris—1. Gr. *Dōris*, genit. *Dōridos*, a sea goddess, mother of half a hundred Nereids. *Ex:* Dorippe (Crust.); Doris (Moll.); Dorid-ella (Moll.); Dorid-opsis (Moll.); Dorid-unculus (Moll.): 2. Gr. *doris*, genit. *doridos*, a sacrificial knife.

dorm—L. *dormio*, to sleep, ppr. *dormiens*, genit. *dormientis*, sleeping; *dormitor* also *dormitator*, a sleeper. *Ex:* dorman-cy<*dormant*; Dormita tor (Pisc.).

dormitat—See dorm.

dors—L. *dorsum*, the back; ML. *dorsalis*, pertaining to the back. *Ex:* dors-ulum; dorsal; dorsi-ferous; dorso-ventral.

dorsat—NL. *dorsatus*, with a back, backed<L. *dorsum*, back.

doru—See dory.

dorx—See dorc.

dory—Gr. *dory*, genit. *doratos*, a spear. *Ex:* Dorato-sepion (Moll.); Doru-don (Mam.); Dory-discus (Prot.); Dory-phora (Ins.); Dory-teuthis (Moll.); Ichthyo-doru-lites (Elasm.).

doryl—NL. *dorylus*, a name for certain ants< Gr. *dorys*, a spear. *Ex:* Dorylo-zelus (Ins.); Dorylus (Ins.).

dorypon—Gr. *doryponos*, toiling with the spear. *Ex:* Doryponus (Av.).

dos—Gr. *dosis*, genit. *doseōs*, a gift.

dosidic—Gr. *dōsydikos*, abiding by the law. *Ex:* Dosidicus (Moll.).

dosin—NL. *dosinia*<Senegal *dosin*, name of a bivalve. *Ex:* Dosin-orbis (Moll.); Dosin-ula (Moll.); Dosin-ia (Moll.); Dosini-opsis (Moll.).

dot—Gr. *dotos*, a gift, *dotēr*, genit. *dotēros*, a giver; *dotikos*, generous. *Ex:* Doter (Ins.); Doticus (Ins.); Doto-cryptus (Ins.); Chiro-dota (Echin.); Olethro-dotis (Ins.). See also doto.

doter—See dot.

dothid—Gr. *dothiēn*, an abscess. *Ex:* Dothid-ella*.

dotic—See dot.

doto—Gr. *Dōtō*, name of a Nereid. *Ex:* Doto (Moll.).

dox—Gr. *doxa*, an opinion; also dignity, glory. *Ex:* A-doxus (Ins.); Hetero-doxus (Ins.); Pro-dox-idae (Ins.).

drab—Gr. *drabē*, a name of uncertain meaning applied by Dioscorides to some cruciferous plant, perh. <*drabē*, acrid, biting, from taste of the leaves. *Ex:* Drab-ella*; Draba*; drabi-folius; Hetero-draba*.

drac—L. *draco*, genit. *draconis*, a dragon<Gr. *drakōn* (fem. *drakaina*), genit. *drakontos*, a dragon, or perh.<*derkomai*, to look terrible, gleam. *Ex:* Dracaena*; Drac-unculus*; Draco (Rept.); Draco-cephalum*; dracont-iasis; Dra contium*; Draconto-myia (Ins.).

dracaen—See drac.

dracon—See drac.

dracont—See drac.

dragm—Gr. *dragma*, genit. *dragmatos*, a handful, a sheaf. *Ex:* Dragmat-ella (Por.).

dram—Gr. *dramein*, inf. aor. of *trechō*, to run; *dramēma*, a race, a running. *Ex:* Ammo-dramus (Av.).

drapet—Gr. *drapetēs*, a fugitive; *drapetidēs* and *drapetiskos*, a small runaway, a slave. *Ex:* Drapet-odes (Ins.); Drapetes (Ins.); Drapetisca (Arach.).

drasm—Gr. *drasmos*, a flight.

drass—Gr. *drassomai*, to grasp. *Ex:* Drass-idae (Arach.); Drass-in-ella (Arach.); Drass-odes (Arach.); Drassus (Arach.).

drast—Gr. *drastas* and *drēstēs*, a servant, worker, agent. *Ex:* Agrio-drastus (Ins.).

drem—See drom.

drep—See drept.

drepan—Gr. *drepanē* and *drapanon*, a sickle> *drepanis*, genit. *drepanidos*, a kind of bird with large sickle-shaped wings. *Ex:* Drepan-aphis (Ins.); Drepana (Ins.); Drepane-phorus (Pisc.); drepani-form; Drepano-ptila (Av.).

drept—Gr. *dreptos*, plucked<*drepō*, to gather, to pluck. *Ex:* Dreptes (Av.); Myceto-drepa (Ins.).

dri—Gr. *drios*, a thicket. *Ex:* driod-ad (Ecol.); Driod-ura (Av.); hero-drium; for Drio-picus (Av.) and Drio-scopus (Av.), see dry.

dril—Gr. *drilos*, a worm, the *membrum virile* or penis; the combining form *dril-* is sometimes used in zoological names in the sense of sperm ducts. *Ex:* Drilo-sphaerus (Prot.); Bdello-drilus (Ann.); Mega-drili (Ann.).

drimy—Gr. *drimys*, pungent, piercing. *Ex:* drim-ium, see -ium 2.; Drimia*; drimy-phyta; Drimys* (Pisc.).

drio—See dry.

driod—See dri.

drom—Gr. *dromos*, a running, a running course< *dramein*, inf. aor. of *trechō*, to run; *dromas*, running; *dromikos*, good at running, swift; *dromaios*, running at full speed. *Ex:* Dremo-therium (Mam.); Dromaeus (Av.); Dromaeo-cercus (Av.); Dromas (Av.); Dromici-ops (Mam.); Dromicia (Mam.); Dromo-gomphus (Ins.); Dromo-mys (Mam.).

dromae—Gr. *dromaios*, swift. *Ex:* Dromaeus (Av.).

dromic—See drom.

dros—Gr. *drosos*, dew; *droseros*, dewy; *drosōdēs*, moist, dewy. *Ex:* Dros-ochrus (Ins.); Drosera*; Droso-phila (Ins.); Droso-phyllum*.

drup—1. Gr. *dryppa*, an overripe olive>NL. *drupa*, a stone fruit, drupe. *Ex:* drup-aceous; drup-ella; drup-etum; drupe; Cymmo-drupa (Prot.): 2. Gr. *drupis*, genit. *drupidos*, a kind of thorn.

drupid—See drup 2.

drus—Ger. *Druse*, pl. *Drusen*, decayed ore, but in the botanical term, druse, it is taken to mean crystals as in the German word *Drusenräume*, cavities in rocks studded with crystals. *Ex:* drusen.

dry—Gr. *drys*, genit. *dryos*, a tree, especially the oak. *Ex:* Drio-picus (Av.); Drio-scopus (Av.); Dryo-balan-ops*; Dryo-bates (Av.); Dryo-scopus (Av.): Ophio-drys (Rept.).

dryad—Gr. *dryas*, genit. *dryados*, a wood nymph; *Dryas*, deity or nymph of the woods. *Ex:* Dryado-blatta (Ins.); Philo-dryas (Rept.).

dryas—See dryad.

dryin—Gr. *dryinos*, oaken < *drys*, an oak. *Ex:* Dryinus (Ins.).

drym—Gr. *drymos* = *drymōn*, a forest, oakwood, coppice; *drymōdēs*, of the woods, woody. *Ex:* Drym-aria*; Drymo-callis*; Drymoda*; Drymonia*.

drynaria—NL. *drynaria*, a fern genus name said to be derived from Gr. *drys*, oak, genit. *dryos*, acc. *dryn* + *aria*. *Ex:* Drynaria*.

dryope—Gr. *Dryopē*, daughter of Dryops and playmate of the wood nymphs. *Ex:* Dryope (Crust.).

drypt—Gr. *dryptō*, to tear. *Ex:* Drypt-odon (Mam.); Drypta (Ins.); Drypto-cephala (Ins.).

drys—See dry.

dubio—L. *dubius*, uncertain. *Ex:* Dubio-teuthis (Moll.).

dubitat—L. *dubitatus*, questioned, pp. of *dubito*, to question.

duct—L. *ductus*, a leading; *ductens*, genit. *ductentis*, drawing, leading, ppr. of *duco*, to lead; *ductor*, a leader. *Ex:* duct; ductus arteriosus; ad-ductor; ovi-duct.

dul—Gr. *doulos*, a slave; *doulōsis*, slavery. *Ex:* Dules (Pisc.); dulosis; Dulus (Av.).

dulc—L. *dulcis*, sweet.

dulich—Gr. *dolichos* = *doulichos*, long; as a substantive, the long course; also a kind of kidney bean. *Ex:* Dulichi-idae (Crust.); Dolichia (Crust.); Dulichium*.

dum—L. *dumus*, a bramble, thorn-bush; *dumetum*, a thicket; *dumosus*, bushy, abounding in bushes. *Ex:* Dumet-ella (Av.); dumet-ose; Dumi-cola (Av.).

dumet—See dum.

dumos—See dum.

duoden—NL. *duodenum*, the first portion of the small intestine (so called because in man it is approximately twelve finger breadths wide) < L. *duodeni*, twelve each. *Ex:* duoden-al; duodenum.

dup—Gr. *doupos*, a rattle, crash, the roar of a torrent, a dull sound, a thud. *Ex:* Dupo-philus (Ins.).

dupetor—Gr. *doupētōr*, a clatterer. *Ex:* Dupetor (Av.).

duplex—See duplic.

duplic—L. *duplex*, genit. *duplicis*, twofold, double; *duplicarius*, a soldier who receives double pay. *Ex:* Duplicaria (Moll.); duplici-dent; Duplici-dentata (Mam.); duplico-dentate.

dur—L. *durus*, hard, durable; NL. *durusculus*, somewhat hard or woody. *Ex:* dura mater; duri-lignosa.

durabil—L. *durabilis*, lasting, durable.

duracin—L. *duracinus*, with hard berries, hard.

duram—L. *duramen*, hardness. *Ex:* duramen.

durat—L. *duratus*, hardened, made callous or insensible; pp. of *duro*, to harden; L. *durateus*, wooden < Gr. *dourateos*.

durio—NL. *durio* < Malay *Durian*, native tree name. *Ex:* Durio*.

duruscul—See dur.

dus—See dys.

dyad—Gr. *dyas*, genit. *dyados*, two. *Ex:* dyad; Dyad-entomum (Ins.); Dyado-zo-arium (Ins.).

dynam—Gr. *dynamis*, power, strength; *dynamikos*, powerful. *Ex:* dynamic murmurs (Med.); Dynamo-saurus (Rept.); Eu-dynamys (Av.).

dynamic—See dynam.

dynat—Gr. *dynatos*, strong, mighty; also productive. *Ex:* Dynato-batis (Elasm.); Dynato-soma (Ins.); Dynatus (Ins.).

dypt—Gr. *dyptēs*, a diver. *Ex:* Eu-dypt-ula (Av.); Eu-dyptes (Av.).

dys—Gr. *dysis*, a dipping; also a setting of the sun, western. *Ex:* Dusi-cyon (Mam.); Cata-dysis (Bry.); Cichla-dusa (Av.); Cymo-dusa (Ins.).

dys-—Gr. *dys-*, inseparable prefix denoting difficulty or trouble, unlucky, bad, ill, hard. *Ex:* Dys-aster (Echin.); Dys-ephyra (Ins.); dys-genic; Dys-odia*; Dys-pteris (Ins.).

dysant—Gr. *dysantēs*, dangerous, cruel. *Ex:* Dy santes (Ins.).

dysarest—Gr. *dysarestos*, hard to please. *Ex:* Dysarestus (Ins.).

dyscol—Gr. *dyskolos*, hard to please, fretful, troublesome. *Ex:* Dyscol-etes (Ins.); Dyscolo-cerus (Ins.).

dyscrit—Gr. *dyskritos*, difficult to distinguish or interpret. *Ex:* Dyscritus (Ins.); Dyscrito-ceras (Moll.).

dysder—Gr. *dysdēris*, quarrelsome, petulant. *Ex:* Dysdera (Arach.).

dyse—Gr. *dysis*, genit. *dyseōs*, a going down, the setting of the sun or stars, the west; also a place of refuge. *Ex:* Dyseo-lemur (Mam.); Dyseo-mermyx (Mam.).

dyseni—Gr. *dysēnios*, stubborn, uneasy. *Ex:* Dysenius (Ins.).

**dysis**—Gr. *dyō*=*dynō*, to put on, clothe, penetrate >NL. *ecdysis*, a getting out of clothing, the act of moulting; *endysis*, a putting on (of clothing).

**dysn**—Gr. *dysnoos*, ill-affected. *Ex.* Dysnocrypta (Ins.).

**dysnoet**—Gr. *dysnoētos*, unintelligible, obscure. *Ex:* Dysnoetus (Ins.); Dysnoeto-pora (Bry.).

**dysnot**—See **dysnoet**.

**dysod**—Gr. *dysōdēs*, ill-smelling; *dysōdia*, stench. *Ex.* Dysodia*; not Dysodus (Mam.) or Dysodonta (Moll.), see dys- and odont; Lygo-dysodia*.

**dysope**—Gr. *dysōpeō*, to shame, to make one change countenance. *Ex:* Dysopes (Mam.).

**dyspet**—Gr. *dyspetēs*, difficult. *Ex:* Dyspet-ornis (Av.).

**dysponet**—Gr. *dysponētos*, bringing trouble. *Ex:* Dysponetus (Ann.).

**dyspros**—Gr. *dysprositos*, hard to get at. *Ex:* dysprosium.

**dystact**—Gr. *dystaktos*, ill-arranged, irregular. *Ex:* Dystact-ella (Moll.).

**dyt**—Gr. *dytēs*, a burrower, diver; *dytikos*, able to dive. *Ex:* Dyticus=Dyt-iscus (Ins.); Acantho-dyta (Ins.); Troglo-dytes (Av.).

**dytic**—See **dyt**.

# E

**e-**—See **ex-**.

**-ea**—NL. *-ea*, ending of generic names taken from personal names ending in *-a*. *Ex:* Juba-ea*. In the case of some of the names made by Linnaeus and other early taxonomists the *-ea* ending was sometimes used in forming generic names ending in vowels other than *a*. The other vowel was first changed to *a* and then the *-ea* added. *Ex:* Brodiaea*<*Brodie;* Jussiaea* <*Jussieu;* Tillaea*<*Tilli.* To avoid confusion these old names are retained. This suffix is also used as an ending for certain zoological names. *Ex:* Asteroid-ea; Echin-oid-ea. Some zoological class names end in *-ae;* as, Calci-spongi-ae; others end in *-a;* as, Tremat-od-a; Insect-a; Onycho-phor-a; Diplo-pod-a. Class names of plants generally end in *-ae;* as Gymno-sperm-ae, or in *-eae,* as, Filincin-eae, Cyano-phyc-eae.

**-eae**—*-eae,* ending of certain sub-class names such as Monocolyledon-eae. In certain other sub-class names the plural ending *-es* or *-i* is used, as in Monocotyledones, Crossopterygii.

**ear**—1. Gr. *ear,* genit. *earos,* the spring; *earinos,* of spring time>L. *earinus,* the color of spring, green. *Ex:* Earina (Ins.); Earinus (Ins.); Earo-phila (Ins.). See also er: 2. Gr. *ear*=*eiar,* blood, juice.

**earin**—See **ear**.

**ebae**—Gr. *ēbaios,* small, poor. *Ex:* Ebae-ides (Ins.); Ebaei-morphus (Ins.); Ebaeus (Ins.).

**eben**—Gr. *ebenos,* the ebony tree, or the wood of this tree>L. *ebeneus,* of ebony, ebon, black. *Ex:* Eben-aceae*; Ebeno-mitra (Moll.); Ebenus*.

**ebor**—L. *ebur,* genit. *eboris,* ivory; *eburneus*= *eburnus,* of ivory. *Ex:* Eboro-ziphius (Mam.); Eburi-fera (Ins.); Eburia (Ins.); Eburio-morpha (Ins.); Eburna (Moll.).

**ebri**—L. *ebrius,* drunk, sated, filled full.

**ebul**—L. *ebulus*=*ebulum,* a name for the dwarf elder.

**ebur**—See **ebor**.

**eburn**—See **ebor**.

**ec-**—Gr. *ek-,* prefix meaning out of. *Ex:* Ec-ballium*, see ball; Ec-copto-cnemis (Ins.); ec-demic (Med.); ec-dysis; Ec-temno-plax (Ins.). See also **eco**.

**ecblastes**—Gr. *ekblastēsis,* a budding. *Ex:* ecblastesis.

**ecclit**—Gr. *ekklitēs,* a drone. *Ex:* Ecclit-ura (Ins.); Ecclites (Ins.).

**eccrem**—Gr. *ekkremēs,* pendent, hanging. *Ex:* Eccremo-carpus*.

**eces**—Gr. *oikēsis,* the act of dwelling, also a place for dwelling, a house. *Ex:* ecesis=oecesis.

**ecetes**—See **oec**.

**ech**—1. Gr. *echis,* genit. *echeōs,* an adder, viper, a serpent. *Ex:* Ech-ites*; Echi-ales*; Echi-aster (Echin.); Echi-ur-oidea (Echin.); Echies

(Rept.); Echis (Rept.); Echio-stoma (Pisc.); Echium*; **2.** Gr. *echō*, to have, hold. *Ex:* ech-ard (Ecol.); Echo-cerus (Ins.); Trich-echidae (Mam.); Tympan-echus (Av.): **3.** Gr. *ēchō*, a sound, echo. *Ex:* Oxy-echus (Mam.): **4**, NL. *echi-*<Gr. *echinos*, a hedge-hog. *Ex:* Echi-mys (Mam.); Pro-echi-mys (Mam.). See also echus.

**echenei**—Gr. *echenēis*, the remora<*echenēis*, holding ships back. *Ex:* Echenei-bothrium (Platy.); Echeneis (Pisc.).

**echet**—Gr. *ēchētikos*, ringing; *ēchetēs*, clear sounding, shrill.

**echidn**—Gr. *echidna*, an adder, viper. *Ex:* Echidna (Mam.), (Rept.); Echidno-cephalus (Pisc.).

**echin**—Gr. *echinos*, a hedge-hog; also a name for the sea-urchin; *echinē*, an urchin's skin; *echinōdēs*, like a hedge-hog>L. *echinatus*, prickly. *Ex:* Echi-mys (Mam.); Echin-anthus (Echin.); Echin-ella (Bry.); Echin-iscus (Arach.); echinate; Echino-dermata; Echino-mastus*; Echinus (Echin.).

**echis**—See ech 1.

**echite**—L. *echite*, a kind of plant. *Ex:* Echites*.

**echm**—Gr. *echma*, genit. *echmatos*, a prop. holdfast, hindrance. *Ex:* Echmat-emys (Rept.); Echmato-phorus (Ins.); Gastr-echmia (Amph.)

**echthist**—See echthr.

**echthodop**—Gr. *echthodopos*, hateful. *Ex:* Echthodopa (Ins.).

**echthr**—Gr. *echthros*, hated, hostile; *echthistos*, most hated. *Ex:* Echthistus (Ins.); Echthrodoca (Ins.); Echthrus (Ins.).

**echus**—Gr. *echusa*, a holding<*echō*, to have. *Ex:* Lom-echusa (Ins.).

**eciton**—NL. *eciton*, name applied to a genus of wandering ants (etym. uncertain). *Ex:* Eciton (Ins.); Eciton-ini (Ins.).

**eclect**—Gr. *eklectos*, picked out; *eklektikos*, selecting. *Ex:* eclectic; Eclectus (Av.).

**eclex**—Gr. *eklexis*, choice, selection. *Ex:* andr-eclexis.

**eclips**—See eclipt.

**eclipt**—Gr. *ekleipō*, to cease, stop, to be deficient, forsake; *ekleipsis*, a failure; *ekleiptikos*, defective, belonging to an eclipse. *Ex:* Eklipea (Ins.); Eclipsis (Ins.); Eclipta*; Eclipto-phanes (Ins.).

**eclosion**—Fr. *éclosion*, an emerging from the egg. *Ex:* eclosion

**ecmel**—Gr. *ekmelēs*, irregular, out of tune. *Ex:* Ecmeles (Av.).

**eco**—Gr. *oikos*, a house. *Ex:* eco-logy=oecology; eco-tome.

**ecpagl**—Gr. *ekpaglos*, terrible. *Ex:* Ecpaglus (Ins.).

**ecphor**—Gr. *ecphoros*, divulged; also violent. *Ex:* Ecphora (Moll.).

**ecphyl**—Gr. *ekphylos*, alien, strange. *Ex:* Ecphylus (Ins.).

**ecphym**—Gr. *ekphyma*, an eruption of pimples. Ecphym-otes (Rept.).

**ecplectic**—Gr. *ekplēktikos*, astounding. *Ex:* Ecplectica (Ins.).

**ecpleo**—**1.** Gr. *ekpleos*, whole, entire, copious. *Ex:* Ecpleo-pus (Rept.). **2.** Gr. *ecpleō*, to sail away.

**ecptom**—Gr. *ekptōma*, a dislocation, falling out. *Ex:* ecptoma.

**ect**—Gr. *ektos*, outside, without. *Ex:* ect-enteron; Ecto-ceras (Arach.); ecto-derm; ecto-parasite; Ecto-procta (Bry.); ecto-therm.

**ectadi**—Gr. *ektadios*, outstretched. *Ex:* Ectadio-phantus (Ins.); Ectadius (Ins.).

**ectas**—Gr. *ektasis*, extension; *ektatos*, capable of extension. *Ex:* Ectasis*; Ectat-omma (Ins.); Ectato-cnemis (Ins.); Cal-ectasia*; par-ectasis (Med.).

**ectat**—See ectas.

**ectemn**—Gr. *ectemnō*, to cut out, to castrate, to weaken. *Ex:* Ectemno-plax (Ins.).

**ecten**—See ectenes.

**ectenes**—Gr. *ektenēs*, drawn-out; also diligent, awake, zealous<*ekteinō*, to extend. *Ex:* Ecten-urus (Platy.); Ecteno-crinus (Echin.).

**ectin**—Gr. *ektinō*, to pay off in full, avenge. *Ex:* Ectinus (Ins.).

**ectop**—Gr. *ectopos*=*ectopios*, displaced, foreign. *Ex:* ectop-ic; Ectopius (Ins.); Ectopio-glossa (Ins.); ectop-y.

**ectopist**—NL. *ektopistes*, a wanderer<Gr. *ektopizō*, to move from a place; *ektopistikos*, migratory. *Ex:* Ectopistes (Av.).

**ectyp**—Gr. *ektypos*, carved, worked in relief. *Ex:* Hol-ectyp-ina (Echin.).

**eczetes**—Gr. *ekzēteō*, to seek out; *ekzētētēs*, an investigation. *Ex:* Eczetesis (Ins.).

**ed**—Gr. *edō*=L. *edo*, to eat; LL. *edibilis*, fit for food. *Ex:* Edo (Ins.); Edo-mya (Ins.); Edo-stoma (Mam.): edible; Ent-edon (Ins.). See also edos.

**edac**—L. *edax*, genit. *edacis*, greedy, devouring.

**edaph**—Gr. *edaphos*, the soil, foundation, bottom *Ex:* edaph-ic (Ecol.); Edaph-odon (Pisc.); edapho-logy; Edapho-saurus (Amph.).

**edax**—See edac.

**edess**—Gr. *Edessa*, city of Macedonia. *Ex:* Edessa (Ins.).

**edest**—Gr. *edestēs*, an eater<*edō*, to eat. *Ex:* Edestes (Pisc.); Edesto-saurus (Rept.); Edestus (Pisc.); Mya-destes (Av.).

**edibl**—See ed.

**edit**—L. *edo*, pp. *editus*, to give out, set high> *editus*, lofty, also to *editus*, excrement.

**-edium**—See aeidi.

**edoli**—Gr. *edōlius*, name of some bird. *Ex:* Edolii-soma (Av.); Edolius (Av.).

edos—Gr. *edōs*, genit. *ēdeōs*, delight, enjoyment.

edr—Gr. *hedra*, dim. *hedrion*, a seat; *hedraios*, sitting much, sedentary; *hedranon*, a seat, dwelling, support. *Ex:* Edri-ophthalma (Crust.); Edrio-aster (Echin.).

edran—See edr.

eduli—L. *edulis*, edible. *Ex:* eduli-ous.

edur—L. *edurus*, very hard.

edy—Gr. *ēdys*, pleasant. *Ex:* Za-edy-us (Mam.).

-ef—See -ex.

effect—L. *efficio*, to complete, accomplish, pp. *effectus*, completed. *Ex:* effector.

effet—L. *effetus*, exhausted.

efficac—L. *efficax*, genit. *efficacis*, effectual, powerful. *Ex:* efficac-y.

efficax—See efficac.

effigur—L. *effiguratio*, genit. *effigurationis*, a representation. *Ex:* effiguration; effigurate.

efflorescen—L. *efflorescens*, genit. *efflorescentis*, flourishing, blooming, ppr. of *effloresco*, to flourish, bloom. *Ex:* efflorescent.

effugi—L. *effugius*, escaping.

effus—L. *effundo*, to spread abroad, pp. *effusus*, spread out, shed. *Ex:* Effusa (Moll.); Effusiana (Moll.).

ega—NL. *Ega*, a geographical name. *Ex:* Ega (Ins.); Ega-droma (Ins.).

egathe—Gr. *ēgatheos*, hallowed, most holy. *Ex:* Egatheus (Av.).

eget—See heget.

eglanter—NL. *eglanterius*, like the briar-rose; OF. *eglentier* = *aiglantier*, etc., the briar-rose.

ego—See aeg.

egregi—L. *egregius*, surprising, excellent.

egrett—Fr. *aigretta*, a tuft of feathers; a kind of heron. *Ex:* Egretta (Av.); egrettus.

-eia—Gr. *-eia* > NL. *-eia*, ending of certain NL. nouns of Greek origin. *Ex:* pharmacopeia < Gr. *pharmakopiia*. For adjectives ending in *-eia*, see -eius.

eid—Gr. *eidos*, form, appearance < *eidō*, to see. *Ex:* Eido-theca (Moll.); Eido-thrips (Ins.); Ido-pompilus (Ins.); Campo-dea (Ins.); Uran-idea (Pisc.).

eidet—NL. *eidetic* < Gr. *eidos*, shape, form. *Ex:* eidetic.

eidolon—Gr. *eidōlon*, an image, phantom. *Ex:* Eidolon (Mam.).

eidos—Gr. *eidos*, a resemblance, form, figure. See -id 2.

eil—Gr. *eilō*, to coil, to roll up, to pack together. *Ex:* eil-oid.

eim—Gr. *eima*, dress, garment > *aneimōn*, without clothing. *Ex:* Aneimia* = Anemia*. For *anemia*, a blood disease, see haem

eir—Gr. *eirō*, to connect, join. *Ex:* Eir-atus (Ins.).

eiren—Gr. *eirenē*, peace, quiet > *Eirenē*, goddess of Peace. *Ex:* Eirenis (Rept.); Irena (Av.); Irene-sauri-pus (Rept.); Ireno-mys (Mam.). See iren.

-eis—L. *-eis*, suffix added to feminine proper names to form adjectives denoting descent or relationship.

eis—1. Gr. *eis*, into. *Ex:* Eis-cladus (Crust.): 2. Gr. *eisō*, = *esō*, within. *Ex:* eso-derm.

eisod—Gr. *esodos* = *eisodos*, an entry < *eis*, into + *hodos*, a way. *Ex:* eisod-ic; esodi-al.

-eius—L. *-eius*, *-a*, *-um*, suffix added to noun stems to form adjectives denoting belonging to. *Ex:* pleb-eius; telmat-eia.

ejuncid—L. *ejuncidus*, lean, slender, growing like a rush.

el—Gr. *helos*, a marsh. *Ex:* El-ornis (Av.). See ele.

elacat—Gr. *ēlakatē*, a spindle, distaff, arrow. *Ex:* Elacates (Pisc.); Elacato-phora (Ins.). See also elagatis.

elach—Gr. *elachys*, comp. *elassōn* (= Attic *elattōn*), superl. *elachistos*, small, mean, little. *Ex:* Elache-soma (Ins.); Elachista*; Elachoceras (Mam.); Elachys (Ins.); Elass-oma (Pisc.); Elasso-chirus (Crust.); Elatto-neura (Ins.); Elatton-istius (Pisc.); Myrm-elachista (Ins.).

elachist—See elach.

elachys—See elach.

elae—Gr. *elaia*, an olive, the olive tree; *elaion*, olive oil, an oily substance; *elaiōdēs*, like an olive, oily; *elaiodochos*, holding oil. *Ex:* Elaeagnus*; Elaeis*; Elaeo-carpus*; Elaeodes = Eleodes (Ins.); Eleodi-phaga (Mam.); elaeodochon; Not-elaea*; Pseud-elaea*.

elaeagn—Gr. *elaiagnos* = *eleagnos*, some Boetian marsh plant < *elaia*, the olive tree + *agnos*, the chaste-tree. *Ex:* Elaeagn-aceae*; see elae, agn; Elaeagnus*.

elaeo—Gr. *elaion*, oil. *Ex:* elaeo-plast.

elaeodoch—See elae.

elagatis—Gr. *ēlakatē*, a spindle. *Ex:* Elagatis (Pisc.), should have been Elakatis.

elain—Gr. *elaineos*, olive-colored. *Ex:* Elain-opsis (Av.). See elae.

elan—L. *elanus*, a kite < Gr. *elanos*, a kite < *elaynō*, to drive. *Ex:* Elan-oides (Av.); Elano-pterus (Av.); Elanus (Av.).

elap—See elop.

elaph—Gr. *elaphos*, a deer, stag. *Ex:* Elaph-idion (Arach.), (Ins.); Elapho-cephalus (Nem.); Elapho-glossum*; Elapho-myces*.

elaphr—Gr. *elaphros*, easy, light, swift, nimble; comp. *elaphroteras*. *Ex:* Elaphrium*; Elaphro-cnemus (Av.); Elaphroterus (Ins.); Elaphrus (Ins.).

elas—Gr. *elasis*, genit. *elaseōs*, a driving, striking. *Ex:* Elasi-poda (Echin.).

**elasm**—Gr. *elasma*=*elasmos*, genit. *elasmatos*, a thin plate, metal plate<*elaynō*, to beat out, strike. *Ex:* Elasma (Coel.); Elasmat-ina (Moll.); Elasmatus (Ins.); Elasmo-branchia; Elasmotherium (Mam.); Elasmus (Ins.).

**elass**—See **elach**.

**elast**—NL. *elastes*, spring, elasticity<Gr. *elatēs*=*elatēr*, a driver, elater. *Ex:* An-elastes (Ins.).

**elat**—1. L. *elatus*, elevating<*effero*, to raise up. *Ex:* Elato-bium (Ins.): 2. Gr. *elatēr*, a driver; *elatērios*, driving away<*elaunō*, to drive forward; *elatos*, driven out; ductile. *Ex:* Elater (Ins.); Elater-idium (Ins.); elaterium; elaterophore: 3. Gr. *elatē*, the silver fir. *Ex:* Elat-omma (Prot.); Cedr-ela*.

**elater**—See **elat** 2.

**elates**—See **elast**.

**elatin**—Gr. *elatinē*, name of a kind of toad-flax. *Ex:* Elatine*.

**elatt**—See **elach**.

**elaut**—L. *elautus*, washed<*elavo*, to wash clean.

**elc**—Gr. *elkō*, to drag. *Ex:* elco-tropism.

**eicobre**—Sp. *El Cobre Canyon*, the copper canyon, from whence specimens were sent. *Ex:* Elcobre-saurus (Rept.).

**ele**—1. Gr. *helos*, genit. *heleos*, a marsh. *Ex:* Eleo-charis*; Eleo-tragus (Mam.); Elo-meryx (Mam.); Elo-saurus (Rept.); Elo-therium (Mam.): 2. Gr. *eleos*, pity, compassion: 3. Gr. *eleos*, a kind of owl: 4. Gr. *ēleos*, astray, foolish: 5. Gr. *elaion*, oil. *Ex:* eleoplast.

**elea**—Gr. *Elea*, name of a town in southern Italy. *Ex:* Elea (Moll.).

**elect**—L. *electus*, chosen<*eligo*, to choose.

**electr**—1. Gr. *Ēlektra*, daughter of Agamemnon. *Ex:* Electra*, (Coel.), (Mam.), (Moll.): 2. Gr. *ēlektron*, amber; also occasionally splendor, something bright (now used in sense of electric or pertaining to electricity or it may even refer to species found as fossils in amber). *Ex:* Electro-phorus (Pisc.); Electro-bisium (Arach.); Electron (Av.).

**eledon**—Gr. *eledōnē*, a kind of octopus. *Ex:* Eledone (Moll.).

**elegan**—L. *elegans*, genit. *elegantis*, neat, elegant; *elegantulus*, somewhat elegant.

**elei**—Gr. *eleios*, a dormouse. *Ex:* Elei-otis*

**elemat**—Gr. *ēlematos*, lazy, trifling.

**elench**—Gr. *elenchos*, an argument of disproof. a cross-examining. *Ex:* Elenchus (Moll.), (Ins.).

**eleod**—See **elae**.

**eleph**—Gr. *elephas*, genit. *elephantos*, an elephant; (the combining form is sometimes used to mean ivory). *Ex:* Elephas (Mam.), (Ins.); Elephant-ichthys (Pisc.); Elephanto-pus*, (Coel.); Elephantus (Mam.); Dolicho-ops-elephus (Ins.).

**elephant**—See **eleph**.

**elephantodet**—Gr. *elephantodetos*, inlaid with ivory. *Ex:* Elephantodeta (Ins.).

**elettaria**—NL. *elettaria* <Sansk. *elā*, a name for cardamon. *Ex:* Elettaria*.

**eleusin**—Gr. *Eleusinē*=*Eleusis*, genit. *Eleusinos* a town in Attica where Ceres was worshipped; also the goddess herself. *Ex:* Eleusina (Ins.); Eleusine*.

**eleuth**—Gr. *eleutheros*, free, not bound. *Ex:* Eleuth-urus (Pisc.); Eleuther-ine*; Eleutherurus (Mam.); Eleuthero-dactylus (Amph.); Eleuthro-zoa (Echin.).

**eli**—Gr. *eleios*, a kind of dormouse. *Ex:* Eliomys (Mam.); Elius (Mam.); See also **hel** 1.

**elic**—Gr. *ēlikia*, stature. *Ex:* Micro-licia*.

**eligm**—Gr. *eligma*, a folding, also a curl; *eligmos*, a winding, sinuosity. *Ex:* Eligm-odon (Mam.); Eligmo-stoma (Moll.); Eligmus (Moll.).

**-elis**—L. *-elis*, suffix added to noun stems to form adjectives meaning, pertaining to.

**elis**—L. *elisus*, forced out, torn out, eradicated; pp. of *elido*, to throw out, shatter.

**eliss**—Gr. *elissō*, to roll, to turn round about. *Ex:* Ex-elissa (Moll.).

**elix**—See **helic**.

**-ell**—L. *-ellus*, *-a*, *-um*, suffix added to noun stems to form diminutives. *Ex:* Euplect-ella (Por.); Pavon-ella (Av.). This suffix is occasionally used in making generic endings for names based on personal names. *Ex:* Parish-ella*; Pary-ella*; Palmer-ella*.

**ellamp**—Gr. *ellampō*, to illuminate, Ellampus (Ins.).

**ellip**—Gr. *ellipēs*, defective. *Ex:* Ellipo-neura (Ins.).

**ellips**—Gr. *elleipsis*, a leaving out, leaving behind; *elleiptikos*, elliptic, defective. *Ex:* Ellips-echinus (Echin.); Ellips-idium (Prot.); Ellipso-bulimina (Prot.); Ellipticus (Ins.); Ellipto-cephalus (Tril.).

**ellipt**—See **ellips**.

**elminth**—See **helminth**.

**elmis**—NL. *elmis*, a generic name of certain clavicorn beetles given by Latreille. Ety. unknown. *Ex:* Elmis (Ins.).

**elo**—See **ele**.

**elod**—Gr. *helōdēs*, marshy. *Ex:* Elodea*; Elodes (Ins.). See **hel** 2.

**elop**—Gr. *elops*, genit. *elopos*, mute, a mute one, the name of a kind of sea-fish; also the name of a kind of serpent>NL. *elaps*. *Ex:* Elap-echis (Rept.); Elapo-cephalus (Rept.); Elaps (Rept.); Elops, (Pisc.); Elop-idae (Pisc.); Elopo-morphus (Pisc.).

**elops**—See **elop**.

elos—See hel 2.

elp—Gr. *elpis*, genit. *elpidos*, hope, expectation; *elpistos*, hoped, to be hoped for; *elpisma*, a thing hoped for. *Ex:* Elp-idium (Crust.); Elpidio-gone (Echin.); Elpido-saurus (Rept.); Elphido-tarsius (Mam.), a misspelling of *elpido*; An-elpistus (Ins.).

elphid—See elp.

elpid—See elp.

elpism—See elp.

elpist—See elp.

eluc—L. *elucus*, a drowsy dreamy one.

elut—L. *elutus*, washed off, pp. of *eluo*, to wash off.

elym—Gr. *elymos*, a case for carrying arrows, a quiver; also an old name for a kind of grass, millet. *Ex:* Elymus*.

elysi—Gr. *ēlysios*, Elysian, i.e., pertaining to Elisium, abode of the blessed dead. *Ex:* Elysia (Moll.).

elytr—Gr. *elytron*, a cover, sheath. *Ex:* Elytr-odon (Ins.); elytri-ger-ous; Elytro-gona (Ins.); elytrum; Hymen-elytra*.

em- —See en-.

emad—See mad.

emarcid—NL. *emarcidus*, withered, flaccid < L. *e*, from + *marcidus*, shrunken.

emarginat—L. *emarginatus*, without a hem or border.

emascul—L. *emasculatus*, castrated, rendered impotent; *emasculator*, one who makes impotent. *Ex:* emascul-ation.

emballo—Gr. *emballō*, to throw in, lay or put in; *embolos*, something that can be thrust in easily, such as a peg, stopper, bolt; *embolion*, a javelin, something thrust in = L. *embolium*. *Ex:* Emballo-theca (Bry.); Emballon-urus (Mam.); Embol-ichthys (Pisc.); Embola (Ins.); embolium; Embolo-branchiata (Arth.).

embaphi—Gr. *embaphion*, a flat vessel. *Ex:* Embaphias (Rept.); Embaphion (Ins.).

embat—Gr. *embatēs*, he that sails in a ship. *Ex:* Embata (Rot.).

emberiz—NL. *emberiza* < Swiss-German *emmeritz*, a bunting. *Ex:* Emberiza (Av.).

embernagr—NL. *embernagra*, a bird name erected by Lesson < *Ember(iza)* + *(Ta)nagra*. *Ex:* Embernagra (Av.).

embi—Gr. *embios*, lively, tenacious of life, long-lived. *Ex:* Embi-idae (Ins.); Embia (Ins.); Embio-myia (Ins.); Embio-toca (Pisc.).

emblem—Gr. *emblēma*, something inserted, a raised ornament, an emblem. *Ex:* emblem-arius; Emblema (Av.), (Moll.); Emblema-soma (Ins.).

embol—See emballo.

embolim—Gr. *embolimus*, inserted, placed between. *Ex:* Embolimus (Ins.).

embrith—Gr. *embrithō*, to be heavy. *Ex:* Em brithes (Ins.); Embritho-saurus (Rept.).

embry—Gr. *embryon*, a foetus. *Ex:* Embryo-crinus (Echin.); embryo-logy; Embryo-phyta*; Embryon (Ins.)*. See also bry 2.

emend—L. *emendō*, to correct, improve; *emendatus*, corrected; *emendatrix*, an improver. *Ex:* emmend-ation.

emers—L. *emersus*, coming forth, emerging.

emes—Gr. *Emesa*, name of a city in Syria. *Ex:* Emesa (Ins.); Emeso-dema (Ins.).

emet—Gr. *emetos*, vomiting; *emetikos*, provoking sickness.

emex—NL. *emex*, a plant name < L. *ex*, out, from + *Rumex*, *Ex:* Emex*.

-emia—NL. *-emia* ( < Gr. *-aimia* < *haima*, blood), suffix indicating a certain blood condition. *Ex:* py-emia.

emin—L. *emineo*, ppr. *eminens*, genit. *eminentis*, to project, stand out. *Ex:* not Eminia (Ann.) nor Emino-drilus (Ann.) which are from a proper name.

eminen—L. *eminens*, genit. *eminentis*, standing high, projecting < *emineo;* Fr. *eminence. Ex:* eminence; eminent.

eminent—See emin.

emiss—L. *emissus*, sending out, letting go; *emissarium*, a drainway, outlet, *Ex:* emiss-ion.

emmal—NL. *emmalus* < Gr. *emmalos*, woolly. *Ex:* Emmalo-chaeta (Ins.); Emmalus (Ins.).

emmel—See emmeles, also en-.

emmeles—Gr. *emmelēs*, sounding in harmony, in tune.

emmen—1. Gr. *emmenō*, to cleave to, to be faithful. *Ex:* Emmen-anthe*; Emmen-odon (Mam.); Emmeno-tarsus (Ins.): 2. Gr. *emmēna*, the menses, *Ex:* emmeno-logy (Med.).

emmest—Gr. *emmestos*, filled.

emmetr—Gr. *emmetros*, suitable, proportioned. *Ex:* Emmetrus (Ins.).

emot—L. *emotus*, removed.

emped—Gr. *empedos*, firm-set, continual. *Ex:* Empedo-psyche (Ins.).

empedocl—L. *Empedocles*, Sicilian philosopher who wrote on plants. *Ex:* Empedoclea*.

emphan—Gr. *emphanēs*, manifest, conspicuous. *Ex:* Emphanes (Ins.).

emphantic—Gr. *emphantikos*, expressive, vivid. *Ex:* Emphantica (Ins.).

empher—Gr. *emphereia*, likeness. *Ex:* Neo-empheria (Ins.).

empheres—Gr. *empherēs*, answering to, like, resembling. *Ex:* Emphere-myia (Ins.); An-empheres (Ins.).

emphrass—Gr. *emphrassō*, to block up. *Ex:* Emphrass-otis (Rept.).

emphrax—Gr. *emphraxis*, a stoppage, obstruction.

**emphyl**—Gr. *emphylos*, in the tribe. *Ex:* Emphylo-coris (Ins.); Emphylus (Ins.).

**emphys**—Gr. *emphysaō*, to inflate; *emphysēma*, an inflation. *Ex:* emphysetus.

**emphysem**—See **emphys**.

**emphyt**—Gr. *emphytos*, implanted, innate. *Ex:* emphyt-ism; Emphyt-oecia (Ins.); emphyto-genesis; Emphytus (Ins.).

**empid**—Gr. *empis*, genit. *empidos*, gnat, mosquito. *Ex.* Emp-idae (Ins.); Empidi-ion (Ins.); Empid-onax (Av.); Empidi-vora (Av.); Em pido-chanes (Av.); Empis (Ins.).

**empis**—See **empid**.

**empodi**—Gr. *empodios*, meeting, impeding. *Ex:* Empodius (Acanth.).

**emprosth**—Gr. *emprosthen*, in front, before, earlier. *Ex:* emprostho-drom-ous.

**empus**—Gr. *empousa*, a ghost, a sprite upon one leg. *Ex:* Empusa* (Ins.).

**emuls**—L. *emulsus*, pp. of *emulgeo*, to milk out. *Ex:* emulsi-fy; emulsion < Fr. *emulsion*.

**emunct**—L. *emunctus*, cleaned < *emungo*, to wipe the nose.

**emyd**—Gr. *emys*, genit. *emydos*, a fresh-water tortoise. *Ex:* Emys; Emy-suchus (Rept.); Emyd-ichnium (Rept.); Emydes (Rept.); Emyd-ura (Rept.); Emydo-sauria (Rept.); Emys (Rept.). See note under my.

**emys**—See **emyd**.

**-en**—Eng. *-en*, suffix, meaning in verbs, to make, as in black-en, in adjectives, made of, as in wood-en.

**en-**—Gr. *en-(em-)*, prefix meaning in, on, among. *Ex:* Em-mel-ichthys (Pisc.); Em-meso-coma. (Ins.); Em-petrium*; em-podium; En-chytraeus (Ann.); en-demic; En-dromis (Ins.); en-tomo-logy: See also oen and enne.

**en**—Gr. *enos*, old, belonging to the former of two periods. *Ex:* Ba-en-odon (Mam.).

**enal**—See **enali**.

**enali**—Gr. *enalios* = *enalos*, in, on, or of the sea. *Ex:* Enali-ornithes (Av.); Enalio-chelys (Rept.); Enalo-droma (Ins.).

**enall**—Gr. *enallos*, changed, contrary. *Ex:* Enall-aster (Echin.); Enalli-pora (Bry.); Enallo-helia (Coel.).

**enant**—Gr. *enantios*, contrary, opposite. *Ex:* Enanti-ulus (Myr.); enantio-blastic; Enantio-sphen (Brach.); Antha-enantia*.

**enarg**—Gr. *enargēs*, visible. *Ex:* Enargo-pelte (Ins.).

**enat**—L. *enatus*, born, arisen < *enascor*, to be born.

**enaul**—1. Gr. *henaulos*, a water course. *Ex:* enaul-ad (Ecol.) Enaulo-spongia (Por.): **2.** Gr. *enaulios*, dwelling in dens < *enaulon*, an abode.

**encars**—Gr. *enkarsios*, oblique. *Ex:* Encarsia (Ins.).

**encaustic**—L. *encausticus*, burned in.

**-ence**—Eng. *-ence* < Fr. *-ence*, L. *-entia*, suffix added to noun stems to signify state, quality or action. *Ex:* somnol-ence.

**encel**—NL. *encelia* < *Christopher Encel*, writer on oak-galls, not from Gr. *encheleion* as given in Webster's New-Int. Dict. *Ex:* Encelia*.

**encephal**—Gr. *enkephalos* the brain. *Ex:* en cephal-ic; di-encephalon.

**ench**—Gr. *enchos*, a spear. *Ex:* Enchei-ziphius (Mam.); Encho-myia (Ins.); Campyl-enchia (Ins.).

**enchel**—See **enchely**.

**enchely**—Gr. *enchelys*, dim. *encheleion*, an eel. *Ex:* Enchel-ina (Prot.); Encheli-ophis (Pisc.); Enchely-core (Pisc.); Enchelyo-lepis (Pisc.); Enchelys (Prot.).

**enchyl**—Gr. *enchylos*, juicy.

**enchym**—Gr. *enchyma*, an infusion; *enchymos*, moistened. *Ex:* Enchymus (Ins.); coen-enchym; par-enchyma. See en- and chym.

**enchytrae**—NL. *enchytrae*, dwelling in an earthen pot < Gr. *en-*, in + *chytra*, a pot. *Ex:* Enchytraeus (Ann.).

**encop**—Gr. *enkopē*, an incision; also a hindrance. *Ex:* Encope (Echin.).

**end**—Gr. *endon*, within. *Ex:* End-amoeba (Prot.); endo-derm; Endo-nychus (Ins.); endo-skeleton; endo-style; endo-therm.

**ende**—Gr. *endeēs*, wanting, in need of. *Ex:* Opis-endea (Ins.).

**endeca**—Gr. *endeka*, eleven. *Ex:* Endeca-pleura (Mam.); Endeca-tomus (Ins.).

**endem**—NL. *endemic*, peculiar or native to a place < a supposed Gr. *endēmikos* < *endēmos*, native. *Ex:* endemic-al; endem-ism.

**endic**—Gr. *endikos*, right, legitimate.

**endiv**—It. *endivia*, endive. *Ex:* endivae-folia.

**endoth**—Gr. *endothi*, within. *Ex:* Endothi-odon (Rept.).

**endox**—Gr. *endoxos*, notable, glorious, esteemed. *Ex:* Endoxo-crinus (Echin.).

**endrom**—Gr. *endromis*, genit. *endromidos*, a garment of fur. *Ex:* Endromis (Ins.).

**endros**—Gr. *endrosos*, dewy. *Ex:* Endrosis (Ins.).

**endym**—Gr. *endyma*, genit. *endymatos*, a garment.

**endymion**—Gr. *Endymion*, name of a beautiful youth with whom Luna fell in love. *Ex:* Endymion*, (Tril.).

**-ene**—Adjectival suffix < L. *-ēnus*, Gr. *-ēnos*, pertaining to. *Ex:* terr-ene, ser-ene. In chemistry used in names of certain hydro-carbons such as benz-ene.

**eneilem**—Gr. *eneilēma*, a wrapper. *Ex:* eneilema.

**eneot**—Gr. *eneotēs*, stupidity.

**enerthe**—Gr. *enerthe*, in the ground, beneath. *Ex:* Enerthe-comal (Prot.).

enes—Gr. *enesis*, an injection.

enet—Gr. *enetos*, injected. *Ex:* Limn-enetes (Mam.).

engoni—Gr. *engōnios*, forming an angle. *Ex:* Engonia (Ins.).

engram—Gr. *en*, in+*gramma*, mark, trace. *Ex:* engram.

engraul—Gr. *engraulis*, genit. *engraulidos*, a small fish, the perch. *Ex:* Engraul-ites (Pisc.); Engrauli-gobius (Pisc.).

engy—Gr. *engys*, near, near at hand but often used in the sense of narrow, straightened. *Ex:* Engy-cystis (Ins.); Engy-ptilla (Av.); Engy-stoma (Amph.).

engyt—Gr. *engytēs*, nearness. *Ex:* Engyt-atus (Ins.).

enhydr—Gr. *enhydris*, an otter; also a water-snake <*enhydros*, living in water. *Ex:* Enhydra (Mam.); Enhydro-cyon (Mam.).

enic—See henic.

enicm—Gr. *enikmos*, "with wet in it," humid. *Ex:* Enicmo-deres (Ins.).

enix—1. L. *enixus*, earnest, with zeal: 2. L. *enixus*, birth; as a passive verb, born.

enizem—Gr. *enizēma*, that on which one sits, a seat. *Ex:* Enizemum (Ins.).

enn—See enne.

enne—Gr. *ennea*, nine. *Ex:* En-arthro-carpus*; Enn-ac-odon (Mam.); Enne-acanthus (Pisc.); Enne-odon (Rept.); Ennea-merus (Ins.); Enneo-conus (Mam.).

enochr—Gr. *enōchros*, palish, rather pale. *Ex:* Enochrus (Ins.).

enod—L. *enodis*, smooth. *Ex:* Enodes (Av.).

enoic—Gr. *enoikeō*, to inhabit. *Ex:* Enoic-yla (Ins.).

enopl—Gr. *enoplos*, armed. *Ex:* Enoplio-morpha (Ins.); Enoplius (Ins.); Enoplo-chiton (Moll.); Enoplo-teuthis (Moll.).

enorm—L. *enormis*, irregular; also enormous <*e-out* of+*norma*, rule.

enos—Gr. *enosis*, a shaking, quaking. *Ex:* enosi-mania (Med.).

enot—Gr. *enotēs*, unity. *Ex:* Enotes (Ins.).

ens—L. *ensis*, a sword >NL. *ensatus*, sword-shaped, ensi-form. *Ex:* Ensi-dens (Moll.); Ensis (Moll.).

ensiger—L. *ensiger*, sword bearing <*ensis*, a sword+*ger*, to bear, carry.

-ensis—L. *-ensis*, adjectival suffix meaning, belonging to. *Ex:* hort-ensis; missouri-ensis; mohav-ensis; praet-ensis.

ent—Gr. *entos*, within, inside. *Ex:* Ent-amoeba (Prot.); ento-derm; Ento-rhipidium (Prot.).

-ent—Eng. *-ent*, adj. suffix <L. *-entem*. *Ex:* depend-ent. See -ant.

ental—See ent and -al.

entalis—NL. *entalis*, a perverted form of *Dentalium*. *Ex:* Entalis (Moll.).

entechn—Gr. *entechnos*, artificial; also artistic, skilled. *Ex:* Entechnia (Ins.).

entel—Gr. *entelēs*, complete, full, perfect. *Ex:* Entel-odon (Mam.); Entelea*; Enteles (Brach.) (Ins.); Entelo-styl-ops (Mam.).

entelech—Gr. *entelecheia*, perfect, perpetual. *Ex:* Entelechia (Ins.).

entell—Gr. *entellō*, to command; *Entellus*, Sicilian hero famous as a pugilist. *Ex:* Entell-ites (Brach.); Entella (Ins.); Entellus (Mam.) <*entellō*, since the animal commands respect of the natives.

enter—Gr. *enteron*, bowel, intestine; *enterikos*, intestinal. *Ex:* Entero-pleura (Moll.); enteron; succus-entericus; mes-enteron.

enthor—Gr. *enthoros*, impregnated. *Ex:* Enthora (Ins.); Enthoro-dera (Arach.).

entin—Gr. *enteinō*, to stretch. *Ex:* Entino-pus (Ins.).

entom—Gr. *entoma*, an insect. *Ex:* Entomo-logy; Eo-entom-idae (Ins.); Acer-entom-idae (Ins.), see a- and cer.

entyp—Gr. *entypoō*, to carve, to cut in relief. *Ex:* Entypo-dera (Ins.); Entypus (Ins.).

entypom—Gr. *entypōma*, an intaglio, a piece of carving. *Ex:* Entypoma (Ins.).

-enus—L. *-enus* >Eng. *-ene*, suffix added to stems of common and proper nouns to form adjectives meaning belonging to. *Ex:* phlo-baph-ene; terr-enus.

enydr—Gr. *enydris*, an otter. *Ex:* Enydris (Mam.); not Enydro-bius (Amph.), see hydr.

enygr—Gr. *enygros*, in the water, damp, watery. *Ex:* Enygro-theres (Av.); Enygrus (Rept.).

eo—See eos.

eolid—Gr. *Ailoidēs*, name of a male descendant of Aiolos. *Ex:* Eolida (Moll.).

eos—Gr. *ēōs*, the dawn, sunrise, the east; *Ēōs*, goddess of dawn corresponding to the Roman *Aurora*. *Ex:* Eo-cene; Eo-hippus (Mam.); Eo-mecon*; Eo-podoces (Av.); Eos (Av.).

eosin—NL. *eosin*, a red stain <Gr. *eos*, dawn. *Ex:* eosino-phil.

ep- —Gr. *epi*, on, upon; also (rarely) to, towards, against, after, etc. *Ex:* ep-harmony (Ecol.); Ep-hydatia (Por.); Ep-hydra (Ins.); Epi-cauta (Ins.); epi-cranium; Epi-gaea*; Epi-lobium*; epi-meron; epi-otic.

epachth—Gr. *epachthēs*, heavy, offensive, annoying, troublesome. *Ex:* Epachthes (Ins.), (Crust.).

epacr—Gr. *epakros*, sharp, pointed at the end. *Ex:* Epacria (Ins.); Epacro-mia (Ins.).

epact—Gr. *epaktos*, imported. *Ex:* Epacto-crinus (Echin.).

epacter—Gr. *epaktēr*, a hunter. *Ex:* Epacter (Ins.).

epacti—Gr. *epaktios*, on the shore. *Ex:* Epactius (Ins.).

epanorth—Gr. *epanorthoō*, to correct, restore; *epanorthēma*, a correction. *Ex:* Epanorthus (Mam.).

eparc—Gr. *eparkēs*, assisting; *eparkesis*, aid, assistance. *Ex:* Eparces (Ins.).

eparm—Gr. *eparma*, genit. *eparmatos*, a swelling. *Ex:* Eparmato-stethus (Ins.).

epaul—Gr. *epaulos*, house. *Ex:* Epaul-oecus (Ins.).

epedan—Gr. *ēpedanos*, frail, weak. *Ex:* Epedanus (Arach.).

epeir—1. Gr. *eperyō*=Ionic *epeiryō*, to pull to. *Ex:* Epeira (Arach.); Epeiro-typus (Arach.); Met-epeira (Arach.): 2. Gr. *ēpeiros*, a continent.

epel—Gr. *epēlis*, a cover, also a freckle. *Ex:* Epel-ichthys (Pisc.); Epelis (Ins.); not Epel-oria (Arth.) which is < Gr. *epēlys*, a stranger.

ependym—Gr. *ependyma*, a tunic, an upper garment. *Ex:* ependyma; ependym-al.

epenthes—Gr. *epenthesis*, an insertion. *Ex:* Epenthesis (Coel.).

epeol—Gr. *ēpiolos*, a moth. *Ex:* Epeolus (Ins.).

eperu—NL. *eperua* < Caribbean *eperu*, name of a fruit. *Ex:* Eperua*.

epheb—Gr. *ephēbos*, a youth; *ephēbikos*, pertaining to youth. *Ex:* ephebic; ephebo-logy.

ephedr—Gr. *ephedra*, a sitting before a place, sitting upon a seat > *ephedra*, the horsetail plant. *Ex:* Ephedra*; Ephedrus (Ins.).

ephel—Gr. *ephēlis*, genit. *ephēlidos*, a freckle.

ephemer—Gr. *ephēmeros*, temporary, living but a day. *Ex:* Ephemer-idae (Ins.); Ephemera (Ins.).

ephesti—Gr. *ephestios*, at home, domestic. *Ex:* Ephestia (Ins.).

ephialt—Gr. *ephialtēs*, the night mare; one who incubates or lies upon. *Ex:* Elphialt-ina (Ins.); Ephialtes (Ins.), (Av.).

ephict—Gr. *ephiktos*, easy to get at, accessible.

ephimer—Gr. *ephimeros*, desired, delightful. *Ex:* Ephimero-pus (Ins.).

ephipp—Gr. *ephippos*, mounted as on a horse; *ephippios*, for putting on a horse. *Ex:* Ephipp-odonta (Moll.); Ephippi-charax (Pisc.); Ephippio-ceras (Moll.). Ephippium (Ins.).

ephistem—Gr. *ephistēmi*, to cause to establish, to give one's attention to. *Ex:* Ephistemus (Ins.).

epholc—Gr. *epholkos*, enticing, alluring.

ephydr—Gr. *ephydros*, wet, living on the water. *Ex:* Ephydra (Ins.); Ephydro-soma (Ins.).

ephyr—Gr. *Ēphyra*, a Greek city, the old name of Corinth. *Ex:* Ephry-ops-idae (Coel.); ephyra; Ephyra (Coel.), etc.

epi—See ep-.

epibat—Gr. *epibatēs*, one who mounts, a stallion; also the head. *Ex:* Epibates (Ins.).

epiblem—Gr. *epiblēma*, that which is thrown over one, a cloak. *Ex:* Epiblema (Ins.); Epiblemum (Arach.).

epibletic—Gr. *epiblētikos*, impinging upon.

epichar—Gr. *epicharis*, beautiful, elegant, pleasing. *Ex:* Epicharis*.

epiclintes—Gr. *epiklintēs*, moving sidewise. *Ex:* Epiclintes (Prot.).

epiclop—Gr. *epiklopos*, thievish, tricky. *Ex:* Epiclopus (Ins.).

epicrat—Gr. *epikratēs*, one with mastery of a thing. *Ex:* Epicrates (Rept.).

epicri—Gr. *epikrion*, the sail-yard of a ship, an antenna. *Ex:* Epicrium (Amph.); Epicrius (Arach.).

epideme—Gr. *epidēmeō*, to live at home, to be prevelant. *Ex:* Epidemia (Ins.); epidemio-logy.

epier—Gr. *epiēros* = *epiēra*, pleasing. *Ex:* Epier-opsis (Ins.); Epierus (Ins.).

epieran—Gr. *epiēranos*, pleasing, acceptable.

epimach—Gr. *epimachos*, that may be easily attacked; also equipped for the fray. *Ex:* Epimachus (Av.).

epimec—Gr. *epimēkēs*, oblong. *Ex:* Epimecis (Ins.).

epio—Gr. *ēpios*, calm, mild, kind. *Ex:* Epio-pelmus (Ins.).

epipact—Gr. *epipaktis*, name of a plant, the hellebore. *Ex:* Epipactis*.

epiped—Gr. *epipedos*, on the ground, level, superficial. *Ex:* Epipeda (Ins.); Epipedo-soma (Ins.).

epiploic—NL. *epiploic* < Gr. *epiploon*, the omentum. *Ex:* an-epiploic; epiploic.

epipolae—Gr. *epipolaios*, superficial, manifest. *Ex:* Epipolaeus (Ins.).

epistax—Gr. *epistaxis*, nose-bleed.

epithem—Gr. *epithēma*, a cover, lid, also a poultice. *Ex:* epitheme; Lin-epithema (Ins.). See -them.

epitimet—Gr. *epitimētēs*, an inspector, avenger. *Ex:* Epitimetes (Ins.).

epiton—Gr. *epitonos*, stretched, intense. *Ex:* Epiton-ites (Echin.).

epod—Gr. *epōdos*, a chorus. *Ex:* Epodus (Ins.).

epomidi—Gr. *epōmidios*, on the shoulder. *Ex:* Epomidio-pteron (Ins.).

epops—Gr. *epops*, a kind of bird, the hoopoo. *Ex:* Epops (Av.).

epset—NL. *epsetus* < *hepsetus*, an old name for anchovies.

ept—NL. *ept* < Gr. *hepta*, seven. *Ex:* Epta-tretus (Elasm.).

epta—See ept.

**eptesicus**—NL. *eptesicus*, a generic name coined by Rafinesque and which he said he derived < Gr. *epiēn*, to fly and *oikos*, house. *Ex:* Eptesicus (Mam.).

**equ–1.** L. *aequus*, equal. *Ex:* equi-rotal; equivalve: **2.** L. *equus*, genit. *equi*, dim. *equuleus*, a horse; *equa*, genit. *equae*, dim. *equula*, a little mare; *equinus*, pertaining to horses; *eques*, a horseman. *Ex:* equ-oid; Eques (Pisc.); Equi-set-ella (Coel.); Equi-setum*; Equula (Pisc.); Equus (Mam.).

**equal**—L. *equalis*, equal.

**equator**—L. *aequator*, an equalizer. *Ex:* equatore-al.

**equet**—NL. *equetus*, a Rafinesque fish name < L. *equus*, a horse.

**equitan**—L. *equitans*, genit. *equitantis*, riding; ppr. of *equito*, to ride.

**er—1.** Gr. *ĕr*, contraction of *ear*, genit. *earos* = L. *ver*, spring. *Ex:* Er-anthemum*; Er-anthis*; Eri-genia*: **2.** Gr. *era*, earth, field. *Ex:* Er-agrostis*: **3.** Gr. *ĕri*, early. *Ex:* Eri-geron*: **4.** Gr. *erion*, wool; *erineos*, woolen; also the wild fig. *Ex:* er-ichthus; Erio-caulon*; erineus; erinose; Erio-dictyon*; Erio-gonum*; Erophila*; Ero-phyllum*; Somat-eria (Av.): **5.** L. *er*, genit. *eris*, a hedgehog. *Ex:* erinus.

**erann**—See erasm.

**eras**—L. *erasus*, scratched out, erased < *erodo*, to gnaw away.

**erasm**—Gr. *erasmios = erannos = erastos = eratos*, amiable, lovely, pleasing. *Ex:* Erasmia (Av.); Erann-ornis (Av.); Eranna (Av.); Eranno (Ann.); Eratio (Ins.).

**erast**—Gr. *erastēs*, fem. *erastria*, a lover. *Ex:* Erastri-facies (Ins.); Erastria (Ins.); porn-erast-ic.

**erastri**—See erast.

**erat**—See erasm.

**erato**—Gr. *Eratō*, the muse of mimicry and love-songs, the Lovely < *eratos*, lovely < *eraō*, to love. *Ex:* Erato (Moll.); Erato-trivia (Moll.); Eratia (Ins.).

**erax**—NL. *erax*, name applied to a genus of insects < Gr. *eraō*, to love. *Ex:* Erax (Ins.); Eraxis (Ins.). See hierax.

**ereb**—Gr. *Erebos*, place of nether darkness; also, a mythical being, the son of Chaos and Darkness; NL. *erebennus*, dark, like *Erebos*. *Ex:* Ereb-ophis (Rept.); Erebia (Ins.); Erebo-thrix (Ins.); Erebus (Ins.).

**erechtites**—Gr. *erechthitēs*, the grounsel < *erechthō*, to rend. *Ex:* Erechtites*.

**erect**—L. *erectus*, erect.

**ereis**—Gr. *ereisis*, a lifting or raising up, a pushing against, a thrusting.

**ereism**—See erism.

**erem—1.** Gr. *ĕrema*, gently, slowly. *Ex:* erema-causis: **2.** Gr. *erēmos*, solitary, uninhabited = L. *eremus*; *erēmitēs*, of the desert, an hermit. *Ex:* erem-ad (Ecol.); Erem-arionta (Moll.); Eremi-acris (Ins.); Eremi-astrum*; Eremi-ornis (Av.); Eremo-bates (Arach.); eremo-philus: **3.** Gr. *erēmia*, a place of solitude; also quietness, stillness, rest. *Ex:* Eremia (Moll.); Eremia-philus (Ins.).

Desert Gilia, *Gilia eremica*, a tiny annual of the desert sands.

**eremic**—L. *eremicus*, of the desert, lonely.

**eremit**—L. *eremitus*, lonely, solitary.

**eremit**—L. *eremitis*, genit. *eremitidis*, alone, solitary.

**eremn**—Gr. *eremnos*, dark, obscure, black, gloomy. *Ex:* Eremno-philus (Ins.).

**eremos**—Gr. *erēmōsis*, a making solitary, desolate. *Ex:* Eremosia (Ins.).

**eremotes**—Gr. *eremōtēs*, a desolator. *Ex:* Eremotes (Ins.).

**erepsin**—NL. *erepsin*, a digestive enzyme < L. *ereptus*, pp. of *eripio*, to pull away. Compare with pepsin, trypsin, etc.

**erept**—Gr. *ereptō*, to crown. *Ex:* Erept-odon (Mam.).

**eret**—See eretm.

**erethistic**—Gr. *erethistikos*, provoking > NL. *erethisticus*; Gr. *erethistēs*, one given to quarreling. *Ex:* Erethistes (Ins.).

**erethiz**—Gr. *erithizō*, Epic impf. *ĕrithizon*, to provoke, vex, excite. *Ex:* Erethizon (Mam.).

**eretm**—Gr. *eretmon*, an oar; *eretēs*, oars, also rowers; *eretikos*, of or for rowing. *Ex:* Eretes (Ins.); Eretm-ichthys (Pisc.); Eretmo-chelys (Ins.); Eretmo-podes (Av.); Pyg-eretmus (Mam.).

**ereunet**—Gr. *ereunētēs*, a searcher, prober. *Ex:* Ereunetes (Av.), etc.

**erg**—Gr. *ergon*, work; *ergasia*, labor, daily business; *ergatēs*, a laborer; *ergatikos*, willing or able to work. *Ex:* Ergas-ilus (Crust.); ergasio-phytes; Ergates (Ins.); Ergaticus (Av.); ergato-gyne; Bathy-erg-oides (Mam.); Poly-ergus (Ins.); syn-erg-id.

**ergasi**—See erg.

ergat—See erg.

ergod—Gr. *ergōdēs*, troublesome.

eri- —Gr. *eri-*, a strengthening prefix meaning very. *Ex:* Eri-glossa (Rept.); Eri-lepis (Pisc.). See also er.

eriarest—NL. *eriarestus* < Gr. *eri*, very + *arestos*, pleasing. *Ex:* Eriaresta (Ins.).

eric—Gr. *ereikē* = *erikē*, heath < *ereikō*, to break, since some ericas were supposed to break bladder stones; NL. *ericetum*, heath > NL. *ericaceus*, pertaining to a heath. *Ex:* Eric-aceae*; Erica*; Erica-meria*; ericet-inus; ericeti-cola.

ericet—See eric.

erichth—Gr. *erechthō*, to break. *Ex:* erichthus; erichth-oid-ina.

ericin—L. *ericinus*, of a hedge-hog.

ericius—L. *ericius*, NL. dim. *ericulus*, a hedge-hog, an urchin. *Ex:* Ericius (Mam.); Ericulus (Mam.).

erict—Gr. *eriktos* = *ereiktos*, bruised, wounded, pounded.

ericul—See ericius.

erigen—L. *erigo*, to raise, ppr. *erigens*, genit. *erigentis*, raising. *Ex:* nervi erigentes.

erin—See er 4.

erinace—L. *erinaceus*, also *ericius*, a hedge-hog. *Ex:* Erinace-olus (Mam.); Erinaceus (Mam.).

erinys—Gr. *Erinys* = *Erynnis*, an avenging deity. *Ex:* Erynnis (Tril.), (Ins.).

erio—See er 4.

eriphia—L. *eriphia* < Gr. *eripheia*, an unknown plant. *Ex:* Eriphia (Crust.).

erism—1. Gr. *ereisma*, genit. *ereismatos*, a prop, support. *Ex:* ereisma; Erismat-ura (Av.); Pachy-erisma (Pisc.):    2. Gr. *erisma*, genit. *erismatos*, a cause of quarrel, dispute; *erismos*, a quarrel, strife. *Ex:* erisma; Erisma*.

erist—Gr. *eristos*, contested; *eristēs*, a contester, wrangler. *Ex:* An-eristus (Ins.).

eristal—L. *eristalis*, name applied by Pliny to an unknown precious stone. *Ex:* Eristalis (Ins.); Eristalio-myia (Ins.); Eristalo-mya (Ins.).

eristic—Gr. *eristikos*, eager for strife. *Ex:* Eristicus (Ins.).

erithac—Gr. *erithakos*, name of some solitary bird. *Ex:* Erithracus (Av.).

erithal—1. Gr. *erithalēs*, much branched. *Ex:* Erithales (Ins.):    2. Gr. *erithalis*, name of some plant, perh. one of the stone crops. *Ex:* Erithalis*.

eritheles—Gr. *erithēlēs*, very flourishing.

erm—Gr. *erma*, a support. *Ex:* Dic-erma*, the first element < *dis*, twice.

ermine—OFr. *ermine*, the ermine > NL. *ermineus*, like the ermine, pertaining to the ermine.

ern—Gr. *ernos*, bud, offshoot, sprout; *ernōdēs*, like a young sprout, branched. *Ex:* Erno-ctona

(Ins.);   Erno-grammus  (Pisc.);   Ernodea*; Ernodes (Ins.).

ernod—See ern.

-ernus—L. *-ernus*, suffix added to noun stems to form adjectives meaning belonging to. *Ex:* hod-ernus.

erod—Gr. *herōdios*, a heron. *Ex:* Erodii; Herodii (Av.); Erodium*.

eros—1. L. *erosus*, gnawed off, consumed, pp. of *erodo*, to gnaw off. *Ex:* Erosa (Pisc.):    2. Gr. *Erōs*, genit. *Erōtos*, god of love < *erōs*, love > *erōtikos*, of or caused by love, pertaining to love. *Ex:* Eros (Ins.); erotic; Eroto-lepsia (Ins.), see er 1.

erot—See eros 2.

erotem—Gr. *erōtēma*, a query, question.

erotyl—Gr. *erōtylos*, a darling. *Ex:* Erotyl-idae (Ins.); Erotylus (Ins.).

erp—See herp.

erpestes—See herp.

erpet—See herp.

errabund—L. *errabundus*, wandering about, inconstant.

erran—L. *errans*, genit. *errantis*, wandering, ppr. of *erro*, to stray, to make a mistake > *erraticus*, wandering to and fro; *erroneus*, wandering, misled; *error*, a wandering, mistake. *Ex:* Errantia (Ann.).

errat—L. *erratus*, a wandering; *errator* a wanderer; *erratilis*, wandering. See also erran.

erratic—See erran.

erromen—Gr. *errōmenos*, stout, vigorous. *Ex:* Erromen-osteus (Pisc.); E· romenus (Ins.).

erron—L. *erro*, genit. *erronis*, a wanderer.

error—L. *error*, a wavering, a wandering from a course.

ersae—Gr. *ersaios* = *ersaēeis*, dewy. *Ex:* ersae-ome; Ersaea (Coel.).

erubescen—L. *erubescens*, genit. *erubescentis*, becoming red, ppr. of *erubesco*, to become red.

eruc—L. *eruca*, a caterpillar; also a kind of crucifer. *Ex:* Eruc-aria*; Eruca (Ins.); eruci-form; eruci-vor-ous.

eruct—1. L. *eructus*, belched forth, impure. *Ex:* eruct-ion:    2. L. *erugo*, to smooth, to make clear of wrinkles, pp. *eructus*, smoothed.

erug—L. *erugo*. to make smooth > NL. *erugosus*, very smooth.

erump—L. *erumpens*, genit. *erumpentis*, breaking out, bursting forth < *erumpo*, to break forth.

ery—Gr. *eryōn*, drawing out, ppr. of *eryō*, to draw, drag. *Ex:* Ery-ops (Amph.).

erycin—L. *Erycina*, a name of Venus > *Erycus*, a high mountain in Sicily and a city near it, famous for its temple of Venus. *Ex:* Erycin-idae (Ins.); Erycina (Ins.).

erymn—Gr. *erymnos*, fenced, made strong, steep. *Ex:* Erymno-chelys (Rept.); Erymnus (Rept.)

eryng—Gr. *ēryngos*, dim. *ēryngion*, a thistle. *Ex:* eryngi-folius; Eryngium*.

eryon—Gr. *eryō*, impf. *eryon*, to drag on the ground, draw. *Ex:* Eryon (Crust.).

erysib—Gr. *erysibē*, mildew, rust. *Ex:* Erysibe*.

erysim—Gr. *erysimon*, the hedge mustard < *eryō*, to draw, to cure. *Ex:* Erysimum*.

erysiph—NL. *erysiphe* < Gr. *erythros*, red+ *siphōn*, a pipe. *Ex:* Erysiphe*.

erythea—Gr. *Erytheia*, an island in the bay of Cadiz where the giant Geryon dwelt. *Ex:* Erythea*.

erythr—Gr. *erythros*, red, reddish; *erythraios*, fem. *erythraia*, red; *erythrotēs*, redness. *Ex:* Erythraea*; Erithina*; erythro-cyte; Erythro-suchus (Rept.); Erythro-xylon*.

erythran—Gr. *erythranos*, red, ruddy > *erythranon*, a kind of ivy with reddish fruits. *Ex:* erythran-ous.

erythron—Gr. *erythronion*, name of a kind of orchidaceous plant. *Ex:* Erythronium*.

eryx—L. *Eryx*, name of an opponent of Hercules > *Eryx*, name of a mountain upon which Eryx was buried, and famous for its temple of Venus. *Ex:* Eryx (Rept.), etc.

-es—1. L. *-es*, plural ending of certain 3rd declension nouns such as *incessor*, pl. *incessores*, a percher: 2. Gr. *-es*, plural ending of certain nouns such as *ortyx*, pl. *ortyges*, a quail: 3. L. and NL. *-es*, plural of nouns used in forming group names especially of birds and mammals. *Ex:* Glir-es (Mam.); Passer-es (Av.); Rasor-es (Av.); Scansor-es (Av.); Steganopod-es (Av.).

-escen—L. *-escens*, genit. *-escentis*, adjectival termination of Latin verbs having usually an inceptive or inchoative force, meaning beginning, beginning to, slightly; corresponding to Eng. *-ish*, *-escence* and *-escent*. *Ex:* adolescense; arbor-escent; pub-escence; pub-escent.

eschar—1. Gr. *eschara*, a fire-place; also a crust, the scab on a wound caused by burning; *escharotikos*, producing a scar. *Ex:* Eschara (Bry.); Eschari-pora (Bry.); Distans-eschar-ella (Bry.);Escharo-pora (Bry.); Rept-eschari-pora (Bry.): 2. Gr. *escharos*, a fish, a kind of sole.

escharion—Gr. *escharion*, a pan of coals; a basis, platform. *Ex:* Escharion (Pisc.).

eschat—Gr. *eschatos*, last in position or in time, the utmost, farthest. *Ex:* Eschat-ura (Ins.); Eschata (Ins.); Eschato-cephalus (Arach.).

esculent—L. *esculentus*, good to eat.

esis—Gr. 1. *ēsis*, genit. *eseōs*, delight. *Ex:* Mon-esis*: 2. Gr. *esis*, a sitting; also a tendency, drive, urging. *Ex:* ec-esis (Ecol.) < *esis*, a sitting, establishing; mon-esis

eso—1. Gr. *esō* = *eisō*, within. *Ex:* eso-derm: 2. *ēso*, to sit, be seated.

esoc—L. *esox*, genit. *esocis*; name of a kind of fish found in the Rhine, a pike. *Ex:* Esoc-idae (Pisc.); esoci-form; Esox (Pisc.).

esod—See eisod.

esop—Gr. *Aisōpos*, a proper name. *Ex:* Esopus (Crust.).

esophag—Gr. *oisophagos*, the gullet, see ois 2. *Ex:* esophagi-al; esophagus.

esoteric—Gr. *esōterikos*, arising within. *Ex:* esoteric.

esox—See esoc.

essenti—L. *essentia*, the essence of anything. *Ex:* essenti-al.

esson—Gr. *ēssōn*, less, weaker. *Ex:* Esson-odontherium (Mam.); Aco-essus (Mam.).

-essus—NL. *essus*, -a, -um, dim. suffix. *Ex:* Vampyr-essa (Mam.). See also esson.

est—See also hest.

esten—See estenomen.

-estes—NL. *-estes*, a contraction of Gr. *edestēs*, an eater. *Ex:* Pyren-estes (Av.).

esth—1. Gr. *esthiō*, to eat. *Ex:* Argyr-esthia (Ins.); Esthio-pterum (Ins.); not Pan-esth-idae (Ins.), see esthes: 2. Gr. *estheō*, to clothe. *Ex:* Esth-onyx (Mam.).

esthes—Gr. *esthēs*, a garment. *Ex:* Lisp(od)-esthes (Moll.); Pan-esth-idae (Moll.); Pan-esthes (Ins.).

esthet—Gr. *esthes*, genit. *esthetos*, a garment. *Ex:* Erythr-esthes (Ins.).

esthl—Gr. *esthlos* = Doric, *eslos*, luckily.

estr—See oestr.

estrait—NL. *estriatus*, without stripes or furrows < L. *e-, ex-*, without+*stria*, a furrow.

esurian—L. *esuriens*, genit. *esurientis*, hungering.

-et—Eng. *-et*, suffix of French derived nouns. *Ex:* sign-et; isl-et.

et—Gr. *etos*, a year. *Ex:* Eto-blattina (Ins.).

etaer—Gr. *etairos*, a companion. *Ex:* Bryss-etaeres (Pisc.); Phil-etaerus (Av.).

etaerio—NL. *etaerio*, a kind of aggregate fruit < Gr. *etaeria*, a companionship, brothership.

etel—Gr. *etelis*, a name of some fish. *Ex:* Etelis.

eteo—Gr. *eteos*, true, genuine. *Ex:* Eteo-philus (Ins.).

eter—Gr. *heteros*, the other, one of two. *Ex:* Etero-di-urus (Ins.); Eteros-onycha (Arach.).

-etes—Gr. *-ētēs*, suffix meaning one who. *Ex:* tryp-etes, a borer < *trypaō*, to bore through; zet-etes, a searcher, seeker < *zēteō*, to seek; etc.

etes—Gr. *etēs*, a neighbor. *Ex:* Aeip-etes (Av.).

etesi—Gr. *etēsios*, annual, lasting a year. *Ex:* Etesius (Ins.).

eth—Gr. *ethos*, genit. *etheos*, character, habit. *Ex:* Simo-ethus (Ins.); Etho-iulus (Arth.).

ethali—Gr. *Aithalion*, a mythological name. *Ex:* Ethalion (Moll.).

ethelast—Gr. *ethelasteios*, conceited. *Ex:* Ethelastia (Ins.).

ethelurg—Gr. *ethelourgos*, willing to work, untiring. *Ex:* Ethelurgus (Ins.).

etheo—1. Gr. *ētheō*, to strain. *Ex:* Etheo-stoma (Pisc.): 2. Gr. *ētheos*, a bachelor. *Ex:* etheogenesis.

ethm—Gr. *ēthmos*, a sieve. *Ex:* ethm-oid; Ethmiopsis (Ins.); Ethmo-sphaera (Prot.); Ethmus (Ins.).

ethn—Gr. *ethnos*, a nation, race; *ethnikos*, of a nation. *Ex:* ethnic; ethno-logy.

etiolat—NL. *eliolatus*, made white < Fr. *étioler*, blanched. *Ex:* etiolat-ed.

etm—Gr. *etmagen* < *temnō*, to cut. *Ex:* Etmopterus (Elasm.).

etr—Gr. *ētron*, the abdomen. *Ex:* Etro-pus (Pisc.); Stir-etrus (Ins.), see steir 1.

-etta—NL. *-etta*, diminutive ending. *Ex:* Ardetta (Av.); Syc-etta (Por.).

-etum—See -etus.

-etus—L. *-etus, -a, -um*, suffix added to noun stems to form nouns designating the place of a thing, or, especially with names of plants, to designate where these grow. *Ex:* arboretum; querc-etus.

eu- —Gr. *eu-* (used before roots beginning with a consonant) = *ev-* (used before roots beginning with a vowel), good, well, true, nice. *Ex:* Eucalypt-us*; Eu-cnida*; Eu-glena (Prot.); Euonymus*; Eu-theria (Mam.); Ev-ento-gnathi (Pisc.); Ev-iuli-soma (Myr.); Ev-odia*; Ev-osmia*; Ev-oto-mys (Mam.).

Beak-fruited Eucalyptus, *Eucalyptus rostrata.*

euag—Gr. *euagōn*, splendid. *Ex:* Euagona (Ins.).

euanth—Gr. *euanthos*, blooming, rich in flowers. gay. *Ex:* Euantha (Ins.).

eudio—Gr. *eudios*, calm, gentle. *Ex:* Eudiocrinus (Echin.).

eudocim—Gr. *eudokimos*, famous, in good standing, well known. *Ex:* Eudocimus (Av.).

eudor—Gr. *eudōros*, generous. *Ex:* Eudor-ina (Av.).

eugnom—Gr. *eugnōmōn*, peaceful, kind-hearted, prudent. *Ex:* Eugnomus (Ins.).

eulab—Gr. *eulabēs*, prudent, wary. *Ex:* Eulabe-ornis (Av.); Eulabea (Av.); Eulabis (Ins.).

-eum—NL *-eum* < Gr. *-eion*, suffix denoting place where. *Ex:* mus-eum, haunt of the Muses.

eumar—Gr. *eumaris*, a thick-soled Asiatic shoe or slipper. *Ex:* Plat-eumaris (Ins.).

eunic—Gr. *Eunikē* = *Euneikē*, name of a Nereid. *Ex:* Eunice (Ann.), etc.; Eunice-opsis (Coel.); Eunicea (Coel.); Eunici-cola (Crust.).

euod—Gr. *eudos*, easy to travel over, easy, favorable, good, successful. *Ex:* euodi-formis.

eupator—Gr. *eupatorion*, agrimony < *Mithridates Eupator*, king of Pontus, said by Pliny to have employed this plant in medicine. *Ex:* Eupatori-eae*; Eupatorium*.

euphorb—Gr. *euphorbion*, an African plant known as spurge < *Euphorbos*, physician to king Jubas. *Ex:* Euphorbia*.

euphrasia—Gr. *euphrasia*, good cheer, delight < *euphrainō*, to delight. *Ex:* Euphrasia*.

euphronid—Gr. *Euphronidēs*, Son of Night. *Ex:* Euphronides (Echin.).

euprepi—Gr. *euprepeia*, good appearance, dignity; *euprepēs*, goodly, comely. *Ex:* Euprepiosaurus (Rept.); Euprepro-cnemis (Ins.).

eur—1. Gr. *Euros*, the east (or east-southeast) wind. *Ex:* Euros (Ins.). See eurot: 2. Gr. *euros*, breadth, width; *eurys*, broad. *Ex:* Eur-illas (Av.).

eurem—Gr. *eurēma*, an invention, discovery. *Ex:* Eurema (Ins.).

eures—Gr. *euresis*, a finding, a discovery, invention.

euret—Gr. *eurētos*, easy to tell. *Ex:* Euret-idae (Por.); Eurete (Por.).

eurin—L. *eurinus*, the east wind.

euro—L. *euros*, eastern. See also eurot.

eurost—Gr. *eurōstos*, strong, vigorous. *Ex:* Eurosto-pus (Av.); Eurostus (Ins.).

eurot—Gr. *eurōs*, genit. *eurōtos*, mold, decay. *Ex:* Euro-phila (Ins.); Eurotia*; euroto-phila (Ecol.).

eury—Gr. *eurys*, wide, broad > *eurynō*, to make wide or broad. *Ex:* Eury-pelma (Arach.); Eury-stomata (Rept.); Eury-taenia*; Eurya*; Euryo-notus (Av.); Euryno-rhynchus (Av.).

euryal—1. L. *Euryalē*, name of one of the Gorgons. *Ex:* Euryal-idae (Echin.); Euryale*, (Echin.): 2. Gr. *euryalōs*, broad. *Ex:* Euryalus (Mam.).

euryn—See eury.

-eus—1. L. *-eus, -ea, -eum*, adjectival terminations added to noun roots to denote the material of which a thing is made, sometimes denoting similarity, as; *aureus*, golden < *aura*, gold; *ligneus*, made of wood, wooden < *lignum*, wood. See -inus: 2. L. *-eus*, termination meaning of or pertaining to, when added to form adjectives from names of Greek men. *Ex:* Achilleus < *Achilles:* 3. Gr. *-eus*, suffix used to designate that a person belongs to some country or town, as; *Eretri-eus*, Eretrian, *Kolon-eus*, of Colonos.

eustath—Gr. *eustathēs*, steadfast, sound, healthy.

eutact—Gr. *eutaktos*, orderly. *Ex:* Eutactus (Ins.).

eutel—Gr. *eutelēs*, cheap, worthless. *Ex:* Eutel-ornis (Av.).

euterpe—Gr. *Euterpē*, one of the Muses < *eu*, well + *terpō*, to delight, please. *Ex:* Euterpe*.

euthem—Gr. *euthēmōn*, neat, pretty. *Ex:* Euthemis*.

euthet—Gr. *euthetos*, well arranged, orderly. *Ex:* Euthetus (Ins.).

euthi—See euthy.

euthy—Gr. *euthys*, straight, direct. *Ex:* Euthi-conus (Ins.); Euthy-neura (Moll.); Euthyo-pter-oma (Ins.).

euthym—Gr. *euthymos*, generous. *Ex:* Euthymus (Ins.).

eutoni—Gr. *eutonia*, vigorous, in a state of vigor. *Ex:* Eutonia (Ins.).

eutych—Gr. *eutychys*, lucky, prosperous. *Ex:* Eutyche-lithos (Pisc.); Eutych-ides (Arach.).

ev- —See eu-.

evagorat—L. *evagoratus*, roaming, extending itself < *evagor*, to wander. *Ex:* Evagora (Ins.)

evan—Gr. *euanios*, taking trouble lightly, easily. *Ex:* Evania (Ins.); Evanio-morpha (Ins.), i.e., of the form of Evania.

evanescen—L. *evanescens*, genit. *evanescentis*, withering away, loosing strength, shrivelling < *evanesco*, to wither, shrink.

evanid—L. *evanidus*, evanescent, feeble, frail.

evect—L. *evectus*, led forth; *eveho*, to lead away, to lift up. *Ex:* evect-ics.

evern—Gr. *evernēs*, sprouting well. *Ex:* Evern-ia*.

evers—L. *eversio*, genit. *eversionis*, an expulsion, an agitation; *eversus*, thrown out, overturned. *Ex:* eversion.

evex—L. *evexus*, rounded at the apex, top.

evict—L. *evictus*, conquered.

evid—Gr. *eueidēs*, comely, well-formed. *Ex:* Evides (Ins.).

evinct—L. *evinctus*, bound, wreathed, crowned.

evod—NL. *evodia* < Gr. *euōdia*, a sweet smell, a pleasant odor. *Ex:* Evodia,* see eu-.

evolans—L. *evolans*, flying, springing forth with speed; ppr. of *evolo*, to fly out.

evoluti—L. *evolutus*, rolled out, pp. of *evolvo*, to unroll; *evolutio*, genit. *evolutionis*, an unrolling or opening of a book. *Ex:* evolution.

ex—Gr. *exis*, a having, a possessing, a habit, that which one has acquired. *Ex:* cach-ex-ia (Med.).

ex- —L. *ex-*, out, beyond, comparable to Gr. *ek-*, out (sometimes contracted to *e-* before roots beginning with b, d, g, h, l, m, n, p, r; usually changed to *ef-* before roots beginning with f). E-nucleator (Av.); E-bapho-chernes (Arach.); ef-ferent; ex-foliate; ex-trinsic; see exo.

exacut—L. *exacutus*, pointed, pp. of *exacuo*, to make sharp, pointed. *Ex:* exacu-ate.

exaeret—Gr. *exairetos*, picked out, selected, taken out, remarkable. *Ex:* Exaerete (Ins.).

exaltat—L. *exaltatus*, very tall, exalted; pp. of *exalto*, to raise up.

exanthem—Gr. *exanthēma*, genit. *exanthematos*, an eruption, a blossoming out < *ex*, out + *antheō*, to put forth flowers, to burst forth from the surface of a thing. *Ex:* exanthema; exanthemat-ous.

exapt—L. *exaptus*, fastened.

exarat—L. *exaratus*, plowed, furrowed, sculptored, pp. of *exaro*, to plow, sculptor. *Ex:* exarate.

exartem—Gr. *exartēma*, an appendage. *Ex:* Exartem-atopus (Ins.).

exasperatul—NL. *exasperatulus*, somewhat irritating.

excell—L. *excello*, to excel; ppr. *excellens*, genit. *excellentis*, excelling.

excels—L. *excelsus*, tall, elevated.

excert—L. *excertus*, projecting, protruding.

excipul—L. *excipula*, a receptacle. *Ex:* excipuli-form.

excis—L. *excisus*, cut out, cut off, castrated, destroyed.

exclam—L. *exclamo*, to call out, make a noise.

excubit—L. *excubitus*, pp. of *excubo*, to keep watch; *excubitor*, a sentinel.

excuss—L. *excussus*, stretched out, extended; also driven out, shaken off < *excutio*, to cast out.

exes—L. *exesus*, devoured, eaten away < *exedo*, to consume.

exigu—L. *exiguus*, short, small, brief; *exiguum*, a trifle, a little.

exil—L. *exilis*, neut. *exile*, small, thin, slender, feeble, dry. *Ex:* exili-ped; Exilia (Moll.); exilis.

**eximi**—L. *eximius*, select, distinguished, uncommon.

**exitial**—L. *exitialis*, destructive, wasting.

**exitios**—L. *exitiosus*, destructive.

**exo**—Gr. *exō*, outside. *Ex:* Exo-chorda (Moll.); Exo-genae\*; exo-skeleton.

**exoch**—Gr. *exochos*, projecting, lofty. *Ex:* Exoch-ura (Mam.); Exocho-blatta (Ins.); Exochus (Ins.).

**exochr**—Gr. *exōchros*, deadly pale.

**exocoet**—Gr. *exōkoitos*, sleeping outside. *Ex:* Exocoet-idae (Pisc.); Exocoetus (Pisc.); Par-exocoetus (Pisc.).

**exoles**—Gr. *exōlēs*, utterly destroyed. *Ex:* Exoles (Pisc.).

**exolesc**—See **exolet**.

**exolet**—L. *exolescus*, matured, pp. of *exolesco*, to grow up; *exoletus*, matured.

**exomal**—Gr. *exomalizō*, to make quite smooth, to form according to law. *Ex:* Exomal-opsis (Ins.); Exomala (Ins.).

**exorat**—L. *exoratus*, persuaded < *exoro*, to persuade by entreaty.

**exorist**—Gr. *exoristos*, banished. *Ex:* Exorista (Ins.).

**exoth**—1. Gr. *exōtheō*, to expel, drive out. *Ex:* Exothea\*: 2. Gr. *exōthō*, to protrude. *Ex:* Exotho-stemon\*.

**exotic**—Gr. *exōtikos*, foreign, not native.

**expans**—L. *expansus*, expanded.

**explanat**—L. *explanatus*, flattened: also made clear, plain.

**explet**—L. *expletus*, perfect, filled to completion.

**explicans**—L. *explicans*, spread out, made plain.

**explicat**—L. *explicatus*, unfolded < *explico*, to unfold.

**explicit**—L. *explicitus*, easy, disentangled.

**explod**—L. *explodo*, to scare away, to cast out; ppr. *explodens*, exploding

**expuls**—L. *expulsus*, driven out.

**exscap**—NL. *exscapus*, without scape < L. *ex*-without + *scapus*, a stalk.

**exsert**—L. *exsertus*, protruding.

**exsili**—L. *exsilio* = *exilio*, to spring out, *exsilium* = *exilium*, a place of retreat.

**exsput**—L. *exsputus*, spit out, expelled.

**exsul**—See **exul**.

**extim**—L. *extimus*, the most remote.

**extra**—L. *extra*, beyond, on the outside, without. *Ex:* extra-embryonic.

**extrane**—L. *extraneus*, strange, from without.

**extrari**—L. *extrarius*, outward.

**extricat**—L. *extricatus*, cleared up, freed, pp. of *extrico*, to free.

**extrors**—L. *extrorsus*, on the outside.

**extrus**—L. *extrusus*, driven out.

**exuber**—L. *exuberans*, genit. *exuberantis*, growing abundantly, superfluous; also extraordinary < *exubero*, to abound.

**exud**—L. *exudo*, to sweat out < *ex*, out + *sudo*, to sweat, pp. *exudatus*. *Ex:* exudate; exudat-oria.

**exul**—L. *exulo* = *exsulo*, to be an exile, ppr. *exu-latus*, banished, exiled.

**exust**—L. *exustus*, burned up, pp. of *exuro*, to burn up.

**exut**—L. *exutus*, drawn off, stripped. *Ex:* exut-ive.

**exuv**—L. *exuviae*, that which is stripped off < *exuo*, to pull or strip off. *Ex:* exuvi-ation.

**exygr**—Gr. *exygros*, watery, liquid, wet.

**exypneustes**—NL. *exypneustes*, an emendation from *exypnestes*, said to be derived from Gr. *exypnizō*, to arouse from sleep. *Ex:* Exypneustes (Rept.).

**eyra**—NL. *eyra* < some S.A. native name; now applied to a wild cat (*Felis eyra*).

**eystath**—Gr. *eustathēs*, durable, stable. *Ex:* Eystathes\*.

# F

**fab**—L. *faba*, a bean < Gr. *phagō*, to eat; *fabarius*, of or belonging to beans. *Ex:* fab-aceus; fabarius.

**facet**—L. *facetus*, well-made, elegant. *Ex:* Facetus (Ins.). See also **faci**.

**facett**—See **faci**.

faci—L. *facies*, form, figure, appearance; also the face, a face>Fr. *face*, dim. *facette*, a face; LL. *facialis*, pertaining to the face, facial. *Ex:* facet; facetted; facial-ium; facio-lingual.

fact—L. *factus*, pp. of *facio*, to make, to bring about; *factor*, a doer. *Ex:* ol-fact-ory.

facul—L. *facula*, a little torch.

facultat—L. *facultas*, genit. *facultatis*, capability, skill. *Ex:* fac-ultat-ive.

faec—L. *faex*, genit, *faecis*; pl. *faeces*, dregs. *Ex:* faeces=feces; fec-ula, de-faec-ation.

faex—See faec.

fag—L. *fagus*, a beech tree. *Ex:* Fag-aceae*; fagion (Ecol.); Fago-pyrum*; Fagus*.

fagine—L. *fagineus*, of the beech tree.

falc—L. *falx*, genit, *falcis*, a sickle, scythe>*falcatus*, sickle-shaped, armed with sickles > *falco*, a falcon, because of its curved talons>Fr. *falcon*, a falcon. *Ex:* Falc-ulea (Av.); Falcunculus (Av.); Falcata*; falcate; Falcator (Av.); falci-form; Falco (Av.); Falco-baenus (Av.), Falcon-idae (Av.); Falx (Pisc.); falx cerebri.

falcon—See falc.

fall—L. *fallo*, to deceive>*fallax*, genit. *fallacis*, deceptive; *falsum*, fraud. *Ex:* Fallaci-turris (Moll.); Falsi-fusus (Moll.); Falso-exo-soma (Ins.).

fallac—See fall.

fallax—See fall.

fals—See fall.

falx—See falc.

familiar—L. *familiarus*, domestic, home-like.

famulator—L. *famulatorius*, slavish.

faranc—NL. *furancia*, a coined name; or perh. from some personal name. *Ex:* Farancia (Rept.).

farcimen—See farcimin.

farcimin—L. *farcimen*, genit. *farciminis*, a sausage. *Ex:* Farcimin-ella (Bry.).

farct—L. *farctus*, filled, pp. of *farcio*, to stuff, fill full. *Ex:* farct-ate.

farin—L. *farina*, flour, meal>*farinosus*, mealy. *Ex:* farin-aceus; farin-alis; Farino-coccus (Ins.); farinose.

fario—See farion.

farion—L. *fario*, genit. *farionis*, the salmon trout.

-farius—NL. *-farius*<L. *farior*, to speak<Gr. *phanai*, to speak; used in sense of occurring in rows. *Ex:* bi-farius; multi-farius.

fasc—L. *fascia*, dim. *fasciola*, a bundle, band; *fasciatus*, bundled. *Ex:* fascia; fasci-ated; Fasciola (Platy.). See also fascis.

fasciari—NL. *fasciarius*, band-like.

fasciat—See fasc.

fascicul—See fascis.

fascinat—L. *fascinator*, a charmer.

fascis—L. *fascis*, dim. *fasciculus*, a bundle. *Ex:* fascicile; fascicul-aris; Fasciculi-pora (Bry.) See also fasc.

Burro-Weed Strangler, *Orobanche fasciculata*, whose fleshy stems occur in "bundles." Redrawn from Desert Wild Flowers—Jaeger. Stanford University Press.

fast—L. *fastus*, pride; *fastosus*, haughty, scornful, full of pride; collateral form, *fastuosus*.

fastidibil—L. *fastidibilis*, loathsome, disagreeable.

fastig—L. *fastigo*, to exalt, pp. *fastigatus*, exalted (cf. L. *fastigium*, a top, summit, slope). *Ex:* fastigi-atus; Fastigi-ceras (Moll.); Fastigi-ella (Moll.).

fastuos—See fast.

fatig—L. *fatigo*, to vex, to make tired; ppr. *fatigans*, genit. *fatigantis*, vexing.

fatisc—*fastisco*, to gape, ppr. *fatiscens*, genit. *fatiscentis*, opening in chinks, gaping, cracking open.

fatu—L. *fatuus*, foolish, simple.

fauc—L. *fauces*, pl. noun meaning the gullet, the hinder parts of the mouth, the passage from the mouth to the pharynx. *Ex:* fauces; fauc-al.

faun—L. *Faunus*, deity of herds and fields>NL. *fauna*, the animal life of a particular period or region. *Ex:* faun-al; fauna; avi-fauna.

faust—L. *faustus*, fortunate, lucky.

fav—L. *favus*, NL. dim. *faveolus*, a honey-comb; NL. *favosus*, like a honey comb. *Ex:* Favispongia (Por.); faveol-ate; Favos-ites (Coel.); Favosi-pora (Bry.); favus: 2. It. *favo*, bean.

faveol—See fav.

favorin—L. *Favorinus*, Roman philosopher of Hadrian's time. *Ex:* Favorinus (Moll.).

favos—See fav.

febr—L. *febris*, fever; *febrilis*, that which produces fever. *Ex:* febri-fugia; febrile.

fec—See faec.

fecul—L. *faecula=fecula*, the lees of wine; *feculentus*, full of dregs or sediment, impure.

fecund—L. *fecundus*, fruitful, rich, abundant.

**fed**—L. *fedus*, an ancient word synonomous with *haedus*, a kid. *Ex:* Fedia*.

**fel**—L. *felis*=*feles*, genit. *felis*, a cat, the prolific one, she that bears young>*felinus*, belonging to a cat. *Ex:* Fel-idae (Mam.); Fel-ichthys (Pisc.); Feli-opsis (Mam.); Felis (Mam.).

**felic**—L. *felix*, genit. *felicis*, fruitful, productive. *Ex:* Felix (Mam.); Felix-astraea (Coel.); Felixi-gyra (Coel.). See also fel.

**feline**—L. *felineus*, of or belonging to a cat. See fel.

**felis**—See **fel.**

**felix**—See **felic.**

**felsin**—L. *Felsina*, name of a town in Tuscany. *Ex:* Felsino-therium (Mam.).

**femin**—L. *femininus*, female<*femina*, woman; *femineus*, feminine. *Ex:* feminine.

**femor**—See **femur.**

**femur**—L. *femur*, the thigh>NL. *femoralis*, pertaining to the thigh. *Ex:* Femori-ferus (Arth.); Femoria (Brach.); femoro-tibial.

**fen**—L. *fenum*=*faenum*, hay, provender. See also **foen.** *Ex:* feno-phil-ous.

**fenestr**—L. *fenestra*, a window. *Ex:* fenestr-alis; Fenestr-ella (Bry.); fenestra ovalis; Fenestra-pora (Bry.); Fenestro-saurus (Rept.).

**fenisec**—L. *fenisex*, genit. *fenisecis*, a mower, a rustic.

**fenisex**—See **fenisec.**

**fer**—1. L. *fero*, to bear, ppr. *ferens*, genit. *ferentis*, bearing; pp. *fertus*, borne>*fertilis*, productive, fertile. *Ex:* fertile; fertilis-ation=fertiliz-ation; fertiliz-in; con-fertum; Pori-fera; vas de-ferens: 2. L. *fera*=*ferus*, a wild animal; *ferus*, wild. *Ex:* fer-al. See rangi.

**ferac**—L. *ferax*, genit. *feracis*, fruitful, rich, fertile.

**ferax**—See **ferac.**

**feroc**—L. *ferox*, genit. *ferocis*, dim. *feroculus*, fierce, wild, bold. *Ex:* Feroculus (Mam.).

**ferox**—See **feroc.**

**ferre**—L. *ferreus*, made of iron, iron-color; also hard, unfeeling.

**ferrugat**—NL. *ferrugatus*, dusky, iron-red.

**ferrugin**—L. *ferrugo*, genit. *ferruginis*, iron rust> *ferruginus*, rusty.

**fert**—See **fer** 1.

**fertil**—See **fer** 1.

**ferula**—L. *ferula*, fennel<*ferula*, a walking stick. *Ex:* Ferula*.

**ferv**—L. *fervens*, genit. *ferventis*, glowing, in earnest, zealous, savage; ppr. of *fervo*, to glow, to boil: *fervidus*, fiery, eager.

**fess**—L. *fessus*, feeble, worn out, decayed.

**festin**—L. *festinus*, quick, hasty<*festino*, to hasten.

**festiv**—L. *festivus*, gay, variegated with bright colors<*festus*, a feast. *Ex:* Festiva (Moll.); Festivus (Ins.).

**festuc**—L. *festuca*, dim. *festucula*, a stem, straw; also a straw-like weed growing among barley. *Ex:* Festuc-aria (Platy.); Festuca*; Festucula (Arach.).

**fet**—L. *fetus*=*foetus*, an offspring, progeny, fruit; *fetus*, pregnant, breeding. *Ex:* fet-al; fet-ation; fetus.

**fetid**—L. *fetidus*=*foetidus*, ill-smelling, fetid; *foetens*, stinking. *Ex:* Foet-orius (Mam.), compr. with Putorius (Mam.); foetid-issimus.

**-fex**—NL. -*fex*, a maker<L. *facio*, to make, do, perform. *Ex:* Tubi-fex (Ann.).

**fiber**—L. *fiber*, the beaver; *fibrinus*, of the beaver. *Ex:* Fiber (Mam.). See also fibr.

**fibr**—L. *fibra*, dim. *fibrilla*, a fiber, filament (either of plant or animal nature). *Ex:* fibr-in; fibr-ino-gen; fibra; fibrilla. See also fiber.

**fibrin**—See **fiber,** also **fibr.**

**fibul**—L. *fibula*, a buckle, clasp, splint. *Ex:* Fibul-aria (Por.); Fibul-aster (Echin.); fibula; Fibulo-ptyxis (Moll.).

**fic**—L. *ficus*, a fig tree, a fig. *Ex:* Fic-aria*; fici-folia; Ficio-myia (Ins.); Fico-phagus (Av.); Ficus*

**-fic**, also -**fice**—L. *fic*, weak root of *facio*, to make, do. *Ex:* chyli-fic; melli-fica; oro-fice.

**-fice**—See -**fic.**

**ficul**—NL. *ficula*, a small fig<L. *ficus*, a fig. *Ex:* Ficula (Moll.); Ficul-ina (Moll.).

**ficuln**—L. *ficulnea*, the fig tree>*ficulnus* fig-like; *ficulneus*, of the fig tree.

**fid**—See **find.**

**fidel**—L. *fidelis*, faithful, true. *Ex:* Fidelis (Moll.).

**fidia**—NL. *fidia*, a non-sense name applied to a genus of beetles. *Ex:* Fidia (Ins.).

**fidicin**—L. *fidicen*, genit. *fidicinis*, a player on the lute or lyre. *Ex:* fidicin-alis; Fidicin-ina (Ins.).

**fidicul**—L. *fidicula*, a small lyre-like instrument.

**fidon**—NL. *fidonia*, a name applied to a genus of geometrid moths<Gr. *pheidos*, sparing, thrifty. *Ex:* Fidonia (Ins.).

**fidus**—L. *fidus*, trustworthy, sure. See also find.

**fierasfer**—NL. *fierasfer*<Fr. (Provencal) *fieras*, a kind of fish +*fer*, fierce, wild. *Ex:* Fierasfer (Pisc.).

**figul**—L. *figulus*, a potter.

**figur**—L. *figuratus*, figured, formed, shaped.

**fil**—L. *filum*, a thread>NL. *filaris*, threadlike; L. *filamentum*, dim. *filamentulum*, a fine untwisted thread. *Ex:* fil-aceus; Filaria (Nemat.); filament-osus; fili-form; filo-plumes; filum terminale.

filag—NL. *filago*, name applied to a genus of plants<L. *filum*, a thread. *Ex:* Filago*.

filament—See fil.

filar—See fil.

filia—L. *filia*, a daughter>NL. *filialis*, pertaining to a son or daughter. *Ex:* filial.

filic—L. *filix*, genit. *filicis*, a fern. *Ex:* filic-ial; filic-inus; filici-form; Filix*.

filix—See filic.

fim—L. *fimus*, dung>*fimetum*, a dunghill. *Ex·* fimet-arius; Fimo-scolex (Ann.).

fimbr—L. *fimbria*>NL. dim. *fimbrilla*, a fringe, border of fibers or threads; *fimbriatus*, fibrous, fringed, bordered with hairs. *Ex:* Fimbri-branchia (Amph.); Fimbri-branchi-ata (Crust.); fimbriate; Fimbrio-torpedo (Pisc.).

fimet—See fim.

fin—1. L. *finis*, boundary, limit. *Ex:* fini-al: 2. AS. *finn*, a fin. *Ex:* fini-form.

find—L. *findo*, (root, *fidi*), to split, pp. *fissus*, cloven, split>*fissilis*, tending to split; *fissio*, genit. *fissionis*, a cleaving; *fissura*; a split, chink, fissure. *Ex:* Fissi-pedia (Mam.); fission; Fissur-ella (Moll.); fissure; Fissuri-rostra (Brach.); bi-fid; pinnati-fid.

firm—L. *firmus*, steadfast, strong. *Ex:* Firmi-termes (Ins.).

firmat—L. *firmatus*, firm, made fast.

firol—NL. *firola*, name applied to a genus of Protozoa (etym. doubtful, but perhaps, as suggested by Agassiz, <Fr. *fiole*, a little bottle, phial. *Ex:* Firol-idae (Prot.); Firola.

fiss—See find.

fissil—L. *fissilis*, tending to split, cleft.

fissur—See find.

fistul—L. *fistula*, a hollow reed-stalk, a pipe, tube >*fistularis*, like a pipe; *fistulatus*, pipe-shaped, furnished with pipes; *fistulosus*, full of pipes. *Ex:* Fistul-aria (Pisc.); fistul-ous; Fistuli-cola (Platy.).

fivor—L. *fivor*, genit. *fivoris*, lead-colored, bluish.

fix—L. *fixus*, attached, fastened, pp. of *figo*, to fasten.

flabell—L. *flabellum*, a small fan. *Ex:* Flabell-ina (Moll.); Flabello-thyris (Brach.); Flabellum (Coel.).

flacc—L. *flaccus*, flabby, hanging down; *flaccidus*, relaxed, flabby, flaccid.

flagell—L. *flagrum*, dim. *flagellum*, a whip; *flagello*, to whip, pp. *flagellatus*, whipped. *Ex:* flagellate; flagelli-form; Dino-flagell-ata (Prot.).

flamm—L. *flamma*, dim. *flammula*, a flame; *flammeus*, flame colored. *Ex:* Flammeo (Pisc.); Flammi-gera (Ins.); flammul-atus; Flammul-ina (Moll.).

flat—1. L. *flatus*, a blowing; *flatulentus*, full of wind<*flo*, to blow. *Ex:* Flat-idae (Ins.); Flata (Ins.); flatulent: 2. NL. *flatus*, flat, akin to Eng. *flat*=Sw. *flat*, flat. *Ex:* Flati-palpus (Ins.); Flato-ptera (Ins.).

flatulent—See flat.

Alabaster Fan-coral, *Flabellum alabastrum*. Redrawn from Voyage of The Challenger—Thomson. Courtesy of Joseph McDonough Co.

flav—L. *flavus*, yellow; *flavicans*, tending to yellow; *flavidus*, of golden yellow, somewhat yellow; *flavesco*, to become yellow, ppr. *flavescens*, genit. *flavescentis*, becoming yellow. *Ex:* Flav-ella (Prot.); Flaveria*.

flavescen—See flav.

flavican—See flav.

flavid—See flav.

flebil—L. *flebilis*, doleful.

flect—L. *flecto*, to bend, pp. *flexus*, bent> *flexibilis*, capable of being bent, pliable; *flexuosus*, winding, bending; NL. *flexor*, a bender. *Ex:* Flecto-notus (Amph.); Flexi-palpus (Arach.); flexible; Flexo-centrus (Ins.); flexor; flexu-ous.

flex—L. *flexus*, bent, turned, curved.

flocc—L. *floccus*, dim. *flocculus*, flock of wool, tuft of wool>*floccosus*, full of flocks of wool. *Ex:* Flocci-fera (Ins.); floccose; floccul-ent; floccule.

flor—L. *flos*, genit. *floris*, dim. *flosculus*, a flower; *floridus*, abounding in flowers; *floresco*, to begin to bloom, ppr. *florescens*, genit. *florescentis*, beginning to bloom; *Flora*, goddess of flowers> *floralis*, of or pertaining to Flora. *Ex:* Flora-conus (Moll.); flori-culture; florid; Florid-ichthys (Pisc.); Floscul-aria (Rot.); Flosculi-pora (Bry.); in-florescence; sparsi-florus.

florens—L. *florens*, genit. *florentis*, glittering

flore pleno—See plen.

florescen—See flor.

florid—See flor.

flos—See flor.

floscul—See flor.

fluct—L. *fluctus*, waving. *Ex:* flucti-fraga.

fluctu—L. *fluctuo*, to waver, to float about, ppr. *fluctuans*, genit. *fluctuantis*, wavering.

**fluit**—L. *fluito*=*fluto*, to float, swim, ppr. *fluitans*, genit. *fluitantis*, floating, swimming, sailing about.

**flumin**—L. *flumen*, genit. *fluminis*, a river. *Ex:* flumen-alis; Flumini-cola (Moll.).

**fluminea**—NL. *fluminea*, a plant name<L. *flumen*, genit. *fluminis*, a river+-*ea*. *Ex:* Fluminea*.

**flustr**—NL. *flustra*, perh.<some Anglo-Saxon **word**, such as *flustrian*, braiding, plaiting. *Ex:* Flustra (Bry.).

**fluv**—L. *fluvius*, a stream, river>*fluviatilis*, of or belonging to a river. *Ex:* Fluvi-cola (Pisc.), etc.; fluviatile; Fluvio-pupa (Moll.).

**flux**—L. *fluxus*, a flowing. *Ex:* ad-flux-ion.

**fod**—L. *fodio*, to dig, dig up, ppr. *fodiens*, genit. *fodientis*, digging, digging up>NL. *fodiator*, a digger. *Ex:* Fodia (Tun.); Fodiator (Pisc.); Ef-fodentia (Mam.).

**foed**—L. *foedus*, ugly, foul.

**foen**—**1.** L. *foenum*=*fenum*, hay, dim. *foeniculum* =*feniculum*. *Ex:* Foeniculum*: **2.** L. *foenus*, that which is produced, such as interest on money. *Ex:* Foeno-morpha (Ins.); Electro-foenus (Ins.); Foenus (Ins.).

**foet**—See **fet.**

**foetid**—See **fetid.**

**foina**—Ital. dial. *foina*, a polecat. *Ex:* Foina (Mam.).

**foli**—L. *folium*, dim. *foliolum*, a leaf>*foliaceus*, leafy; *foliosus*, full of leaves, leafy. *Ex:* foli-fer-ous; foliaceous; Folio-seris (Coel.); foliol-ate; Foliol-ina (Por.); foliose; ex-foli-ation.

Small-leaved Amsonia, *Amsonia brevifolia*. The genus *Amsonia* was named after Charles Amson, physician of Colonial Virginia. Re-drawn from Desert Wild Flowers—Jaeger. Stanford University Press.

**foll**—L. *follis*, a bag or sac>dim. *folliculus*, a small sac, the bladder, the scrotum. *Ex:* Folli-crinus (Echin.); follicle; follicul-ate; Follicul-ina (Moll.).

**fomes**—See **fomit.**

**fomit**—L. *fomes*, genit. *fomitis*, tinder. *Ex:* Fomes*.

**font**—L. *fons*, genit. *fontis*, dim. *fonticulus*, a fountain, spring>*fontanus*, genit. *fontanalis*, pertaining to a fountain or spring. *Ex:* Fontan-iana (Moll.); Fonti-cola (Platy.); Fonto-nema (Nem.).

**fontan**—See **font.**

**fontanell**—Fr. *fontanelle*, a little fountain.

**fonticul**—See **font.**

**for**—L. *foris*, a gate, door; as adj. out-of-doors, abroad. *Ex:* Tri-foris (Moll.).

**foramen**—See **foramin.**

**foramin**—L. *foramen*, genit. *foraminis*, a hole, a perforation. *Ex:* foramen; Foramin-ites (Ann.); Foramini-fera (Prot.).

**forat**—L. *foro*, to perforate, bore through, pp. *foratus*, bored through; *foratus*, a boring. *Ex:* Forat-ella (Bry.); Forati-dolium (Moll.); im-per-forate.

**forcep**—See **forcip.**

**forcip**—L. *forceps*, genit. *forcipis*, forceps, nipper <*formus*, warm, hot+*capio*, to take. *Ex:* Forcep-ina (Por.); forceps; forcip-ate; Forcip-ula (Ins.); Forcip-ul-ata (Echin.); Forcipo-myia (Ins.).

**forfex**—See **forfic.**

**forfic**—L. *forfex*, genit. *forficis*, dim. *forficula*, scissors, shears>NL. *forficatus*, forked. *Ex:* forficate; Forficula (Ins.), etc.; Forficulo-tarpa (Ins.).

**form**—L. *forma*, shape, figure, appearance, nature>dim. *formula*, a small pattern, rule, principle, formula; *formo*, to form, pp. *formatus*, formed, fashioned, molded; *formosus*, finely formed, beautiful>dim. *formosulus*, pretty. *Ex:* format-ive; Formi-coris (Ins.); Formo-cryptus (Ins.); formul-ate.

**format**—See **form.**

**formic**—L. *formica*, an ant>*formicinus*, of or like ants. *Ex:* Formic-oma (Ins.), see homo 2.; Formica (Ins.); Formica-leon (Ins.); Formici-capa (Av.); Formico-thrips (Ins.).

**formos**—*formosa*, name of an island in the Pacific Ocean<L. *formosus*, beautiful. *Ex:* Formos-ania (Pisc.); Formos-aphis (Ins.); Formoso-sena (Ins.). See also form.

**formul**—See **form.**

**fornac**—L. *fornax*, genit. *fornacis*, a furnace. *Ex:* Fornax (Ins.).

**fornax**—See **fornac.**

**fornic**—L. *fornix*, genit. *fornicis*, a vault; also a brothel>*fornicatus*, arched over. *Ex:* Fornicia (Ins.); Fornico-cassis (Ins.); fornix.

**fornix**—See **fornic.**

**fort**—L. *fortis*, strong, brave, powerful.

**fortunat**—L. *fortunatus*, lucky, doing well, prosperous.

**foss**—L. *fossa*, a ditch, dim. *fossula*; *fossilis*, dug up, 'dug out; *fossor*, a digger, grave-digger<

**fodo,** to dig, pp. *fossus,* dug. *Ex:* fossa; Fossa (Mam.); Fosso-pora (Coel.); fossil; Fossulaster (Echin.); Fossor (Mam.); fossori-al; infoss-ate.

**fossat**—L. *fossatus,* dug around, pierced; as a noun, a boundary, i.e., something with a ditch dug about it <*fossa,* a trench, a hole.

**fot**—L. *fotus,* warmed, pp. of *foveo,* to warm.

**fov**—L. *foveo,* to keep warm, to cherish, nourish, support. *Ex:* fov-illa.

**fove**—L. *fovea,* dim. *foveola,* a pit >*foveatus,* dim. *foveolatus,* pitted. *Ex:* fovei-form.

**foveat**—See **fove.**

**foveol**—See **fove.**

**fracid**—L. *fracidus,* mellow, soft.

**fract**—L. *frango,* to break, pp. *fractus,* broken. *Ex:* Fract-armilla (Moll.); Fracto-phloeus (Ins.); in-fracted; re-fractus.

**fraen**—See **fren.**

**frag**—1. L. *frag,* the root of *frango,* to break, *fragilis,* brittle and *fragmentum,* a piece. *Ex:* Frag-arium (Tun.); Fragilo-cyathus (Coel.); Ossi-fraga (Av.); Saxi-fraga*;   2. L. *fragum,* a strawberry plant. *Ex:* Frag-aria*.

**fragil**—See **frag 1.**

**fragos**—L. *fragosus,* breakable, fragile.

**francolin**—Pg. *francolim* = Sp. *francolin,* a partridge; *francolin,* dim. of Port. *frango,* a hen. *Ex:* Francolinus (Av.).

**frater**—L. *frater,* a brother <Gr. *phratēr.*

**fraud**—L. *fraudo,* to cheat; ppr. *fraudans,* genit. *fraudantis,* cheating, defrauding; *fraus,* genit. *fraudis,* a cheat.

**fraus**—See **fraud.**

**fraxin**—L. *fraxinus,* the ash tree <Gr. *phrassō,* to hedge or inclose, since the ash was formerly used for hedges. *Ex:* Fraxinus*.

**fregat**—It. *fregata,* a frigate. *Ex:* Fregatt-ornis (Av.); Fregata = Fregatta (Av.).

**fremit**—L. *fremitus,* a dull, roaring or murmuring sound <*fremo,* to hum, growl, murmur; pp. *fremitus.*

**fren**—L. *frenum* = *fraenum,* NL. dim. *frenulum* = *fraenulum,* a bridle, curb. *Ex:* fren-ate; Bi-fren-aria*; bi-fren-atus.

**fres**—L. *fresus,* crushed, bruised, pp. of *frendo,* to grind to bits.

**fret**—L. *fretus,* a channel, a straight; *fretensis,* inhabiting straights.

**frig**—L. *frigor,* genit. *frigoris,* coldness; *figidus,* cold. *Ex:* Frigidi-lacuna (Moll.); frigori-deserta (Ecol.).

**frigid**—See **frig.**

**frigor**—See **frig.**

**fringill**—L. *fringilla,* name of a kind of small bird. *Ex:* Fringilla (Av.); Fringilli-parus (Av.).

**fritill**—L. *fritillus,* a dice-box. *Ex:* Fritill-aria*, (Tun.); Fritillum (Tun.).

**frond**—L. *frons,* genit. *frondis,* a leaf, frond > *frondeus,* leafy, covered with leaves; *frondosus,* full of leaves, leafy; *frondator,* one who prunes trees; NL. dim. *frondiculus,* a small frond, small leaf. *Ex:* Frondi-pora (Bry.); Frondiculina (Prot.); Frondo-vagin-ul-ina (Prot.); albi-frons.

**frondator**—See **frond.**

**frons**—See **frond** and **front.**

**front**—L. *frons,* genit. *frontis,* forehead, brow; *frontosus,* many-browed, shameless. *Ex:* frons; front-adi-form; front-al; fronto-clypeal; Secti-frontes (Ins.).

**frontat**—NL. *frontatus,* with forehead <*frons,* genit. *frontis,* the forehead.

**fruct**—L. *fructus,* a fruit. *Ex:* Fructi-cantor (Av.); fructi-form.

**frug**—L. *frux,* genit. *frugis,* fruits of the earth; *frugilegus,* gathering fruit. *Ex:* frugi-fer-ous; frugi-vor-ous; Frugilegus (Av.).

**frument**—L. *frumentum,* corn, grain >*frumentarius,* of or belonging to grain; *frumentor,* to forage. *Ex:* frument-aceus.

**frust**—L. *frustum,* LL. dim. *frustulum,* piece, part. *Ex:* frustul-ent; frustul-ose.

**frustr**—1. L. *frustra,* uselessly, to no purpose > NL. *frustraneous,* useless. *Ex:* Frustranea*: 2. L. *frustrulum,* a small piece. *Ex:* frustul-ose. 3. L. *frustrator,* a deceiver: *frustratus,* deceived.

**frux**—See **frug.**

**frutesc**—See **frutic.**

**frutex**—See **frutic.**

**frutic**—L. *frutex,* genit. *fruticis,* a bush, shrub > *fruticesco,* to become bushy, ppr. *fruticescens,* genit. *fruticescentis* = *frutescens,* genit. *frutescentis,* becoming bushy; *fruticosus,* shrubby. *Ex:* Fruti-cola (Av.); frutic-ul-ose; Frutico-trochus (Moll.); fruticose.

**fuc**—L. *fucus,* a kind of rock-lichen from which was extracted a red dye <Gr. *phykos,* seaweed, sea-wrack. *Ex:* fuci-vorous; Fuco-myia (Ins.); Fucu-therium (Mam.); Fucus*.

**fucat**—L. *fuco,* to paint, falsify, pp. *fucatus,* painted <*fucus,* a kind of lichen, see fuc.

**fug**—L. *fugio,* to flee >*fugax,* genit. *fugacis,* swift, fleeting. *Ex:* fugaci-ous; Cimici-fuga*; nidi-fug-ous.

**fugac**—See **fug.**

**fugax**—See **fug.**

**fugit**—L. *fugitus,* escaped, departed <*fugio,* to flee.

**fulcr**—L. *fulcrum,* a prop <*fulcio,* to support. *Ex:* fulcr-atus; Fulcr-ella (Moll.); fulcrum.

**fulg**—L. *fulgeo,* to shine; ppr. *fulgens,* genit. *fulgentis,* shining, glowing >*fulgor,* a flash of lightning, glitter; *fulgur,* flashing lightning. *Ex:* Fulgia (Prot.); Fulgor-idium (Ins.); Fulgoro-thrips (Ins.); Fulgur-odes (Ins.); Fulguro-fusus (Moll.); ex-fulgent.

**fulgen**—See **fulg.**

**fulgor**—See **fulg.**

**fulgur**—See **fulg.**

**fulic**—L. *fulica*, dim. *fulicula*, a coot; also *fulix*, genit. *fulicis*, a coot. *Ex:* fulic-arius; Fulica (Av.); Fulix (Av.).

**fulig**—L. *fuligo*, genit. *fuliginis*, soot. *Ex:* Fuligo* (Prot.); fuligin-ous.

**fuligul**—NL. *fuligula*, name applied to a genus of sea-ducks <*fulica*, a coot. *Ex:* Fuligula (Av.).

**fulix**—See **fulic.**

**fullo**—L. *fullo*, a beetle with white spots; also a cloth dresser.

**fulmar**—ME. *fulmar*, the pole-cat (perhaps< Icel. *full*, foul +*mar*, the sea-mew). *Ex:* Fulmarus (Av.).

**fultur**—L. *fultura*, a prop, support, suspensor.

**fulv**—L. *fulvus*, reddish yellow, tawny, gold-colored. *Ex:* fulv-escent; Fulvia (Moll.); Fulvius (Ins.).

**fulvast**—L. *fulvaster, -tra, -trum*, yellowish.

**fum**—L. *fumus*, smoke >*fumidus*, smoky. *Ex:* Fum-aria*; fumi-frons.

**fumator**—L. *fumator*, one who deceives, one who throws up a smoke screen; also a ripener of wine <*fumo*, to smoke.

**fumifer**—L. *fumifer*, smoking, steaming.

**fun**—L. *funis*, dim. *funiculus*, a cord, rope >NL, *funilis*, rope-like. *Ex:* Fun-ambul-us (Mam.); funi-form; Funi-sciurus (Mam.); Funicul-ina (Coel.); funiculus.

**funal**—L. *funalis*, made of rope.

**funambul**—L. *funambulus*, a rope-dancer.

**fund**—L. *fundus*, bottom, foundation; dim. *fundulus*, a kind of sausage; in mechanics, a piston; *fundamentum*, foundation. *Ex:* Fund-aspis (Ins.); fundament; Fundul-ichthys (Pisc.); Fundulo-bombus (Ins.); fundus.

**fundat**—L. *fundatus*, firm, established <*fundo*, to fasten.

**fundul**—See **fund.**

**funebr**—See **funer.**

**funer**—L. *funus*, genit. *funeris*, a funeral, burial >*funebris*, of or belonging to a funeral; *funereus*, of or belonging to a burial; *funestus*, deadly, destructive; LL. *funerarius*, pertaining

to a funeral; ML. *funeralis*, belonging to a burial. *Ex:* Funebri-bombus (Ins.); funere-al.

**funest**—L. *funestus*, deadly.

**fung**—L. *fungus*, dim. *fungulus*, a mushroom, fungus. *Ex:* fungi-form; Fungia (Coel.); fungi-vor-ous; Fungulus (Tun.); Fungus*.

**funicul**—See **fun.**

**fur**—L. *furo*, to rage, be furious; ppr. *furens*, genit. *furentis*, raging. *Ex:* Fur-unculus (Mam.); Furi-ella (Mam.); Furi-pterus (Mam.); Furia (Nemat.).

**furac**—L. *furax*, genit. *furacis*, inclined to pilfer, steal.

**furax**—See **furac.**

**furc**—L. *furca*, dim. *furcilla*=*furcula*, a fork> *furcatus*, dim. *furcillatus*, forked. *Ex:* Furci-pus (Ins.); Furcilli-ger (Ins.); Furco-myia (Ins.).

**furcans**—NL. *furcans*, forking <a supposed L. *furco*, to fork.

**furfur**—L. *furfur*, bran. *Ex:* furfur-aceus.

**furios**—L. *furiosus*, full of fury, raging.

**furn**—L. *furnus*, an oven; *furnorum*, of ovens.

**furnari**—L. *furnarius*, a baker. *Ex:* Furnarius (Av.).

**furtiv**—L. *furtivus*, stolen, concealed.

**furv**—L. *furvus*, swarthy.

**furvescen**—L. *furvescens*, genit. *furvescentis*, growing dark in color.

**fus**—1. L. *fusus*, dim. *fusulus*, a spindle. *Ex:* Fusi-fer (Por.); fusi-form; Fusi-mitra (Moll.); Fuso-spirula (Moll.); Fusul-ina (Prot.); Fusus (Moll.): 2. L. *fusus*, spread out, broad, copious <*fundo*, to spread. *Ex:* humi-fusus.

**fusan**—Fr. *fusain*, the spindle tree<L. *fusus*, a spindle. *Ex:* Fusanus*.

**fusc**—L. *fuscus*, brown, dark, dusky >*fuscatus*, somewhat dusky. *Ex:* Fusco-mitra (Moll.); Fusco-naia (Moll.).

**fuscul**—NL. *fusculus*, somewhat dark <*fuscus*, dark.

**fust**—L. *fustis*, dim. *fusticulus*, a knobbed stick, club. *Ex:* Fusti-ger (Ins.); Fusticul-aria (Coel.); Fustis (Pisc.).

**-fy**—Eng. *-fy*, suffix, meaning to make. *Ex:* saponi-fy, scari-fy.

# G

gad—Gr. *gados*, name of a kind of fish. *Ex:* Gadidae (Pisc.); Gad-ulus (Pisc.); Gadus (Pisc.).

gae—See ge.

gaes—Gr. *gaison = gaisos*, a spear, javelin. *Ex:* Gaesa (Ins.); Gaeso-myrmex (Ins.).

gagat—Gr. *gagatēs*, jet, also black, velvet black. *Ex:* Gagat-ellus (Ins.); Gagati-ceras (Moll.).

gaia—See ge.

gal—Gr. *gala*, genit. *galaktos*, milk > *galaxaios*, milky. *Ex:* Gal-anthus*; galact-urus; Galactodendrum*; Galax*; Gale-go*, see ag < *agō*; galo-chrous; Poly-gala*; Ornitho-galum*. See also gale.

galact—See gal.

galangal—Chinese *galangal*, a plant name.

galanth—L. *Galanthis*, a female attendant of Alcemone, changed into a weasel by Lucina. *Ex:* Galanth-ula (Coel.); Galanthia (Ins.); Galanthis (Crust.); not Galanthus*, see gal.

galapag—Sp. *Galapagos*, name of a group of islands in the Pacific Ocean, characterized by the presence of a large number of tortoises < *galapago*, a tortoise. *Ex:* ?Galapag-urus (perhaps Gala-pagurus) (Crust.); Galapago-myia (Ins.); Galapagos-ia (Ins.).

galat—Gr. *Galatea*, name of a sea-nymph. *Ex:* Galatea = Galathea (Moll.); Galath-odes (Crust.); Galathe-ascus (Crust.); Galathea (Crust.).

galath—See galat.

galax—Gr. *galaxaios*, milky. *Ex:* Galax*; Galaxias (Pisc.).

galb—1. L. *galba*, a small worm, the ash-borer; also a personal name. *Ex:* Galba (Ins.): 2. L. *galbus*, yellow > *galbinus*, greenish yellow.

galban—L. *galbanum*, the greenish yellow resin of a certain plant of Syria. *Ex:* galban-us.

galbe—L. *galbeus = galbeum*, an arm band a ribbon used as an ornament.

galbin—See galb.

galbul—1. L. *galbula*, dim. of *galbina*, name of a kind of small bird, perhaps the female yellow oriole < *galbus*, yellow. *Ex:* Galbul-idae (Av.); Galbula (Av.): 2. L. *galbulus*, the nut of the cypress tree. *Ex:* galbulus.

gale—1. L. *galea*, a helmet > *galeola*, a helmet-shaped vessel; *galeo*, to cover with a helmet; pp. *galeatus*, covered with a helmet. *Ex:* Galeruca (Ins.); Gale-andra*; Gale-orchis*; galea; galeate; galei-form; galeo-theca; Galeol-ella (Moll.): 2. Gr. *galeē = galē*, an animal of the weasel kind, a "cat." *Ex:* Gal-ictis (Mam.); Gale-opsis*; Galeo-bdolon*; Spilo-galia (Mam.); Spilo-gale (Mam.): 3. Gr. *galeos*, a shark; *galeōdēs*, like a shark. *Ex:* Gale-saurus (Rept.); Galeo-cerdo (Elasm.); Galeodes (Arach.); *Ex:* Galeus (Elasm.): 5. Dutch *Gagel* > NL.

gale, a name for sweet willow or *gaule*, or, as the Scotch say, *gall-bush*.

galeat—See gale.

galega—Sp. *galega*, a plant name < Gr. *gala*, milk. *Ex:* Galega*.

galen—Gr. *galēnos*, calm. *Ex:* A-galena (Arach.).

galeol—See gale 1.

galer—L. *galerum = galerus*, dim. *galericulum*, a cap or helmet-like covering for the head < *galea*, a helmet. *Ex:* Galer-aster (Echin.); Galer-ita (Moll.), (Ins.); Galleri-pitta (Av.); Galeria = Galleria (Ins.); Galero-clypeus (Echin.); Galerus (Moll.); galericul-ate.

galeruc—NL. *galeruca*, name applied to a genus of beetles, perhaps < L. *galerum*, a helmet + *eruca*, a caterpillar. *Ex:* Galeruc-idae (Ins.); Galeruca (Ins.).

galgul—L. *galgulus*, name of a kind of bird, the witwall or great spotted woodpecker. *Ex:* Galgul-idae (Ins.); Galgulus (Ins.), (Av.).

gali—Gr. *galion*, a plant called bed-straw. *Ex:* Galium*.

galid—Gr. *galideus*, a young marten. *Ex:* Galidia (Mam.); Galid-ictis (Mam.).

gall—1. L. *galla*, the oak-apple, gall-nut. *Ex:* gall-ic; Galli-cola (Ins.): 2. L. *gallus*, a cock; *gallina*, dim. *gallinula*, a hen > *gallinaceus*, pertaining to poultry; fowl-like. *Ex:* Galliformes (Av.); Gallinul-ops (Av.); Gallo-perdix (Av.); Gallus (Av.).

galler—See galer.

gallin—See gall 2.

gallit—Sp. *El Gallito*, a town of Porto Rica. *Ex:* Gallito-bius (Myr.).

galpinsia—Anagram of *Salpingia*. *Ex:* Galpinsia*.

galumm—L. *galumma*, genit. *galummatis*, a cover. *Ex:* galumm-ate.

gam—Gr. *gamos*, a marriage; *gametē*, a wife; *gametēs*, a husband. *Ex:* gamete; gametophyte; gamo-genesis; A-gama (Moll.); a-gamous.

gamas—NL. *gamasus*, name applied to a genus of mites (etym. uncertain). *Ex:* Gamas-oidea (Arach.); Gamaso-morpha (Arach.); Gamasus (Arach.).

gamb—Indian *gamba*, "covered breasts." *Ex:* Gamba-therium (Mam.). See also gambos.

gambos—L. *gambosus*, with a swelling near the hoof. *Ex:* Gambo (Av.).

gambr—Gr. *gambros*, adjacent, related by marriage. *Ex:* Gambro-stola (Ins.); Gambrus (Ins.); Homo-gambrus (Ins.).

gambusi—NL. *gambusia* < Cuban *gambusino*, a word signifying absence of, nothing, a mere farce. *Ex:* Gambusia (Pisc.).

**gamet**—See **gam.**

**gammar**—L. *gammarus* = *cammarus*, a kind of lobster < Gr. *kammaros*, a kind of lobster. *Ex:* Gammar-idae (Crust.); Gammaro-tettix (Ins.); Gammarus (Crust.).

**gamph**—See **gomph**, of which it is probably a corruption.

**gamps**—Gr. *gampsos*, bowed, curved, crooked. *Ex:* Gamps-acanthus (Pisc.); Gampsorhynchus (Av.).

**-gamy**—Gr. *-gamia* (< *gamos*, marriage), suffix denoting reproduction, marriage. *Ex:* polygamy.

**gan**—Gr. *ganos*, genit. *ganeos*, beauty, lustre. *Ex:* gan-oid; Ganeo (Platy.); Gano-cephala (Amph.); Ecto-ganus (Mam.); Ori-ganum*; Pro-gano-saurus (Rept.).

**gangli**—Gr. *ganglion*, a swelling, a tumor under the skin. *Ex:* ganglio-cyte; Ganglio-pus (Crust.); ganglion.

**gangren**—Gr. *gangraina*, an eating ulcer. *Ex:* gangren-ous; gangrene.

**ganymed**—Gr. *Ganymēdēs*, the cup-bearer of Zeus or of the Olympian gods. *Ex:* Ganymeda (Echin.); Ganymede-bdella (Ann.).

**gapo**—Tupi Indian *ygapó*, a forest island covered with water. *Ex:* gapo (Ecol.).

**gargal**—Gr. *gargalos*, a tickling. *Ex:* gargalesthesia.

**gargar**—Gr. *gargara*, lots of, plenty. *Ex:* gar gara-phylla; Gargara (Ins.).

**garrul**—L. *garrulus*, chattering. *Ex:* Garrulus (Av.).

**garrup**—Pg. *garoupa*, name of a kind of rockfish. *Ex:* Garrupa (Pisc.).

**garumn**—L. *Garumna*, name of a river in southwestern France. *Ex:* Garumn-aster (Echin.).

**garzett**—It. *garza*, dim. *garzetta*, a heron < Sp. *garza*, a heron. *Ex:* Garzetta (Av.).

**gas**—Eng. *gas*, < Gr. *chaos*, the boundless immeasurable space. *Ex:* gaso-plankton; gaso-vacuoles.

**gaster**—Gr. *gastēr*, genit. *gasteros*, by syncope, *gastros*; NL. dim. *gastrula*, the belly, stomach. *Ex:* Gaster-osteus (Pisc.); Gasteria*; Gasterosyphon (Moll.); gastraea; Gastr-idium*; Gastrimargos (Mam.); Gastro-poda (Moll.); gastrulation; gastrula; soleno-gaster.

**gastr**—See **gaster.**

**gaul**—Gr. *gaulos*, a pail, a round-bottomed vessel. *Ex:* Myla-gaulus (Mam.).

**gaur**—**1.** Gr. *gauros*, most elegant, haughty, majestic. *Ex:* Gaur-ella*; Gaur-ambe (Ins.); Gaura*; Gauro-myrmex (Ins.): **2.** Hindu *gaur* = *gour*, a bovine animal of Central India. *Ex:* Bos-gaurus (Mam.).

**gaus**—Gr. *gausos*, crooked, bent outwards. *Ex:* Gauso-centrus (Ins.).

**gavia**—L. *gavia*, name of a kind of bird, perhaps the sea-mew. *Ex:* Gavia (Av.); Gavii-formes (Av.).

**gavial**—NL. *gavialis*, name applied to a genus of crocodiles < Hind. *ghariyāl*, name of the Gangetic crocodile. *Ex:* Gavialis (Rept.); Gavialosuchus (Rept.).

**gavis**—L. *gavisus*, delighting < *gaudeo*, to rejoice.

**gaz**—Gr. *gaza*, treasure. *Ex:* Gaza (Moll.).

**ge**—Gr. *gē* = *ga* = *gaia*, the earth, land > *geios*, of the earth. *Ex:* Gaea (Ins.); Gaia-dendron*; Ge-atractus (Rept.); Ge-oica (Ins.); Geosaurus (Rept.); geo-tropic; bio-geo-graphy; epi-gae-ic; epi-gei-ous; Noto-gaea (Zoo-geo.).

**gecc**—NL. *gecco* = *gecko* = *gekko*, name applied to a genus of lizards, certain species of which make a croaking or chirping noise, whence the name "gecko" < Malay *gēkok*. *Ex:* Gecco = Gecko = Gekko (Rept.); Gecco(n)-idae (Rept.); Gekk-ota (Rept.).

**geck**—See **gecc.**

**geios**—See **ge.**

**geison**—See **geiss.**

**geiss**—Gr. *geison* = *geisson*, a hem, the eaves of a house, border. *Ex:* Geisso-rhiza*; Geissospermum*; Geissois*; Geissono-ceras (Av.).

**geit**—Gr. *geitōn*, genit. *geitonos*, a neighbor. *Ex:* geitono-gamy; Gito-gnathus (Ins.); Gitonischius (Ins.); Potamo-geton*.

**gekk**—See **gecc.**

**gel**—**1.** L. *gelo*, to freeze, congeal, pp. *gelatus*, frozen > *gelidus*, icy, frosty, stiff. *Ex:* gel; gelineae; gelat-in; Gel-idium*: **2.** Gr. *gelōs*, genit. *gelōtos*, laughter; *gelastos* and *gelasimos*, laughable; *gelastēs* = *gelasinos*, a laugher. *Ex:* Gelasimus (Crust.); Gelasinus (Mam.); Ge lastes (Av.); Gelasto-coris (Ins.); Gelo-chelidon (Av.); Geloto-labis (Ins.): **3.** Gr. *gela*, the light of the sun < *geleō*, to shine. *Ex:* Gela*.

**gelasim**—See **gel 2.**

**gelasin**—See **gel 2.**

**gelast**—See **gel 2.**

**gelat**—See **gel 1.**

**gelechi**—Gr. *gēlechēs*, sleeping on the ground < *gē*, earth + *lechos*, a bed. *Ex:* Gelechi-idae (Ins.); Gelech-ia (Ins.).

**gelid**—See **gel 1.**

**gelli**—L. *Gellius*, Roman family name. *Ex:* Gellius (Por.).

**gelot**—See **gel 2.**

**gelsem**—It. *gelsomino*, jessamine. *Ex:* Gelsemium*.

**gem**—Gr. *gemō*, to be full; *gemos*, a load. *Ex:* Gemo-phaga (Ins.).

**gemell**—See **gemin.**

**gemin**—L. *geminus*, dim. *gemmula* = NL. dim. *gemellus*, a twin, one born at the same time; *gemino*, to double, pp. *geminatus*, doubled, paired. *Ex:* Gemelli-pora (Bry.); gemellus; Gemini-dens (Moll.); Gemino-ropa (Moll.).

gemm—L. *gemma*, a bud, gem < *gemmo*, pp. *gemmatus*, to put forth buds, set with gems. *Ex:* gemm-ula; Gemm-oliva (Moll.); Gemma (Moll.); Gemmi-pora (Coel.); Gemmato-phora (Coel.); Phys-gemm-aria (Coel.).

gemmat—See gemm.

gen—1. (a). Gr. *genos*, genit. *geneos*, a race, kind, descent < *gignesthai*, second aor. *genesthai*, to be produced, be born, become > *genesis*, origin, descent; *genelēs*, an ancestor. *Ex:* Geno-cidaris (Echin.); Geno-mys (Mam.); geno-type; Genos-iris*; genesio-logy; genet-ic; Di-genea (Platy.); hetero-gene-ous; Pletho-genesia (Ins.): (b). L. *genus*, genit. *generis*, pl. *genera*, a race, generation, stock, etc.; *gens*, a race, people, clan, etc.; *gigno* = OL. *geno*, pp. *genitus*, to beget, produce > *genitalis*, of or belonging to birth; *genero*, pp. *generatus*, to generate > *generatio*, genit. *generationis*, generation, the act of generating or begetting. *Ex:* genera; genital; genito-urinary; genus; Gigni-mentum (Pisc.); Gigno-peltis (Tril.): 2. L. *genu*, dim. *geniculum*, a knee > *geniculatus*, having a knot or protuberance like a knee or elbow, knotted, kneed. *Ex:* geni-al; geniculate; genuflexu-ous: 3. Gr. *genys*, the cheek, under jaw (L. *gena*, a chin) > *geneion*, a chin; *geneias*, genit. *geneiados*, beard; *geneiatēs*, bearded. *Ex:* gena; Geneiado-laelaps (Ins.); Geneion (Pisc.); Geneo-glossa (Ins.); Geniates (Ins.); genioglossal; Geny-ornis (Av.); Geny-pterus (Pisc.); A-genei-osus (Pisc.); A-geni-aspis (Ins.); Coelo-genys (Mam.); Dicro-genium (Ins.).

geneiad—See gen 3.

geneiat—See gen 3.

geneo—See gen 3.

generat—See gen 1 (b).

genes—See gen 1 (a).

genet—See gen 1 (a)

genett—O. Fr. *genette*, a civet cat. *Ex:* Genetta (Mam.).

-genic—NL. *-genic*, adj. combining form meaning giving rise to, originating. *Ex:* pyro-genic.

genicul—See gen 2.

genist—L. *genista* = *genesta*, name applied to several kinds of plants, especially the broom < Celtic *gen*, a bush. *Ex:* Genista*; Genisti-fex (Ins.).

genit—See gen 1 (b).

gennad—Gr. *gennadas*, of noble birth. *Ex:* Gennadas (Av.).

gennae—Gr. *gennaios*, genuine, good. *Ex:* Gen naeo-crinus (Echin.).

gens—L. *gens*, a tribe, race, nation. *Ex:* gens.

gentian—Gr. *gentianē*, name of a kind of plant, the gentian < *Gentios*, name of an Illyrian king who was said to have first discovered the properties of the gentian. *Ex:* Gentiana*; Gentian-aceae*.

genys—See gen 3.

geometr—Gr. *geōmetrēs*, a land-measurer. *Ex:* Geometr-odes (Ins.); Geometra (Ins.).

geonom—Gr. *geōnomos*, a colonist. *Ex:* Geonoma*.

georych—Gr. *geōrychos*, throwing up the earth. *Ex:* Georychus (Mam.).

gephur—See gephyr.

gephyr—Gr. *gephyra*, a bridge. *Ex:* Gephyr-anodus (Mam.); Gephyr-ina (Arach.); Gephyr-rhina (Mam.); Gephyrea; Gephyro-phora (Bry.).

ger—1. L. *gero*, to bear, carry. *Ex:* Globi-ger-ina (Prot.); seti-ger-ous: 2. Gr. *gerōn*, genit. *gerontos*, an old man; *gerontikos*, pertaining to an old man; *gēras*, old age > *agēraton*, a plant, "not growing old" < *a*, not + *gēras*, old. *Ex:* gero-morph-ism; geron-ic; geront-ism; geronto-ge-ous; Ageratum*; Thalasso-geron (Av.).

geran—Gr. *geranos*, a kind of bird, the crane > *geranion*, a plant called the crane's-bill. *Ex:* Geranium*; Gerano-aetus (Av.); Gerano-myia (Ins.); Limno-geranus (Av.).

gerb—See gerbill.

gerbill—Fr. *gerbille*, dim. form < Ar. *gerbo*, name of a kind of small rodent, the gerbil = jerboa. *Ex:* Gerbill-iscus (Mam.); Gerbillo-philus (Ins.); Gerbillus (Mam.); Gerbo-ides (Mam.).

germ—L. *germen*, genit. *germinis*, a bud, off-shoot > *germino*, to sprout, bud, pp. *germinatus*, sprouted, budded, germinated > *germinatio*, genit. *germinationis*, a sprouting. *Ex:* germ; germination.

geron—See ger 2.

-gerous—Eng. suffix < L. *-ger*, to bear, carry. *Ex:* seti-gerous.

gerres—L. *gerres*, a kind of salted fish. *Ex:* Gerres (Pisc.).

gerrh—Gr. *gerrhon*, anything made of wicker-work such as a screen or shield. *Ex:* Gerrho-notus (Rept.); Gerrho-saurus (Rept.); Platy-gerrhus (Ins.).

gerul—L. *gerulus*, a bearer, one who carries.

gerygone—Gr. *gērygonē*, born of sound < *gērys*, voice, speech + *gonē*, progeny. *Ex:* Gerygone = Gerigone (Av.); Eu-gerygone (Av.).

geryon—Gr. *Geryōnē*, name of a three-bodied monster < *geryō*, to shout. *Ex:* Geryon (Coel.), etc.; Geryon-idae (Coel.); Geryonia (Coel.).

gest—L. *gero*, to carry, bear, pp. *gestus*, carried, borne > *gestio*, genit. *gestionis*, a managing, doing, performing. *Ex:* di-gestion; in-gestion.

gestat—L. *gesto*, to bear, carry, pp. *gestatus*, borne, carried > *gestatio*, genit. *gestationis*, a carrying, bearing; *gestator*, a bearer. *Ex:* gestation.

get—See geit.

gethe—Gr. *gētheō*, to rejoice. *Ex:* Meli-gethes (Ins.), (Av.).

gethosyn—Gr. *gēthosynos*, glad. *Ex:* Gethosynus (Ins.).

getul—L. *Getulus*, belonging to the Getulians, a people of Morocco, i.e. of the African coast. *Ex:* Getulus (Mam.).

geum—L. *geum*, name of a kind of plant, the herb-bennet or avens < Gr. *geuō*, to have a taste, to give a relish. *Ex:* Geum*.

geust—NL. *geustia* < Gr. *geusis*, taste. *Ex:* ambly-geustia; allotrio-geustia.

giand—Hindustani, *gianda*, name for the large Asiatic rhinoceras. *Ex:* Gianda-therium (Mam.).

gibb—L. *gibbus*, bent, hunched > *gibber*, a hunch, hump; *gibberosus* = *gibbosus*, hunched, humped. *Ex:* Giber-ella*; Gibbi-rhynchia (Arach.); Gibbi-um (Ins.); Gibbo-discus (Prot.); gibbose; gibbous.

gigant—See gigas.

giganth—NL. *giganthes*, large flowered < Gr. *gigas*, a giant + *anthos*, a flower.

gigart—Gr. *gigarton*, a grape seed. *Ex:* Gigartina*.

gigas—Gr. *gigas*, genit. *gigantos*, a giant; also mighty. *Ex.* Gigant-ostraca (Crust.); Giganteo-trochus (Moll.); Giganti-pitta (Av.); Giganto-monas (Prot.); gigas.

gign—See gen 1 (b).

gilv—L. *gilvus*, pale yellow; yellowish.

gimn—See gymn.

gingiv—L. *gingiva*, the gum. *Ex:* gingiv-al; gingivo-labial.

gingko—Jap. *gingko*, vernacular name of the maiden-hair fern tree. *Ex:* Gingko*; Ginko-ales*.

ginglim—See ginglym.

ginglym—Gr. *ginglymos*, a hinge-joint. *Ex:* Ginglym-acarus (Arach.); Ginglymo-stoma (Elasm.); ginglimus.

giraff—NL. *giraffa* (Sp., Pg. *girafa*), a giraffe < Ar. *zarāf* = *zarāfa* = *zorāfa*, a giraffe. *Ex:* Giraff-oidea (Mam.); Giraffa (Mam.); Giraffo-myia (Ins.).

giss—Gr. *geisson*, a hem, border. *Ex:* Gisso-crinus (Echin.).

git—See geit.

gith—L. *gith*, name of a kind of plant with black aromatic seeds, the corn-cockle or Roman coriander. *Ex:* Gith-opsis*; Gith-ago*.

giton—See geit.

glab—L. *glaber*, smooth > *glabellus*, hairless; *glabro*, pp. *glabratus*, to make smooth, deprive of hair and bristles. *Ex:* glab-ellum; glabr-ous; glabrate; Glabri-scala (Moll.); Glabro-pecten (Moll.).

glabr—See glab.

glacial—L. *glacialis*, frozen.

glad—L. *gladius*, dim. *gladiolus*, a sword; *gladiator*, one who handles the sword. *Ex:* gladii-

formis; Gladio-graptis (Coel.); Gladiolus*; gladius.

gland—See glans.

glani—Gr. *glanis*, name of a kind of fish. *Ex:* Glanio-stomi (Pisc.).

glans—L. *glans*, genit. *glandis*, an acorn, dim. *glandula*, a gland; *glandulosus*, glandulous, glandular. *Ex:* glandi-fer-ous; Glandulo-nodosaria (Prot.); glandulose; Glans (Moll.); Juglans*.

glaph—Gr. *glaphō*, to make hollow, carve, adorn. *Ex:* Glaphi-urus (Mam.); Glapho-stoma (Ins.).

glaphyr—Gr. *glaphyros*, hollow; also neat, elegant, polished, well finished < *glaphō*, to hew, carve > *glaphyria* smoothness, polish. *Ex:* Glaphyra (Ins.); Glaphryia*, because of its elegance; Glaphyro-cystis (Echin.).

glare—L. *glarea*, NL. dim. *glareola*, gravel > *glareosus*, full of gravel, gravelly. *Ex:* Glareola (Av.); glareosus; glareous.

glarid—Gr. *glaris*, genit. *glaridos*, a chisel. *Ex:* Glarid-odon (Rept.); Glarido-glanis (Pisc.).

glauc—Gr. *glaukos*, silvery, gleaming; also bluish-green or grey > *glaukos*, name of a kind of gray-colored fish; *glaukion*, the juice of a plant similar to the horned poppy, so-called because the plant from which it is extracted has glaucous foliage; *glaux*, dim. *glaukidion*, name of a kind of owl, so called because of its glaring eyes; *Glaukonomē*, name of a Nereid. *Ex:* glauc-escent; Glauco-theca*; glaucous; Glaucus (Moll.), (Pisc.); Glaucidium (Av.); Glaucium*; Glauconome (Coel.), etc.; Crypto-glaux (Av.). See also glaux.

glauce—Gr. *Glaukē*, a mythological name. *Ex:* Glauce-a (Moll.).

glaux—Gr. *glaux* = *glax*, name of a kind of plant, the milk vetch. *Ex:* Glaux*; not Crypto-glaux (Av.), see glauc.

glea—See gli.

gleb—L. *gleba*, dim. *glebula*, a clod. *Ex:* gleba; Gleba (Moll.); glebula.

glechoma—Gr. *glēchōn*, pennyroyal. *Ex:* Glechoma* = Glecoma*.

glen—1. Gr. *glēnē*, a cavity, socket for a bone. *Ex:* glen-oid; Glen-urus (Ins.); Gleno-gnatha (Arach.); Gleno-tremites (Echin.): 2. Gr. *glēnē*, genit. *glēnēs*, the eyeball. *Ex:* Actino-glena (Prot.); Eu-glena (Prot.); Haplo-glenius (Ins.); Macro-glenes (Ins.): 3. Gr. *glēnos*, a star, radiance, things to stare at, curiosities, wonders. *Ex:* Tricho-glenus (Ins.).

gli—Gr. *glia*, glue. *Ex:* gli-oma (Path.); gli-osa; meso-glea; neuro-glia.

glin—Gr. *glinos* = *gleinos*, name applied by Theophrastus to the maple. *Ex:* Glinus*.

glir—L. *glis*, genit. *gliris*, a dormouse. *Ex:* Gliridae (Mam.); Glir-iscus (Mam.); Gliri-sorex (Mam.); Glis (Mam.); Glis-cebus (Mam.); Clavi-glis (Mam.).

**glis**—See **glir**.

**glischr**—Gr. *glischros*, glutinous, sticky; also greedy. *Ex:* glischr-in; Glischro-pus (Mam.); Glischrus (Moll.).

**glob**—L. *globus*, dim. *globulus*, a globe, ball > *globosus*, round as a ball; *globo*, pp. *globatus*, to make into a ball; *globator*, one who uses a globe. *Ex:* globator; Globi-ger-ina (Prot.); Globo-monas (Prot.); globose; Globul-aria*; Globulea*; haemo-glob-in.

Jaw of Globe-toothed Mososaurian, *Globidens*. Redrawn from Volume 41, Proceedings of the United States National Museum.

**gloch**—Gr. *glōchin*, genit. *glōchinos* = *glōchis*, a projecting point. *Ex:* gloch-idium; Glochi-cerus (Ins.); Glochino-myia (Ins.); Tri-glochin*; Tri-glochino-pholis (Mam.).

**gloe**—See **gloi**.

**gloi**—Gr. *gloios*, any glutinous substance; as adj., gelatinous, viscid, sticky. *Ex:* gloea; Gloeo-capsa*; Gloeo-dinium (Prot.); gloeo-spore; Gloio-nycteris (Mam.); Eri-gloea (Ins.).

**glom**—L. *glomus*, genit. *glomeris*, a ball, round body, a clue of yarn > *glomero*, to wind, pp. *glomeratus*, wound. *Ex:* Glomeris (Myr.); glomer-ites; glomer-ulus; glomerate; Glomo-spira (Prot.); Glomus (Ins.).

**glomer**—See **glom**.

**glori**—L. *gloria*, honor, praise, glory; *gloriosus*, superb, full of glory, renowned.

**gloss**—Gr. *glōssa* = Attic. *glōtta*, the tongue; *glōttikos*, of the tongue. *Ex:* Glossi-ptela (Av.); Glosso-petalon*; Glott-ella (Moll.); Glotto-ceras (Moll.); Rhipido-glossa (Moll.).

**glott**—See **gloss**.

**glottic**—See **gloss**.

**glottid**—See **glottis**.

**glottis**—Gr. *glōttis*, genit. *glōttidos*, the mouth of the windpipe, glottis < *glōtta* = *glōssa*, the tongue. *Ex:* Glottis (Av.); Glottidia (Brach.); epi-glottis; pro-glottis = pro-glottid.

**glum**—L. *gluma*, a hull, husk. *Ex:* glum-aceous; glume.

**glut**—1. Gr. *gloutos*, the rump > NL. *gluteus*. *Ex:* glute-al; gluteus; Lio-gluta (Ins.): 2. L. *gluto*, to swallow. *Ex:* de-glut-ition: 3. L. *gluta*, glue. *Ex:* Gluta*.

**glutin**—L. *gluten*, genit. *glutinis*, glue; *glutineus*, gluey. *Ex:* glutin.

**glyc**—Gr. *glykys*, sweet, pleasant > *glykeros*, sweet > *Glykera*, a feminine proper name. *Ex:* Glyc-ine*; Glyc-osma*; Glycer-ella (Ann.); Glycera (Ann.); Glyceria*; Glyce-lima (Moll.); glyco-gen; Glycy-meris (Moll.).

**glymm**—Gr. *glymma*, genit. *glymmatos*, an engraved or carved figure. *Ex:* Glymmat-acanthus (Pisc.); Epi-glymma (Ins.).

**glyph**—Gr. *glyphē*, a carving < *glyphō*, to carve. *Ex:* Glyph-alimus (Por.); Glyph-odon (Rept.); Glyphi-pteryx (Ins.); Glyphio-ceras (Moll.); siphono-glyph.

**glyphid**—Gr. *glyphis*, genit. *glyphidos*, the notched end of an arrow < *glyphō*, to carve, notch. *Ex:* Glyphid-odon = Glyphis-odon (Pisc.); Glyphis (Pisc.).

**glyphis**—See **glyphid**.

**glypt**—Gr. *glyptos*, good for carving, carved; *glyptēs*, a carver, sculptor; *glyptēr*, a chisel. *Ex:* Glypt-icus (Echin.); Glypt-odon (Mam.); Glypto-cranium (Arachn.); Glypto-pleura*; Helmintho-glypta (Moll.).

**gnampt**—Gr. *gnamptos*, curved. *Ex:* Gnampt-odon (Ins.).

**gnaph**—Gr. *gnaphos*, prickly teasel, a wool-comber's card; also an instrument of torture. *Ex:* Gnaph-osa (Arach.); Gnapho-cerus (Av.).

**gnaphal**—Gr. *gnaphalion*, name of a kind of down-covered plant used in stuffing cushions, the cud-weed. *Ex:* Gnaphal-odes*; Gnaphalium*; Gnaphalo-cera (Ins.).

**gnapt**—Gr. *gnaptōr*, a cloth-dresser. *Ex:* Gnaptor (Ins.).

**gnar**—L. *gnarus*, skillful.

**gnat**—See **nasc**.

**gnath**—Gr. *gnathos*, the jaw. *Ex:* gnath-ite; Gnath-odon (Moll.); gnatho-pod; Gnatho-stomata; Chaeto-gnatha.

**gnaur**—Eng. *gnaur*, also *knar* and *knur* < ME. *knarre* < Dan. *knor*, a knob. *Ex:* gnaur.

**gnesi**—Gr. *gnēsios*, genuine, legitimate, real. *Ex:* Gnesia (Ins.); Gnesio-ceros (Platy.); gnesio-gamy.

**gnet**—NL. *gnetum*, name applied to a genus of plants < a native name, *gnemon* = *gnemo*, a gymnosperm species growing on the island of Ternate in the Malay Archipelago. *Ex:* Gnet-acea*; Gnet-ales*; Gnetum*.

**gnom**—Gr. *gnōma*, a mark, sign, opinion. *Ex:* Gnoma (Ins.).

**gnomon**—Gr. *gnōmōn*, discerning, judicious; as subst., a judge; inspector, a carpenter's square, the index of a sun-dial. *Ex:* Gnomon (Ins.); Gnomon-iscus (Crust.); Meli-gnomon (Av.).

**gnoph**—Gr. *gnophos* = *dnophos*, darkness, gloom; *gnopheros* = *dnopheros*, dusky, gloomy. *Ex:* Gnopho-myia (Ins.); Gnophria (Ins.).

**gnorim**—Gr. *gnōrimos*, known, notable. *Ex:* Gnorimo-crinus (Echin.); Gnorimus (Ins.).

gnorism—Gr. *gnōrismos*, a recognizing, a making evident or known. *Ex:* Gnorismo-schema (Ins.).

gnost—Gr. *gnōstos*, well-known. *Ex:* Gnostus (Ins.).

gnot—Gr. *gnōtos*, understood, well-known. *Ex:* Gnoti-carina (Ins.); Gnotus (Ins.).

gnypet—Gr. *gnypetos*, falling down on the knees, weak. *Ex:* Gnypet-ella (Ins.); Gnypeta (Ins.); Gnypeto-morpha (Ins.).

gobi—L. *gobio*=*gobius*, a fish of small value, the gudgeon. *Ex:* Gobi-esox (Pisc.); Gobio-soma (Pisc.).

goer—Gr. *goeros*, mournful, with distress. *Ex:* Goeri-us (Ins.).

goet—Gr. *goēs*, genit, *goētos*, a howler, a wizard.

goliath—Heb. *Goleath*, giant of Gath slain by David. *Ex:* Goliathi-ceras (Moll.); Goliatho-cera (Ins.).

gomph—Gr. *gomphos*, a club, a bolt, a bond or fastening. *Ex:* Gompho-carpus*; Gompho-gnathus (Rept.); Gompho-therium=Gampho-therium (Mam.); Gomphus (Ins.):    2. Gr. *gomphios*, a molar tooth. *Ex:* Oxy-gomphius (Mam.); Poly-gomphius (Mam.).

gomphrena—NL. *gomphrena*<L. *gromphaena*, name of an amaranth. *Ex:* Gomphrena*.

gon—1. Gr. *gōnia*, a corner, joint, a knee, an angle>L. *goniatus*, angled. *Ex:* Goniat-ites (Moll.)<Gr. *gōnia*+*lithos*, a stone; Gonio-nemus (Coel.); Gono-ptera (Ins.); Anti-gonon*; Poly-gonum*:    2. Gr. *gonos*, also *gonē*, progeny, generation, seed>*goneuō*, to generate; *goneus*, a father; *gonimos*, able to

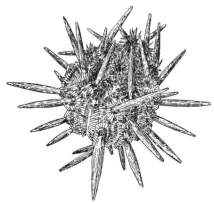

Tube-spined Angle-crown Urchin, *Goniocidaris canaliculata*. Redrawn from Voyage of The Challenger—Thomson. Courtesy of Joseph McDonough Co.

produce, productive; NL. *gonas*, genit. *gonadis*, a germ gland, reproductive gland or organ, gonad. *Ex:* gon-angium; gonad; goneo-clinic; Gonimo-phyllum*; gono-duct; gono-theca; **meso-gonim-icus**; Peri-gonimus (Coel.).

gonad—See gon 2.

gonat—Gr. *gony*, genit. *gonatos*, the knee; *gonation*, the knee-joint. *Ex:* Gonato-zygon*; Gonato-pus (Ins.); Gonato-nema*; Gonatium (Arach.); Gony-aulax (Prot.); Gony-stylus*; Gonyo-stomum (Prot.); gony-oncus (Med.); gony-theca; gonys. See also gon.

goneus—See gon 2.

gongyl—Gr. *gongylos*, round. *Ex:* Gongylus (Ins.), (Rept.); Gonglyo-spermae*.

goniat—See gon 1.

gonim—See gon 2.

-gony—M. -*gony*<L. -*gonia* (<Gr. *gigyesthai*, to be born), suffix denoting production, generation. *Ex:* sporo-gony. See also gonat.

gony—See gonat.

gordi—Gr. *Gordios*, name of a king of Phrygia> L. *nodus Gordius*, an inextricable knot, the Gordian knot. *Ex:* Gordi-acea (Nemat.); Gordi-ichthys (Pisc.); Poly-gordius (Ann.); Gordio-drilus (Ann.).

gorg—L. *Gorgo*, genit. *Gorgonis*=Gr. *Gorgō*, name of a female monster of terrible aspect< Gr. *gorgos*, fierce. *Ex:* Gorgo-dera (Platy.); Gorgo-saurus (Rept.); Gorgonia (Coel.); Gorgono-cephalus (Echin.); Eu-gorgia (Coel.).

gorget—OFr. *gorgete*, dim. of *gorge*, the throat, neck. *Ex:* gorget.

gortyn—Gr. *Gortynē*, an ancient city in Crete. *Ex:* Gortyna (Ins.).

goryt—Gr. *gōrytos*, a quiver. *Ex:* Goryt-odes (Ins.); Gorytes (Ins.).

gos—Gosiute Indian, *gosi*, desert. *Ex:* Gos-iulus (Arth.).

gossip—See gossyp.

gossyp—L.    *gossipion*=*gossypion*=*gossypinus*, name of a kind of plant, the cotton-tree< Arabic *gothn*, *qothn*, a soft substance. *Ex:* gossip-inus; Gossyp-aria (Ins.); Gossypium*.

gracil—L. *gracilis*, comp. *gracilior*, superl. *gracil-limus* (see lim 1.), slender, thin, simple. *Ex:* Gracil-aria*.

gracul—L. *graculus*, the jackdaw; also the cormorant. *Ex:* Graculi-pica (Av.): Gracul-avus (Av.); Gracula (Av.); Graculus (Av.).

grad—L. *gradior*, to step, walk. *Ex:* Tardi-grada (Arach.).

gradat—L. *gradatus*, having steps, step-by-step, by degrees.

graec—L. *Graecus*, Graecian, of Greece.

graeciz—NL. *graecizans*, speaking Greek; i.e., Graecian<Gr. *Graekizō*, to speak Greek.

grai—Gr. *graia*, gray.

grall—L. *grallae*, stilts>*grallator*, one who walks on stilts. *Ex:* Grallato-termes (Ins.); Grallator (Pisc.); Gralli-pes (Av.); Grallo-desmus (Myr.); Grallus (Av.).

grallat—See grall.

gramin—L. *gramen*, genit. *graminis*, grass>
*gramineus*, of or pertaining to grass. *Ex:*
Gramin-eae*; Gramini-cola (Av.); gramineous;
gramino-logy.

gramm—Gr. *grammē*, a mark, line, a writing;
*grammikos*, linear, lined; *gramma*, genit. *grammatos*, a letter, that which is drawn>*graphō*, to
draw, write. *Ex:* Gramm-anthes*; Grammato-
phyllum*; Grammo-psittaca (Av.); Crypto-
gramma*; Melano-grammus (Pisc.).

grammac—Gr. *grammakos*, streaked>NL. *grammacus*, streaked.

grammate—Gr. *grammateus*, a writer. *Ex:*
Grammateus (Pisc.).

gran—L. *granum*, a seed, grain>*granatus*, many
seeded; *granosus*, full of seeds or grains. *Ex:*
Granat-ellus (Av.); Granat-ina (Av.); Granati-
vora (Av.); Granato-crinus (Echin.); Grani-
fera (Moll.); grani-vor-ous; Grano-cardium
(Moll.).

granit—Eng. *granite*<It. *granito*, granite<L.
*granum*, grain: NL. *graniticus*, granite-loving,
of granite. *Ex:* granit-icola.

granul—L. *granulus*, a small grain. *Ex:* granul-
ation; granule.

graph—Gr. *graphē*, a drawing, painting, representation by means of lines, description;
*grapheion*, a style, pencil; *graphis*, genit.
*graphidos*, pencil, a graving tool; also a draw-
ing<*graphō*, to scratch, represent by means of
lines, draw, write. *Ex:* Graphi-urus (Mam.);
graphio-hex-aster; Graphis*; Auto-grapha
(Ins.).

graps—Gr. *grapsaios*, a crab. *Ex:* graps-oid;
Grapsi-cepon (Crust.); Grapso-branchus (Ann.);
Pachy grapsus (Crust.).

grapt—Gr. *graptos*, inscribed, painted. *Ex:*
Grapt-emys (Rept.); Grapto-lith-ina (Coel.);
Allo-graptus (Ins.); grapto-lite.

grassator—L. *grassator*, an idler, a lazy roamer.

gratiola—NL. *gratiola*, a plant name<L. *gratia*,
grace. *Ex:* Gratiola*.

gratus—L. *gratus*, pleasing, acceptable.

grav—L. *gravis*, heavy, burdened>*gravidus*,
pregnant or laden with young. *Ex:* grave-
olens; Gravi-ceps (Pisc.); gravid; Gravid-iana
(Moll.).

graveolens—L. *graveolens*, strong-scented.

gravid—See grav.

greg—L. *grex*, genit. *gregis*, a flock>*gregalis*=
*gregarius*, of or belonging to a herd or flock.
*Ex:* gregal-oid; Gregar-ina (Prot.); Gregari-
ella (Moll.); gregarious; loco-gregi-form.

gregal—See greg.

gregar—See greg.

gress—L. *gressus*, a step, course<*gradior*, to
walk>NL. *gressorius;* L. *gressor*, a walker.
*Ex:* Gress-lyo-saurus (Rept.); gressori-al;
Gressoria (Ins.).

grex—See greg.

grill—See gryll.

gripe—Gr. *gripeus*, a fisherman. *Ex:* Gripeus
(Av.).

griph—Gr. *griphos*, a woven fish basket, anything intricate or puzzling, a riddle. *Ex:* Griph-
apex (Ins.); Gripho-dictya (Por.); not Griphus
(Brach.), see gryph.

gris—ML. *griseus*=*griseis*, gray<Ger. *greis*=
OHG. *gris*, gray; Fr. *grison*, grey-headed. *Ex:*
Griseo-gaurena (Ins.); gris-escent; Grison
(Mam.).

grom—L. *groma*=*gruma*, a surveyor's measuring
rod. *Ex:* Gromia (Prot.); Allo-gromia (Prot.).

gromph—Gr. *gromphas*, genit. *gromphados*, an
old sow. *Ex:* Gromphado-rhina (Ins.).

The Hideous Sow-nose, *Grompha-
dorhina portentosa*, a wingless cockroach from Madagascar. Redrawn
from Cambridge Natural History.
The Macmillan Co.

gron—Gr. *grōnē*, a cavern; *grōnos*, eaten out. *Ex:*
Gron-ops (Ins.); Grona*; Gronias (Pisc.);
Grono-therium (Mam.).

gross—L. *grossus*, thick; also a name applied to
a green fig. *Ex:* Grosso-pterus (Tril.).

grossul—Fr. *grosielle*, a gooseberry>NL. *gros-
sula*, a gooseberry>*grossularia*, a gooseberry
bush. *Ex:* Grossulari-aceae*.

grossular—See grossul.

gru—L. *grus*, genit. *gruis*, name of a kind of bird,
the crane. *Ex:* Gru-idae (Av.); Grui-formes
(Av.); Grus (Av.).

grum—L. *grumus*, dim. *grumula*, a hillock, heap.
*Ex:* grum-ose; Grumilea*.

grumil—See grum.

grunn—L. *grunnio*, to grunt; *grunniens*, grunt-
ing. See also gry.

grup—See gryp.

grus—See gru.

gry—Gr. *gry*, a grunt, anything utterly insignificant, a minute particle of anything. *Ex:* gryo-
chrome; Gryon (Ins.); Hoplo-gryon (Ins.).

gryll—L. *gryllus*=*grillus*, a cricket, grasshopper
<Gr. *gryllos*. *Ex:* Grilli-vora (Av.); Gryllo-
blatta (Ins.); Gryllo-talpa (Ins.); Gryllus
(Ins.).

grymae—Gr. *grymea*, bag, a chest. *Ex:* Grymaeo-
mys (Mam.).

grymea—See grymae.

gryn—Gr. *grynos*, a fagot, dry wood for kindling.
*Ex:* Gryno-bius (Ins.); Gryno-charis (Ins.).

**gryon**—See **gry.**

**gryp**—*grypos*, hook-nosed, curved > *grypanios*, bent with age; *grypōsis*, a curving. *Ex:* Grypidius (Ins.); Gryp-odon (Pisc.); Grypo-suchus (Rept.); gryposis; Grypus (Av.); Andragrupes (Ins.).

**gryph**—LL. *gryphus*, a griffin < L. *gryps*, a griffin < Gr. *gryps*, a griffin < *grypos*, hook-nosed, curved. *Ex:* Gryphaea (Moll.); Gryphochiton (Moll.); Gryphus (Av.); Pseudogryphus (Av.).

**guaiac**—Abor. West Indian *guaiac*, name for lignum-vitae. *Ex:* Guaiacum*.

**guan**—NL. *guano* < Sp. *guano* < Peruv. *huanu*, ordure, dung. *Ex:* guano; guano-phore.

**guarea**—Vernacular name, prob. West Indian *guarea*, name of some tropical American plant. *Ex:* Guarea*.

**gubernacul**—L. *gubernaculum*, a rudder. *Ex:* gubernaculum.

**guirac**—Mex. *guiraca*, name for a kind of bird. *Ex.* Guiraca (Av.).

**gul**—1. L. *gula*, the throat. *Ex:* gul-ar; gulomental; albi-gulus; atri-gul-aris:    2. L. *gulo*, a glutton; *gulosus*, gluttonous < *gula*, the throat. *Ex:* Gulo (Mam.); Gulosus (Av.):    3. Gr. *gaulos*, any round vessel. *Ex:* Tetragulus (Verm.).

**gumm**—Gr. *kommi*, gum > L. *gummi* = *gummis* = *cummi*, etc., gum > *gummosus*, gummy; ML. *gumma*, genit. *gummatis*, gum. *Ex:* gummatous; gummi-fera; gummose.

**gurg**—L. *gurgito*, to engulf, pp. *gurgitatus*, engulfed, flooded. *Ex:* re-gurgitate.

**gustat**—L. *gusto*, to taste, pp. *gustatus*, tasted. *Ex:* gustat-orius.

**gut**—L. *gutus* = *guttus*, a narrow-necked vessel such as a flask. *Ex:* guti-form.

**gutt**—L. *gutta*, dim. *guttula*, a drop, drop-like spot > *guttatus*, containing drops or drop-like masses; spotted, speckled; *gutturosus*, that has a tumor in the throat. *Ex:* gutta; guttate; gutti-form; gutto-lineata; guttul-ate; gutturose.

**guttur**—L. *guttur*, genit. *gutteris*, the throat. *Ex:* guttur-al; gutturo-nasal.

**gyal**—See **gyalect.**

**gyalect**—NL. *gyalecta*, name applied to a genus of lichens < Gr. *gyalon*, a hollow; hollow vessel. *Ex:* Gyalecta*; gyalecti-form.

**gyascut**—NL. *gyascutus*, name applied to a genus of beetles (etym. uncertain). *Ex:* Gyascutus (Ins.).

**gyg**—1. Gr. *gygēs*, a water-bird. *Ex:* Gyges (Prot.); Gygis (Av.); not Gygo-geo-mys (Mam.) which involves a corruption of *zygos*, a yoke:    2. Gr. *Gygēs*, son of Dascylus.

**gyi**—1. Gr. *gyion*, limb, foot > *dexiogyios*, nimbleness of foot. *Ex:* Dexiogyia (Ins.):    2. Gr. *gyios*, lame.

**gyminda**—Anagram of *Myginda*, an allied plant genus. *Ex:* Gyminda*.

**gymn**—Gr. *gymnos*, naked, lightly clad > *gymnastikos*, pertaining to athletic exercises, gymnastic; *gymnastēs*, a trainer of athletes; *gymnoō*, to strip naked; L. *gymnicus*, gymnastic. *Ex:* Gimno-mera (Ins.); Gymn-(n)ema*; Gymnusa (Ins.); Gymn-ophiona (Amph.); gymnoblast; Gymno-gramme*; A-gymnastus (Ins.).

**gyn**—Gr. *gynē*, genit. *gynaikos*, a woman, wife > *gynaikeion*, feminine, the women's division of a house. *Ex:* gyn-andro-morph; Gyn-erium*; gyneco-logy; gyneceum = gynecium; Gynocardia*; Coelo-gyne*; poly-gyny.

**gyneco**—See **gyn.**

**gyp**—1. Gr. *gyps*, genit. *gypos*, a hawk, vulture. *Ex:* Gyp-aetus (Av.); Gypo-psittacus (Av.); Gymno-gyps (Av.):    2. Gr. *gypsos*, chalk, gypsum. *Ex:* Gyps-ornis (Av.); Gypso-phila*; gypso-phytes.

**gypag**—NL. *gypagus*, name applied to a genus of birds < Gr. *gyps*, a vulture + *arpagē*, a hook. *Ex:* Gypagus (Av.).

**gyps**—See **gyp.**

**gypsat**—L. *gypsatus*, covered with gypsum.

**gyr**—Gr. *gyros*, round; a circle. *Ex:* Gyr-

Gyrophragmium, a conspicuous fungus of deserts. The name refers to the radially arranged lamellae of the gleba.

encephala (Mam.); Gyra-thrix (Platy.); Gyropus (Ins.); Ana-gyrus (Ins.); Plagio-gyra (Moll.); Poly-gyra (Moll.).

**gyrans**—L. *gyrans*, whirling, going in a circle < *gyro*, to turn round in a circle.

**gyrat**—L. *gyratus*, made in a circular form, rounded. *Ex:* acro-gyratus.

**gyrin**—Gr. *gyrinos*, a tadpole > *gyros*, round, spiral. The combining form *gyrin*- often refers only to *Gyrinus*, a genus of water-beetles. *Ex:* Gyrin-odon (Mam.); Gyrin-urus (Pisc.); Gyrineum (Moll.); Gyrino-philus (Amph.); Gyrinus (Ins.).

# H

haban—Sp. *Habana*= *Havana*, name of the capital city of Cuba. *Ex.* Haban-aster (Echin.).

haben—L. *habena*, dim. *habenula*, a rein, strap. *Ex:* Haben-aria*.

habit—L. *habeo*, to have, hold, keep>*habito*, to dwell, reside>*habitus*, condition, appearance, attire, nature; *habitatio*, genit. *habitationis*, a dwelling. *Ex:* habit; habitat.

habr—See abr.

had—Gr. *Hadēs*, the underworld. *Ex:* Had-en-oecus (Ins.).

hadr—Gr. *hadros*= *hathros*, thick, stout; *hadrotēs*, thickness, strength, vigor. *Ex:* Hadr-urus (Arach.); Hadro-saurus (Rept.); Hadrotes (Ins.); Hathro-metra (Echin.); Eu-hadrus (Moll.).

hadrot—See hadr.

hadryn—Gr. *hadrynō*, to ripen; *hadrynsis*, a ripening, coming to maturity; *hadryntikos*, ripening, making stout, strengthening.

haed—L. *haedus*, a young goat. *Ex:* Nemor-haedus (Mam.); Nemor-hedus (Mam.), see nem 2.; Haedus (Ins.).

haem—Gr. *haima*, genit. *haimatos*, blood; *haimōnios*, blood-red; *haimateros*, bloody, hence red; *haimēros*, bloody. *Ex:* Haem-ulon (Pisc.); Haemat-ornis (Av.); Haematera (Ins.); Haemato-pinus (Ins.); Haemeria (Av.); Haemo-dorum*; haemo-globin=hemo-globin; Haemonia (Ins.).

haemat—See haem.

haesitat—L. *haesitatus*, fixed, made fast; *haesitator*, one who does not move.

hagi—Gr. *hagios*, holy, devoted to the Gods. *Ex:* Hagio-mantis (Ins.); Hagio-psar (Av.); Elae-agia*.

hal—1. Gr. *hals*, genit. *halos*, the sea>*halimos*, belonging to the sea>Gr. *halimon*, name of a sea-shore plant, the orach. *Ex:* Hal-campa (Coel.); Hal-ocy-ptena (Av.); Hal-saurus (Pisc.); Hali-physema (Prot.); halimi-folium; Halimo-dendron*; halo-limnic: 2. L. *halo*, to breathe, ppr. *halens*, genit. *halentis*, breathing>*halitus*, a breath. *Ex:* in-halent; halituous.

halcy—Gr. *halkyōn*, the kingfisher. *Ex:* Halcy-ornis (Av.); Halcyon (Av.). See also alcyon.

halec—L. *halec*=*alec*, a herring. *Ex:* Haleco-morphi (Pisc.).

halepensis—NL. *halepensis*, of or belonging to *Aleppo*= *Alepo*, in Asia<*Alep*+*ensis*, belonging to.

hales—L. *Halesus*, son of Agamemnon. *Ex:* Halesus (Ins.).

halict—NL. *halictus*, name applied to a genus of bees, perh.<Gr. *halizō*, to gather together, to be massed together into a ball. *Ex:* Halicto-phagus (Ins.); Halictus (Ins.).

halieut—Gr. *halieutēs*, a fisher, a seaman. *Ex:* Halieut-ichthys (Pisc.).

halim—See hal 1.

halipl—Gr. *haliploos*, sailing on the sea, a sailor; also covered with water. *Ex:* Halipl-idae (Ins.); Haliplus (Ins.).

halit—See hal 1.

hall—Gr. *hallomai*, to leap; *hallomēnos*, leaping. *Ex:* Hallo-pus (Rept.); Hallomenus (Ins.); Allo-saurus (Rept.).

hallomen—See hall.

hallux—NL. *hallux* from L. *hallex* (*allex*), genit. *hallicis*, the great toe. *Ex:* hallux.

halm—Gr. *halma*, genit. *halmatos*, a leap, spring. *Ex:*Halma-tettix (Ins.); Halmat-urus (Mam.); Halmato-rhagada (Moll.).

halosim—Gr. *halōsimos*, easily caught. *Ex:* Halosimus (Ins.).

halter—Gr. *haltēres*, weights held in the hands to give momentum in leaping. *Ex:* Halter-idium (Prot.); halteres; Halteri-phorus (Ins.); Haltero-phora (Prot.).

haltic—Gr. *haltikos*, good at leaping, swift. *Ex:* Halti-cor-idae (Ins.); Haltic-ella (Ins.); Haltica (Ins.); Haltico-ptera (Ins.); Halticus (Ins.).

halys—Gr. *halysis*, a chain>*halysidōtos*, wrought like a chain. *Ex:* Halys-ites (Coel.); Halysi-astraea (Coel.); Halysidota (Ins.); Halysio-rhynchus (Platy.); Halysis (Platy.); ?Halizia (Ins.).

halysidot—See halys.

ham—L. *hamus*, dim. *hamulus*, a hook>*hamatus*, hooked. *Ex:* Ham-ites (Moll.); hamate; Hami-termes=Ami-termes (Ins.); Rostro-hamus (Av.). See also hama.

hama—Gr. *hama*, all together, at the same time. *Ex:* Hama-melis*; according to Don it is derived from Gr. *omos*, like+*mēlea*, an apple>*homomēlis* of Athenaeus.

hamadry—Gr. *Hamadryas*, name of a wood nymph. *Ex:* Hamadry-opsis (Ins.); Hamadryas (Ins.).

hamart—Gr. *hamartia*, error. *Ex:* hamarto-plasia.

hamat—See ham.

hamax—Gr. *hamaxa*, a wagon. *Ex:* Hamaxo-bium (Ins.).

hamm—Gr. *hamma*, genit. *hammatos*, a knot, a noose. *Ex:* Hamma-cerus (Ins.); Hammato-cerus (Ins.); Hammato-cherus and Hammato-chaerus (Ins.), see chaer.

hana—Jap. *hana*, flower.

hapal—Gr. *hapalos*, soft, tender. *Ex:* Hapal-idae (Mam.); Hapal-otis (Mam.); Hapales (Mam.); Hapalis (Av.); Hapalo-carcinus (Crust.).

haph—Gr. *haphē*, a touching, grasping. *Ex:* haph-algesia (Med.); Hapho-helix (Moll.), an-aphia (Med.); An-apho-thrips (Ins.).

**hapl**—Gr. *haploos*, single, simple. *Ex:* Aplodontia (Mam.); Aplo-pappus*; hapl-odont; hapl-oid; Hapl-omi (Pisc.); Haplo-doci (Pisc.); Haplo-tax-idae (Ann.).

**hapt**—Gr. *haptō*, to fasten, bind, reach, overtake; *haptos*, fastened, fixed. *Ex:* hapto-phore; synapsis; Syn-apta (Echin.).

**haren**—L. *harena = arena*, sand. *Ex:* Haren-actis (Coel.). See also aren.

**hareng**—ML. *harengus*, a herring. *Ex:* Harengula (Pisc.); harengi-form.

**harm**—Gr. *harmos*, a joint. *Ex:* Harm-idium (Ins.); Harmo-stomium (Platy.); Harmo-thoe (Ann.).

**harmos**—Gr. *harmosis*, an adapting, a fitting together.

**harmoz**—Gr. *harmozō* = Attic *harmottō*, Doric *harmozdō*, to regulate. *Ex:* harmozone.

**harp**—Gr. *harpē*, a sickle, a hook; also a rapacious sea-fish; *harpē*, a bird of prey, the kite > NL. *harpes*, genit. *harpedis*. *Ex:* Harpe (Pisc.); Harpe-phyllum*; Harped-idae (Tril.); Harpes (Tril.); Harpia (Av.); Harpium (Ins.).

**harpact**—See **harpag**.

**harpag**—Gr. *harpagē*, a hook for seizing, a robbery, rape, seizure; *harpax*, genit. *harpagos*, robbing, rapacious; *harpaktēr*, a robber; *harpaktikos*, robberish; *harpazō*, to seize. *Ex:* Harpactes (Pisc.); Harpacticus (Ins.); Harpacto-stigma (Ins.); Harpactor (Pisc.); Harpago-ornis (Av.); Harpago-phytum*; Harpago-xenus (Ins.); Harpax (Moll.).

**harpal**—Gr. *harpaleos*, greedy; also attractive, alluring. *Ex:* Harpal-iscus (Ins.); Harpalo-bius (Ins.); Harpalus (Ins.).

**harped**—See **harp**.

**harpy**—Gr. *Harpyiai*, "The Snatchers," winged monsters, harpies < *harpazō*, to seize. *Ex:* Harpyia (Av.); Harpyo-nycteris (Mam.).

**hast**—L. *hasta*, dim. *hastula*, a spear > *hastatus*, spear-shaped. *Ex:* Hast-ina (Ins.); Hastatella (Prot.); Hasti-ger-ina (Prot.); Hasto-spiculum (Nem.); Hastula (Moll.).

**hastilis**—L. *hastilis*, of a javelin or spear.

**hathro**—Gr. *hathroos*, assembled in crowds, heaps. *Ex:* Hathro-metra (Echin.). See also hadr.

**hatter**—NL. *hatteria* < Maori *tuatera*, name of a kind of lizard-like reptile. *Ex:* Hatteri-idae (Rept.); Hatteria (Rept.).

**haust**—L. *haurio*, to draw up, pp. *haustus*, drawn up > *haustor*, a drawer of water; *haustrum*, NL. dim. *haustellum*, a pump, a machine for drawing water. *Ex:* haust-orium; Haustell-ata (Ins.), (Crust.).

**heaut**—Gr. *heautou*, fem. *heautēs*, neut. *heautou*, of himself, herself, itself. *Ex:* heauto-type.

**hebe**—**1.** Gr. *hēbē*, youth, puberty; pubescence > *Hēbē*, personification of youth; *hēbētikos*,

pertaining to puberty. *Ex:* Hebe-cephalus (Ins.); hebe-gynum; Hebe-loma*; hebe-petalous; hebetic: **2.** L. *hebes*, genit. *hebetis*, blunt, dull, dim, stupid; *hebeto*, pp. *hebetatus*, to make blunt, to weaken; *hebetor*, one who weakens, makes dull. *Ex:* hebetate.

**hebet**—See hebe 2.

**hecat**—**1.** Gr. *hekaton*, a hundred > NL. *hecto-*. *Ex:* Hecata-saurus (Rept.); hecato-phyll-ous; hecto-cotylus; Hecto-phyll-idae (Ins.): **2.** Gr. *hekatē*, far-shooting > *Hekatē*, an epithet of Artemis (Diana) and *Hekatos*, an epithet of Apollo. *Ex:* Hecate (Nem.).

**hecist**—Gr. *hēkistos*, least. *Ex:* Hecisto-cyphus (Echin.); hecisto-thermic.

**hect**—See hecat 1.

**hecyr**—Gr. *hekyra*, a mother-in-law; *hekyros*, father-in-law. *Ex:* Hecyr-ida (Ins.).

**hed**—Gr. *hedos*, genit. *hedeos*, dim. *hedion*, a seat, foundation, dwelling place. *Ex:* hedium = hedion (Ecol.); Hedo-bia (Ins.); Hedo-tettix (Ins.). See also hedy.

**hedeom**—NL. *hedeoma*, name applied to a genus of plants < Gr. *hēdys*, sweet + *osmē*, smell. *Ex:* Hedeoma*.

**heder**—L. *hedera*, ivy < Celtic *hedra*, ivy; *hederaceus*, of ivy, ivy-green; *hederatus*, covered with ivy. *Ex:* Heder-ella (Bry.); Hedera*; hederae-folium.

**hedon**—Gr. *hēdonē*, pleasure > *hēdonikos*, pleasurable. *Ex:* Hedone (Ins.); hedonic.

**hedy**—Gr. *hēdys*, dim. *hēdylos*, sweet. *Ex:* Hed-osmum*; Hedy-carpus*; Hedy-chium*; Hedylus (Ins.); Hedys-arum*, the last element < *arōma*, perfume.

**hegemon**—Gr. *hēgemōn*, a leader.

**heget**—Gr. *hēgētōr*, a leader, chief; *hēgētēs*, a leader. *Ex:* Hegeto-therium (Mam.); Hegetor (Ins.).

**hegeter**—Gr. *hēgētēr*, a guide. *Ex:* Hegetero-cara (Ins.).

**heil**—Gr. *heilō*, to pack closely, to shrink up, to wind or twine around; *heilipous*, to walk with rolling gait, to have legs which roll when walking. *Ex:* Heilipus (Ins.).

**hel**—**1.** Gr. *hēlios*, the sun. *Ex:* Elio-cidaris (Echin.); Hel-arctes (Mam.); Heli-chrysum*; Helio-zoa (Prot.); Helo-chara (Ins.); Ichthhelis (Pisc.); **2.** Gr. *helos*, genit. *heleos*, a marsh; *helōdēs*, frequenting marshes; marshy. *Ex:* Elodea*; Elodes (Ins.); Elosia (Amph.); Heli-naia (Av.); Helio-chloa*; helo-bius; Helohyas (Mam.); Helodes (Ins.); Helonias*. **3.** Gr. *hēlos*, a nail. *Ex:* Hel-odus (Pisc.); Helo-derma (Ins.); Helo-pus (Ins.).

**hela**—NL. *hela* < Gr. *hellos*, a young deer. *Ex:* Hela-mys (Mam.).

**helc**—**1.** Gr. *helkō*, to drag, draw, attract. *Ex:* helco-dermat-ous; Helco-soma (Prot.); helco-tropism; Helcon (Ins.): **2.** Gr. *helkos*, a wound, an ulcer. *Ex:* helco-plasty (Med.); Helco-stizus (Ins.).

**helenium**—NL. *helenium* < Gr. *helenion*, name of a plant. *Ex:* Helenium\*.

**helic**—1. Gr. *helix*, genit. *helikos*, anything twisted or wound; also a tendril, a kind of ivy. *Ex:* Helic-ina (Moll.); Helico-phanta (Moll.); Helix (Moll.); Is-helix (Ins.); Oreo-helix (Moll.): 2. Gr. *hēlix*, genit. *hēlikos*, of the same age, a comrade: 3. Gr. *hēlikos*, as large as, as strong as. *Ex:* Helico-olenus (Pisc.).

**helict**—Gr. *heliktos*, rolled, twisted; *helictēr*, anything twisted. *Ex:* Helicteres\*; Helicto-stylus (Moll.).

**heligm**—1. Gr. *heligmos*, a winding. *Ex:* Heligma (Ins.); Heligmo-merus (Arach.); Heligmus (Moll.): 2. Gr. *heligma*, genit. *heligmatos*, a fold, wrapper; also a curl of hair.

**helix**—See helic 1, 2.

**hellad**—Gr. *Hellas*, genit. *Hellados*, Hellas, Greece. *Ex:* Hellado-therium (Mam.).

**hellebor**—Gr. *helleboros*, name of a kind of plant, the hellebore. *Ex:* Helleborus\*.

**hellu**—L. *helluo* = *heluo*, a glutton. *Ex:* Helluo = Heluo (Ann.); Helluo (Ins.); Helluo-gaster (Ins.).

**helmins**—See helminth.

**helminth**—Gr. *helmins*, genit. *helminthos*, a bug, a worm either flat or round. *Ex:* Helmins (Nem.); Helmintho-glypta (Moll.); Platy-helminthes; Ster-clmintha (Entozoa).

**helod**—See hel 2.

**helonias**—See hel 2.

**helot**—Gr. *hēlōtos*, nailed, nail-shaped. See hel 3.

**heluo**—See hellu.

**helv**—L. *helvus*, honey yellow > *helveolus* = *helvolus*, yellowish.

**helvell**—L. *helvella*, a kind of pot-herb. *Ex:* Helvella\*; Helvell-ales\*.

The Helvella Fungus, *Helvella crispa*. Note saddle-shaped cap with drooping lobes. Redrawn from The Romance of The Fungus World—Rolf. J. B. Lippincott Co., Chapman & Hall, Ltd.

**helveol**—See helv.

**helvol**—See helv.

**helxin**—Gr. *helxine*, a kind of plant with woody capsules. *Ex:* Helxine\*.

**hem**—See haem.

**hemato**—See haem.

**hemer**—1. Gr. *hēmera*, day. *Ex:* Hemer-arachne (Arach.); Hemero-bius (Ins.); Hemero-callis\*; Hemero-campa (Ins.): 2. Gr. *hēmeros*, cultivated, tamed. *Ex:* Hemero-dromus (Av.); hemero-phytes.

**hemi-** —Gr. *hēmi*-, inseparable prefix < *hēmisys*, half. *Ex:* Hemi-carpha\*; Hemi-chorda; Hemigalus (Mam.); Hemi-ptera (Ins.); Hemy-trypa (Bry.).

**hemion**—Gr. *hēmionos*, a mule. *Ex:* Hemion-iscus (Crust.).

**hemionit**—Gr. *hēmionitis*, genit. *hēmionitidos*, name of a kind of fern. *Ex:* Hemionitis\*.

**hemo**—See haem.

**hemy**—See hemi.

**henic**—Gr. *henikos*, single, singular. *Ex:* Enicodes (Ins.); Enico-cephal-idae (Ins.); Henic-urus (Ins.); Henico-phatnus (Ins.).

**henicm**—Gr. *henikmos*, humid, "with wet in it."

**henotic**—Gr. *henōtikos*, serving to join. *Ex:* Henoticus (Ins.).

**heo**—See heos.

**heos**—Gr. Attic. *heōs* = dawn, east. *Ex:* Heo-anthropus (Mam.); Heos-emys (Rept.).

**hepat**—Gr. *hēpar*, genit. *hēpatos*, the liver > *hēpatikos*, pertaining to or affecting the liver. *Ex:* Hepatica\*; hepato-lith; Hepato-zoon. (Prot.).

**hephiast**—Gr. *Hephiastos*, a name for Vulcan. *Ex:* hephiast-ic.

**hephth**—Gr. *hephthos*, boiled, dressed; also languid. *Ex:* Hephtho-pelta (Crust.).

**hepial**—Gr. *hēpialēs*, a nightmare, shivering fit. *Ex:* Hepialus (Ins.).

**hept**—Gr. *hepta*, seven. *Ex:* Hept-actis (Echin.); Hepta-melus (Ins.); Hepto-stomum (Platy.).

**her**—1. Gr. *hērōs*, a hero. *Ex:* Heros (Pisc.): 2. Gr. *heros*, wool. *Ex:* Hero-phila (Ins.).

**heracl**—Gr. *Heraklēs* = L. Hercules, god of physical strength. *Ex:* Heraclea (Arach.); Heracleum\*; Heraclia (Ins.).

**herb**—L. *herba*, grass, green crops, an herb; *herbaceus*, grassy, grass-colored. *Ex:* herbacous; herbi-vorous.

**herbul**—L. *herba*, green crops, herbs > *herbula*, a little herb. *Ex:* Passer-herbula (Av.).

**herc**—Gr. *herkos*, genit. *herkeos*, a wall, fence. Herc-odon (Moll.); Herco-ceras (Moll.); herco-gamic.

**hercyn**—L. *Hercyna*, female attendant of Prosperina; *Hercynia silva*, forest of ancient Germany. *Ex:* Hercyna (Ins.); Hercyn-ella (Ins.), Hercyno-saurus (Rept.).

**heredit**—L. *heres*, genit. *heredis*, an heir>*hereditas*, heirship, inheritance. *Ex:* heredit-y.

**heren**—L. *haereo*, to hang, to adhere, ppr. *haerens*, genit. *haerentis*, hanging, cleaving. *Ex:* ad-herent; co-herent.

**heriad**—NL. *heriades*, a word irregularly formed <Gr. *erion*, wool. *Ex:* Heriades (Ins.).

**herm**—1. Gr. *herma*, genit. *hermatos*, a prop, support; also a mound; *hermin*, genit. *herminos*, a prop. *Ex:* Hermato-stroma (Prot.); Hermin-ium*: 2. Gr. *Hermēs*, name of a god; *hermaios*, named after Hermes. *Ex:* herm-aphrodit-ic; Herm-inea (Ins.); Herm in-idae (Ins.); Hermaeo-phaga (Ins.); Hermodactylus*.

**hermae**—See **herm** 2.

**hermat**—See **herm** 1.

**hermin**—See **herm** 1.

**hermos**—Sp. *hermosa*, beautiful> *Hermosa*, a place name. *Ex:* Hermosa-illa (Pisc.); Hermosa (Arach.); Hermosi-ornis (Av.).

**hern**—Gr. *hernos*, genit. *herneos*, a sprout, perh. >L. *hernia*, a rupture. *Ex:* Herni-aria*; hernia (Med.).

**herod**—See **erod**.

**herp**—Gr. *herpō*, to creep>*herpēs*, a creeper; *herpeton*, a reptile; *herpēstēs*, a creeping thing. *Ex:* Herpestes (Mam.); herpeto-logy; Herpetomonas (Prot.); Cath-erpes (Av.); Hyl-erpetum (Amph.).

**herpet**—See **herp**.

**hesper**—Gr. *hesperos*, of or at evening, the west, western> *Hesperos*, the evening star, Hesperus; *Hesperia*, land of the west, i.e., Italy and Spain; *hesperis*, name of a kind of plant, the night-scented gilly-flower. *Ex:* Hesper-ornis (Av.); Hesperi-phona (Av.); Hesperia (Ins.); Hesperis*; Hespero-callis*.

**hest**—Gr. *hēstos*, glad, agreeable. *Ex:* Thallestus (Ins.).

**hestern**—L. *hesternus*, of yesterday.

**hetaer**—Gr. *hetaireia*, companionship; *hetairos*, a comrade. *Ex:* Hetaer-ina (Ins.); Hetaeriobius (Ins.); Hetaero-dipsas (Rept.).

**hetero**—Gr. *heteros*, other, different. *Ex:* hetero-cercal; Hetero-japyx (Ins.); Hetero-teuthis (Moll.).

**hevea**—Native name *heve*, a plant of northern South America. *Ex:* Hevea*.

**hex**—1. Gr. *hex*, six. *Ex:* Hex-actin-ell-idae (Por.): Hexa-poda: 2. Gr. *hexis*, habit. *Ex:* hexo-logy.

**hexeris**—L. *hexeris*, a kind of small boat with six banks of oars.

**hian**—L. *hio*, to open, stand open, pp. *hiatus*, opened, ppr. *hians*, genit. *hiantis*, opening, gaping. *Ex:* Hians (Av.).

**hiascens**—L. *hiascens*, opening; ppr. of *hiasco*, to open.

**hiat**—L. *hiatus*, a gap, cleft. *Ex:* Hiat-ella (Moll.); hiati-cola; hiatus. See also hian.

**hibern**—L. *hibernus*, pertaining to winter; *hibernaculum* a winter residence. *Ex:* Hibernula (Echin.); Hibernia=Hybernia (Ins.); hibernaculum.

**hibisc**—Gr. *hibiskos*, the marsh mallow. *Ex:* Hibiscus*.

**hidr**—Gr. *hidrōs*, sweat>*hidrōtikos*, causing to perspire, apt to sweat. *Ex:* hidro-plankton (Ecol.); Hidroticus (Coel.).

**hidrotic**—See **hidr**.

**hidrys**—See **hidryt**.

**hidryt**—Gr. *hidrytos*, firmly fixed; *hidrysis*, a seat, a foundation, something settled. *Ex:* Hidryta (Ins.).

**hiem**—L. *hiems*, genit. *hiememis*, the winter> *hiemalis=hyemalis*, of or belonging to winter, wintry.

**hiemal**—See **hiem**.

**hier**—Gr. *hieros*, sacred. *Ex:* Hier-aetus (Av.); Heir-emys (Rept.); Hiero- ceryx (Ins.); Hiero-chloe*; Hiero-falco (Av.).

**hierac**—Gr. *hierax*, genit. *hierakos*, a falcon, hawk. *Ex:* Hierac-idea (Av.); Hieracium*; Hieraco-glaux (Av.); Meri-erax (Av.); Microhierax (Av.).

**hierax**—See **hierac**.

**hilar**—Gr. *hilaros*, cheerful, gay. *Ex:* Hilarempis (Ins.); Hilara (Ins.); not Hilaria*, named after A. St. Hilaire, French botanist.

**hilum**—L. *hilum*, a trifle, little thing, now used to designate the "eye" of a bean, etc. *Ex:* hilum=hilus.

**himant**—Gr. *himas*, genit. *himantos*, a leather strap or thong. *Ex:* Himan-thalia*; Himantopus (Av.).

**himat**—Gr. *himation*, a cloak, covering. *Ex:* Himat-anthus*; Himat-idium (Ins.); Himatiopetalum (Myr.); Himato-labus (Ins.); Leuchimatium (Ins.).

**himer**—Gr. *himeros*, a yearning, desire; *himero-*, in compounds means lovely, beautiful. *Ex:* Himero-crinus (Echin.); Himero-metra (Echin.).

**himert**—Gr. *himertos*, longed for, desired, beautiful. *Ex:* Himerta (Ins.); Himerto-soma (Ins.); Himertus.

**hinn**—L. *hinnus*, dim. *hinnulus*, a mule< Gr. *hinnos*, a mule. *Ex:* Hinn-ites (Moll.) Hinnulus (Moll.); Hinnus (Moll.); hinny.

**hipp**—Gr. *hippos*, dim. *hipparion*, a horse> *hippelatēs*, one who drives horses. *Ex:* Hipparionyx (Brach.); Hipparion (Mam.); Hipparium (Mam.); Hippe-astrum*; Hippelates (Ins.); Hippo-potamus (Mam.); Hypp-onyx (Moll.); Eo-hippus (Mam.); epi-hippium; Pleo-ippus (Mam.).

**hippar**—See **hipp**.

**hippelat**—See **hipp**.

hippocrat—Gr. *Hippokratēs*, father of medicine. *Ex:* Hypocratea*.

hippodam—Gr. *Hippodameia*, wife of Pirithous. *Ex:* Hippodamia (Ins.).

hippolais—NL *hippolais* < Gr. *hypolais*, the singing hedge-sparrow. *Ex:* Hippolais (Av.).

hippolyt—Gr. *Hippolytē*, queen of the Amazons; *Hippolytos*, son of Theseus. *Ex:* Hippolytes (Crust.); Hippolytus (Crust.).

hippomed—Gr. *Hippomedōn*, one of the seven heroes against Thebes. *Ex:* Hippomedon (Crust.).

hippur—Gr. *hippouris*, the plant called horsetail. *Ex:* Hippuris*; Hippuri-phila (Ins.).

hircin—L. *hircus*, dim. *hirculus*, a goat > *hircinus*, pertaining to or smelling like a goat. *Ex:* Hirculus*; hircinous; Hircinia (Por.); Hircinothrips (Ins.).

hirn—L. *hirnea*, dim. *hirnula*, a jug. *Ex:* Hirneola*.

hirsut—See hirt.

hirt—L. *hirtus*, rough, hairy > *hirsutus*, hairy, shaggy, bristly. *Ex:* Hirsut-ina (Ins.); Hirsuto-palpus (Ins.); Hirti-termes (Ins.); Hirtodroso-phila (Ins.).

hirud—L. *hirudo*, genit. *hirudinis*, a leech. *Ex:* Hirud-ella (Platy.); Hirudi-soma (Myr.); Hirudinea (Ann.); Hirudo (Platy.).

hirund—L. *hirundo*, a swallow. *Ex:* Hirund-oecus (Ins.); Hirundi-napus (Av.); Hirundo (Av.); Hirundo-lanius (Av.).

hisp—NL. *hispa*, name applied to a genus of bees < L. *hispidus*, shaggy, rough. *Ex:* Hispidae (Ins.); Hispa (Ins.).

hispan—L. *Hispania*, Spain > *Hispanicus*, of Spain, Spanish. *Ex:* Hispanicus (Pisc.); Hispanio-desmus (Myr.), Hispano-mydas (Ins.).

hispid—L. *hispidus*, spiny, shaggy, rough. *Ex:* hispid-ulous. See also hisp.

hispo—L. *Hispo*, a Roman surname. *Ex:* Hispo (Arach.).

hist—Gr. *histos*, a web; tissue. *Ex:* hist-amine; histo-genic; histo-logy; histo-lysis; met-ist-oid.

hister—1. Etruscan *hister*, an actor > L. *histrio*, genit. *histrionis*, a stage player > *histrionicus*, relating to an actor. *Ex:* Hister (Ins.); Histeridae (Ins.); Histrionicus (Av.): 2. Gr. *hysteros*, coming after, behind. *Ex:* Histeromerus (Ins.).

histio—Gr. *histion*, a sheet, sail, web. *Ex:* Histioteuthis (Moll.); Histiona (Prot.); Istio-phorus (Pisc.); Elatton-istius (Pisc.).

histrionic—See hister 1.

hiulc—L. *hiulcus*, gaping, split wide open.

hod—Gr. *hodos*, a path. *Ex:* Hod-urus (Ins.); Hodo-mys (Mam.); es-odic.

hodiern—L. *hodiernus*, of this day, of the present time.

hodoipor—Gr. *hodoiporos*, a traveller. *Ex:* Hodoiporus (Av.).

hol—Gr. *holos*, whole, entire. *Ex:* hol-ard (Ecol.); Hol-ectyp-ina (Echin.); hol-endobiotic; holo-blastic; holo-centric; Holo-chilus (Mam.); Holo-tricha (Prot.).

The Black Sail-bearer, *Istiophorus nigricans*, one of the pelagic sailfishes with habits much like the swordfish. Redrawn from Fishes of North and Middle America—Jordan.

holc—1. Gr. *holkos*, a track, furrow, groove. *Ex:* holc-odont; Holco-notus (Pisc.); Olco-steph anus (Moll.); Adel-olcus (Ins.); An-olc-ites (Moll.); Odont-olcae (Av.): 2. Gr. *holkos*, a kind of grain. *Ex:* Holcus*: 3. Gr. *holkos*, attractive, greedy: 4. *hoklē*, an attraction, drawing.

holothur—Gr. *holothourion*, a sort of water polyp, sea-cucumber. *Ex:* Holothuria (Echin.); Holothurio-philus (Crust.).

holover—L. *holoverus*, wholly purple.

hom—See homo.

homaem—Gr. *homaimos*, related by blood. *Ex:* Homaemus (Ins.).

homal—Gr. *homalos*, smooth, even; *homalotēs*, level ground, equilibrium, evenness. *Ex:* Homali-notus (Ins.); Homal-oid; Homal-onychus (Arach.); Homalo-ptera (Ins.); Homolo-saurus (Rept.); Homalota (Ins.).

homalia—Gr. *homalia*, evenness of surface. *Ex:* Homalia (Moll.).

homalot—See homal.

homar—OFr. *homar*, a lobster. *Ex:* Homar-idae (Crust.); Homarus (Crust.).

homelys—Gr. *homēlys*, genit. *homēlydos*, a companion. *Ex:* Homelys (Crust.), etc.

homer—Gr. *homēreō*, to meet, to agree. *Ex:* Homeria*.

homil—Gr. *homilos*, a being together. *Ex:* Homilo-stola (Ins.). See also omil.

homin—See homo.

homo—1. L. *homo*, genit. *hominis*, dim. *homunculus*, a man. *Ex:* Homin-idae (Mam.); Homo (Mam.); Homunculus (Mam.): 2. Gr. *homos*, common, joint, equal, alike, likewise. *Ex:* Hom-onyx (Ins.); Homo-ptera (Ins.); homo-zygote; Formic-oma (Ins.).

**homoe**—Gr. *homoios*, like, resembling; *homoiōsis*, a becoming like, a likeness; *homoiousios*, of like nature; *homoiō*, to make like, to be like. *Ex:* Homoeo-sauria (Rept.); homoeo-type; homoeozoic; homoeosis; Homoeusa (Ins.); homoiotherm; Amph-omoea (Moll.); Ip-omoea*.

**homoio**—See homoe.

**homolo**—See homal.

**homolog**—Gr. *homologia*, conformity; *homologos*, agreeing, conforming. *Ex:* homolog-ous; homology.

**homor**—Gr. *homoros*, neighboring. *Ex:* Homorocerus (Ins.); homorus; Homorus (Ins.). See omor.

**hoots**—Indian *hoots*, name for the big brown and grizzly bears of the north Pacific coast of N. America.

**hopl**—1. Gr. *hoplon*, a tool, weapon. *Ex:* Hoplarchus (Pisc.); Hoplio-cnema (Ins.); Hoplocampa (Ins.): 2. *hoplē*, a hoof. *Ex:* Hoplia (Ins.).

**hoplist**—Gr. *hoplistēs*, armed; also a warrior's dress. *Ex:* Hoplisto-cerus (Ins.).

**hoplit**—Gr. *hoplitēs*, heavily armed. *Ex:* Hoplitus (Ins.).

**hor**—1. L. *hora*, an hour. *Ex:* hor-arius: 2. Gr. *horion*, a boundary; *horios*, of boundaries, of limits, of bounds; *horos*, a boundary, limit. *Ex:* Horio-cerus (Moll.); Horo-genus (Ins.).

**horae**—L. *horae*, a dial, clock > *Horae*, goddesses of the seasons. *Ex:* Horaeo-cera (Ins.); Horaeometra (Echin.).

**horam**—Gr. *horama*, that which is seen. *Ex:* Cat-horama (Ins.).

**horde**—L. *hordeum*, barley > *hordeaceus*, of or pertaining to barley; L. *hordeolus*, a sty. *Ex:* hordeaceus; hordeolum (Med.); Hordeum*.

**horia**—L. *horia*, dim. *horiola*, a small vessel, fishing boat. *Ex:* Horia (Ins.).

**horio**—Gr. *horios*, of boundaries. *Ex:* Horiopleura (Echin.).

**horism**—Gr. *horismos*, a marking by bounds. *Ex:* horismo-logy; Horismus (Ins.).

**horiz**—Gr. *horizōn*, the boundary line of a circle, horizon. *Ex:* Horizo-ceras (Av.); Horizo-stoma (Moll.).

**horm**—1. Gr. *hormos*, a cord, chain. *Ex:* Hormiphora (Cten.); Hormio-pterus (Ins.); Hormius (Ins.); hormo-gonium: 2. Gr. *hormē*, an attack, the point of starting. *Ex:* hormion.

**hormen**—Gr. *hormenos*, a stem, sprout.

**hormin**—Gr. *horminon*, a kind of sage which was supposed to have aphrodisiac properties < *hormaō*, to excite. *Ex:* Horminum*.

**hormon**—Gr. *hormaō*, to excite, ppr. *hormōn. Ex:* hormone.

**horn**—L. *hornus*, of the present year; *hornotinus*, of this year.

**horp**—Gr. *horpex*, genit. *horpēkos*, a sapling, a branch of a tree; anything made from a branch of a tree such as a point or spike. *Ex:* Horpoceras (Moll.).

**horre**—L. *horreum*, genit. pl. *horreorum*, a barn or storehouse, a granary.

**horren**—L. *horrens*, genit. *horrentis*, bristly, rough, trembling, creating confusion or horror, ppr. of *horreo*, to stand on end, to stand the hair on end.

**horrid**—L. *horridus*, standing on end, projecting, rough, prickly < *horreo*, to stand on end. *Ex:* Horrido-labis (Ins.).

**hort**—L. *hortus*, dim. *hortulus*, a garden > *hortensis*, of or for a garden; *hortulanus*, a gardener. *Ex:* horti-culture; Horto-bombus (Ins.); Hortulanus (Av.); Hortulia (Rept.); hortus; not Hortia*, which is derived from a personal name.

**horus**—L. *horus*. < Egyptian name for the sun.

**hosp**—See hospit.

**hospit**—L. *hospes*, genit. *hospitis*, a host; also a visitor.

**host**—ME. host(e) < L. *hospes*, one who entertains a guest, a host. *Ex:* host.

**houbar**—Ar. *hubara*, a bustard. *Ex:* Houbara (Av.); Houbar-opsis (Av.).

**huanac**—NL. *huanacos* < Peruv. *huanacu*, name for the llama.

**hum**—L. *humus*, the ground, soil. *Ex:* humi-fuse; humi-stratus, see stern; Hum-ulus*; humus.

**humat**—L. *humatus*, buried, covered with earth.

**humer**—L. *humerus*, the upper bone of the arm, shoulder > *humerale*, a cape for the shoulders. *Ex:* humer-al; humerale; humero-ulnar; humerus.

**humil**—L. *humilis*, small, dwarfish, on the ground. *Ex:* Humil-aria (Moll.); humilis.

**humor**—L. *humor*, moisture, fluid. *Ex:* humor; humor-al.

**humos**—NL. *humosus*, full of earth > *humus*, soil, earth + -*osus*, full of.

**huperuthrus**—See hyperythr.

**hy**—1. Gr. *hys*, genit. *hyos* = L. *sus*, genit. *suis*, a hog. *Ex:* Hyo-phorbe*; Hyo-therium (Mam.): 2. Gr. *hyoeides* (< the Gr. letter Υ + *eidos*, form), shaped like the letter upsilon (Υ). *Ex:* Hi-odon = Hy-odon (Pisc.); hyo-gloss-al; hyoid.

**hyacinth**—Gr. *hyakinthos*, name of a kind of plant (not the modern hyacinth) < Gr. *Hyakinthos*, youth beloved of Apollo and unluckily killed by him. From his blood Apollo caused the hyacinth to grow. *Ex:* Hyacinthus*.

**hyaen**—Gr. *hyaina*, a hyena < *hys*, a hog, because of the bristly mane. *Ex:* Hyaen-anche*; Hyaen-odon (Mam.); Hyaena = Hyena (Mam.) Hyaeno-gnathus (Mam.).

**hyal**—Gr. *hyaleos* = *hyalinos*, glassy, shining. *Ex:* hyal-escent; Hyalea (Moll.); hyalo-plasm;

Hyalo-nema (Por.); Hyalo-pterus (Ins.); hya line.

**hyas**—Gr. *Hyas*, pl. *Hyades*, daughters of Atlas. *Ex:* Hyas (Amph.).

**hyb**—Gr. *hybos*, hump-backed; also a tuber. *Ex:* Hyb-anthus*; Hyb-odus (Elasm.); Hybophorus (Ins.).

**hybern**—See hibern.

**hybrid**—L. *hybrida*, a mongrel, the progeny of a tame sow and a wild boar. *Ex:* hybrid; hybridi-form; Hybrido-neura (Ins.).

**hybridis**—NL. *hybridis* < L. *hybrida*, a mongrel.

**hybrist**—Gr. *hybristēs*, a licentious, insolent person; *hybristos*, unrestrained, insolent. *Ex:* Hybristes (Ins.).

**hyda**—NL. *hyda* < Gr. *hydor*, water. *Ex:* hyda-thode. See -thode.

**hydat**—Gr. *hydatis*, genit. *hydatidos*, a watery vesicle; *hydatikos*, living in water. *Ex:* Hydatina (Rot.); hydati-form; Hydaticus (Ins.); hydatid; Hydato-gena (Platy.).

**hydatic**—See hydat.

**hydn**—Gr. *hydnon*, a tuber, also the name of kind of edible fungus. *Ex:* Hydno-carpus*; Hydno-cerina (Por.); Hydno-pora (Coel.); Hydnum*.

**hydr**—Gr. *hydōr*, genit. *hydatos*, water > *hydra* = Ionic *hydrē*, a kind of water serpent; *hydrainō*, to water, to wash; *hydrochoos*, a water pourer; in composition the Gr. *hydor* appears as a prefix *hydro-*. *Ex:* Hydat-ina (Rot.); Hydr-achna (Ins.), see arachn.; Hydractinea (Coel.); Hydr-angea*; hydr-anth; Hydra (Coel.); Hydraena (Ins.); hydro-theca; Hydrochous (Ins.); En-ydro-bius (Amph.).

**hydrast**—NL. *hydrastis* < Gr. *hydōr*, water + *draō*, to act; *drastēs*, an agent, performer. *Ex:* Hydrastis*.

**hyemal**—See hiem.

**hyen**—See hyaen.

**hyet**—Gr. *hyetos*, rain; *hyetios*, rainy, belonging to rain. *Ex:* Hyet-ornis (Av.); Hyeto-ceryx (Av.); Hyeto-mantis (Av.).

**hyg**—Gr. *hygiēs*, healthy > *hygiainō*, to be in good health; *Hygeia*, goddess of health < *hygeia* health. *Ex:* hygiene.

**hygr**—Gr. *hygros*, wet, moist; *hygrotēs*, wetness. *Ex:* Hygr-oecia (Ins.); Hygrio-bia (Ins.); Hygro-bia (Ins.); Hygro-ge-us (Ins.); Hygrotophila (Ins.); Hygrotus (Ins.).

**hygrot**—See hygr.

**hyl**—Gr. *hylē* = Doric *hyla*, a wood; *hylōdēs*, woody, bushy; *hylaios*, belonging to the forest, savage. *Ex:* Hyl-acantha (Moll.); Hyl-onax (Av.); Hyla (Amph.), or perhaps < Gr. *hylaō*, to bark, or < *Hyla*, vocative of Hylas, friend of Hercules; Hylaeo-saurus (Rept.); Hylaeus (Ins.); Hylaia (Ins.); Hyle-myia (Ins.); Hylobia (Ins.); Hylo-cerus*; Hylodes (Av.); Agalm-yla*; prot-yle; pter-ylae.

**hylact**—Gr. *hylactēs* = *hylaktētēs*, a barker; *hylacteō*, to bark; *hylaktikos*, given to barking. *Ex:* Hylactes, a name given to a genus of certain S.A. birds, some of which "yelp like dogs."

**hylae**—See hyl.

**hylaeus**—L. *Hylaeus*, one of the hounds of Actaeon, the hunter, who, found spying on Diana while bathing, was changed by the goddess into a stag. He was afterwards devoured by his own dogs.

**hylast**—Gr. *hylastēs*, a tree feller. *Ex:* Hylastinus (Ins.); Hylastes (Ins.).

**hylecoet**—Gr. *hylēkoitēs*, a forest dweller. *Ex:* Hylecoetus (Ins.).

**hylurg**—Gr. *hylourgos*, a carpenter. *Ex:* Hylurgops (Ins.); Hylurgus (Ins.).

**hymen**—Gr. *hymēn*, genit. *hymenos*, a parchment, a membrane, in medicine referring to the vaginal membrane; *hymenōdēs*, membra-

Caverned Membrane-head Fish, *Hymenocephalus cavernosus*, a deep-sea macruroid fish with large eyes. Redrawn from Fishes of North and Middle America—Jordan.

nous; *Hymēn*, god of marriages. *Ex:* Hymenaster (Echin.); hymen-alis; Hymen-aea* < *Hymēn*; Hymeno-callis*; Hymeno-clea*; Hymeno-ptera (Ins.); Hymenodes (Ins.).

**hyn**—See hynn.

**hynn**—Gr. *hynnē* = *hynis*, a plowshare. *Ex:* Hyni desmus (Myr.); Hynnis (Pisc.); Hynobius (Rept.).

**hyp**—Gr. *hypo*, under, beneath, less than usual. *Ex:* hyp-axial; Hyp-ochro-taenia (Ins.); Hypoxis*; Hyp-udaeus (Mam.) < *hypo* + *oudas*, the ground; hypo-cotyl; Hypo-pitys*.

**hypag**—Gr. *hypagō*, to lead or bring under. *Ex:* Hypago-ptera (Ins.).

**hypeco**—Gr. *hypēkoon*, a plant with leaves like rue < *hypecheō*, to rattle. *Ex:* Hypecoum*.

**hypelat**—Gr. *hypelatē*, Pliny's name for the butchers-broom < *hypo*, under + *elatē*, pine or fir. *Ex:* Hypelate*.

**hypen**—Gr. *hypēnē*, the hair on the upper lip, the moustache; also the under part of the face. *Ex:* Hypena (Ins.); Hypena-gonia (Ins.); Hypeno-rhynchus (Ins.).

**hyper**—1. Gr. *hyper* = *hypeir*, above, beyond, over. *Ex:* Hyper-anthus (Av.); hyper-trophy; Hypero-cerus (Moll.): 2. Gr. *hyperos*, a pestle. *Ex:* Hypera (Ins.); Hypero-dapedon (Rept.); Hypero-morpha (Ins.): 3. Gr. *hyperōē*, the palate. *Ex:* Hypero-odon (Mam.); Hypero-treta (Cycl.).

**hyperbor**—Gr. *hyperboreos*, beyond the north wind, of the extreme north. *Ex:* Hyperborea (Ins.).

**hyperic**—Gr. *hyperikon*, name of a kind of plant, the St. John's wort <*hypo*, under+*erikē*, heath; Linnaeus says it is derived from Gr. *yper*, upper and *eikōn*, an image. *Ex:* Hypericum*.

**hyperion**—Gr. *Hyperiōn*, the Sun-god. *Ex:* Hyperion (Ins.).

**hypermegeth**—Gr. *hypermegethēs*, exceedingly difficult. *Ex:* Hypermegethes (Av.).

**hyperythr**—Gr. *hyperythros*, somewhat red.

**hyph**—Gr. *hyphē*, a web>*hyphasma*, something woven; *hyphantos*, woven; *hyphainō*, to weave; *hyphantēs*, fem. *hyphantria*, a weaver. *Ex:* hypha; Hyphaene*; Hyphantria (Ins.); Hyphantro-phaga (Ins.); Hyphasma*.

**hyphaen**—See **hyph**.

**hyphal**. Gr. *hyphalos*, under the sea. *Ex:* Hyphal-aster (Echin.).

**hyphant**—See **hyph**.

**hyphasm**—See **hyph**.

**hyphydr**—Gr. *hyphydros*, found in water. *Ex:* Hyphydrus (Ins.).

**hypn**—1. Gr. *hypnos*, sleep>*hypnōtikos*, drowsy; putting to sleep, narcotic; *hypnōdia*, sleepiness. *Ex:* Hypnea*; hypnody; Hypnota (Ins.); Hypnoticus (Crust.); Phyl-ypnus (Pisc.): 2. Gr. *hypnon*, tree-moss. *Ex:* Hypno-phila (Ins.).

**hypo**—See **hyp**.

**hypochoer**—Gr. *hypochoiris*, a cichoriaceous plant, the cat's-ear, perh. <*hypo*, under+ *choiros*, pig, these animals being fond of its roots. *Ex:* Hypochoeris*.

**hypoloep**—Gr. *hypoloipos*, surviving; also abandoned, defective. *Ex:* Hypoloepus (Ins.).

**hypomen**—Gr. *hypomenō*, to stay behind, to lag, to survive. *Ex:* hypomen-ous.

**hyponom**—Gr. *hyponomeuō*, to undermine; *hyponomē*, an underground passage. *Ex:* Hyponomeut-idae=Yponomeut-idae (Ins.); Hyponomeuta (Ins.); Hyponome (Echin).

**hyps**—Gr. *hypsos*, high, height; *hypsi*, on high, aloft. *Ex:* hyps-odont; Hypsi-camera (Ins.); hypsi-steno-cephalic.

**hypsil**—Gr. *hypsilon*, upsilon, the Gr. letter Υ. hypsil-oid; hypsili-form.

**hypsist**—Gr. *hypsistos*, highest.

**hypti**—Gr. *hyptios*, laid back. *Ex:* Hyptio-crinus (Echin.); Hyptis*.

**hyptiot**—Gr. *hyptiotēs*, flatness. *Ex:* Hyptiotes (Arach.).

**hyptis**—See **hypti**.

**hypyl**—Gr. *hypylos*, unsound beneath. *Ex:* Hypylus (Ins.).

**hysgin**—Gr. *hysginon*, a bright crimson vegetable dye. *Ex:* Hysginum (Prot.).

**hyssop**—Heb. *ēzōph*>Gr. *hyssōpos*, an aromatic plant. *Ex:* Hyssopus*.

**hystat**—Gr. *hystatos*, last.

**hyster**—1. Gr. *hystera*, the womb; *hysterikos*, suffering in the womb, hysterical. *Ex:* Non-hystera (Nem.): 2. Gr. *hysteros*, later, behind, coming after. *Ex:* Hystero-neura (Ins.).

**hystric**—Gr. *hystrix*, genit. *hystrichos*, a porcupine. *Ex:* Hystric-idae (Mam.); Hystrico-thrips (Ins.); Hystrix (Mam.).

**hystrix**—See **hystric**.

## I

(Consonant I=J and the two are often Latinized interchangeably).

**-ia**—Gr. and L. *-ia*, ending of Gr. and L. nouns denoting quality of or state of being; often changed to *y* in English derivatives, e.g. Gr. *philosophia*=Eng. philosophy; if a Latin word ends in *ia* preceded by the root-ending *t* the *tia* is often changed to *cy* in English words, e.g., clemency<L. *clementia*. This same euph- onious ending, formed from the stem vowel *i* or connecting-vowel with *a* added, is often found in the generic names of plants and animals, especially those derived from personal names, as well as in class names of animals (Amphibia, Reptilia, Mammalia) and in other scientific classificatory terms, (bacteria). In medicine it is used to terminate the names of diseases or diseased states (dementia, pneumonia, melancholia).

iach—1. Gr. *iacheō*, to cry out. *Ex:* Iache (Av.): 2. Gr. *Iakchos*, mystic name of Bacchus, also a song sung in his honor. *Ex:* Iachus (Mam.).

iachr—Gr. *iachros*, softened, melted; also quiet.

-iacus—L. *-iacus*, adj. termination of Greek names ending in *ias*, as; Archiacus < *Archias*.

-iades—L. *-iades*, masculine patronymic. *Ex:* Anchis-iades, son of Anchises.

ialtr—NL. *ialtris*, name applied to a genus of snakes < Gr. *iallō* (verbal adj. *ialtos*), to send forth; also to bind. *Ex:* Ialtris (Rept.).

iamat—Gr. *iama*, genit. *iamatos*, medicine, the art of healing. *Ex:* iamato-logy.

-ian—NL. *-ian*, a variant of *-an*, suffix of L. origin meaning related to, connected with, belonging to. *Ex:* mammal-ian; nemert-ian; reptil-ian.

ianth—Gr. Gr. *ianthinos*, violet-colored. *Ex:* Iantho-cicla (Av.); Ianthina = Janthina (Moll.) Ianthin-opsis (Moll.); Janthino-soma (Ins.).

iapet—Gr. *Iapetos*, a Titan, son of Uranus and Ge and father of Atlas. *Ex:* Iapetus (Ins.); not Japet-ella (Moll.); named after Japetus Steenstrupp, brilliant Danish conchologist.

iapyg—See iapyx.

iapyx—Gr. *Iapyx*, genit. *Iapygos*, son of Daedalus; also the west northwest wind. *Ex:* Iapygidae = Japyg-idae (Ins.); Iapyx = Japyx (Ins.); Hetero-japyx (Ins.).

-ias—1. L. *-ias*, fem. patronymic ending. *Ex:* Thest-ias, daughter of Thestius: 2. NL. *-ias*, suffix denoting possession or special characteristic. *Ex:* Ul-ias (Mam.); Arct-ias (Mam.).

-iasis—Gr. *-iasis*, suffix terminating names of diseases. *Ex:* amoeb-iasis: elephant-iasis; filariasis; my-iasis. It may also denote an action or process. *Ex:* odont-iasis.

iaspid—Gr. *Iaspis*, genit. *Iaspidos*, jasper; L. *iaspideus = jaspideus*, jasper-like. *Ex:* iaspideus; Iaspis (Ins.).

iaspis—See iaspid.

iatr—Gr. *iatros*, a physician; *iatreia*, medicine, healing, *iaktrikos*, able to cure, of or for a physician; *iatrikē*, the art of healing. *Ex:* pediatrics.

iatrics—See iatr.

iber—Gr. *Ibēria*, Spain, the northeast part of Spain. The modern Georgia in the Caucasus was also called Iberia by the ancients. *Ex:* Iberia (Ins.); Ibero-gallus (Mys.).

iberid—Gr. *ibēris*, genit. *ibēridos*, a kind of crucifer, the pepperwort. *Ex:* Iberis*.

iberis—See iberid.

ibex—L. *ibex*, name of a kind of goat, the chamois. *Ex:* Ibex (Mam.).

ibid—Gr. *ibis* > L. *ibis*, genit. *ibidis*, name of a sacred Egyptian bird that subsisted upon water animals, the ibis. *Ex:* Ibidium*; Ibidopodia (Av.); Ibis (Av.).

ibis—See ibid.

ibolium—NL. *ibolium* < Jap. *ibola*, the wax tree. *Ex:* ibolium.

ibycter—Gr. *ibyctēr*, genit. *ibyctēros*, one who begins a war song. *Ex:* Ibycter = Ibicter (Av.).

ic—1. Gr. *eikos*, like truth; i.e., likely, probable. *Ex:* ico-type: 2. Gr. *eikō*, to yield, give way. *Ex:* Ic-osteus (Pisc.); Ico-chilus (Mam.).

-ic—Eng. *-ic*, suffix added to nouns to form adjectives < L *-icus*, Gr. *-ikos*, denoting belonging to, relating to. *Ex:* cosm-ic, relating to the cosmos.

-ica—See icus.

icac—Sp. *icacō*, the coco plum. *Ex:* Icac-ina*; Icacin-aceae*.

-ical—See tic.

ican—Gr. *ikanos*, becoming, competent, sufficing; *ikanotēs*, fitness, sufficiency. *Ex:* Ican-odus (Pisc.); Icanotia (Moll.).

icar—Gr. *Ikaros*, son of Daedalus; *Icaria*, an island in the Aegean Sea near which Icarus was drowned. *Ex:* Icar-idion (Ins.); Icari-astrum (Ins.); Icaria (Ins.).

icel—1. Gr. *ikelos*, like, resembling: 2. Gr. *Ikelos*, son of Hypnos, god of sleep. *Ex:* Icel-ichthys (Pisc.); Icel-inus (Pisc.); Icelus (Arach.), (Pisc.).

icery—NL. *icerya*, name given to a genus of scale insects, (origin uncertain). *Ex:* Icerya (Ins.).

ichn—Gr. *ichnos*, a track, trace, *Ex:* ichno-logy; Ichno-carpus*; Asaph-oid-ichnus (Tril.); par-ichnos.

ichneum—Gr. *ichneumōn*, the tracker; an Egyptian animal which hunts out the eggs of crocodiles; also the name of a spider-hunting wasp. *Ex:* Ichneumia (Mam.); Ichneumon (Ins.).

chthy—Gr. *ichthys*, genit. *ichthyos*, a fish. *Ex:* Ichthy-ornis (Av.); ichthyo-logy; Ichthyo-sauria (Rept.); Sebast-ichthys (Pisc.).

-icius—L. *-icius*, suffix signifying made of or belonging to or indicating likeness or possession of a character.

icle—See -cle.

-icola—NL. *-icola*, an inhabitant < L. *incola*, an inhabitant. *Ex:* granit-icola, etc. See col. Some etymologists would derive such words as graniticola < *granite* + NL. *cola*, to inhabit < L. *colo*, to inhabit, the *-i-* being only a connecting vowel.

icon—L. *icon*, genit. *iconis*, a figure, image < Gr. *eikōn*, a figure, image, likeness. *Ex:* Icon-aster (Echin.); Icones.

icos—Gr. *eikosi*, twenty. *Ex:* Icosi-dactylo-crinus (Echin.).

-icos—Gr. *-ikos*, suffix denoting ability or fitness when added to Gr. verbals, as in *archikos*, fit to rule; when added to Gr. nouns it denotes relation as in *basilikos*, kingly. See tic.

**icot**—Gr. *eikotōs*, suitable, reasonable. *Ex:* Icot-urus (Av.). Icoto-pus (Crust.).

**ict**—1. *ict*, contraction of Gr. *ichthys*, a fish. *Ex:* Ict-aleurus (Pisc.); see aelur; Ictio-bus (Pisc.): 2. Gr. *iktis*, genit. *iktidos*, a weasel, the yellow-breasted marten. *Ex:* Ict-ides (Mam.); Icti-cyon (Mam.); Icti-therium (Mam.); Ictido-gnathus (Rept.); Ictido-mys (Mam.); Gal-ictis (Mam.), see gale 2.

**icter**—Gr. *ikteros*, jaundice; hence yellowness; also a bird of yellowish-green color, the oriole, which upon being seen was said to cure jaundice. *Ex:* Icter-idae (Av.); Icteria (Av.); Ictero-cephalus (Ins.); Icterus (Av.).

**icteric**—Gr. *ikterikos*, yellowish. *Ex:* Icterica (Ins.).

**icthy**—See **ichthy.**

**ictid**—See **ict** 2.

**ictin**—Gr. *iktinos*, a kind of bird, the kite. *Ex:* Ictinia (Av.); Ictini-astur (Av.); Ictino-aetus (Av.); Ictinus (Av.).

**-icus**—1. L. *-icus, -a, -um*, suffix added to noun stems to form adjectives meaning belonging to. *Ex:* californ-icus: 2. NL. *-icus, -a, -um*, commonly used ending employed to emphasize or intensify a certain character. *Ex:* nar-ica, here calling attention to the long nose<L. *naris*, the nose.

**id**—Gr. *idios*, distinct. *Ex:* id; id-ant. See idio.

**-id**—1. Eng. *-id*,<L. *idēs*, patronymic termination meaning son of. Compare with -ides below. *Ex:* ir-id*; nere-id; sipuncul-id: 2. Eng. *-id*, termination of nouns derived from zoological family names; thus: fringillid< Fringill-idae and psocid<Psoc-idae: 3. Gr. *eidos*, resemblance. *Ex:* trache-id; broch-ido-dromus, etc: 4. Eng. *-id*, having the quality of, that which. *Ex:* flu-id>L. *fluo*, to flow.

**-ida**—NL. *-ida* (assumed neuter of Gr. *-ides*), used to terminate zoological group names of various classificatory ranks. *Ex:* Arachn-ida; Phoron-ida; Tri-clad-ida (Platy.).

**-idae**—L. *-idae*, feminine plural adjectival suffix, added to stems of generic names to form family names. *Ex:* Nymphal-idae (Ins.). See also eid.

**idaeus**—L. *Idaeus*, belonging to Mt. Ida.

**idal**—L. *Idalia*, mountain city in Cyprus, sacred to Venus. *Ex:* Idalia (Ins.), etc.

**idant**—See **id.**

**-ide**—See **-ite.** In chemistry *-ide* is used in making names of compounds. *Ex:* sulph-ide.

**idem**—Gr. *eidemōn*, with knowledge, expert in a thing. *Ex:* Idemum (Ins.).

**ideo**—Gr. *idea*, an idea, the appearance of a thing, a class, sort. *Ex:* Ideo-blothrus (Arach.); not Ideodelphys (Mam.), which is based on an anagram of Eo-di-delphys (Mam.).

**-ides**—Gr. *-idēs*, patronymic termination meaning son of, often added to proper and common names to indicate descent or relationship. *Ex:* Potam-ides (Moll.). See -id.

**-ideus**—NL. *-ideus*<Gr. *eidos*, adj. suffix denoting similarity.

**-idia**—See **-idius**, also **-idium.**

**idiast**—Gr. *idiastēs*, a hermit. *Ex:* Idiasta (Ins.).

**idic**—Gr. *idikos*, own's own, special.

**idio**—Gr. *idios*, one's own, distinct, peculiar. *Ex:* idio-blast; Idio-cerus (Ins.); idio-morphous; Acte-dium (Ins.).

**idiomat**—Gr. *idiōmatikos*, peculiar, characteristic. *Ex:* idiomatic.

**-idion**—Gr. *-idion*=*-ion*, diminutive suffix giving rise to Eng. *-idium*. *Ex:* anther-idium; Mys-idion (Moll.); pyg-idium; ryncho-teuth-ion.

**-idium**—See **-idion.**

**-idius**—NL. *-idius, -a, -um*<Gr. *idion*, a diminutive ending. *Ex:* Crith-idia (Prot.) Balanat-idium (Prot.).

**idmon**—Gr. *idmōn*, skillful>*Idmōn*, father of Arachne; also the name of the son of Apollo, an Argonaut. *Ex:* Idmonea (Bry.); Idmoma (Arach.).

**ido**—See **eid.**

**idol**—Gr. *eidōlon*, an image, phantom. *Ex:* Idolo-thrips (Ins.); Idolus (Ins.).

**idolon**—See **idol.**

**idomen**—Gr. *Idomeneus*, chief of the Cretans. *Ex:* Idomeneus (Mam.).

**idone**—L. *idoneus*, proper, suitable, satisfactory.

**idotea**—See **idothea.**

**idothea**—Gr. *Eidothea*, name of a sea nymph. *Ex:* Idothea (Crust.); Idotea (Crust ).

**idr**—Gr. *hidrōs*, sweat, perspiration; also gum, resin. *Ex:* brom-idrosis (Med.); osm-idrosis (Med.).

**idris**—Gr. *idris*, genit. *idrios*, knowing, skilled. *Ex:* Idris (Ins.); Idriso-bdella (Ann.).

**idros**—See **idr.**

**-idus**—L. *-idus*, adjectival termination sometimes added to the root of neuter verbs to denote quality or state. See -id 4. *Ex:* alg-idus, cold<*algeo*, to be cold; mad-idus, moist< *madeo*, to be wet.

**-iensis**—NL. *-iensis*, adj. ending meaning of or connected with; similar to *-ensis*, which see. *Ex:* timor-iensis.

**ign**—See **igneus.**

**ignar**—L. *ignarus*, inexperienced, unaware.

**ignav**—L. *ignavus*, inactive, without spirit; *ignavia*, idleness. *Ex:* Ignavus (Mam.).

**ignescens**—L. *ignescens*, taking fire, burning.

**igneus**—L. *igneus*, of fire, fiery; *igniarius*, pertaining to fire; *ignis*, fire.

**igniar**—See **igneus.**

**ignican**—L. *ignicans*, genit. *ignicantis*, flaming, fire-colored.

ignit—L. *ignitus*, glowing, fiery.

ignot—L. *ignotus*, unknown, ignored. *Ex:* Ignotornis (Av.); Ignotus (Ins.).

iguan—Sp. *iguana* < Carib. *yuana* = *igoana*, name of a kind of large lizard, the iguana. *Ex:* iguan-id; Iguan-odon (Rept.); Iguano-gnathus (Rept.).

il- —See in-.

ilapinast—Gr. *eilapinastēs*, a guest, one who feasts. *Ex:* Ilapinastes (Ins.).

-ile—L. *-ile*, suffix indicating a place where things are kept or stored. *Ex:* ov-ile, a sheepfold > *ovis*, sheep. In ecological terminology, a suffix denoting a society. *Ex:* irid-ile (Ecol.).

ile—1. L. *ile* = *ileum* = *ilium*, pl. *ilia*, the groin, small intestine; *iliacus*, relating to colic. *Ex:* ile-al; ileo-caecal; iliacus; ilio-caecal; ilium: 2. Gr. *eileō* = *eilō*, to collect. *Ex:* Ile-omus (Ins.).

ilet—Gr. *eilētos* arched, wound, wreathed. *Ex:* Oph-ileta (Moll.).

ilex—See ilic.

ili—See ily.

iliac—See ile.

ilic—L. *ilex*, genit. *ilicis*, ancient name of a plant, the holm oak (Quercus ilex). *Ex:* Ilex*; ilici-folia.

iling—Gr. *illingos*, a whirling or spinning round, dizziness < *illō*, to roll. *Ex:* Ilingo-ceras (Mam.).

ilio- —See ile; also ily.

-ilis—1. L. *-ilis*, and *-bilis*, adjectival terminations added to the root of Latin words to denote capability, worth, of the character of. *Ex: amabilis*, worthy to be adored or loved < *amo*, to love; *ductilis*, ductile < *duco*, to lead; *horribilis*, dreadful < *horreo*, to be horrid; *hostilis*, hostile < *hostis*, an enemy; *fertilis*, fertile < *fero*, to produce, *parti-bilis*, divisible < *pars*, genit. *partis*, part: 2. *-ilis*, adj. ending meaning related to, derived from. *Ex: civilis* < *civis*, a citizen.

ilium—See ile.

ill—1. Gr. *illō*, to roll, turn; also to roll the eyes, look askance; *illōdēs*, squinting, distorted. *Ex:* Hydr-illa*; not Illigera*, named after C. W. Illiger, naturalist: 2. Gr. *illos*, the eye; also as an adj., squinting. *Ex:* Illo-sporium*: 3. Gr. *illas*, a thrush. *Ex:* Diaphor-illas (Av.).

illad—Gr. *illas*, genit. *illados*, a rope, band.

illaen—Gr. *illainō*, to look crosswise, to wink at, squint. *Ex:* Illaen-urus (Tril.); Illaenus (Tril.).

illaquaet—L. *illaquaetus*, trapped, entangled.

illas—See ill 3 and illad.

illecebros—L. *illecebrosus*, alluring.

illect—L. *illectus*, not collected, unread.

illep—L. *illepidus*, impolite, unpleasant.

illici—L. *illicium*, enticing, seductive. *Ex:* Illicium*.

illigat—L. *illigatus*, tied on, fastened.

illin—L. *illino*, to spread over, anoint; pp. *illitus*, smeared. *Ex:* Illino-dendron (Coel.).

Holly-leaved Burrobush, *Franseria ilici-folia*. The genus Fanseria was named after Ant. Franser, 18th century physician and botanist of Madrid. Redrawn from Desert Wild Flowers—Jaeger. Stanford University Press.

illit—See illin.

illod—See ill 1.

illot—L. *illotus*, dirty, unwashed.

illuden—L. *illudo*, to ridicule, deceive; ppr. *illudens*, genit. *illudentis*, mocking, deceiving.

-illus—L. *-illus*, *-a*, *-um*, diminutive suffix. *Ex:* Spong-illa (Por.).

illusori—L. *illusorius*, of a mocking character, illusionary.

illustr—L. *illustris*, bright, lustrous, illuminated; *illustratus*, pictured, made clear; pp. of *illustro*, to illuminate.

ily—Gr. *ilys*, genit. *ilyos*, mud, slime. *Ex:* Iliosuchus (Rept.); Ily-bius (Ins.); Ilyo-cryptus (Crust.); Ilys-anthes*; Ilysia (Rept.).

ilys—See ily.

im- —See in-.

im—L. *imus*, lowest. *Ex:* imi-ceps.

-im—L. *-im* > NL. *-ima*, superlative ending. *Ex:* Penth-ima*; pulchr-ima.

-ima—See -im.

imag—L. *imago*, genit. *imaginis*, likeness, image. *Ex:* imagin-al disk; imago.

imbecill—L. *imbecillus*, feeble.

imberb—L. *imberbis*, beardless.

imbib—L. *imbibo*, to drink in > Fr. *imbibition*. *Ex:* imbibition.

imbric—L. *imbrico*, to cover with tiles or scales; pp. *imbricatus*, covered with tiles or scales.

imbricat—See imbric.

imbut—L. *imbutus*, wetted, stained.

Hawskbill Turtle, *Chelone imbricata*. Redrawn from Cambridge Natural History. The Macmillan Co.

*Ex:* Imbric-aster (Echin.); Imbric-ina (Ann.); imbricate.

imit—L. *imitor*, to imitate; ɒp. *imitatus*, copied >*imitator*, one who imitates; *imitatio*, genit. *imitationis*, an imitation. *Ex:* Imitato-crinus (Echin.); Imitator (Ins.); Imito-ceras (Moll.).

imitat—See imit.

immers—L. *immersio*, to immerse; pp. *immersus*, submerged, immersed. *Ex:* Immersi-dens (Moll.).

immit—L. *immitis*, rough, savage, sour.

immixt—1. L. *immixtus*, not mixed: 2. *immixtus*, blended<*immisceo*, to intermix, blend.

immot—L. *immotus*=*inmotus*, immovable, motionless.

immun—L. *immunis*, free, exempt from public service; *immunitas*, freedom from public service or state burdens. *Ex:* immune; immun-ization, immunit-y, immuno-protein.

immund—L. *immundus*, foul, unclean.

immut—L. *immuto*, to change, alter; *immutator*, one who changes.

impact—L. *impactus*, driven in, pushed in; pp. of *impingo*, to push.

impar—L. *impar*, unequal. *Ex:* impari-pinnate.

imparat—L. *imparatus*, unfurnished, not made ready.

impatien—L. *impatiens*, genit. *impatientis*, without feeling, impatient. *Ex:* Impatiens*.

impavid—L. *impavidus*, fearless.

impens—L. *impensus*, large, ample, strong.

imperat—L. *imperator*, a leader, chief.

imperme—NL. *imperme-*<L. *in*, not+*per*, through+*meo* to go. *Ex:* imperme-ate: imperme-able.

impetigin—NL. *impetiginosus*, full of pustular eruptions<impetigo, a skin disease which is in turn derived from L. *impetere*, to attack, rush upon violently.

impiger—L. *impiger*, diligent, unwearied.

implant—NL. *implant*<L. *in*, in+*planto*, to plant, set. *Ex:* implant-ation.

implet—L. *impletus*, filled, made fat, full.

implex—L. *implecto*, to twist, entwine; pp. *implexus*, twisted.

impolit—L. *impolitus*, unpolished, rough.

imponen—L. *imponens*, genit. *imponentis*, deceiving; ppr. of *impono*, to impose, cheat.

impress—1. L. *impressus*, pressed into, impressed; pp. of *imprimo*: 2. L. *impressus*, not milked: 3. L. *impressus*, a pressing upon, impressing<*imprimo*.

impur—L. *impurus*, unclean, impure.

in-—1. L. *in-* (assimilated forms, *-il, -im, -in, -ir*), prefix meaning in, into, within, upon, against, towards. *Ex:* il-lude; im-bibe; im-pregnate; In-fusoria (Prot.); in-cisor; in-sect; ir-roratus: 2. L. *in-* (assimilated forms *il-, im-, ir-*, etc.), inseparable prefix, akin to Eng. *un-*, meaning not, without. *Ex:* Il-loricata (Rot.); im-berbis; Im-pennes (Av.): Im-placent-alia (Mam.); im-pudicus; in-ermis; ir-regularis.

-in—1. NL. *-in*, a suffix used in chemistry to denote an activator (pancreat-in, secret-in), also a derivative of (glycer-in). In some cases it is used in the making of names of elements (chlor-in): 2. Fr. and Eng. *-in*, suffix<L. *-inus, -a, -um*, belonging to, like. See -ine.

-ina—L. *-ina*, suffix denoting likeness, derived from L. neut. pl. of the adjectival ending *inus*. *Ex:* Anser-ina (Arth.); Tigr-ina (Mam.): 2. *-ina*, dim. suffix. *Ex:* Granat-ina (Av.).

inach—Gr. *Inachus*, a river god, son of Oceanus. *Ex:* inach-ides; Inach-oides (Crust.); Inachus (Crust.).

-inae—NL. *-inae*, suffix forming New-Latin names of sub-families of animals. *Ex:* Nymphal-inae (Ins.).

inaequal—L. *inaequalis*, unequal, uneven.

inan—L. *inanis*, empty<*inano*, to make empty; pp. *inanitus*, emptied; *inanitio*, genit. *inanitionis*, emptiness. *Ex:* inanition.

inanit—See inan.

inapert—L. *inapertus*, closed, not open.

inaurat—L. *inauratus*, covered with gold.

incan—L. *incanus*, hoary. *Ex:* incan-ous; Incania (Moll.).

incautus—L. *incautus*, heedless, improvident, wanting thrift.

incert—L. *incertus*, uncertain, doubtful.

incest—L. *incestus*, defiled.

inciden—1. L. *incidens*, genit. *incidentis*, ppr. of *incido*, to fall upon, to happen: 2. L. *incidens*, genit. *incidentis*, ppr. of *incido*, to cut through or into, to cut off.

incil—L. *incilis*, pertaining to a ditch<*incile*, a ditch.

incis—L. *incido*, to cut into, cut open, cut up; pp. *incisus*, cut. *Ex:* incisi-form; inciso-dentatus; Inciso-lema (Ins.); incisor<NL. *incisor*, a cutter.

incisiv—NL. *incisivus* <L. *incido*, to cut into+ *-ivus*, suffix forming verbal adj. of quality or tendency. *Ex:* foramen incisivum. See -ivus.

inclaud—NL. *inclaudens*, genit. *inclaudentis*, not closing; ppr. of *inclaudo*(<L. *in*, not+*claudo*, to close).

inclin—L. *inclinus*, bending, leaning over; also unbending.

inclus—L. *inclusus*, confined, shut up, included.

incol—L. *incola*, an inhabitant. *Ex:* Incol-acris (Ins.); prat-incol-ous.

incolorat—NL. *incoloratus* <L. *incolor*, without color+*-atus*, provided with, lit., not provided with color.

incommod—L. *incommodus*, troublesome.

incompt—L. *incomptus*, unadorned, rude.

inconstant—L. *inconstans*, genit. *inconstantis*, inconstant, fickle, unsteady, wavering.

incrassat—L. *incrassatus*, thickened, made stout.

increbescent—L. *increbesco*=*increbresco*, to become frequent, to increase; ppr. *increbescens*, genit. *increbescentis*, increasing.

increment—L. *incrementum*, an increase.

incrustat—L. *incrustatus*, covered with mud, smeared over, cloaked, pp. of *incrusto*; inf *incrustare*, to incrust.

incub—ML. *Incubus*, a demon supposed to be the cause of nightmare. *Ex:* Incubus (Ins.). S.

incud—L. *incudo*, to forge with a hammer; pp. *incudatus*, forged>*incus*, genit. *incudis*, an anvil. *Ex:* incudate; Incudi-fera (Ins.); incus.

incult—L. *incultus*, unadorned, rough.

incumb—L. *incumbo*, to lean, recline; ppr. *incumbens*, genit. *incumbentis*, leaning, reclining.

incus—See incud.

ind—Gr. *Indos*, name of a river in Asia, the *Indus*> India, name of a country in Asia, India>*Indikos*, of India, Indian>L. *indicum* >Sp. *indigo*, a dark-blue dye, indigo. *Ex:* Ind-agrion (Ins.); Ind-arctus (Mam.); Indigofera*; Indo-blastus (Echin.); Sap-indus*.

indagat—L. *indagator*, a tracker, searcher; *indagatrix*, a female searcher.

indentat—L. *indentatus*, indented, pp. of LL. *indento*, to notch, to cut into (<L. *in*, in+*dens*, genit. *dentis*, tooth).

indic—L. *Indicus*, of India.

indicat—L. *indico*, to point out; pp. *indicatus*, pointed out, designated; *indicator*, one who points out. *Ex:* Indicator (Av.).

indict—L. *indictus*, announced, fixed.

indigen—L. *indigenus*, native; *indigena*, a native. *Ex:* indigenous.

indigo—See ind.

individu—L. *individuus*, inseparable, not divided. *Ex:* individu-al.

indr.—Sanskr. *Indra*, god of thunderstorms. *Ex:* Indra-bovis (Mam.); Indra-therium (Mam.).

indris—Malagasay *indris*, name of a kind of animal, the lemur. *Ex:* Indris (Mam.).

indument—See indut.

indurat—L. *induratus*, hardened; pp. of *induro*, to make hard. *Ex:* indurat-ion.

indus—See indut.

indut—L. *induo*, to put on clothes; pp. *indutus*, clothed>*indumentus*, a garment; *indusium*, a tunic; *induviae*, clothes. *Ex:* indument; indusiate; indusium; induvi-al; induviae, pl. of *induvium*.

induv—See indut.

-ine—L. *-inus*, *-a*, *-um*, suffix denoting of or pertaining to, like, characterized by, as in can-ine, lacustr-ine; palustr-ine, rupestr-ine; *-in* and *ine* are used in forming names of elements or compounds as in brom-ine, lecith-in.

-ineae—NL. fem. pl. suffix used in forming suborder or sub-group names in botanical nomenclature. *Ex:* Querc-ineae*.

inebrians—L. *inebrians*, making drunk.

ineffect—L. *ineffectus*, not completed.

inept—L. *ineptus*, not suitable or fit, unbecoming, out of place.

inerm—L. *inermis*, unarmed, defenseless. *Ex:* Inermi-costa (Moll.); Inermia (Pisc.); inermis.

infan—L. *infans*, genit. *infantis*, an infant; speechless, mute>*infantilis*, of or belonging to infants.

infarct—L. *infarctus*, stuffed; pp. of *infarcio*, to stuff. *Ex:* infarct.

infaust—L. *infaustus*, unfortunate.

infect—1. L. *infectus*, unfinished: 2. L. *infectus*, stained, injected; *infectivus*, having to do with dyeing. *Ex:* infect-ion.

infectiv—See infect 2.

infens—L. *infensus*, hostile, bitter.

infer—L. *inferus*, underneath, low. *Ex:* Infer-arctic Zone; Inferi-cornia (Ins.); Infero-clypeus (Echin.).

infernal—L. *infernalis*, of the lower world.

infest—L. *infesto*, to attack, molest; ppr. *infestans*, genit. *infestantis*, molesting, disturbing.

infid—L. *infidus*, not trustworthy.

inflat—L. *inflo*, to blow into, inflate; pp. *inflatus*, inflated. *Ex:* Inflat-aster (Echin.); Inflati-ceras (Moll.).

inflex—L. *inflecto*, to bend, curve; pp. *inflexus*, bent.

infloresc—L. *infloresco*, to begin to bloom; ppr. *inflorescens* genit. *inflorescentis*. *Ex:* inflorescence.

inform—L. *informis*, deformed.

infortunat—L. *infortunatus*, unfortunate.

infra- —L. *infra-*, prefix indicating below, beneath, on the under side, etc. *Ex:* Infra-clypeus (Echin.); infra-red.

**infract**—L. *infractus*, broken, bent.

**infucat**—L. *infucatus*, painted, bedaubed, smeared.

**inful**—L. *infula*, a band, bandage>*infulatus*, adorned with a band. *Ex:* Insuful-aster is an error for Inful-aster (Echin.).

**infumat**—L. *infumatus*, dried in smoke, smoked.

**infundibul**—L. *infundibulum*, a funnel<*infundo*, to pour into. *Ex:* Infundibul-ops (Moll.); Infundibula (Ann.); infundibuli-form.

**infus**—L. *infundo*, to pour in; pp. *infusus*, poured in; *infusio*, genit. *infusionis*, a pouring in; LL. *infusor*, one who pours in>*infusorium*, a vessel for pouring. *Ex:* infusion; Infusoria (Prot.).

**infuscat**—L. *infuscatus*, made dark.

**inga**—NL. *inga*<some South American plant name. *Ex:* Inga*.

**ingen**—L. *ingens*, genit. *ingentis*, great, remarkable, large. *Ex:* ingens.

**ingest**—L. *ingestus*, pp. of *ingero*, to carry in> *ingestio*, genit. *ingestionis*, a pouring in, an introduction, ingestion.

**inglis**—*inglis*, an obsolete variation of *English*. *Ex:* Crypt-inglisia (Ins.).

**ingluv**—L. *ingluvies*, a crop. *Ex:* ingluvies.

**ingrat**—L. *ingratus*, unpleasant.

**inguin**—L. *inguen*=*inguina*, the groin>*inguinalis*, pertaining to the groin. *Ex:* inguinal; in guino-labial.

**inhaeren**—L. *inhaereo*, to adhere, stick fast; ppr. *inhaerens*, genit. *inhaerentis*, adhering.

**inhaes**—L. *inhaesus*, attached, made to adhere <*inhaereo*, to cleave to, hang to, adhere.

**inhibit**—L. *inhibitus*, kept back; *inhibitio*, genit. *inhibitionis*, a curbing. *Ex:* inhibit-ion; inhibit-ory.

**ini**—Gr. *inion*, the occiput, the muscle in the back of the neck. *Ex:* ini-ad; ini-encephalus; Ini-omi (Pisc.); Ini-ophthalma (Moll.); inio-glabell-ar; inion. See also inia.

**inia**—Bolivian *inia*, name of a kind of dolphin. *Ex:* Ini-idae (Mam.); Ini-opsis (Mam.); Inia (Mam.).

**inimic**—L. *inimicus*, hostile. *Ex:* inimic-al.

**initial**—L. *initialis*, original.

**injust**—L. *injustus*, excessive, wrongful.

**innat**—L. *innatus*, natural, inborn. *Ex:* innate.

**innix**—L. *innixus*, leaning upon; also ended, terminated.

**innoxi**—L. *innoxius*, not harmful.

**innuit**—L. *innuitus*, nodding, pp. of *innuo*, to nod.

**ino**—**1.** Gr. *is*, genit. *inos*, a fiber, muscle, nerve; also strength, force; *inōdēs*, fibrous. *Ex:* Ino-carpus*; Ino-ceramus (Moll.); ino-phyllus; ino-tropic: **2.** Gr. *Inō*, name of a sea goddess, the daughter of Cadmus. *Ex:* Ino (Moll.), etc.

**inoculat**—L. *inoculatus*, implanting, ingrafting; *inoculatio*, genit. inoculationis, an inoculating. *Ex:* inoculat-ion.

**inod**—See ino 1.

**inop**—L. *inops*, genit. *inopis*, poor, helpless, weak. *Ex:* inopi-folia.

**inopinat**—L. *inopinatus*, unexpected.

**inornat**—L. *inornatus*, undecorated, not beautiful.

**-inos**—Gr. *-inos*, suffix added to noun stems to form adjectives denoting the material or source of a thing. *Ex:* lith-inos, made of stone.

**-inous**—Eng. *-inous*<L. *-inus*, *-a*, *-um*, adj. ending meaning, belonging to. *Ex:* er-inous.

**inquiet**—L. *inquies*, genit. *inquietis*, restless.

**inquilin**—L. *inquilinus*, a tenant. *Ex:* inquiline; Inquilinium (Prot.).

**inquinan**—L. *inquinans*, genit. *inquinantis*, staining, making foul or impure; ppr. of *inquino*, to pollute, stain, discolor.

**inquinat**—L. *inquinatus*, befouled, made impure.

**inscript**—L. *inscriptus*, inscribed, written upon; also unwritten.

**insculpt**—L. *insculpo*, to cut in, engrave, imprint; pp. *insculptus*, etched, carved.

**inseren**—L. *inserens*, genit. *inserentis*, introducing; ppr. of *insero*, to insert, to introduce.

**insess**—L. *insideo*, to sit in, sit upon; pp. *insessus*, sat upon; *insessor*, a sitter. *Ex:* Insessores (Av.), the plural of *incessor*. See -es.

**insidians**—L. *insidians*, sitting in or upon; ppr. of *insideo*, to be seated, to be fixed.

**insign**—L. *insignis*, unique, well-marked, extraordinary. *Ex:* Insigni-pitta (Av.).

**insit**—L. *insitus*, grafted; *insitio*, a grafting. *Ex:* insit-ion.

**insolent**—L. *insolens*, genit. *insolentis*, different, unusual, arrogant. *Ex:* Insolenti-palpus (Ins.); Insolentia (Moll.).

**insolit**—L. *insolitus*, unaccustomed, uncommon.

**insonus**—L. *insonus*, silent.

**insperat**—L. *insperatus*, unexpected, not even hoped for.

**inspissat**—L. *inspissatus*, thickened.

**inspurcat**—L. *inspurcatus*, defiled.

**instar**—L. *instar*, form, figure. *Ex:* instar.

**institut**—L. *institutus*, grafted, planted, fixed, established, also begun, undertaken; pp. of *instituo*, to place, to found, establish.

**insuet**—L. *insuetus*, accustomed, habituated.

**insuful**—See **inful.**

**insul**—L. *insula*, an island. *Ex:* Insula-saurus (Rept.); Insuli-cola (Ins.); insul-in.

**insuls**—L. *insulsus*, tasteless, absurd.

**insurg**—L. *insurgo*, to rise up, rise upon; ppr. *insurgens*, genit. *insurgentis*, rising up, arising. *Ex:* Insurgus (Ins.).

**intact**—L. *intactus*, untouched, uninjured, without blemish, intangible.

**integer**—L. *integer*, entire, perfect, spotless; *integerrimus*, very perfect or complete.

**integr**—L. *integro*, to repair, renew; pp. *integratus*, renewed; *integer*, fem. *integra*, whole, unchanged. *Ex:* Integer-aster (Echin.); Integricardium (Moll.); integro-pallial; integrum.

**integument**—L. *integumentum* (<*in-*, in+ *tegumen* a cover), a covering. *Ex:* integument; integument-ary.

**inter**—L. *inter*, preposition meaning between, among. *Ex:* inter-costal; inter-osseous.

**intercal**—L. *intercalo*, to insert; pp. *intercalatus*, inserted; *intercalaris*=*intercalarius*, of or for insertion. *Ex:* intercalare; intercalate.

**interced**—L. *intercedo*, to be or go between, ppr. *intercedens*, going between.

**interfect**—L. *interfectus*, slaying, pp. of *interficio*, to destroy, slay; *interfector*, a murderer.

**interius**—See **interior**.

**internat**—L. *internasco*, to grow between; pp. *internatus*, grown between.

**interpres**—L. *interpres*, genit. *interpretis*, an explainer, translator, go-between.

**interrit**—L. *interritus*, not afraid, not frightened.

**interstinct**—L. *interstinctus*, separated, marked off in squares.

**intertext**—L. *intertextus*, interwoven.

**intestin**—L. *intestinus*, internal>*intestinum*, an intestine. *Ex:* intestin-al.

**intim**—L. *intimus*, innermost. *Ex:* intima.

**intort**—L. *intortus*, twisted, distorted.

**intr**—L. *intra*, inside>*intrinsecus*, on the inside. *Ex:* intr-oxyl-ic; intrinsic; intro-venous.

**intribulat**—L. *intribulatus*, pressed, pressed out.

**intrinsic**—See **intr**.

**introit**—L. *introitus*, entered; as a noun, an entrance.

**intrud**—L. *intrudo*, to thrust in; pp. *intrudus*, intruded.

**intrus**—L. *intrusus*, thrust in, intruded; pp. of *intrudo*, to thrust.

**intus**—L. *intus*, within, on the inside. *Ex:* Intus-plicata (Moll.); intus-susception.

**intut**—L. *intutus*, without guard, insecure, unsafe.

**intyb**—L. *intybus*, chicory<Gr. *entybon*, chicory. *Ex:* intyb-aceus.

**inul**—L. *inula*, the plant called elecampane. *Ex:* Inula*.

**inult**—L. *inultus*, unhurt, not revenged.

**inuncan**—L. *inuncans*, genit. *inuncantis*, hooked, covered with hooks, ppr. of *inunco*, to hook.

**inuncat**—L *inuncatus*, seized, caught on a hook.

**inunct**—L. *inunctus*, smeared, covered with grease or oil<*inunguo*, to smear. *Ex:* inunction.

**inundat**—L. *inundatus*, overflowed, submerged.

**-inus**—L. *-inus, -a, -um*, suffix added to noun stems to form adjectives meaning belonging to, like. *Ex:* Abr-inus (Ins.); manat-inus; panamint-inus; radul-inus.

**inust**—L. *inustus*, burned.

**inutil**—L. *inutilis*, without use; also injurious.

**invers**—L. *inversus*, inverted, upset; pp. of *inverto*, to put upside down, invert.

**invis**—L. *invisus*, overlooked, unseen; also hated, detested.

**involucr**—NL. *involucrum*, a wrapper, envelope <*involvo*, to roll up, wrap up. *Ex:* involucratus; involucre.

**io**—1. Gr. *Iō*, daughter of the river god, Inachus. *Ex:* Io (Ins.): 2. Gr. *ios*, an arrow; also poison. rust >*iodokos*, holding poisoned arrows. *Ex:* Io-ctonus (Arach.); Io-glossus (Pisc.); io-terium; iodocus.

**-io**—See **-ion** 2.

**iod**—See **ion**.

**iodoc**—See **io**.

**iole**—L. *Iolē*, daughter of Eurytus. *Ex:* Iol-ella (Crust.); Iole (Crust.).

**-ion**—1. Gr. *iōn*, neut. of *ion*, ppr. of *eimi*, to go, come; to enter, penetrate, to enter with violence. *Ex:* ion; cat-ion. 2. L. *-io, -ionis*, noun suffix denoting thing acted upon, state, result of. *Ex:* castrat-ion; solut-ion; summation. See also -idion and ion: 3. Gr. *-ion* or *-iōn*, occasionally used patronymic suffix; as Kron-ion, son of Cronos: 4. Gr. *-ion*, dim. ending. Ex: Stephan-ion*.

**ion**—Gr. *ion*, genit. *iontos*, a violet>*ioeidēs*, like a violet, violet-colored. *Ex:* Iod-amoeba (Prot.); Ioda-monas (Prot.); Iodo-pleura (Av.); Ion-idium*; Ion-opsis*; Ion-oxalis*; Calyptr-ion*; Erpet-ion*; Leuco-jum*. See also -ion.

**ionth**—1. Gr. *ionthas*, genit. *ionthados*, downy, soft, shaggy, speckled: 2. Gr. *ionthos*, young hair, the root of a hair.

**-ior**—L. *-ior* (*neut. -ius*), masculine and feminine ending added to adjectival stems to form the comparative; as, *hebetior* from *hebes*, dull (stem, *hebet-*), etc. *Ex:* angust-ior: brev-ior, vivid-ior. This same ending *-ior* is sometimes used by modern word coiners as an ending for other adjective stems.

**ior**—Gr. *iōros*, a watchman. *Ex:* Ioro-pus (Av.).

**iph**—1. Gr. *Iphis*, one of the Argonauts; also a Cretan girl who was changed into a man. *Ex:* Iphis (Ins.), (Crust.); Iphisa (Rept.): 2. Gr. *iphi*, stoutly, mightily. *Ex:* Iphi-crates (Ins.); Iphi-pus (Ins.).

**iphigen**—Gr. *Iphigeneia*, daughter of Agamemnon; *iphigeneia*, most valiant, mighty. *Ex:* Iphigenia (Moll.).

iphion—Gr. *iphion*, name of some herb. *Ex:* Iphion*.

iphthim—Gr. *iphthimos*, spirited, strong. *Ex:* Iphthimo-rhinus (Ins.); Iphthimus (Ins.).

ipn—Gr. *ipnos*, an oven, furnace, lantern. *Ex:* Ipn-ops (Pisc.); Ipno-domus (Av.).

ipomoea—See ips.

ippus—See hipp.

ips—1. Gr. *ips*, genit. *ipos*, a worm that eats vines, also one that eats wood <*iptō*, to injure. *Ex:* Ips (Ins.): 2. Gr. *ipsos*, ivy. *Ex:* Ipomoea*, see homoe.

ipsi—NL. *ipsi-* <Gr. *ipos*, same. *Ex:* ipsi-lateral.

iracund—L. *iracundus*, red with anger, angry.

iren—Gr. *Eirēnē*, goddess of peace <*eirēnē*, peace, a time of peace. *Ex:* Irena (Av.). See eiren.

iresin—Gr. *eiresiōnē*, a harvest garland wound with wool, a crown. *Ex:* Iresine*.

irid—Gr. *iris*, genit. *iridos*, the rainbow, the iris of the eye, name of a kind of lily >NL. *irideus*, rainbow-like. *Ex:* irid-escent; Iridi-pitta (Av.); Irido-myrmex (Ins.); Iris*.

irio—L. *irio*, genit. *irionis*, a kind of cress.

iris—See irid.

irpex—See irpicin.

irpicin—NL. *irpicinum* <L. *irpex*, genit. *irpicinis*, a large rake with iron teeth, a harrow.

irremot—L. *irremotus*, unmoved.

irretit—L. *irretitus*, caught in a net.

irrigu—L. *irriguus*, supplied with water.

irris—L. *irrisor*, one who derides, a mocker; *irrisus*, a scoffing, mocking. *Ex:* Irrisor (Av.).

irritabilit—L. *irritabilitas*, irritability; *irritabilis*, easily excited. *Ex:* irritabilit-y.

irrorat—L. *irroratus*, moistened with dew <*in,* upon+*roro*, to distil dew <*ros*, genit. *roris,* dew. The past participle *irroratus* is sometimes erroneously used to mean freckled or mottled.

is—Gr. *isos*, equal, similar. *Ex:* is-anthous; Is-anthus*; Is-aria*; Is-urus (Elasm.); Iso-lepis*; Iso-poda (Crust.); iso-tropic.

-is—Gr. *-is* genit. *-idos*, patronymic suffix meaning daughter of; as Priam-is, daughter of Priam. See isid.

isat—Gr. *isatis*, an herb supplying dye. *Ex:* Isatis*, perh. <Gr. *isazō*, to rend equal.

-isc—NL. *-iscus*, *-a*, *-um*, dim. suffix <Gr. *-iskos*. *Ex:* Bassar-iscus (Mam.); Bemat-iscus (Mam.); Centr-iscus (Pisc.); Centr-isc-idae (Pisc.); Troch-iscus (Moll.).

isch—Gr. *ischō*, to restrain, check. *Ex:* isch-ury (Med.); isch-aemia (Med.).

ischi—Gr. *ischion*, the hip-joint, hip. *Ex:* ischio-cele (Med.); ischium.

ischn—Gr. *ischnos*, slender, withered, weak. *Ex:* Ischn-ura (Ins.); Ischno-chiton (Moll.).

ischy—Gr. *ischys*, strength. *Ex:* Ischy-odus (Pisc.); Ischy-pterus (Pisc.); Ischyo-pteron (Ins.).

ischyr—Gr. *ischyros*, strong. *Ex:* Ischyro-mys (Mam.); Ischyro-psal-idae (Arach.).

-iscus—See -isc.

isdrom—Gr. *eisdromē*, an attack, assault. *Ex:* Isdromas (Ins.).

isid—Gr. *Isis*, genit. *Isidos*, Egyptian goddess of fecundity. *Ex:* Isid-ium; Isis (Coel.); Isis-ina (Coel.).

isis—See isid.

island—NL. *islandus* <Dan. *Island*, Iceland > NL. *islandicus*, of or pertaining to Iceland.

-ism—Eng. suffix *-ism* (<Gr. *-ismos*, L. *-ismus*), often meaning, when added to nouns, a state or a condition, a doctrine or practice of. *Ex:* Darwin-ism.

iso—See is.

isoet—Gr. *isoetēs*, name of a kind of plant, the small houseleek <*isoetēs*, equal in years <*isos*, equal+*etos*, year. *Ex:* Isoetes*.

-issimus—L. *-issimus*, *-a*, *-um*, superlative suffix denoting very much, most. The superlative of Latin adjectives is regularly formed by adding this suffix to the stem of the positive which then loses its final ending. *Ex:* ramos-issimus; virid-issimus.

-ist—L. *-ista*, suffix denoting one who practices, one who is skilled in. *Ex:* systemat-ist.

istem—Gr. *istēmi*, to make to stand, to place. *Ex:* end-istem.

ister—See hist.

isthm—Gr. *isthmos*, any narrow passage such as a neck of land between two seas; *isthmikos*, like an isthmus. *Ex:* isthmi-an; Isthmio-phora (Platy.); Isthmo-coris (Ins.).

-istic—NL. *-istic*, adj. suffix, meaning pertaining to as agent <Gr. *-istes* +Eng. *-ic*. *Ex:* pan-o-istic, see -oistic; mer-o-istic, hol-o-istic.

istio—See histio.

-istos—Gr. *-istos*, suffix added to certain adjectives to form superlatives; exactly similar to L. *-issimus*, see above. *Ex:* brach-istos; tach-istos.

istos—Gr. *histos*, a web, a tissue. *Ex:* Micr-ist-odus (Elasm.). See also hist.

-istus—NL. *-istus* <Gr. *-istos*, superlative suffix. *Ex:* call-istus, cf. Gr. *kallistos*, most beautiful.

-ita—Sp. *-ita*, dim suffix. *Ex:* Nomad-ita (Ins.).

itam—Gr. *itamos*, hasty, reckless; *itamotēs*, boldness. *Ex:* Itamo-plex (Ins.); Itamus (Ins.).

ite—Gr. *itea*, a willow; also a wicker shield. *Ex:* Itea*; Itea-crinus = Iteo-crinus (Echin.); Itea-phila (Ins.); iteo-logy; Iteo-myia (Ins.).

-ite—1. Eng. *-ite* (<Gr. *-itēs*), suffix denoting one of a group, a native of, etc. *Ex:* polyp-ite = polyp-ide: 2. Eng. *-ite* (<Gr. *-itēs*), suffix used to denote that a thing is closely joined to

or is a part of something. *Ex:* exo-pod-ite, stern-ite:    **3.** Eng. *-ite* (<Gr. *-itēs*), suffix denoting the fossil nature of a thing. *Ex:* trilob-ite.

**iter**—L. *iter*, a passage. *Ex:* iter.

**-ites**—Gr. *-itēs*, suffix denoting having to do with, of the nature of, like, belonging to; also denoting agent or doer. *Ex:* Ocean-ites (Av.); Tring-ites (Av.). It is often used arbitrarily to indicate the fossil character of a genus. *Ex:* Bactr-ites (Moll.); Bacul-ites (Moll.); Cupres-ites*; Eu-cet-ites (Mam.); Pithecul-ites (Mam.).

**ith**—Gr. *ithys*, erect, straight. *Ex:* Itha-genes (Av.), etc.; Itho-mia (Ins.); Ithys (Av.); Ithy-phallus (Ins.); Ithyo-clino-stomum (Platy.).

**ithys**—See **ith.**

**-itic**—NL. *-itic*, suffix < Gr. *-itikos*, pertaining to, of the nature of. *Ex:* dendr-itic; arthr-itic.

**itin**—Gr. *iteinos*, made of willow. *Ex:* itinus.

**-ition**—Eng. *-ition*, suffix of compound nouns, meaning act of. *Ex:* duglut-ition; imbib-ition.

**-itis**—**1.** L. *-itis*, adj. termination meaning like, of the nature of (Gr. *-itēs*, masc.; *-itis*, fem.) > Eng. *-ite*. See -ite 3:    **2.** Gr. *-itis* (fem.) > NL. *-itis*, suffix used in medical terminology to indicate inflammation of a part. *Ex:* tonsil-itis, rhin-itis.

**itonid**—NL. *itonid* < lepidopteran genus *Itonia*. *Ex:* Itonid-idae (Ins.).

**-itus**—L. *-itus*, *-a*, *-um*, adjectival ending, meaning provided with, having. *Ex:* piper-itus.

**ity**—Gr. *itys*, genit. *ityos*, an embrace, a garland; also the rim of a shield or the shield itself. *Ex:* Ityo-cara (Ins.); An-itys (Ins.).

**-ity**—Eng. *-ity* < thematic vowel *-i-*+*-ty*, suffix used in forming abstract nouns. *Ex:* poros-ity; variabil-ity.

**itys**—See **ity.**

**iul**—**1.** L. *iulus*, a catkin, plant-down < Gr. *ioulos*, down, the down of plants. *Ex:* Iulo-tricha (Ins.); juli-flora:    **2.** L. *Iulus*, son of Aeneas. *Ex:* Iulus (Moll.):    **3.** Gr. *ioulos*, a centipede. *Ex:* Jul-idae (Myr.); jul-oid; Julus (Myr.); Acanth-iulus (Myr.).

**-ium**—**1.** L. *-ium*, suffix added to noun and verb-stems denoting offices and groups. *Ex:* sensor-ium:    **2.** NL. *-ium*, in ecology used as a loca-

tive suffix denoting a formation. *Ex:* driod-ium (Ecol.):    **3.** NL. *-ium*, suffix used in forming names of chemical elements. *Ex:* sod-ium, stront-ium:    **4.** NL. *-ium* < Gr. *-ion*, dim. ending. *Ex:* pyren-ium; pyxid-ium:    **5.** NL. *-ium*, ending of generic names, consisting of euphonic *i*+L. neut. ending, *-um*, *Ex:* Centaur-ium*; Chelidon-ium*. See also -ia.

**-ius**—**1.** L. *-ius*, suffix expressing passive but occasionally active qualities. *Ex:* *eximius* excepted, exempt; also select, distinquished, extraordinary:    **2.** NL. *-ius*, suffix used in Latinizing personal names. *Ex:* Brown-ius: **3.** L. *-ius*, neuter singular comparitive ending added to adjectival stems. See -ior:    **4.** L. *-ius*, adj. suffix signifying belonging to, derived from, as *regius* < *rex*, the king.

**-ive**—Eng. *-ive*, adj. suffix < L. *-ivus*, *-a*, *-um*. See -ivus. *Ex:* disjunct-ive; capt-ive; abort-ive.

**-ivus**—L. *-ivus*, *-a*, *-um* (*-vus*), suffix added to verb stems to form verbal adjectives expressing quality or tendency, fitness, pertaining to, as; *proter-vus*, violent < *protero*, to trample; *incis-ivus* < *incido* to cut into; *recid-ivus*, restored < *recido*, to fall back.

**ix**—Gr. *ixos*, the mistletoe berry or the mistletoe plant; also birdlime, prepared from mistletoe; sticky, a miserly fellow; *ixōdēs*, like birdlime. *Ex:* Ix-oreus (Av.); Ixia*; Ixio-lirion*; Ixo-brychus (Av.), here *ixos* is taken to mean "reed"; see Jordan, Manual of Vertebrate Animals, p. 285; Ixon-anthes*; Ixodes (Av.).

**ixal**—Gr. *ixalos*, jumping, darting. *Ex:* Ixal-idium (Ins.); Ixalis (Amph.).

**ixod**—See **ix.**

**ixor**—NL. *Ixora*, name applied to a genus of plants < Sanskr. *icvara*, a master, lord; cited as the name of a Malabar deity to whom flowers are offered. *Ex:* Ixor-eae*; Ixora*.

**ixys**—Gr. *ixys*, the loins, the small of the back. *Ex:* Cin-ixys (Rept.).

**iyng**—Gr. *iynx*, genit. *iyngos*, name of a kind of bird, the wryneck. *Ex:* Iyngi-picus (Av.); Iynx (Av.).

**iynx**—See **iyng.**

**-iza**—NL. *-iza*, a mutilated form of Gr. *spiza* a finch. *Ex:* Acanth-iza (Av.); Aphr-iza (Av.)

# J

(J = consonant I and the two are often
Latinized interchangeably.)

**jacamar**—Tupi *jacama-ciri*, name of a kind of
bird, the jacamar. *Ex:* Jacamar-alcyon (Av.).

**jacarand**—Tupi *jacaranda*, name of a kind of
tree. *Ex:* Jacaranda*.

**jactat**—L. *jactatus*, tossed about, scattered;
*jactator*, a boaster, a person who makes a
display of himself.

**jacul**—L. *jaculor*, to throw > *jaculator*, a thrower.
*Ex:* Jaculus (Mam.); e-jaculator.

**jambosa**—Hindu *jambu*, vernacular name of the
rose-apple. *Ex:* Jambosa*.

**jan**—L. *Janus*, Roman god with two opposite
faces. *Ex:* jani-ceps; Janus (Ins.); Janus-ia*.

**janth**—See ianth.

**japyg**—Gr. *Iapyx*, genit. *Iapygis*, a native of the
south of Italy. *Ex:* Japyg-idae (Ins.); Iapyx
(Ins.).

**japyx**—See iapyx.

**jasion**—Gr. *iaisōnē*, some wild pot-herb, now unknown. *Ex:* Jaisone*.

**jasmin**—Ar. *yāsmīn*, name of a kind of shrub.
*Ex:* Jasminum*.

**jasp**—Eng. *jasper* < L. *iasper*, genit. *iaspides*,
jasper; NL. *jaspideus*, like jasper.

**jass**—NL. *jassus*, name applied to a genus of
homopterous insects, perh. < L. *Iassus*, name of
a town on the coast of Caria. *Ex:* Jass-idae
(Ins.); Jassus (Ins.).

**jat**—Gr. *iatēr* = *iatēs* = *iatros*, a physician. *Ex:*
Jat-ropha*, see troph; Jateo-rhiza*.

**jejun**—L. *jejunus*, empty, hungry, fasting. *Ex:*
jejunum.

**jocinor**—L. *jocinor*, genit. *jocinoris*, the liver.

**jocos**—L. *jocosus*, full of jest or fun.

**jub**—1. L. *juba*, a mane > *jubatus*, having a mane,
maned, crested. *Ex:* Jub-ella (Bry.); juba;
jubate:    2. L. *Juba*, king of Numidia. *Ex:*
Juba-ea*.

**jucund**—L. *jucundus*, pleasant, agreeable. *Ex:*
Jucund-acris (Ins.); Jucundus (Ins.).

**jug**—L. *jugo*, to join; marry < *jugum*, a yoke >
*jugalis*, pertaining to a yoke. *Ex:* jugum; A-
juga*; con-jug-ants; con-jug-ation.

**jugat**—L. *jugatus*, joined, connected, pp. of *jugo*,
to join, marry.

**jugland**—L. *juglans*, genit. *juglandis*, a walnut,
walnut tree < *Jovis*, Jove, Jupiter + *glans*, an
acorn or any acorn-shaped fruit. *Ex:* Juglando-
crinus (Echin.); Juglans*.

**juglans**—See jugland.

**jugular**—NL. *jugularis* < L. *jugulum*, the collar-
bone. *Ex:* jugular; Jugulares (Pisc.).

**jugulat**—L. *jugulator*, a slayer, cut throat.

**juic**—Eng. *juice* < ME. *juce, juse* < L. *jus*, genit.
*juris*, soup, broth, gravy. *Ex:* juic-y; juice.

**jul**—See iul.

**julis**—L. *julis*, a kind of rock-dwelling fish.

**jum**—NL. *-jum* < Gr. *ion*, violet. *Ex:* Leuco-
jum*.

**junc**—L. *juncus*, a rush > *junceus*, made of
rushes, rush-like. *Ex:* Junco (Av.); Juncus*.

**juncag**—NL. *juncago*, genit. *juncaginis*, name applied to a genus of rush-like plants < L. *juncus*,
a rush. *Ex:* Juncagin-aceae*; Juncago*.

**junct**—L. *junctus*, joined; pp. of *jungo*, to join,
unite, fasten. *Ex:* dis-junctus.

**juniper**—L. *juniperus*, the juniper tree. *Ex:*
juniper-inus; Juniperi-fex (Ins.); Juniperus*.

**jurass**—Fr. *Jurassique*, pertaining to the Jura
Mountains, also to the Jurassic period.
*Ex:* Jurassi-cardium (Moll.); Jurrasi-phorus
(Moll.), probably an error for Jurassi-phorus.

**jurgios**—L. *jurgiosus*, quarrelsome, full of dispute
< *jurgo*, to dispute.

**jurras**—See jurass.

**juven**—L. *juvenis*, young > *juvenilis* also *juvenalis*
youthful; *juvenesco*, to grow up, ppr. *juvenes-
cens*, genit. *juvenescentis*, growing up. *Ex:*
juvenal; juvenile; re-juvenescence.

**juvenc**—L. *juvencus*, youthful, young; also a
young bullock.

**juxta**—L. *juxta*, near to, nigh. *Ex:* juxta-posi-
tion; Juxta-pulex (Ins.); juxta-spinal.

# K

(The letter K is often used interchangeably with the letter C, hence many combining forms in K are listed under C. A great many of the generic and specific names in K are derived from personal and geographic names. S.

**kal**—1. Gr. *kalia*, dim. *kalidion*, a granary, hut, bird's nest. *Ex:* kalidium:   **2.** Gr. *kalon*, wood. *Ex:* Kalo-termes (Ins.):   **3.** Gr. *kalos* = poetical *kallimos*, beautiful; *kallos*, beauty. *Ex:* Kalli-trichia (Ins.); Kallima (Ins.); Kalio-sphinga (Ins.); Kallo-bombus (Ins.), see cal.

**kall**—See **kal 3.**

**kallim**—See **kal 3.**

**kann**—Gr. *kanna*, reed. *Ex:* Kanna-bateo-mys (Mam.).

**kar**—Austrian *kar*, hollows dug out by glaciers. *Ex:* kar-herbage (Ecol.).

**kary**—See **cary.**

**kat**—See **cat.**

**kathet**—Gr. *kathetos*, vertical. *Ex:* Katheto-stoma (Pisc.).

**keitloa**—East African *keitloa*, name of the two-horned rhinoceros. *Ex:* Keitloa (Mam.).

**kelaen**—Gr. *kelainos*, black. *Ex:* Kelaena-nes-ian (Ethn.).

**kelea**—See **cele.**

**ken**—Gr. *kenos* = *keneos*, empty. *Ex:* ken-apo-phytes; ken-enchyma; Ken-odon (Mam.); Keno-dactylus (Ins.).

**kentr**—See **cent.**

**ker**—See **cor 1.**

**kerat**—See **cer 1.**

**kerm**—See **cherm.**

**kin**—Gr. *kineō*, verbal adj. *kinētos*, to move > *kinēsis*, movement; *kinētikos*, causing motion. *Ex:* kineto-nucleus; Kino-thorax (Rept.); A-cinet-actis (Prot.); karyo-kinesis; Poly-cinetis (Ins.).

**kinesis**—See **kin.**

**kinet**—See **kin.**

**kinkl**—See **cincl 1.**

**kiss**—Gr. *kissos*, ivy. *Ex:* Kisso-phagus (Ins.).

**kitt**—See **citt.**

**kleo**—Gr. *kleos*, a rumor, report; also fame, glory. *Ex:* Kleo-thrips (Ins.).

**klin**—Gr. *klinō*, to turn aside, bend. *Ex:* klino-morphy; Klino-thrips (Ins.).

**knem**—See **cnem.**

**koal**—Native Australian *koolah*, name of the kangaroo > Eng. *koala*, name of a kind of marsupial mammal, the "native bear" of Australia. *Ex:* Koala (Mam.); Koale-mus (Mam.).

**kogia**—NL. *kogia*, "a barbarous and unmeaning name" but perh. < "*Cogia Effendi*, who observed whales in the Mediterranean." *Ex:* Kogia (Mam.).

**koir**—See **choer.**

**kole**—See **cole.**

**kolen**—See **colen.**

**koll**—See **coll 2.**

**kont**—Gr. *kontos*, a pole. *Ex:* chondrio-kont.

**korem**—See **corem.**

**krik**—See **cric.**

**krimn**—Gr. *krimnon*, a kind of coarse meal. *Ex:* Krimno-chelidon (Av.).

**krit**—Gr. *kritos*, chosen, picked out, separated. *Ex:* krito-chyma; Krito-saurus (Rept.).

**krumm**—Ger. *krumm*, crooked. *Ex:* krumm-holz (Ger. *Holz*, wood.).

**kumb**—Gr. *kumbē*, a boat, cup, bowl; also the head. *Ex:* kumbe-cephalic (Anthrop.).

**kurt**—Gr. *kurtos*, curved, arched, humped; also a bird-cage, a fishing basket. *Ex:* Kurt-odon (Mam.); Kurti-formes (Pisc.); Kurtus (Pisc.).

**kym**—See **cym.**

**kyn**—Gr. *kynos*, dog. *Ex:* Kynos (Mam.).

**kyph**—Gr. *kyphos*, humpbacked, gibbous. *Ex:* kyph-osis; Kypho-balaena (Mam.); Kypho-clon-ella (Por.); Kyphus (Moll.).

**kyrt**—See **cyrt 2.**

**kyt**—Gr. *kytos*, a hollow, anything that contains something. *Ex:* Kyto-rrhinus (Ins.).

# L

la—1. Gr. *las*, genit. *laos*, stone. *Ex:* La-ornis (Av.); Lao-phonte (Crust.); Lao-pithecus (Mam.); Strepsi-las (Av.): 2. Gr. *laos*, people. *Ex:* la-rithmics, Gr. *arithmos*, numbers.

lab—1. *labia*=*labium*, dim. *labiellum*, a lip; *labiatus*, lipped; Low L. *labialis*, pertaining to the lips; *labeo*, one who has large lips. *Ex:* Labeo (Pisc.); Labeo-scala (Moll.); Labi-atae*; labi-ose; labi-palpi; labia cerebri; labial; labio-dental; Labo-chirus (Arach.).

labe—1. Gr. *labē*, a handle, a taking hold or accepting. *Ex:* Cerco-labes (Mam.); Oncino-labes (Echin.): 2. L. *labes*, a spot, defect: 3. L. *labes*, genit. *labis*, a falling down, sinking in.

labecul—L. *labecula*, a stain, disgrace.

labell—See labr.

labeo—See lab.

labes—See labe.

labid—1. Gr. *labis*, genit. *labidos*, a pair of forceps, a handle, clasp. *Ex:* Labid-esthes (Pisc.); Labid-ura (Ins.); Labidia (Ins.); labido-phorous; Labido-saurus (Rept.); Za-labis (Mam.): 2. L. *labidus*, slippery.

labil—L. *labilis*, neut. *labile*, slipping, gliding. *Ex:* labile.

labis—See labid.

labium—See lab.

laboratory—LL. *laboratorium*, a place for work, especially of a scientific nature. Cf. L. -*orium*> Eng. -*ory*, a suffix added to noun stems meaning place of or for, as in ambulatory, a place for walking.

labori—L. *labor*, toil; *laboriosus*, laborious. *Ex:* Labori-ops-thyrus (Ins.).

labr—1. L. *labrum*, dim. *labellum*, a lip; *labrosus*, thick-lipped. *Ex:* Labelli-nacra (Moll.); label lum; Labri-stomus (Pisc.): 2. Gr. *labros*, fierce, greedy, boistrous. *Ex:* Labr-odon (Pisc.); Labro-phagus (Pisc.); Labro-saurus (Rept.); Scissi-labra (Moll.): 3. L. *labrus*, a kind of fish. *Ex:* Labr-idae (Pisc.); Labrus (Pisc.).

labrac—Gr. *labrax*, genit. *labrakos*, the sea-wolf or bass<*labros*, greedy=NL. *labrax*, genit. *labracinis*, a genus of fishes. *Ex:* Labr-oides (Pisc.); Labrac-opsis (Pisc.); Labracinus (Pisc.); Labrax (Pisc.).

labrax—See labrac.

labrossyt—NL. *labrossyta*<Gr. *labrosytos*, rushing furiously. *Ex:* Labrossyta (Ins.).

laburn—L. *laburnum*, the bean-trefoil. *Ex:* Laburnum*.

labyrinth—Gr. *labyrinthos*, a tortuous passage, any coiled up body. *Ex:* Labyrinth-odon (Amph.); Labyrinth-ula (Prot.); labyrinthi-form; Labyrintho-myxa (Prot.).

lac—1. L. *lacus*, a basin, a lake, pond; originally anything hollow<Gr. *lakkos*, a cistern. *Ex:* Laco-somat-idae (Ins.); Lacus (Ins.): 2. Gr. *lakkos*, a hole or pit. *Ex:* Bathy-laca (Pisc.): 3. Gr. *lakis*, genit. *lakidos*, a tear, rending. *Ex:* Laci-phorus (Ins.); Laci-stema*. See also lacc 1.

lacathea—See cathem.

lacc—1. Fr. *lac*=Ital. *lacca*, varnish; related to Pers. *laka* and Hind. *lakh*, to dye. *Ex:* lac-scale; lacc-ase; Phyto-lacca*: 2. Gr. *lakkos*, a pond, pit. *Ex:* Lacco-philum (Coel.); Lacco-saurus (Amph.); Cato-laccus (Ins.); see lac.

lacer—L. *lacero*, to tear, pp. *laceratus*, torn to pieces, mangled; *lacerus*, mangled, torn; Fr. *laceration*<L. *laceratio*, genit. *lacerationis*, a tearing. *Ex:* laceration.

lacert—1. L. *lacerta*, a lizard; ML. *lacertilis*, of or pertaining to a lizard. *Ex:* Lacerta (Rept.); lacerti-form; Lacertilia (Rept.); Lacertina (Rept.): 2. L. *lacertosus*, strong, powerful <*lacertus*, the muscular part of the arm. *Ex:* Lacerto-belus (Ins.).

laceryz—Gr. *lakeryza*, one that cries. *Ex:* Laceryzon (Av.).

lachan—Gr. *lachanon*, a vegetable, garden herb. *Ex:* Lachana (Ins.).

lache—Gr. *lachos*, share, portion; *Lachesis*, one of the three Fates, Disposer of lots<*lacheō*, to apportion by lot. *Ex:* Laches-ana (Arach.); Lachesis (Rept.); not Lachenalia*, named after W. de Lachenal, Swiss botanist.

laches—See lache.

lachn—Gr. *lachnē*=*lachnos*, woolly hair, down; *lachnaios*=*lachnēeis*, woolly. *Ex:* Lachn-anthes*; Lachn-aphis (Ins.); Lachnaea*; Lachneis (Ins.); Lachno-campa (Ins.); Lachno-mys (Mam.); Lacno-desmus (Myr.) for Lachno-desmus (Myr.); Di-lachnus (Ins.).

lachos—See lache.

lachrim—L. *lachrima*, an old form of *lacrima*, pl. *lacrimae* >ML. *lachrymalis*, of or pertaining to tears. *Ex:* lachrimae-form; lachrimal= lacrimal=lachrymal.

lachrym—See lachrim.

lacid—Gr. *lakis*, genit. *lakidos*, a rent; *lakistos*, torn. *Ex:* Lacist-odes (Ins.); Lacisto-rhynchus (Platy.).

lacin—L. *lacinia*, a thing torn, the edge of a garment. *Ex:* Lacin-aria*; lacini-ate; Lacini-orbis (Moll.); lacinia; laciniol-ate, dim. of laciniate; lacinul-ate; lacinula.

lacis—See lacid.

lacn—See lachn.

lacrim—See lachrim.

lact—L. *lac*, genit. *lactis*, milk, see gala; *lacteus*, milky; *lactescens*, genit. *lactescentis*, becoming milky, ppr. of *lacteo*, to milk; *lactarius*, belonging to milk; *lactuca*, lettuce, so called because of its milky juice. *Ex:* Lactarius*; lacte-al; lactescent; lacti-vorous; Lacto-bacillus*; Lactuca*.

**lactic**—Gr. *laktikos*, stubborn, calcitrant. *Ex:* Lactica (Ins.). See also lact.

**lactuc**—See lact.

**lacun**—L. *lacuna*, ditch, pit; *lacunosus*, full of pits, pitted; *lacuno*, to hollow out. *Ex:* Lacunella (Moll.); lacun-ule; Lacuna (Moll.); Lacunos-ella (Brach.); lacunose. See also lacunar.

**lacunar**—L. *lacunar*, genit. *lacunaris*, a paneled ceiling, so called from its sunken spaces or *lacunae*.

**acustr**—NL. *lacuster*, genit. *lacustris*, pertaining to a lake<*lacus*, a lake. *Ex:* lacustr-al; lacustrine; Lacustri-cola (Pisc.).

**lacydes**—L. *Lacydes*, Academician of Cyrene, pupil of Arcesilas. *Ex:* Lacydes (Ann.).

**ladan**—Gr. *lēdanon* or *ladanon*, a gummy substance gathered from Cistus creticus, or *lada*. *Ex:* Ladanum*; ladani-folia.

**ladas**—Gr. *Ladas*, one of Alexander the Great's runners whose name became a proverb for speed. *Ex:* Ladas (Moll.).

**ladon**—Gr. *Ladōn*, one of Actaeon's hounds; also mythical father of Daphne.

**lae**—Gr. *laios*, left. *Ex:* Laeo-cochlis (Moll.); laeo-torma; laeo-tropic.

**laedor**—See loidor.

**laelaps**—Gr. *lailaps*, genit. *lailapos*, a hurricane, a dark furious storm. *Ex:* Laelaps (Rept.); Genciado-laelaps (Ins.); Longo-laelaps (Arach.).

**laeli**—1. L. *Laelia*, name of a vestal virgin. *Ex:* Laeli-eae*:   2. L. *Lailia*, a Roman woman of culture: *Ex:* Laelia (Ins.); Laeli-opsis (Ins.), i.e., of the appearance of the insect, Laelia.

**laem**—See laim.

**laemarg**—Gr. *laimargos*, greedy. *Ex:* Laemargus (Elasm.).

**laemat**—See laim.

**laena**—L. *laena* = Gr. *laina* = *chlainē*, a cloak, garment > NL. *laenatus*, cloaked. *Ex:* Laena (Ins.); Diplo-laena*; Notho-laena* = Nothochlaena*.

**laenat**—See laena.

**laeo**—See lae.

**laeph**—Gr. *laiphos*, a shabby torn garment, a sail. *Ex:* Laeph-otis (Mam.).

**laes**—L. *laesus*, injured, damaged, harmed, pp. of *laedo*, to wound.

**laestrygon**—Gr. *Laistrygonēs*, a race of fierce giants who murdered the comrades of Odysseus; also an ancient people of Italy. *Ex:* Laestry gonus (Arach.).

**laet**—L. *laetus*, gay, pleasing, abundant; *laetabilis*, joyful. *Ex:* Laeti-acantha (Ins.); not Laetia*, named after Jan de Laet, Belgian botanical patron.

**laetamin**—L. *laetamen*, genit. *laetaminis*, dung, manure.

**laetabil**—See laet.

**laetm**—Gr. *laitma*, genit. *laitmatos*, the depth of the sea. *Ex:* Laetm-aster (Echin.); Laetmogone (Echin.); Laetmo-nice (Ann.); see nic, or perhaps it should be Laetm-onice<Gr. *onikē*, asinine or<*Laetmonica*, a mythological name.

**laetmat**—See laetm.

**laetmonice**—See laetm.

**laev**—1. L. *laevis* = *levis*, smooth; *laevigatus* = *levigatus*, slippery, smooth. *Ex:* Laevi-cardium*; Laevo-zebrinus (Moll.); Levi-pali-fer (Coel.):   2. L. *levis* = *laevis*, light, nimble, small:   3. L. *laevus*, to the left; also unsuitable, unfavorable. *Ex:* laevu-l-ose (the *l* is a connective).

**laevigat**—See laev.

**lag**—Gr. *lagōs*, dim. *lagidion*, a hare. *Ex:* Lagorchestes (Mam.); Lag-urus*; Lagidium (Mam.); Lago-morpha (Mam.); Lago-mys (Mam.); Lago-thrix (Mam.).

**lagar**—Gr. *lagaros*, lax, empty. *Ex:* Lagarista (Ins.); Lagar-otis (Ins.); Lagaro-crinus (Echin.); Lagarus (Ins.).

**lagen**—L. *lagena* = *lagaena*, a flask < Gr. *lagēnos* = *lagynos*, a flask. *Ex:* Lagen-aria*; Lagen-ella (Prot.); lageni-form; Lageno-rhynchus (Mam.).

**lagetta**—Native Jamaican *lagetto*, name for a tree. *Ex:* Lagetta*.

**lagid**—See lag.

**lagn**—Gr. *lagnos*, lascivious, lewd. *Ex:* Lagnus (Arach.).

**laguncul**—L. *laguncula*, a small jug or bottle. *Ex:* Laguncul-aria*.

**laim**—Gr. *laimos*, the throat > L. *lamium*, the dead-nettle, named because of the throated flowers. *Ex:* Laemo-bothrion (Ins.); Laimodon (Av.); Lamium*; Lemo-phoeus (Ins.) for Laemo-phloeus (Ins.); Lemo-sthena (Myr.); Crypto-laemus (Ins.); Gymno-laem-ata (Bry.).

**lais**—1. Gr. *Lais*, name of two Greek courtesans celebrated for their beauty. *Ex:* Lais (Arach.): 2. Gr. *laios*, a kind of thrush. *Ex:* Helio-lais (Av.); Uro-lais (Av.).

**lal**—Gr. *lalō*, to speak; *lalos*, said, spoken; *eulalos*, well spoken, well-said. *Ex:* Eu-lalia (Ins.).

**lam**—See lamb, also laim and lami.

**lama**—Peruvian *lama* = *llama*, name for *Lama peruvina*. *Ex:* Lama (Mam.); Llama (Mam.).

**lamachus**—Gr. *Lamachos*, name of an Athenian, lit. one eager for fight. *Ex:* Lamach-elia (Ins.); Lamachus (Ins.).

lamb—L. *lambo*, to lick or lap up, to bathe, pp. *lambitus*, lapped, bathed. *Ex:* Lam-petra (Cycl.).

lamban—Gr. *lambanō*, to grasp, to apprehend. *Ex:* Lambana (Ins.).

lambd—Gr. *lambda*, the Greek letter λ. *Ex:* lambdo-id-al, see eido; Lambdo-therium (Mam.).

lambit—See lamb.

lamell—See lamin.

lami—Gr. *Lamia*, mythical monster said to feed on human flesh. *Ex:* Lam-ictis (Mam.); Lami-idae (Ins.); Lamia (Ins.); Lamia-saurus (Rept.).

lamin—L. *lamina*, dim. *lamella*, a thin plate, leaf, layer; *lamellatus*, thinly layered; NL. *laminatus*, layered. *Ex:* lamin-ar; Lamin-aria*; lamina; laminate; lamini-form; Lamell-aria (Moll.); lamellate; Lamelli-cornia (Ins.); Lamello-copt-urus (Ins.).

lamium—See laim.

lamn—Gr. *lamna*, a fish of prey < *Lamia*, name of a horrible man-eating monster. *Ex:* Lamn-odus (Pisc.); Lamna (Elasm.); Lamni-ceps (Ins.); Lamno-stoma (Pisc.). See also lami.

lamp—Gr. *lampas*, genit. *lampados*, a lamp, a torch; *lampē*, a torch; *lampetēs*, the shining one. *Ex:* Lamp-ornis (Av.); Lampadio-teuthis (Moll.); Lampas-opsis (Moll.); Lampo-desmus (Myr.); Lampo-soma (Dipt.); A-lampetis (Ins.); Nemato-lampas (Moll.). For Lampsilis (Moll.) see lampr.

lampabil—L. *lampabilis*, shining.

lampad—See lamp.

lamper—Gr. *lampēros*, covered with slime. *Ex:* Lamperos (Ins.).

lampet—See lamp.

lampetra—See lamb.

lampr—Gr. *lampros*, shining, beautiful; *lamprotēs*, brightness; also clear, sonorous. *Ex:* Lampr-empis (Ins.); Lampra (Ins.); Lampri-gera (Ins.); Lamprias (Ins.); Lampro-peltis (Rept.); Lampro-phonus (Av.); Lamprot-ornis (Av.); Lamprotes (Ins.); Lamprotis*; Lampsilis (Moll.) < *lampros* + *psilos*, smooth.

lamprim—Gr. *lampreimōn*, clad in fine robes. *Ex:* Lamprima (Ins.).

lamprot—See lampr.

lampyr—Gr. *lampyris*, genit. *lampyridos*, a glow worm. *Ex:* Lampyr-idae (Ins.); Lampyris (Ins.).

lan—L. *lana*, wool; *lanatus*, wooly; *lanuginosus*, downy; *lanosus*, full of wool; *lanugo*, woolly substance, down. *Ex:* Lan-orus (Ins.); lanate; lani-fer-ous; lanugo.

lanc—L. *lancea*, a small light spear; *lanceolatus*, lance-like, armed with a pointed weapon.

languid—L. *languidus*, faint; also sluggish, dull, slow.

langur—L. *languria*, from Celtic *langa*, a lizard from whose urine a stone called langurium was obtained. *Ex:* Languria (Ins.).

lani—L. *lanius*, a butcher < *lanio*, to tear in pieces. *Ex:* Lani-odon (Mam.); Lanio-vires (Av.); Lanius (Av.); Lani-idae (Av.); Mio-lania (Rept.).

laniat—L. *laniator*, a butcher. *Ex:* Laniator-es (Arach.).

lanici—L. *lanicius*, woolen, woolly, fleecy.

lanos—See lan.

lantan—NL. *lantana* < an old Italian name for Viburnum which it somewhat resembles in foliage. *Ex:* Lantana*. See also lanthan.

lantern—L. *lanterna*, NL. dim. *lanternul*, a lantern, lamp. *Ex:* Lanterna (Prot.).

lanthan—Gr. *lanthanō*, to escape notice, to be unknown, unseen; related to *lathanō*, to make to forget. *Ex:* Lanthan-otis (Rept.); Lanthano-therium (Mam.) = Lantano-therium.

lanug—See lan.

lanul—L. *lanula*, a tiny lock of wool.

lao—See la.

laodic—1. *Laodikē*, a nymph:  2. Gr. *Laodikea*, a city of Phrygia. *Ex:* Laodicea (Coel.).

laomed—Gr. *laomedōn*, ruler of the people > *Laomedōn*, King of Troy. *Ex:* Laomedea (Coel.); Laomedes (Coel.).

lapar—Gr. *lapara*, the flank, loin, the soft part of the body between the ribs and hip. *Ex:* lapar-ect-omy (Surg.); Laparo-myrmex (Ins.); laparo-tomy (Surg.); Laparus (Pisc.).

lapath—L. *lapathium*, sorrel. *Ex:* lapathi-folium.

laphyct—Gr. *laphyktēs*, a heavy eater. *Ex:* Laphyctes (Ins.), (Av.), etc.

laphyr—Gr. *laphyra*, plunder, booty. *Ex:* Laphyr-agogus (Ins.); Laphyra (Ins.); Laphryo-scopus (Ins.).

lapid—L. *lapis*, genit. *lapidis*, dim. *lapillus*, stone; *lapidosus*, stony, full of stones. *Ex:* lapidi-col-ous; Lapidosus (Moll.); Lapillo-cystis (Echin.).

lapillo—See lapid.

lapith—Gr. *lapithēs*, a swaggerer. *Ex:* Lapithes (Ins.).

lapp—L. *lappa*, a burr, NL. dim. *lappula*; *lappaceus*, burr-shaped, burr-like. *Ex:* lappaceous; Lappula*.

lappet—Sw. *lapp*, a patch; A.S. *laeppa*, a loosely hanging portion; Icel. *lapa*, to hang down; *lappet* < *lapp* + dim. ending -*et*. *Ex:* lappet.

lapponic—L. *lapponicus*, of Lapland.

lappul—See lapp.

laps—L. *lapsus*, slipping, falling, ruined, dead < *labor*, to sink, slip, glide.

lapsan—Gr. *lapsanē*, a kind of edible plant, probably a crucifer. *Ex:* Lapsana*.

lapt—Gr. *laptō*, to lick, lap, touch. *Ex:* Lapto-trachelus (Ins.); Lapton (Ins.).

lar—1. Gr. *laros*, a ravenous sea-bird=L. *larus*, a gull. *Ex:* Lar-idae (Av.); Lario-saurus (Rept.); Larus (Av.):　2. Gr. *laros*, dainty, sweet:　3. L. *Lar*, tutelary god of field and house. *Ex:* Lari-idae (Ins.); Laria (Ins.).

larc—Gr. *larkos*, a charcoal-basket. *Ex:* Larc-idium (Prot.); Larco-pyle (Prot.).

lardace—Fr. *lardace*, having the appearance of lard. *Ex:* lardaceus.

larent—L. *Larentia*, a name of Flora or perhaps< *Larentia*, nurse of Romulus and Remus. *Ex:* Larenti-oides (Ins.); Larentia (Ins.).

larg—L. *largus*, large, abundant.

laria—See lar 3.

laric—L. *larix*, genit. *laricis*, the larch-tree; NL. *laricinus*, pertaining to the larch. *Ex:* Larico-bius (Ins.); Larix*.

larifug—L. *larifuga*, a wanderer. *Ex:* Larifuga (Arach.).

larim—L. *larimos*=*larinos*, name of some fish. *Ex:* Larim-ichthys (Pisc.); Larimus (Pisc.).

larin—Gr. *larinos*, fatted, fat. *Ex:* Larino-poda (Ins.); Larinus (Ins.). See larim.

larix—See laric.

larmier—Fr. *larmier*<*larme*, a tear. *Ex:* larmier.

larnac—Gr. *larnax*, genit. *larnakos*, a box, chest. *Ex:* Larn-acantha (Prot.); Larnac-idium (Prot.); Larnaco-spongus (Por.).

larnax—See larnac.

larv—NL. *larva*, the immature form of an animal undergoing metamorphosis, often worm-like in form; <L. *larva*, a mask, spectre, scarecrow; *larvatus*, masked. *Ex:* Larv-ulina (Prot.); larva; larvi-form; Larvi-vora (Av.).

laryng—Gr. *larynx*, genit. *laryngos*, the larynx, gullet. *Ex:* Laryng-odus (Ins.); Laryngo-gramma (Av.); larynx.

larynx—See laryng.

las—See la.

lasan—Gr. *lasana*, a gridiron. *Ex:* Lasanius (Pisc.).

lasciv—L. *lascivulus*, a little playful, frisky, dim. of *lascivus*, playful, frolicsome.

lasi—Gr. *lasios*, hairy, woolly, shaggy. *Ex:* Lasi-andra*; Lasi-urus (Mam.); Lasia* (Ins.); Lasio-nycteris (Mam.); Lasius (Ins.); Gyro-lasia (Ins.).

-lasm—See elasm.

lass—L. *lassus*, faint, drooping.

lat—1. L. *latus*, genit. *lateris*, the side, flank. *Ex:* Lateri-branchi-aea (Moll.); latero-version; lati-folia, lati-rostrate; lati-sternal:　2. L. *latus* =Gr. *latos*, a fish of the Nile. *Ex:* Lates (Pisc.).

latag—See latax.

latani—NL. *latania*<West Indian *allatani*, name of a palm. *Ex:* Latania*.

latax—Gr. *latax*, genit. *latagos*, a kind of water animal, probably a beaver. *Ex:* Latax (Mam.); Latax-ina (Mam.); Lataxia (Mam.).

latebr—L. *latebra*, a hiding place<*lateo*, to lurk; *latebrosus*, full of holes, hidden, obscure. *Ex:* latebra;　Latebri-cola　(Arach.);　Latebrus (Pisc.).

laten—L. *latens*, genit. *latentis*, ppr. of *lateo*, to lurk, lie hid, hidden; *latescens*, genit. *latescentis*, concealed, ppr. of *latesco*, to be concealed. *Ex:* latent; latescent.

later—L. *later*, genit. *lateris*, brick, tile; *latericius* =*lateritius*, built of bricks; sometimes used in sense of brick-red. See also lat. S.

lateric—See later.

laterici—L. *latericius*, made of bricks.

latern—L. *laterna*=*lanterna*, a lantern.

lates—See lat 2.

latesc—See laten.

latex—See latic.

lathan—See lanthan.

lathetic—Gr. *lathetikos*, likely to escape notice. *Ex:* Latheticus (Ins.).

lathr—Gr. *lathrē*=*lathra*, secretly; *lathraios*, hidden; *lathridios*, secret; *lathrimaios*, secretly. *Ex:* Lathr-aea*; Lathrid-ulus (Ins.); Lathridius (Ins.);　Lathrio-soma　(Av.);　Lathrimaeum (Ins.);　Lathro-plex　(Ins.);　Anchy-lathron (Ins.).

lathrid—See lathr.

lathrimae—See lathr.

lathyr—Gr. *lathyros*, a kind of plant, vetchling. *Ex:* Lathyrus*.

latic—L. *latex*, genit. *laticis*, a liquid, fluid. *Ex:* latex; latici-fer-ous.

latil—NL. *latilus* (<L. *latus*, broad, wide), generic name of certain fishes. *Ex:* Latil-idae (Pisc.); Latilus (Pisc.); Lopho-latilus (Pisc.); Caulo-latilus (Pisc.).

latir—NL. *latirus*, name for a genus of snails, said to be derived from Gr. *lathyros*, a plant name. *Ex:* Latirus (Moll.).

latitat—L. *latitatus*, concealed, hidden, pp. of *latito*, to hide.

latr—1. L. *latro*, a robber; *latrunculus*, a robber, free booter. *Ex:* Latro-dectus (Arach.), see dect. 2; Latrunculus (Pisc.):　2. L. *latro*, to bark; *latrans*, a barker, a ppr. used as a noun: 3. Gr. *latron*, pay, hire; *latris*, a hand-maid>L. *Latris*, genit. *Latridis*, a proper name. *Ex:* Latrid-idae (Pisc.); Latris (Pisc.).

latrans—See latr.

latrat—L. *latratus*, pp. of *latro*, to bark, roar, rage; *latrator*, a barker. See also latr.

latreut—Gr. *latreutēs*, a hireling, a hired servant. *Ex:* Latreutes (Crust.).

laud—L. *laudo*, to praise; *laudatus*, worthy, praised.

**lauda**—See **alauda**.

**laur**—1. L. *laurus*, the laurel; *laurinus*, of laurel. *Ex:* Laurelia\*, Latinized from the Eng. *laurel*; lauri-folius; laurino-xylon; Lauro-cerasus\*; Laurus\*: 2. Gr. *laura*, a drain, passage. *Ex:* lauro-philus; lauro-phyta.

**laut**—L. *lautus*, washed; also clean, neat, splendid <*lavo*, to wash. *Ex:* Lauto-conus (Moll.).

**lavandula**—ML. *lavandula*, the lavender<L. *lavo*, to wash. *Ex:* Lavendula\*.

**lax**—L. *laxus*, wide, loose, spacious. *Ex:* Laxi-spira (Moll.); Laxo-phyllum (Prot.).

**lazul**—Low L. *lazulum*, *lazurius*, *lazur*=Sp. *azul*, blue.

**leaena**—L. *leaena*, a lioness<Gr. *leaina*, a lioness; *Leaena*, heroic courtesan in Athens. *Ex:* Leaena (Ann.).

**leb**—Gr. *lebēs*, genit. *lebētos*, a kettle; *lebias*, a kind of fish suitable to be cooked in a kettle; *lebistēs*, a kind of fish<*lebēs*. *Ex:* Lebe-dieropsis\*; Lebistes (Pisc.); Oxy-lebius (Pisc.).

**leberid**—Gr. *lebēris*, genit. *lebēridos*, a serpent's skin, a girdle resembling one made of serpent's skin; the shell of beans, a husk. *Ex:* Acantholeberis (Arth.); Grapto-leberis (Arth.).

**leberis**—See **leberid**.

**lebist**—See **leb**.

**lec**—Gr. *lekos*, genit. *lekeos*; dim. *lekis*, genit. *lekidos*, a dish, plate, pot. *Ex:* Lecidea\*; Leco-myia (Ins.); leco-trop-al; glypho-lec-ine.

**lecan**—Gr. *lekanē*, dim. *lekanion*, a dish, pot. *Ex:* Lecan-ora\*, see or 9.; Lecani-cephalus (Platy.); Lecanium (Ins.); Lecano-bius (Ins.).

**lecher**—OFr. *lecheros*<OFr. *lecheor*, a gormand. *Ex:* lecher-ous.

**lechr**—Gr. *lechrios*, slanting, crosswise. *Ex:* Lechri-odonta (Amph.); Lechri-orchis (Platy.); Lechrio-pyla (Prot.).

**lecid**—See **lec**.

**lecith**—Gr. *lekithos*, the yolk of an egg. *Ex:* lecith-in; Lecitho-phora (Platy.); centro-lecith-al.

**lect**—1. Gr. *lektos*, selected, chosen; *lektēs*, a speaker, one chosen. *Ex:* Lecto-somus (Av.); lecto-type; Amphi-lectus (Por.): 2. L. *lectus*, a couch>LL. *lectualis*, pertaining to a bed. *Ex:* lectual: 3. L. *lectus*, selected, choice, ppr. of *lego*, to select.

**lectuari**—L. *lectuarius*, belonging to a bed; also the bed itself.

**lectul**—L. *lectulus*, a bed.

**lecyth**—Gr. *lēkythos*, an oil-jar. *Ex:* Lecythi-era (Ins.); Lecythio-crinus (Echin.); Lecythis\*; Lecytho-plastes (Av.).

**led**—1. Gr. *lēdon*, mastic, an Oriental shrub. *Ex:* ledi-tannic; Ledum\*: 2. Gr. *Lēda*, mother of Pollux and Castor. *Ex:* Led-idae (Moll.); Leda (Moll.).

**leg**—1. L. *lego*, to bring together, collect. *Ex:* ostra-legus; podi-leg-ous: 2. Gr. *legō*, to lie down. *Ex:* Thryo-legus (Av.): 3. Gr. *legō*, to choose, pick out; also to say, speak. *Ex:* Hydro-legus (Av.).

**legitim**—L. *legitimus*, fertilized by its own sperm or pollen.

**legn**—Gr. *legnon*, a border or colored edge; *legnotos*, with a colored border. *Ex:* Legno-notus (Pisc.); Legnotis\*; Sapro-legnia\*.

**legnot**—See **legn**.

**legum**—L. *legumen*, genit. *leguminis*, a leguminous plant; ML. *leguminosus*, leguminous. *Ex:* legume; Leguminos-ae\*; Leguminos-ites\*.

**lei**—Gr. *leios*, smooth; *leiotēs*, smoothness; *leioō*, to make smooth, pound fine; *leiōsis*, a polishing. *Ex:* Lei-urus (Pisc.); Leio-bunum (Arach.); Lejo-pyge (Tri.); Li-odon (Rept.); Lio-saurus (Rept.); Lion-urus (Pisc.), the *n* is superfluous; Liota (Ins.). See also leip.

**leich**—Gr. *leichō*, to lick up, to play with the tongue. *Ex:* Leicho-myle (Platy.); Cyto-leich-idae (Arach.).

**leimon**—See **limon**.

**leip**—Gr. *leipō*, to leave, quit, to die. *Ex:* Leip-oa (Av.), lit. egg-deserter<*leipō*+*ōa*, pl. of *ōon*, an egg; Leip-onyx (Mam.); Leipo-ceros (Ann.).

**leir**—Gr. *leiros*, pale. See also liri.

**leist**—Gr. *lēistēs*, a robber, plunderer; *lēistos*, to be stolen. *Ex:* Leist-arches (Ins.); Leistes (Av.); Leisto-phorus (Ins.).

**lejo**—See **lei**.

**lem**—1. Gr. *lēma*, proudness, will, desire. *Ex:* Lema (Ins.); see also laim: 2. Gr. *lēmē*, gum, rheum: 3. Gr. *loimos*, plague. *Ex:* lemo-logy. See also laim.

**lemb**—Gr. *lembos*, a little boat; *lembōdēs*, boat-shaped. *Ex:* Lemb-ulus (Moll.); Lembo-ides (Arth.); Lembo-pteris (Ins.); Lembus (Prot.).

**lemm**—1. Gr. *lemma*, genit. *lemmatos*, a bark, peel, rind; now denoting a sheath. *Ex:* Lemma-myia (Ins.); Lemmato-pora (Bry.); neuro-lemma; sarco-lemma: 2. NL. *lemmus*, the lemming<Norw. *lemja*, to maim, strike. *Ex:* Lemmo-mys (Mam.); Lemmus (Mam.): 3. Gr. *lēmma*, an assumption, something taken for granted.

**lemn**—Gr. *lemna*, a kind of water plant. *Ex:* Lemna\*; Lemna-phila (Ins.); not Lemnus (Mam.) which is an error for Lemmus (Mam.).

**lemnisc**—L. *lemniscus*, a ribbon; *lemniscatus*, adorned with ribbons. *Ex:* lemniscate; Lem niscia (Moll.); Lemnisco-mys (Mam.); lem niscus.

**lemo**—See **laim**.

**lemon**—Gr. *leimōn*, a meadow; *leimōnias*, a meadow nymph. *Ex:* Lemoni-idae (Ins.); Lemonias (Ins.); Scoto-lemon (Arach.).

**lemph**—Gr. *lemphos*, slime, snivel. *Ex:* Lempho-lemma\*; Lemphus (Ins.).

**lemur**—L. *lemures*, shades, ghosts, cf. Gr. *Lamia*, a devouring monster. *Ex:* Lemur (Mam.); Lemur-avus (Mam.); Lemur-phthirus (Ins.); Lemuro-limnas (Av.).

**len**—1. Gr. *lēnos*, wool. *Ex:* Leno-thrix (Mam.): 2. L. *lenis*, soft, mild. *Ex:* leni-fy.

**lend**—L. *lens*, genit. *lendis*, a nit. *Ex:* lendi-gerum.

**lens**—See lent.

**lent**—1. L. *lens*, genit. *lentis*, dim. *lenticula*, a lentil; *lenticularis*, of or pertaining to a lentil. *Ex:* Lens\*; lenti-form; Lentibularia\* the c altered to b by Gesner; lenticle; lenticular; Lenticula; lenticul-ate (Moll.); Lento-spora (Prot.): 2. L. *lentus*, slow, pliant, tough, tenacious, viscous. *Ex:* lentous.

**-lent**—L. -*lentus*, suffix denoting fullness, proness to. *Ex:* pesti-lentus; succu-lent; puru-lent.

**lentig**—L. *lentigo*, genit. *lentiginis*, a freckle, a lentil-shaped spot; *lentiginosus*, freckled. *Ex:* lentigin-ose; lentigin-ous; Lentigo (Moll.).

**lentin**—L. *Lentinus*, a Roman proper name. *Ex:* Lentinus\*.

**lentisc**—L. *lentiscus*, the mastic-tree. *Ex:* Lentiscus\*.

**leo**—Gr. *leōn*, genit. *leontos*, a lion. *Ex:* leochromous; Leon-otis\*; Leon-urus\*; Leontodon\*; Leonto-podium\*; Myrme-leon (Ins.); not Leonia\* which was named after D. Francisco Leon, promotor of "Flora Peruviana et Chilensis." S.

**leon**—See leo.

**leonin**—L. *leoninus*, of or belonging to a lion.

**leont**—See leo.

**leot**—Gr. *leiotēs*, smoothness. *Ex:* Leotia\*.

**lep**—Gr. *lepis*, genit. *lepidos*, dim. *lepion*, also *lepidion*, a scale; *lepidōtos*, scaly; *lepos*, a scale. *Ex:* Lep-idium\*; Lep-omis, see pom; Lepido-ptera (Ins.); Lepido-teuthis (Moll.); Lepidota (Amph.); Lepo-derma (Platy.); not Lepi-lemur (Mam.) which is in part from L. *lepidus*, pleasing; Lepti-ota\*; Calli-lepis (Arach.).

**lepachys**—See pachy.

**lepad**—Gr. *lepas*, genit. *lepados*, a limpet. *Ex:* Lepad-ella (Rot.); Lepad-idae (Crust.); Lepado-crinus (Echin.); Lepado-gaster (Pisc.); Lepas (Moll.); Concho-lepas (Moll.).

**leparg**—Gr. *lepargos*, with white skin or feathers. *Ex:* Lepargus (Ins.).

**lepas**—See lepad.

**lepid**—L. *lepidus*, pretty, neat, graceful. *Ex:* Lepi-lemur; not Lepido-ptera (Ins.); see lep.

**lepism**—Gr. *lepisma*, genit. *lepismatos*, scale, rind, something scaled off. *Ex:* Lepisma (Ins.); Lepismato-phila (Prot.); Lepism-ina (Ins.); Lepism-odes (Ins.).

**lepismat**—See lepism.

**lepist**—L. *lepista*, a goblet < Gr. *lepastē*, a goblet. *Ex:* Lepista (Prot.).

**lepor**—L. *lepus*, genit. *leporis*, a hare. *Ex:* Lepor-idae (Mam.); Lepori-conus (Moll.); Lepus (Mam.).

**lepr**—Gr. *lepra*, leprosy; *lepras*, rough; *lepros* scaly > L. *leprosus*, scurfy. *Ex:* Lepr-alia (Bry.); lepra; Lepro-nyssus (Arach.).

**leprad**—Gr. *lepras*, genit. *leprados*, rough. *Ex:* leprado-folia.

**leps**—Gr. *lēpsis*, a seizing, catching. *Ex:* Eroto-lepsia (Ins.); photo-lepsy.

**lept**—1. Gr. *leptos*, slender, thin, small, weak > *lepton*, the small gut; also a minute piece of money. *Ex:* Lept-an-odonta (Moll.); Lept-

California Prickly-Phlox, *Leptodactylon californicum.*

echinus (Echin.); Lept-ino-tarsa (Ins.), see inos; Lept-urus; Lept-aena\*; ? Leptilon\*; Leptinus (Ins.); Lepto-chloa\*; Lepto-chiton (Moll.); lepto-nema; Lepton (Moll.); Eury-lepta (Platy.): 2. Gr. *lēptēs*, one who accepts or takes. *Ex:* Cerco-leptes (Mam.), -*leptes*, here in sense of takes hold; Eu-leptes (Rept.): 3. NL. *lept-* from a supposed Gr. *leptos*, solid, filled. *Ex:* leptom; lepto-id.

**leptacin**—See leptale.

**leptale**—Gr. *leptaleos*, slender, delicate = poet. *leptakinos*. *Ex:* Leptacinus (Ins.); Leptalea (Ins.); Leptaleo-ceras (Moll.); Leptaleum\*.

**lepteces**—Gr. *leptēkēs*, fine, pointed, delicate. *Ex:* Lepteces (Crust.).

**leptes**—See lept 2.

**leptic**—Gr. *lēptikos*, given to accepting. *Ex:* Lepticus (Ins.).

**leptin**—See lept 1.

leptom—See lept 3.

leptosyn—Gr. *leptosynē,* slenderness. *Ex:* Leptosyna (Ins.); Leptosyne*.

leptot—Gr. *leptotēs,* thinness. *Ex:* Leptotes (Ins.).

leptyn—Gr. *leptynō,* to make thin; in passive, to be reduced. *Ex:* Leptyno-concha (Moll.).

leptysm—Gr. *leptysmos,* a thinning. *Ex:* Leptysm-ina (Ins.); Leptysma (Ins.).

lepur—See lepyr.

lepus—See lepor.

lepyr—Gr. *lepyron,* a shell, husk; *lepyros,* in a shell or rind. *Ex:* Lepur-andra*; Lepyri-actis (Echin.); Lepyro-lobus (Moll.); lepyro-phylly; Lepyrus (Ins.).

lerem—Gr. *lērēma,* silly talk. *Ex:* Lerema (Ins.).

lern—Gr. *Lerna,* a marsh in Argolis in which the Hydra dwelled. *Ex:* Lern-anthro-pus (Crust.); Lernaeo-poda (Crust.); Lerneo-myzon (Crust.).

lerod—Gr. *lērōdēs,* silly. *Ex:* Lerodea (Ins.).

lerwa—NL. *lerwa* < Nepalese *larwā,* a partridge. *Ex:* Lerwa (Av.).

lesbia—L. *lesbias* = *lesbia,* a precious brilliant colored stone found in Lesbos; Gr. *Lesbos,* an island in the Aegean Sea the inhabitants of which were reputed to be exceedingly sensual. *Ex:* Lesbia (Ins.), (Av.); Lesbian-ism.

lest—Gr. *lēstēs,* also *lēstēr,* a robber; *lēstikos,* piratical; *lēsteuō,* to rob. *Ex:* Lesteva (Ins.); lesto-biot-ic; Archi-lestes (Ins.); Caeno-lestes (Mam.); Ornitho-lestes (Rept.).

lesteu—See lest.

lestev—See lest.

lestic—See lest.

lestis—Gr. *lēstis* = *lēsthē,* a forgetting, forgetfulness. *Ex:* Lestis (Ins.).

lestr—Gr. *lēstris,* piratical; *lēstrikos,* inclined to rob. *Ex:* Lestri-melitta (Ins.); Lestrico-thynnus (Ins.); Lestris (Av.).

lestrigon—Gr. *Laistrygonēs,* a legendary cannibalistic people of giant size. *Ex:* Lestrigonus (Crust.).

let—L. *letum,* death > *letifer,* death-dealing, fatal, lethal.

leth—Gr. *lēthos* = *lēthē,* a forgetting, escaping notice. *Ex:* Leth-enteron (Pisc.); Letho-cerus (Ins.).

lethae—Gr. *lēthaios,* oblivious. *Ex:* Lethaeus (Ins.).

lethal—L. *lethalis,* deadly. *Ex:* lethal.

ethargic—Gr. *lēthargikos,* drowsy, pertaining to drowsiness. *Ex:* lethargic.

letif—See let.

leuc—Gr. *leukos,* white, bright, light; *leukon,* white; *leukainō,* to whiten. *Ex:* Leuc-andra (Por.); Leuca-dendron*; Leucaena*; Leucas*; Leuci-corus (Pisc.); leuco-cytes; Leuconostoc*; Leuco-solenia (Por.); leucon; ochro-leuca.

leucani—Gr. *laukaniē* = *leukaniē,* the throat. *Ex:* Leucania (Ins.).

leucipp—1. Gr. *Leukippē,* daughter of Thestor. *Ex:* Leucippe (Crust.): 2. Gr. *Leukippos,* father of Phoebe and Hilaira carried off by Castor and Pollux. *Ex:* Leucippus (Av.).

leucisc—NL. *leuciscus,* < Gr. *leukiskos,* the white mullet. *Ex:* Leuciscus (Pisc.).

leucoj—NL. *leucojum* < Gr. *leukos,* white + *ion* (NL. *jon*), the violet. *Ex:* Leucojon*.

leucon—1. Gr. *leukon,* white; *leukon,* pl. *leukōnes,* a grove of white poplars. *Ex:* Leucones (Por.): 2. L. *Leucon,* genit. *Leuconis,* one of Actaeon's hounds. See also leuc.

leucophaet—L. *leucophaetus,* with dark gray garments < Gr. *leucophaios,* ash-colored.

leucoth—1. Gr. *Leukothoē,* daughter of Orchamus, king of Babylonia. *Ex:* Leucotho-ella (Crust.); Leucothoe* (Crust.): 2. Gr. *Leukothea,* a sea nymph. *Ex:* Leucothea (Moll.).

leur—Gr. *leuros,* smooth, even. *Ex:* leur-odont; Leuro-gnathus (Elasm.).

lev—See laev.

levator—L. *levator,* a lifter < *levo,* to lift up, to raise. *Ex:* levator.

levidens—L. *levidensus,* thin, slight.

levigat—L. *levigatus* = *laevigatus,* made smooth, polished, pp. of *levigo,* to make smooth.

levir—L. *levir,* brother-in-law.

li—See lei.

liass—Fr. *lias,* a sort of limestone; in geology, the lower division of the Jurassic; *liassique,* of the lias. *Ex:* Liasso-tipula (Ins.).

liatri—NL. *liatris* (origin unknown), name for a genus of herbs. *Ex:* Liatris*.

lib—Gr. *libas,* anything that drops or trickles, a spring; *libos,* tears. *Ex:* Libo-cedrus*.

liban—1. Gr. *libanos,* incense. *Ex:* Liban-otis*, libani-fer-ous: 2. L. *Libani,* of Mt. Lebanon.

libat—L. *libator,* one who makes a drink offering.

libell—L. *libellus,* a little book, dim. of *liber;* *libellulus,* a very little book. *Ex:* Libellula (Ins.); Libellulo-soma (Ins.).

liber—1. L. *liber,* the inner bark. *Ex:* libero-ligneous: 2. L. *libero,* to set free. *Ex:* libero-motor; libro-plast: 3. L. *liber,* genit. *liberi,* a child.

libid—L. *libido,* genit. *libidinis,* pleasure, desire. *Ex:* libido.

libr—L. *liber,* genit. *libri,* a book, dim. *libellus.* *Ex:* libri-form. See also liber 2.

liby—Gr. *Libys,* genit. *Libyos,* a Lybian. *Ex:* Liby-pithecus (Mam.); Liby-thea (Ins.); Libyo-drilus (Ann.).

libyss—Gr. *Libys,* fem. *Libyssa,* a Lybian. *Ex:* Libyssa (Ins.).

**lic**—L. *licium*, thread, the end of a thread. *Ex:* Licea*. See also lix.

**lica**—See **lich**.

**licani**—NL. *licania*, anagram of S.A. Indian *calignia*, a plant name. *Ex:* Licania*.

**lich**—Gr. *lichas*, genit. *lichados*, a steep cliff; also the space between the fore-fingers. *Ex:* Licaphrium (Mam.), see phrix; Lichas (Tri.), some would derive it from Gr. *Lichas*, a personal name, which seems more probable.

**lichan**—Gr. *lichanos*, the fore-finger<*leichō*, to lick, from its use in licking up food. *Ex:* Lichan-otus (Mam.); Lichan-ura (Rept.).

**lichas**—See **lich**.

**lichen**—Gr. *leichēn*, a lichen. *Ex:* Lichen-aria (Coel.); licheni-vor-ous; Licheno-phagus (Ins.); Licheno-pora (Bry.); Lichina*.

**lichn**—Gr. *lichnos*, dainty; also greedy. *Ex:* Lichno-ptera (Ins.). See also lychn.

**licia**—See **elic**.

**licin**—L. *licinus*, bent, reflexed or twined upward. *Ex:* Licinus (Ins.).

**licmet**—Gr. *likmētos*, winnowing. *Ex:* Licmetis (Av.).

**licn**—Gr. *liknon*, a winnowing fan: also a cradle in which the infant Bacchus was carried. *Ex:* Likno-daemus (Arach.); Oto-licnus (Mam.).

**lien**—L. *lien*, genit. *lienis*, the spleen *Ex:* lien-al; lieno-gastric; lieno-renal.

**lig**—1. L. *ligo*, to bind; pp. *ligatus*, bound; *ligamentum*, a band, tie; *ligatura*, a band. *Ex:* ligamentus; ligature; ad-ligant.     2. Gr. *Ligeia*, name of a water nymph. *Ex:* Ligidium (Crust.); Ligi-idae (Crust.); Ligia= Ligyda (Crust.); Ligia (Ins.):     3. Gr. *liga*, in clear loud tone. See also ligaen.

**ligaen**—Gr. *ligainō*, to cry with a loud, clear voice.

**ligament**—See **lig 1**.

**lign**—L. *lignum*, wood; *lignosus*, woody; *ligneus*, wooden. *Ex:* Ligni-cola (Ann.); ligni-fic-ation; ligni-vor-ous; ligno-cellulose; lignum-vitae; libero-ligne-ous.

**lignator**—L. *lignator*, a wood-cutter.

**ligon**—L. *ligo*, genit. *ligonis*, a grub-axe. *Ex:* Ligoni-pes (Arach.); Ligono-dina (Ann.).

**ligul**—L. *ligula*, a little tongue, see lingu. *Ex:* Ligul-aria*; ligul-ate; Ligul-ops (Brach.); Ligula (Platy.); liguli-flor-ous.

**ligustic**—L. *Ligusticus*, of or from Liguria> *ligusticum*, a plant name. *Ex:* Ligusticum*.

**ligustr**—L. *ligustrum*, a plant named privet. *Ex:* Ligustrum*.

**ligy-**—Gr. *ligy-* in compounds, meaning clear, shrill, loud; *ligyphonos*, clear-voiced. *Ex:* Ligy-molpa (Ins.); Ligy-pterus (Ins.).

**ligyd**—See **lig 2**.

**ligyr**—Gr. *ligyros*, clear, shrill. *Ex:* Ligyrus (Ins.).

**lili**—L. *lilium* (<Gr. *leirion*, a lily); L. *liliaceus*, of or from lilies. *Ex:* Lili-ales*; Lilium*.

**lim**—1. L. *limus*, mud; *limosus*, fem. *limosa*, full of mud, slime. *Ex:* Limi-cola (Av.); Limosella*; Limosa (Av.):     2. L. *limus*, bandage or apron trimmed with purple:     3. L. *lima*, a rasp, file, dim. *limula*; *limatus*, filed, rasped. *Ex:* Lima (Moll.); Lima-pontia (Moll.); Limat-ula (Moll.):     4. L. *limus*, sidelong, askew, aslant; *limulus*, a little askew. *Ex:* Limul-ites (Crust.); Limulus (Crust.):     5. Gr. *limos*, hunger. *Ex:* Dendro-limus (Ins.); Eu-lima (Moll.).

**limac**—L. *limax*, genit. *limacis*, a slug; kindred to *limus*, slime, mud; *limaceus*, of mud; slime. *Ex:* Lima-pontia (Moll.); Limac-arion (Moll.); Limac-idae (Moll.); limaci-formis; Limas-ella (Moll.); Limax (Moll.); Ario-limax (Moll.).

**limand**—NL. *limanda*<L. *limus*, mud. *Ex:* Limanda (Pisc.).

**limat**—L. *limatus*, polished; *limatulus*, somewhat polished or filed. *Ex:* Limato-gaster (Ins.).

**limax**—See **limac**.

**limb**—L. *limbus*, an edge; *limbatus*, bordered. *Ex:* limb-ic; Limb-oria*; Limbato-chlamys (Ins.); bi-limb-ose.

**limbat**—See **limb**.

**limen**—Gr. *limēn*, genit. *limenos*, a harbor; *Limenitēs*, god of the harbor of Priapus. *Ex:* Limen-archis (Ins.); Limenitis (Ins.).

**limer**—Gr. *limēros*, hungry. *Ex:* Limer-odes (Ins.).

**limit**—L. *limes*, genit. *limitis*, the border, limit; *limitaneus*, that is on the border; *limitatus*, bounded.

**limn**—Gr. *limnē*, marsh, pond; *limnētēs*, living in marshes; *limnas*, genit. *limnados*, poet. fem. of *limnaios*, of or from the marsh>*Limnōreia*, a Nereid. *Ex:* Limn-anthes*; Limn-erium (Ins.), the last element of uncertain meaning; Limnatornis (Av.); Limnetis (Crust.), fem. of *limnētēs*; Limnias (Rot.); Limno-bium*; limno-logy; Limnoria (Crust.); Lymn-aea (Moll.); Amphilimna (Echin.).

**limnad**—See **limn**.

**limnet**—See **limn**.

**limodes**—Gr. *limōdēs*, famished, hungry.

**limodoron**—Gr. *limodoron*, name of some wild plant. *Ex:* Limodoron*.

**limogn**—Fr. *Limogne*, a place in France. *Ex:* Limogni-therium (Mam.).

**limon**—Gr. *leimōn*, a meadow, any bright or flowery surface; *leimōnērēs*, belonging to a meadow; *leimōnion*, limonium, sea-lavender or snake-weed. *Ex:* leimon-apo-phyte; Leimoniptera (Av.); Limoneres (Av.); Limoni-dromus (Av.); Limonium*; Acantho-limon*.

**limos**—See **lim 1**.

**limul**—See **lim 4**.

**-limus**—Six Latin adjectives ending in *-ilis* form their superlative by adding *-limus* to the stem. These are: *difficilis*, difficult; *similis*, like; *gracilis*, slender; *humilis*, low; *dissimilis*, like; *facilis*, easy. Thus: *similis*, superl. *simillimus*.

**limus**—See lim.

**lin**—1. L. *linea*, dim. *lineola*, a line; *linearis*, pertaining to a line or lines; *lineatus*, streaked, marked with lines<*lineo*, to make straight. *Ex:* linea alba; Linea-dinium (Prot.); linear; lineo-polar; lino-spor-ous:    2. L. *linum*, thread, flax, rope, cable; *lineus*, flaxen, of flax; *linarius*, a linen weaver. *Ex:* Lin-anthus*; Linaria*, fem. of *linarius*; lin-oides; Lineo-palpa (Ins.); Lineus (Platy.); linin; Lin osyris*; Linum*; Came-lina*:    3. Gr. *linon*, net. *Ex:* Lino-pteris*; linon fibers.

**linct**—L. *linctus*, licked<*lingo*, to lick.

**lind**—L. *lindus*, a town of Rhodes. *Ex:* lindiform, i.e. of the form of Lindia (Rot.).

**linear**—See lin.

**lineat**—See lin.

**linga**—Sanskr. *linga*, the penis, symbol of Siva, Hindu deity. *Ex:* Linga (Moll.).

**lingo**—Malay *lingoa*, a Malayan tree furnishing a valuable hardwood. *Ex:* Lingoum*.

**lingu**—L. *lingua*, dim. *lingula*, tongue; *linguatus*, gifted with a tongue; *lingulatus*, tongue-shaped. *Ex:* Linguat-ul-ina (Arach.); Linguata (Amph.); Lingula (Brach.); Linguli-pora (Brach.); Lingulo-cystis (Echin.).

**lingul**—See ligul, also lingu.

**linit**—L. *linitus*, smeared, bedaubed.

**linyph**—Mod. Gr. *linyphos*, linen for wearing. *Ex:* Linyphia (Arach.); Linyphi-idae (Arach.).

**lio**—See lei.

**lion**—See lei.

**lip 1.** Gr. *lipos*, fat, lard, tallow; *liparos*, sleek, oily, shiny with oil. *Ex:* lip-oid; Liparia*; Liparis*; Liparo-crinus (Echin.); Lipe-urus (Arth.); Neo-liparis (Pisc.):    2. Gr. *leipō*, to be wanting, to quit, to fall from. *Ex:* Lip-otus (Mam.); Lip-urus (Mam.); Lipo-branchia (Arach.); Lipo-lexis (Ins.); Lipo-rhynchia (Platy.); Lipo-stoma*; lipo-xen-ous; Axono-lipa (Coel.):    3. Gr. *liparia*, perseverance.

**lipar**—See lip 1 and 3.

**lipasm**—Gr. *lipasma*, genit. *lipasmatos*, fatness, a fattening substance.

**lipaug**—Gr. *lipaugēs*, having lost its splendor or brilliance. *Ex:* Lipaugus*.

**lippul**—L. *lippulus*, somewhat blear-eyed.

**lir**—1. L. *lira*, dim. *lirella*, a ridge; *lirulatus*, ridged; NL. *liratus*, bearing ridges. *Ex:* Lirat-ella (Ins.); lirate; Lirator (Moll.); lirelli-form; Liri-odon (Moll.). See also liri:    2. Gr. *liros*, bold, lewd. *Ex:* Liro-pecten (Moll.).

**lirat**—See lir.

**lirell**—See lir.

**liri**—Gr. *leiron*, a lily>*leiros*, lily-white, delicate, pale. *Ex:* Leiro-notus (Ins.); Lirio-dendron*; Dasy-lirion*. See also lir.

**lirion**—See liri.

**liriope**—L. *Liriope*, a fountain-nymph, mother of Narcissus. *Ex:* Liriope*, (Coel.).

**lisp**—Gr. *lispos*, smooth, polished. *Ex:* Lisp(od)-esthes (Moll.); Lispo-gnathus (Crust.); Lispo-thrips (Ins.).

**liss**—Gr. *lissos*=*lissē*, smooth. *Ex:* Liss-amphibia (Amph.); Liss-encephala (Mam.); Lisso-delphis (Mam.); Lisso-flagellata (Prot.).

**listr**—Gr. *listron*, dim. *listrion*, a tool for smoothing, a shovel, hoe; *listrōtos*, leveled, polished. *Ex:* Listrio-therium (Mam.); Listro-poda (Ins.); Listro-stachys*; Listrota (Ins.).

**lit**—1. L. *litus*, besmeared; pp. of *lino*, to befoul, smear over:    2. L. *litus*, the sea shore:    3. L. *litus*, a shearing:    4. Gr. *litos*, simple, small, smooth, slender; *litotēs*, plainness, simplicity. *Ex:* Liti-opa (Moll.); Lito-pterna (Mam.); Lito-siphon*; Lito-tarsus (Mam.); Litos anthes*; not Lito-cranius (Mam.), which comes in part from Gr. *lithos*, stone.

**litarg**—Gr. *litargos*, running quick. *Ex:* Litarg-ellus (Ins.); Litargo-somus (Ins.); Litargus (Ins.).

**-lite**—NL. *-lite*, combining form<Gr. *lithos*, a stone. Used in names of minerals, rocks and stony objects such as fossils. *Ex:* Nummu-lites (Prot.), nummu-lite.

**-lites**—See -lite.

**lith**—Gr. *lithos*, stone; *lithōsis*, petrifying, turning into stone; *lithōdēs*, like stone; ML. *lithistes*, a kind of sponge<*lithizō*, to look like a stone; *lithax*, genit. *lithakos*, stony. *Ex:* lith-ichno-zoa; Lith-urgus (Ins.); Lithio-phanes (Av.); Lithistes (Por.); Lithist-ida (Por.); lithizo; Litho-carpus*; litho-desma; Lithodes (Crust.); Lithosia (Ins.); Lito-cranius (Mam.); Lytho-glyptus (Moll.); Neo-lithic (Anthr.); Paleo-lithic (Archeo.).

**lithac**—See lith.

**lithad**—Gr. *lithas*, genit. *lithados*, a stone. *Ex:* Lithado-thrips (Ins.)

**lithas**—See lithad.

**lithodom**—Gr. *lithodomos*, a stone-mason. *Ex:* Lithodomus (Moll.).

**litigios**—L. *litigiosus*, quarrelsome.

**litor**—See littor.

**litotes**—Gr. *litotēs*, plainness. *Ex:* Litotes (Nemat.).

**litsea**—NL. *litsea*<the Chinese *li tsai*, a little plum. *Ex:* Litsea*.

**litterat**—L. *litteratus*, branded, marked with letters.

littor—NL. *littus* <L. *litus,* the sea shore; *litoralis,* improperly *littoralis,* belonging to the sea shore; Fr. *littoral,* the sea shore. *Ex:* Littorella*; Littor-ina (Moll.); littoral; littorideserta.

litu—L. *lituus;* a trumpet or staff with curved end; NL. *lituatus,* forked and with points turned a little downwards. *Ex:* Litu-ites (Moll.); Litu-ola (Prot.); litui-form.

litur—L. *lituro,* to erase, pp. *lituratus,* erased.

litus—L. *litus,* the coast, shore of a lake.

liv—NL. *livius,* lead-colored, bluish-gray <L. *liveo,* to be of a blue-black color, livid. *Ex:* livia.

liven—L. *livens,* genit. *leventis,* bluish, black and blue, of the color of lead.

livescen—L. *livescens,* genit. *livescentis,* turning black and blue, blue <*livesco.* See liven.

livid—L. *lividus,* blue, bluish, leaden color.

livon—Russian *Livonia,* Baltic province of Russia. *Ex:* Livonia (Moll.).

livor—L. *livor,* genit. *livoris,* lead colored, leadenblue.

lix—L. *lix,* genit. *licis,* ashes, lye; *lixivius* <*lixivium,* containing lye, alkaline salts. *Ex:* lixivi-al; Lixo-somus (Ins.); Lixus (Ins.).

ixiv—See lix.

loas—South American *loasa,* a plant name of unknown origin given by Adanson. *Ex:* Loasaccac*; Loasa.*

lob—Gr. *lobos,* a lobe; also a capsule or pod >NL. *lobulus,* a small lobe, a lobule; NL. *lobosus,* full of lobes, ragged, tattered; NL. *lobatus,* lobed. *Ex:* Lob-actis (Coel.); Lobata (Cten.); Lobato-mixis (Ins.); Lobi-pes (Av.); Lobiophasis (Av.); Lobo-ceras (Ins.); Lobosa (Prot.); Lobotes (Pisc.); Lobul-aria*; lobule.

lobat—See lob.

lobot—Gr. *lobotēs,* lobed. *Ex:* Lobotes (Pisc.).

oc—L. *locus,* dim. *locellus,* place. *Ex:* locellus.

loch—Gr. *lochos,* an ambush; also a company or a body of people; *lochētikos,* lying in ambush; *lochites,* one who lives in ambush, a recluse, a fellow soldier. *Ex:* lochetic; Lochites (Ins.); Lochito-myia (Ins.); Amphi-loch-oides (Arth.); Archi-lochus (Av.).

locheum—See lochi

lochi—Gr. *locheia,* childbirth; *locheuma,* genit. *locheumatos,* a child, that which is born; *lochios,* pertaining to childbirth. *Ex:* Locheuma (Ins.); lochia; lochio-rrhagia (Med.); Aristolochia*.

lochit—See loch.

lochm—Gr. *lochmē,* a thicket, bush; *lochmaios,* of the bushes; *lochmōdēs,* bushy. *Ex:* lochmocola; Lochmo-phasis (Av.); lochmo-phyta.

lochmod—See lochm.

locul—L. *loculus,* a small place, a cell, dim. of *locus,* a place. *Ex:* locul-ar; Loculi-pora (Bry.).

loculament—L. *loculamentum,* a box, case. *Ex:* loculament-ose; loculamentum.

locust—L. *locusta,* locust, grasshopper. *Ex:* Locust-idae (Ins.); Locusta (Ins.); Locustivora (Ins.).

lodic—L. *lodix,* genit. *lodicis,* dim. *lodicula,* a coverlet, blanket. *Ex:* lodicule.

lodoicea—NL. *lodoicea,* a modification of Gr. *Laodikē,* daughter of Priam. *Ex:* Lodoicea*.

loech—NL. *loechus* <a supposed Gr. *loechos,* a licking, erroneously derived from Gr. *leicho,* to lick. *Ex:* Haemato-loechus (Platy.).

loem—See loim.

loesth—Gr. *loisthos,* left behind. *Ex:* Loesthia (Ins.).

log—Gr. *logos,* word or discourse >L. *lego,* to speak, also >NL. *-logia* and *-logy. Ex:* analog-ous; ana-logue; histo-logy; homo-log-ous; zoo-logy.

-logy—See log.

loidor—Gr. *loidoros,* abusive as subst., a railer. *Ex:* Loidor-usa (Av.).

loim—Gr. *loimos,* a plague. *Ex:* Loemo-psylla (Ins.); Loimos (Platy.).

loip—Gr. *loipos,* the remaining. *Ex:* Loipophyllum (Coel.).

loli—L. *lolium,* an old name for darnel. *Ex:* Lolium*.

lolig—L. *loligo,* genit. *loliginis,* a cuttle fish. *Ex:* Loligo (Moll.); Loligo-sepia (Moll.); Lolliguncula (Moll.). When the ending unculus, -a, -um is added to a noun stem it is sometimes the custom to double a consonant in the stem, with a sense of humor involved.

lollig—See lolig.

lom—Gr. *lōma,* genit. *lōmatos,* fringe, border of a robe. *Ex:* Lom-aria*; Loma-myia (Ins.); loma-stome; Lomat-ium*; Lomato-stoma (Moll.) Lomo-mus (Mam.); A-lomia*; Cyclo-loma*.

lomat—See lom.

loment—L. *lomentum,* bean meal. *Ex:* loment; loment-aceous; Loment-aria*.

lomvia—Faeroëse *lomvia,* a bird name. *Ex:* Lomvia (Av.).

lonch—Gr. *lonchē,* a spear; *lonchitis,* a plant with spear-shaped seeds; *lonchimos,* of a spear; *lonchērēs,* armed with a spear; *lonchōtos,* furnished with a point. *Ex:* Lonch-idia (Ins.); Loncheres (Mam.); Lonch-aea (Ins.); Lonchitis*; Loncho-carpus*; Eu-lonchus (Ins.); Steno-lonche (Ins.).

lonchot—See lonch.

long—L. *longus,* long. *Ex:* Long-ichneumon (Ins.); Longi-cornia (Ins.); Longo-laelaps (Arach.).

longaev—L. *longaevus*, ancient.

longchit—L. *longchitis*, genit. *longchitidis*, name of some plant < Gr. *longchitis*, genit. *longchitidos*. *Ex:* Longchitis*.

longe—See long.

longuri—L. *longurio*, a slender youth, sapling; *longurius*, a long pole.

Long-flowered Acleisanthes, *Acleisanthes longiflora*. A rare plant of the S. W. American deserts. The generic name means without closing flowers. Redrawn from Desert Wild Flowers—Jaeger. Stanford University Press.

lop—Gr. *lopos* = *lopisma*, shell, husk, peel, bark. *Ex:* lopo-lith; Lopo-poma (Moll.); Lopus (Ins.); A-lopo-glossus (Rept.); Leio-lopisma (Rept.).

lopad—1. Gr. *lopas*, genit. *lopados*, a flat dish. Lopadio-crinus (Echin.): 2. Gr. *lōpas*, genit. *lōpados*, a garment, tunic. *Ex:* Lopadophorus (Por.).

loph—Gr. *lophos*, the crest; also the nape of the neck; *lophōdēs*, like a ridge; *lophōtos*, crested. *Ex:* Loph-ortyx (Av.); Lophi-odon (Mam.); Lophio-mys (Mam.); Lopho-chiton (Moll.); lopho-phore; lopho-trich-ous; Lophotocarpus*; Stomo-lophus (Coel.).

lophyr—Gr. *lophouros*, with tufted tail. *Ex:* Lophyr-iscus (Moll.); Lophyro-plectus (Ins.); Lophyrus (Av.).

lopim—Gr. *lopimos*, easy of decortication. *Ex:* Lopimia*.

lopisma—See lop.

loquac—L. *loquax*, genit. *loquacis*, chattering, talkative.

loquax—See loquac.

lor—L. *lorum*, a thong, strap; *loreus*, made of thongs; *loratus*, bound with thongs, now taken to mean strap-shaped. *Ex:* lor-al = lore-al; Lor-anthus*; Lora (Moll.); lore = lora; Loripes (Moll.); Loro-petalum*; lorum; not Lor-idium (Mam.) which is Latinized from Fr. *loris*, see loris.

lord—Gr. *lordos*, bent backward; *lordōsis*, a spinal curvature with convexity in front. *Ex:* Lordo-phlips (Ins.); lordosis.

lordot—NL. *lordot* < Gr. *lordos*, bending back. *Ex:* lordot-ic.

lori—Malay, *lūrī*, *nūrī* = *lory* or *lori*, a lory or parrot. *Ex:* Lori-inae (Av.); Loriculus (Av.), NL. dim. of *Lorius*; Lorius (Av.).

loric—L. *lorica*, a corselet made of thongs; *loricatus*, clad in mail. *Ex:* lorica; Loricata (Moll.); loricate; not Lor-iculus (Av.), see lori.

Polylophodont Molar of Stegodon. Redrawn from Textbook of Palaeontology—Zittel. The Macmillan Co.

loris—NL. *loris* < Dutch *loeris*, clown, booby or perh. < Flemish *lorrias*, lazy. It also is commonly said to be a native East Indian name signifying "bashful cat" or "bashful monkey." *Ex:* Lor-idium (Mam.); Loris (Mam.).

lot—1. L. *lotus*, a washing. *Ex:* lot-ic (Ecol.): 2. L. *lotus* < Gr. *lōtos*, a sweet, juicy-fruited shrub; also the name of several other plants. *Ex:* loti-form; Lotus*.

lota—OFr. *lote*, some ganoid fish, the pout. *Ex:* Lota (Pisc.).

lotor—NL. *lotor*, a washer < participial form, *lotus*, of the verb *lavo*, to wash + -*or*, a doer. See lot and lutor.

lotus—See lot 2.

lox—Gr. *loxos*, crooked, slanting. *Ex:* Lox-omma (Amph.); Lox-ops (Av.); Loxia (Av.); Loxo-donta (Mam.); loxo-loph-odont; Loxo-(n)ia*; Pyrrhu-loxia (Av.).

lubric—L. *lubricus*, smooth, slippery. *Ex:* Lubrico-gobius (Pisc.); lubricous.

luc—L. *lux*, genit. *lucis*, light; *lucus*, dim. *luciola*, at daybreak; also a wood thicket sacred to the deities; *lucens*, shining, conspicuous. *Ex:* Lucilia (Ins.); Luci-dota (Ins.); luci-fer-ase; luciferin; Luciola (Ins.); Lux-ilus (Pisc.). See also luci.

lucan—M.L. *lucanus*, a beetle mentioned by Pliny < *luca*, the stag beetle. *Ex:* Lucan-idae (Ins.); Lucanus (Ins.).

lucania—NL. *lucania*, a coined name. *Ex:* Lucania (Pisc.).

lucar—L. *lucaris*, of or belonging to a grove.

lucen—L. *lucens*, genit. *lucentis*, shining, bright, visible < *luceo*, to shine < Gr. *leukos*, white, bright. *Ex:* trans-lucent.

lucerna—L. *lucerna*, a lamp. *Ex:* Lucern-aria (Coel.); Lucern-ula (Ins.); Lucerna (Moll.).

luci—L. *lucius*, the pike. *Ex:* Lucio-brama (Pisc.); Lucio-gobius (Pisc.).

**lucid**—L. *lucidus*, clear, bright, shining. *Ex:* Lucid-ella (Moll.).

**lucin**—1. L. *lucinus*, light-bringing: 2. L. Lucina, goddess of childbirth, also a name of Diana. *Ex:* Lucina (Moll.); Lucin-aea*.

**lucipet**—L. *lucipetus*, light-seeking.

**lucisat**—L. *lucisator*, a producer of light.

**luctan**—L. *luctans*, genit. *luctantis*, struggling, ppr. of *luctor*, to wrestle, struggle.

**luctat**—L. *luctator*, a wrestler.

**luctuos**—L. *luctuosus*, causing grief, sorrow.

**lucubrans**—L. *lucubrans*, genit. *lucubrantis*, working at night, ppr. of *lucubro*, to work by lamp-light.

**lucum**—Quechan Indian *lucuma*, a plant name. *Ex:* Lucuma*.

**lud**—L. *ludo*, to play, sport, pp. *lusus*, playing; *ludor*, a player.

**luden**—L. *ludens*, genit. *ludentis*, playing; ppr. of *ludo*, to play, sport.

**ludi**—L. *ludius*, a stage-player. *Ex:* Ludius (Ins.).

**ludibund**—L. *ludibundus*, playful.

**ludificat**—L. *ludificatus*, a mocking.

**luff**—Ar. *lūfah*, a plant name>NL. *luffa*. *Ex:* Luffa*.

**lugent**—L. *lugens*, genit. *lugentis*, ppr. of *lugeo*, to mourn, to wear mourning apparel.

**lugubr**—L. *lugubris*, sorrowful, dark, gloomy.

**lumb**—L. *lumbus*, loin; *lumbaris*, of the loins. *Ex:* lumbo-abdominal; lumbar.

**lumbric**—L. *lumbricus*, an earthworm. *Ex:* Lumbrici-cola (Nemat.); Lumbrici-nereis (Ann.); Lumbricus (Ann.).

**lumen**—See lumin.

**lumin**—L. *lumen*, genit. *luminis*, light; also a window, brightness, splendor; *luminosus*, full of light; *luminatus*, lighted. *Ex:* lumen; lumin-escence.

**lumpen**—Dan. *lumpen*, a name for *Zoarces viviparus* with which these fishes were confounded. *Ex:* Lumpen-ella (Pisc.); Lumpenus (Pisc.).

**lun**—L. *luna*, the moon; *lunatus*, crescent-shaped <*luno*, to crook like a sickle, to bend like a half-moon. *Ex:* Lun-aria*; lun-ule; lunate; luni-frons.

**lup**—1. L. *lupus*, dim. *lupulus*, a wolf. *Ex:* Lupulus (Mam.); Lupus (Mam.): 2. L. *lupus*, dim. *lupululus*, the hop plant (Humulus lupulus). *Ex:* lupul-inus.

**luper**—See lyper.

**lupin**—1. L. *lupinus*, belonging to a wolf <*lupus*, a wolf: 2. L. *lupinus*, a lupine. *Ex:* Lupinus*.

**lurid**—L. *luridus*, pale yellow.

**luror**—L. *luror*, genit. *luroris*, paleness, yellowish colored.

**-lus**—L. *-lus*, *-la*, *-lum*, dim. added to certain noun stems, as in *alveo-lus*, a small cavity < *alveus; virg-ula*, a small branch <*virga*, branch; *atrio-lum*, a small vestibule <*atrium*, a hall.

**lus**—L. *lusus*, a game; *lusor*, genit. *lusoris*, one who plays, a sport <*ludo*, to sport, play. *Ex:* lusi-form; lusus.

**lusc**—L. *luscus*, one-eyed. *Ex:* not Luscus (Mam.) which is an error for Cuscus (Mam.).

**luscin**—L. *luscinia*, a nightingale. *Ex:* Lusciniopsis (Av.); Luscinia (Av.).

**lusitanic**—L. *lusitanicus*, of Portugal.

**lusori**—L. *lusorius*, belonging to a player.

**lustrabil**—L. *lustrabilis*, conspicuous.

**lustran**—L. *lustrans*, genit. *lustrantis*, going around, encircling, also examining <*lustro*, to go around.

**lut**—1. L. *lutum*, yellow-weed; as a color, clay-yellow>NL. *lutescens*, genit. *lutescentis*, becoming or appearing to be clay-yellow, and *luteus*, golden-yellow. *Ex:* lutei-ventris; lutein; luteo-fulv-ous; lutescent; corpus luteum: 2. L. *lutum*=*lutus*, mud, mire; *luteus*, muddy; *lutarius*, belonging to mud>*lutensis*, living in mud; *lutosus*, full of mud. *Ex:* luti-col-ous; luti-fer-ous: 3. Gr. *loutēs*, a bather. *Ex:* sigo-lutes.

**luteol**—L. *luteolus*, yellowish. <*luteus*, yellow+ *-olus*, a dim. ending.

**lutes**—See lut 3.

**lutesc**—See lut 1.

**lutian**—NL. *lutianus*, a compound word derived <a Malayan name for fishes. *Ex:* Lutianus (Pisc.).

**lutit**—L. *lutitus*, bedaubed or befouled with mud.

**lutor**—L. *lutor*, genit. *lutoris*, a washer.

**lutr**—L. *lutra*, dim. *lutreola*, otter. *Ex:* Lutraria (Moll.); Lutr-avus (Mam.); Lutra (Mam.); Lutreola (Mam.); Lutrix (Mam.). See also lytros.

**lutulent**—L. *lutulentus*, muddy, impure, filthy.

**lux**—See luc.

**luxat**—L. *luxatus*, put out of joint, pp. of *luxo*, to dislocate.

**luxil**—NL. *luxilus*, a name coined by the eccentric Rafinesque <L. *lux*, light. *Ex:* Luxilus (Pisc.).

**luxuos**—NL. *luxuosus*, luxurious <L. *luxus*, excess, luxury.

**luxurian**—L. *luxurians*, genit. *luxuriantis*, growing rankly, abounding in richness.

**luxus**—L. *luxus*, a dislocation. See also luxuos.

**luzul**—NL. *luzula* (<It. *luzziola*, *lucciola*, a glow worm <L. *luceo*, to shine), name for a plant called adder's tongue. *Ex:* luzul-ina; Luzula*; luzuli-folia.

**ly**—Gr. *lyō*, to loose>*lysis*, a loosing. *Ex:* Ly-encephala (Mam.); Lyo-sphaera (Pisc.); dialysis; haemo-lysis. See also lys.

**-ly**—Eng. *-ly*, suffix meaning like. *Ex:* seriati-ly.

**lyc**—**1.** Gr. *lykos*, a wolf; also a kind of noose, a flesh-hook; *lykōdēs*, wolfish; *lykaina*, a she-wolf; *lykaon*, a wolf-like animal. *Ex:* Ly-cyon (Mam.), *c* omitted; Lyc-alopex (Mam.); Lyc-iscus (Mam.); Lyc-orus (Mam.); Lycaena (Ins.); Lycaon (Mam.); Lyco-gala*; Lyco-perdon*; Lyco-podium*; Lycodo-nus (Pisc.), *nus* without meaning; Lycus (Ins.): **2.** Gr. *lykos*, a kind of spider<*lykos*, a wolf, so named because of its predatory habits. *Ex:* Lycosa (Arach.): **3.** Gr. *lykos*, a hood or door-knocker. *Ex:* lyco-trop-ous.

**lychn**—**1.** Gr. *lychnos*, a lamp. *Ex:* Lychni-fugus (Ins.); Lichno-canium (Prot.); Lychno-gaster (Ins.): **2.** Gr. *lychnis*, genit. *lynchni-dos*, a plant with bright scarlet flower. *Ex:* Lychnido-spiza (Av.); Lychnis*.

**lyci**—Gr. *lykion*, a kind of thorny Lycian shrub. *Ex:* Lycium*.

**lycoperd**—NL. *lykoperdon*, name applied to a genus of puffballs<Gr. *lykos*, wolf+*perdomai*, to break wind. *Ex:* Lycoperd-ina (Ins.); Lycoperdon*.

**lycor**—L. *Lycoris*, a Roman actress mentioned by Virgil. *Ex:* Lycoris*, (Ann.); not Lycorus (Mam.) which comes from Gr. *lykos*, wolf +*oros*, mountain.

**lyct**—NL. *lyctus*<Gr. *Lyctos*, founder of a Cretan city. *Ex:* Lycto-pholis (Ins.); Lyctus (Ins.).

**lyg**—Gr. *lygos*, a pliant twig, willow-like tree; *lygōdēs*, flexible, pliant, like a willow twig. *Ex:* Lygeum*; Lygo-cerus (Ins.); Lygo-desmia*; Lygo-soma (Rept.); Lygod-ium*.

**lygae**—Gr. *lygaios*, shadowy, gloomy. *Ex:* Lygae-idae (Ins.); Lygaeo-scytus (Ins.); Lygeo-morphus (Ins.); Lygaeus (Ins.); not Lygeum* which is derived from *lygoō*, to tie, bind.

**lygar**—NL. *lygaria*<Gr. *lygē*, twilight, something obscure. *Ex:* Lygaria (Ins.).

**lygdin**—Gr. *lygdinos*, of white marble, marble-white. *Ex:* Lygdinus (Arach.).

**lygeo**—See lygae.

**lygin**—Gr. *lyginos*, bent together, shrivelled, wrinkled. *Ex:* Lygino-pteris*; Lyginus (Ins.).

**lygism**—Gr. *lygisma*, genit. *lygismatos*, a twist; *lygismos*, a bending, twisting. *Ex:* Lygisma (Pisc.).

**lygist**—Gr. *lygistos*, bent, pliant; *lygistēs*, a basket-maker; *lygistikos*, readily bending. *Ex:* Lygisto-pterus (Ins.).

**lygr**—Gr. *lygros*, baneful, cowardly. *Ex:* Lygr-ommat-oides (Arach.); Lygro-charis (Ins.).

**lym**—**1.** Gr. *lyma*, genit. *lymatos*, filth: **2.** Gr. *lymē*, maltreatment, ruin. *Ex:* Lyme-xylon (Ins.); not Lym-odon (Mam.) which is probably an anagram of Myl-odon (Mam.).

**lymant**—Gr. *lymantēr=lymantōr*, a destroyer; *lymantērios*, injurious, destructive. *Ex:.* Ly mantes (Ins.); Lymantor (Ins.); Lymantria (Ins.).

**lymat**—See lym.

**lymn**—See limn.

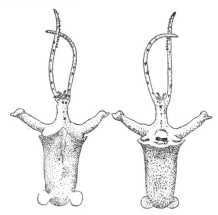

Lyre-eyed Marvel-of-the-deep, *Bathothauma lyromma*, a squid. Redrawn from Natural History Magazine.

**lymph**—L. *lympha*, water. *Ex:* lymph; lympho-genesis.

**lync**—Gr. *lynx*, genit. *lynkos*, the lynx. *Ex:* Lync-odon (Mem.); Lynchus (Mam.); Lynx (Mam.).

**lynceus**—L. *Lynceus*, one of the Argonauts famed for his sharp sight. *Ex:* Lynceus (Mam.).

**lynx**—See lync.

**lyp**—Gr. *lypē*, pain of body or mind, a sad condition. *Ex:* Lype (Ins.).

**lypem**—Gr. *lypēma*, genit. *lypēmatos*, pain.

**lyper**—Gr. *lypēros*, painful, sad, sorrowful. *Ex:* Lyper-anthus*; Lyperia*; Lupero-saurus (Rept.); Lyperus (Ins.).

**lypr**—Gr. *lypros*, wretched, poor; *lyprotēs*, wretchedness. *Ex:* Lypro-corrhe (Ins.); Lyprus (Ins.).

**lyr**—Gr. *lyra*=L. *lyra*, a lyre; NL. *lyratus*, lyre-like. *Ex:* Lyr-oda (Ins.); Lyr-urus (Av.); lyrate; lyri-form; Lyro-derma (Mam.).

**lyricen**—L. *lyricen*, genit. *lyricinis*, a lute player. *Ex:* Lyricen (Ins.).

**lys**—Gr. *lysis*, a loosing<*lyō*, to loose; *lysios*, releasing, loosening. *Ex:* Lisi-anthus*; lyse-gen-ic=lysi-gen-ic; Lysi-chitum*; lysi-genetic; Lysi-loma*; Lysi-urus; Lysio-gnatha (Ins.); Lysio-notus*; Lysio-squilla (Arthr.); haemo-lysis; para-lysis.

**lysimachia**—Gr. *lysimachion*, a kind of herb used in medicine. *Ex:* Lysimachia*.

**lysinoe**—Gr. *Lysinoē*, city in Pisidia. *Ex:* Lysinoe (Moll.).

yss—Gr. *lyssa*=Attic *lytta*, madness. *Ex:* Lyssodes (Mam.); Antho-lyza*; not Lyssakina (Por.) which is poorly made from Gr. *lysis*, a loosing.

lyt—Gr. *lytos*, dissolvable, broken; *lytērios*, loosening, releasing. *Ex:* Lyt-acra (Ins.); Lytechinus (Echin.); A-lytes (Amph.); Autolytus (Ins.); caryo-lytes; Hippo-lyte (Crust.).

lyth—See lith.

lythr—Gr. *lythron*, blood, gore, hence sometimes used in sense of red; *lythrōdēs*, defiled with blood. *Ex:* Lithr-ichthys (Pisc.); Lythr-aria (Ins.); Lythr-ulon (Pisc.); Lythrodes (Ins.); Lythrum*.

lytros—Gr. *lytrōsis*, a freeing. *Ex:* Lutro-stylis*.

lytta—Gr. *lytta*=*lyssa*, fury, canine madness; also the worm under the tongue of dogs which was removed because of the belief that it produced madness. *Ex:* Lytta (Ins.).

-lyze—NL. *-lyze*, an ending used to form transitive verbs from noun stems to which the ending *-lysis* has been added; thus, from catalysis we have cata-lyze and from paralysis, paralyze.

# M

-ma—Gr. *-ma*, neuter suffix denoting thing that, that which. *Ex:* aro-ma, stig-ma.

macacus—NL. *macacus* < Fr. *macaque* < a native name of a monkey in the Congo. *Ex:* Macacus (Mam.).

macar—Gr. *makaria*, happiness; *makarios*=*makaros*, blessed, happy. *Ex:* Macaria (Ins.); Macaro-crinus (Echin.).

macell—1. Gr. *makella*, a miner's pick-axe with one point. *Ex:* Macell-odon (Rept.); Macellodous (Rept.); Macella (Ins.); Macello-menia (Moll.): 2. Gr. *makellon*, a slaughter house, an enclosure.

macer—See macr.

macerat—L. *maceratio*, a steeping, soaking *Ex:* macerat-ion.

mach—Gr. *machē*, battle, combat; *machētēs*, a fighter. *Ex:* Mach-odon (Ins.); Machetes (Av.); macho-polyp; Lysi-machia*; Odonto-machus (Ins.).

machaer—Gr. *machaira*, a saber, a dagger; *machairion*, a surgeon's knife. *Ex:* Machaer-odus = Machair-odus (Mam.); Machaer-ites (Ins.); Machaera (Moll.); Machaer-ium*; Machaeropus (Crust.); Machetes (Av.); Makaira (Pisc.).

machair—See machaer.

machet—See mach.

machil—NL. *machilis*, name for a genus of insects. *Ex:* Machilis (Ins.).

machl—Gr. *machlos*, lustful. *Ex:* Machlo-stomus (Av.).

machlyd—Anagram of *chlamydo*. *Ex:* Machlydotherium (Mam.).

machomen—Gr. *Machomenē*, the militant one *Ex:* Machomena (Ins.).

Jamaican Great-armed Stalk-eyed Crustacean, *Macrobrachium jamaicense*. Redrawn from Volume 38, Proceedings of the United States National Museum.

machrinus—NL. *machrinus* < Gr. *makros*, large +*rhinos*, nose, a fabrication of the eccentric Rafinesque. *Ex:* Machrinus (Pisc.).

macilent—L. *macilentus*, thin, lean < *macies*, leanness.

macr—1. Gr. *makros*, long, large. *Ex:* Macr-acantha (Arach.); macr-aden-ous; macr-andr-ous; Macr-oto-lagus (Mam.); Macr-ura (Crust.); Macro-dactyla (Coel.); macro-scopic: 2. L. *macer*, fem. *macra*, neut. *macrum*, lean.

macrescen—L. *macrescens*, growing lean.

mact—L. *mactus*, honored, adored. *Ex:* Macta (Ins.). See mactat.

mactan—L. *mactans*, genit. *mactantis*, deadly, killing < *macto*, to afflict, slaughter, punish; pp. *mactatus*, punished.

mactat—L. *mactatus*, killed, pp. of *macto*, to kill; *mactator*, a killer; *mactans*, killing, ruining.

mactr—Gr. *maktra*, a kneading trough. *Ex:* Mactra (Moll.); Mactro-desma (Moll.).

macul—L. *macula*, spot, stain, mark; *maculosus*, speckled, mottled, full of spots, blotted, stained; *maculo* to make spotted; ppr. *maculans*, spotting; pp. *maculatus*, spotted, variegated, full of spots. *Ex:* maculated; maculiformis; bi-maculate.

mad—1. L. *madidus*, moist; *madens*, genit. *madentis*, moistening, ppr. of *madeo*, to be wet, to drip; *madefactus*, moistened. *Ex:* E-madus (Ins.): 2. L. *madidans*, weeping, giving rise to moisture, ppr. of *madido*, to moisten: 3. Gr. *mados*, smooth>NL. *anomados*, not smooth, uneven<Gr. *aneu*, without+*mados*. *Ex:* Ano-madus (Ins.).

madar—Gr. *madaros*, flaccid, bald. *Ex:* Madarus (Ins.).

madefact—See mad.

madia—Chilean, *madi*, the name of some plant. *Ex:* Madia*.

madid—See mad.

madon—Gr. *madōnia*, a name of the water-lily. *Ex:* Madoni-actis (Coel.).

madre—Ital. *madre*, mother. *Ex:* Madre-myia (Ins.); madre-pore.

maeandr—Gr. *maiandros*, a winding<Gr. *Maiandros*, winding river of Phrygia. *Ex:* Maeandrina (Coel.); Maeandro-seris (Coel.); Maeandrusa (Ins.).

maen—Gr. *mainē*, a small fish which was salted. *Ex:* Maena (Pisc.); Archeo-maene (Pisc.); Pisco-maena (Pisc.).

maenad—L. *Maenas*, genit. *Maenadis*, priestesses who worshiped at the altar of Priapus. *Ex:* Maenas (Ins.).

maenas—See maenad.

maeren—L. *maereo*, to be sad, ppr. *maerens*, genit. *maerentis*, saddening.

maest—L. *maestus*, sad, dejected.

magdal—L. *magdalia*, waltzing figures. *Ex:* Magdal-inus (Ins.); Magdalis (Ins.).

mag—Gr. *Magos*, one of the Magi or priests of Persia, a magician. *Ex:* Magus (Mam.); Paulo-magus (Av.).

magestus—NL. *magestus*, anagram of *Megastus*. *Ex:* Magestus (Mam.).

magic—L. *magicus*, magical, mysterious.

magilus—NL. *magilus*, from a native name for a mollusk. *Ex:* Magilus (Moll.).

magister—L. *magister*, a magistrate, chief.

magm—Gr. *magma*, genit. *magmatos*, a kneaded mass, salve<*massō*, to knead. *Ex:* magm-oid; magmat-ic.

magn—L. *magnus*, great. *Ex:* magna-glans; magni-fi-er; magnum.

mai—1. Gr. *maia*, a large kind of crab; also a good mother; *Maia*, daughter of Atlas and mother of Mercury. *Ex:* Mai-idae (Crust.); Mai-opsis (Crust.); Maia (Crust.): 2. Gr. *Maios*, May. *Ex:* Mai-anthemum*.

maieu—Gr. *maieusis*, childbirth; *maieutikos*, of or pertaining to obstetrics. *Ex:* maieutic, maieusio-phobia.

maior—See majuscul.

majal—1. L. *majalis*, a castrated boar: 2. NL. *majalis*=L. *maialis*, pertaining to or of May.

major—L. *major*, greater, larger; compar. of *magnus*, great; *majorinus*, of a larger kind or form.

majus—L. *majus*, great<the root *mag* from which comes *magis* and *magnus*. See major.

majuscul—L. *majusculus*, somewhat greater, larger or older, a dim. from *major*=*maior*, greater.

makaira—See machaer.

mal—1. L. *malum*<Gr. *mēlon*=Doric *malon*, an apple; L. *malus*, an apple-tree. *Ex:* Malus*; Micro-melum*: 2. L. *malus*, a mast, a beam. *Ex:* mali-ger: 3. L. *malus*, bad, ugly. *Ex:* mal-form-ation; for Sauromalus, see omal: 4 Gr. *malos*, woolly, soft. *Ex:* Malurus (Av.), "here taken to mean slender"; Mala-pter-urus (Pisc.); for Malaclemmys (Rept.) see malac; Mero-malus (Ins.): 5. L. *mala*, cheek, jaw. *Ex:* mala; mali-pedes; malo-plasty (Surg.); deuto-malae; proto-mala: 6. Gr. *malos*=*mallos*, dim. *malion*, a lock of hair. *Ex:* Cori-malia (Ins.). See also malac and omal.

malac—Gr. *malakos*, soft, gentle; Gr. *malaxis*, a softening; L. *malacissans*, genit. *malacissantis*, ppr. of *malacisso*, to make soft. *Ex:* Malaclemmys (Rept.); Mala-pter-ursus (Pisc.); malacissant; Malaco-bdella (Nemert.); Malaco-edus (Av.); malaco-logy, science of soft animals, i.e. mollusks; Malaco-thrix*. Malaxa(Ins.); malax-ation.

malach—Gr. *malachē*, mallow; *malachion*, a woman's dress of mallow color. *Ex:* Malachius (Ins.); Malacho-dendron*.

malari—Eng. *malaria*, a disease<It. *malaria*, bad air. *Ex:* Malari-ology.

malax—See malac.

malign—L. *malignans*, genit. *malignantis*, ppr. of *maligno*, to treat maliciously; *malignus*, wicked, evil. *Ex:* malignant; malignan(t)-cy.

malit—L. *malitia*, badness>*malitiosus*, wicked, full of tricks, knavish.

mall—Gr. *mallos*=*malos*, a lock of wool; *mallōtos*, fleecy; *mallōsis*, a being dressed with wool. *Ex:* Mallo-mys (Mam.); Mallotus (Pisc.); homo-malia; Strepsi-mallus (Ins.).

malle—L. *malleus*, a hammer<*malleo*, to hammer. malle-able; Malleus (Moll.).

mallot—See mall.

malperi—NL. *malperia*, an anagram of *Palmeri*. *Ex:* Malperia*.

malt—AS. *mealt*=Ger. *malz*=Eng. *malt*. *Ex:* malt-ase; malt-ose.

**malth**—Gr. *malthē*, soft wax; *malthōdēs*, pliant, adhesive, after the manner of soft wax. *Ex:* Malth-aster (Ins.); Malth-opsis (Pisc.), like Malthe; Malth-inus (Ins.); Malthe (Pisc.); Malthodes (Ins.).

**malthac**—Gr. *malthakos*, soft. *Ex:* Malthaco-soma (Ins.).

**malthon**—Gr. *malthōn*, effeminate, weak. *Ex:* Malthon-ea (Ins.).

**malva**—Anc. L. *malva*, name of some plant < Gr. *malachē*, the mallow < *malassō*, to soften, because of its soft leaves or from its reputed relaxing powers. *Ex:* Malv-aceae*; Malva*; Malv-astrum*.

**mam**—NL. *mam-* prefix, indicating a mammal. *Ex:* Mam-ateles-us, *Ateles*, with *mam*, prefixed, indicating a mammal, +*us* added to give it the form of a Latin masculine noun (Latin tree names ending in *us* are feminine.). In the same way *a* is often used as a plant-name ending.

**mamm**—L. *mamma*, dim. *mammilla*, breast, pap, teat; *mammalis*, of or for the breasts; among modern naturalists taken to mean having breasts. *Ex:* Mamm-aster (Echin.); mammalogy; mammal; Mammalia; Mammill-aria*; mammilla.

**mammut**—Tartar *mammantu*, a ground dweller. *Ex:* Mammut (Mam.); Mammut-idae (Mam.).

**man**—1. Gr. *manos*, wide, loose, roomy, thin, rare. *Ex:* Manis-uris*; Mano-lepis (Rept.); Mano-mera (Ins.); mano-meter; Mano-pora (Coel.): 2. L. *manus*, a hand. *Ex:* Manul-ea* < NL. *manul* + euphonius ending ea; manual: 3. Tartar *manul*, name of a cat.

**manac**—NL. *manacus* < Dutch *manneken*, dim. of *man*, a dwarf. *Ex:* Manacus (Av.); manikin.

**manat**—Haitian, *manati*, big beaver; NL. *manatinus*, like the manatee or sea-cow. *Ex:* Manat-(t)herium (Mam.); Manat-us (Mam.); manatee.

**manc**—L. *mancus*, maimed, infirm, imperfect. *Ex:* Manc-alla (Av.), Sw. *alla*, a kind of auk; Manca-lias (Pisc.), the ending -lias is a defective diminutive; Manca-sellus (Arthr.)

**mandibul**—L. *mandibula*, a mandible < *mando*, to chew. *Ex:* mandible; Mandibul-arca (Pisc.); mandibuli-form; cerato-mandibul-ar.

**mandr**—Gr. *mandra*, an enclosed place, a stable. *Ex:* Mandro-spongia (Por.); Chlorito-mandra (Ins.).

**mandragor**—Gr. *mandragoras*, the mandrake. *Ex:* Mandragora*, (Ins.).

**manduc**—L. *manducor*, to chew; *manduco*, genit. *manduconis*, a chewer, glutton; *manducus*, a glutton, a queer chewing figure. *Ex:* Manducus (Pisc.).

**manes**—Gr. *manēs*, a cup; also a slave. *Ex:* Scoto-manes (Mam.); Thryo-manes (Av.), David Starr Jordan would derive *manes* from Gr. *mainomai*, to rage, to be furious, but Coues (p. 295 Key to N. A. Birds) would derive it from Gr. *manēs*, a kind of cup.

**mang**—Pg. *manga*, Tamil *mankay*, mango, a tropical fruit. *Ex:* Mangi-fera*; mango.

**mania**—Gr. *mania* = *manē*, madness, frenzy, passion. *Ex:* Hippo-mane* < Gr. *hippomanēs*, as an adjective (said of mares), mad after the horse; as a substantive, an aphrodisiac made from spurge; maniac-al, as if > L. *maniacus* < Gr. *mania*, madness.

**maniac**—See mania.

**manic**—L. *manica*, a sleeve; *manicatus*, sleeved. Manic-aria*; Manic-ellus (Ins.); Manico-coris (Ins.). See also manac.

**maniculat**—NL. *maniculatus*, small handed < L. *manicula*, a little hand.

**manihot**—Tupi *mandihoca*, manioc, cassava. *Ex:* Manihot*.

**manipul**—L. *manipulus*, a handful, a bundle. *Ex:* manipul-ation.

**manis**—NL. *manis*, assumed singular of L. *Manes*, gods of the underworld and spirits of the dead, ghosts. *Ex:* Mani-dae (Mam.); Manis (Mam.). See also man.

**mann**—1. Gr. *mannos* = *manos*, a collar, necklace. *Ex:* Mann-odon (Mam.); Manno-charax (Pisc.): 2. Gr. *manna*, a morsel, grain of mica.

**mansuet**—L. *mansuetus*, tame, also tamed.

**mant**—Gr. *mantis*, genit. *manteōs*, a soothsayer, a kind of grasshopper with thin forefeet; *mantikos*, prophetic (the poetic form is *manteios*). *Ex:* Mante-odon (Mam.); Manteo-ceras (Mam.); Mantico-ceras (Moll.); Mantido-phaga (Ins.); Mantis; Mantis-atta (Arach.); Mantio p-id (Ins.), the *p*, a contraction of *ops*, face, or perh. < *Mantis pagana* Fabricius; Mantisia*; Rhampho-mantis (Av.).

**mante**—See mant.

**mantell**—L. *mantellum*, a cloth or cloak. *Ex:* Mantelli-ceras (Moll.); Mantellum (Moll.).

**mantic**—See mant.

**mantichora**—Gr. *mantichōras* = *martichōras*, a fabulous beast compounded of a lion, scorpion and porcupine and having a human head. *Ex:* Mantichora (Ins.); Montichor-ula (Ins.).

**manticora**—See mantichora.

**mantid**—See mant.

**mantisp**—See mant.

**mantiss**—L. *mantissa*, an addition of no value. *Ex:* Mantissa (Moll.).

**manubrium**—L. *manubrium*, a handle or haft. *Ex:* manubrium.

**manucod**—NL. *manucodia* < Malay name for the bird of paradise, *manuk-dewata*, bird of the gods. *Ex:* Manucodia (Av.).

**manul**—See man.

maori—New Zealand *Maori*, primitive inhabitants of the islands. *Ex:* Maori-ana (Moll.); Maori-drilus (Ann.).

mar—1. Gr. *maron*, a bitter herb. *Ex:* Petromar-ula\*: 2. L. *mare*, genit. *maris*, the sea. *Ex:* mari-col-ous. See also mas.

maraen—Gr. *marainō*, to die away, go out slowly, to quench. *Ex:* Maraen-aspis (Ins.); Maraeno-biot-us (Crust.).

marant—Gr. *marantos*, withering<*marainō*, to wither, to fade. *Ex:* A-maranthus\*; A-marantus\*; not Maranta\* which was named after B. Maranti, Venetian botanist.

marasm—Gr. *marasmos*, a wasting. *Ex:* Marasmarcha (Ins.); Marasmius\*; marasmus (Med.).

marc—1. L. *marcus*, a hammer; *marculus*, a small hammer: 2. L. *marcidus*, flaccid, shrunken, withered; *marcens*, genit. *marcentis*, withering, feeble; ppr. of *marcio*, to wither, to be weak. *Ex:* e-marcidus.

marcesc—L. *marcesco*, to fade, ppr. *marcescens*, genit. *marcescentis*, withering, fading. *Ex:* marcescent.

marcid—See marc 2.

mardon—Gr. *Mardonios*, a Persian general. *Ex:* Mardonius (Myr.).

marec—Pg. *marreca*, fem. of *marreco*, the widgeon. *Ex:* Mareca (Av.). See also marica.

marg—1. L. *margo*, genit. *marginis*, edge, border. *Ex:* Margin-aster (Echin.); Marginis (Ins.); Margino-spongia (Por.); E-margin-ula (Moll.); E-margin-ata (Av.); E-margin-ella (Moll.): 2. Gr. *margos*, furious, greedy. *Ex:* Margus (Ins.); Gastri-margos (Mam.).

margar—Gr. *margaritēs*, a pearl; *margarōdēs*, pearl-like. *Ex:* Margar-iscus (Pisc.); Margar-ops (Ins.); Margari-ona (Moll.); Margarites (Moll.); Margaris\*; Margariti-fera (Moll.); Margaro-perdix (Av.); Margarodes (Ins.).

margel—Gr. *margelis*, a pearl. *Ex:* Margelis (Coel.); Margel-opsis (Coel.).

margin—See marg.

marian—NL. *marianus* (<*Mary*, Virgin mother of Jesus), new specific name of the Carduus Mariae of Fuchs.

marica—L. *Marica*, a nymph, mother of the Latins. *Ex:* Marica=Mareca (Av.).

maril—Gr. *marilē*, live embers or sparks. *Ex:* Marila\* (Av.); Marilo-chen (Av.).

marin—1. L. *marinus* of the sea, marine<*mare*, the sea: 2. Gr. *marinos*, a kind of marine fish.

marisc—L. *mariscus*, a kind of rush. *Ex:* Mariscus\*.

marit—L. *maritus*, married. *Ex:* marit-al.

maritim—L. *maritimus*, of or belonging to the sea; also inconstant.

marmar—Gr. *marmaros*, marble, any rock of crystalline structure. *Ex:* Marmaro-netta (Av.); Marmaro-saurus (Rept.).

marmor—L. *marmor*, marble; *marmoratus*, marbled, pp. of *marmoro*, to overlay with marble.

marmos—Fr. *marmose*, a name of undetermined origin (not from OF. *marmoset*, a puppet; later, an ape) applied to certain S. Amer. opossums. *Ex:* Marmosa (Mam.).

marmota—Fr. *marmotte*<L. *Mus montis*, lit., mountain mouse. *Ex:* Marmota (Mam.).

marpt—Gr. *marptō*, to seize. *Ex:* Marpt-usa (Arach.): Marpto-desmus (Arth.).

marrub—NL. *marrubium*<Heb. *marrob*, bitter juice. *Ex:* marrubi-folium; Marrubium\*.

marsip—Gr. *marsypos*=*marsipos*, a purse, bag; dim. *marsypion*=L. *marsupium*. *Ex:* Marsipo-laemus (Mam.); marsupi-al; Marsupio-crinus (Echin.); marsupium; Marsypi-anthus\*.

marsup—See marsip.

marsyp—See marsip.

martes—L. *martes*, a marten. *Ex:* Martes (Mam.).

mas—L. *mas* genit. *maris*, a male.

masar—NL. *marsaris*<Gr. *masaomai*, to chew, thrust out the lip. *Ex:* Masaris (Ins.); Masarididae (Ins.).

maschal—Gr. *maschalē*, the armpit, a bay, hollow. *Ex:* Maschal-odonta (Ins.); Maschalo-stachys\*.

mascul—L. *masculus*, male, vigorous, strong. *Ex:* e-mascul-ate.

mass—1. L. *massa*, a lump, *mass*. *Ex:* mass-ula: 2. Gr. *massōn*, larger, comp. of *makros*, large. *Ex:* Mass-ornis (Av.); Masso-spondyl-us (Rept.).

mast—Gr. *mastos*, a breast. *Ex:* Mast-odon (Mam.); masta-trophia (Med.); Masto-therium (Mam.); Mastodon-saurus (Rept.).

mastac—Gr. *mastax*, genit. *mastakos*, mouth, jaw. *Ex:* Mastac-ides (Ins.); Mastaco-mys (Mam.); mastax; Psilo-mastax (Ins.).

mastax—See mastac.

master—See mastr.

mastes—See mastr.

mastich—Gr. *mastichē*=L. *mastic*, gum of the mastic-tree. *Ex:* masticho-dendron.

mastig—1. Gr. *mastix*, genit. *mastigos*, a whip; NL. *mastic*, a whip, Gr. *mastigion*, a small whip. *Ex:* Mastic-ophis (Rept.); Mastig-urus (Rept.); mastigium; Mastigo-phora (Prot.); Mastigo-proctus (Arach.); Mastigus (Ins.): 2. L. *mastigia*, a rascal, rogue. *Ex:* Mastig-ella (Prot.).

mastiges—NL. *mastiges*<Gr. *mastix*, pl. *mastiges*—Cyclo-mastiges (Prot.). See mastig.

mastiv—NL. *mastivus*, pertaining to or of the mastiff (bat) <Eng. *mastiff*<OF. *mestif*, mongrel.

mastix—See mastig 1

mastr—Gr. *mastēr*, genit. *mastēros*, a searcher; *mastēs*, in compounds, a seeker, searcher. *Ex:* Mastrus (Ins.); Syro-mastes (Ins.).

mastrucat—L. *mastrucatus*, covered with wool, clothed in skin; L. *mastruca=mastruga*, a sheep-skin.

masturb—Eng. *masturbate* <L. *masturbor*, to cause orgasm by hand, perh. <*manus* hand +*stuprare*, to defile, to rape. *Ex:* masturbation.

matae—Gr. *mataios*, empty, idle, foolish. *Ex:* Mataeo-cephalus (Pisc.); Mataeo-mera (Ins.); Mataeo-ptera (Av.); Mataeus (Ins.).

matar—L. *materis=mataris*, a Celtic javelin. *Ex:* Matara (Ins.); Mataris (Ins.).

matax—L. *mataxa=metaxa*, raw silk. *Ex:* Mataxa (Moll.).

matern—See matr.

matr—L. *mater*, genit. *matris*, mother; *maternus*, motherly. *Ex:* matern-al; matro-clinous; pia mater >matr-al.

matric—L. *matrix*, genit. *matricis*, the womb or uterus; often used in the sense of a place where anything is generated <*mater*, mother. *Ex:* Matric-aria*; matrix.

matrix—See matric.

matron—L. *matrona*, a wife, matron; *matronalis*, belonging to a matron.

matt—L. *matta*, mat. *Ex:* matt-ula; Matta (Arach.).

matur—L. *maturus*, ripe; *maturatus*, ripened, pp. of *maturo*, to ripen. *Ex:* matur-ation.

matutin—L. *matutinus*, of the dawn, morning. *Ex:* matutin-al.

maur—Gr. *mauros*, dark, obscure. *Ex:* Maurachelia (Ins.); Mauro-myia (Ins.).

maxill—L. *maxilla*, the jawbone, jaw; *maxillaris* of or belonging to the jaw. *Ex:* Maxillaria*; maxilli-ped; maxillo-jugal; pre-maxilla.

maxim—L. *maximus-a-um*, superl. of *magnus*, large. *Ex:* maximum.

mayac—NL. *mayaca*, from a native plant name. *Ex:* Mayaca*.

mayten—Sp. *mayten* <Araucan *mantun*, a plant name. *Ex:* Maytenus*.

maz—1. Gr. *mazos*, one of the breasts, a teat. *Ex:* Maz-odus (Pisc.); Mazus*: 2. Gr. *maza*, a barley cake. *Ex:* maz-ic; mazo-lysis.

mazama—Mex. *mazame*, name for some Mexican ungulate. *Ex:* Mazama (Mam.).

me- —Gr. *mē*, prefix meaning not. *Ex:* Meryhnchites (Ins.). See also meat.

meandr—See maeandr.

meant—L. *meantes*, going, pl. ppr. of *meo*, to go. *Ex:* Meantes (Amph.).

meat—L. *meatus*, a passage <*meo*, to go, to pass. *Ex:* meato-meter; meatus.

mec—Gr. *mēkos*, length, height. *Ex:* Mec-odonta (Amph.); Eu-meces (Rept.); Hypo-mecus (Ins.); Pro-meco-gnathus (Ins.).

mechan—Gr. *mechanē*, a machine; *mechanikos*, of or pertaining to machines. *Ex:* mechanotropism.

mecist—Gr. *mēkistos*, longest, tallest, greatest; irregular superl. of *makros*, long. *Ex:* Mecistura (Av.); mecisto-cephalic; Mecisto-scelis (Ins.).

mecon—Gr. *mēkōn*, a poppy; *mēkōnion=*L. *meconium*, poppy juice; also the fecal matter discharged by new-born children. *Ex:* meconidium; Mecon-opsis*; meconium; Dendromecon*.

mecyn—Gr. *mēkynō*, to prolong, extend. *Ex:* Mecyn-odes (Ins.); Mecyna (Ins.); Mecynocera (Crust.).

mecysm—Gr. *mēkysmos*, a lengthening. *Ex:* Mecysm-auchenius (Arach.).

med—1. L. *medius*, middle. *Ex:* medi-cisterna; medi-thorax; medio-cortex; Medio-cricetus (Mam.): 2. Gr. *medōn*, a protector, a guardian <*medō*, to guard, protect. *Ex:* Medon (Ins.): 3. Gr. *mēdos*, bladder, pl. *mēdea*, the genitals: 4. Gr. *medion*, a plant name. *Ex:* Hali-meda*: 5. Gr. *mēdos*, a plan; *mēdomai*, to plan.

medeola—NL. *medeola* <Gr. *Mēdeia*, the sorceress. *Ex:* Medeola*.

mediastin—1. L. *mediastinus*, a helper, drudge: 2. ML. *mediastinus*, medial. *Ex:* mediastinum.

medic—L. *medicus*, healing, curative; also a physician. *Ex:* medic-al; medicine; medicodental.

medicag—Gr. *mēdikē*, a kind of clover from Media. *Ex:* Medicago*.

medin—*Medina*, a region of W. Africa; NL. *medinensis*, belonging to or of Medina.

mediocr—L. *mediocris*, ordinary, moderate, not important.

medon—See med.

medull—L. *medulla*, marrow, pith; *medullosus*, marrowy, pithy; *medullatus*, fat, abounding in richness; *medullaris*, situated in the marrow or inmost part. *Ex:* medulla oblongata; medullar-y; medullat-ed.

medus—L. *Medusa*, daughter of Phorcus, who lured Neptune with her golden hair and became by him the mother of Pegasus. *Ex:* medu(s)-some; Medus-aster (Echin.); medusa; medusi-form.

meg—Gr. *megas*, great; fem. *megalē*; *megistos*, fem. *megista*, greatest. *Ex:* Meg-encephalum (Mam.); Mega-chile (Ins.; mega-nucleus; Mega-scops (Av.); megalo-cyte; Megalosaurus (Rept.); megalo-sphere; Megista (Ins.): Megisto-pus (Ins.).

megal—See meg.

megist—See meg.

mei—Gr. *meiōn*, less, smaller, fewer, comp. of *mikros*, small; *meiōsis*, reduction. *Ex:* Meioceras (Moll.); meio-phylly; meio-taxy; meiotherm; meiosis; miotic; mi-urus; Mio-corax (Av.); Mion-ornis (Av.); Miono-meris (Ins.); A-mei-urus (Pisc.).

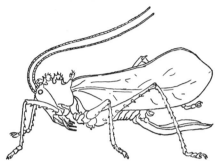

Large-toothed Locust, *Eumegalodon*.

meiz—Gr. *meizōn*, greater. *Ex:* Meiz-odon (Rept.); Meizo-glossa (Ins.).

mel—1. L. *mel*, genit. *mellis*, honey>*mellita*, honey cake; Gr. *melleis*, pertaining to honey< Gr. *meli*, genit. *melitos*, honey. *Ex:* Mel-ursus (Mam.); Meli-gethes (Ins.); Meli-lotus*; Melitea (Ins.); Melli-vora (Mam.); Mellinus (Ins.); Mellita (Echin.): 2. Gr. *melos*, song, melody. *Ex:* mel-ic; Meli-erax Melo-pelia (Av.); Melo-spiza (Av.): 3. Gr. *mēlon*, a tree-fruit, an apple=L. *malum. Ex:* Melodinus*; Melo-nycteris (Mam.); Chamae-melum*; Chaeno-meles*; Chryso-melus (Ins.); Micro-melum*: 4. L. *meles*, badger. *Ex:* Mel-ictis (Mam.); Meles (Mam.); Melo-gale (Mam.): 5. Gr. *mēlē*, a probe. *Ex:* Meloe (Ins.), etym. doubtful: 6. Gr. *melos*, a limb, *Ex:* Aniso-meles*; Aniso-melus (Ann.): 7. Gr. *melia*, the ash-tree. *Ex:* Meli-aceae*; Melia*.

mela—Gr. *melas*, fem. *melaina*, black; *melania*, blackness; *melaneimōn*, clothed in black. *Ex:* Mel-anthium*; Mela-leuca*; Mela-netta (Av.); Melan-erpes (Av.); melanin; Melanimon (Ins.); Melano-chroi (Ethnol.); Melano-dendron*; Cori-melaena (Ins.).

melaen—See mela.

melamp—Gr. *melampous*, black-footed. *Ex:* Melampus (Mam.).

melampsor—NL. *melampsora*, a word badly formed<Gr. *melas*, genit. *melanos*, black+ *psora*, a scale. *Ex:* Melampsora*.

melan—See mela.

melas—See mela.

melasm—Gr. *melasma*, a black spot. *Ex:* Mel asm-ia*.

meleagr—Gr. *meleagris*, a guinea-fowl, hence used in sense of speckled. *Ex:* Meleagr-ina (Moll.); Meleagris (Av.); Meleagro-teuthis (Moll.).

melem—Gr. *melēma*, genit. *melēmatos*, a beloved object; also care, anxiety.

meles—See mel 3, 4, 5.

melet—Gr. *meletē*, care, attention.

meliboea—Gr. *Meliboia*, a feminine personal name; also a sea shore town of Thessaly. *Ex:* Meliboea (Moll.).

melic—1. NL. *melica*<Ital. *meliga*, maize, millet. *Ex:* Melica*: 2. L. *melicus*, tuneful.

melicarius—L. *melicarius*, full of music <*melicus*, tuneful.

melicert—Gr. *Melikertēs*, son of Athamas and Ino. He was changed into a marine deity. *Ex:* Melicerta (Rot.); Melicertum (Coel.).

melichr—Gr. *melichros*, honey-colored. *Ex:* Melichrus*.

melin—1. Gr. *mēlinos*, made of apples or quinces: 2. Gr. *melinos*, ashen: 3. L. *melinus*, quince colored, like quinces: 4. Gr. *melinē*, millet: 5. L. *melinus*, a marten.

melism—1. Gr. *melisma*, genit. *melismatos*, a song: 2. Gr. *melismos*, a dismembering, a dividing.

meliss—Gr. *melissa*=Attic *melitta*, the honey-bee. *Ex:* Melissa*; meliss-aceus; Melisso-pus (Ins.); Melitta (Ins.); Melittis*.

melit—Gr. *Melitaios*=L. *Melitaeus*, of Malta, Melitan; *melit-ensis*, of Malta. *Ex:* Meletaea (Coel.), (Ins.), fem. of L. *Melitaeus*. See also mel 1.

melitt—see meliss.

melizo—Gr. *melizō*, to sing. *Ex:* Melizo-philus (Av.).

mell—See mel 1.

melle—L. *melleus*, belonging to honey.

melletes—Gr. *mellētēs*, a delayer, one who loiters. *Ex:* Melletes (Pisc.).

mellit—See mel 1.

meloch—Arab. *meluchia*, borrowed through the Syriac from Gr. *malachē*=*molochē*, the mallow. *Ex:* Melochia*.

melod—Gr. *melōdia*, a melody, a melodious song; *melōdos*=L. *melodus*, pleasantly singing.

melolonth—Gr. *mēlolonthē*, the cockchafer< *mēloō*, to explore +*onthos*, dung. *Ex:* Melolontha (Ins.).

melon—Gr. *mēlon*, apple. *Ex:* Melon-echinus (Echin.).

melothr—Gr. *mēlothron*, a wild vine. *Ex:* Melothria*.

melpomene—Gr. *Melpomenē*, muse of tragic and lyric poetry. *Ex:* Melpomene (Arach.).

melyr—Gr. *melouris*, genit. *melouridos*, a maker of songs, a kind of locust. *Ex:* Melyr-odes (Ins.); Melyrid-ella (Ins.); Melyris (Ins.).

melyrid—See melyr.

memat—Gr. *mematon*, desired, sought for< *maō*, to desire, seek. *Ex:* Memato-ptychias (Pisc.).

**membrac**—Gr. *membrax*, genit. *membrakos*, a kind of cicada. *Ex:* Membrac-idae (Ins.); Membracis (Ins.).

**membran**—L. *membrana*, the skin of the body; also a parchment. *Ex:* membran-ula; Membrani-pora (Coel.).

**memecyl**—L. *memecylon*, edible fruit of the strawberry-tree. *Ex:* Memecylon*.

**memina**—Ceylonese *memina*, name of a deerlet. *Ex:* Memina (Mam.).

**memnon**—Gr. *Memnōn*, king of the Ethiopians, slain by Achilles and changed by Aurora into a bird. *Ex:* Memnon (Arach.).

**memnoni**—L. *memnonius*, black, brownish-black.

**men**—1. Gr. *menos*, disposition, force, vigor, courage. *Ex:* Men-ac-odon (Mam.); Meno dora*; Eu-men-idae (Ins.); Eu-menes (Ins.), good disposition: 2. Gr. *mēnē=menōs*, genit. *mēnados*, moon; *mēniskos*, a little moon or crescent; *noumēnios*, used at the new moon. *Ex:* Men-aspis (Elasm.); Men-epites (Bry.); Men-ura (Av.); Mene (Pisc.); Meni-spermum*; Meniscium*, L. dim of *meniscus* < Gr. *mēniskos*; Menisco-essus (Mam.); meniscus; Lepido-menia (Moll.); Neo-menia (Moll.); Numenius (Av.), from its crescent beak: 3. Gr. *mēn*, genit. *mēnos* > L. *mensis*, month. *Ex:* meno-pause (Med.); mens-al; menses: 4. Gr. *menō*, to remain, abide, to be unchanged. *Ex:* Meno-poma (Amph.); Meno-rhyncha (Ins.); Meno-typhla (Mam.); Chrysi-menia*; epimenus; Nycti-mene (Av.): 5. Gr. *mēnis*, genit. *mēnidos*, later *mēnidos*, ire, wrath. *Ex:* Tachy-menis (Rept.).

**menad**—See **men** 2.

**mendac**—L. *mendax*, genit. *mendacis*, a liar.

**mendax**—See **mendac**.

**mendic**—L. *mendicus*, beggarly, needy, mean.

**mendos**—L. *mendosus*, full of faults, blemishes.

**menidi**—NL. *menidia* < an old fish name < Gr. *mēnē*, moon. *Ex:* Menidia (Pisc.).

**mening**—Gr. *mēninx*, genit. *mēningos*, membrane > NL. *meninx*, pl. *meninges*. *Ex:* mening-itis (Med.); meninges.

**meninx**—See **mening**.

**menisc**—See **men** 2.

**menopon**—NL. *menopon*, name of some insect (derivation uncertain). *Ex:* Menopon (Ins.).

**mens**—L. *mensa*, table; *mensalis*, of or belonging to a table. *Ex:* mensa; com-mensal.

**menstru**—L. *menstrualis*, monthly, the monthly course of women; *menstruo*, to discharge the menses. *Ex:* menstru-ation; menstrual.

**ment**—1. L. *mentum*, chin, beard; *mento*, one with a long chin. *Ex:* mental foramen and nerve; Menti-cirrhus (Pisc.); menti-ger-ous, bearing a mentum; mento-labial; mentum: 2. L. *mens*, genit. *mentis*, the mind. *Ex:* ment-al.

**mentagr**—L. *mentagra*, a skin eruption of the chin. *Ex:* mentagra-phyte.

**menth**—See **minth**.

**mentul**—L. *mentula*, membrum virile, the penis. *Ex:* mentula.

**meny**—Gr. *mēnyō*, to disclose. *Ex:* Meny-anthes*. Some would derive it from Gr. *mēn*, month but this does not account for the *y*.

**meotic**—Gr. *meiōtikos*, small, mean. *Ex:* Meotica (Ins.).

**mephit**—L. *mephitis*, bad odor, damp of the earth. *Ex:* Mephit-inae (Mam.); mephit-ic; Mephit-idia*; Mephitis (Mam.).

**mer**—1. Gr. *meros*, a part. *Ex:* Mer-aphodius (Ins.); mer-o-istic, see oo; mer-ont; meri-carp; meri-stele; meri-thallus; mero-blastic; mero-genesis; mero-plankton; Mero-tettix (Ins.); mero-zo-ite; blasto-mere; epi-mer-ite; hypomeron; myo-mere; Tri-mero-cerat-idae (Moll.): 2. Gr. *meros*, the thigh. *Ex:* Mer-anoplus (Ins.); mero-cer-ite; mero-gnathite; meron; Mero-stomata (Crust); meros-thenic; Oedemerus (Ins.); Hetero-mera (Ins.): 3. L. *merus*, pure, genuine.

**mercat**—L. *mercator*, a trader.

**mercenari**—L. *mercenarius*, hired for wages; also one hired.

**mercuri**—L. *Mercurius*, Roman god of commerce, messenger of the gods; ML. *mercurialis*, pertaining to mercury. *Ex:* Mercurialis*.

**merd**—L. *merda*, dung, excrement. *Ex:* merdi-vor-ous.

**merg**—L. *mergus*, a diver < *mergo*, to immerse; ppr. *mergens*, genit. *mergentis*, diving, dipping. *Ex:* Merg-anser (Av.); Merg-ulus (Av.); Mergus (Av.).

**mergen**—L. *mergens*, genit. *mergentis*, diving, ppr. of *mergo*, to plunge, dive.

**meridional**—L. *meridionalis*, southern.

**merimn**—Gr. *merimna*, care, thought, solicitude. *Ex:* A-merimnon*, the house leek which requires no care in cultivation.

**mering**—Gr. *mērinx*, genit. *mēringos = smērinx*, genit. *smēringos*, a bristle. *Ex:* Smeringo-pus (Arach.); A-smerinx (Ins.).

**merinx**—See **mering**.

**meriones**—Gr. *Mēriōnēs*, companion of Idomeneus. *Ex:* Meriones (Mam.).

**merism**—Gr. *merisma*, genit. *merismatos*, a part; *merismos*, a dividing, division. *Ex:* merismatic; Merismo-pedia*; Merismo-pteria (Moll.).

**merist**—Gr. *meristos*, divisible, divided; *meristēs*, a divider. *Ex:* meristem, Merist-odon (Pisc.); meristic; Meristo-belus (Ins.); meristo-genetic; Iso-merista (Ins.).

**merit**—L. *meritus*, deserving; also fit, proper.

**merluci**—NL. *merlucius = merluccius* < Fr. *merluche* < L. *mar*, the sea+*lucius*, some fish, prob. the pike. *Ex:* Merlucius (Pisc.); also Merluccius (Pisc.).

mermis—See mermith.

mermith—Gr. *mermis*, genit. *mermithos*, a cord, string. *Ex:* Mermis (Nemat.); mermitho-gyne.

meron—See mer 1.

merop—See merops.

merope—Gr. *Meropē*, one of the Pleiades. *Ex:* Merope*.

merops—Gr. *merops*, genit. *meropos*, a bird, the bee-eater. *Ex:* Merops (Av.); Merop-idae (Av.).

meros—L. *merosus*, pure, unmixed. See also mer 2.

merul—L. *merula*, the European black bird. *Ex:* Merul-idae (Av.); Merula (Av.).

meryc—Gr. *mēryx*, genit. *mērykos*, a ruminating animal. *Ex:* Meryc-hyus (Mam.) see hy<*hys*, a pig; Meryc-oid-odon (Mam.); Meryco-hippus (Mam.); Amphi-meryx (Mam.).

meryx—See meryc.

mes—Gr. *mesos*, middle, the half; *mesaios*, intermediate. *Ex:* Mes-embry-anthemum*. According to Dr. Asa Gray the original spelling was Mesembrianthemum (Gr. *mesembria*, midday+*anthos*, a flower); later Linneaus changed it to its present form, deriving it from *mesos*, middle and *embryon*, embryo; mes-enchyme; Mes-ore-odon (Mam.); Mes-pilus*; Mesaeus; mesio-labial; meso-glea; meso-nephros; Meso-odon (Mam.); Macro-mesus (Ins.).

mesae—NL. *mesaeus*<Gr. *mesaios*, intermediate, between; L. *mesa*, middle.

mesaul—Gr. *mesaulos*, belonging to a hall, the door between two halls, a passage between cattle. *Ex:* Mesaulus (Ins.).

mesio—See mes.

mesit—Gr. *mesitēs*, an arbitrator. *Ex:* Mesites (Av.); Mesitus (Ins.).

mespil—Gr. *mespilē*, the medlar tree. *Ex:* Mes pilia (Echin.); Mespilus*.

messapus—L. *Messapus*, mythical prince of Messapia. *Ex:* Messapus (Arach.).

messor—L. *messor*, a reaper. *Ex:* Messor (Ins.).

mest—Gr. *mestos*, filled, full, stuffed. *Ex:* Mest-urus (Pisc.); Mesto-charis (Ins.); Mesto-soma (Myr.); hadro-mest-ome.

mesua—Arab. *Mesue*=*Mesua*, Arabian physician of the eighth and ninth centuries. *Ex:* Mesua*.

met—See meta.

meta—1. Gr. *meta*, next to, in company with, among, in the midst of, between, after, behind, later in time; in composition it may denote change. *Ex:* Met-arctos (Mam.); met-encephalon; met-enteron; meta-carpal; Meta-chirus (Mam.); Meta-crinus (Echin.); meta-genesis; meta-mer-ic; Methy-lobates (Mam.): 2. L. *meta*, dim. *metula*, goal, extremity; also a conical column.

metacin—Gr. *metakineō*, to shift, alter. *Ex:* Metacin-ops (Ins.).

metall—Gr. *metallon*, a cave where metals are found, then later the metal itself; *metallitēs*, metallic. *Ex:* Metall-ina (Ins.); Metallo-psar (Av.); Metallo-timarcha (Ins.); Metallon (Ins.).

metastas—Gr. *metastasis*, a removal, a change.

metax—Gr. *metaxa*, raw silk.

metaxy—Gr. *metaxy*, found in the middle, between. *Ex:* Metaxy-blatta (Ins.); Metaxy-therium (Mam.); Metaxya (Ins.); Metaxys (Ins.).

metel—Gr. *metel*, a plant, the mad-solanum. *Ex:* metel-oides.

meteor—1. Gr. *meteōros*, lofty, high; *meteōrōs*, unsettled, in suspense, high in air. *Ex:* Meteor-idae (Ins.); Meteorus (Ins.): 2. Gr. *meteōria*, forgetfulness.

metheis—See methes.

methes—Gr. *mētheis*=*mēdeis*, nobody, naught. *Ex:* Metheisa (Ins.); Pro-methes (Ins.).

methidrys—Gr. *methidrysis*, change, removal. *Ex:* Methidrysis (Ins.).

metholc—Gr. *metholkē*, a drawing away. *Ex:* Metholcus (Ins.).

methori—Gr. *methorios*, bordering, bounding, adjacent. *Ex:* Methorio-pterus (Av.); Metho rius (Ins.).

methy-—In Gr. compounds sometimes=*meta*-, which see.

methyl—1. Gr. *methy*, genit. *methyos*, wine, mead. *Ex:* methyl<*methy*, wine+*hylē*, wood, i.e., wine of wood.

meticulos—L. *meticulosus*, fearful, timid<*metus*, fear.

metis—1. Gr. *mētis*, wisdom, craft. *Ex:* A-poly-metis (Moll.): 2. Gr. *Mētis*, daughter of Oceanus, personification of prudence. *Ex:* Metis (Echin.); ?Ancylo-metis (Ins.).

meto—Gr. *Metōn*=L. *Meto*, celebrated Athenian astronomer, discoverer of the Metonic cycle. *Ex:* Meto (Coel.).

metoch—Gr. *metochē*, a sharing. *Ex:* metoch-ous; metoch-y.

metoec—Gr. *metoikos*, a settler, emigrant, stranger, one who changed his abode. *Ex:* Metoecus (Ins.); Metoico-ceras (Moll.).

metoic—See metoec.

metop—Gr. *metōpon*=*metōpion*, the brow, countenance; *metōpias*, with broad or high forehead. *Ex:* Metop-acanthus (Pisc.); Metopias (Amph.); Metopo-ceros (Rept.); Metopo-ctea (Arach.); Metoponium (Ins.); Anco-metopius (Ins.); Cyclo-metopon (Crust.).

metopias—See metop.

metopidi—Gr. *metōpidios*, on the fore-head. *Ex:* Metopidius (Av.).

metopor—Gr. *metopōron*, the harvest time, late autumn.

metr—1. Gr. *mētra*, the uterus; also the heart of a tree; *mētēr*, genit. *mētros*, a mother = Doric *matēr*; *mētridios*, having a womb, fruitful. *Ex:* Metridium (Coel.); Metro-xylon*; Metro-sideros*, in sense of the heart of a tree; A-metro-somus (Ins.); endo-metrium: 2. Gr. *metron*, a measure. Used as a combining form in word-building *metro-* often means by measure, with measure. *Ex:* Metro-nectes (Ins.); Metron (Ins.): 3. Gr. *metreō*, in active voice, to measure; in passive, to be measured round, to be surrounded, and in this sense is often used in making generic names of echinoderms and other radiate animals. *Ex:* Acantho-metra (Prot.); Astero-metra (Echin.); Dactylo-metra (Coel.). See also metrio.

metrio—Gr. *metriōs*, within measure, moderately; *metriotēs*, moderation, the middle course < *metron*, a measure, limit. *Ex:* Metri-inae (Ins.); Metri-omphalus (Moll.); Metrio-dromus (Mam.); Metrio-therium (Mam.); Metriotes (Ins.); Metrius (Ins.).

metul—See meta 2.

meum—L. *meum*, an umbelliferous plant < Gr. *mēon*, a plant called spignel. *Ex:* Meum*.

mez—Gr. *meizōn*, stronger, greater, comp. of *megas*, great. *Ex:* Mezo-neurum*; Mezo-treron (Av.).

mezentia—L. *Mezentia*, mythical Etruscan king. *Ex:* Mezentia (Arach.).

mezereum—Persian *Madzaryoun*, the plant we call daphne, and from which Linnaeus made the name *Mezereum*. *Ex:* Mezereum*.

mi—See mei; also myi.

miaephon—Gr. *miaiphonōs*, stained with blood. *Ex:* Miaephonus (Arach.).

miant—Gr. *miantos*, dyed, stained. *Ex:* Mianto-nota (Ins.).

miar—Gr. *miaros*, stained with blood, defiled. *Ex:* Miaro-blatta (Ins.); Miarus (Ins.).

miasm—Gr. *miasma*, pollution, defilment. *Ex:* miasma.

miastor—Gr. *miastōr*, a guilty wretch, one who brings pollution. *Ex:* Miastor (Ins.).

micans—L. *micans*, twinkling, glowing < *mico*, to shine.

micari—L. *micarius*, frugal, a crumb gatherer < *mica*, a crumb, morsel > NL. *mica*, now applied to a group of minerals. *Ex:* Micaria (Arach.); Micario-soma (Arach.).

micc—Gr. *mikkos*, Doric for *mikros*, little. *Ex:* Micco-cephal-ites (Ins.); Miccus (Ins.).

micid—L. *micidus*, thin, poor, lean < *mica*. See also micari.

micr—Gr. *mikros* = *smikros*, small. *Ex:* Micr-aster*; Micr-elasma (Moll.); Micr-otus (Mam.); Micro-di-pod-ops (Mam.); Micro-licia*, see elic; micro-scope; micro-spore; Micry-phantes (Arach.).

micry—See micr.

mict—Gr. *miktos*, mixed, thrown together, blended. *Ex:* mict-ic; mict-ium (Ecol.); Micto-mys (Mam.); a-mictic.

micul—L. *micula*, a small crumb < *mica*, a crumb; NL. *miculatus*, made of small bits or crumbs.

mid- —AS. *midd-* > Eng. *mid-*, a prefix meaning middle, intermediate. *Ex:* mid-autumn; mid-intestine.

midamus—L. *Midamus*, one of the fifty sons of Aegyptus. *Ex:* Midamus (Arach.).

midas—1. Gr. *Midas*, well-known ancient king: 2. Gr. *midas*, a destructive insect in pulse. *Ex:* Midas = Mydas (Ins.); Myda-idae (Ins.).

migad—Gr. *migas*, genit. *migados*, promiscuous, mixed. *Ex:* Migado-phonus (Ins.); Ara-migas (Ins.); Thrico-migas (Ins.).

migas—See migad.

migr—L. *migrans*, genit. *migrantis*, wandering < *migro*, to wander abroad, to go away; *migratus*, wandering. *Ex:* migrant; migrat-ion.

migrat—L. *migratus*, pp. of *migro*, to wander; *migratio*, a removal, migration > Fr. *migration*; LL. *migrator*, a wanderer. *Ex:* migration; migrator-y.

mikro—See micr.

mili—L. *milium*, millet > NL. *miliola*; L. *miliarius*, belonging to millet. *Ex:* miliarius; milio-lite; Miliol-idae (Prot.); Miliola (Prot.); Milium*.

milic—See mylic.

milich—Gr. *meilichos* = *meilichios*, gentle, kind. *Ex:* Milichi-linus (Ins.).

militar—L. *militaris*, belonging to a soldier, war-like.

mille—L. *mille*, a thousand. *Ex:* mille-ped = mille-pede; Mille-pora (Coel.).

milt—Gr. *miltos*, red chalk, ochre; *miltōtos*, painted red. *Ex:* Milto-trogus (Ins.); Milto-gramma (Ins.).

milv—L. *milvus*, dim. *milvulus*, a bird known as the kite; *milvinus*, of or belonging to a kite, rapacious. *Ex:* Milv-aquila (Av.); Milv-ago (Av.); Milvulus (Av.); Milvus (Av.).

milyas—NL. *milyas* < L. *Milyas*, a district in Lycia. *Ex:* Milyas (Ins.).

mim—L. *mimus*, dim. *mimulus*, a mimic < Gr *mimō*, genit. *mimous*, an ape; Gr. *mimos* = *mimētēs*, an imitator; *mimētikos*, imitative. *Ex:* Mim-aster (Echin.); Mim-ulus*; Mimesa (Ins.); Mimet-anthe*; Mimetes (Mam.); mimetic; Mimo-nectes (Crust.); mimo-type; Mimon (Mam.); Mimos-ina (Prot.); Mimosa*; Mimus (Av.); Mimus-ops*; Astro-mimus (Por.).

mimallon—Gr. *Mimmalōn*, one of the Bacchantes. *Ex:* Mimmalon-ides (Ins.).

mimer—Gr. *mēimerēs*, halved. *Ex:* Mimer-astria*.

mimet—See mim.

minac—L. *minax*, genit. *minacis*, jutting out, projecting; also full of threats.

minaret—Sp. *minarete* < Ar. *manārat*, a lighthouse turret, minaret. *Ex:* Minaretus (Moll.).

minax—See minac.

ming—L. *mingo*, to urinate, ppr. *mingens*, genit. *mingentis*, urinating. *Ex:* retro-mingent.

mini—See miny.

miniat—L. *miniatus*, colored with cinnabar or vermillion.

minim—L. *minimus* = *minumus*, least, smallest.

minor—L. *minor*, smaller, inferior.

minth—1. Gr. *mintha*, mint = L. *menta* or *mentha*. *Ex:* Mentha*; Mintho-dexia (Ins.); Acantho-mintha*: 2. Gr. *minthos*, human ordure < *minthoō*, to cover with dung, to defile.

minum—See minim.

minur—See minyr.

minus—L. *minus*, less, subtracting (neuter of the comparative form *minor*).

minuscul—L. *minusculus*, very small, trifling < *mimus*, small.

minut—L. *minutus*, minute < *minuo*, to make small.

miny—Gr. *minys*, small. *Ex:* Minio-pteris (Mam.); Miny-tragus (Mam.); Miny-trema (Pisc.).

minyad—L. *minyas*, genit. *minyadis*, an herb fabled to have magical qualities. *Ex:* Minyadinae (Coel.); Minyas (Coel.).

minyas—See minyad.

minyr—Gr. *minyros*, whining, complaining. *Ex:* Minurus (Ins.).

mio—Gr. *meiōn*, less, smaller; the form *mio-* is sometimes used by paleontologists to denote a genus of Miocene Age as in Mio-hippus (Mam.). *Ex:* mi-urus; Mio-gryllus (Ins.); Mio-lania (Rept.); Mio-pithecus (Mam.); Miocene; Mion-ornis (Av.).

mion—See mei.

mir—1. L. *mirus*, wonderful < *miror*, to wonder at; *mirabilis*, marvelous, strange. *Ex:* Mirablilis*; Miro-phasma (Ins.); Miro-termes (Ins.); rete mirabile: 2. NL. *miris*, generic name for the oval leaf-bugs (etym. uncertain). Mir-idae (Ins.); Miris (Ins.).

mirab—See mir.

mirac—Gr. *meirax*, genit. *meirakos*, a young girl. Miraces (Ins.); Mirax (Ins.). See also miracidi.

miracidi—Gr. *meirakidion*, also *meirakion*, a boy, a youthful person, a stripling. *Ex:* miracidium.

miranda—L. *miranda* (feminine form of the gerundive of the verb *miror*), wonderful, admirable. *Ex:* Miranda (Arach.).

mirax—See mirac.

mirza—Pers. *mirza*, prince. *Ex:* Mirza (Mam.).

mis—Gr. *miseō*, to hate; *misētēs*, a hater;

misēma, an object of hate. *Ex:* mis-andry; mis-anthrop-ous; Mis-erythrus (Av.); Misetus (Ins.); Miso-lampra (Ins.); Miso-phthora (Ins.); Myio-misa (Ins.).

mis- —*mis-*, Eng. prefix meaning ill, wrong or wrongly. *Ex:* mis-graft; mis-identify.

misc—See misch.

misch—Gr. *mischos*, a pedicel, the stalk of a leaf; also a husk. *Ex:* Misc-anthus*; Miscodera (Ins.); Misco-gaster (Ins.); Mischo-cyttarus (Ins.); Mischo-phloeus*; Macro-mischa (Ins.); Platy-mischus (Ins.).

miscic—L. *miscix*, genit. *miscicis*, changeable inconstant.

miscix—See miscic.

misell—L. *misellus*, poor, wretched.

misem—See mis.

miser—L. *miser*, fem. *misera*, neut. *miserum*, wretched, worthless, vile.

miset—See mis.

mist—1. L. *mistus*, a mixing, mingling. *Ex:* Mist-echinus (Echin.); misto-form: 2. Gr. *meistos*, most. *Ex:* Mist-ichthys (Pisc.), here taken in the sense of most small.

misth—Gr. *misthos*, wages, hire > *misthodotēs*, one who hires. *Ex:* Misthodotes (Ins.).

mit—Gr. *mitos*, thread. *Ex:* mit-apsis; mito-chondria; Mito-pus (Arach.); mit-osis; mitotic, as if from a Gr. *mitotikos*. See also mitis.

mitell—L. *mitella*, dim. of *mitre*, a mitre. *Ex:* Mitella*.

mithrac—L. *mithrax*, genit. *mithracis*, a Persian precious stone. *Ex:* Mithrac-ulus (Crust.).

mithrax—See mithrac.

mitis—L. *mitis*, harmless, mild, without spines.

mitiuscul—L. *mitiusculus*, rather mild.

mitotic—NL. *mitotic* (as if < a Gr. *mitotikos*), pertaining to mitosis.

mitr—Gr. *mitra*, a head-band, head dress, cap > L. *mitratus*, wearing a mitre or other headdress; ML. *mitralis*, pertaining to a mitre. *Ex:* mitr-al valve of the heart; Mitr-aria*; Mitr-aster (Echin.); Mitr-ella*; Mitr-idae (Moll.); Mitra (Moll.); Mitra(s)-acme*; Mitre-phanes (Av.); mitri-form; Mitr-acea (Mol.); Mitro-crinus (Echin.); Mitro-morpha (Moll.).

mitral—See mitr.

mitri—Gr. *mitrion*, a small cap, dim. of *mitra*, a cap. *Ex:* Physco-mitrium*; Mitrio-stigma*.

mitrul—L. *mitrula*, a small head-band or bandage; dim. of *mitra*, a head-band *Ex:* Mitrul-aria (Moll.): Mitrula (Moll.).

mitten—L. *mitto*, to send, to cause to go, ppr. *mittens*, genit. *mittentis*. *Ex:* intro-mittent.

mix—Gr. *mixis*, mingling. *Ex:* mixi-pterygium; Mixo-dectes (Mam.); Mixo-saurus (Rept.); mixo-trophic; endo-mixis.

**mixt**—L. *mixtus*, blended, united <*misceo*, to mix. *Ex:* Mixto-therium (Mam.); com-mixtus.

**mnem**—Gr. *mnēma*, genit. *mnēmatos*, a memorial; *mnēstis* also *mnēmē*, memory; *mnēmōn*, mindful; *mnēmonikos*, of or for memory or remembrance. *Ex:* Mnemat-idium (Ins.); Mnemi-opsis (Cten.); mnemonic; mnest-ic.

**mnesi**—Gr. *mnēsi-*, in composition meaning re-collecting, mindful; *mnēsios*, of memory > *amnesia*, without memory. *Ex:* Mnesi-dacus (Ins.); Mnesi-theus (Arach.); Mnesio-lochus (Ins.); amnesia.

**mnest**—Gr. *mnēstēr*, genit. *mnēstēros*, a wooer, suitor; also calling to mind. *Ex:* Mnester-odes (Ins.). See also mnem.

**mnestr**—Gr. *Mnēstra*, a mythological name. *Ex:* Mnestra (Coel.).

**mni**—Gr. *mnion*, moss. *Ex:* Mni-erpes (Pisc.); Mnio-bia (Rot.); Mnio-tilta (Av.); Mnium*; Em-mnion (Pisc.).

**mniar**—Gr. *mniaros*, mossy, soft as moss. *Ex:* Mniarum*.

**mobil**—L. *mobilis*, neut. *mobile*, loose, moveable.

**mobul**—NL. *mobula* (etym. uncertain), a genus of rays. *Ex:* Mobula (Elasm.).

**mochl**—Gr. *mochlos*, a lever. *Ex:* Mochl-odon (Rept.).

**mochth**—Gr. *mochthos*, hard work, toil; *mochthēros*, suffering. *Ex:* Mochtherus (Ins.); Mochtho-picus (Av.).

**modest**—L. *modestus*, calm, unassuming, modest.

**modic**—L. *modicus*, moderate, ordinary.

**modiol**—L. *modiolus*, a small measure; also the nave of a waterwheel <*modius*, a measure. *Ex:* Modiol-aria (Moll.); Modiol-opsis (Moll.); Modiola* (Moll.); Modioli-cola (Crust.); modiolus; Modiolus (Moll.).

**modul**—L. *modulus*, a small measure. *Ex:* Modulus (Moll.). See also modiol.

**moech**—Gr. *moichos*, an adulterer. *Ex:* Moecha (Ins.).

**moer**—Gr. *Moiris* = L. *Moeris*, an ancient lake in Egypt. *Ex:* Moeri-therium (Mam.). See also moir.

**moeren**—L. *moerens*, genit. *moerentis*, mourning, full of sorrow, sad; ppr. of *moeror* = *maeror*.

**moest**—L. *moestus*, sorrowful.

**mog**—Gr. *mogos*, trouble, travail. *Ex:* Mogo-plistus (Ins.).

**moho**—Hawaiian *moho*, a bird name. *Ex:* Moho (Av.).

**moiety**—Fr. *moitie*, a half <L. *medius*, middle, half. *Ex:* moiety (Anthro.).

**moir**—Gr. *moira*, a part or portion, lot; *Moira*, goddess of fate. *Ex:* Moera-phora (Ins.); Mio-moera (Ins.); Moir-aster (Echin.); Moira (Echin.).

**mol**—L. *mola*, genit. *molaris*, a millstone. *Ex:* mol-ula; Mola (Pisc.); molar, molari-form.

**molar**—See mol.

**mole**—L. *moles*, a mass. *Ex:* mole-cule.

**molest**—L. *molestus*, disturbed.

**molg**—Gr. *molgos*, skin, hide. *Ex:* Molg-ophis (Amph.); Molg-ula (Tun.); Molg-ul-idae (Tun.).

**molge**—Ger. *Molge*, name of the triton, a sala-mander. *Ex:* Typhlo-molge (Amph.).

**molim**—L. *molimen*, genit. *moliminis*, an effort. *Ex:* molimen; molimin-ous.

**molitor**—L. *molitor*, genit. *molitoris*, a miller, grinder; also one who attempts, contrives < *molo*, to grind.

**moll**—L. *mollis*, soft; *molluscus*, soft. *Ex:* Moll-ugo*; molli-fy; molli-pilose; Mollis-ia*; Mollo-crinus (Echin.); Mollusca, neut. pl. of *molluscus*, soft.

**molle**—1. Peruvian *molli* <Western S.A. *molle*, name for the pepper tree, *Schinus molle*: 2. L. *molle*, neut. sing. of *mollis*, soft, see moll.

**mollesc**—L. *mollesco*, to become soft; ppr. *mollescens*, genit. *mollescentis*, becoming soft. *Ex:* mollescent.

**mollit**—L. *mollitus*, made soft, flexible.

**mollitia**—L. *mollitia*, tenderness, softness.

**mollug**—L. *mollugo*, genit. *molluginis*, a plant name <L. *mollis*, soft. *Ex:* Mollugo*.

**mollusc**—See moll.

**molop**—Gr. *mōlops*, genit. *mōlōpos*, the mark of a wound. *Ex:* Molopo-spermum*; Molops (Ins.).

**moloss**—Gr. *Molossos*, the Molossus hound or mastiff <*Molossis*, a district in Epirus famous for its hounds. *Ex:* Moloss-ops (Mam.); Molossus (Mam.).

**molothr**—Gr. *molothros*, an error for *molobros*, a parasite, a greedy fellow, a vagabond. *Ex:* Molothrus (Av.); Call-othrus (Av.).

**molp**—Gr. *molpē*, a song; *molpastēs*, a dancer. *Ex:* Molpastes (Av.); Molpe-myia (Ins.); Eu-molpe (Ins.); Eu-molpo-crio-geton (Ins.); Eu-molpus (Ins.).

**molpadi**—Gr. *Molpadia*, a feminine name. *Ex:* Molpadia (Echin.).

**molpast**—See molp.

**molybd**—Gr. *molybdos*, lead. *Ex:* Molybd-ichthys (Pisc.); Molybdo-phanes (Av.).

**molyc**—See molyx.

**molyn**—Gr. *molynō*, to defile, stain; *molynsis*, pollution. *Ex:* molyns-ic.

**molyns**—See molyn.

**molyt**—Gr. *mōlytēs*, soft, feeble. *Ex:* Molytes (Ins.).

**molyx**—Gr. *molyx*, genit. *molykos*, soft, feeble. *Ex:* Hypo-molyx (Ins.).

**mom**—Gr. *momos*, ridicule, disgrace. *Ex:* Momus (Av.); Calli-momus (Ins.).

**momot**—NL. *momot(us)* from *motmot*, a Mexican name; derived through a misspelling. *Ex:* Momot-idae (Av.); Momotus (Av.).

**momph**—Gr. *momphos*, blame, reproof, a ground for complaint.

**mon**—Gr. *monas*, genit. *monados* = L. *monas*, genit. *monadis*, single, a unit. *Ex:* Mon-anthes*; Mon-esis*; Mon-olene (Pisc.); Mon-omphalus (Moll.); Mon-oo-phorum (Platy.); monad; Mono-tremata (Mam.); A-monas-therium (Ins.); Herpeto-monas (Prot.); Proto-monad-ina (Prot.).

**monac**—See monach.

**monach**—1. Gr. *monachos*, solitary; also a monk = L. *monax*, genit. *monacis*, a monk. *Ex:* monacus; Monacha (Av.); Monachus (Mam.); Eri-monax (Pisc.): 2. Amer. Indian *monax*, the marmot. *Ex:* Monax (Mam.).

**monax**—See monach.

**monedul**—L. *monedula*, a jackdaw. *Ex:* Monedula (Av.).

**monel**—L. *monela*, a reminding, admonishing.

**moner**—Gr. *monērēs*, single. *Ex:* Moner-ula; Monera (Prot.); Monere-bia (Ins.).

**mones**—NL. *moneses* (<Gr. *monos*, alone), a plant name. *Ex:* Moneses*.

**monet**—L. *moneta*, a stamp or die for coining money < *Moneta*, a name for Juno in whose temple money was coined.

**mongo**—See mungos.

**monias**—Gr. *monias*, solitary. *Ex:* Monias (Av.).

**monil**—L. *monile*, genit. *monilis*, a necklace. *Ex:* Monil-ispis (Moll.); Monili-cornes (Ins.); monili-form; Monilia*; Monilo-pora (Coel.).

**monim**—Gr. *monimos*, stable, fixed. *Ex:* monimo-stylic; Monimo-topus (Pisc.).

**monit**—L. *monitus*, advised, put in mind of < *moneo*, to remind. *Ex:* Moniti-lora (Moll.).

**monitor**—L. *monitor*, a reminder. *Ex:* Monitor (Rept.).

**monomach**—Gr. *monomachos*, a gladiator. *Ex:* Monomachus (Ins.).

**monos**—Gr. *monōsis*, singleness. *Ex:* monos-y.

**mons**—See mont.

**monster**—L. *monstrum*, a divine omen indicating misfortune > NL. *monster*, an animal of grotesque form. *Ex:* ?Monstera*; Monstr-illa (Crust.); monstrosity < Fr. *monstruosité*.

**monstr**—See monster.

**mont**—L. *mons*, genit. *montis*, dim. *monticulus*, a mountain; *montanus*, belonging to a mountain, of mountains. *Ex:* mons pubis; montane; monti-colus; monti-genum; Monticuli-pora (Bry.); cis-montane; ultra-montane.

**mops**—Malay *mops*, name of a bat. *Ex:* Mops (Mam.); Eu-mops (Mam.).

**mor**—1. L. *morus*, a mulberry tree < Gr. *morea*, a mulberry. *Ex:* Mor-aceae; Mor-inda*, see indic; Morus*; syca-more: 2. L. *morus* < Gr. *mōros*, foolish, sluggish. *Ex:* Moris (Av.); Moro-pus, Moro-pod-idae (Mam.); Moro-saurus (Rept.): 3. Gr. *moros*, destiny, death, the son of night: 4. L. *mora*, a hindrance, drag. See also mores. 5. L. *morio*, a monster, deformed person.

**morb**—L. *morbus*, disease; *morbidus*, sickly.

**morchell**—Ger. *Morchel*, a morel, an edible fungus. *Ex:* Morchella*.

**mord**—L. *morsus*, a biting; ML. *morsitans*, etching; *mordicus*, biting, *mordax*, genit. *mordacis*, biting, pungent. *Ex:* Mord-ella (Ins.); Mordica*; prae-morsus.

**mordax**—See mord.

**mores**—L. *mos*, genit. *moris*, pl. *mores*, custom, manner, behavior. *Ex:* mores (Ethn.).

**morhua**—ML. *morhua*, the cod. *Ex:* Morhua (Pisc.).

**morim**—Gr. *morimos* = *morsimos*, doomed by fate, destined. *Ex:* Morimus (Ins.); Morsimus (Ins.).

**morind**—NL. *morinda* (contraction of *Morus indica*), a plant genus. *Ex:* Morinda*.

**morinell**—NL. *morinella*, an old name for the turnstone. *Ex:* Morinella (Av.).

**moringu**—Tamil *malangu*, eel. *Ex:* Moringu-idae (Pisc.); Moringua (Pisc.).

**morio**—See morion.

**morion**—1. L. *morio*, genit. *morionis*, a fool, an arrogant person. *Ex:* Morio (Ins.); *Morion-ella* (Moll.): 2. Gr. *morion*, dim. of *moror*, a part, piece. *Ex:* Mono-morium (Ins.). See also mor. 5: 3. Gr. *mōrion*, a narcotic plant, a nightshade.

**morium**—See morion.

**morm**—Gr. *mormō*, genit. *marmoos*, a bugbear; *mormolykeion*, a hobgoblin. *Ex:* Morm-odes*; Mormo-myia (Ins.); Mormo-ops (Mam.); Mormolyce* (Ins.).

**mormolyc**—See morm.

**mormyr**—Gr. *mormyros*, a kind of sea fish mentioned by Aristotle. *Ex:* Mormyrus (Pisc.)

**moron**—1. Gr. *mōros*, neut. *mōron*, stupid. *Ex:* moron; moron-ism: 2. Gr. *moros*, death, fate: 3. NL. *morone*, name for a genus of fishes (etym. unknown.). *Ex:* Moron-idae (Pisc.); Morone (Pisc.).

**morone**—See moron.

**moros**—L. *morosus*, fretful, also slow, late in appearing.

**morph**—1. Gr. *morphē*, form, shape; *morphōsis*, a shaping. *Ex:* Morph-elaphus (Mam.); Morph-ippus (Mam.), see hipp; morpho-logy; morphosis; gyn-andro-morph; poly-morphic: 2. Gr. *Morphō*, a name of Venus, bestower of beauty. *Ex:* Morpho (Ins.).

morphn—Gr. *morphnos*, epithet of an eagle, expressing color, hence dusky, dark. *Ex:* Morphn-aetos (Av.); Morphnus (Av.).

mors—See mord.

morsim—See morim.

morsitan—NL. *morsitans*, genit. *morsitantis*, biting (irregular, ppr. of L. *mordeo*, to bite).

mortu—L. *mortuus*, dead.

morul—1. L. *morulus*, dark-colored, black. *Ex:* Morul-ina (Moll.); morul-ose; Morula (Moll.); Morulus (Moll.): 2. Mod. L. *morula*, dim. of L. *morus*, a mulberry. *Ex:* morul-ation; morula.

mosa—L. *Mosa*, the river Meuse. *Ex:* Mosasaurus (Rept.).

mosaic—LL. *mosaicus*, parti-colored, spotted, mosaic-like.

mosch—1. Gr. *moschos*, musk; NL. *moschatus*, musky. *Ex:* Mosch-osma*; Moscha (Av.); Moscho-mys (Mam.); Moschus (Mam.): 2. Gr. *moschos*, a calf, a heifer; also, rarely, a young shoot or twig. *Ex:* Moschi-ola (Mam.); Hyae-moschus (Mam.).

mosell—L. *Mosella*, a river of Belgic Gaul, the modern Moselle. *Ex:* Mosello-saurus (Rept.).

moso—NL. *moso* < L. *Mosa*, the river Meuse. *Ex:* Moso-saurus (Rept.).

mot—L. *motus*, a moving; *motor*, a mover < *moveo*, to move; *motatio*, motion. *Ex:* mot-ile; motor; im-motate.

motacill—L. *motacilla*, the wagtail. *Ex:* Motacill-idae (Av.); Motacilla.

motmot—See momot.

mox—See myx.

muc—L. *mucus*, mucus, snivel; *muccosus*, full of mucous; *muculentus*, snivelling; NL. *mucin*, a constituent of mucus < L. *muc(us)* +*in*. *Ex:* muc-oid; muci-fic; muci-par-ous; mucino-gen; muco-duct; mucosa; mucous; muculent.

muced—L. *mucedus*, mouldy < *muceo*, to mould. *Ex:* muced-in-ous.

mucid—L. *mucidus*, musty, mouldy < *mucus*.

mucin—See muc.

mucor—L. *mucor*, genit. *mucoris*, mould, mouldiness. *Ex:* Mucor*; Mucor-ales*.

mucro—L. *mucro*, genit. *mucronis*, a sharp point, hence a sword; *mucronatus*, pointed. *Ex:* mucronate.

muculent—See muc.

mucun—Braz. *mucuna*, a plant name. *Ex:* Mucuna*.

mugil—L. *mugil* = *mugilis*, a sea fish, perhaps the mullet. *Ex:* Mugil (Pisc.); Mugil-oides (Pisc.); Mugilis-gobius (Pisc.); Mugilo-stoma (Pisc.).

muilla—NL. anagram of Allium. *Ex:* Muilla*.

mulg—L. *mulgeo*, to milk, to suck > NL. *mulgedium* one with milk. *Ex:* Mulgedium*. Capri-mulg-idae (Av.).

mulin—L. *mulinus*, pertaining to a mule. *Ex:* Mulin-idae*; Mulinum*.

mull—L. *mullus*, a kind of fish, the red mullet. *Ex:* Mull-idae (Pisc.); Mullus (Pisc.). See also myll.

multi—L. *multus*, many > *multi-*, prefix meaning many, much. *Ex:* Multi-ceps (Platy.); multi-fari-ous; multi-fid-ous.

mund—L. *mundus*, cleanly, neat. *Ex:* Mundia*.

mundul—L. *mundulus*, neat, trim < *mundus*, adorned, clean.

mungos—Tamil, *monegos*, a mongoose. *Ex:* Mongo (Mam.); Mungos (Mam.).

munitus—L. *munitus*, armed, protected < L. *munio*, to fortify.

mur—1. L. *mus*, genit. *muris*, dim. *musculus*, a mouse; *murinus*, belonging to mice, mouse-like. *Ex:* Mur-idae (Mam.); Muri-lemus (Mam.); murine; Murinus (Mam.); Mus (Mam.): 2. L. *murus*, a wall. *Ex:* muri-form.

muraen—L. *muraena*, murena, a fish of which the ancients were very fond; also a black-stripe in the form of this fish. *Ex:* Muraen-esox (Pisc.); Muraen-idae (Pisc.); Muraena (Pisc.); Muraeno-saurus (Pisc.).

mural—L. *muralis*, growing on walls, belonging to walls.

murex—See muric.

muric—L. *murex*, genit. *muricis*, the purple fish. *Ex:* Murex (Moll.); Muric-idae (Moll.); Muric-opsis (Moll.); Murici-arius (Moll.); murici-form. See also muricat.

muricat—L. *muricatus*, pointed, full of sharp points < *murex*, genit. *muricis*, a pointed rock. *Ex:* muricate; Muric-aria*.

murin—See mur 1.

mus—1. L. *mus*, a mouse < Gr. *mys*, a mouse. *Ex:* Mus (Mam.); Para-mus (Mam.): 2. NL. *musa* < Ar. *muze*, the banana, the plantain. *Ex:* Musa*; Muso-phaga (Av.); Muso-vora (Av.).

musc—1. L. *musca*, a fly; *muscarius*, belonging to flies; *muscarium*, a fly brush. *Ex:* Musca (Ins.); muscari-form; muscarium; Musci-capa (Av.); Musci-vora (Av.); Musco-pteryx (Ins.): 2. L. *muscus* = Gr. *moschos*, musk. *Ex:* Musc-ari*, but should it not be Muscarium < L. *muscarium*, a fly-brush? 3. L. *muscus*, moss. *Ex:* musc-oid; musco-logy.

muscad—Fr. *muscade*, a nutmeg. *Ex:* Muscadi-vores (Av.).

muscid—L. *muscidus*, mossy.

muscipul—L. *muscipula*, a mouse trap.

muscul—L. *musculus*, a muscle. *Ex:* muscul-ar; musculo-cutaneous. See also mur 1.

music—Gr. *mousikē*, music; *musikos* = L. *musicus*, a musician; also as an adj. of or pertaining to music. *Ex:* Musico-derus (Ins.); Musicus (Av.).

**musiv**—L. *musivus*, of or belonging to a muse.

**must**—L. *mustus*, new, fresh.

**mustel**—L. *mustela*, a weasel; *mustelinus*, weasel-colored, i.e, tawny; also weasel-odored. *Ex:* Mustela (Mam.); Mustelina (Mam.).

**mutab**—See mutat.

The Dog-phallus Stinkhorn Fungus, *Mutinus caninus*. Redrawn from The Romance of The Fungus World — Rolfe. J. B. Lippincott Co., Chapman & Hall, Ltd.

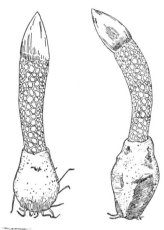

**mutat**—L. *mutatus*, change, alteration; *mutabilis*, changeable. *Ex:* Mutato-copt-ops (Ins.); mutat-ion; per-mutate.

**mutic**—L. *muticus*, curtailed, cut off. *Ex:* mutic; Mutic-aria (Moll.).

**mutilat**—See mutill.

**mutill**—NL. *mutilla*, perh.<L. *mutilus*, maimed; *mutilo*, pp. *mutilatus*, to cut off, cut short. *Ex:* Mutill-oides (Ins.); Mutilla (Ins.).

**mutin**—L. *Mutinus*, a name for Priapus, hence by transference, the penis. *Ex:* Mutinus*.

**my**—1. Gr. *mys*, genit. *myos*, mouse; dim. *myskos*; *myōdēs*, like a mouse. *Ex:* My-arion (Mam.); My-odes (Mam.); My-otis (Mam.); Myo-gale (Mam.); Mys-ateles (Mam.); Myscebus (Mam.); Mys-lemur (Mam.); Mys-idion (Moll.); Aconae-mys (Mam.), when *mys*, as here, is preceded by e of the previous combining form, it may be confused with Gr. *emys*, a tortoise; Geo-mys (Mam.); Pity-mys (Mam.); Pero-myscus; Thomo-mys (Mam.): 2. Gr. *mys*, muscle. *Ex:* My-aria (Moll.); myo-coele; myo-commata; myo-logy; myo-mere; Myotome; Desmo-my-aria (Hemichorda): 3. L. *mya*, a sea-mussel<Gr. *myax*, genit. *myakos*, the sea-mussel. *Ex:* My-opsis (Moll.); Mya (Moll.); Myo-concha (Moll.); Myo-dora (Moll.); Soleno-mya (Moll.): 4. Gr. *myō*, to close. *Ex:* Myo-coryna (Ins.); Myo-por-um*; Myo-rrhinus (Ins.): 5. Gr. *myia*, a fly. *Ex:* my-iasis.

**myac**—Gr. *myax*, genit. *myakos*, the sea-mussel. *Ex:* Myac-idae (Moll.); Myax (Moll.). See also my 3.

**myagr**—1. L. *Myagrus*, Egyptian divinity: 2. Gr. *myagra*, a mouse-trap; *myagros*, a mouser, a kind of snake; also a kind of plant. *Ex:* Myagra (Av.); Myagro-stoma (Moll.); not Myagrion (Ins.), see agrio; Myagrus (Ins.), (Av.).

**myaria**—See my 2.

Merriam Two-footed Mouse or Kangaroo Rat, *Dipodomys merriami*.

**myax**—See myac.

**myc**—Gr. *mykēs*, genit. *mykētos*, a fungus; also any knobbed body shaped like a fungus; *mykētinos*, made of mushrooms>NL. *mycelium*. *Ex:* mycel-oid; myc-elium, see hel. 3; Mycet-ales*; Mycetina (Ins.); Myceto-coccus (Ins.); Myceto-zoa; Myco-mya (Ins.); Myxomycetes*; Schizo-mycetes*.

**mycal**—Gr. *Mykalē*, mountain of ancient Greece. Mycale (Por.).

**mycel**—See myc.

**mycen**—NL. *mycena*<Gr. *mykēs*, a fungus. *Ex:* Mycena*.

**mycet**—See myc.

**mycetes**—Gr. *mykētēs*, a bellower. *Ex:* Mycetes (Mam.). See also myc.

**mych**—Gr. *mychios*=*mychos*, inward, inmost. *Ex:* Myceto-mychus (Ins.); mycho-gemia; Mycho-thorax (Ins.); not Endo-mychus (Ins.) which is said to be<Gr. *endon*, within+*mykēs*, a fungus.

**myct**—Gr. *myktēr*, genit. *myktēros*, the nose, snout. *Ex:* Mycter-ella (Ins.); Mycter-odes (Ins.); Mycteria (Av.); Myctero-mys (Mam.); Mycterus (Ins.); Mycto-dera (Ins.).

**myd**—Gr. *mydos*, decay, wetness, clamminess< *mydaō*, to be damp or wet. *Ex:* mydas; Mydaus (Mam.).<*myda-*+L. masc. ending *us*. See also midas.

**mydale**—Gr. *mydaleos*, wet, damp, mouldy.

**mydaus**—See myd.

**mydr**—Gr. *mydros*, red-hot metal, any red-hot mass. *Ex:* Mydro-soma*; Mydro-thauma (Ins.).

myel—Gr. *myelos*, marrow; often now taken to mean the spinal cord or some connection with it. *Ex:* myelin; Myelo-dactylus (Echin.); myelo-coel; myelo-mere; Myelon*.

mygal—Gr. *mygalē*, a field mouse. *Ex:* mygaloid; Mygal-odonta (Arach.); Mygale (Arach.).

mygm—Gr. *mygmos*, a moaning, muttering, a strange noise.

myi—Gr. *myia*=Attic *mya*, a fly. *Ex:* Myadestes (Av.), see edest; my-iasis=my-iosis; Myio-chanes (Av.); Myio-psittacus (Av.); Myo-phthiria (Ins.); Rhopalo-myia (Ins.).

myl—Gr. *mylos*, akin to *mylē*, a mill, a molar tooth, a grinder. *Ex:* Myl-odon (Mam.); Mylagaulus (Mam.); Myle-stoma (Pisc.); mylo-glossus; mylo-hyoid; Tri-mylus (Mam.).

myla—See myl.

mylabris—See mylacr.

mylac—Gr. *mylax*, genit. *mylakos*, a millstone. *Ex:* Mylacus (Ins.); Mylax (Pisc.).

mylacr—Gr. *mylakris*, a millstone; also a kind of cockroach in flour-mills=*mylagris*=*mylabris*. *Ex:* Mylabris (Ins.); Mylacris (Ins.).

mylagr—See mylacr.

mylax—See mylac.

myli—Gr. *mylias*, a millstone. *Ex:* Mylio-batis (Elasm.).

mylic—Gr. *mylikos*, belonging to a mill. *Ex:* Milico-trochus (Moll.).

myll—1. Gr. *myllainō*, to change the form of the mouth. *Ex:* Myllaena (Ins.):    2. Gr. *myllon*, a lip:    3. Gr. *myllos*, bent, awry, crooked. *Ex:* Myllo-cer-ops (Ins.):    4. Gr. *myllos*, a salted sea-fish. *Ex:* Mullus (Pisc.):    5. Gr. *myllas*, genit. *myllados*, a prostitute.

mymar—Gr. *mymar*, ridicule, blame, mockery. *Ex:* Mymar (Ins.); Mymar-illa (Ins.); Mymaromma (Ins.); Mymaro-thrips (Ins.).

myo—See myi.

myodoch—Gr. *myodochos*, harboring mice. *Ex:* Myodocha (Ins.); Myodochus (Ins.).

myop—Gr. *myops*, the horsefly. *Ex:* Myopornis (Av.).

myox—Gr. *myōxos*, the dormouse. *Ex:* Myoxoides (Mam.); Myoxo-mys (Mam.); Myoxus (Mam.).

myr—1. Gr. *myrizō*, to rub with ointment < *myron*, a balsamic juice, sweet oil; Gr. *myristikos*, fit for anointing. *Ex:* Myr-odia*; Myro-phylla (Ins.); Myro-xylon*; Myristica*; Myristici-vora (Av.); A-myris*:    2. Gr. *myros*, a kind of sea-eel. *Ex:* Myr-ophis (Pisc.).

myri—Gr. *myrias*, genit. *myriados*, the number 10,000; akin to *myrios*, myriad, numberless, infinite, immense. *Ex:* Myria-n ites (Ann.), the *n* used as a connective; Myrio-lithes (Bry.); Myria-poda (Myr.); myriad; Myrio-pora (Bry.); Myrio-zoum (Myr.).

Myrica - leaved Bernardia, *Bernardia myricafolia*, a shrubby spurge from the S. W. American deserts. Redrawn from Desert Wild Flowers—Jaeger. Stanford University Press.

myric—Gr. *myrikē*, ancient name of the tamarisk < *myrō*, to flow, since it grows on banks of running streams. *Ex:* Myric-aria*; Myrica, Myric-aceae*.

myristic—See myr.

myrm—1. Gr. *myrmēx*, genit. *myrmēkos*, also *myrmos*, an ant; *myrmedōn*, an ant hill. *Ex:* Myrm-oecia (Ins.); Myrmeco-phila (Ins.); myrmeco-phag-ous; Myrmedon-ia (Ins.); Myrmex (Ins.); Myrmic-inae (Ins.); Myrmo-phila (Av.); Pogono-myrmex (Ins.):    2. Gr. *myrmēkion*, a kind of spider. *Ex:* Mymec-isca (Arach.); Myrmecium (Arach.); Megamermecium (Arach.).

myrmec—See myrm.

myrmic—See myrm.

myrsi—NL. *myrsi*<Gr. *myrsinē*, myrtle. *Ex:* Myrsi-phyllum*.

myrsil—Gr. *Myrsilos*, king of Lydia. *Ex:* Myrsilus (Mam.).

myrsin—Gr. *myrsinē*, a name of myrrh. *Ex:* Myrsine*.

myrt—L. *myrtus*, the myrtle. *Ex:* Myrt-aceae*; Myrtus*.

myrtea—L. *Myrtea*, an epithet of Venus. *Ex:* Myrtea (Moll.).

myrtil—Gr. *Myrtilos*, son of Mercury, charioteer of Oenomaus. *Ex:* Myrtilus (Moll.).

mys—See my.

mysi—Gr. *mysis*, a closing of the lips or eyes. *Ex:* Mys-idae (Crust.); Mysis (Crust.).

myst—Gr. *mystis*, genit. *mystidos*, a mystic. *Ex:* Mysti-cetus (Mam.); Mysto-mys (Mam.).

mystac—Gr. *mystax*, genit. *mystakos*, upper lip, mustache. *Ex:* Mystac-ides (Ins.); mystac-inous; Myataco-cete (Mam.); mystax; Mystaxus (Ins.).

mystax—See mystac.

**mysteri**—L. *mysterium*, a mystery < Gr. *mystērion*. *Ex:* Mysteria (Ins.).

**mystes**—Gr. *mystēs*, a priest, sometimes taken to mean a dark color. *Ex:* Mystes (Ins.).

**mysti**—Gr. *mystis*, one initiated. *Ex:* Oreomystis (Av.).

**mystico**—Gr. *mystikos*, mystic, secret. *Ex:* Mystico-crinus (Echin.).

**mystr**—Gr. *mystrion*, a spoon; dim. of *mystron*, a spoon. *Ex:* Mystri-ophis (Rept.); Mystriosaurus (Rept.); Mystrium (Ins.); Mystro-mys (Mam.); Mystro-petalon*; Mystro-phorus (Ins.); Mystro-thamnus*.

**mytho**—Gr. *mythos*, myth, fable. *Ex:* Mytho-mantis (Ins.); Mytho-mys (Mam.).

**mytil**—Gr. *mytilos*, a sea mussel. *Ex:* Mytil-idae (Moll.); Mytil-aspis (Ins.); Mytil-oides (Moll.); Mytili-meria (Moll.); Mytilo-phagus (Pisc);. Mytilus (Moll.).

**myx**—1. Gr. *myxa*, slime, mucus; *myxinos*, a slime fish. *Ex:* Myx-aster (Echin.); Myxa (Moll.); Myxin-idae (Cycl.); Myxine (Cycl.); Myxo-bolus (Prot.); Myxo-dictum (Prot.): 2. Gr. *myxa*, a kind of plum.

**myxin**—See **myx.**

**myz**—1. Gr. *myzō*, to suck in; *myzaō*, to suck. *Ex:* Moxo-stoma (Pisc.); an error for Myzostoma; Myz-ine (Ins.); Myzo-bdella (Ann.); Myzo-stoma (Ann.); Myzus (Ins.); Antho-myza (Ins.); Malaco-myza (Ins.): 2. Gr. *myzō*, to mutter. *Ex:* Myzo-mela (Av.).

# N

**nab**—NL. *nabis*, an insect genus, perh. < L. *nabis*, a giraffe, but it may be derived from L. *Nabis*, genit. *Nabidis*, name of a Spartan king. *Ex:* Nab-icula (Ins.); Nab-idae (Ins.); Nabido-morpha (Ins.); Nabis (Ins.).

**nabid**—See **nab.**

**nabl**—1. Gr. *nabla*, a musical instrument of ten strings; *nablistēs*, one who plays the nabla. *Ex:* Nabla (Ins.); Nablistes (Ins.): 2. Gr. *nablē*, a ship.

**nac**—Gr. *nakos*, fleece, a skin with fleece on. *Ex:* Naco-phora (Ins.); Naco-spatangus (Echin.).

**nacc**—L. *nacca*, a fuller. *Ex:* Nacc-ula (Moll.); Nacca (Moll.); Labelli-nacca (Moll.).

**nacre**—Fr. *nacre*, Sp. *nacar*, mother of pearl < Pers. *nakar*. *Ex:* nacre; nacre-ous.

**nact**—1. Gr. *naktos*, full pressed, solid: 2. L. *nactus*, obtained, pp. of *nanciscor*, to get, obtain, to meet with, stumble on.

**naem**—See **nem 2.**

**naenia**—See **nenia.**

**naet**—Gr. *naetēs*, an inhabitant. *Ex:* Naetes (Ins.).

**naeus**—See **nai.**

**naev**—L. *naevus*, a spot, a birth mark; *naevia*, spotted, varied. *Ex:* Naev-osa (Moll.); naev-ose; nevius.

**nai**—Gr. *naiō*, to dwell, to inhabit. *Ex:* Naja (Ins.); Epi-nae-us (Ins.); Heli-naia (Av.); Pelo-naia (Hemichorda).

**naiad**—See **naias.**

**naias**—L. *nais*, genit. *naidis* = *naias*, genit. *naiadis*, a water nymph. *Ex:* Naiad-idae*; Naiad-ites (Moll.); Naid-idae (Ann.); Naiado-chelys (Rept.); Naias*; Nais (Ann.); Najado-chelys (Rept.).

Small Dormouse Phalanger, *Dromicia nana.* Redrawn from The Wild Animals of Australia—Le Souef and Byrrell.

**nais**—See **naias.**

**naja**—NL. *naja* < Skr. *nāga*, a snake. *Ex:* Naja (Rept.).

najad—See naias.

nama—Gr. *nama*, genit. *namatos*, a stream, anything flowing. *Ex:* Nama*; Nama-myia (Ins.); namat-ium (Ecol.); namato-philus.

namat—See nama.

nan—See nann.

nandina—1. Jap. *nandin*, a plant name. *Ex:* Nandina*: 2. NL. *nand(us)*, etym. doubtful. *Ex:* Nand-ina (Pisc.) <*nand*+L. neut. pl. adj. suff. -*ina*.

nandinia—West African *nandinia*, native name for the palm civet. *Ex:* Nandinia (Mam.).

nann—Gr. *nanos* = *nannos* = L. *nanus*, dwarf; Gr. *nanōdēs*, dwarfish, small. *Ex:* Nan-elaphus (Mam.); Nann-ugo (Mam.); Nanno-batrachus (Amph.); nanno-plankton; Nanno-suchus (Rept.); Nannus (Av.); Nano-hy-us (Mam.); Nano-drilus (Platy.); Nanodes (Av.).

nao—Gr. *naos*, temple. *Ex:* Nao-saurus (Rept.); not Nao-morpha (Ins.) which comes in part from Gr. *naus*, Attic. genit. *naos*, a ship.

nap—1. L. *napus*, a kind of turnip. *Ex:* nap-ellus; napi-formis: 2. Gr. *napē*, a glade, woody dell; *napaios*, of a wooded dale >L. *Napaea*, a wood nymph. *Ex:* Nap-ochus (Ins.); Nap-odon-ictis (Mam.); Napaea*; Napaeo-zapus (Mam.); Napo-myia (Ins.); Napo-phila (Av.).

napae—See nap 2.

nar—L. *naris*, pl. *nares*, nostril. *Ex:* Nar-ica, see icus 2.; nares; nari-al; nari-corn; not Naresia (Bry.); named after Capt. Geo. Nares of the Challenger Expedition.

narc—Gr. *narkē*, numbness, stupor; *narkōdēs*, stupid, dull; *narkōtikos*, benumbing; *narkaō*, to grow numb. *Ex:* Narc-aci-ontes (Elasm.); Narc-inc (Elasm); Narco-batus (Elasm.); Narco-medusae (Coel.); narcot-ize; Ptcro-narc-ella (Ins.); Tetro-narce (Elasm.).

narciss—Gr. *narkissos*, the plant narcissus, named for its narcotic properties (see narc); also the youth Narcissus. *Ex:* Narciss-astrea (Coel.); narcissi-form; Narcissus*.

narcot—See narc.

nard—Gr. *nardos*, nard, the Indian spikenard; also an ointment prepared from it. *Ex:* Nard-osmia*; Nardo-stachys*; Nardus*; nard-inus.

nares—See nar.

narinari—Brazilian *narinari*, name of some fish.

narinos—L. *narinosus*, broad-nosed.

narthec—Gr. *narthēx*, genit. *narthēkos*, ferula, an umbelliferous plant, the stems of which were used as rods. *Ex:* Nartheci-um*; Nartheci-us (Ins.).

nas—L. *nasus*, the nose, a nozzle; *nasutus*, large-nosed; ML. *nasalis*, of or pertaining to the nose. *Ex:* Nas-turtium*, see tort; nasal; Nasalis (Mam.); nasi-form, naso-graph; naso-frontal; Nasu-inae (Mam.); Nasua (Mam.); nasus; nasute: atro-nasus.

nasc—L. *nascor*, pp. *natus* = *gnatus*, to be born, spring forth, ppr. *nascens*, genit. *nascentis*, arising; *natalis*, of or belonging to one's birth. *Ex:* nascent; natal; in-nate.

nasic—NL. *nasica*, with large or pointed nose < L. *nasus*, nose.

nasio—NL. *nasion*, the middle point of the fronto-nasal suture. *Ex:* nasio-mental; nasion.

nasiterna—L. *nasiterna* = *nassiterna*, a watering-pot with a large nose or spout. *Ex:* Nasiterna (Av.).

nasm—Gr. *nasmos*, a stream. *Ex:* nasmo-philous.

nass—L. *nassa* = *naxa*, a narrow-necked fish basket. *Ex:* Nass-arius (Moll.); Nass-ell-aria (Prot.); Nassa (Moll.); Callia-nassa (Arthr.). See also anass.

nast—Gr. *nastos*, crammed full, pressed close < *nassō*, to press, to cram. *Ex:* nast-ic; Nasto-cerus (Ins.); chemo-nast-ic; epi-nast-ic.

nastes—Gr. *nastēs* an occupant. *Ex:* Campo-nastes (Ins.); Crypto-nastes (Ins.).

nasu—See nas.

nasut—See nas.

nat—1. L. *nato*, to swim, ppr. *natans*, genit. *natantis*, floating, swimming; *natator*, a swimmer >*natrix*, genit. *natricis*, a water-snake. *Ex:* nata-podium; natant; Natantia (Por.); natatori-al; Natric-idae (Rept.); Natrix (Rept.): 2. L. *natis*, pl. *nates*, the rump, buttocks = ML. *natica*, pl. *naticae*, nates. *Ex:* Natic-idae (Moll.); Natica (Moll.). See also nasc.

natal—NL. *natalus*, name for a genus of bats. *Ex:* Natalus (Mam.). See also nasc.

natic—See nat 2.

natric—See nat 1.

natrix—See nat 1.

nau—Gr. *naus*, a ship; *nuubatēs*, a seaman; *naukratēs*, the master of a ship; also the name of a fish. *Ex:* Nau-cor-idae (Ins.); Naucrates (Pisc.); Nausi-thoe (Coel.).

naubat—See nau.

naucrat—See nau.

naumach—Gr. *naumachos*, fit for sea fighting. *Ex:* Naumacho-crinus (Echin.).

naupact—Gr. *Naupaktos*, the name of a town. *Ex:* Naupactus (Ins.); Naupact-opsis (Ins.).

naupli—Gr. *nauplios*, a kind of mollusk that sails in its shell as in a ship (Pliny). *Ex:* nauplius.

naus—See nau.

nause—Gr. *nausia*, ship-sickness < *naus*, a ship; L. *nauseosus*, that produces nausea, nauseous.

naut—Gr. *nautēs* (poet. *nautilos*) dim. *nautiskos*, a sailor. *Ex:* Naut-ichthys (Pisc.); Nautil-oidea (Moll.); Nautilo-ceras (Moll.); Nautiscus (Pisc.); Exo-nautes (Pisc.).

nautil—See naut.

**nautisc**—See **naut.**

**nav**—1. L. *navis*, dim. *navicella=navicula*, a ship. *Ex:* Navicella (Moll.); Navicula (Prot.); pseudo-naviculae:   2. L. *navus*, diligent, active.

**navig**—L. *navigo*, to sail; *navigator*, a sailor.

**-ndi**—L. *-ndi*, genit. gerund ending used in an adjectival sense after nouns. It has a meaning corresponding to the English verbal in *-ing*. *Ex:* potentia audiendi, the power of hearing; potentia gererandi, ability to procreate.

**-ndus**—L. *-ndus*, gerundive ending. The gerundive may be used as an adjective implying obligation. See -ndi.

**ne**—Gr. *neō*, to spin, to swim, to go. *Ex:* Nephila (Arach.).

**ne-** —Gr. *nē*, poetic particle used as a prefix meaning not. *Ex:* Ne-cremnus (Ins.); Ne-elaps (Rept.); Ne-litris*, Gr. *elytron*, a cover; Ne-mertes (Crust.); Ne-penthes*; Ne-pus (Mam.). See also neo.

**neaer**—Gr. *Neaira*, a girl mentioned in the writings of Horace. *Ex:* Neaera (Moll.); Neaero-mya (Moll.).

**neal**—See **near.**

**neanic**—See **near.**

**neanthes**—Gr. *neanthēs*, a new-blown bud. *Ex:* Neanthes (Ins.).

**near**—Gr. *nearos=nealēs*, youthful; also fresh, as of fish; *neanikos*, youthful, *nealōtos*, newly caught; *neazō*, to be young. *Ex:* Neaera (Moll.); Neaerus (Ins.); Neaero-mya (Moll.); Nealotos (Pisc.).

**neas**—1. Gr. *nēas*, Ionic of *naus*, a ship. *Ex:* Neas-ellus (Crust.); Neas-ura (Ins.):   2. Gr. *neasis*, newness, freshness.

**neat**—1. Gr. *neatos*, the last, lowest. *Ex:* Neatus (Ins.):   2. Gr. *neatos*, renewed.

**nebali**—NL. *nebalia*, from a proper name. *Ex:* Nebali-opsis (Crust.); Nebalia (Crust.).

**nebrac**—Gr. *nebrax*, genit. *nebrakos*, a young deer, a young animal. *Ex:* Lago-nebrax (Mam.).

**nebrax**—See **nebrac.**

**nebri**—Gr. *nebris*, a fawn skin>*nebrias*, dappled like a fawn. *Ex:* Nebria (Ins.); Nebrio-soma (Ins.).

**nebul**—L. *nebula*, vapor, smoke; *nebulosus*, dark, clouded. *Ex:* Nebuli-pora (Coel.); Nebulosicus (Echin.).

**necan**—L. *necans*, genit. *necantis*, killing, destroying <*neco*, to put to death.

**necator**—LL. *necator*, a slayer. *Ex:* Necator (Nemat.), (Av.).

**necatrix**—L. *necatrix*, she who slays.

**necopin**—L. *necopinus*, not expecting.

**necr**—Gr. *nekros*, a dead body, the dead, dead; *nekrōsis*, a killing, mortification; *nekrotēs*, deadness, a state of death, mortification. *Ex:* Necr-astur (Av.); Necr-odes (Ins.); Necrodasy-pus (Mam.); necro-phagus; necrosis; necrot-ic.

**nect**—1. Gr. *nēktos*, neut. *nēkton*, swimming; *nēktēr*, fem. *nēktris*, a swimmer. *Ex:* Nect-urus (Amph.); not Nect-andra*, see nectar; necto-pod; Necto-saurus (Rept.); necto-some; nek ton; Nectria*; Calo-nectris (Av.); Chiro-nectes (Mam.); Cysto-nectae (Coel.):   2. L. *necto*, to join, ppr. *nectens*, genit. *nectentis*, joined, bound. *Ex:* an-nectens. See also nectar.

**nectar**—Gr. *nektar*, drink of the gods; *nektareos*, scented; also beautiful, divine. *Ex:* Nect-andra*; Nectar-inia (Av.); nectari-fer-ous; nectaro-theca; nectar-y.

**nectr**—See **nect.**

**necy**—Gr. *nekys*, a dead body, the spirits of the departed. *Ex:* Necy-mylacris (Ins.); Necy-opa (Ins.); Necyo-mantes (Ins.).

**necydal**—Gr. *nekydalos*, the silkworm larva. *Ex:* Necydalis (Ins.).

**neda**—NL. *neda*, an arbitrary combination of letters used to designate a genus of beetles. *Ex:* Neda (Ins.).

**nedy**—Gr. *nēdys*, the belly, any of the large cavi ties in the body. *Ex:* Nedy-stoma (Pisc.) Nedyo-pus (Myr.).

**nedym**—Gr. *nēdymos*, sweet, delightful. *Ex:* Nedymo-serica (Ins.).

**neel**—Gr. *neēlyx*, a new comer. *Ex:* Neel-idae (Ins.).

**neg**—L. *nego*, to refuse, to deny. *Ex:* Securi-nega*.

**negan**—L. *negans*, genit. *negantis*, denying, ppr. of *nego*, to refuse, deny.

**neglect**—L. *neglectus*, neglected, not chosen< *nec*, not+*lego*, to choose, gather. *Ex:* Neglecta (Moll.); Neglecti-ana (Moll.).

**negret**—Gr. *nēgretos*, unawakened, sound in slumber. *Ex:* Negretus (Ins.).

**negrit**—Sp. *negrito*, dim. of *negro*, black. *Ex:* Negrito-myia (Ins.); Negritius (Ins.).

**negundo**—NL. *negundo*<Malayalan name of a tree, a name now applied to the maple. *Ex:* Negundo*.

**neid**—Gr. *nēis*, genit. *nēidos*, unknowing, feeble. *Ex:* Neid-idae (Ins.); Neidos (Ins.); neidio-plankton (Ecol.).

**neilo**—Gr. *Neilos*, the Nile. *Ex:* Neilo-carus (Ins.).

**neis**—See **neid.**

**nekt**—See **nect.**

**nel**—Gr. *nēlēs*, merciless. *Ex:* Neleo-mys (Mam.) Neleo-thymus (Ins.); Neli-opisth-us (Ins.); Nelo-mys (Mam.).

**neleges**—Gr. *nēlegēs=anēlegēs*, reckless. *Ex:* Neleges (Ins.).

**neleo**—See **nel.**

nelio—Gr. *Neilos*, the Nile. *Ex:* Neilo-carus (Ins.).

nelip—Gr. *nēlipos*, bare-footed. *Ex:* Nelipo-phygus (Ins.).

nelumb—NL. *nelumbo* < Ceylonese name for the lotus lily. *Ex:* Nelumbi-um*; Nelumbo*.

nem.—1. Gr. *nēma*, genit. *nēmatos*, thread. *Ex:* Nema-caulis*; Nema-lion*, see leo; Nema-palpus (Ins.); Nema-therium (Mam.); Nemat-helminthes; Nemat-oda (Nemat.); Nemat-ura (Moll.); nemato-cyst; Nematus (Ins.); Nemo-cera (Ins.); Nemo-glossa (Ins.); Nemo-p-anthus*, see pod; Nemo-seris*; Cali-nem-urus (Ins.); Hetero-nema (Nemat.): 2. L. *nemus*, genit. *nemoris* = Gr. *nemos*, genit. *nemeos*, a pasture, a woodland glade; L. *nemoralis*, of a wood or grove, sylvan. *Ex:* Naemor-haedus (Mam.); Nemeo-bi-idae (Ins.); Nemeo-bius (Ins.); Nemio-blastus (Ins.); Nemo-lestes (Mam.); Nemo-phila*; Nemo-spiza (Av.); Nemor-aea (Ins.); Nemori-cola (Av.): 3. Gr. *nemō*, to distribute, to dwell; also to es-teem, govern, regulate.

nemert—Gr. *Nēmertēs*, name of a Nereid < *nēmertēs*, the unerring one. *Ex:* nemert-ian; Nemert-inea (Ann.); Nemertes (Ann.); Ne-merto-drilus (Ann.).

nemes—Gr. *nemesis*, the snap-dragon. *Ex:* Nemesia*.

nemestrin—L. *Nemestrinus*, god of groves. *Ex:* Nemestrinus (Ins.).

nemor—See nem 2.

nenia—L. *nenia* = *naenia*, a funeral song. *Ex:* Naenia (Av.); Nenia (Moll.).

neo—Gr. *neos*, new, recent, young. *Ex:* Ne-aco-mys (Mam.); Ne-arctos (Mam.); ne-enceph alon; Neo-fiber (Mam.); Neo-Lamarck-ism; Meo-meris (Mam.), a misprint for Neo-meris; Neo-rnithes (Av.); Neo-sorex (Mam.); Nu-menius (Av.).

neoss—See neott.

neoter—L. *neotericus*, new, modern < Gr. *neōterikos*, new.

neotes—Gr. *neotēs*, youth; also rashness.

neotrot—Gr. *neotrōtos*, lately wounded, with fresh sores. *Ex:* Neotrotus (Ins.).

neott—Gr. *neossia* = Attic *neottia*, a bird's nest. *Ex:* Neotti-coris (Ins.); Neottia*; Neotto-pteris*.

nep—L. *nepa*, a scorpion. *Ex:* Nep-idae (Ins.); Nep-idium (Ins.); Nepa (Ins.); Nepeta*, a named used by Pliny; not Nepus (Mam.), see ne.

nepenth—See ne.

neph—Gr. *nephos* = *nephelē*, a cloud; *nephelion*, a cloud-like spot. *Ex:* Nepha (Ins.); Nephel-opsis (Ann.); Nephelium*; Nepho-ect-etes (Av.); Nepho-therium (Mam.); Anti-nephele (Ins.); Epi-nephelus (Pisc.).

nephel—See neph.

nephr—Gr. *nephros*, the kidneys. *Ex:* nephr-idium; Nephr-osteon (Mam.); Nephro-lepis*; nephro-stome; meso-nephros.

nephthy—Gr. *Nephthys*, Egyptian goddess, wife of Typhon. *Ex:* Nephthya (Coel.); Nephthys (Ann.); Nephthytis*.

nepio—Gr. *nēpios* = *nēpion*, an infant. *Ex:* Nepio-blatta (Ins.); Nepio-teuthion (Moll.); nepion-ic; Calli-nepion (Moll.).

nepion—See nepio.

nepos—See nepot.

nepot—L. *nepos*, genit. *nepotis*, a descendant, grandson.

nept—L. *neptis*, a granddaughter, dim. *nepticula*. *Ex:* Nepticula (Ins.); Neptis (Ins.).

neptun—L. *Neptune*, fabled god of the sea. *Ex:* Neptun-ella (Moll.); Neptunea (Moll.); Nep tunia*.

ner—Gr. *nēros* = *aros*, humid, fluid; also used in the sense of a swimmer. *Ex:* Ner-ophis (Pisc.)· Nero-philus (Ins.).

nere—Gr. *Nēreis*, genit. *Nēreidos*, name of a sea-nymph > L. *Nerine*, a nereid. *Ex:* Nereid-aster (Echin.); Nereis (Ann.); Nereo-cystis*; Nereo-graptus (Coel.); Nerine*; Nerinea (Moll.).

nereid—See nere.

nereis—See nere.

nerine—See nere.

nerit—1. Gr. *nēritēs* = *nēreitēs*, a kind of shell-fish that swims in the water. *Ex:* Nerit-ina (Moll.); Nerit-onyx (Moll.); Nerita (Moll.); Neritae-formis (Moll.); Neriti-conus (Moll.); Nerito-dryas (Moll.); Nerito-globus (Moll.); Amphi-nerita (Moll.): 2. Gr. *nēritos*, undisputed; also immense, vast, grand.

neritic—NL. *noritic* < *nerita*, the sea mussel +adj. ending -*ic*, pertaining to. *Ex:* neritic, an ecological term applied to flowing waters.

nerium—Gr. *nērion*, the oleander. *Ex:* Nerium*.

nert—Gr. *nertos*, some bird of prey. *Ex:* Hali-nertus (Av.).

nerter—Gr. *nerteros*, lower, inferior, the world be-low, the dead. *Ex:* Nertera*, because of its creeping habits.

nerthr—NL. *nerthrus*, a generic name, prob. < L. *Nerthus*, earth-goddess. *Ex:* Nerthrus (Ins.).

nerv—L. *nervus*, a sinew, tendon, nerve < Gr. *neuron*, a sinew, tendon; nerve; NL. *nervillus*, a fine nerve. *Ex:* Neur-ada*, see aden; neuri-lemma; neuro-pore; Neuro-trichus (Mam.); neurone; nervi-nervorum; nerville; Nevr-omus (Ins.); Poly-nevra (Ins.).

nes—Gr. *nēsos*, island, dim. *nēsion*; *nēsiōtēs*, an islander > NL. *nesioticus*, belonging to an is-land. *Ex:* Nes-oryzo-mys (Mam.); Nesio-phasma (Ins.); nesiote; nesioticus; nesium; Neso-draba*; Neso-pithecus (Mam.); Neso-spiza (Av.); Nesos-us (Mam.), see sus.

nesaea—Gr. *Nēsaiē*, a sea-nymph, one of the Nereids. *Ex:* Nesaea*.

nesc—L. *nescio*, to be ignorant, ppr. *nesciens*, genit. *nescientis*, ignoring; *nescentia*, ignorance. *Ex:* Nesc-idium*; Nescio-therium (Mam.).

nesiot—See nes.

ness—See nett.

nestis—Gr. *nēstis*, hungry, destitute. *Ex:* Odonestis (Ins.).

nestor—Gr. *Nestōr*, King of Pylos, who took part in the seige of Troy. *Ex:* Nestor (Av.); Nestori-therium (Mam.).

net—Gr. *nētos*, heaped, piled up; also spun and in this sense used in spider names. *Ex:* Argyro-neta (Arach.); Lepto-neta (Arach.); Micro-neta (Arach.).

neth—Gr. *nēthō*, to spin. *Ex:* Neth-ea (Por.); Calli-neth-is (Arth.).

nethr—L. *Nethrus*, goddess of Earth. *Ex:* Nethrus*.

netr—Gr. *nētron*, a spindle. *Ex:* Netro-cera (Ins.); Netro-stoma (Moll.).

nett—Gr. Attic *nētta* = Ionic *nēssa*, duck; *nēttion*, a duckling. *Ex:* Nesso-rhinus (Ins.); Nett-arion (Av.); Netta-rrhinus (Ins.); Nettium (Av.); Aristo-netta (Av.); Charito-netta (Av.); Glaucio-netta (Av.); Melan-netta (Av.).

neu—Gr. *neuō*, to incline, to nod. *Ex:* Neuglenes (Ins.); Neu-raphes (Ins.).

neur—See nerv.

neust—Gr. *neustos* and *neustikos*, able to swim; also inclining, nodding, assenting; *neustēr*, a swimmer. *Ex:* Neustic-urus (Rept.); Neusticosaurus (Rept.); Neusto-saurus (Rept.).

nevr—See nerv.

nex—1. Gr. *nēxis*, a swimming. *Ex:* Nexi-spongia (Por.); Philo-nexis (Moll.): **2.** L. *nexus*, bound, fastened together, pp. of *necto*, to bind. *Ex:* nexus: **3.** L. *nex*, genit. *necis*, death.

nexil—L. *nexilis*, tied together < *necto*, to join. *Ex:* Nexil-arius (Pisc.); Nexil-osus (Pisc.).

nic—1. Gr. *neikos*, strife. *Ex:* Nico-bium (Ins.): 2. Gr. *nikē*, conquest, ascendency. *Ex:* Laetmo-nice (Ann.), see laetm.

nices—Gr. *nikēeis*, conquering. *Ex:* Tachy-nices (Mam.).

nicothoe—Gr. *Nicothoē*, the name of a Harpy. *Ex:* Nicothoe (Crust.).

nict—L. *nicto*, to wink, blink; *nictitatio*, genit. *nictitationis*, a winking. *Ex:* nictitating membrane. See also nyct.

nictitat—See nict.

nid—L. *nidus*, dim. *nidulus*, nest; *nidulatus*, nestled, pp. of *nidulor*, to nestle. *Ex:* nidi-fic-ate; Nidi-valvata (Arach.); Nidul-arium*; nidulatus; nidus.

nidament—L. *nidamentum*, the materials of which a nest is made. *Ex:* nidament-al.

nidoros—L. *nidorosus*, steaming, reeking as with a bad odor.

nidul—See nid.

nigell—NL. *nigellus*, dim. of L. *niger*, black. *Ex:* Nigell-astrum (Coel.); Nigella* (Coel.).

Whisk Broom, *Eriogonum nidularium*, a dainty little buckwheat with specific name referring to the nest-like appearance of the old plants. Redrawn from Desert Wild Flowers—Jaeger. Stanford University Press.

nigr—L. *niger*, dark, black, deathly; *nigrescens*, genit. *nigrescentis*, blacking, ppr. of *nigresco*, to grow or become black; *nigricans*, black, swarthy; NL. *nigritus*, blackened. *Ex:* niger; Nigr-avis (Av.); nigrescent; Nigri-lauda (Av.); nigri-pinnis; Nigrit-ella (Moll.); Nigrito-myia (Ins.).

nilio—L. *nilios*, genit. *nilionis* = Gr. *neilios*, a precious stone. *Ex:* Nilio (Ins.); Nilion-idae (Ins.).

nilotic—L. *Niloticus*, of the Nile < *Nilus*, the Nile.

nimbat—L. *nimbatus*, frivolous.

nimbos—L. *nimbosus*, rainy, stormy.

nimi—L. *nimius*, excessive, beyond reason, much used. *Ex:* Byrso-nima*.

nimr—Biblical *Nimrod*, name of a hunter. *Ex:* Nimr-avus (Mam.).

nin—*Nina*, goddess of the sea-depths. *Ex:* Nin-ella (Moll.); Nina (Moll.).

ninox—NL. *ninox*, name for a genus of Old World owls. *Ex:* Ninox (Av.).

niob—Gr. *Niobē*, daughter of Tantalus. *Ex:* Niob-ella (Tri.); Niobe* (Tri.).

niph—Gr. *nipha*, snow; *niphas*, genit. *niphados*, a snowflake; *niphetōdēs*, like snow; *niphoblēs* = *niphobolos*, snowclad. *Ex:* Niph-argus (Crust.); Niphad-onyx (Ins.); Niphado-lepis (Ins.); Niphaea*; Niphetodes (Ins.); Nipho-grapta (Ins.); Nipho-stola (Ins.).

niphad—See niph.

nipt—Gr. *niptēr*, genit. *niptēros*, a basin. *Ex:* Niptera*; Nipter-axis (Moll.); Niptero-crinus (Echin.).

**nis**—1. L. *nisus*, striving, pp. of *nitor* to strive: 2. Gr. *Nisus*, king of Megara, fabled to have been changed into a sparrow-hawk. *Ex:* Nis-aetus (Av.); Nisu-ella (Av.); Nisus (Av.).

**nisu**—See nis 2.

**nit**—L. *nitidus*, dim. *nitidulus*, bright, trim, glittering < *niteo*, to shine; ppr. *nitens*, shining. *Ex:* Nitid-ella (Moll.); nitid-ous; nitidi-folius; Nitido-pecten (Ins.); Nitidul-idae (Ins.); Nitidula (Ins.). See also nitel.

**nitel**—L. *nitella*=*nitela*, splendour, brightness. *Ex:* Nitela (Ins.); Nitella*; Nitelo-pterus (Ins.).

**nitell**—See nitel.

**nitid**—See nit.

**nitr**—L. *nitrum*, nitron, native soda. *Ex:* Nitr-aria*; nitro-bacteria; Nitro-phyla*.

**nitroso-**—NL. *nitroso-*, a combining form used to designate certain nitrifying bacteria: also used by chemists to designate the radical NO. *Ex:* Nitroso-bacter*.

**niv**—L. *nix*, genit. *nivis*, snow; *niveus*, snowy; *nivosus*, full of snow, snowy. *Ex:* eu-niv-al (Ecol.).

**nix**—L. *nixus*, similar to *nisus*, an effort, pressure. *Ex:* nixus formativus. See also niv.

**nixan**—L. *nixans*, leaning, resting, pp. of *nixor*, to lean.

**nobil**—L. *nobilis*, known, famous. *Ex:* Nobilis (Ins.).

**noc**—L. *noceo*, to injure; ppr. *nocens*, genit. *nocentis*, injuring. *Ex:* noci-ceptor.

**nochel**—Gr. *nōchelēs*, slow, sluggish. *Ex:* Nocheles (Ins.).

**nocit**—L. *nocitus*, harmed < *noceo*, to injure.

**nociv**—L. *nocivus*, hurtful < *noceo*, to hurt.

**noct**—L. *nox*, genit. *noctis*, night; *nocturnalis*, nocturnal; *nocturnus*, of the night; LL. *noctulus*, nocturnal > Fr. *noctule*, name of a bat. *Ex:* Nocti-lio (Mam.), see leo; Nocti-luca (Prot.); Noctu-idae (Ins.); Noctua (Ins.); Noctul-inia (Mam.); Noctula (Mam.); nocturnal.

**nod**—1. L. *nodus*, dim. *nodulus*, knotty, nobby; *nodosus*, full of knots, knobs. *Ex:* nod-al; Node-pus (Ins.); Nodi-scala (Moll.); Nodi-termes (Ins.); Nodo-cephalus (Myr.); Nodos-aria (Prot.); Nodos-ina (Por.); Nodul-aria*: 2. Gr. *nōdos*, toothless. *Ex:* Nodus (Mam.).

**nodul**—See nod.

**nodyn**—Gr. *nōdynos*, alleviating pain, without pain.

**noem**—Gr. *noēma*, the understanding. *Ex:* noema-tacho-meter; Noemon (Ins.); A-noema (Mam.).

**nola**—L. *nola*, a little bell, from Nola in Campania where bells were said to have been first made (cf. L. *campana*, bell); LL. dim. *nolana*. *Ex:* Nol-idae (Ins.); Nola (Ins.); Nolana*;

Nolan-ea*; not Nolin-a*, a plant name honoring P. C. Nolin, French agricultural writer.

**nom**—1. Gr. *nomos*, law, order, custom. *Ex:* Nom-arthra (Mam.); Nom-onyx (Av.): 2. Gr. *nomas*, genit. *nomados*, roaming about for pasture, nomads < *nomos*, a meadow, pasture, abode. *Ex:* Nomad-ita (Ins.); Nomada (Ins.); Nomia (Ins.): 3. Gr. *nomos*, a pasture, an abode; *nomē*, a feeding. *Ex:* Nome-bius (Ins.); Nomo-tettix (Ins.); Chreo-noma (Ins.); Eremo-nomus; My-nomes (Mam.); Nycti-nomus (Mam.); Nycti-nom-ops (Mam.).

**nomad**—See nom 2.

**nome**—Gr. *nomeus*, a herdsman. *Ex:* Nome-idae (Pisc.); Nomeus (Pisc.).

**non**—L. *nonus*, ninth.

**nonn**—Gr. *nonnos*, a monk; *nonna*, a nun. *Ex:* Nonn-ula (Av.); Nonnus (Ins.).

**nopacht**—NL. *nopachtus*, anagram of Panochthus. *Ex:* Nopachtus (Mam.).

**nopal**—Sp. *nopal*, a cactus name. *Ex:* Nopalea*.

**nops**—Gr. *nōps*, genit. *nōpos*, wholly blind. *Ex:* Nops (Arach.); Oo-nops (Arach.).

**norm**—L. *norma*, rule; *normalis*, according to rule, in a straight way or line. *Ex:* norm-al normo-cyte.

**nos**—Gr. *noseō*, to be sick, *nosēma*, sickness; *nosōdēs*, making sick; *noseros*, unhealthy. *Ex:* Nosema (Prot.); Noserus (Ins.); Noso-dendr-idae (Ins.); Noso-xylon (Ins.).

**noser**—See nos.

**nost**—Gr. *nostos*, a return home; pleasantness; *nostimos*, that will return, desirable; also productive. *Ex:* nost-algia (Med.); Nostima (Ins.); Nosto-ceras (Moll.); nosto-logy.

**nostim**—See nost.

**nostoc**—NL. *nostoc*, name of a plant, prob. an invented name. *Ex:* Nostoc*.

**nostr**—L. *nostras*, genit. *nostratis*, of our land. *Ex:* nostrat-ic.

**not**—1. Gr. *nōtos*, the back. *Ex:* Not-ommata (Rot.); noto-chord; Noto-phorus (Mam.); Noto-lophus (Ins.); Noto-necta (Ins.); noto-podi-um; Noto-pteris (Mam.); notum; Chaeto-not-oidea (Rot.); Scaphi-notus (Ins.): 2. Gr. *notos*, the south wind, the south quarter; *notothen*, from the south. *Ex:* Not-elaea*; Not-elephus (Mam.); Not-ictus (Mam.); Not-ornis (Av.); Notohippus (Mam.); Noto-pithecus (Mam.); Notothen-idae (Pisc.); Noto-therium (Mam.).

**not-**—L. *not-*, prefix meaning not. *Ex:* Not-acantha (Ins.).

**notabil**—L. *notabilis*, noteworthy. *Ex:* Notabilia (Ins.).

**notacm**—Gr. *notakmon*, with armored back.

**notat**—L. *notatus*, marked, distinguished, perceptible < *noto*, to mark. *Ex:* Notata (Ins.); in-notatus.

**noteo**—Gr. *notios*, southern. *Ex:* Noteo-saurus (Rept.); Noteo-suchus (Rept.). See also notio.

**noter**—Gr. *noteros*, moist, damp. *Ex:* notero-phil-ous; Noterus (Ins.).

**noteus**—Gr. *nōteus*, one who carries on his back. *Ex:* Noteus (Rot.).

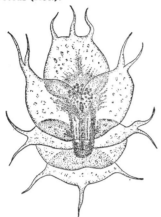

Elated Back-wing Bearer, *Notoptero-phorus elatus*, a copepod. Redrawn from Smithsonian Scientific Series.

**noth**—Gr. *nothos*, spurious, bastard. *Ex:* Noth-arctus (Mam.); Noth-olca (Rot.); Nothi-erax (Av.); Notho-cyon (Mam.); Notho-laena*.

**nothr**—Gr. *nōthros*=*nōthēs*, sluggish, slothful. *Ex:* Nothro-phontes (Av.); Nothro-pus (Mam.); Nothro-therium (Mam.).

**nothus**—L. *nothus*, false, counterfeit. See noth.

**notial**—L. *notialis*, southern.

**notid**—Gr. *notis*, genit. *notidos*, moisture, rain. *Ex:* Notido-bia (Ins.).

**notidan**—Gr. *nōtidanos*, with pointed dorsal fin. *Ex:* Notidan-oidea (Elas.); Notidanus (Elasm.).

**notio**—1. Gr. *notios*, wet, damp, moist; *notiōdēs*, wet, moist. *Ex:* Notio-myia (Ins.); Notio-philus (Ins.): 2. Gr. *notios*, southern, from the south. *Ex:* Notio-cetus (Mam.); Notio-myia (Ins.); Notio-sorex (Mam.).

**notothen**—See not 2.

**nous**—Gr. *noos*, contr. *nous*, mind. *Ex:* A-nous (Av.).

**nov**—L. *novus*, new.

**novacul**—L. *novacula*, a razor. *Ex:* Novacula (Pisc.).

**novem**—L. *novem*, nine. *Ex:* novem-fida.

**noverc**—L. *noverca*, a stepmother.

**noxios**—L. *noxiosus*, injurious

**nu**—See neo.

**nub**—1. L. *nubes*=*nubis*, a cloud; *nubilus*, cloudy: 2. L. *nubo*, to marry. *Ex:* Pro-nuba (Ins.).

**nubecula**—L. *nubecula*, a small cloud, dim. of *nubes*. *Ex:* Nubecula (Moll.); Nubeculo-spira (Prot.).

**nubil**—See nub.

**nuc**—L. *nux*, genit. *nucis*, dim. *nucella*, nut, kernal. *Ex:* Nuc-ula (Moll.); Nucella (Moll.); Nuces (Echin.); Nuci-fraga (Av.).

**nuch**—ML. *nucha*, neck; *nuchalis*, of the neck. *Ex:* nuchal; ligamentum nuchae.

**nucle**—L. *nucleus*, a little nut, the kernel, the inner part. *Ex:* Nucle-ar-ina (Prot.); Nucle-aria (Moll.); nucle-olus; Nucleo-spira (Brach.); nucleus.

**nud**—L. *nudus*, naked; *nudatus*, stripped of its cover, uncovered. *Ex:* Nuda (Cten.); Nudi-branchiata (Moll.); nudi-caulis; Nudi-spongia (Por.); Nudo-bius (Ins.); Nudo-chernes (Arach.); de-nudatus.

**nugac**—See nugator.

**nugator**—L. *nugator*, a jester; *nugax*, genit. *nugacis*, jesting, joking. *Ex:* Nugator (Ins.).

**nugax**—See nugator.

**null**—L. *nullus*, not any, none. *Ex:* nulli-caulus; Nulli-crinis (Crust.); Nulli-pora (Coel.).

**numen**—Gr. *noumēnios*, happening at the time of the new moon; also a kind of curlew. *Ex:* Numenes (Ins.); Numenia (Moll.); Numenius (Av.).

**numid**—L. *numida*, a Numidian. *Ex:* Numidi-cola (Ins.), lit. living on Numida; Numida (Av.).

**numin**—L. *numen* genit. *numinis*, a nodding of the head, a command. See also numen.

**nummul**—L. *nummulus*, money<*nummus*, a coin. *Ex:* Nummu-lites (Prot.), see -lite; Nummul-aria (Prot.).

**nunc**—L. *nuncia*=*nuntius*, a messenger. *Ex:* inter-nunci-al.

**nuneche**—Gr. *nounecheia*, discretion; *nounechēs*, with good sense, discreet, wise. *Ex:* Nuneches (Ins.).

**nunt**—See nunc.

**nuphar**—Gr. *nouphar*, name of a medicinal plant, perh. a water-lily. *Ex:* Nuphar*.

**nuptial**—L. *nuptialis*, pertaining to a marriage.

**nut**—L. *nutans*, genit. *nutantis*, nodding, ppr. of *nuto*, to nod; *nutatio*, a nodding. *Ex:* nutation.

**nutan**—See nut.

**nutr**—1. L. *nutrix*, dim. *nutricula*, she who nourishes, a nurse: 2. Sp. *nutra*=*nutria*, an otter <L. *lutra*. *Ex:* Nutria (Mam.).

**nutrit**—L. *nutrio*, to feed, rear, nourish; *nutritus*, nourishing. *Ex:* nutrit-ion.

**nych**—See onych.

**nyct**—Gr. *nyx*, genit. *nyktos*, night; *nykteus*, nocturnal; *nyctalops*, that sees at night; *nyktalos*, sleepy; *nyktios*, nightly. *Ex:* Nyct-ago*, -ago as in Plantago*; Nyct-anassa (Av.); Nyctala (Av.); Nyctalops (Arach.); Nyctea

(Av.); Nycti-corax (Av.); Nycti-ornis (Av.); Nycti-ellus (Mam.); nycti-pelagic; Nyctophilus (Mam.); Calo-nyction*.

**nyctal**—See nyct.

**nycter**—Gr. *nykteros*, nocturnal>*nykteris*, genit. *nykteridos*, a bat; *nyktereutēs*, one who hunts by night; *nykterinos*, belonging to the night; *nykterios*, nocturnal. *Ex:* Nycter-bia (Ins.); Nyctereutes (Mam.); Nycteri-bia (Ins.); Nycterinia*; Nyctero-bius (Mam.); Balionycterus (Mam.).

**nygm**—Gr. *nygma*, genit. *nygmatos*, a puncture, a prick, a sting. *Ex:* Nygmatia (Dipt.); Nygmat-onchus (Nemat.); Philo-nygmus (Ins.).

**nym**—See onym.

**nymph**—L. *Nympha*, goddess of waters, meadows and forest<*nymphē*, a bride, a nymph; Gr. *nymphaios*, sacred to the nymphs; *nymphaia*, the water-lily; L. *nymphalis* of or pertaining to a fountain. *Ex:* nymph; nymphal; Nymph-aster (Echin.); Nymph-ophidium (Rept.); Nymphaea*; Nymphal-ites (Ins.); Nympho-troctes (Ins.); Nympho-stola (Ins.).

**nymphon**—Gr. *nymphōn*, the bridechamber. *Ex:* Nymphon (Arach.); Nymphon-ella (Ins.).

**nyroca**—NL. *nyroca*<Russ. *nirok*, a goosander, merganser. *Ex:* Nyroca (Av.).

**nysi**—Gr. *Nysios*=*Nysaios*, of Nysa, name of a number of sites sacred to Bacchus. *Ex:* Nysius (Ins.).

**nyss**—Gr. *nyssō*=Attic. *nyttō*, to prick, stab, afflict. *Ex:* Nyss-odon (Mam.); Nysso-notus (Ins.); Nysson (Ins.); Nyttum (Ins.): Liponyssus (Arth.).

**nyssa**—1. Gr. *nyssa*, "name of some tree"; also a goal, starting post. *Ex:* Nyssa*: 2. L. *Nyssa*=*Nysa*, the nurse of Bacchus. *Ex:* Nyssa (Arach.).

**nystactes**—Gr. *nystaktēs*, a nodder, sleeper< *nystazō*, to nod. *Ex:* Nystactes (Mam.), (Av.).

**nystal**—Gr. *nystalos*, nodding, drowsy; *nystaleos*, drowsy.

**nyth**—Gr. *nythos*, dumb. *Ex:* Nytho-phona (Ins.); Nytho-saurus (Rept.).

**nytt**—See nyss.

**nyx**—Gr. *nyxis*, genit. *nixeōs*, a pricking, puncture. *Ex:* Nyxeo-philus (Ins.).

**nyxetes**—NL. *nyxetes*, a piercer, one who punctures<Gr. *nyxis*, a piercing, +-*ētēs*, one who. *Ex:* Nyxetes (Ins.).

**nyxis**—See nyx.

# O

**o**—Gr. *ōion*=*ōon*, egg. *Ex:* O-idium*; o-oecia. See also oo; pan-o-istic; hol-o-istic.

**oa**—Gr. *oa*=*ōa*, a border, fringe. *Ex:* Dactyl-oa (Rept.).

**oari**—1. Gr. *ōiarion*, a small egg. *Ex:* Oariopherus (Ins.); Oario-stylus (Ins.): 2. NL. *oario-*<*ovario-*<NL. *ovarium*. *Ex:* oario-cele; oario-tomy (Surg.).

**oarism**—Gr. *oarisma*, genit. *oarismatos*, familiar discourse. *Ex:* Oarisma (Ins.).

**ob-**—L. *ob*, over, against, toward, facing (usually changed before *c*, *f*, *g* and *p* to *oc*, *of*, *og* and *op*). *Ex:* Ob-ovo-thyris (Brach.); Ob-ovatus (Moll.); ob-plete; ob-tect; Ob-tortio (Moll.); oc-cipital; op-plete, see pleo.

**obbat**—NL. *obbatus*, beaker-shaped<*obba*, a beaker.

**obel**—1. Gr. *obelias*, a round cake. *Ex:* Obel-aria (Coel).; Obelia (Coel.): 2. Gr. *obelos*, a spit,

dagger; also a horizontal line; *obeliskos*, an obelisk, a dagger, blade. *Ex:* Obel-ura (Ins.); Obeliscus (Prot.); Obelo-streptus (Myr.); Obelo-phorus (Ins.).

**oberon**—O.Ger. *Oberon*, king of the fairies. *Ex:* Oberonia*; Oberonus (Ins.).

**obes**—L. *obesus*, stout, fat, pp. of *obedo*, to eat one's self fat. *Ex:* Obes-ula (Moll.); obese; Obeso-mon-omma (Ins.).

**obex**—See obic.

**obfuscat**—L. *obfuscatus*, darkened.

**obic**—L. *obex*, genit. *obicis*, a barrier, bar, wall. *Ex:* Obex (Moll.); obices (Ecol.).

**obion**—NL. *obione*, a plant name. "Origin of the word unknown unless from the river Obi, in Siberia, whence the original species came." *Ex:* Obione*.

**obis**—NL. *obisium*, name for a genus of spiders. *Ex:* Obisi-phaga (Ins.); Obisium (Arach.).

objicien—L. *objiciens*, objecting.

obligat—L. *obligatus*, obliged. *Ex:* obligate; obligat-ive.

obliqu—L. *obliquus*, slanting sidewise. *Ex:* Obliqu-aria (Moll.); Obliqui-pecten (Moll.).

oblit—1. L. *oblitus*, forgotten, pp. of *obliviscor*, to forget: 2. L. *oblitus*, besmeared, pp. of *oblino*, to smear over.

oblitterat—L. *oblitteratus*, erased, obliterated, forgotten.

oblong—L. *oblongus*, somewhat long, oblong.

obmutescens—L. *obmutescens*, keeping silence.

obnupt—L. *obnuptus*, covered, veiled <*obnubo*, to cover with a veil.

obol—Gr. *obolos*, a small coin; worthless. *Ex:* Obol-aria*; Obol-ella (Brach.); Obolo-cera (Ins.); Obolus (Brach.).

obpallens—NL. *obpallens*, tending towards yellow, yellowish, fading.

obri—Gr. *obria=obrikala*; the young of animals. *Ex:* Obricala (Ins.); Obrio-morpha (Ins.); Obrium (Ins.); Piez-obria (Ins.).

obrical—See obri.

obrim—Gr. *obrimos*, strong. *Ex:* Obrimus (Ins.).

obruss—L. *obrussa*, a testing of gold by fire, a proving.

Long-leaved Piñon Brickellia, *Brickellia oblongifolia*, of the desert mountains of the S. W. United States. The genus *Brickellia* was named after Dr. J. Brickell, early botanist of Savannah, Georgia. Redrawn from Desert Wild Flowers—Jaeger. Stanford University Press.

obrut—L. *obrutus*, thrown down, pp. of *obruo*, to throw or cast down. *Ex:* obrute.

obscur—L. *obscurus*, dusky. *Ex:* Obscura (Moll.); Obscur-ella (Moll.).

obscurat—L. *obscuratus*, darkened, obscured; pp. of *obscuro*, to make dark, to obscure. *Ex:* obscurat-or. See -or.

obsit—L. *obsitus*, barred, pp. of *obsero*, to make secure.

obsolesc—L. *obsolescens*, genit. *obsolescentis*, thrown out, worn out; ppr. of *obsolesco*, to wear out gradually.

obsolet—L. *obsoletus*, effaced, worn out, indistinct, without clear markings.

obstetric—L. *obstetrix*, genit. *obstetricis*, midwife < *obsto*, to stand facing or before. *Ex:* obstetric; Obstetricans (Amph.); Obstetrix-ella (Ins.).

obtect—L. *obtectus*, covered over.

obturat—L. *obturatus*, closed up, pp. of *obturo*, to stop up. *Ex:* obturat-or foramen.

obtus—L. *obtusus*, blunt, pp. of *obtundo*, to make dull. *Ex:* Obtus-ella (Moll.); Obtusi-palpus (Ins.); Obtusi-termes (Ins.).

obvelat—L. *obvelatus*, hidden, covered <*obvelo*, to conceal.

oc—See ob.

occ—1. L. *occa*, a harrow. *Ex:* Occ-ella (Pisc.); Occa (Pisc.): 2. Gr. *okkos*, an eye. *Ex:* Meni-ocus*.

occidental—L. *occidentalis*, western, westerly.

occidu—L. *occiduus*, setting, sinking, falling below.

occipit—L. *occiput*, genit. *occipitis*, the back part of the head. *Ex:* occipit-al; Occipit-odontus (Nemat.); occiput.

occisor—L. *occisor*, a murderer. *Ex:* Occisor (Ins.).

occlud—See occlus.

occlus—L. *occlusus*, shut up, pp. of *occludo*, to shut, close; *occlusio*, genit. *occlusionis*, a closing, shutting. *Ex:* occlusion.

ocean—L. *oceanus* <Gr. *ōkeanos*, ocean; in Homer, the great river supposed to encompass the earth. *Ex:* Oceania (Coel.); Oceano-droma (Av.); Oceanus (Av.).

ocell—L. *ocellus*, a little eye, dim. of *oculus*, eye; *ocellatus*, spotted as with little eyes. *Ex:* Ocellat-aria (Ins.); ocellat-ed; Ocellata (Ins.); ocellus.

ocellat—See ocell.

*Euphorbia occelata*, Little-eyed Spurge, showing nectiferous glands set like eyes on the calyxlike nvolucre.

och—Gr. *ochos*, anything which bears, a wagon. Och-odontus (Ins.); Ocho-gona (Myr.); ocho-petal-ous Styl-ochus (Platy.).

ochem—Gr. *ochēma*, anything that bears or supports.

oches—Gr. *ocheō*, to continue, to endure, to carry, to sustain. *Ex:* Chelis-oches (Ins.).

ochet—Gr. *ochetos*, a leather water pipe, drain, channel. *Ex:* ochet-ium (Ecol.); Ochet-odon (Mam.); Ocheto-myrmex (Ins.); Ochetus (Echin.); An-ochetus (Ins.).

ochl—Gr. *ochlos*, a moving crowd, a mob; *ochlōdēs*, turbulent, unruly: *ochlēsis*, disturbance. *Ex:* Ochle-rotatus (Ins.); Ochlo-chaete*; Ochlo-genes (Ins.); Ochlodes (Ins.); Ochlodus (Pisc.).

ochler—Gr. *ochlēros*, troublesome, turbulent. *Ex:* Ochlero-ptera (Ins.); Ochlerus (Ins.).

ochm—Gr. *ochma*, genit. *ochmatos*, that which holds, a band. *Ex:* Ochm-acanthus (Pisc.).

ochn—Gr. *ochnē* = *onchnē*, a pear-tree, a pear. *Ex:* Ochna*.

ochotona—NL. *ochotona* < Mongol name of the pika. *Ex:* Ochotona (Mam.).

ochraceus—NL. *ochraceus*, like ocher, pale yellow.

ochro—Gr. *ōchra*, yellow-ochre; *ōchros*, pale. *Ex:* ochro-leucus; Ochro-sidia (Ins.); Ochro-tettix (Ins.); Ochros-ia*.

ochrom—Gr. *ōchrōma*, paleness. *Ex:* Ochroma*.

ochth—Gr. *ochthos* = *ochthē*, rising ground, hill, hump, a mud bank, a sea-shore dune; *ochthōdēs*, humped, warted. *Ex:* Ochthe-dromus (Ins.); Ochthe-phila (Ins.); Ochthera (Ins.); Ochtho-dromus (Av.); ochtho-philus; ochtho-phyta; Ochthod-ium*; Olci-ochthes (Arth.); Pan-ochthus (Mam.).

ochyr—Gr. *ochyros*, firm, stout, strong; *ochyrotēs*, firmness; *ochyrōtikos*, serving to strengthen. *Ex:* Ochyro-cera (Arach.); Ochyrotica (Ins.).

ocim—L. *ocimum* < Gr. *ōkimon*, a sort of clover, an aromatic plant, basil. *Ex:* Ocimum*.

ocn—Gr. *oknos*, sluggish, lazy; also a bittern; *oknēros*, hesitating, timid. *Ex:* Ocner-odes (Ins.); Ocnero-drilus (Ann.); Ocno-therium (Mam.); Heter-ocnus (Av.).

ocner—See ocn.

ocotea—S. Amer. Indian *ocotea*, a plant name. *Ex:* Ocotea*.

ocr—Gr. *okris*, a ridge, summit, a jagged point, any roughness. *Ex:* Ocr-odon (Mam.).

ocrea—L. *ocrea*, a greave or legging; *ocreatus*, wearing leggings, booted. *Ex:* ocreate; Ocreatus (Av.).

oct—L. *octo*, eight. *Ex:* Octa-odon (Mam.); Octo-bunus (Arach.); Octo-pus (Moll.); Cun-oct-antha (Coel.).

octav—L. *octavus*, the eighth.

octonari—L. *octonarius*, consisting of eight parts.

ocul—L. *oculus*, eye. *Ex:* Ocul-ina (Prot.); Oculo-spongia (Por.); Oculus (Echin.).

ocus—See occ.

ocy—Gr. *ōkys*, swift, quick; *ōkyporos*, quick-going; *ōkypetēs*, swift flying; *okyrhoēs*, swift-flowing, swift-moving. *Ex:* Ocy-dromus (Av.); Ocy-phaps (Av.); Ocypetes (Mam.); Ocy-poda (Crust.); Ocyroe = Ocyrhoe (Cten.); Hal-ocy-ptena (Av.).

ocym—NL. *ocymum*, said by Mathiolus to be from *ozō*, to smell. *Ex:* Ocymo-ideae*; Ocymum*.

ocypetes—See ocy.

ocypor—See ocy.

ocyr—See ocy.

-od—NL. *-od-* < Gr. *eidos*, form. *Ex:* Phae-od-aria (Prot.); phae-od-ellum; phae-od-ium.

-oda—See -ode.

-ode—1. Eng. suffix *-ode* < Gr. *-ōdēs* (< *o*+*eides*, like; *eidos*, form), meaning like, a thing like. *Ex:* Camp-odea (Ins.); Luc-odes*; Nemat-oda; Phloe-odes (Ins.); Sarc-odes*. See also oid: 2. *-ode*, suffix < Gr. *hodos*, a way, path. *Ex:* cath-ode; electr-ode. See also odo.

odea—See -ode.

odeg—Gr. *odēgos*, a guide, a teacher.

-odes—See -ode.

odeum—See deum.

odin—Gr. *ōdis*, genit. *odinos*, the pains of travail. *Ex:* par-odinia (Med.).

odites—Gr. *oditēs*, a traveller. *Ex:* Odites (Ins.); Porrh-odites (Ins.).

odm—See osm.

odo—Gr. *hodos*, a way. *Ex:* Odo-nestis (Ins.); Odo-stemon*; Odo-stomi-opsis (Moll.); Di-odia*; not Odobaenus nor Odocerus, see odont.

odoi—Gr. *hodois*, belonging to a way or journey; *hoditēs* = *hodoiporos*, a traveller. *Ex:* Odoiporus (Ins.).

odon—See odont.

odont—Gr. *odous*, genit. *odontos*, tooth. *Ex:* Odo-benus = Odo-baenus (Mam.); Odo-coileus (Mam.); Odon-ata (Ins.); Odont-ites*; Odonto-ceti (Mam.); Cerat-odus (Pisc.); Dys-odonta (Moll.); Dys-odus (Mam.); loph-odont; Tri-odia*.

odor—L. *odorus*, fragrant; *odoratus*, sweet-smelling < *odoro*, to give off fragrance.

odot—Gr. *odōtos*, practicable, feasible.

odur—Gr. *odouros*, a conductor; a pirate. *Ex:* Odur-ella (Av.); Odura (Av.); not Ichthyo-doru-lites (Elasm.), see dory.

odus—See odon.

odyn—Gr. *odynē*, pain; *odynēros*, painful. *Ex:* Odyner-opsis (Ins.); Odynero-myia (Ins.); Odynerus (Ins.).

odyner—See odyn.

oe—Latin *oe* equivalent of Greek *oi* and often rendered as simple *e* in English. *Ex:* oesophagus = esophagus; oestral = estral; oecology = ecology.

**oec**—Gr. *oikos*, house; *oiketēs*, an inhabitant. *Ex:* Oecet-ina (Ins.); Oeco-bius (Arach.); oeco-logy; Oeco-phylla (Ins.); Oiko-micron (Hemichorda); andr-oecium; Heter-oicus (Ins.); mon-oeci-ous; Pedi-oecetes (Av.); syn-oecy; zo-ecium.

**oecet**—See oec.

**oecot**—Gr. *oikotōs*, reasonably, probably. *Ex:* Oecoto-peria (Ins.).

**oed**—Gr. *oidēma*, genit. *oidēmatos*, a swelling, a tumor; *oidos*, a swelling. *Ex:* Oede-cnema (Ins.); Oede-machilis (Ins.); Oedema-peza (Ins.); Oedemat-archa (Ins.); Oedemato-cera (Ins.); Oidemia (Av.); Oedi-cephalus (Ins.); Oedi-cnemus (Av.); Oedo-gonium*; Oedo-thorax (Arach.); oedema=edema.

**oedem**—See oed.

**oedip**—Gr. *Oidipous*, swollen-footed, Greek hero who solved the riddle of the Sphinx. *Ex:* Oedip-midas (Mam.); Oedipus (Mam.).

**oeg**—Gr. *oigō*, open. *Ex:* Oeg-ops-id (Moll.); Oego-conia (Ins.); Oego-phymia (Por.); not Oegoceros (Mam.) nor Oegocera (Moll.) nor Oegocera (Ins.), see aeg.

**oen**—**1.** Gr. *oinos*, wine-colored; *oinopoieō*, to make wine; *oinanthē*, name of a kind of plant; *oinothēras*, a kind of willow-herb, the root of which smelled like wine. *Ex:* Oenanthe*; oeno-cytes; oeno-logy=eno-logy; Oenopo-pelia (Av.); Oenothera*: **2.** Gr. *oinas*, a wild pigeon. *Ex:* Oena (Av.); Cal-oenas (Av.); Cal-oenad-idae (Av.); Mes-oenas (Av.).

**oenothera**—See oen.

**oes**—Gr. *oisos*, the willow. *Ex:* Oeso-cerus (Ins.). See also ois 2.

**oesophagus**—See ois 2.

**oest**—Gr. *oistos*, an arrow, shaft. *Ex:* Oesto-phora (Moll.).

**oestr**—Gr. *oistros*, a strong desire; orgasm; the sting of a gadfly, anything that drives one mad; *oistrēlatos*, goaded on as by a gadfly. *Ex:* estral; Oestrelata (Av.); Oestro-phasia (Ins.); Oestrus (Ins.); met-estrum.

**of**—See ob.

**officin**—L. *officinalis*, of or pertaining to an office, shop; *officina*, an office.

**og**—See ob.

**ogc**—See onc.

**oglin**—NL. *oglinum*, an unexplained name applied to the thread herring by Le Sueur.

**ogm**—Gr. *ogmos*, straight line, a furrow. *Ex:* Ogmo-balaena (Mam.); Ogmo-rhinus (Mam.); Dis-ogmus (Ins.); Tele-ogmus (Ins.).

**ogyg**—**1.** *Ōgygēs*, a legendary king of Athens in whose reign there was a flood. *Ex:* Ogyg-opsis (Arach.): **2.** Gr. *Ōgygia*, mythical island home of Calypso. *Ex:* Ogygia (Tri.).

**oic**—Gr. *oikos*, house, home. *Ex:* Oico-monas, (Prot.). See also oec.

**-oid**—NL. *-oid*=*-oides*, a contraction of Gr. *-o+eidos*, denoting likeness of form, a thing that is like. *Ex:* dipl-oid; disc-oid; Erio-camp-oides (Ins.); odont-oid. See also ode and -oed.

**-oidea**—**1.** Gr. *-ōideos*=L. *-oideus*, adj. suffix meaning form of, type of. *Ex:* Sepiol-oidea (Moll.): **2.** Gr. *-oidea*>NL. *-oidea*, suffix used in making a super-family name from a family name, as Ammon-oidea (Moll.); Sipuncul-oidea (Moll.).

**-oides**—See -oid.

**oig**—See oeg.

**oiko**—See oec.

**oin**—See oen.

**oio**—Gr. *oios*, alone, only, one. *Ex:* Oio-rrhinus (Ins.); Oio-zona (Ins.). See ois, also o.

**ois**—**1.** Gr. *ois*, genit. *oios*, sheep. *Ex:* Oio-ceros (Mam.); Pseud-ois (Mam.): **2.** Gr. *oisō*, I shall carry. *Ex:* oesophagus<*oisō*+*phagēton*, food; Oesophagi-cola (Platy.); Stom-oisia*.

**oist**—Gr. *oistos*, an arrow. *Ex:* Oisto-phora (Ins.); Oistus (Ins.).

**oistic**—NL. *oistic*, pertaining to an egg<Gr. *ōon*, an egg +NL. *-istic*, pertaining to as agent. *Ex:* pan-oistic, producing only eggs.

**ol**—**1.** L. *olor*, a smell, odor<*oleo*, to smell. *Ex:* ol-factory: **2.** Gr. *holos*, whole, entire. *Ex:* olo-petal-arius; Olo-phrinus (Ins.); Olo-stylia*; Anis-ol-ornis (Av.): **3.** Gr. *olos*, mud, also the ink of Sepia. See also -olus.

**-ola**—See olus.

**olax**—L.L. *olax*, odorous. *Ex:* Olax*.

**olbi**—**1.** Gr. *olbios*, blessed, happy, wealthy. *Ex:* Olbi-orchilus (Av.); Olbius (Ins.): **2.** *Olbia*, a name for several cities. *Ex:* Olbia (Crust.).

**olbodot**—Gr. *olbodotēs*, a giver of bliss. *Ex:* Olbodotes (Mam.).

**olc**—See holc.

**ole-**—See olus, also olen 2.

**olea**—L. *olea*, olive<Gr. *elaia*, the olive<*leios*, smooth; *oleaginus*, of the olive. *Ex:* Ole-acea*; Olea*; not Olearia*, prob. named after Adam Olearius, German traveller; oleaginous.

**oleagin**—See olea.

**oleaster**—L. *oleaster*, the wild olive tree. *Ex:* Oleaster*.

**olecran**—Gr. *ōlekranon*, the elbow. *Ex:* olecran-al; olecranon.

**olen**—**1.** L. *Olenus*=Gr. *Ōlenos*, husband of Lethaea who with her was changed into a stone. *Ex:* Olen-ellus (Tri.); Olenus (Tri.): **2.** Gr. *ōlenē*, arm, elbow. *Ex:* ole-cranon; Olen-opsis (Mam.); Dicro-olene (Pisc.); Helico-olenus (Pisc.); Sarc-olene (Moll.); Tri-olena (Prot.).

**olens**—L. *olens*, genit. *olentis*, smelling, sweet-smelling, ppr. of *oleo*, to smell. *Ex:* grave-olens; red-olent, see redolens.

olent—See olens.

oleos—L. *oleosus*, oily, full of fat.

oler—Gr. *oleros*, impure, turbid.

oleraceus—L. *oleraceus*, resembling herbs, vegetable.

oles—Gr. *olesai*, to destroy, ruin; *ōlesis*, destruction. *Ex:* Olesi-campe (Ins.); Olesi-coccus (Ins.).

olethr—Gr. *olethros*, death; *olethrios*, deadly, destructive. *Ex:* Olethria (Ins.); Olethro-dotis (Ins.); Olethrus (Ins.).

olfact—See ol.

olibr—Gr. *olibros*=*olisthēros*, slippery, hard to catch. *Ex:* Olibro-porus (Ins.); Olibro-soma (Ins.); Olibrus (Ins.).

olic—Gr. *olikos*, universal, general.

olid—L. *olidus*, emitting a smell, stinking. *Ex:* Olido-sus (Mam.).

olig—Gr. *oligos*, few, small; as a prefix often used to denote Oligocene Age or derivation. *Ex:* Olig-odon (Rept.); Oligo-bunus (Mam.); Oligo-

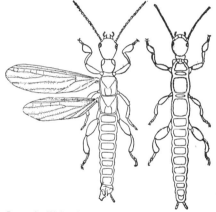

Gurney's Web-spinner, *Oligotoma Gurneyi*, male and female, a New Zealand Embiarid.

cene; Oligo-chiton (Moll.); oligo-tropic; Olyg-odon (Mam.); Mer-oligon (Ins.); Zo-oligus (Mam.).

oligor—Gr. *oligōreō*, to esteem lightly. *Ex:* Oligoria (Ins.); Oligorus (Pisc.).

olinx—NL. *olinx*, a furrow<a supposed Gr. *olinx*. *Ex:* Olinx (Ins.); Para-olinx (Ins.).

olisb—Gr. *olisbos*, penis coriaceus. *Ex:* Olisbea*.

olisth—Gr. *olisthos*, slipperiness; *olisthēros*, slippery, tottering; *olisthanō*, to slip. *Ex:* olisthium (Ecol.); Olisthan-ella (Platy.); Olistherus (Ins.); Olistho-pus (Ins.).

olitori—L. *olitorius*, of or pertaining to vegetables.

oliv—L. *oliva*, an olive; ML. *olivascens*, growing olivaceous. *Ex:* Oliv-ancillaria (Moll.) Oliv-ella (Moll.); Oliv-ina (Prot.); Oliva (Moll.)· oliv-ary body; Spong-oliva (Prot.).

olivari—L. *olivarius*, pertaining to the olive.

olix—Gr. *olixoō*, to make a little less. *Ex:* Olixon (Ins.).

oll—1. L. *olla*,dim. *ollicula*,a pot or jar. *Ex:*Ollicula (Arth.):      2. Gr. *ollos*, Ionic crasis for *o allos* =*allos*, other, of a different sort. *Ex:* Oll-otis (Amph.); Ollo-chirus (Rept.).

ollicul—See oll.

ollul—L. *ollula*, a little pot <*olla*, a pot.

ololyg—Gr. *ololygōn*, the croaking of the male frog. *Ex:* Ololygon (Av.).

olor—L. *olor*, genit. *oloris*, a swan. *Ex:* Olor (Av.).

-olum—See -olus.

-olus—L. *-olus*, *-a*, *-um*, diminutive suffix=Eng. *-ole*. *Ex:* atri-olum; aure-olum; Corrigi-ola*; Gladi-olus*; nucle-olus; oste-ole, peti-ole; Pyrola*; Sepi-ola (Moll.).

olyg--See olig.

olynth—Gr. *olynthos*, a fig that remains unripe. *Ex:* Olynth-ella (Por.); Olyntha (Por.); Olyn tho-scelis (Por.); Olynthus (Por.).

olyr—Gr. *olyra*, name for some grass. *Ex:* Olyra*; olyr-oides.

om—1. Gr. *ōmos*, raw, crude, savage. *Ex:* Om-orgus (Ins.); omo-phagic:      2. Gr. *ōmos*, the shoulder. *Ex:* omo-hyoid; Omo-mys (Mam.); Omo-saurus (Rept.); Opisth-ome (Pisc.); for somewhat similar ending see homo:      3. Gr. *omos*, one and the same, common:      4. Gr. *omōs*, likewise, alike.

-oma—Gr. *-ōma*, suffix used in medicine to denote morbid condition of some part; usually used in ref. to a tumor. *Ex:* carcin-oma; fibr-oma; glauc-oma.

omad—Gr. *omados*, noise.

omal—Gr. *homalos*, level, flat; *homalēs*, level, even. *Ex:* Omalo-pleurus (Pisc.); Ant-omalus (Ins.); Pter-omalus (Ins.); Saur-omalus (Rept.). See also homal.

omasum—L. *omasum*, stomach of a bullock. *Ex:* ab-omasum.

ombr—Gr. *ombros*, a rain storm, rain. *Ex:* ombro-phile; ombro-phob-ous; Pter-ombrus (Ins.).

omeg—Gr. *ō mega*, last letter of the Greek alphabet. *Ex:* Omeg-odus (Mam.); Omega-syrphus (Ins.).

oment—L. *omentum*, the fat skin, membrane. *Ex:* omento-pexy (Med.); omentum.

omil—Gr. *homilos*, a being together, intercourse; *homilos*, a crowd; *homileō*, to associate with. *Ex:* Omil-urus (Ann.).

omm—Gr. *omma*, genit. *ommatos*, eye; also an appearance, aspect; *ommatosterēs*, blind, robbed of eyes. *Ex:* Omma-strephes (Moll.); ommat-idium; Ommato-istius; Ommato-phoca (Mam.); Ommatoster-g-us (Mam.), see erg; A-stiphr-omma (Ins.); ari-ommus; Hali-omm-ura (Prot.); Lox-omma (Amph.); Pachyl-omma (Ins.).

**ommat**—See **omm.**

**omni**—L. *omnis*, all; generally used in sense of unrestricted. *Ex:* Omni-rete-pora (Bry.); omni-vor-ous.

**omoea**—See **homoe.**

**omoi**—See **homoe.**

**omophr**—Gr. *omophrōn*, in agreement, united. *Ex:* Omophron (Ins.).

**omor**—Gr. *homoros*, having the same borders with. *Ex:* Omor-ophius (Ins.); Di-omorus (Ins.). See also homor.

**omot**—Gr. *omotēs*, rudeness, fierceness, crudity. *Ex:* Omotes (Ins.).

**omphal**—Gr. *omphalos*, the navel; *omphalōdēs*, like a navel. *Ex:* Omphal-ina (Moll.); Omphalea*; Omphalius (Moll.); Omphalophora (Ins.); Omphalodes*; omphalod-ium; Chrys-omphalus (Ins.); Eu-der-omphalus (Ins.).

**on**—See **ono.**

**onan**—*Onan*, the son of Judah. See Gen. XXXVIII, 9. *Ex:* onan-ism.

**onagr**—1. Gr. *onagros*=L. *onager*=*onagrus*, the wild ass. *Ex:* Onagr-odes (Ins.): 2. Gr. *onagra*, a kind of plant. *Ex:* Onagr-aceae*; Onagra*. See anogra, an anagram of Onagra.

**onax**—Gr. *anax* (=*onax*), genit. *anaktos*, a king. *Ex:* Cyno-my-onax (Mam.); Empid-onax (Av.).

**onc**—1. Gr. *onkos*, (sometimes wrongly transliterated *ogkos*), a protuberance, tubercle; *onkēros*, tumid, swollen; *onkinos*, a hook. *Ex:*

Batlike Barb-head, *Ogcocephalus vespertilio*, a batfish from the warm seas of Florida. Redrawn from Fishes of North and Middle America—Jordan.

Onc-idium* (Moll.); Onchi-doris (Moll.); Onchus (Pisc.); Onci-deres (Ins.); Oncino-labes (Echin.); Oncinus*; Onco-ceras (Moll.); Ogcocephalus (Pisc.): 2. NL. *onca* (>some native name), specific name of the spotted cat, *Felis onca* of tropical America. *Ex:* Onci-felis (Mam.); Onco-ides (Mam.).

**onch**—See **onc.**

**oncin**—See **onc.**

**ondatr**—N.A. Indian *ondatra*, name for the muskrat. *Ex:* Ondatra (Mam.).

**oneir**—Gr. *oneiros*, a dream; *oneirōdēs*, dreamlike. *Ex:* oneiro-dynia; Oneiro-phantus (Echin.); Oneirodes (Pisc.).

**oneista**—Gr. *onēistos*, most useful, superl. of *oneios*, useful. *Ex:* Oneista (Ins.).

**onik**—See **onych.**

**onisc**—See **ono.**

**ono**—Gr. *onos*, dim. *oniskos*, an ass; also a name applied to several kinds of insects. *Ex:* On-osma*; Oniscus (Crust.); Onisco-myia (Ins.); Ono-brychis*; Ono-hipp-idium (Mam.); Onotragus (Mam.); Camp-oniscus (Ins.); hemionus.

**onoclea**—Gr. *onokleia*, a plant name<*onos*, a vessel+*kleō*, to close. *Ex:* Onoclea*.

**onon**—Gr. *onōnis*, name of a leguminous plant, the rest-harrow. *Ex:* Ononis*.

**onopordon**—Gr. *onopordon*, the cotton-thistle. *Ex:* Onopordon*.

**ont**—1. Gr. *on*, genit. *ontos*, a being. *Ex:* ontogony; Onto-cetus (Mam.); spor-ont: 2. Gr. *ōntos*, actually, verily.

**onth**—Gr. *onthos*, manure. *Ex:* Onth-oecus (Ins.); Ontho-lestes (Ins.); Ontho-philus (Ins.).

**onust**—L. *onustus*, loaded down, burdened, pp. of *onusto*, to load.

**onych**—Gr. *onyx*, genit. *onychos*, nail, talon, claw. *Ex:* Onychi-urus (Ins.); Onycho-mys (Mam.); Onycho-tes (Av.); Onikia (Moll.); Dicrost-onyx (Mam.); Homal-onychus (Arach.); Petal-onyx*; Tri-onyx (Rept.).

**onym**—Gr. *onyma*=*onoma*, a name; also of good name, lucky. *Ex:* Onym-acris (Ins.); Callionymus (Pisc.); Eu-onymus*.

**onyx**—See **onych.**

**oo**—Gr. *ōon*=*ōion*, an egg. *Ex:* o-oecia; O-idium*; Oo-dectes (Mam.); oo-genesis, oo-sperm; Oon-ops (Arach.); Mon-oo-phorum (Platy.).

**oon**—See **oo.**

**ootoc**—Gr. *ōotokos*, laying eggs. *Ex:* Diplootocus (Av.).

**op-**—See **ob-.**

**op**—1. Gr. *ops*, genit. *opos*, voice>Gr. *Kalliopē*, the beautiful voiced, the Muse presiding over eloquence. *Ex:* Calli-ope (Av.); see calliope: 2. Gr. *opē*, a hole, chink, opening. *Ex:* Opegrapha*; Opo-crates (Ins.); Eri-ope*; Liti-opa (Moll.); Heter-opia (Por.).

**opac**—L. *opacus*, shaded, shady.

**opacat**—L. *opacatus*, covered, shaded.

**opad**—Gr. *opados*, an attendant; *opadēsis*, a following after, pursuit. *Ex:* Opado-thrips (Ins.).

**opalin**—ML. *opalinus*<L. *opalus*, an opal. *Ex:* Opalin-opsis (Prot.); Opalina (Prot.).

**-ope**—See **op.**

**opeas**—See **opeat.**

**opeat**—Gr. *opeas*, genit. *opeatos*, dim. *opētion*, an awl. *Ex:* Opeas (Moll.); Opeti-odon (Rept.); Opetio-ptila (Av.); Syn-opeas (Ins.); Tom-opeas (Mam.).

**operan**—L. *operans*, genit. *operantis*, active, efficient, ppr. of *operor*, to work.

operari—L. *operarius*, a laborer<*opera*, work, toil.

operat—L. *operator*, a worker.

opercul—L *operculum*, a cover, lid<*operio*, to conceal, to shut up. *Ex:* opercul-ate; Opercul-ina*; Operculi-phorus (Ins.); operculum.

opert—L. *opertus*, hidden<*operio*, to hide.

opeti—See opeat.

ophel—Gr. *ophelos*, use, help; *ōpheleia*, service; *ōphelimos*, useful. *Ex:* Opheli-mimus (Ins.), mimic of Ophelimus (Ins.); Ophelia*; Ophelimus (Ins.); Ophelos-ia (Ins.); Ophelus*.

ophelim—See ophel.

opheo—NL. *opheo-*<Gr. *ophis*, a snake. *Ex:* Opheo-drys (Rept.).

ophi—Gr. *ophis*, genit. *opheōs*, dim. *ophidion*, a snake, a serpent; *ophioneos*, belonging to, or like a serpent; *ophiōn*, name of a fabulous animal. *Ex:* Ophi-acantha (Echin.); Ophi-saurus (Rept.); Ophi-ur-oidea (Echin.); Ophid-ascaris (Nemat.); Ophidia (Rept.); Ophidio-cephalus (Rept.); Ophio-glossum*; Ophion (Ins.); Agath-ophiona (Ins.); Gymn-ophiona (Ins.); Gymn-ophiona (Rept.); Pitu-ophis (Rept.);=Pity-ophis (Rept.); Thamn-ophis (Rept.).

ophid—See ophi.

ophion—See ophi.

ophiur—See ophi.

ophius—Gr. *Ophioussa*, Serpent-Island, a name of Cythnos. *Ex:* Ophiusa (Ins.).

ophrus—See ophry.

ophry—Gr. *ophrys*, brow, eyebrow>L. *ophrys*, a plant with two leaves, bifoil. *Ex:* Ophry-aster (Echin.); ophryas; Ophrys*; Ophrysia (Av.); Eu-ophrys (Arach.); Staur-ophrya (Prot.); Semn-ophrys (Ins.).

ophrys—See ophry.

ophthalm—Gr. *ophthalmos*, the eye. *Ex:* oph thalm-ic; Opthalmo-myia (Ins.); Tetr-ophthalmus (Ins.).

opia—See opys.

opic—L. *opicus*, uncouth, rude. *Ex:* Opica (Moll.).

opidn—Gr. *opidnos*, dreaded. *Ex:* Opidnus (Ins.).

opifer—L. *opifer*, helpful.

opilio—L. *opilio*, a shepherd, also a kind of bird >NL. *opiliones*. *Ex:* Opilio (Arach.); Opilio-acarus (Arach.); Opiliones (Arach.).

opim—L. *opimus*, rich, filled, fat, fruitful< *opimo*, to fill full, to make fat.

opimia—L. *Opimia*, unfaithful vestal virgin who was burned alive. *Ex:* Opimia (Crust.).

opinat—L. *opinatus*, conjecture, supposition< *opinor*, to imagine, suppose.

opipar—L. *opiparus*, rich, beautiful.

opis—1. Gr. *opisō*, backwards. *Ex:* Opiso-cardium (Moll.): 2. Gr. *Opis*, a name of Artemis. *Ex:* Opis (Crust.).

opisth—Gr. *opisthen*, behind, at the back; *opisthios*, neut. *opisthion*, the hinder part. *Ex:* Opisth-arthri (Elasm.); opisth-otic; opisthen-ar, see thenar; opisthion; Opistho-branchia (Moll.); opistho-coelous; Mon-opistho-discinea (Platy.).

opisto—See opisth.

opius—See ops.

opl—1. Gr. *hoplon*, genit. *hoplontos*, armor; also a tool; *hoplitēs*, armed; *hōplismenos*, armed. *Ex:* Opla-cerus (Mam.); Ople-gnathus (Pisc.); Oplismenus*; Oplo-therium (Mam.); Oplon-aeschna (Ins.); An-opl-ura (Ins.); Amb-oplites (Pisc.); Anis-oplia (Ins.): 2. Gr. *hoplē*, hoof. See also hopl.

ople—1. Gr. *hoplē*, a hoof. See hopl.: 2. Gr. *hopleō*, to make ready. *Ex:* Sphaer-oplea*.

oplismen—See opl 1.

oplit—See opl.

oplon—See opl.

opo—Gr. *opos*, vegetable juice. *Ex:* Opo-balsamum*; Opo-lemur (Mam.), here *opos*, is used in the sense of "fat"; Opo-panax*; opo-therapy (Med.).

opor—Gr. *opōra*, the end of summer, the time of fruits, also the fruit itself; *opōrinos*, autumnal, belonging to the end of summer. *Ex:* Oporanthus*; Opor-ornis (Av.); Oporinos*; Geopora*.

oporin—See opor.

ops—1. Gr. *opsis*, genit. *opseōs*, aspect, view, appearance. *Ex:* Castan-opsis*; Ichthy-opsida; Oryz-opsis*; Tri-cerat-ops (Rept.): 2. Gr. *ops*, genit. *opos*, the eye, the face. *Ex:* Opsi-ceros (Mam.); Megal-ops (Crust.); Mer mar-opus (Ins.); Sandal-ops (Moll.); Trach-ur-ops (Pisc.): 3. Gr. *opse*=*opsios*, late, after a time. *Ex:* Opseo-trophus (Ins.); opsi-gamy; opsi-mathy: Opsio-myia (Ins.): 4. Gr. *opson*, genit. *opsontos*, cooked meat; also seasoning, sauce, dainties. *Ex:* opsi-uria (Med.); opso-mania (Med.).

opseo—Gr. *opsis*, genit. *opseōs*, a sight, appearance. *Ex:* opseo-spermata. See also ops 3.

opsigon—Gr. *opsigonos*, late-born. *Ex:* opsigon-y.

opsio—See ops 3.

opsopoe—Gr. *opsopoieō*, to feed on dainties. *Ex:* Opsopoe-odus (Pisc.); Opsopoea (Ins.).

opt—Gr. *optēr*, genit. *optēros*, an explorer, a spy. *Ex:* Megal-opta (Ins.).

optabil—L. *optabilis*, wished for, desired.

optat—L. *optatus*, pleasing, desired.

optes—See copt.

optic—Gr. *optikos*, of or for sight>*opto-*, a combining form. *Ex:* optic; optico-cilliary; Optico-pteryx (Ins.); opto-gram; opto-metry.

optim—L. *optimus*, superl. of *bonus*, good, i.e. best. *Ex:* optimum.

opto—See optic.

opul—1. L. *opulus*, a kind of maple. *Ex:* Opulus*: 2. *opulens*, genit. *opulentis*, becoming rich; also fine, splendid; *opulentus*, rich, wealthy.

opunti—NL. *opuntia*, name of a cactus, said to be derived from Gr. *Opous*, genit. *Opountos*, a town in Greece, where a cactus-like plant "herba Opuntia" grew. *Ex:* Opunti-ales*; Opunti-aspis (Ins.); Opuntia*; Opuntia-phila (Ins.).

opys—Gr. *opys*, juice. *Ex:* Chrys-opia*.

-or—1. L. *-or*, noun suffix denoting state or quality, as in *pallor:* agent or doer, as in *captor* and *obturator*; akin to Eng. *-er:* 2. NL. *-or* < L. *-orium*, suffix meaning place where, as in *mirror*.

or—1. Gr. *oros*, genit. *oreos*, a mountain; *oreitēs*, a mountaineer; *Orestēs*, son of Agamemnon, lit., a mountaineer. *Ex:* Or-ac-odon (Mam.); Or-odus (Mam.); Ore-amnos (Mam.); Ore-gaster (Echin.); Oreo-mys (Mam.); Oreo-carya*; Oreo-helix (Moll.); Oreo-spiza (Av.); Orestes (Moll.); Ori-bates (Arach.); Ori-ganum*, see gan; Oro-hippus (Mam.). See also ur: 2. Gr. *ōra*, dawn, spring. *Ex:* Prot-oro-saurus (Rept.): 3. Gr. *oraō*, to see. *Ex:* Or-acanthus (Pisc.): 4. L. *os*, genit. *oris*, mouth, oral, pertaining to the mouth > *orificium*, an opening. *Ex:* os-culum (See os); orifice (*os+facio*, to make); or-ad; oro-nasal: 5. Gr. *ōra*, care, concern. *Ex:* Pneum-ora (Ins.): 6. Gr. *horos*, margin, limit. *Ex:* aethal-orus; Pent-horum*: 7. Gr. *oros*, whey of milk. *Ex:* oro-therapy (Med.): 8. Gr. *hōros*, year, season: 9. Gr. *hōra*, beauty. *Ex:* Lecan-ora*: 10. Gr. *orō*, to excite. *Ex:* Oro-bus*; oro-phallic.

orari—L. *orarius*, belonging to the coast.

orat—Gr. *oratos*, visibl∴. *Ex:* Orato-stylum (Ins.).

oratori—L. *oratorius*, belonging to an orator.

orb—L. *orbis*, dim. *orbulina*, circle, ring; *orbita*, orbit; *orbitus*, circular; *orbiculatus*, circular. *Ex:* Orbea*; Orbi-cella (Coel.); orbit-al; Orbit-oides (Prot.); Orbito-lites (Ann.); Orbulina (Prot.); Plan-orbis (Moll.).

orbicul—See orb.

orbit—See orb.

orbon—L. *Orbona*, goddess sacred to parents bereft of children. See orbus.

orbus—L. *orbus*, devoid of children, without parents, empty. See orbon.

orc—1. L. *orca*, a kind of whale, the great killer. *Ex:* Orca (Mam); Orca-ella (Mam.): 2. L. *orca*, a dice box: 3. L. *Orcinus*, belonging to Hades < *Orcus*, the Lower World. *Ex:* Orcinus (Mam.), or perh. it should be Orc-inus (L. *orca*, a whale+-inus, like.).

orch—Gr. *orchis*, a testicle > *orchis*, a kind of plant, the orchid, so named because of the form of its root > L. *orchis*, NL. genit. *orchidis*. *Ex:* orcheo-bius (Prot.); Orchio-mys (Mam.); Orichid-aceae*; crypt-orchid-ism; Orchis*, mes-orchium.

orches—Gr. *orchēsis*, the art of dancing; *orchēstēs*. a dancer. *Ex:* Orches-ella (Ins.); Orchest-ina (Arach.); Orchestes (Ins.) Orchestia (Arth.); Orchesto-merus (Ins.).

orchest—See orches.

orchil—Gr. *orchilos*, a wren. *Ex:* Olbi-orchilus (Av.).

orcin—See orca.

orcul—L. *orcula*, a cask. *Ex:* Orcul-ella (Moll.); Orcula (Prot.); orculi-form.

orcyn—L. *orcynus*, the name of kind of sea fish < Gr. *orkynos*, a fish name. *Ex:* Orcynus (Pisc.).

ordi—L. *ordior*, to begin, to begin a web, to spin. *Ex:* prim-ordi-al.

ordovic—L. *Ordovices*, a people of early Britain. *Ex:* Ordovici-an.

ore—See or 1.

oreas—Gr. *oreias*, an Oread or mountain nymph. *Ex:* Oreas (Mam.); Orias (Mam.).

orect—Gr. *orektos*, stretched out. *Ex:* Orecto-gnathis (Ins.); Orecto-lobus (Pisc.); Orectis (Ins.).

orecto—Gr. *orektikos*, appetitive < *orexis*, desire, appetite. *Ex:* Orecto-gnathus (Ins.); Orecto-lab-idae (Pisc.).

oreg—Gr. *oregō*, to stretch, to stretch out, to reach. *Ex:* Orego-cera (Ins.).

oregm—Gr. *oregma*, genit. *oregmatos*, a stretching out. *Ex:* Oregma (Ins.).

orein—See or 1.

-oreios—NL. *-oreios*, adj. ending of Gr. origin meaning "of mountain" as in *cyanoreios*, of blue mountain.

oreo—See or 1.

oresci—Gr. *oreskios*, overshadowed by mountains. *Ex:* Orescius (Ins.); Oreskios (Av.).

oress—Gr. *orres*- inseparable stem used in forming words referring to mountains. *Ex:* Oress-aula (Ins.); Oressi-noma (Ins.); Oreso-chen (Av.).

orest—See or 1.

orgad—Gr. *orgas*, genit. *orgados*, a meadow, any well-watered, fertile spot of land. *Ex:* orgado-philus.

organ—Gr. *organon*, an organ. *Ex:* organ-ic; organo-poda (Ins.); organo-trophic.

orgasm—Fr. *orgasme*, orgasm, the climax of sexual excitement, derived either < Gr. *orgasmos*, a kneading, a softening through kneading or < *orgaō*, to swell, especially with lust. *Ex:* orgasm.

orgil—Gr. *orgilos*, inclined to be angry, irritable. *Ex:* Orgilo-morpha (Ins.), i.e., shaped like Orgilus; Orgilus (Ins.).

orgyia—Gr. *orgyia*, the length of the outstretched arms. *Ex:* Orgyia (Ins.).

orias—See oreas.

oribas—Gr. *oreibasos*, a mountain-climber. *Ex:* Oribasus (Ins.).

oribat—Gr. *oreibatēs*, mountain-ranging. *Ex:* Oribat-ella (Arach.); Oribat-odes (Arach.); Oribata (Arach.).

orien—L. *oriens*, genit. *orientis*, rising; ppr. of *orior*, to rise, bring forth; Fr. *orientation. Ex:* orientation.

orim—Gr. *orimos*, mature, ripe. *Ex:* Orimo-dema (Ins.).

orin—1. Gr. *orinō*, to excite. *Ex:* Orino-dromus (Ins.): 2. Gr. *oreinos*, from the mountains, mountain dwelling.

oriol—OFr. *oriol*, the oriole. *Ex:* Oriolus (Av.).

orism—Gr. *orismos*, finite, the end, marked out by boundaries; *oristikos*, for defining. *Ex:* Meg-orismus (Ins.). See also horism.

orit—L. *oritis*, a precious stone. *Ex:* Orit-iscus (Av.); Orit-urus (Av.).

-orius—1. L. *-orius, -a, -um*, adj. suffix, referring to the place of a thing. *Ex:* audit-orium; tent-orium. See Sept-oria* under sept: 2. Gr. *oreios*, of or pertaining to mountains.

orm—Gr. *ormos*, a cord, chain. *Ex:* Ormo-car pum*; Ormo-xylon*; Ormos-ella (Prot.); Ormosia*. See also horm.

ormen—Gr. *hormenos*, a shoot, stalk. *Ex:* Ormen-alurus (Mam.); Ormeno-flata (Ins.).

orn—L. *ornus*, the wild ash. *Ex:* Ornis*.

ornan—L. *ornans*, genit. *ornantis*, ornamenting; ppr. of *orno*, to fit out, adorn.

ornat—L. *ornatus*, decorated, adorned. *Ex:* Ornat-ella (Bry.); Ornati-pitta (Av.).

ornis—See ornith.

ornith—Gr. *ornis*, genit. *ornithos*, dim. *ornithion*, bird. *Ex:* ornis; Ornith-ischia (Rept.); Ornithion (Av.); Ornitho-logy; Ornitho-rhynchus (Mam.); Ornitho-sauria (Rept.); Hesperornis (Av.).

oro—See or 1.

orobanch—Gr. *orobanchē*, the broom-rape, a parasitic plant. *Ex:* Orobanch-aceae*; Orobanche*.

orobus—Gr. *orobos*, the bitter vetch. *Ex:* Orobus*.

oront—Gr. *Orontēs*, name of a Syrian River. *Ex:* Oront-ium*.

oroph—Gr. *orophē*, roof, the top of anything. *Ex:* Oroph-odon (Mam.); Orophea*; Oropho-crinus (Echin.); Ag-orophius (Mam.).

orpa—See orpac.

orpac—Gr. *orpēx*=Dor. *orpax*, genit. *orpakos*, a sapling, young shoot; also a lance or spike. *Ex:* Orpaco-phora (Ins.); Pan-orpa (Ins.).

orphn—Gr. *orphnos*, dark, dusky; *orphnē*, night. *Ex:* Orphn-oecus (Arach.); Orphn-urgus (Echin.); Orphne-biota (Ins.); Orphno-xanthus (Crust.); Orphnus (Moll.).

orress—See oress.

orrh—1. Gr. *orrhos*, whey, serum. *Ex:* orrho-rrhea (Med.); orrho-therapy (Med.): 2. Gr. *orrhos*, the rump.

orrhod—Gr. *orrhōdia*, terror, fright.

ors-—Gr. *orsi-*, exciting; *orsō*, fut. of *ornymi*, to rouse, excite. *Ex:* Orsi-macha (Ins.).

orsodacn—Gr. *orsodaknē*, a bud-eating insect; *daknō*, to bite. *Ex:* Orsodacne (Ins.).

ort—L. *ortus*, born, pp. of *orior*, to be born, to come forth. *Ex:* ab-ortion.

ortal—Gr. *ortalis*, young bird, the young of any family. *Ex:* Ortal-idae (Ins.); Ortal-istes (Ins.); Ortalis (Av.), (Ins.); Ortalo-ptera (Ins.).

orth—Gr. *orthos*, straight. *Ex:* Orth-echinus (Echin.); Orthis (Brach.); Ortho-ceras (Moll.); ortho-genesis; Ortho-ptera (Ins.); An-orth-ura (Av.).

orthagorisc—Gr. *orthagoriskos*, a suckling pig. *Ex:* Orthagoriscus (Pisc.).

orthrio—Gr. *orthrios*, early. *Ex:* Orthrio-mys (Mam.).

orthros—Gr. *orthros*, dawn, about day-break. *Ex:* Orthros-anthus*.

ortig—See ortyg.

ortyg—Gr. *ortyx*, genit. *ortygos*, the quail. *Ex:* Oritigi-ornis (Av.); Ortigo-spiza (Av.); Ortyg-onax (Av.); Ortygio-metra (Av.); Ortygo-cichla (Av.); Ortyx-elus (Av.), meaning of the last element uncertain; Loph-ortyx (Av.); Ore-ortyx (Av.).

ortyx—See ortyg.

-orum—L. *-orum*, genit. plural ending of second declension masculine and neuter. Latin nouns and first and second declension masculine and neuter Latin adjectives. *Ex:* dumet-orum.

orus—Gr. *ouros*, a watcher, warden. *Ex:* pyl-orus. See also or 6.

oruss—Gr. *orussō*, to dig through, perforate. *Ex:* Orussus (Ins.).

-ory—Eng. *-ory*<L. *-orius, -a, -um*, place where, place of, that which is suitable for or serves for. *Ex:* fact-ory; audit-ory (noun); ambulat-ory.

orych—Gr. *orychō*, to dig. *Ex:* Orycho-teuthis (Moll.).

oryct—Gr. *oryktēr*, a tool for digging; *oryktēs*, one who digs, a digger; in paleontological terms used in the sense of fossil. *Ex:* Oryctero-pus (Mam.); Oryctes (Ins.); Orycto-lagus (Mam.); Heli-oryctes (Ins.); Phyllon-orycter (Ins.).

oryg—Gr. *oryx*, genit. *orygos*, a sharp tool for digging; also a kind of gazelle or antelope. *Ex:* Orygo-cera (Ins.); Orygo-therium (Mam.); Oryx (Mam.); Lisso-dend-oryx (Por.).

orygm—Gr. *orygma*, genit. *orygmatos*, a pit, trench. *Ex:* orygma; Orygmato-bothrium (Platy.); Orygmus (Ins.).

oryss—Gr. *oryssō*, to dig, pierce. *Ex:* Oryssus (Ins.); Ge-oryssus (Ins.).

oryx—See oryg.

oryz—Gr. *oryza*, rice. *Ex:* Oryz-oryctes (Mam.); Oryz-opsis*; Oryza*; Oryzae-philus (Ins.); Oryzo-mys (Mam.).

-os—Gr. *-os*, ending of many Greek nouns and adjectives. It corresponds to the Latin ending *-us*. In making Latin and New Latin words from Greek words ending in *-os*, the *-us* ending is used. Thus: *Sphaerocarpos*>*Sphaerocarpus*; Gr. *nanos*, a dwarf>L. *nanus*, a dwarf; Gr. *olynthos*, a fig.>NL. *olynthus*, a fig.

os—See oss.

-osa—See -osus.

osch—1. Gr. *oschē*, the scrotum. *Ex:* oscheo-cele (Med.); osche-al: 2. *oschos*, a young branch, shoot.

oscill—L. *oscillo*, to swing>NL. *oscillator*, a swinger. *Ex:* Oscill-aria (Prot.); Oscillator-ia*.

oscin—L. *oscen*, genit. *oscinis*, pl. *oscines*, a singing bird, a divining bird from whose notes auguries were taken. *Ex:* Oscines (Av.); oscini-an; Oscini-soma (Ins.); Oscinis (Ins.); Oscino-mima (Ins.).

oscitan—L. *oscitans*, genit. *oscitantis*, listless, sluggish, ppr. of <*oscito*, to gape. *Ex:* oscitant.

oscul—1. L. *osculor*, to kiss; pp. *osculatus*, kissed. *Ex:* Osculatia (Av.); in-osculat-ion: 2. L. *osculum*, a little mouth. *Ex:* Osculi-gera (Moll.); Osculi-pora (Bry.).

osculat—See oscul.

-ose—See -osus.

osiris—Gr. *Osiris*, an Egyptian deity, husband of Isis. *Ex:* Osiris (Ins.).

-osis—Gr. *-osis*, suffix indicating, esp. in disease, a state of, or an increase in production: Often used in a similar sense to -iasis, which see. *Ex:* acid-osis; melan-osis; osm-osis; phlog-osis; sten-osis; trichin-osis.

osm—1. Gr. *osmē*=*odmē*, smell, scent; *osmērēs*, emitting an odor; *osmēsis*, a smelling; *osmētos*, that can be smelled. *Ex:* Osm-aelurus (Mam.); Osm-anthus*; osme-terium; Osmerus (Pisc.); osmesis; Osmet-ectis (Mam.); Osmia (Ins.); Osmo-phila (Ins.); Croc-osmia*; Di-osma*; Thamn-osma*: 2. Gr. *ōsmos*, a thrusting,

impulse. *Ex:* osm-osis; osmo-meter; osmo-tropism.

osmer—See osm 1.

osmund—*Osmunder*, a Saxon god. *Ex:* Osmunda*.

osphr—Gr. *osphrainomai*, aor. *osphromenos*, to smell, track by smelling; *osphradion*, a strong scent; *osphrantērios*, able to smell; *osphrantikos*, quick of scent. *Ex:* osphradium; Osphranter (Mam.); Osphrantic-um (Crust.); Osphromenus (Pisc.).

osphrant—See osphr.

osphy—Gr. *osphys*, the lower part of the back. *Ex:* Osphy-olax (Pisc.); Tret-osphys (Mam.).

oss—L. *os*, genit, *ossis*, dim. *ossiculum*, bone. *Ex:* os-calcis; oss-icle; osse-in; Ossi-fraga (Av.); ossi-fragus; ossi-fy; Ossiculum (Coel.).

ost—See oste.

ostar—Gr. *ostarion*, a little bone. *Ex:* ostari-phytum; Ostario-physi (Pisc.).

oste—Gr. *osteon*, bone. *Ex:* Ost-ichthys (Pisc.); osteo-logy; Osteo-pera (Mam.); Lepis-osteus (Pisc.); tele-ost; Tri-osteum*.

ostent—L. *ostentus*, a specter, a showing, a display; *ostentator*, a displayer. *Ex:* Ostentator (Ins.).

oster—Gr. *ostēros*, quick, nimble.

osti—L. *ostium*, dim. *ostiolum*, a door. *Ex:* osti-ate; ostiole; ostium.

ostin—Gr. *ostinos*, of bone. *Ex:* Ostin-ops (Av.).

ostrac—Gr. *ostrakon*, dim. *ostrakion*, a shell; *ostrakōdēs*, testaceous. *Ex:* Ostracion (Crust.); Ostraco-derm (Pisc.); Malac-ostraca (Crust.); peri-ostracum.

ostre—L. *ostrea*=Gr. *ostreon*, an oyster. *Ex:* Ostrea (Moll.); ostrea-culture; Ostrei-genus (Moll.); Ostreo-bium*.

ostreat—L. *ostreatus*, rough, covered with scales.

ostry—Gr. *ostrya*, a kind of hard-wood tree. *Ex:* Ostrya*.

-osus—L. *-osus, -a, -um*, termination denoting full of, augmented, prone to. *Ex:* dum-osa; filament-osus; filament-ose; ramul-osus, sax-osa; scoli-osus.

osyr—L. *osyris*, a plant name<Gr. *osyris*. *Ex:* Lin-osyris*.

ot—Gr. *ous*, genit. *ōtos*, dim. *ōtion*, nom. plu. *ōta*, the ear; *ōtikos*, of the ear; *ot-* when used as a prefix sometimes refers to a large ear, or at least a larger ear; *ōtōeis*, eared. *Ex:* Ot-elaphus (Mam.); Oti-phoca (Mam.); Otio-rynchus (Ins.); Oto-coris (Av.); oto-conium; oto-porpae; Oto-spermo-philus (Mam.); otic; Otoes (Mam.); A-otus*; Micr-otus (Mam.): par-otic. See also otid and parot.

-ota—NL. *-ota*, suffix meaning having, as in Amniota, i.e. having an amnion. Sometimes used to end a group name as in Gekkota.

otacust—Gr. *ōtakousteō*, to listen to. *Ex:* Ota custes (Ins.).

otar—Gr. *ōtaros*, large-eared; *ōtarion*, a little ear, dim. from *ous*, genit. *ōtos*, an ear. *Ex:* Otar-idae (Mam.); Otaria (Mam.), both from *otaros*.

oter—NL. *otero*, from a supposed Gr. *oteros*, the other. *Ex:* Otero-gnathus (Rept.).

Desert Horned Lark, *Otocoris alpestris*

othe—Gr. *ōtheō*, to burst forth, to thrust. *Ex:* Otheo-stethus (Ins.); Heli-othis (Ins.).

othis—See **othe**.

othn—Gr. *othneios*, strange. *Ex:* Othnio-cryptus (Ins.); Othnius (Ins.).

othon—Gr. *othonē*, fine-linen, sail-cloth, a sail. *Ex:* Cycl-othone (Pisc.).

othrus—See **molothrus**.

otid—Gr. *ōtis*, genlt. *ōtidos*, a bustard. *Ex:* Otid-idae (Av.); Otid-us (Av.); Otis (Av.); Syphe-otis (Av.).

otio—See **ot**.

otios—L. *otiosus*, idle.

otis—See **otid**.

otl—Gr. *otlos*, suffering, distress. *Ex:* Otlo-phorus (Ins.).

otob—Gr. *otobos*, a shrill noise, a loud sound.

otrynter—Gr. *otryntēr*, one who excites, < *otrynō* to excite, instigate. *Ex:* Otrynter (Pisc.).

otus—L. *otus*, a horned owl. *Ex:* Otus (Av.).

ouden—Gr. *ouden* < *oude*, neither, not+*eis*, one, i.e. not one, not any, none. *Ex:* Ouden-odon (Rept.).

ouistit—Brazilian *ouistiti*, name of a monkey. *Ex:* Ouistitis (Mam.).

oul—See **ul**.

ouran—See **uran**.

ourax—Gr. *ourax*, Attic name of a gallinaceous bird. *Ex:* Ourax (Av.).

ouro—See **ur** 1 and 3.

-ous—1. OFr. *-ous*, *-eus* < L. *-osus*, full of > Eng. *-ose*, suffix meaning full of, abounding in, pos-sessing the qualities of. *Ex:* bulb-ous, fibr-ose: 2. *-ous*, suffix used in chemistry to denote a valence lower than that of a similar substance bearing the ending *-ic*; thus ferr-ous and ferr-ic; nitr-ous and nitr-ic.

ous—Gr. *ous*, ear. *Ex:* Ous-tropis*.

ov—L. *ovum*, dim. *ovulum*, egg. *Ex:* ov-enchyma; ovi-duct; ovi-gerous; ovary; Ovi-clypeus (Echin.); ovi-parous; ovi-positor; Ovul-aster (Echin.).

oval—Fr. *oval*, *ovale*, elliptical, oval. *Ex:* oval; Oval-astrea (Coel.); ovali-form.

ovari—NL. *ovarium*, ovary < L. *ovum*, egg. *Ex:* ovario-tomy (Surg.); Ovarium (Echin.).

ovat—L. *ovatus*, egg-shaped, having oval spots. *Ex:* Ovati-cella (Bry.); Ovati-ceras (Moll.).

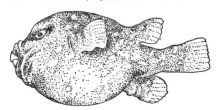

The Bristly Egg-shaped Fish, *Ovoides setosus*. Redrawn from Fishes of North and Middle America—Jordan

ovi—L. *ovis*, sheep; *ovinus*, pertaining to or of sheep. *Ex:* Ovi-bos (Mam.); ovine; Ovis (Mam.).

ovill—L. *ovillus*, belonging to sheep.

ovin—See **ovi**.

ovis—See **ovi**.

ovul—See **ov**.

ox—Gr. *oxys*, sharp, keen, quick, also sometimes in sense of acid and shrill > *oxalis*, sorrel; *oxytenēs*, sharp, pointed; *oxyntos*, making sharp, making acid, verb. adj. of *oxynō*, to make acid. *Ex:* Oxalis*; oxe-ate; Oxy-coccus*; oxy-diactis; Oxy-echus (Av.); Oxy-opes (Arach.); Oxy-rhino (Ins.); Oxy-soma (Ins.); Oxy-stomata; Oxyia*; oxynt-ic; Oxytenia*; Amphi-oxus (Cephalochorda); Pom-oxis (Pisc.).

oxalis—See **ox**.

oxy—See **ox**.

oxydo—Fr. *oxyde* < *oxide*, an oxide. *Ex:* Carb-oxydo-monas*.

oxygo—Gr. *oxygoos*, shrill-wailing. *Ex:* Oxygous (Mam.).

oxynoe—NL. *oxynoe*, a name without meaning devised by Rafinesque. *Ex:* Oxynoe (Moll.).

oxyntic—See **ox**.

oz—1. Gr. *ozō*, to smell; *ozaina* = *ozolis*, an offen-sive polypus of the nose. *Ex:* Ozaena (Ins.); Ozo-gnathus (Ins.); Ozol-ictis (Mam.); Rhin-ozolis (Mam.):    2. Gr. *ozos*, a branch, twig. *Ex:* Caly-oza (Ins.); Goni-ozus (Ins.).

ozanna—NL. *ozanna*, name for an antelope, ori-gin uncertain. *Ex:* Ozanna (Mam.).

ozol—See **oz**.

ozot—Gr. *ozōtos*, branched. *Ex:* Ozoto-ceras (Mam.).

# P

pa—Gr. masc. *pas*, neut. *pan*, genit. *pantos*, all. *Ex:* Pa-lophus (Ins.); Pa-mmelas (Pisc.), see mela; Pa-phagus (Ins.).

pabul—L. *pabulum*, nourishment < *pabulo*, to feed; *pabularis*, of or fit for fodder. *Ex:* pabulous.

pac—Peruvian *paca, pacos*, name for the *llama*. *Ex:* Pacos (Mam.).

pach—See pachy.

pachn—Gr. *pachnē*, hoar-frost < Gr. *pachnoō*, to thicken, congeal. *Ex:* Pachne-phorus (Ins.).

pachy—Gr. *pachys*, thick; *pachos*, thickness, *pachylos*; thickish, fat; *pachetos*, thick or stout; *pachistikos*, very thick; *pachytēs*, thickness. *Ex:* Pach-astr-ella (Por.); Pach-echinus (Echin.); Pachi-gaster (Ins.); Pacho-choerus (Mam.); Pachy-dermata (Mam.); Pachygnatha (Arach.); Pachy-onus (Mam.); Pachystima*, see stigm; pachy-tene; Pachyl-ister (Ins.) Pachyl-ota (Ins.); Pachys-andra*; Gymno-paches (Ins.); Lep-pachys*, involving Gr. *lepis*, a scale; Para-pachyta (Ins.).

pachyl—See pachy.

pachyma—NL. *pachyma* < Gr. *pachys*, thick. *Ex:* Pachyma*.

pacific—L. *pacificus*, peace-making, peaceable > *Pacific Ocean*. *Ex:* Pacific-agrion (Ins.); Pacifico-droma (Av.).

pact—L. *pactus*, agreeing, made fast, settled < Gr. *pactos*, solid, firm, coagulated < *pactoō*, to bind. *Ex:* Pact-ilia*, the last element seemingly without meaning; Pacto-pus (Ins.); Pacto-stoma (Ins.).

pactil—L. *pactilis*, wreathed, woven together < *pango*, to make fast, set together.

pad—Gr. *pados*, a kind of tree, perh. *Prunus padus*. *Ex:* Padus*.

padin—Gr. *padinos*, of the tree, *pados*. *Ex:* Padina*.

paect—Gr. *paiktos*, joked or sported with; *paiktēs*, fem. *paikteira*, a dancer, player < *paizō*, to sport, joke. *Ex:* Paectes (Ins.); Paectira (Ins.); Paecto-phyllum (Myr.).

paed—Gr. *pais*, genit. *paidos*, child. *Ex:* paed-iatrices = ped-eatrics (Med.); paedo-genesis; Paedo-therium (Mam.); Paido-pithex (Mam.); dasy-paedes; Helio-pais (Av.).

paeder—Gr. *paiderōs*, rouge; a kind of opal; a kind of plant with rosy flowers. *Ex:* Paederia*; Paederus (Ins.).

paedid—L. *paedidus*, filthy, stinking < *paedor*, a bad smell.

paegm—Gr. *paigma*, genit. *paigmatos*, play, sport. *Ex:* Anemo-paegma*,

paenulat—L. *paenulatus*, wearing a cloak of wool.

paeonia—Gr. *paiōnia*, a kind of plant, the peony < *Paiōn*, a mythical physician for whom the plant was named. *Ex:* Paeonia*.

paet—L. *paetus*, with leering eyes.

paeus—Gr. *poieō*, to make > *pēlopoios*, a potter, one who fashions with mud. *Ex:* Astero-paeus (Ins.); Pelo-paeus (Ins.).

pag—1. Gr. *pagos*, that which is fixed or firmly set; *pagios*, solid. *Ex:* Pagi-odon (Mam.); Pagio-poda (Ins.); sterno-pagus:    2. Gr. *pagos*, ice, frost. *Ex:* Pago-mys (Mam ); Pago-phila (Av.); Eu-pago-deres (Ins.).

pagan—L. *paganus*, of the country or village, rustic. *Ex:* Pagana (Moll.).

pagell—See pagr.

pagin—L. *pagina*, a leaf, a page. *Ex:* Pagin-ula (Mam.); pagina.

pagio—See pag 1.

pagod—Pg. *pagode* < Tamil *pagavadi*, a pagoda. *Ex.* Pagodi-spira (Moll.).

pagoll—NL. *pagolla* < Lapp. *pago*, a name applied to the plover. *Ex:* Pagolla (Av.).

pagr—Gr. *pagros* = *phagros*, a kind of fish, the sea-bream; ML. dim. *pagellus*. *Ex:* Pagellus (Pisc.); Pagrus (Pisc.); Holo-pagrus (Pisc.).

pagum—NL. *paguma*, "a coined word, evidently modelled after *Puma*" *Ex:* Paguma (Mam.).

pagur—Gr. *pagouros*, a crab. *Ex:* Pagur-oides (Crust.); Pagur-opsis (Crust.): Paguro-dacty lus (Ins.); Pagurus (Crust.).

paid—See paed.

pais—See paed.

pal—1. Gr. *palē*, fine meal. *Ex:* Pal-orus (Ins.): 2. Gr. *palē*, fight, battle:    3. Gr. *palos*, a shaking, casting of lots. *Ex:* Erythro-palum*: 4. L. *palus*, a stake. See also palae and palus.

palae—Gr. *palaios*, ancient, old in years; as a prefix it indicates an ancient or extinct type or form. *Ex:* Pal-oplo-therium (Mam.); Pal-orchestes (Mam.); Palae-ac-odon (Mam.); Palae-an-odonta (Moll.); palae-onto-logy; Palaeo-balaena (Mam.); palaeo-lithic; Palaeozoic; Palaio-trogos (Mam.); Paleo-lemur (Mam.).

palaio—See palae.

palam—Gr. *palamē*, a web, the palm of the hand; also a device, method. *Ex:* Palam-phora (Echin.); Palamo-pus (Rept.); Micro-palama (Av.).

palamed—Gr. *Palamēdēs*, mythical hero of the Trojan war. *Ex:* Palamedea (Av.).

palan—L. *palans*, genit. *palantis*, supporting, ppr. of *palo*, to prop up, support.

palaquim—NL. *palaquim* < Tagalog *palac*, the gutta-percha tree. *Ex:* Palaquim*.

palar—See palus.

palass—Gr. *palassō*, to besprinkle, spot, defile. *Ex:* Palasso-pora (Coel.).

palat—L. *palatum*, the roof of the mouth, the palate. *Ex:* palat-al; palat-ine; palate; palato-dent-al.

pale—Gr. *palē*, a fight; *palaiō*, to wrestle, fight. *Ex:* Paleo-pragma (Ins.); Cero-pales (Ins.). See palae.

palea—L. *palea*, chaff, straw; *palearis*, belonging to chaff. *Ex:* pale-aceous; paleae-formis; palei-form; not Palea-crita (Ins.), which is derived from Gr. *palaios*, ancient.

paleat—L. *paleatus*, mixed with chaff, chaffy.

palei—See palea.

paleo—See palae, also pale.

palimbol—Gr. *palimbolos*, thrown back, reversed; also uncertain. *Ex:* Palimbola (Ins.).

palin—Gr. *palin*, again, once more; also back, backwards. *Ex:* palin-drom-ous; palin-genesis.

palingen—Gr. *palingenesia*, new birth. *Ex:* Palingeni-idae (Ins.).

palinur—L. *Palinurus*, pilot of Aeneas. *Ex:* Palinur-ichthys (Pisc.); Palinurus (Crust.).

palisad—Fr. *palissade*, a fence of stakes<L. palus, a stake. *Ex:* palisade cells.

palitans—L. *palitans*, wandering about.

paliur—Gr. *paliouros*, a kind of shrubby plant. *Ex:* Paliurus*.

pall—Gr. *pallō*, to quiver, to toss. *Ex:* Pallopteras (Ins.); Rhipi-pallus (Ins.).

pallac—1. Gr. *pallax*, genit. *pallakos*, a youth. *Ex:* Pallaco-coris (Ins.): 2. Gr. *pallakis*, a concubine.

pallad—See pallas.

pallant—Gr. *pallas*, genit. *pallantos*, one young in years, a youth. See also pallas.

pallas—Gr. *Pallas*, genit. *Pallados*, goddess of wisdom to whom the owl was sacred. *Ex:* Micro-pallas (Av.); not Pallasi-carbo (Av.) named in honor of P. S. Pallas, ornithologist.

pallene—Gr. *Pallēnē*, peninsula of Macedonia where the battle between the gods and giants took place. *Ex:* Pallene (Ins.).

pallens—L. *pallens*, wan, pale, a faint color, greenish, yellowish<*palleo*, to be pale; *pallescens*, genit. *pallescentis*, becoming pale, fading, withering, ppr. of *pallesco*, to fade.

pallescen—See pallens.

palli—L. *pallium*, a mantle, cover; *palliatus*, cloaked. *Ex:* palli-al sinus; Palli-fera (Moll.); Pallio-branchiata (Brach.); pallium; Pallium (Moll.).

pallid—L. *pallidus*, pale, pallid. *Ex:* pallidi-florus.

palliolat—L. *palliolatus*, hooded, covered<*pallium*, a cover.

palm—1. L. *palma*, the palm of the hand, sometimes used in the sense of broad, as in L.

palmipes, broad-footed; *palmatus*, marked like the palm of the hand; *palmaris*, pertaining to the palm of the hand. *Ex:* Palm-asterias (Echin.); Palma-corixa (Ins.); Palma-cystes (Echin.); palmar arch; palmati-fid; Palmato-gecko (Rept.); Palmatus (Mam.); Palmipes (Mam.): 2. Gr. *palmos*, a quivering motion, pulsation. *Ex:* Palm-ella*; palmo-spasmus: 3. L. *palma*, a palm-tree>Fr. *palm-*

Palmate-leaved Gourd, *Cucurbita palmata*. Redrawn from Desert Wild Flowers—Jaeger. Stanford University Press.

*iste*, a palm dweller. *Ex:* Palmista (Mam.); palmo-grapher.

palmyr—Gr. *Palmyra*, city of Syria. *Ex:* Palmyr-ides (Ann.); Palmyr-opsis (Ann.); Palymyra (Ann.).

palo—See palae.

palp—L. *palpo* and *palpor*, to stroke, to feel one's way>NL. *palpus*, a feeler. *Ex:* palp; palp-ulus; palpi-formis; palpo-cil; palpon.

palpebr—L. *palpebro*, to wink frequently, to blink; *palpebratio*, genit. *palpebrationis*, a winking. *Ex:* palpebrate.

palt—Gr. *paltos*, a dart. *Ex:* Palt-odus (Ann.); Palto-themis (Ins.); Palto-thyreus (Ins.); Pros-palt-ella (Ins.).

palud—L. *palus*, genit. *paludis*, a marsh; *paludosus*, swampy. *Ex:* Palud-estrina (Moll.); Palud-omus (Moll.), see dom 2; palud-ous; Paludi-cella (Bry.).

paludat—L. *paludatus*, dressed in a soldier's cloak.

palumb—L. *palumbes*, a wood-pigeon, ring-dove; *palumbinus*, of wood-pigeons. *Ex:* Palum bina*.

palus—L. *palus*, a stake, prop, NL. dim. *palulus*; *palaris*, pertaining to stakes. *Ex:* palari-ramosus; pali-form; palule, palus; See also palud.

palustr—L. *paluster*, fem. *palustris*, neut. *palustre*, marshy. *Ex:* palustr-al; palustr-ine.

pam—See pan.

pambasil—Gr. *pambasileus*, an absolute monarch.

pamm— —Gr. prefix *pamm-*, in composition denoting all, very much, exceeding. See pan.

pampa—Peruv. *pampa*, a plain. *Ex:* Pampatherium (Mam.).

pampin—L. *pampinus*, the young twining shoot of a vine, a tendril. *Ex:* pampini-form; pam pino-cele (Med.).

pan—**1.** Gr. masc. *pas*; fem. *pasa*, genit. *pases*; neuter *pan.* genit. *pantos*, all; (*pan* becomes *pam* before radicals beginning with *b* or *p*). *Ex:* P-antholops (Mam.); Pam-machus (Ins.); Pam-micrus (Ins.); Pam-phagus (Prot.); Pam-philius (Ins.); pam-pro-dactyl; Pan-cratium,* see crat 2; pan-creas· pan-genesis; pan-oistic; Pan-orpa (Ins.); Pano-therium (Mam.); Panto-lestes (Mam.); Pasi-lobus (Ins.); Pasi-peda (Ins.) **2.** Gr. *panos*, a torch. *Ex:* Panus (Ins.): **3.** Gr. *Pan*, rural god of Arcadia. *Ex:* Pan (Mam.); Pan-isco-mima (Ins.); Pan-iscus (Mam.); Pan-iscus (Ins.). **4.** L. *panis*, bread: *Ex:* com-pan-ion.

panag—Gr. *panagēs*, wholly hallowed. *Ex:* Panag-aeus (Ins.).

panax—L. *panax* = *panacea*, an herb which was supposed to heal all diseases; Gr. *panakēs*, all healing. *Ex:* Panax*; Acantho-panax; Opo-panax*.

pancreas—See pan 1.

pand—L. *pandus*, bent, crooked. *Ex:* re-pand.

pandan—NL. *pandanus* < Malay *pandang*, conspicuous. *Ex:* Pandanus*.

pandar—Gr. *Pandaros*, son of Lycaon. *Ex:* Pandarus (Ins.).

pandion—Gr. *Pandiōn*, a king of Athens. *Ex:* Pandion (Av.).

Pandor—Gr. *Pandōra*, name of the first mortal woman on whom all the gods bestowed gifts; < *pas* (*pan*-), all + *doron*, gift. *Ex:* Pandor-ina*; Pandora (Cten.), (Moll.); Pandorea*.

pandur—L. *pandura*, a musical instrument of three strings, invented by Pan; *panduratus*, fiddle-shaped. *Ex:* panduri-form.

pang— L. *pango*, to fasten, drive in. *Ex:* Pangus (Ins.).

panice—L. *paniceus*, made of bread. *Ex:* Paniceum (Prot.).

panicl—L. *panicula*, a tuft, dim of *panus*, a swelling, an ear of millet. *Ex:* panicl-ed; panicle; Panicul-aria*; panicul-ate.

panicul—See panicl.

panicum—L. *panicum*, an old name for millet, panic grass. *Ex:* Panicum*.

pann—L. *pannus*, dim. *panniculus*, a piece of cloth, rags, a garment; *panneus*, ragged, tattered; *pannosus*, full of rags. *Ex:* panniculus muscle; pannose.

pannicul—See pann.

pannon—Gr. *Pannonia*, a geographical name. *Ex:* Pannonia (Moll.).

panope—L. *Panope*, a sea-nymph. *Ex:* Panope (Moll.); Panope-us (Crust.); Neo-panope (Arth.).

pans—L. *pansus*, expanded, stretched open < *pando*, to spread out, extend.

panshura—Bengalese *panshura*, name for a chelonian. *Ex:* Panshura (Rept.).

pansyd—Gr. *pansydiē*, fast, with full speed. *Ex:* Pansydia (Ins.).

pant—See pan.

pantel—Gr. *pantelēs*, all-complete, absolute. *Ex:* Pantele-ella (Ins.) ; Panteles (Ins.); A-panteles (Ins.).

pantex—See pantic.

panther—Gr. *panthēr*, a panther. *Ex:* Panthera (Mam.).

pantherin—L. *pantherinus*, with spots like a panther.

pantic—L. *pantex*, genit. *panticis*, the belly and its contents.

panto—See pan.

panulir—NL. *panulirus*, anagram of Palinurus, an arthropod genus. *Ex:* Panulir-idae (Crust.); Panulirus (Crust.).

panurg—Gr. *panourgos*, ready to do anything, wicked, knavish. *Ex:* Panurgus (Ins.).

pao—Gr. *paos*, Dor. of *pēos* (the older form), one related by marriage. *Ex:* Pao-phil-us (Ins.).

papaver—L. *papaver*, a poppy < *papa*, pap, thick milk. *Ex:* Papaver*; Papaver-aceae*; Papaver-ales*.

paphinia—Gr. *Paphos*, city of Cyprus, sacred to Venus. *Ex:* Paphinia*.

papilio—L. *papilio*, genit. *papilionis*, a butterfly, moth. *Ex:* Papilio (Ins.); papilion-aceous; Papilion-idae (Ins.).

papill—L. *papilla*, a nipple, pimple. *Ex:* papillary; papill-ate; papilli-form; Papillo-porus (Arach.).

papio—NL. *papio* < Fr. *papion*, a baboon. *Ex:* Papio (Mam.).

papp—Gr. *pappos*, a grandfather, also the first down on the chin; L. *pappus*, plant down. *Ex:* papp-escent; Papp-ichthys (Pisc.); Pappi-pappus (Ins.); Pappo-geo-mys (Mam.); Pappo-saurus (Rept.); pappus.

papul—L. *papula*, pl. *papulae*, a pimple. *Ex:* papulae; Papuli-scala (Moll.).

papyr—Gr. *papyros*, the paper-reed. *Ex:* papyraceous; papyri-fera; Papyri-scala (Moll.); Papyro-theca (Moll.).

par—**1.** Gr. *para*, beside, near > *para-*, a prefix meaning near, beside, parallel; also beyond; in the names of extinct genera *para-* is used to denote relationship. *Ex:* Par-apsida (Rept.); par-enchyma; par-enter-al; Par-eu-menes (Ins.); par-eunia; Par-otia (Av.); par-otid, comp. with L. *parotid*, genit. *parotidis*, a tumor near the ear; para-basal; Para-bates (Ins.); para-blast; Para-cyn-odon (Mam.); para-glossa; para-lysis; para-physis; para-podia; para-site, see sit; para-type; Paro-ceras (Mam.); Paro-nychia*: **2.** L. *pareo*, to bring forth, to beget. *Ex:* par-ous; acantho-paria; ovi-par-ous; vivi-par-ous; for Para-galia (Mam.) see per 2: **3.** L. *par*, genit. *paris*, equal. *Ex:* pari-digitate. See also parus: **4.** Gr. *paros*, before time, early. *Ex:* Paro-lamia (Ins.). **S.**

**para**—See par.

**paracm**—Gr. *parakmē*, the point where decay sets in. *Ex:* paracme.

**parad-** —NL. *paradi-*, a mutilated form of *paradisi-*<Gr. *paradeisos*, a park, paradise. *Ex:* Paradi-galla (Av.).

**paradis**—LL. *paradisus*<Gr. *paradeisos*, an enclosed park, a pleasure ground. *Ex:* Paradisea (Av.); Paradisia*.

**paradox**—Gr. *paradoxos*>L. *paradoxus*, strange, contrary to all expectation, unaccountable. *Ex:* Paradox-ides (Tri.); Paradox-urus (Mam.); Paradoxo-mys (Mam.).

**paral**—Gr. *paralios*=*paralos*, maritime; sometimes used in the sense of blue, i.e. like the sea. *Ex:* paral-inus.

**parallel**—Gr. *parallēlos*, parallel. *Ex:* parallelo-drome.

**paralysis**—See par 1.

**paramec**—Gr. *paramēkēs*, oblong, rather long, oval. *Ex:* Parameci-um (Prot.); Parameco-soma (Ins.); Paramek-odon (Crust.).

**paramyth**—Gr. *paramythia*, encouragement, consolation. *Ex:* Paramythia (Av.).

**paranom**—Gr. *paranomos*, lawless, contrary to law. *Ex:* Paranomus (Ins.).

**parasite**—See par 1.

**parastat**—Gr. *parastatēs*, one who aids, stands by, a comrade; *oi parastatai*, the testicles. *Ex:* para-parastat-ic; anti-parastata.

**parat**—L. *paratus*, prepared, ready. *Ex:* pre-parat-ion; Parato-carpus*.

**paratin**—Gr. *parateinō*, to stretch out. *Ex:* Paratinus (Ins.).

**parc**—L. *parcus*, sparing, frugal, poor; *parce*, sparingly. *Ex:* parci-costata.

**pard**—**1.** L. *pardus*<Gr. *pardos*=*pardalis*, genit. *pardaleōs*, the pard, leopard or panther; Gr. *pardalōtos*, spotted like the pard. *Ex:* Pard-anthus*; Pardal-ina (Mam.), dim. of *pardalis*; pardal-ine; Pardalotus (Av.); Pardo-felis (Mam.); Camelo-pardus (Mam.): **2.** Gr. *pardalos*, a kind of bird, the starling.

**pardal**—See pard.

**paredr**—Gr. *paredros*, near. *Ex:* Paredro-coris (Ins.).

**parei**—Gr. *pareia*, cheek. *Ex:* Parei-oplit-ae (Pisc.); Pareia-sauria (Rept.).

**pares**—Gr. *paresis*, a letting go, a weakening, paralysis. *Ex:* Pares-urus (Ins.); paresis (Med.).

**pareudiast**—Gr. *pareudiastēs*, name of some water bird frequenting land in good weather. *Ex:* Pareudiastes (Av.).

**pareun**—Gr. *pareunos*, lying beside or with a bedfellow (*par*, beside+*eunē*, a couch), sexual intercourse. *Ex:* pareunia.

**paria**—NL. *paria*, derived from some geographical name. *Ex:* Paria (Ins.).

**paries**—See pariet.

**pariet**—L. *paries*, gen. *parietis*, a wall; *parietalis*, pertaining to walls. *Ex:* Pariet-aria*; parietal; parieto-jugal; not Parietis (Mam.) which is curiously derived from *para*, near+*iktes*, weasel.

**paril**—L. *parilis*, equal, like.

**parinari**—Brazilian *parinari*, the gingerbread tree. *Ex:* Parinari-um*.

**paris**—Gr. *parisos*, almost equal, evenly balanced. *Ex:* Paris*; Pariso-crinus (Echin.); Pariso-morphus (Ins.).

**parm**—L. *parma*, a small round shield, a target; *parmatus*, shielded. *Ex:* Parm-elia*, the last element perh.<Gr. *heileō*, to roll up; Parm-unculus (Moll.); Parma-cochilea (Moll.); Parmi-palpus (Ins.); Parmo-phorus (Moll.).

**parmul**—L. *parmula*, a small shield<*parma*, a shield.

**parn**—Gr. *Parnēs*, a mountain in Greece. *Ex:* Parn-idae (Ins.); Parnus (Ins.).

**parnass**—Gr. *Parnasos*=L. *Parnassus*, celebrated mountain of Greece. *Ex:* Parnassia*; Parnassius (Ins.).

**parnop**—Gr. *parnops*, a kind of locust. *Ex:* Parnopes (Ins.).

**paro**—See par.

**parod**—Gr. *parodos*, a way, passage. *Ex:* Parodo-ceras (Moll.).

**paromal**—Gr. *parōmalos*, nearly even or equal. *Ex:* Paromalus (Ins.).

**paropi**—Gr. *parōpia*, an eye-shade. *Ex:* A-paropion (Ins.).

**parosela**—Anagram of *Psoralea*. *Ex:* Parosela*.

**parot**—Gr. *parōtis*, genit. *parōtidos*, the gland beside (*par-*) the ear (*ous*), also a curl of hair beside the ear. *Ex:* parot-itis (Med.): Parot-ia (Av.); parotido-auricularis.

**-parous**—Eng. *-parous*, suffix<L. *pario*, to bring forth, to produce, give birth to. *Ex:* ovi-parous; vivi-parous; Vivi-parous (Moll.).

**parra**—L. *parra*, a bird of evil omen, perhaps the wheat-ear. *Ex:* Parr-inae (Av.); Parra (Av.).

**pars**—L. *pars*, genit. *partis*, pl. *partes*, a part; *partile*, divisible. *Ex:* pars petrosa.

**parthen**—Gr. *parthenos*, virgin; *parthenios*, maidenly, pure. *Ex:* Parthenia (Moll.); Parthenice*.

poetical form of the following related generic name; Parthenium*; partheno-genesis; Par theno-merus (Crust.); partheno-spore.

parthenop—Gr. *Parthenopē*, a siren said to have been cast up and drowned on the shore at Naples. *Ex:* Parthenopa (Mam.); Parthenopea (Crust.).

parti—Fr. *parti*, divided. *Ex:* parti-colored.

partial—LL. *partialis*, not general or entire <L. *pars*, genit. *partis*, a part.

partibil—L. *partibilis*, divisible <*pars*, genit. *partis*, a part.

particeps—L. *particeps*, a sharer, partaker.

partim—L. *partim*, partly <*pars*, genit. *partis*, a part.

partit—L. *partitus*, divided. *Ex:* partite; pluri-partite.

parturien—L. *parturiens*, genit. *parturientis*, desiring to bring forth young, ppr. of *parturio*. *Ex:* parturient.

parturit—L. *parturitus*, produced, pp. of *parturion* to desire to bring forth, to produce. *Ex:* parturit-ion.

paru—Brazilian *paru*, name of the harvest fish.

parus—L. *parus*, a titmouse, tomtit. *Ex:* Par-idae (Av.); Parus (Av.); Auri-parus (Av.).

parv—L. *parvus*, little, small, petty; *parvitas*, smallness; *parvulus*, small. *Ex:* Parv-oliva (Moll.); Parvi-cardium (Moll.); Parvo-chlamys (Moll.); Parvulus (Amph.); Parvus (Av.).

parvit—See parv.

paryph—Gr. *paryphē*, a hem, border. *Ex:* Paryphe-phorus (Av.); Parypho-ceras (Moll.); Echino-paryphi-um (Platy.); Eu-parypha (Moll.); Eu-paryphi-um (Platy.); Leuco-paryphus (Ins.).

pas—See pan.

pascal—See pascu.

pasch—Gr. *paschō*, to be in passion. *Ex:* Pascho-anthus*.

paschal—L. *paschalis*, belonging to Easter.

pascu—L. *pascuum*, a pasture; *pascor*, to feed; *pascualis* = *pascalis*, grazing. *Ex:* pascu-al; pascu-ous.

pasiphae—Gr. *Pasiphaē*, sister of Circe, wife of Minos, mother of the Minotaur. *Ex:* Pasiphae (Ins.).

paspal—Gr. *paspalos*, millet. *Ex:* Paspal-idium*; Paspalum*.

pass—L. *passio*, a passion, also a suffering. *Ex:* Passi-flora*.

passal—Gr. *passalos* = Attic *pattalos*, a peg. *Ex:* Passal-ac-odon (Mam.); Passal-aster (Echin.); Passal-ites (Mam.); Passal-oecus (Ins.); Passalo-teuthis (Moll.); Passalus (Ins.).

passer—L. *passer*, pl. *passeres*, a sparrow; *passerinus*, sparrow-like, of or fit for a spar-

row. *Ex:* Passeri-formes (Av.); Passerina*; passerine; Passero-myia (Ins.).

past—L. *pastus*, pasture, food.

pastic—L. *pasticus*, fattened, well fed.

pastin—L. *pastinaca*, the carrot, also the parsnip >*pastinum*, a dibble with form like that of a carrot. *Ex:* Pastinaca*.

pastor—L. *pastor*, a herdsman, a shepherd. *Ex:* Sturno-pastor (Av.).

pat—Gr. *patos*, a beaten way, path <*pateō*, to tread, walk. *Ex:* Pat-uron; ichthyo-pato-lite; Peri-patus (Arth.).

pataec—NL. *pataecus* <Gr. *Pataikoi*, Phoenician deities of strange dwarfish form. *Ex:* Pataecus (Pisc.).

patag—1. L. *patagium*, the gold border of a garment. *Ex:* patagium: 2. Gr. *patagos*, a clatter, a crash, a chattering. *Ex:* Patag-ornis (Av.); ?Patago-spirio (Prot.): 3. *Patagonia*, an area of South America. *Ex:* Patagon-ula*.

patan—Gr. *patanē*, a flat dish. *Ex:* Patano-phyma (Por.).

patell—See patin.

paten—L. *patens*, genit. *patentis*, standing open, spreading, extending, accessible. *Ex:* patent.

pater—1. L. *pater*, genit. *patris*, a father; *paternus*, fatherly; *patrius*, belonging to one's father; Gr. *pater*, Attic *patros*, father. *Ex:* patern-al; patri-clinous; patri-lineal; Patrio-felis (Mam.); Patro-therium (Mam.): 2. L. *patera*, a dish, a saucer. *Ex:* pateri-form.

patern—See pater.

path—Gr. *pathos*, suffering; also accident, experience; *pathētikos*, sensitive, sensuous, pathetic. *Ex:* patheticus; patho-logy; Anti-path-aria (Coel.); Cirri-pathes (Coel.); Osteo-pathy; Sticho-pathes (Coel.); sym-pathetic.

-pathy—See path.

patibul—L. *patibulus*, yoked.

patienti—L. *patientia*, enduring, patient <*patior*, to bear, support.

patin—L. *patina*, dim. *patella*, a dish, stew-pan; Gr. *patella*, dim. *patellion*, a flat dish, pan. *Ex:* patella; Patellio-crinus (Echin.); patello-femoral; patina; Patino-pecten (Moll.).

patr—See pater.

patri—See pater.

patrio—Gr. *patrios*, belonging to one's father. *Ex:* Patrio-felis (Mam.).

patro—See pater 1.

patruel—L. *patruelis*, descended from an uncle.

pattal—See passal.

patul—L. *patulus*, spread out, extended, standing open; *patulens*, genit. *patulentis*, spreading out >*propatulus*, uncovered. *Ex:* Patul-axis (Moll.); Patul-oscula (Por.); Patula (Moll.); patulent; Patuli-zon-ites (Moll.); Propatulus*.

patulen—See **patul.**

**pauc**—L. *paucus*, few. *Ex:* Pauci-dentata (Moll.).

**paul**—**1.** Gr. *paula*, a pause, rest. *Ex:* Paulo-magus (Av.); paulo-spore: **2.** L. *paulus*, little. *Ex:* Paulo-crinus (Echin.); Paulo-scirtes (Pisc.).

**paunch**—OFr. *panche* < L. *pantex*, the belly. *Ex:* paunch.

**paupercul**—L. *pauperculus*, poor, dim. of *pauper*, not wealthy, of small income.

**paur**—Gr. *pauros*, little, small. *Ex:* Paur-odon (Mam.); Paur-urus (Ins.); Pauro-poda (Myr.); Pauro-tylus (Arach.).

**paus**—Gr. *pausis*, cessation. *Ex:* dia-pause; meno-pause.

**pauss**—NL. *paussos*, perhaps from Gr. *Pausos*, a mountain in Greece. *Ex:* Pauss-idae (Ins.); Pausso-brenthus (Ins.); Pausso-tropus (Ins.); Paussus (Ins.).

**pauxi**—*puoxi*, native bird name on the coast of Paria in Venezuela. *Ex:* Pauxia (Av.); Pauxis (Av.).

**pav**—See **pavon.**

**pavett**—NL. *pavetta* < Singhalese *pāvattā*, a plant name. *Ex:* Pavetta*.

**pavid**—L. *pavidus*, timid, filled with fear, trembling < *paveo*, to fear, tremble.

**pavon**—L. *pavo*, genit. *pavonis*, dim. *pavoncella*, a peacock. *Ex:* Pav-uncula (Av.); Pavo (Av.); pavon-ine, peacock blue; Pavonium (Coel.); not Pavon-ia*, named after Dr. Josef Pavon, co-author with Ruez of a Flora Peruviana; Galo-pavo (Av.).

**paxill**—L. *paxillus*, a peg, small stake. *Ex:* paxilla, paxilli form; Paxillo-somus (Ins.); Paxillus (Ins.).

**pe**—Gr. *peos*, the penis. *Ex:* peo-phobia (Med.); Tatarto-peus (Ins.).

**pebrin**—Fr. *pebrine*, a disease of silkworms accompanied with the appearance of black pepper-like spots < Pg. *pebre*, pepper. *Ex:* peb rin-ous; pebrine.

**pec**—Gr. *pekō*, to comb. *Ex:* Peco-pteris*. See also pecor and pectin.

**pecor**—L. *pecus*, genit. *pecoris*, pl. *pecora*, cattle, animals, beasts. *Ex:* Pecora (Mam.).

**pect**—**1.** Gr. *pēktos*, fixed, compact, congealed; related to Gr. *pēxis*, a coagulation, fastening. *Ex:* pect-ase; Pect-ostraca (Crust.); pect-in; pect-iz-ing; pecto-cellul-osis; Pecto-glossa (Amph.); pectose: **2.** L. *pectus*, genit. *pectoris*, breast; *pectoralis*, belonging to or pertaining to the breast. *Ex:* pectoral; pectus. See also pectin.

**pectin**—L. *pecten*, genit. *pectinis*, pl. *pectines*, a comb; also a shell-fish, the scallop; *pectinator*, one who combs < Gr. *pektiō*, to comb. *Ex:* Pect-unculus (Moll.); Pecten (Moll.); pectin-

eus; Pectin-idae (Moll.); pectinate; pectinato-pinnate; Pectinator (Mam.); pectines; Pectini-branchia (Moll.); Pectis*; Pecto-carya*.

**pecuar**—L. *pecuarius*, relating to cattle, a grazier. *Ex:* Pecuarius (Av.).

**ped**—**1.** L. *pes*, genit. *pedis*; dim. *pediculus*, a foot, the base of anything; *pedalis*, of or belonging to the foot. *Ex:* ped-uncle; Ped-unculata (Arth.); pedi-ger-ous; pedi-palp; pedicle; pedo-motor; bi-pes; centi-pede; Cirri-pedia (Arth.); Cypri-pedium*; Podice-ps (Av.), see podic: **2.** Gr. *pedon*, ground, earth. *Ex:* Ped-ina (Arach.); Pedo-mys (Mam.). See also paed: **3.** Gr. *pedē*, a fetter. *Ex:* sym-peda: **4.** *pēdon*, oar, related to *pous*, foot. *Ex:* ? Boo-pedon (Ins.), or it may be from *pedon*, earth, soil in allusion to its grassland habitat.

**pedalio**—Gr. *pēdalion*, a kind of plant; also a prop, a rudder. *Ex:* Pedalion (Rot.), (Moll.); Pedalium*.

**pedamin**—L. *pedamen*, genit. *pedaminis*, a stake, prop, tree; *pedaminus*, staked, propped up.

**pedan**—Gr. *pedanos*, short. *Ex:* Pedano-stethus (Arach.); Pedanus (Ins.).

**pedat**—L. *pedatus*, having feet < *pedo*, to foot, to furnish with feet. *Ex:* Pedata (Echin.); pedati-formis; pedati-sect.

**pedes**—NL. *pedes* < Gr. *pedaō*, to shackle, restrain. *Ex:* Apteno-pedes (Ins.); Atalo-pedes (Ins.).

**pedesis**—Gr. *pēdēsis*, a leaping < *pedaō*, to leap; to throb. *Ex:* dia-pedesis.

**pedet**—**1.** Gr. *pēdētēs*, a leaper, a dancer. *Ex:* Pedetes (Mam.); Poly-pedetes (Amph.): **2.** Gr. *pedētēs*, a prisoner, one chained.

**pedi**—**1.** Gr. *pedion*, a plain, level country; *pedios*, dweller of the plains. *Ex:* Pedio-cetes (Av.); Pedio-nomus (Av.); pedio phytes (Ecol.); ?Cypri-pedium*; Dia-pedium*; Merismo-pedia*. See also ped: **2.** Gr. *pedion*, dim. of *pedē*, a fetter, anklet, instep. *Ex:* Pedio-coccus*.

**pediac**—Gr. *pediakos*, even, smooth. *Ex:* Pedia cus (Ins.).

**pedicel**—See **pedicul 2.**

**pedicell**—See **pedicul 1.**

**pedicul**—**1.** L. *pediculus*, dim. *pedicellus*, a louse; *pediculous*, infested with lice; *pedicularis*, pertaining to lice. *Ex:* Pedicularia (Moll.); Pedicularis*; Pediculus (Ins.): **2.** L. *pediculus* = NL. *pedicellus*, a little foot. *Ex:* pedicel; pedicell-aria; pedicul-ate; Pedicul-ati (Pisc.).

**pedil**—Gr. *pedilon*, a sandal. *Ex:* Pedil-anthus*; Pedil-idae (Ins.); Pedilo-phorus (Ins.); Pedilus (Ins.).

**pedin**—Gr. *pedinos*, flat, found on the plain. *Ex:* Pedino-pelta (Ins.); Pedino-thuria (Echin.); Pedinus (Ins.).

**pedipes**—NL. *pedipes* < L. *pes*, genit. *pedis*, a foot + *pes*, a foot. *Ex:* Pedipes (Moll.).

peduncul—NL. *pedunculus*, dim. of L. *pes*, genit. *pedis* foot. *Ex:* peduncle; peduncul-ar.

peg—1. Gr. *pēgos*, well put together, solid. *Ex:* Pego-pus (Ins.); Pego-somum (Platy.); ptero-pegum: 2. Gr. *pēgē*, well, fountain, stream. *Ex:* Peg-antha (Coel.); Pego-myia (Ins.); Cero-pegia*. See also pegaz.

pegan—Gr. *pēganon*, rue<*pēgos*, solid. *Ex:* Peganum*.

pegas—Gr. *Pēgasos*, Pegasus, a winged horse sprung from the blood of Medusa. *Ex:* Pegasus (Pisc.).

pegaz—Gr. *pēgazō*, to spring forth. *Ex:* Geo-pega (Av.).

pegm—Gr. *pēgma*, genit. *pēgmatos*, anything joined, a frame-work. *Ex:* Pegma-crinus (Echin.); Arachno-pegma (Prot.).

pegrid—L. *pegris*, genit. *pegridis*, a sea-mussel.

pegris—See pegrid.

pel—1. Gr. *pēlos*, clay, mud. *Ex:* pel-ochtho-phyta; Pela-mys (Mam.); Pelo-bates (Amph.); Pelo-dryas (Amph.); Pelo-medusa (Rept.); Pelo-mys (Mam.); pelo-phile: 2. Gr. *pelos*, brown, dusky: 3. Gr. *pelas*, close, handy.

pelag—Gr. *pelagos*, the sea; *pelagios*, of the sea> L. *pelagicus*, of or pertaining to the ocean, marine. *Ex:* Pelag-ornis (Av.); Pelagia (Moll.); Pelagios (Mam.); pelago-phyta*; Pelago-saurus (Rept.); Pelagus (Moll.).

pelam—Gr. *pēlamis*=*pēlamys*, a young tunny fish. *Ex:* Pelam-ichthys (Pisc.); Pelamis (Pisc.).

pelan—Gr. *pelanos*, a thick liquid, a sacrifice offered to the gods. *Ex:* Pelan-echinus (Echin.).

pelarg—Gr. *pelargos*, a stork. *Ex:* Pelarg-anax (Av.); Pelarg-opsis (Av.); Pelargo-crex (Av.); Pelargoni-um*<Fr. *pelargon*, name for the storkbill; Pelargos (Av.).

pelasg—Gr. *Pelasgos*, mythical founder of the Pelasgian race; a dweller of the Peloponnesus >L. *Pelasgia*, the Peloponnesus. *Ex:* Pelasgi-ella (Moll.); Pelasgia (Moll.).

pelec—See pelecy.

pelecan—Gr. *pelekan*, genit. *pelekanos*, a pelican. *Ex:* Pelecan-ichthys (Pisc.); Pelecanus (Av.).

pelecy—Gr. *pelekys*, genit. *pelykeos*, a hatchet; related to the barbarous form *pelyx*. *Ex:* pelec-oid; Peleco-ceras (Moll.); Pelecy-clus (Ins.), see clei; Pelecy-phora*; Pelecy-poda (Moll.); Pelecy-stoma (Ins.); Argyro-pelecus (Pisc.).

peli—Gr. *pelios*, black, black and blue, dark. *Ex:* Peleo-poda (Ins.); Peli-perdix (Av.); Pelio-cypas (Ins.); Pelio-derma (Ins.); pelion; Pelios-anthes*; pelious.

pelia—Gr. *peleia*, a dove. *Ex:* Chaeme-pelia (Av.); Melo-pelia (Av.).

pelichn—Gr. *pelichnē*=*pella*, a wooden bowl, a drinking cup. *Ex:* Pelichni-bothrium (Platy.).

pelidn—Gr. *pelidnos*, livid. *Ex:* Pelidna (Av.); Pelidno-coris (Ins.); Pelidn-ota (Ins.); Pelidnus (Ins.).

pelin—Gr. *pēlinos*, of clay. *Ex:* Pelino-bius (Arach.).

pell—1. Gr. *pellos*=*pelos*, dusky, dark-colored. *Ex:* Pell-aea*; Pello-bunus (Arach.); Pello-chromis (Pisc.); Pello-psyche (Ins.): 2. Gr. *pella*, a cup, a wooden bowl, related to *pelyx*, a basin. *Ex:* platy-pell-ic: 3. L. *pellis*, dim. *pellicula*, a skin, hide; *pellitus*, covered with skin. *Ex:* pell-agra; pelli-branchiate; pellicle; Pellicul-ites (Coel.); Pellicula (Moll.); a-pell-ous.

pellac—L. *pellax*, genit. *pellacis*, deceitful, seductive. *Ex:* Pellax (Moll.).

pellax—See pellac.

pellegrin—It. *Monte Pellegrina*, mountain in Sicily. *Ex:* Pellegrina (Mam.).

pellex—See pellici.

pellici—L. *pellex* (=*paelex*), genit. *pellicis*, a male concubine, a substitute.

pellion—L. *pellio*, genit. *pellionis*, a dealer in furs.

pellit—See pell. 3.

pelluc—L. *pelluceo*=*perluceo*, to shine through< *per*, through+*luceo*, to shine; *pellucidatus*, transparent; *pellucidus*, clear, transparent. *Ex:* pellucid.

pellucen—L. *pellucens*, genit. *pellucentis*, shining; ppr. of *pelluceo*=*perluceo*, to shine, to be transparent, to be clear.

pelm—Gr. *pelma*, genit. *pelmatos*, the sole of the foot; also a stalk. *Ex:* Pelmat-ellus (Ins.); Pelmato-zoa (Echin.); Pelmo-poda (Arach.); Steno-pelmatus (Ins.).

pelop—Gr. *pēlopoios*=*pēloplathos*, a potter. *Ex:* Pelopaeus, also Pelopoeus (Ins.).

pelops—Gr. *Pelops*, son of Tantalus. *Ex:* Pelops (Arach.); Pelopsis (Arach.).

pelor—Gr. *pelōr*, a monster, prodigy; *pelōros*, monstrous. *Ex:* Pelor (Rept.); Pelori-adapis (Mam.); Peloria*; Peloro-rrhinus (Ins.); Pelorus (Moll.).

pelt—Gr. *peltē*, dim. *peltarion*, a small shield; L. *peltatus*, armed with a shield. *Ex:* Pelt-andra*; Pelt-aria*; Pelt-ella (Moll.); peltate; Pelte-philus (Mam.); pelti-form; Pelto-chelys (Rept.); Pelto-phorum*; Holco-pelte (Ins.).

pelurg—Gr. *pēlourgos*, working in mud. *Ex:* Pelurga (Ins.).

pelusi—NL. *pelusios*<Gr. *pēlos*, mud, clay. *Ex:* Pelusios (Rept.).

pelv—L. *pelvis*, a basin. *Ex:* pelvi-form; pelvis.

pelyc—Gr. *pelyx*, genit. *pelykos*, a basin; also very rarely, an axe. See pelecy. *Ex:* Pelyc-ictis (Mam.); Pelyc-odus (Mam.); Pelyc-rhamphus (Mam.); Pelyco-sauria (Rept.).

pelyx—Gr. *pēlyx*, a rent, cleft. See also pelyc.

pemm—Gr. *pemma*, genit. *pemmatos*, a cake. *Ex:* Pemmat-ites (Por.); Pemmato-discus (Platy.).

pemmat—See pemm.

pempher—Gr. *pemphēris*, a kind of fish. *Ex:* Pempheris (Pisc.); Pemphero-mima (Ins.).

pemphig—Gr. *pemphix*, genit. *pemphigos*, a bubble, blister, pustule=*pemphis*. *Ex:* Pemphigaspis (Crust.); Pemphigo-notus (Ins.); Pemphigus (Ins.); Pemphis*; Pemphix (Crust.).

pemphis—See pemphig.

pemphix—See pemphig.

pemphredon—Gr. *pemphrēdōn*, a kind of wasp. *Ex:* Pemphredon (Ins.).

pempt—1. Gr. *pemptos*, the fifth, oneself and four others. *Ex:* Pempto-porus (Myr.): 2. Gr. *pemptō*, to send; *pemptos*, sent.

pendul—L. *pendulus*, hanging down; also doubtful, uncertain<*pendeo*, to hang down. *Ex:* pendul-inus; Fili-pendula*.

pene—L. *paene* (less correctly *pene*), almost. *Ex:* pene-plain.

penelop—1. Gr. *Pēnelopē*, a fem. proper name, lit., a weaver. *Ex:* Penelop-ides (Av.); Penelopina (Av.); Penelope (Av.): 2. Gr. *penelops*, a kind of duck.

penes—Gr. *penēs*, genit. *penētos*, a poor man, a day laborer; *penestēs*, a laborer. *Ex:* Penetoblatta (Ins.); Penestes (Ins.); Penesto-glossa (Ins.); Neuro-penes (Ins.); Phlebo-penes (Ins.).

penest—See penes.

penet—See penes.

peneus—L. *Peneus*<Gr. *Pēneios*, father of Daphne. *Ex:* Peneus (Crust.).

peni—See penis.

penia—Gr. *penia*, poverty, need. *Ex:* Penia-gone (Echin.). See also penis.

penichr—Gr. *penichros*, poor, needy; *penichrotēs*, poverty. *Ex:* Penichro-dema (Ins.); Penichroa (Ins.); Penichrotes (Ins.); Penichrus (Ins.).

penicill—L. *penicillus*=*penicillum*, a painter's brush or pencil of hairs; *penicillatus*, having the form of a pencil, ending in a tuft of fine hairs. *Ex:* Penicill-aria (Por.); Penicill-idia (Ins.); penicill-ate; Penicilli-um*; Penicillus (Moll.); penicill-in.

penis—L. *penis*, the penis, a tail>*penitus*, furnished with a tail; see penit. *Ex:* pen-ile=penial; Penio-cereus*; Penis-comus (Ins.); penis-filum; penisi-form.

penit—L. *penitus*, inward, interior. See also penis.

penn—L. *penn*=*pinna*, feather, pen, wing; *pennatus*=LL. *pennatulus*, winged. *Ex:* Penn-aria (Coel.); Penn-ula (Av.); Penna-poda (Ins.); Pennatula (Coel.); penni-ger-ous; Pennisetum*; penno-pluma; Pinna (Moll.); pinniform. S.

pennat—See penn.

pensia—See pent.

pensil—L. *pensilis*, hanging down. *Ex:* pensile.

pent—Gr. *pente* (often in compounds, *penta-* or *pental-*), five. *Ex:* Pent-horum*; Pent-rem-ites

(Blast.), acc. to some it should be Pen-tremites; Penta-chaeta*; Penta-clethra*; pentadactyl; penta-mer-ous; Penta(s)-acme*; Pental-astrum (Prot.); Acro-pentias (Ins.); Dia-pensia*; Pent-stemon*=Pen-stemon*.

pental—See pent.

pentapet—Gr. *pentapetes*, a name for cinquefoil. *Ex:* Pentapetes*.

Showy Beardtongue, *Penstemon speciosus*, a handsome species of high desert mountains of S. W. United States.

penth—Gr. *penthos*, sorrow, mournfulness; *penthikos*, mournful; *penthētēr*=*penthētōr*, a mourner; *pentheres*, mourning; *penthētrios*, in sign of mourning. *Ex:* Penth-estes (Av.), Gr. *esthēs*, a garment; Penth-ina (Ins.); Pentherichthys (Pisc.); Pentheria (Av.); Penthetor (Mam.); Penthetria (Av.); Penthic-odes (Ins.); Pentho-phlebia (Ins.); Ne-penthes*.

pentheus—Gr. *Pentheus*, legendary king of Thebes. *Ex:* Pentheus (Crust.).

peper—Gr. *peperi*, genit. *pepereōs*, pepper. *Ex:* Peper-omia*, the ending omia of uncertain meaning, but probably meaning like.

pepl—1. Gr. *peplos*, a robe or coat. *Ex:* Peploglyptus (Ins.); Calli-pepla (Av.); Cosmo-pepla (Ins.); Phaino-pepla (Av.): 2. Gr. *peplis*, a plant name for one of the spurges. *Ex:* Peplidium*.

pepo—See pepon.

pepon—Gr. *pepōn*, genit. *peponos*, ripe, mellow> L. *pepo*, genit. *peponis*, a kind of melon. *Ex:* pepo; pepo-ponon-idium; peponium; Peponocranium (Arach.); Pepono-cyathis (Coel.).

peps—Gr. *pepsis*, digestion>Ger. *Pepsin*. *Ex:* pepsin; Pepsis (Ins.); Pepso-nema (Platy.).

pept—Gr. *peptō*=*pessō*, to soften, ripen, digest. *Ex:* pept-ic; pept-ose; pepto-nephr-idia. See also peps and pepton.

pepton—Gr. *pepton*, neut. of *peptos*, cooked, digested. *Ex:* peptone. See also pept.

**per-** —L. *per-* prefix denoting through as in per-oral; all over, during, throughout, by, as in per-ennial; an intensive meaning well, very much, completely as in per-fasciatus; per-niger. See also peri.

**per—1.** Gr. *pĕros*, maimed. *Ex:* Per-otis*, not Perotis (Ins.) which is derived from Gr. *peirō*, to bore through; Per-osis (Ins.); Pera-ceras (Mam.); Pero-dict-icus (Mam.); Pero-pteryx (Av.):    **2.** Gr. *pĕra*, dim. *pĕridion*, a pouch. *Ex:* Pera-cardia (Crust.); Pera-ceras (Mam.); Pera-meles (Mam.); Pera-mus (Mam.); Pera-thereutes (Mam.); Pero-phora (Coel.); Pero-stylus (Moll.); Osteo-pera (Mam.).

**peramium**—NL. *peramium*, a plant name, etym. unexplained. *Ex:* Peramium*.

**peran—**Gr. *perainō*, to finish, conclude, also to penetrate. *Ex:* peran-osis.

**peras—**See perat.

**perat—**Gr. *peras*, genit. *peratos*, limit, boundary; *perātos*, passable; *peratos*, on the opposite side. *Ex:* Perat-odonta (Ins.); Perato-stoma (Moll.) Peratus (Av.).

**perates—**Gr. *peratēs*, a migrant. See also perat.

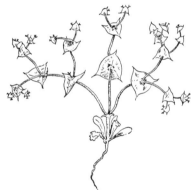

Punctured Bract, *Oxytheca perfoliata*, a queer little plant of the deserts of the S. W. United States. The name *Oxytheca*, lit. spiny cup, refers to the spiny involucre. Redrawn from Desert Wild Flowers—Jaeger. Stanford University Press.

**perc—1.** Gr. *perkē*, the perch, a kind of fish < *perknos*, dark-colored. *Ex:* Perc-esoces (Pisc.); Perc-ichthys (Pisc.); Perc-ophis (Pisc.); Perca (Pisc.); perci-form:    **2.** Gr. *perkos* = *perknos*, a kind of hawk < *perknos*, dark colored.

**percellen—**L. *percellens*, genit. *percellentis*, smiting.

**percit—**L. *percitus*, excited, aroused.

**percn—**Gr. *perknos*, dark-colored. *Ex:* Percno-bapta (Ins.); Percno-pterus (Av.); percno-some.

**percontat—**L. *percontator*, an inquirer, questioner; *percontatrix*, a female questioner; *percontativus*, universally known.

**percurs—**L. *percursus*, passed over; pp. of *percurro*, to run through, to pass over.

**perd—**Gr. *perdomai*, to break wind. *Ex:* Lyco-perdon*. See also perdit.

**perdic—**L. *perdix*, genit. *perdicis*, partridge. *Ex:* Per-ortyx (Av.); Perdic-idae (Av.); Perdix (Av.); Ammo-perdix (Av.); Margaro-perdix (Av.).

**perdit—**L. *perditus*, lost, hopeless, ruined < *perdo*, to destroy; *perditor*, a destroyer. Fr. *perdu*, lost, undone; *Perdita*, disowned, and eventually lost, child of Leontes and Hermione. A character in Shakespeare's Winter's Tale. *Ex:* Perdita (Ins.).

**perditrix—**L. *perditrix*, she that makes ruin < *perditor*, a destroyer.

**perdix—**See perdic.

**perdon—**See perd.

**peregrin—**L. *peregrinus*, strange, foreign. *Ex:* peregrin-oid; Peregrinus (Ins.).

**pereio—**Gr. *peraioō*, inf. *peraioun*, to transport, carry to the other side; *peraios*, beyond the sea, on the other side. *Ex:* Pereio-notus (Crust.); pereio-pod; pereion.

**perempt—**L. *peremptor*, a destroyer. *Ex:* Peremptor (Ins.).

**perenn—**L. *perennis*, throughout the year, through the year. *Ex:* perenni-al; Perenni-branchi-ata (Amph.).

**perennans—**L. *perennans*, continuing, lasting through the years; ppr. of *perenno*, to endure, to last a long time.

**perennit—**L. *perennitas*, continuance.

**peres—**L. *peresus*, worn, wasted, ruined, pp. of *peredo*, to waste, ruin.

**perfect—**L. *perfectus*, complete.

**perforat—**L. *perforatus*, perforated, pp. of *perforo*, to bore through. *Ex:* perforat-orium; Perforata (Coel.).

**perfus—**L. *perfusus*, poured over, moistened, wetted.

**pergamen—**L. *pergamena*, parchment, an invention of Eumenes, King of Pergamon. *Ex:* pergamene-ous.

**pergul—**L. *pergula*, an arbor. *Ex:* Pergul-aria*.

**peri-** —NL. *peri-*, prefix < Gr. *peri*, near, around, roundabout, all around; comparable to L. *circum*, near. *Ex:* Per-chroerus (Mam.); per-embryium; peri-andricus; peri-anth; peri-blem; Peri-bos (Mam.); peri-cardium; peri-carp; Peri-come*; peri-cycle; Peri-dermium*; peri-gynous; peri-osteum; peri-ostracum; peri-patetic; Peri-patus (Arth.); peri-phery; peri-toneum = peri-tonaeum < Gr. *peritonaion*, extension all around.

**perichar—**Gr. *pericharēs*, very glad. *Ex:* Perichares (Ins.).

**periculos—**L. *periculosus*, threatening, full of danger.

perider—Gr. *perideris*, a necklace. *Ex:* Perideropsis (Moll.); Perideris (Moll.).

peridi—Gr. *pēridion*, dim. of *pēra*, a wallet. *Ex:* Peridio-mys; peridium; Peridium (Prot.).

peridin—Gr. *peridineō*, to whirl or wheel around. *Ex:* Peridini-ella (Prot.); Peridinium (Prot.).

perimec—Gr. *perimēkēs*, very long. *Ex.* Perimecoceras (Moll.); Perimekes (Ins.).

perimek—See perimec.

perine—Gr. *perineos*=*perinaion*, the space between the anus and the sexual parts. *Ex:* perineo-scrotal; perineum.

perio—See pereio.

period—Gr. *periodos*, a going around, a period of time. *Ex:* photo-period-ism.

periorg—Gr. *periorgēs*, wrathful. *Ex:* Periorges (Ins.).

peripat—Gr. *peripatos*, a walking about<*peri*, about+*patos*, a way. *Ex:* Peripatus (Arth.).

peripher—Gr. *periphereia*, the circumference of a circle. *Ex:* peripher-al.

periploc—Gr. *periplokē*, a twining, entangling. *Ex:* Periploca*.

perisem—Gr. *perisēmos*, handsome, remarkable. *Ex:* Perisemus (Ins.).

perisoreus—Gr. *perisōreuō*, to heap up all around. *Ex:* Perisoreus (Av.).

periss—Gr. *perisso*=Att. *perittos*, uneven, extraordinary, odd. *Ex:* Periss-odon (Moll.); Perisso-dactyla (Mam.); Perisso-netta (Av.); Perisso-pterus (Ins.); Perissus (Ins.); Perittocrinus (Echin.).

peristalt—Gr. *peristaltikos*, grasping and compressing. *Ex:* peristaltic.

perister—1. Gr. *peristera*, a pigeon, dove. *Ex:* Perister-inae (Av.); Peristero-podes (Av.); 2. Gr. *peristereōn*=*peristerion*, a kind of verbena. *Ex:* Peristeria*; Peristerion*.

peristict—Gr. *peristiktos*, spotted all over, dappled. *Ex:* Peristicta (Ins.).

peritel—Gr. *periteleō*, to finish all round. *Ex:* Peritel-inus (Ins.); Peritelus (Ins.).

peritone—L. *peritoneum*, the membrane surrounding the intestines=Gr. *peritonaion*. *Ex:* peritone-al; peritoneum.

peritt—See periss.

perjur—L. *perjurus*, false, lying.

perl—NL. *perla*, an insect name<Fr. *perla*< Low L. *perula*, a pearl; given in reference to the rounded head "which rivals a pearl in brilliance." *Ex:* Perl-idae (Ins.); Perla (Ins.); Chloro-perla (Ins.).

perlat—L. *perlatus*, carried through, completed, pp. of *perfero*, to carry through.

perm—Russ. *Perm*, a province of Eastern Russia. This combining form is frequently used to denote Permian age. *Ex:* Permi-an; Permocidaris (Echin.).

permeabl—L. *permeabilis*, passable. *Ex:* permeable.

permist—L. *permistus*, mixed, mingled, confused.

permutat—L. *permutatus*, completed, completely changed.

pern—Gr. *perna*, a ham, femur; also a mollusk. *Ex:* Pern-aria (Moll.); Perna (Moll.); Pernatherium (Mam.); Perno-pecten (Moll.); not Pern-opsis (Av.); see pernis.

pernic—L. *pernix*, genit. *pernicis*, nimble, swift.

pernis—NL. *pernis*, a bird of prey, etym. uncertain. *Ex:* Pernis (Av.).

pernix—See pernic.

pernox—L. *pernox*, continuing through the night.

peron—Gr. *peronē*, a brooch, the fibula; also something pointed. *Ex:* Peron-ella (Por.); perone-al; Perone-cera (Ins.); Peroni-aster (Echin.); peroni-um; Perono-scelis (Ins.); Perono-spora* (Coel.); Belo-perone*.

peronat—L. *peronatus*, rough-booted. *Ex:* peronate.

perper—Gr. *perperos*, vain-glorious, boasting. *Ex:* Perperus (Ins.).

perpes—See perpet.

perpet—L. *perpes*, genit. *perpetis*, perpetual, lasting throughout.

perpusill—L. *perpusillus*, very small.

persa—L. *Persa*, daughter of Oceanus. *Ex:* Persa (Ann.).

perscit—L. *perscitus*, very clever, most fine.

persea—Gr. *persea*, name of some fruit-bearing tree. *Ex:* Persea*.

persephon—Gr. *Persephonē*, Proserpine, daughter of Ceres and Jupiter. *Ex:* Persephona (Crust.).

persic—1. L. *persica*, peach. *Ex:* Persic-ula (Moll.); persici-folium; Lyco-persicon*: 2. L. *persicus*, belonging to Persia.

persiccat—L. *persiccatus*, dried, perfectly dry.

persisten—L. *persistens*, genit. *persistentis*, continuing, persevering, persisting; ppr. of *persisto*, to continue.

person—L. *persona*, a mask; *personatus*, wearing a shield.

personat—L. *personatus*, masked. *Ex:* Personatus (Moll.).

perspicillat—L. *perspicillatus*, conspicuous, spectacular.

perstan—L. *perstans*, genit. *perstantis*, fixed, standing firm<*persto*, to stand firm.

persuas—L. *persuasor*, a convincer.

pertenu—L. *pertenuis*, very small or thin<*tenuis*, thin, small.

perth—Gr. *perthō*, to waste, to get by plunder. *Ex:* Pertho-stoma (Ins.); Blito-pertha (Ins.); Rhizo-pertha (Ins.).

perthecat—L. *perthecatus*, well sheathed.

**pertic**—L. *pertica*, a pole, a measuring rod. *Ex:* Pertica (Pisc.).

**pertinac**—L. *pertinax*, genit. *pertinacis*, pertinacious. *Ex:* Pertinac-ides (Ins.); Pertinax (Ins.).

**pertinax**—See **pertinac**.

**pertrit**—L. *pertritus*, worn out, bruised, rubbed to bits.

**pertus**—L. *pertusus*, perforated, punctured < *pertundo*, to pierce. *Ex:* pertusate.

**perul**—L. *perula*, a little pouch, dim. of *pera*, a pouch; NL. *perulatus*, pouch-like; in botany used in sense of furnished with protecting scales or perula. *Ex:* Perul-aria*; perulate; perule.

**perus**—1. NL. *perus* < Gr. *peirō*, to pierce, bore through. *Ex:* Ans-perus (Rot.); Chiro-perus (Rept.): 2. Gr. *pēros*, maimed.

**perv**—L. *pervius*, passable.

**pervalid**—L. *pervalidus*, very strong.

**pervers**—L. *perversus*, overthrown, put to silence, turned the wrong way.

**pes**—See **ped**.

**pessim**—L. *pessimus* (superl. of *malus*, bad), wicked, destructive, ill-formed.

**pessul**—L. *pessulus*, a bolt. *Ex:* pessul-ar; pessulus.

**pet**—L. *peto*, to seek; also to demand, to desire. *Ex:* acro-pet-al.

**petac**—L. *petax*, genit. *petacis*, greedy. *Ex:* Petac-ula (Av.).

**petal**—Gr. *petalos*, outspread, flat; *petalon*, a leaf, flower leaf; also a leaf of metal. *Ex:* petal; Petal-odes (Ins.); Petal-odus (Elasm.); Petal-onyx*; petali-ferous; Petalia (Mam.); Petalo-conchus (Moll.); Petalo-crinus (Echin.); Petalo-stemon*; Loro-petalum*.

**petas**—Gr. *petasos*, a broad-brimmed felt hat; a broad umbellated leaf. *Ex:* Petas-idae (Coel.); Petas-ites*; Petas-unculus (Moll.); Petasi-ger (Platy.); Petaso-phora (Av.); petaso-spores; Petasus (Coel.).

**petasat**—L. *petasatus*, prepared for a journey, having a cap on.

**petasm**—Gr. *petasma*, something spread out. *Ex:* petasma.

**petaur**—Gr. *petauron*, a perch, spring board > *petauristēs*, a tumbler, a rope dancer, a vaulter. *Ex:* Petaur-ula (Mam.); Petaurista (Mam.); Petauristes (Ins.); Eu-petaurus (Mam.).

**petax**—See **petac**.

**peten**—L. *petens*, genit. *petentis*, going, seeking < *peto*, to assail, to go, to seek, also to inquire. *Ex:* alti-petens.

**petes**—NL. *petes*, a flier < Gr. *petomai*, to fly. *Ex:* Acro-petes (Mam.).

**petig**—L. *petigo*, genit. *petiginis*, a scab. *Ex:* Petig-urus (Tri.); Petigo-pora (Bry.).

**petil**—L. *petilus*, slender, thin.

**petin**—Gr. *peteinos*, = *petēnos*, winged. *Ex:* Petino-crinus (Echin.); Petino-pus (Ins.).

**petiol**—L. *petiolus*, a little foot, little leg; stem or stalk < *pes*; NL. *petiolatus*, stalked. *Ex:* petiol-aris; Petiol-urus (Arach.); Petiolata (Ins.).

**petiolan**—NL. *petiolans*, genit. *petiolantis*, producing petioles.

**petr**—Gr. *petra*, a rock; ledge or shelf of rock; *petros*, a stone > *petrōdēs*, like a stone, stony; > L. *petra*, a rock; *petronius*, of or belonging to a mountain, rock; *petrosus*, full of rocks, rocky. *Ex:* Petr-aster (Echin.); Petra-doria*; Petri-cola (Moll.); petri-fy; Petro-dromus (Mam.); Petro-selinum*; petrodo-phyta; Petronius (Av.); petros-al; Em-petrum*, not Petrea*, named after Lord Petre, botanical patron of England.

**petrae**—L. *petraeus*, growing among rocks < *petra*, a rock.

**petrod**—See **petr**.

**petron**—See **petr**.

**petulan**—L. *petulans*, genit. *petulantis*, freakish, impudent.

**petun**—NL. *petunia* < Fr. *petun*, tobacco. *Ex:* Petunia*.

**peuc**—Gr. *peukē*, a pine, fir. *Ex:* Peuc-estes (Ins.); Peuc-aea (Av.); Peuce-dramus (Av.); Peuce-phyllum*; Peuco-glyphus (Ins.).

**peucedan**—Gr. *peukedanon*, name for hog's fennel, an umbelliferous plant < *peukē*, pine + *danos*, parched. *Ex:* Peucedanum*.

**peus**—See **pe**.

**pex**—1. L. *pexus*, wooly < *pecto*, to comb. *Ex:* Pexo-cnem-idae (Ins.); Pexo-mya (Ins.); Pexo-tettix (Ins.): 2. Gr. *pexis*, a fixing, fastening.

**pexat**—L. *pexatus*, clothed in a wooly garment < *pexus*, wooly.

**pez**—1. Gr. *pezos*, on foot, walking. *Ex:* Pez-erpes (Ins.); Pezo-phaps (Av.); Pezo-porus (Av.); Pezo-thrips (Ins.); Pezus (Av.); Platy-pez-idae (Ins.); Tany-peza (Ins.): 2. Gr. *pezis*, a stalkless fungus.

**peziz**—L. *pezica*, a sessile mushroom < Gr. *pezis*, a sessile mushroom, puffball; *pezikēs*, mushrooms without foot or stalk. *Ex:* Peziza*. See pez.

**pezopor**—Gr. *pezoporos*, going by land passage. *Ex:* Pezoporus (Av.).

**phab**—Gr. *phabo-*, in compounds meaning a dove < *phaps*, genit. *phabos*, a wild pigeon. *Ex:* Phab-alectryo (Av.); Phabo-typus (Av.).

**phac**—Gr. *phakos*, a lentil, lentil-shaped, a seed; in modern technical terms it often means a lens. *Ex:* Phac-idia*; Phac-ops (Tri.); Phac-ota (Ins.); Phaca*; Phace-phorus (Ins.); phaco-cyst; Phaco-choerus (Mam.); Phacus (Prot.); Calo-phaca*.

phacel—Gr. *phakelos=phakellos*, a cluster, a bundle. *Ex:* Phacelia*; Phacelo-domus (Av.); Phacello-pegma (Por.); Phakelia (Por.).

phae—Gr. *phaios*, dusky, dark, gray. *Ex:* Phae-aster (Prot.); phaeo-chrous; Phaeo-laema (Av.); Phaeo-phyc-eae*; phaeo-plast; Phaeo-pus (Av.).

phaec—Gr. *phaikos*, splendid. *Ex:* Phaeca (Av.).

phaedr—Gr. *phaidros*, bright, beaming>*phaidrotēs*, sprightliness. *Ex:* Phaedra-nassa*; Phaedro-ctonus (Ins.); Phaedro-tettix (Ins.); Phaedrotes (Ins.); Phaedrus (Av.).

phaedrom—See phaidrom.

phaen—See phaner.

phaenolis—Gr. *phainolus*, light-giving. *Ex:* Phaenolis (Ins.).

phaeo—See phaio.

phaeth—Gr. *Phaethōn*, the sun god, Helios; *phaethontis*, shining. *Ex:* Phaeth-ornis (Av.); Phaethon (Av.); Phaethont-idae (Av.).

phaeton—Fr. *phaeton*, lit. chariot of the sun<Gr. *Phaethōn*, the sun god, Helios. *Ex:* Phaeton (Ins.).

phag—Gr. *phagō*, to eat. *Ex:* phago-cyte; cyto-phage; eso-phagus=oeso-phagus; Eu-phagus (Av.); macro-phage; Xylo-phaga (Ins.).

phaget—Gr. *phagēton=phagēma*, food.

phaidr—See phaedr.

phaidrom—Gr. *phaidromos*, neat, bright, agile. *Ex:* Phaedromus (Av.).

phain—See phaner.

phaio—Gr. *phaios*, dark, dusky. *Ex:* Phaeo-chroa (Av.); Phaeo-gryllus (Av.); Phaio (Ins.); Phaio-mys (Mam.); Phaio-spor-ales*; Phajus*.

phaj—See phaio.

phak—See phac.

phal—1. Gr. *phalos*, light, shining, white. *Ex:* Spheno-phalos (Mam.): 2. Gr. *phalēs=phallos*, the penis. *Ex:* Tetra-phalus (Ins.). See phall.

phalacr—Gr. *phalakros*, bald. *Ex:* Phalacro-corax (Av.); Phalacro-seris*; Phalacrus (Ins.).

phalaen—Gr. *phalaina=phallaina*, a devouring monster, a whale; also a moth. *Ex:* Phalaen-opsis*; Phalaena (Ins.); Phalaeni-vora (Av.); Phalaeno-ptilus (Av.); Phalain-esthes (Ins.); Phalaino-soma (Ins.).

phalain—See phalaen.

phalang—1. Gr. *phalanx*, genit. *phalangos*, a line of battle, soldiers formed in ranks; also the bone between the joints of the finger. *Ex:* Phalang-acris (Ins.); phalange; Phalanger (Mam.); phalanges, Eng. pl. of phalanx or phalange; Phalango-pora (Coel.): 2. Gr. *phalangion*, a spider. *Ex:* Phalang-idae (Arach.); Phalangium (Arach.).

phalanth—Gr. *phalanthos*, bald in front. *Ex:* Phalantho-rrynchus (Ins.).

phalar—Gr. *phalaris*, old name for a grass; also a coot<*phalaros*, having a patch of white (on its forehead.). *Ex:* Phalar-idion (Av.); Phalaris*; Phalaro-pus (Av.).

phaler—Gr. *phalēros*, Ionic for *phalaros*, with a patch of white; L. *phalero*, to adorn, display. *Ex:* Phaleria (Ins.); Phalero-pygus (Echin.).

phall—Gr. *phallos*, a stick of wood cut as a symbol of the male organ and carried in processions, the penis. *Ex:* phall-ic; Phallo-soma (Geph.); Phallus*; Phallusi-opsis (Tun.); A-phall-arion (Moll.); Acantho-phallus (Rept.); Amorpho-phallus*. See also phal.

phan—See phaner.

phanae—Gr. *phanaios*, light bringing, emitting brightness. *Ex:* Phanaeus (Ins.).

phaner—Gr. *phaneros*, visible, open, evident<*phainō*, to show; *phanos*, light; a torch. *Ex:* Phaeno-coma*; Phaino-pepla (Av.); Phaner-aster (Echin.); phanero-gam-ous; Phanero-pter-inae (Ins.); Phanero-zonia (Echin.); pheno-type; A-phanisma*; A-phano-trochus (Moll.); Nycti-phanes (Arth.).

phanogen—Gr. *phanogeneia*, born of a sea-god. *Ex:* Phanogenia (Echin.).

phant—Gr. *phantos*, visible. *Ex:* A-phant-aphis

Stinkhorn Fungus, *Phallus impudicus*. Redrawn from The Romance of The Fungus World —Rolfe. J. B. Lippincott Co., Chapman & Hall, Ltd.

(Ins.); A-phanto-loba (Ins.); Bathy-phantes (Arth.).

phantas—Gr. *phantasis*, a show, appearance. *Ex:* Phantasio-myia (Ins.); Phantasis (Ins.).

phantasm—Gr. *phantasma*, an appearance, image, vision. *Ex:* Phantasma (Ins.); Phantasmi-ella (Ins.); Phantasmo-coris (Ins.).

phao—Gr. *phaos*, *phaeos*, Att. contr. *phōs*, genit. *phōtos*, light. *Ex:* phao-meter; Phaos (Ins.).

phaon—*Phaōn*, legendary youth of Lesbos, beloved by Sappho. *Ex:* Phaon (Ins.).

phaps—Gr. *phaps*, genit. *phabos*, a wild pigeon. *Ex:* Phaps (Av.); Phapsi-treron (Av.); Lopho-phaps (Av.); Odidi-phaps (Av.). See also phab.

phar—1. Gr. *pharos*, a web, piece of cloth, a mantle, a garment > *aphares*, without covering, unclad. *Ex:* Phare-odus (Pisc.); Pharo-macrus (Av.); Pharus*; Aphare-us (Crust.); Aphareo-caris (Crust.): **2.** NL. *phar*, having < Gr. *phareō*, to have. *Ex:* Calli-phari-xen-idae (Ins.); **3.** Gr. *pharos*, a lighthouse: **4.** Gr. *pharos*, a plow.

pharang—Gr. *pharanx*, genit. *pharangos*, a cleft, chasm. *Ex:* Pharang-ispa (Ins.).

pharaon—L. *Pharao*, genit. *Pharaonis*, Pharaoh, title of Egyptian kings. *Ex:* Pharaon-aster; (Echin.); Pharaon-ella (Moll.); Pharaonis (Moll.).

pharc—Gr. *pharkis*, genit. *pharkidos*, a wrinkle. *Ex:* Pharci-ceras (Moll.); Pharcid-ella (Moll.); Pharcido-notus (Moll.); Pharkido-notus (Av.).

pharetr—Gr. *pharetra* = *pharetreōn*, a quiver for arrows. *Ex:* Pharetra (Moll.); Pharetria (Prot.); Pharetro-spongia (Por.); pharetron-id.

pharmac—Gr. *pharmakon*, a healing drug, poison; *pharmakos*, a poisoner, magician (the combining form *pharmac-* denotes a poison). *Ex:* pharmaco-logy; Pharmaco-phagus (Ins.); Pharmacus (Ins.).

pharnac—L. *Pharnaces*, king of Pontus. *Ex:* Pharnaceum*.

phars—Gr. *pharsos*, a piece torn off > *pharsophorus*, a standard-bearer. *Ex:* Pharsophorus (Mam.).

pharyng—Gr. *pharynx*, genit. *pharyngos*, the pharynx. *Ex:* Pharyng-ella (Prot.); pharyngobranchial; Pharyngo-dictyon (Tun.).

pharynx—See pharyng.

phas—Gr. *phasis*, an appearance aspect; also rarely, an assertion. *Ex:* Di-phasia (Coel.); Myio-phasia (Ins.); telo-phase; Xylo-phasia (Ins.). See phasian.

phasc—Gr. *phaskon* = *phaskos* = *sphagnos*, a kind of moss growing in trees. *Ex:* Phasc-aceae*; Phascum*.

phasco—See phascol.

phascol—Gr. *phaskōlos*, a leather bag, pouch. *Ex:* Phasco-gale (Mam.); Phascolo-mys (Mam.); Phascolo-soma (Ann.); Phascolo-therium (Mam.).

phasel—See phaseol.

phaseol—Gr. *phasēlos* = *phaseolos* = *phasiolos*, a kind of bean with an edible pod, kidney bean, > *phasēlos*, a little boat. *Ex:* phaseol-ite; Phaseolus*, (Moll.).

phasgan—Gr. *phasganon*, a sword. *Ex:* Phasgan-odus (Pisc.); Phasgan-ura (Ins.); Phasganophora (Ins.); Phasgania (Ins.); Phasganus (Pisc.).

phasian—Gr. *Phasis* = L. *Phasis*, genit. *Phasidis* or *Phasidos*, the river Phasis in Colchis, which flows into the Black Sea > L. *phasianus* = Gr. *phasianos*, a pheasant. *Ex:* Phasian-ella (Moll.); Phasian-urus (Av.); Phasiano-trochus (Moll.); Phasidus (Av.); Oreo-phasis (Av.).

phasid—See phasian.

phasm—Gr. *phasma*, genit. *phasmatos*, an apparition, vision, a monster. *Ex:* Phasm-ichthys (Pisc.); Phasma (Ins.); Phasma-gyps (Av.); Phasmat-opsis (Moll.); Phasmo-mantis (Ins.).

phasmat—See phasm.

phassa—Gr. *phassa*, the ring dove, a wild pigeon. *Ex:* Phassa (Av.); Petro-phassa (Av.).

phatag—East Indian *phatagin* = *phatagen*, name for the scaly ant-eater. *Ex:* Phatages (Mam.); Phatagin-us (Mam.).

phatn—Gr. *phatnē*, a feeding trough, a panel. *Ex:* Phatn-acra (Ins.); Phatn-aspis (Prot.); Archiphatna (Prot.); Henico-phatnus (Ins.)

phaul—Gr. *phaulos* = collat. form *phlauros*, little, easy, light, simple, despised, rude. *Ex:* Phaulacridium (Ins.); Phaulo-mys (Mam.); Phaulotettix (Ins.); Phlauro-lestes (Ins.); Phaylopsis*.

phaus—Gr. *phausis*, a lighting, shining bright. *Ex:* Phaus-ina (Arach.); Phausis (Ins.); Euphausia (Crust.).

phayl—See phaul.

phebal—See phibal 2.

pheg—Gr. *phēgos*, a kind of oak with edible nut, the beech tree. *Ex:* Pheg-ornis (Av.); Phegomyia (Ins.); Phego-pteris*; Epi-phegus*.

phegg—Gr. *pheggos*, genit. *pheggeos*, splendor, lustre. *Ex:* Pheggo-misetes (Ins.); Argyrophegges (Ins.).

pheid—Gr. *pheidos*, thrifty > *pheidōlos*, as substantive, a miser; as adjective, sparing. *Ex:* Pheidole (Ins.); Pheidoli-phila (Ins.); Pheidologeton (Ins.).

pheidol—See pheid.

phele—Gr. *phēloō*, to deceive, ensnare; *phēlōma*, a deception; *phelētēs*, a cheat, knave. *Ex:* Pheles (Ins.); Pheletes (Ins.); Chondro-phelia (Ins.); Demo-pheles (Ins.).

phelet—See phele.

phell—1. Gr. *phelleus*, stony ground. *Ex:* phellophilous; phello-phyta: **2.** Gr. *phellos*, cork > *phellinos*, made of cork. *Ex:* Phelli-opsis (Coel.); Phellin-odes (Ins.); Phelline*; Phello-dendron*; Phello-derma (Por.); phello-gen.

phellin—See phell.

phelom—See phele.

phem—Gr. *phēmē*, a voice, report, fame. *Ex:* Pheme(r)-anthus*; brady-phemia; Poly-phemus (Ins.) < *Polyphēmos*, the many-voiced giant.

phemonoe—L. *Phemonoe* < Gr. *Phēmonon*, daughter of Apollo. *Ex:* Phemonoe (Coel.), (Av.), (Ins.).

phenac—Gr. *phenax*, genit. *phenakos*, a cheat, imposter. *Ex:* Phenac-arion (Moll.); Phenacodus (Mam.); Phenaco-bius (Pisc.); Phenacomys (Mam.); Phenax (Rept.); Plectro-phenax (Av.).

**phenax**—See phenac.

**phenes**—Gr. *phēnē*, a sea-eagle, the osprey. *Ex:* Phene (Av.).

**pheng**—Gr. *phengō*, to make bright, shine. *Ex:* Pheng-odes (Ins.); Phengus (Moll.).

**pheno**—See phaner, also phenomen.

**phenomen**—Gr. *phainomenon*, that which is seen, a phenomenon. *Ex:* pheno-logy = phenomenology.

**pheo**—See phaio.

**pher**—Gr. *pherō*, to bear, to carry. *Ex:* Pherocladus (Ins.); Phero-trichis*; peri-pher-y.

**pherecyd**—Gr. *Pherekydēs*, celebrated philosopher from Syros, instructor of Pythagoras. *Ex:* Pherecydes (Arach.) = Pherecides (Arach.).

**pherus**—Gr. *Pherousa*, daughter of Nereus and Doris. *Ex:* Pherusa (Moll.), etc.; Pherusi-dae (Moll.).

**pheug**—Gr. *pheugō*, to flee. *Ex:* Pheug-inus (Ins.); Pheugo-podius (Av.).

**phial**—1. Gr. *phialē*, a broad flat vessel or bowl. *Ex:* Phial-actis (Coel.); phiali-form; Phialosphaera (Ins.); phialo-spore: 2 L. *Phiala* = Gr. *Phialē*, companion of Diana. *Ex:* Phiala (Ins.); Phiale (Arach.).

**phiar**—Gr. *phiaros*, gleaming, shining, bright. *Ex:* Phiara (Ins.); Phiarus (Ins.).

**phibal**—1. Gr. *phibaleōs*, a kind of fig. *Ex:* Phibalo-thrips (Ins.): 2. Gr. *phibaleē*, a kind of myrtle. *Ex:* Phibalium*.

**phidipp**—Gr. *Pheidippos*, grandson of Hercules. *Ex:* Phidippia (Arach.); Phidippus (Arach.).

**phidol**—See pheid.

**phil**—Gr. *philos*, loving, fond of, having affinity for; also a friend > *philesis*, a loving; *philētōr*, a lover; *philikos*, amiable, friendly. *Ex:* Philacarus (Arach.); Phil-etaerus (Av.); Philesia*; Philetor (Mam.); Philo-dina (Prot.); Philo-hela (Av.); Philo-pterus (Ins.); Philo-scia (Arth.); not Philo-tria* see phyll and tria; Ammo-phila*; Dino-philus (Ann.).

**philadelph**—Gr. *philadelphon*, a sweet-flowering shrub, perhaps the jasmine. *Ex:* Philadelphus*.

**philaen**—NL. *philaenus* < L. Philaeni, patriots of Carthage who, because of love of country, submitted to living burial. *Ex:* Philaenus (Ins.).

**philantomb**—Liberian *filintongue*, name for an antelope. *Ex:* Philantomba (Mam.).

**phile**—Gr. *phileō*, to love. *Ex:* Phile-pitta (Av.).

**philedon**—Gr. *philēdonos*, fond of pleasure. *Ex:* Philedon (Av.).

**philes**—See phil.

**philet**—See phil.

**phill**—See phyll.

**phillyr**—NL. *phillyrea* < Gr. *phillyrea*, a plant name used by Theophrastus < Gr. *philyna*, the linden tree. *Ex:* Phillyrea*.

**philomides**—Gr. *philomeidēs* = *philommeidēs*, glad, cheering. *Ex:* Philomides (Ins.).

**philydr**—Gr. *philydros*, water-loving. *Ex:* Philydra*.

**philyn**—See phillyr.

**philypn**—Gr. *philypnos*, sleep-loving. *Ex:* Philypnus (Pisc.).

**philyre**—Gr. *philyrea*, a kind of shrub, philyrea. *Ex:* Philyrea*.

**phim**—Gr. *phimos*, a muzzle > *phimosis*, a muzzling, that which stops an opening. *Ex:* Phimocrinus (Echin.); Phimo-phorus (Ins.); phimosis; Phimosus (Av.); Dicto-phimus (Prot.).

**phiomia**—*Faiyum*, lit. lake province, a district of Egypt. *Ex:* Phiomia (Mam.).

**phitecus**—See pithec.

**phlao**—Gr. *phlaō*, to crush, to eat greedily. *Ex.* Phlao-cyon (Mam.).

**phlattothrat**—Gr. *phlattothrat*, "sound and fury without sense." *Ex:* Phlattothrata (Arach.).

**phlaur**—Gr. *phlauros*, trivial, useless. *Ex:* Phlauro-centrum (Prot.).

**phle**—1. Gr. *phleōs*, an old name for a marsh reed. *Ex:* Phleo-cryptes (Av.); phleo-ides; Phleum*: 2. Gr. *phleō*, to teem with abundance, to babble.

**phleb**—Gr. *phleps*, genit. *phlebos*, vein, blood vessels; *phlebōdēs*, with large veins, full of veins. *Ex:* phleb-itis (Med.); Phleb-odium*; phlebo-edesis; Phlebo-tomus (Ins.); Lepto-phlebia (Ins.); Lysi-phlebus (Ins.); Oligo-phlebodes (Ins.).

**phleg**—Gr. *phlegōn*, flame < *phlegō*, to kindle, burn up. *Ex:* Phlegon (Ins.).

**phlegeth**—Gr. *phlegethō*, to burn, scorch; *Phlegethōn*, fiery river of the underworld. *Ex:* Plegethontia (Amph.).

**phlegm**—1. Gr. *phlegma*, genit. *phlegmatos*, heat, later meaning phlegm; *phlegmatikos*, like phlegm, dull, stupid, heavy; *phlegmatōdēs*, inflammatory. *Ex:* Phlegma-cera (Arach.); phlegmatic; Phlegmo-doris (Moll.); Chryso-phlegma (Av.): 2. *phlegmos* = *phlogmos*, blood.

**phlegmat**—See phlegm 1.

**phlegr**—Gr. *Phlegra*, a city in Macedonia. *Ex:* Phlegra (Arth.).

**phleum**—Gr. *phleōs*, a water plant, a kind of flowering rush or reed. *Ex:* Phleum*.

**phlex**—Gr. *phlexis*, an unknown bird. *Ex:* Uro-phlexis (Av.).

**phli**—Gr. *phlia*, doorposts. *Ex:* Titano-phlium (Coel.).

**phlib**—Gr. *phlibō* = *thlibō*, to press, squeeze. *Ex:* Phlibo-stoma (Ins.).

**phlict**—See phylkt.

**phlictaen**—See phylkt.

**phlips**—Gr. *phlipsis*, pressure < *phlebō*, to squeeze, straighten. *Ex:* Lordo-phlips (Ins.).

**phloe**—Gr. *phloios*, bark of tree, rind. *Ex:* Phloe-odes (Ins.); phloem; Phloeo-mys (Mam.); Phloeo-spora*; phloeo-terma; Drymo-phloeus*; meso-phloem.

**phloem**—See phloe.

**phlog**—Gr. *phlox*, genit. *phlogos*, a flame, hence reddish. *Ex:* Phlog-acanthus*; Phlog-monas (Av.); phlogi-folius; Phlogo-thauma (Ins.); Phlox*; Hydro-phlox (Pisc.).

**phloi**—See phloe.

**phlom**—Gr. *phlomos*, mullein. *Ex:* Phlomis*.

**phlox**—See phlog.

**phlyar**—Gr. *phlyaros* = *phlyaria*, foolish silly talk, nonsense. *Ex:* Phlyaria (Ins.).

**phlyct**—See phlykt.

**phlyctaen**—See phlykt.

**phlyctain**—See phlykt.

**phlycten**—See phlykt.

**phlyda**—Gr. *phlydaō*, to have an excess of moisture > *phlydaros*, flabby. *Ex:* Phlyda (Ins.).

**phlykt**—Gr. *phlyktis*, genit. *phlyktidos*; also *phlyktaina*, dim. *phlyzakion*, a blister. *Ex:* Phlictaeno-pora (Por.); Phlyctaeno-gastra (Ins.); Phlycti-ceras (Moll.); phlyktio-plankton (Ecol.); Phlysacium (Tri.); Phlyzakium (Myr.); Eu-phlictis (Amph.).

**phlys**—Gr. *phylsis*, an eruption.

**phlysac**—See phlykt.

**phlyzac**—See phlykt.

**phob**—1. Gr. *phobos*, fright, panic, terror. *Ex:* phob-ic; Phobo-campe (Ins.):   2. *phobē*, a lock or curl of hair.

**phober**—Gr. *phoberos*, formidable. *Ex:* Phobereo-therium (Mam.); Phobero-mys (Mam.).

**phobetic**—Gr. *phobetikos*, timid. *Ex:* Phobetico-myia (Ins.).

**phobetr**—Gr. *phobētron*, a scare-crow, bug-bear. *Ex:* Phobetro-mimus (Ins.); Phobetron (Ins.); Phobetrum (Ins.).

**phoc**—Gr. *phōkē*, a kind of mammal, a seal. *Ex:* Phoc-ascaris (Nemat.); phoc-ine; Phoc-odontia (Mam.); Phoca (Mam.); Phoca-geneus (Mam.); Phocan-ella (Mam.); Phoco-cetus (Mam.).

**phocaen**—Gr. *phōkaina*, porpoise. *Ex:* Phocaen-opsis (Mam.); Phocaena (Mam.).

**phod**—Gr. *phōs*, genit. *phōdos* or *phōis*, genit. *phōidos*, a blister. *Ex:* Phodo-mys (Mam.).

**phoeb**—Gr. *phoibos*, pure, radiant. *Ex:* Phoeb-odus (Pisc.); Phoebo-philus (Ins.).

**phoebetr**—Gr. *phoibētria*, a soothsayer, purifier. *Ex:* Phoebetrea (Av.).

**phoenic**—Gr. *phoinix*, genit. *phoinikos*, purple-red, crimson; also the date palm; a fabulous bird. *Ex:* Phoenic-ulus (Av.); Phoenico-pterus (Av.); Phoenix*; Acantho-phoenix*; Archonto-phoenix*.

**phoenix**—See phoenic.

**phoeno**—See phaen, also phon 2.

**phoin**—See phon 2.

**phol**—1. Gr. *pholis*, genit. *pholidos*, a horny scale, particularly of reptiles; also a kind of fish; *pholidōtos*, armed with scales. *Ex:* Pholi-urus*; Pholid-a-pus (Pisc.); pholid-osis; Pholido-carpus*; Pholidota*; Pholisma*; A-pholid-emys (Rept.); Di-pholis*; Ophio-pholis (Echin.):   2. Gr. *phōlas*, genit. *phōlados*, lurking in a hole; also a mollusc that makes holes in stones < *phōleuō*, to lurk in a hole. *Ex:* Phola-meria (Moll.); Pholad-idae (Moll.); Pholado-blatta (Ins.); Pholas (Moll.); Pholia (Av.); Para-pholas (Moll.).

**pholad**—See phol 2.

**pholc**—Gr. *pholkos*, squint-eyed. *Ex:* Pholc-idae (Arach.); Pholco-phora (Arach.); Pholcus (Arach.).

**phole**—Gr. *phōleos*, a hollow, lurking place. *Ex:* Pholeo-philus (Rept.); Pholeo-ptynx (Av.).

**pholet**—Gr. *phōlētēr*, one who hides in a den or tends to live in one place.

**pholeuon**—Gr. *phōleuon*, lurking in a den. *Ex:* Pholeuon (Arach.); Pholeuon-opsis (Ins.).

**pholid**—See phol 1.

**phollicodes**—Gr. *phollikōdēs*, full of cavities, spongy. *Ex:* Phollicodes (Ins.).

**pholoe**—Gr. *Pholoē*, mountain in Thessaly where the Centaurs lived. *Ex:* Pholoe (Ann.).

**phom**—NL. *phoma*, name of a fungus < Gr. *phōis*, blister, burn. *Ex:* Phom-opsis*; Phoma*.

**phon**—1. Gr. *phōnē*, sound, voice. *Ex:* phon-ation; A-phonia (Ins.); A-phono-gryllus (Ins.):   2. Gr. *phonos* = *phoinos*, murderous, blood-red; also, murder, homicide; *phonōdēs*, like blood; *phoneus*, a murderer; *phoneutēs*, fem. *phoneutria*, a murderer; *phonoktonos*, murdering. *Ex:* Phoneutria (Av.); Phoneyusa (Arach.); Phonio-cercus (Arach.); Phono-rhynchus (Platy.); Phonoc-dromus (Mam.); Phonoctonus (Ins.); Myio-phoneus (Av.); Thely-phon-idae (Arach.).

**phonac**—Gr. *phonax*, genit. *phonakos*, eager for blood.

**phonax**—See phonac.

**phone**—See phon.

**phoneus**—See phon 2.

**phonio**—See phon 2.

**phonoc**—See phon 2.

**phont**—Gr. *phontēs*, a murderer. *Ex:* Clisio-phontes (Prot.).

**phor**—1. Gr. *phōr*, thief; also, a kind of bee, the robber bee. *Ex:* Phora (Ins.); Phor-ella (Dipt.); Phora-dendron*; Phora-rhacos (Av.):   2. Gr. *phora*, motion, movement. *Ex:* phoro-zoo-id; phoro-zoon:   3. Gr. *phoros*, a bearing; *phoreus*, a bearer < *pherō*, to bear. *Ex:* phor-anthium; Phor-odon (Ins.); Phoreo-tropis (Tri.); chromato-phore; gono-phore; photo-phore; not Tri-phoris (Moll.), nor Tri-phorus (Moll.), both of which are badly formed in part from L. *foris*, a gate or door: **4.** Gr. *phŏros*, a detector.

**phorb**—Gr. *phorbē*, pasture, fodder; *phorbas*, genit. *phorbados*, giving pasture or food; *euphorbos*, well fed. *Ex:* Phorb-antus (Mam.), see anta; Phorbas (Ins.); Phorbia (Ins.); Hyo-phorbe*; not Euphorb-ia*, which was named after Euphorbus, ancient physician; meso-phorbium.

**phorc**—Gr. *phorkos*, white, gray. *Ex:* Phorco-rraphis (Crust.); Phorcus (Moll.).

**phores**—Gr. *phorēsis*, a being borne, a wearing. *Ex:* phoresy; cata-phoresis.

**phoret**—Gr. *phorētos*, borne, carried, moveable. *Ex:* Phoreto-phyra (Prot.).

**phorm**—Gr. *phormos*, dim. *phormis*, genit. *phormidos*, a woven basket, a mat; *phormion*, a mat, something plaited; also a kind of plant. *Ex:* Phormium*; Phormo-spyris (Prot.); Arto-phormis (Prot.).

**phorminct**—Gr. *phorminktēr*, a harpist. *Ex:* Phormincter (Ins.).

**phorming**—Gr. *phorminx*, genit. *phormingos*, an ornamented lyre. *Ex:* Phormingo-chilus (Arach.).

**phorminx**—See **phorming**.

**phoron**—L. *Phoronis*, surname of Io, daughter of Inachus who was changed into a white heifer and afterwards wandered all over the earth before being restored to her original form. *Ex:* Phoron-ida (Phor.); Phoronis (Phor.).

**phort**—Gr. *phortis*, a small ship, a ship of burden; *phortion*, a load, burden. *Ex:* Phortion (Moll.); Spongo-phortis (Prot.).

**phortic**—Gr. *phortikos*, fit for carrying; also vulgar, common. *Ex:* Phortic-ella (Ins.); Phortica (Ins.); Phortico-somus (Ins.); Phorticus (Ins.).

**phos**—See **phot**.

**phot**—Gr. *phōs*, genit. *phōtos*, light; *phōteinos*, shining, light. *Ex:* Phos (Moll.); Phos-phorus (Ins.); phos-phoreo-centa; Photinia*; photon; photo-genic; photo-phore; photo-trop-ism; Tri-phosa (Ins.).

**photin**—See **phot**.

**phox**—Gr. *phoxos*, tapering, pointed. *Ex:* Phox-aster (Echin.); Phoxi-chilus (Pantopoda); Phoxo-cephalus (Arth.).

**phoxin**—Gr. *phoxinos*, an unknown river-fish. *Ex:* Phoxinus (Pisc.).

**phoyx**—Gr. *phōyx*, a species of heron. *Ex:* Meso-phoyx (Av.).

**phract**—Gr. *phraktos*, fenced in, protected. *Ex:* Phract-amphibia (Amph.); Phract-ura (Pisc.); Phracto-mys (Mam.); Caro-phractus (Ins.).

**phrad**—Gr. *phradē*, a hint, understanding; *phradēs*, prudent, skillful. *Ex:* Phradis (Ins.); Phrado-noma (Ins.); Caeno-phrada (Ins.).

**phragm**—Gr. *phragma*, genit. *phragmatos*, hedge, fence; *phragmos*, a hedge, barricade; *phragmitēs*, growing in hedges. *Ex:* Phragmato-pora (Coel.); Phragmites*; phragmo-cone; Phragmo-pedilum*; dia-phragm.

**phrass**—See **phrax**.

**phrast**—Gr. *phrastōr*, a guide; *phrastēr*, an informer, expounder; *phrastikos*, suited for guiding. *Ex:* Phrastero-thrips (Ins.); A-phrast-ura (Av.); A-phrasto-bracon (Ins.).

**phrat**—Gr. *phratēr*, genit. *phrateros*, brothers of a tribe; *phratoria = phatra*, a clan, tribe. *Ex:* Phratora (Ins.); phratry. See also frater.

**phrax**—NL. *phrax* < Gr. *phrassō*, to fence in, to stop up.

**phreat**—Gr. *phrear*, genit. *phreatos*, a well, cistern, pit. *Ex:* Phreat-ichthys (Pisc.); Phraeto-bius (Pisc.); Phreato-thrix (Ann.).

**phren**—Gr. *phrēn*, diaphragm; also the heart or mind, the brain. *Ex:* phren-ic; phreno-gastric; Phreno-picus (Av.); Za-phrentis (Coel.).

**phreoryct**—Gr. *phreōryktēs*, a well digger. *Ex:* Phreoryctes (Ann.); Phreoryctus (Ann.).

**phret**—Gr. *phreatia*, a tank, basin, well. *Ex:* phreto-philus (Ecol.).

**phric**—Gr. *phrix*, genit. *phrikos*, a ripple; *phrikōdēs*, causing fits, shivering, horrible. *Ex:* Phrico-carabus (Ins.); Phricodo-ceras (Moll.); Lica-phrium (Mam.), err. for Lica-phricum; Ophio-phrixus (Echin.).

**phrict**—Gr. *phriktos*, causing one to shudder, horrible. *Ex:* Phricta (Ins.); Phrictus (Ins.).

**phriss**—Gr. *phrissō = Att. *phrittō*, to be rough on the surface, to stand on end, to feel chill, to shudder *Ex:* Phrisso-trich-ium (Ins.); Phrysso-cystis (Echin.); Cocco-phrissus (Ins.).

**phrix**—Gr. *phrixos*, standing on end, bristling. *Ex:* Phrixo-cephalus (Crust.); Phryxo-trichus (Arach.). See also phric.

**phronim**—Gr. *phronimos*, clever, thoughtful, sensible. *Ex:* Phronima (Crust.), so named because many of them cleverly protect themselves by living inside the transparent cases of tunicates.

**phront**—Gr. *phrontis*, genit. *phrontidos*, thought, care, attention. *Ex:* Prontis (Coel.); Phronto-soma (Ins.).

**phrontid**—See **phront**.

**phrud**—Gr. *phroudos*, gone away, vanished, ruined. *Ex:* Phrudo-phleps (Ins.); Phrudus (Ins.).

**phrur**—Gr. *phrouros*, a guard; *phrourion*, a prison, fort. *Ex:* Phruro-lithus (Arach.); Calli-phruria = Cali-phruria*; Calli-phrurus (Ins.); Xylo-phrurus (Ins.).

**phrya**—Gr. *phryas*, root, sucker. *Ex:* Podo-phrya (Prot.).

**phryct**—**1.** Gr. *phryktos*, a burning torch, fire-signal: **2.** Gr. *phryctē*, a kind of gum, resin.

phryg—1. Gr. *phrygios*, dry:    2. Gr. *Phrygia*, ancient land of Asia Minor; *Phrygios*, of Phrygia. *Ex:* Phrygi-anus (Coel.); Phrygio-murex (Moll.), probably from its likeness to the conical Phrygian cap.

phrygan—Gr. *phryganon*, a dry stick, in pl. a bundle of brush wood, dry sticks. *Ex:* Phrygan-opsis (Ins.); Phryganea (Ins.); Phrygano-philus (Ins.).

phrygil—Gr. *phrygilos*, a finch. *Ex:* Phrygilus (Av.).

phrygm—NL. *phrygma*, etym, unknown. *Ex:* Phrygma*.

phryn—Gr. *phrynos* = *phrynē*, a toad. *Ex:* Phryn-ichthys (Pisc.); Phryne (Amph.); Phryno-soma (Rept.); Acantho-phrynus (Arach.).

phrysso—See phriss.

phryx—Gr. *Phryx*, genit. *Phrygos*, a Phrygian; among the Greeks, a bye-word for cowardice. *Ex:* Phryx (Moll.); Phryxo-lestes (Moll.). See also phrix.

phthan—Gr. *phthanō*, to overtake, to arrive first, to anticipate. *Ex:* Phthano-coris (Ins.).

phthart—Gr. *phthartos*, destructible, perishable. *Ex:* Ptharto-micrus (Ins.); Phthartus (Ins.).

phtheir—See phthir.

phthin—Gr. *phthinas*, genit. *phthinados*, waning, wasting. *Ex:* Phthino-branchii (Pisc.); Phthino-stoma (Ins.).

phthir—1. Gr. *phtheir*, a louse > *phtheirōdēs*, in-fested by lice. *Ex:* Phtheir-ichthys (Pisc.); Phthirius (Ins.); Phthiro-thrips (Ins.); Echino-phthirius (Ins.); Phyto-phthires (Ins.):    2. Gr. *phtheirō*, to corrupt, to destroy. *Ex:* Phtheiro-spermum*; Phthirusa*.

phthis—Gr. *phthinō*, fut. *phthisō*, to waste away. *Ex:* phthis-aner; phthiso-gyne.

phtho—Gr. *phthoē*, corruption. *Ex:* Dendro-phthoe*.

phthon—Gr. *phthonos*, malice, envy. *Ex:*Pthon-andria (Ins.); Pthono-sema (Ins.); A-phthona (Ins.).

phthoner—Gr. *phthoneros*, envious. *Ex:* Pthoner-odes (Ins.).

phthong—Gr. *phthongos*, the voice, a sound. *Ex:* phthongo-meter; a-phthongia.

phthor—Gr. *phthora*, destruction; *phthorimos*, destructive, corrupting. *Ex:* Phthora-mys (Mam.); Phthorima (Ins.); Phthoro-blastis (Ins.); Miso-phthora (Ins.).

phthorim—See phthor.

phug—See phyg.

phy—Gr. *phyō*, to grow, to generate, to be born; *phyē*, growth, stature. *Ex:* Erio-phyes (Arach.); Eu-phy-dryas (Ins.); Hetero-phy-idae (Platy.); Macro-phya (Ins.); mono-phy-odont; Pycno-phyes (Rot.); Xeno-phya.

phyc—1. Gr. *phykos*, painted, colored with cos-metics. *Ex:* Phyco-tettix (Ins.); a-phyca (Ins.):

2. Gr. *phykos* also *phykion*, see weed > *phyki-ōdēs*, covered with sea-weed; *phykes*, fem. *phykis*, a kind of fish living in a sea-weed. *Ex:* Phycis (Pisc.); Phycita (Ins.); Phyciodes (Ins.); Phyco-drom-idae (Ins.); Phyco-mycetes*; A-phyco-morpha (Ins.).

phyct—Gr. *phyktos*, to be eluded, avoidable.

phyg—Gr. *phygē*, flight; *phygas*, genit. *phygados*, a fugitive; *phygadeyō*, to exile, banish. *Ex:* Phuga-therium (Mam.); Phygadeuon (Ins.); Phygas (Moll.); phygo-blastema; Phos-phuga (Ins.); Semio-phygas (Ins.).

phygad—See phyg.

phyl—Gr *phylē*, *phylon*, a tribe, race *Ex:* Phyl-archus (Ins.); phylo-geny; phylum

phylact—Gr. *phylax*, genit. *phylaktos* = *phylaktēr*, genit, *phylaktēros*, a guard; *phylaktikos*, vigil-ant, cautious, guarding < *phylassō*, to guard. *Ex:* Phylacter (Ins.); Phylactero-poda (Ins.); Phylacto-laemata (Bry.); Allo-phylax (Ins.); ana-phylaxis; Hydro-philax*; Xylo-phylax (Ins.).

phylax—See phylact.

phylet—Gr. *phyletēs*, a tribesman, one of the same tribe; *phyletikos*, of or for a tribe. *Ex:* phyletic; Phileto-bius (Ins.); Phyletus (Ins.).

phyll—Gr. *phyllon*, leaf > *phyllōdēs*, like leaves, rich in leaves; *phylleion*, green-stuff, green-herbs. *Ex:* Phyll-acanthus (Echin.); phyll-ode; Phyllo-gonum*; Phyllo-spadix*; A-phyllon*; chloro-phyll; Di-phylleia*; Peri-phylla (Coel.).

phyllodoc—L. *Phyllodoce*, a sea-nymph, daughter of Nereus and Doris. *Ex:* Phyllodoce*, (Ann.).

phylyctaen—Gr. *phylyktaina*, a blister, bubble. *Ex:* Phlyctaena*.

phym—Gr. *phyma*, genit. *phymatos*, a swelling, tumor. *Ex:* Phym-echinus (Echin.); Phyma-phora (Ins.); Phymat-odes*; Phymato-cera (Ins.); phymato-d-eus; Phymo-soma; Di-octo-phyme (Nemat.).

phymat—See phym.

phyrt—Gr. *phyrtos*, mixed, worked up. *Ex:* Phrytus (Ins.).

phys—1. Gr. *physis*, growth. *Ex:* para-physes; An-ec-physis (Ins.):    2.. *phykes*, fem. *phy-kis*, ancient name of some fish living in Fucus *Ex:* Phys-ic-ulus (Pisc.):    3. Gr. *physa*, a pair of bellows, an air bubble, bladder. *Ex:* Phys-aria*; Physa (Moll.); Physo-gaster (Ins.); Physo-phora (Coel.); Physo-stegia*. See also physet.

physal—NL. *physalus*, the rorqual whale < Gr. *physalos*, a kind of whale. *Ex:* Physalus (Mam.).

physal—1. Gr. *physalis* = *physallis*, genit. *physall-lidos*, a bladder, bubble. *Ex:* Phys-aria*; Physal-odes*; Physalia (Coel.); Physalis*; Physalido-notus (Ann.); Physalo-ptera (Ins.):    2. Gr. *physalos*, a kind of whale; also a kind of toad which puffs itself up. *Ex:* Physalus (Mam.).

physalid—See physal.

physc—Gr. *physkē*, a sausage, the large intestine, a blister. *Ex:* Physca-phora (Por.); Physcia*; Physcius (Ins.); Physco-mitrium*; Physco-soma (Geph.); Physcus (Ins.).

physem—Gr. *physēma*, a bubble, a snorting; dim. *physēmation. Ex:* Physema*; Physematium (Prot.); Physemo-phorus (Ins.).

Thick-leaved Ground-cherry, *Physalis crassi-folia.* Redrawn from Desert Wild Flowers—Jaeger. Stanford University Press.

physet—Gr. *physētēr*, a blower, also an instrument for blowing; also the blow-hole of whales. *Ex:* Physet-odon (Mam.); Physeter (Mam.); Physo-therium (Mam.).

physi—Gr. *physis*, genit. *physeos* = Ion. *physios*, nature, natural qualities; *physikos*, natural. *Ex:* physic-al; physio-logy; apo-physis.

physic—See physi.

physis—See physi.

phyt—1. *phyton*, tree, plant, slip. *Ex:* Phyt-elaphas*; Phyt-helios (Prot.); phyto-bio-logy; Phyto-lacca*; Phyto-mastig-ina (Prot.); phyto-phagus; Petro-phytum*; sapro-phyta; xero-phyte; zoo-phyte:  2. Gr. *phyton*, a creature, animal. *Ex:* Phyto-crinus (Echin.); Astero-phyton (Echin.): Crota-phytus (Rept.).

phyteum—Gr. *phyteuma*, name of a plant, perhaps a species of Reseda. *Ex:* Phyteuma*.

phyxi—Gr. *phyxion*, a place of refuge; *phyxios*, causing or belonging to flight. *Ex:* Phyxio-schema (Arach.).

phyzel—Gr. *phyzēlos*, shy, cowardly. *Ex:* Phyzelus (Ins.).

pia—L. *pia*, fem. of *pius*, tender, kind. *Ex:* pia mater; Pia (Ins.).

piab—Tupi Indian *piaba*, name of a fresh-water fish. *Ex:* Piab-archus (Pisc.); Piabucus (Pisc.).

piale—Gr. *pialios*, poet. for *piōn*, fat. *Ex:* Piale-oidea (Ins.).

piantic—Gr. *piantikos*, falling. *Ex:* piantic.

piar—Gr. *piar*, fat, the thick juice of certain trees, cream; *piaros*, fat, rich. *Ex:* Piar-anthus*; Piaro-pus*; Piaro-soma (Ins.); Piarus (Ins.).

piaz—Gr. *piazō*, Doric for *piezō*, to press tight, to oppress, damage. *Ex:* Piaz-urus (Ins.); Piazo-mias (Ins.); Piazo-rhinus (Ins.).

pic—1. L. *pica*, a magpie. *Ex:* Pica (Av.):  2. L. *picus*, a woodpecker. *Ex:* Pic-ulus (Av.); Pici-corvus (Av.); pici-form; Pic-oides (Av.); Pico-laptes (Av.), see colaptes; Picus (Av.). See also picea.

picea—L. *picea*, the pitch-pine < *pix*, genit. *picis*, pitch; *piceus*, pitch-black or pitch-brown. *Ex:* Picea*; Picea-cantor (Av.); pici-pes.

piceat—L. *piceatus*, smeared with pitch; also thievish as with sticky hands < *picea*, the pitch pine.

picin—L. *picinus*, pitch black < *pix*, genit. *picis*, pitch.

picn— See pycn.

picr—Gr. *pikros*, bitter, pungent; also rough; *pikrasmos*, bitterness. *Ex:* Picr-adenia*; Picr-amnia* (*pikros+thamnos*, a bush); Picr-odon (Mam.); Picrasma*; Picris*; Picro-dendron*; picro-toxin; A-picra*.

pict—L. *pictus*, painted, embroidered, pp. of *pingo*, to paint; *pictilis*, embroidered; *pictor*, a painter. *Ex:* Picti-blennius (Pisc.); Picto-palpus (Ins.); Pictus (Moll.).

picumn—L. *Picumnus*, deity of fertility. *Ex:* Picumna (Ins.); Picumnus (Av.).

pidac—Gr. *pidax*, genit. *pidakos*, a fountain, a spring. *Ex:* Di-padax*.

pidax—See pidac.

piedraia— Sp. *piedraia*, name of some bacterial disease. *Ex:* Piedraia*.

pieris—Gr. *Pieris*, a muse from the district of Pieria in Thessaly. *Ex:* Pier-colias (Ins.); Pier-idae (Ins.); Calo-pieris (Ins.); Pieris (Ins.).

pies—See piest.

piest—Gr. *piestēr*, a squeezer; *piestos*, compressible, depressed; *piesis*, compressing < *piezō*, to press. *Ex:* Pies-cephalus (Av.); Piest-chilus (Moll.); Piesto-pleura (Ins.); Piestus (Ins.); Piez-ata (Ins.); Pieza; Piezo-bria (Ins.); Piczo-gaster (Ins.); Ano-piesta (Ins.).

piez—See piest.

pig—See pyg.

pigm—See pygm.

pigment—L. *pigmentum*, paint, color. *Ex:* pigmento-phage.

pigr—L. *pigrus*, lazy < *pigreo*, to be indolent, dilatory; *pigrescens*, genit. *pigrescentis*, becoming slow, ppr. of *pigresco*, to become sluggish, dilatory.

pil—1. L. *pilus*, a hair; *pilosus*, hairy < Gr. *pilos*, dim. *pilion*, a hair > Gr. *pilinos*, made of felt.

*Ex:* pili-fer-ous; pili-ger-ous; Pili-psocus (Ins.); Pilio-colobus (Mam.); Pilo-carpus*; Pilo-cereus*; pilo-gland-ul-ose; pilos-ellus; pilose. See also pile:   2. L. *pila* dim. *pilula,* a ball; also a pillar = Gr. *pilos,* a ball, bullet. *Ex:* Pilo-bolus*; Pilul-aria*; Mes-pilus*, Gr. *mesos,* the half; Thallasso-pila (Prot.).

**pilat**—L. *pilatus,* thick, dense; also armed with a javelin.

**pile**—Gr. *pileos,* a cap; *pilos,* dim. *pilidion* (=L. *pileus,* dim. *pileolus*), a felt cap; L. *pileatus,* covered with a cap. *Ex:* Pile-opsis (Moll.); Pilea*; pileate; pilei-form; pileo-rhiza; pile-olat-ed; Pileolus (Moll.); pileus; Pilo-carpus; pilidium.

**pilem**—Gr. *pilēma,* genit. *pilēmatos,* felt or anything made of it, such as a hat. *Ex:* Pilemia (Ins.).

**pileol**—See **pile.**

**pilidi**—Gr. *pilidion,* a small felt night-cap. *Ex:* Pilidio-nota (Ins.); pilidium.

**pilin**—See **pil.**

**pilos**—See **pil.**

**pilul**—L. *pilula,* a globule. *Ex:* Pilul-aria*; Pilula (Moll.).

**pilumn**—L. *Pilumnus,* Roman deity, personification of the pestle < *pilum,* a pestle. *Ex:* Pilumno-plax (Crust.); Pilumnus (Crust.).

**pimel**—Gr. *pimelē,* soft fat, lard; *pimelēs,* fat. *Ex:* Pimel-odus (Pisc.); Pimele-pterus (Pisc.); Pimelea*; Pimeli-pterus (Pisc.); Pimelo-metopon (Pisc.); A-pimeles (Ins.).

**piment**—Sp. *pimienta;* Port. *pimenta,* pepper. *Ex:* Pimenta*.

**pimpinella**—It. *pimpinella,* pimpernel. *Ex:* Pimpinella*.

**pimpla**—Gr. *Pimpla,* a mountain in Thrace, the site of a sacred spring. *Ex:* Pimpla (Ins.); Caeno-pimpla (Ins.); Xantho-pimpla (Ins.).

**pin**—1. L. *pinus,* a pine, a tree producing pitch > L. *pineus,* relating to a pine. *Ex:* Pin-aceae*; pin-etum; Pin-ites*; pin-oid* pine-al; Pine-mys (Mam.); Pini-cola (Av.);   2. Gr. *pinō,* to drink. *Ex:* pino-meter; Haemato-pinus (Ins.); not Xylo-pinus (Ins.), see pin 4: 3. Gr. *pinos,* dirt, filth:   4. Gr. *peinaō,* to be hungry. *Ex:* Xylo-pinus (Ins.).

**pinac**—Gr. *pinax,* genit. *pinakos,* a plank, tablet, a platter. *Ex:* pinac-enchyma; pinac-oid; Pinacio-phora (Prot.); Pinaco-ceras (Moll.); pinaco-cyte; Pinax (Ins.).

**pinar**—Gr. *pinaros,* dirty. *Ex:* Pinar-ornis (Av.); Pinaro-lestes (Av.); Pinaro-pappus*.

**pinax**—See **pinac.**

**pinet**—L. *pinetum,* a pine-wood; pl. genit. *pinetorum* of pine-woods. *Ex:* pinetum.

**pingu**—L. *pinguis,* fat, strong, stout; *pinguitia,* fatness; *pinguiculus,* somewhat fat. *Ex:* pingu-id; Pingui-pes (Pisc.); Pinguicula*; Pinguitia (Av.).

**pinn**—1. Gr. *pinnē = pinna,* genit. *pinnatos,* a kind of bivalve. *Ex:* Pinna (Moll.); Pinnato-pora (Bry.); Pinno-theres (Crust.). See pinnother also penn.   2. L. *pinna = penna,* a wing, also a feather; *pinnatus,* plumed, with feathers, winged; *pinniger,* bearing wings or feathers. *Ex:* pinn-ule; pinna, pinnate.

**pinnother**—Gr. *pinnothērēs,* the crab that lives in and guards the *pinna's* shell (*therein,* to guard). *Ex:* Pinnotheres (Crust.).

**pintad**—Sp. *pintado,* painted, mottled. *Ex:* Pintado (Av.).

**pinul**—NL. *pinula* < L. *pinnula,* a little plume or fin. *Ex:* pinulus.

**pio**—Gr. *piōn,* fat, rich milk; also, plump, sleek; *piotēs,* fatness. *Ex:* Pio-phila (Ins.); Pio-proso-pus (Ins.); Pion-ites (Av.); pion-notes, see noth; Piono-lema (Ins.); Pionus (Av.); Piote-palpus (Ins.); Piotes (Ins.).

**pion**—See **pio.**

**piot**—See **pio.**

**pip**—Galibi *pipa,* native name of the Surinam toad. *Ex:* Pip-idae (Amph.); Pipa (Amph.).

**pipatio**—L. *pipatio,* genit. *pipationis,* a chirping.

**piper**—L. *piper,* pepper; *piperatus,* peppery. *Ex:* Piper*; Piper-ales*; Piperi-vor-ous (Av.).

**pipien**—L. *pipiens,* genit. *pipientis,* peeping, ppr. of *pipio,* to peep.

**pipil**—NL. *pipilo* < L. *pipo,* to chirp, to peep. *Ex:* Pipile (Av.); Pipilo (Av.).

**pipistrell**—It. *pipistrello,* also *vespitrello,* dim. from L. *vespertilio,* a bat. *Ex:* Pipistrellus (Mam.).

**pipr**—Gr. *pipra = pipō,* the woodpecker. *Ex:* Pipr-idae (Av.); Pipra (Av.); Pipri-picus (Av.); Pipro-morpha (Av.); Musci-pipra (Av.).

**pipt**—Gr. *piptō,* to fall. *Ex:* Pipt-adenia*; Pipt-anthus*; Pipt-urus*; Pipto-meris*; Pipto-merus (Rept.).

**pipuncul**—NL. *pipunculus,* a dim. of uncertain meaning applied to a genus of flies. *Ex:* Pipunculus (Ins.).

**pir**—L. *pirum,* a pear > NL. dim. *pirula. Ex:* Piro-plasma (Prot.), a Latin-Greek barbarism; Pirula (Moll.). See also pyr.

**pirang**—*piranga,* native name of some South American bird. *Ex:* Piranga (Av.).

**pirat**—L. *pirata = Gr. peiratēs,* one who attacks ships. *Ex:* Pirat-ella (Crust.); Pirata (Arach.); Pirates (Ins.); Pirato-saurus (Rept.).

**piricul**—NL. *piricula,* dim. of L. *pirum,* pear. *Ex:* Piricul-aria*.

**pirr**—See **pyrrh.**

**pirul**—See **pir.**

**pis**—1. L. *pisum = Gr. pisos = pison,* a pea; Gr. *pisinos,* made of peas. *Ex:* Pis-aster (Echin.); Pis-idium (Moll.); pisi-form; piso-uncinat-us; Pison (Ins.); Pisum*:   2. Gr. *pisos,* a humid place, meadow. *Ex:* Piso-bia (Av.). Not Pisonia* named after Dr. William Pison, traveller in Brazil.

**pisan**—NL. *pisania,* a native of, or belonging to Pisa in Tuscanny. *Ex:* Pisan-ella (Moll.); Pisani-ana (Moll.); Pisania (Moll.).

**pisaur**—L. *Pisaurum,* an old Italian town. *Ex:* Pisaurus (Arach.).

**pisc**—L. *piscis,* a fish; *picsator,* a fisherman, fem. *piscatrix. Ex:* Pisc-idia*, the last element <L. *caedere,* to kill; piscatori-al; Piscatrix (Av.); Pisces; Pisci-cola (Ann.); pisci-vorous; Chloro-pisca (Ins.).

**pisculent**—L. *pisculentus,* abounding in fish.

**piss**—Gr. *pissa,* pitch>*pissōdēs,* like pitch, yielding pitch. *Ex:* Pisso-notus (Ins.); Pissodes (Ins.).

**pist**—1. Gr. *pistos,* authentic, pure, genuine; related to *pistis,* confidence, faith; *pistikos,* credible. *Ex:* Pisto-saurus (Rept.); A-lyto-pistis (Ins.): 2. Gr. *pistos,* liquid, watery. *Ex:* Pistia*.

**pistacia**—Gr. *pistakē,* a kind of tree, the pistachio <anc. Persian *pistah,* name for a kind of nut, the pistachio nut. *Ex:* Pistacia*=Pistachio*.

**pistic**—See **pist.**

**pistil**—See **pistill.**

**pistill**—L. *pistillum,* pestle. *Ex:* pistil; pistili-fer-ous; pistill-ate; pistill-id; pistillo-ger; Pistillum (Moll.).

**pisyn**—Gr. *pisynos,* confiding in. *Ex:* Pisyno-limnas (Av.).

**pitang**—Tupi *pitangus,* name of a large-billed flycatcher of S.A., so named because of its call<*pitangua,* suddenly murmuring. *Ex:* Pitangus (Av.).

**pith**—Gr. *pithos,* dim. *pithiskos,* a large jar with a wide mouth. *Ex:* Pithiscus (Ins.); Pitho-phora*; Pith-gaster (Ins.); Pitho-helix (Moll.) =Pytho-helix (Moll.). Pith, the botanical term, comes from an Anglo-Saxon word, *pidka,* pith, marrow.

**pithan**—Gr. *pithanos,* probable, persuasive, plausible; *pithanotēs,* plausibility. *Ex:* Pithan-otaria (Mam.); Pithan-oto-mys (Mam.); pithano-logy; Pithanotes (Ins.).

**pithe**—See **pithec.**

**pithec**—Gr. *pithēkos=pithex,* one who plays tricks, an ape. *Ex:* Pithe-sciurus (Mam.); Pithec-anthro-pus (Mam); Pithec-ulus (Mam.); Pithecia (Mam.); Pitheco-ctenium*; Pithex (Mam.); Cerco-pithecus (Mam.), erroneously Certo-phitecus (Mam.); Paido-pithex (Mam.).

**pithex**—See **pithec.**

**pithod**—NL. *pithode,* the nuclear barrel-shaped figure formed in cytokinesis. *Ex:* pith-ode. See **-ode 1.**

**pithon**—Gr. *pithōn,* a little ape, flatterer; also a cellar. *Ex:* Pithon-ella (Prot.).

**pitt**—1. Gr. *pitta,* resin. *Ex:* Pitto-notus (Ins.); Pitto-sporum*: 2. NL. *pitta*<Tamil *ponunki pitta,* a small jay. *Ex:* Pitt-idae (Av.)· Pitta (Av.); Coraco-pitta (Av.).

**pitu**—See **pity.**

**pituit**—L. *pituita,* phlegm, slime. *Ex:* pituit-ary; pituit-ous.

**pity**—Gr. *pitys,* genit. *pityos,* the pine. *Ex:* Pity-mys (Mam.); Pity-omma (Prot.); Pity-ophis (Rept.); Pityo-philus (Ins.); Pitys*; Sciado-pitys*.

**pityr**—Gr. *pityron,* bran, husks of grain, scurf. *Ex:* pityri-asis (Med.); Pityri-opsis (Av.); Pityro-gramma*; Pityro-sporum*.

**pix**—See **picea.**

**piz**—Gr. *piezō,* to press, squeeze, grasp. *Ex:* Piz-onyx (Mam.). See also **piest.**

**plac**—Gr. *plax,* genit. *plakos,* a flat round plate, tablet, broad surface>*plakōdēs,* tabular, foliated. *Ex:* plac-oid; Plac-odus (Rept.); plac odont; plac-ula; Placo-pharynx (Pisc.); Pla cus*; Plax-haplos; Plaxi-phora (Moll.); Di-placus*; Steno-plax (Moll.). See also **plakin.**

**placam**—See **plocam.**

**placent**—L. *placenta*=Gr. *plakous,* a small round, flat cake. *Ex:* placent-ary; Placent-alia (Mam.); placent-ation; placenta; Placenti-ceras (Moll.).

**placid**—L. *placidus,* mild, gentle. *Ex:* Placida (Moll.); Placidus (Ins.).

**placin**—Gr. *plakinos,* made of boards. *Ex:* Placino-lopha (Por.).

**placit**—L. *placitus,* pleasing<*placeo,* to please.

**placun**—NL. *placuna,* name for a genus of mol-lusks<Gr. *plax,* anything flat. *Ex:* Placuna (Moll.).

**plad**—Gr. *plados,* abundance of fluids. *Ex:* plado-boles.

**pladar**—Gr. *pladaros,* wet, damp. *Ex:* pladar-osis (Med.).

**pladasm**—Gr. *pladasmos,* wettness.

**plaes**—Gr. *plaisos=blaisos,* crooked, bent. *Ex:* Plaesio-mys (Brach.); Plaesio-rhinus (Ins.); Plaesius (Ins.).

**plag**—1. L. *plaga,* a snare, a wound, blow<Gr. *plēgē;* NL. *plagatus,* striped, wounded. *Ex:* plaga; Plaga (Ins.); plagate; Plago-pterus (Pisc.), here used in sense of wound: 2. Gr. *plagos,* side (seldom used in compounds.)> *plagios,* see next entry.

**plagiat**—L. *plagiator,* a thief, kidnapper; *plagiatus,* stolen<*plagio,* to steal.

**plagio**—Gr. *plagios,* oblique, transverse, slanting; also the sides, flanks<*plagos,* side. *Ex:* Plagi-anthus*; Plagi-echinus (Echin.); plagi-odont; Plagio-bothrys*; plagio-dromous; Plagio-lepis (Ins.); plagio-tropic.

**plagius**—NL. *plagiusus*<Gr. *plagios,* oblique.

**plakin**—Gr. *plakinos*, made of boards. *Ex:* Placino-lopha (Por.); Plakin-astr-ella (Por.); Plakina (Por.).

**plan**—1. Gr. *planos*, roaming; *planē*, a wandering; *planēs*, also *planētēs*, a roamer. *Ex:* not Planera*, named after J. J. Planer of Erfurt; Planes (Ins.); Plano-sarcina*; plano-spore; Eremio-planis (Rept.); Nycti-planus (Mam.); Peri-planeta (Ins.): 2. *planos*, deceiving. *Ex:* Plan-odos (Mam.); Plan-ops (Mam.): 3. L. *planus*, dim. *planula*, flat, level, even; *planarius*, level, flat. *Ex:* plan-aea, the last element of uncertain meaning; Plan-orbis (Moll.); Planaria (Platy.); Planarium*; Planiceros (Mam.); Plani-pennia (Ins.); Planispiral; plano-lite; planula; Typhlo-plana (Platy.).

**planari**—See **plan 3.**

**planc**—L. *plancus*, flat-footed. *Ex:* Plancus (Av.).

**planct**—L. *planctus*, a wailing, a beating of arms, a rustling>NL. *plancterus*, a fish name. *Ex:* Plancterus (Pisc.). See also plankt.

**planest**—NL. *planesticus*, generic name of the American robin, appar.<Gr. *planētēs*, a wanderer. *Ex:* Plenesticus (Av.).

**planet**—Gr. *planētēs*, a wandering star, planet; *planētikos*, disposed to wander. *Ex:* Planetatherium (Mam.); Planetica (Ins.); Planetolestes (Ins.).

**plankt**—Gr. *plankton*, neut. of *planktos*, wandering, roaming. *Ex:* Plancto-plana (Platy.); plankto-phyte; plankton; Thallaso-plancta (Prot.).

**plant**—1. L. *planta*, the sole of the foot>*plantaris*, pertaining to the sole of the foot. *Ex:* plantar; planti-grade: 2. Eng. *plant*, a vegetable organism<L. *planta*, lit. a flat, spreading shoot<L. root from which *platus*, flat, is derived: 3. L. *planto*, to set, plant. *Ex:* im-plant-ation.

**plantag**—L. *plantago*, genit. *plantaginis*, plantain. *Ex:* Plantagin-ales*; Plantago*.

**planula**—See **plan 3.**

**plas**—Gr. *plasis*, a moulding. *Ex:* macro-plasia; ana-plasis.

**plasm**—Gr. *plasma*, genit. *plasmatos*, anything formed or moulded, an image. *Ex:* pla-some, a crude butchery of plasmato-some; plasmodium; Plasmato-ceras (Moll.); plasmo-gamy; plasmo-lysis; Plasmo-para*; proto-plasm.

**plasmod**—NL. *plasmodium* (<Gr. *plasma+eidos*), a multi-nucleated mass of naked protoplasm. *Ex:* plasmodio-carpus; plasmodium. See plasm.

**plasome**—See **plasm.**

**plasson**—Gr. *plassō*, ppr. *plasson*, to form or mold. *Ex:* plasson.

**plast**—Gr. *plastos*, molded, formed; *plastēs*, fem. *plastis*, a builder, a moulder, modeller,

creator. *Ex:* Plast-andrena (Ins.); plastid; plastid-ule; plasto-gamy; Plasto-phora (Ins.); plasto-some; bio-plast.

**plastid**—See **plast.**

**plastr**—Fr. *plastron*, a breastplate. *Ex:* plastr-al; Plastr-emys (Rept.); plastron; Plastron (Ins.).

**-plasty**—Eng. *-plasty*, suffix indicating act of forming, also development, growth<Gr. *plastos*, formed. *Ex:* auto-plasty; ecto-plasty.

**plat**—See **platy.**

**platac**—Gr. *platax*, genit. *platakos*, a kind of fish. *Ex:* Platac-idae (Pisc.); Platax (Pisc.).

**platale**—L. *platalea*, the spoon-bill. *Ex:* Platalea (Av.).

**platam**—Gr. *platamōn*, anything broad, flat. *Ex:* Platam-ops (Ins.); Platamo-myia (Ins.): Platamon (Crust.).

**platamod**—Gr. *platamōdēs*, flat, broad. *Ex:* Platomodes (Ins.).

**platan**—L. *platanus*<Gr. *platanos*, the oriental plane-tree. *Ex:* Platanus*.

**platanist**—Gr. *platanistēs*, "a fish," apparently a dolphin. *Ex:* Platanista (Mam.).

**platax**—See **platac.**

**platess**—L. *platessa*, the plaice. *Ex:* Platessa (Pisc.).

**platy**—1. Gr. *platys*>*platos*, flat, broad, wide. *Ex:* Plat-ac-odon (Mam.); plat-ur-ous; Platigonus (Mam.); Plato-pterus (Pisc.); Platycercus (Av.); Platy-helminthes; Platy-pus (Ins.); Platy-rrhina (Mam.); Platy-stemon*: 2. Sp. *Rio de La Plata*, the Argentine river La Plata. *Ex:* Plata-eo-mys (Mam.); Platatherium (Mam.).

**platyc**—Gr. *platykos*, broad. *Ex:* Platycis (Ins.).

**platyn**—Gr. *platynō*, to make broad. *Ex:* Platynaspis (Ins.); Platynus (Ins.).

**platysm**—Gr. *platysma*, genit. *platysmatos*, anything broad, a plate. *Ex:* platysma muscle.

**plaus**—L. *plausor*, a clapper of hands, an applauder.

**plaut**—L. *plautus*, flat-footed. *Ex:* Plautus (Av.).

**plax**—See **plac.**

**plaz**—Gr. *plazō*, to wander, rove.

**ple**—1. Gr. *pleōs*, full>L. *pleo*, to fill, pp. *pletus*, filled. *Ex:* ple-odont; Pleo-pus; op-plete: 2. Gr. *pleō*, to sail>NL. *pleon*, a sail; pleo-pod =pleo-pod-ite; pleon (a term used in connection with crustaceans), see also pleon under pleio.

**plebei**—L. *plebeius*, of common people, common, vulgar. *Ex:* Plebeius (Ins.); Plebejus (Av.); Plebi-donax (Moll.).

**plebi**—See **plebei.**

**pleco**—Gr. *plekō*, to weave, twine, twist. *Ex:* Plec-otus (Mam.); Pleco-glossus (Pisc.). Pleco-ptera (Ins.). See plect.

**plect**—Gr. *plektos*, plaited, twisted>*plektanē*,

anything twisted or wreathed; L. *plectilis,* woven, plaited. *Ex:* Plect-aster (Echin.)

European Long-eared Woven-ear Bat, *Plecotus auritus.* Redrawn from Bats—Allen. By permission of the President and Fellows of Harvard College.

Plectan-ella (Prot.); Plecto-gnathi (Pisc.); Plecto-spondyli (Pisc.); Eu-plect-ella (Por.).

**plectes**—Gr. *plēktēs,* a striker, a brawler. *Ex:* Plectes (Ins.).

**plectil**—See **plect.**

**plectr**—Gr. *plēktron,* a thing to strike with; also a cock's spur. *Ex:* Plectr-eurys (Arach.); plectr-idium; Plectr-ites*; Plectro-mantis (Amph.); Plectro-phenax (Av.); A-plectrum*.

**pleg**—1. Gr. *plēgas,* genit. *plēgados,* a sickle, a scythe. *Ex:* Plegad-is (Av.) the last element of the word is evidently without meaning; Plegad-ornis (Av.): 2. Gr. *plēgē,* a blow, stroke. *Ex:* plega-phonia; Plega-derus (Ins.); para-plegia = para-plegy.

**plegad**—See **pleg 1.**

**plegm**—Gr. *plegma,* genit. *plegmatos,* anything twined, wicker-work. *Ex:* Plegma-pterus (Ins.); Plegmato-ptera (Ins.); Plegmo sphaera (Prot.); Oro-plegma (Prot.).

**plegy**—See **pleg.**

**plein**—Gr. *plein,* Attic for *pleon,* more, full. *Ex:* Pleino-peltus.

**pleio**—Gr. *pleiōn,* more>*pleio-,* a prefix used commonly in generic names to indicate Pliocene occurrence or age. *Ex:* pleio-morphy; pleio-nexy; pleo-chroic; pleo-phagus; pleon (botanical term), see pleon under ple; Pleiono-gaster (Ann.); Plio-cene; Plio-hippus (Mam.). See also ple.

**pleion**—See **pleio.**

**pleist**—Gr. *pleistos,* most, superl. of *polys,* much; also used as a prefix to indicate Pleistocene (<*pleistos+kainos,* recent) occurrence or age. *Ex:* Pleist-odontes (Ins.); Pleisto-cene; Plisto-trichia (Ins.); Careno-plistus (Ins.).

**plemmyr**—Gr. *plēmmyra,* the high tide. *Ex:* Plemmyra (Ins.).

**plen**—L. *plenus,* full. In the specific name, *flore pleno,* the masculine ablative, *pleno*

agrees with the masc. ablative *flore.* Thus *flore pleno* means "with full flower" or "full flowered": a good example of the ablative of quality having the force of an adjective.

**pleo**—See **ple,** also **pleio.**

**pleon**—See **ple 2,** also **pleio.**

**pleonaz**—Gr. *pleonazō,* to be more than enough. *Ex:* Pleonazo-neura (Ins.).

**pleonect**—Gr. *pleonektēs,* one who is grasping; *pleonexia,* grasping. *Ex:* Pleonecto-poda (Ins.); Pleonexes (Crust.).

**pleonex**—See **pleonect.**

**pler**—Gr. *plērēs,* full; *plērōma,* completion, that which fills; *plērōsis,* a filling up. *Ex:* Pler-odon (Rept.); Plere-gnathus (Mam.); plero-cerc-oid; plero-morph; Pleroma*; plerome; plerosis (Med.); Aspido-pleres (Arth.).

**plerom**—See **pler.**

**ples**—Gr. *plēsios,* near, recent. *Ex:* Ples-ictis (Mam.); Ples-ippus (Mam.); Plesi-arcto-mys (Mam.); Plesia (Ins.); Plesio-meta (Arach.); Plesio-saurus (Rept.); plesius; Di-plesion (Pisc.); Geitono-plesium*.

**plesio**—See **ples.**

**plesion**—Gr. *plēsion,* adverb meaning near, hard by. *Ex:* Di-plesion (Pisc.).

**pless**—Gr. *plēssō,* to smite. *Ex:* plessi-meter.

**plet**—See **ple.**

**pleth**—Gr. *plēthō,* to be full, to become full, complete; *plēthys,* fullness. *Ex:* Pleth-aelurus (Mam.); Pleth-odon (Amph.); Pletho-spira (Moll.); Plethus (Ins.).

**plethor**—Gr. *plethōrē,* increase, fullness. *Ex:* plethora.

**plethyn**—Gr. *plēthynō,* to increase.

**plethysm**—Gr. *plēthysmos,* an enlargement. *Ex:* Plethysmo-chaeta (Ins.); plethysmo-graph.

**pleur**—Gr. *pleura = pleuron,* a rib, the side. *Ex:* pleur-al; pleur-odont; Pleura-pogon*; Pleuro-branchia (Cten.); Pleuro-gynia*; Pleuro-nectes (Pisc.).

**pleuric**—Gr. *pleurikos,* at the side. *Ex:* Pleurico-spora*.

**pleurothen**—Gr. *pleurothen,* from the side.

**pleust**—Gr. *pleustēs,* a sailor; *pleustikos,* ready for sailing <*pleō,* to sail. *Ex:* Pleustes (Crust.); pleuston.

**plex**—1. L. *plexus,* a knitting, interweaving <*plecto,* to weave. *Ex:* Plex-aur-ella (Coel.); plexeo-blastus; plexi-form; Plexo-cera (Crust.); solar-plexus: 2. Gr. *plēxis,* a stroke. *Ex:* plexi-meter.

**plexipp**—Gr. *Plēxippos,* one of the numerous sons of Aegyptus. *Ex:* Plexippus (Arach.).

**plic**—L. *plico,* to fold, to braid; *plicatus,* braided, doubled up; *plicatura,* a folding; *plicatilis,* capable of being folded. *Ex:* plica; Plica-phora (Moll.); Plicat-odon (Mam.); Plicatula (Moll.); plicature; plici-form; Plici-pennes (Ins.); Plico-bulla (Moll.); ad-plictus; bi-plicate.

**plicat**—See **plic.**

Coldenia plicata, a perennial plant with gray-green "plaited" leaves. Coldenia was named after Dr. C. Colden, colonial botanist. Redrawn from Desert Wild Flowers—Jaeger. Stanford University Press.

**plinth**—Gr. *plinthos*, dim. *plinthis*, a brick. *Ex:* Plinth-aster (Echin.); plinthi-form; Plintho-lepis (Pisc.); Plinthus (Ins.).

**plio**—See **pleio.**

**ploc**—Gr. *plokeus*, a braider; *plokos*, a lock of hair, curl, wreath; *plokion*, a necklace or chain; *plokē*, a complication, a twisting together; *plokios*, twined. *Ex:* Ploce-alauda (Av.); plocei-form; Ploceus (Av.); Plocio-cerus (Ins.); Plocion-us (Ins.); ploco-carpium; Argyro-ploce (Ins.); Peri-ploce*; Sym-plocos*.

**plocam**—Gr. *plokamos* = poet. *plokamis*, genit.

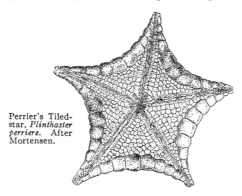

Perrier's Tiled-star, *Plinthaster perriere.* After Mortensen.

*plokamidos*, tresses, a curl of hair, twisted hairs. *Ex:* Plocama*; Plocamia (Por.); Plo camium*; Placamo-ceras (Moll.); Plocamus (Ins.).

**ploe**—Gr. *pleō*, to sail > *ploion*, a floating vessel.

*Ex:* Ploe-soma; Ploeo-physa (Coel.); Ploio-chiton (Moll.).

**ploho**—anagram of *hoplo.* *Ex:* Ploho-phorus (Mam.).

**ploiari**—Gr. *ploiarion*, a small boat—Ploiari-idae (Ins.).

**-ploid**—NL. *-ploid*, suffix < Gr. *-ploos*, as in *diploos*, double. *Ex:* poly-ploid; compare with diploid (derived from Gr. *diploos*, double + *oid* )

**ploim**—Gr. *plōimos*, sea-worthy. *Ex:* Ploima (Rot.).

**ploio**—See **ploe.**

**ploran**—L. *ploro*, to wail, cry out; ppr. *plorans.* genit. *plorantis*, crying, wailing.

**plos-** —Gr. *plos*, that from which something arises, that which is beside, near, besides. This prefix enters into many Greek compounds.

**plot**—Gr. *plōtēr*, genit. *plōtēros*, a swimmer, a sailor; *plōtikos*, skilled in seamanship. *Ex:*

Ossicles from body wall of the sea-cucum-ber, Bathyplotes. From Fishes of North and Mid-dle America—Jordan.

Plot-actis (Coel.); Ploteres (Ins.); Ploteris (Ins.); Ploto-nemertes (Nemert.); Plotus (Av.); Eu-plotes (Prot.).

**ploter**—See **plot.**

**plotes**—See **plot.**

**plum**—L. *pluma*, dim. *plumula*, a soft feather; *plumatus*, feathered. *Ex:* Plum-aster (Echin.); Plumat-ella (Bry.); plumi-corn; Plumo-hali-chondria (Por.); Plumul-aria (Coel.); plumule; filo-plume.

**plumb**—L. *plumbum*, lead; *plumbeus*, leaden, lead-colored.

**plumbag**—L. *plumbago*, genit. *plumbaginis*, a plant called leadwort < *plumbum*, lead. *Ex:* Plumbagin-aceae*; Plumbago*.

**plumul**—See **plum.**

**plur**—L. *plus*, genit. *pluris*, more; when used as a combining form, *pluri-* means several. *Ex:* plur-al; pluri-locul-ar; pluri-par-ous; pluri-vor-ous.

**plus**—Gr. *plousios*, wealthy. *Ex:* Plusi-odonta (Ins.); Plusi-otis (Ins.); Plusia (Ins.); Plusio-tricha (Ins.). See also plur.

**plut**—Gr. *ploutos*, wealth. *Ex:* Plut-ella (Ins.); Pluto-thrix (Ins.).

**plute**—L. *pluteus*, a shed, a breast-work, a painter's easel. *Ex:* plutei-form; pluteus; echino-pluteus.

**pluton**—NL. *plutonius*, dusky < *Pluto* = Gr. *Ploutōn*, God of the dark underworld. *Ex:* Pluton-aster (Echin.); Plutono-thrips (Ins.).

**pluv**—L. *pluvia*, rain, a shower>*pluvialis*, pertaining to rain. *Ex:* Pluvi-anus (Av.); pluvial; Pluviales (Av.); Pluviali-formes (Av.).

**pluvial**—See **pluv.**

**plynteri**—Gr. *plyntērios*, of or for washing. *Ex:* Plynteria (Ins.).

**plysi**—Gr. *plytos*, washed; *plysis*, a washing. *Ex:* A-plysia (Moll.).

**plyti**—See **plysi.**

**pne**—Gr. *pneō*, to breathe; *pneustikos*, of, or for breathing. *Ex:* pneo-gaster; Entero-pneusta (Hemichorda); Tri-pneustes (Echin.).

**pneum**—Gr. *pneuma*, genit. *pneumatos*, air, wind, breathing; *pneumonikos*, of the lungs; *pneumatikos*, inflated, of or caused by the wind. *Ex:* Pneum-aria*; pneumato-phore; Pneumobranchus (Pisc.); Pneumo-dermon (Moll.); Pneumon-oeces (Platy.); pneumonic.

**pneumat**—See **pneum.**

**pneus**—Gr. *pneusis*, a blowing. *Ex:* pneusis.

**pneust**—See **pne.**

**pnig**—Gr. *pnigos*, genit. *pnigotos*, a choking; *pnigaliōn*, a night-mare. *Ex:* Pnig-odes (Ins.); Pnigalion (Rept.); Pnige-acanthus (Pisc.); Pnigo-cichla (Av.).

**pnigal**—See **pnig.**

**pnigot**—See **pnig.**

**pno**—Gr. *pnoē*, breath; *pnoos*, breathing. *Ex:* Pnoc pyga (Av.); Di-pnoi (Pisc.); dys-pnoe.

**po**—Gr. *poa*, in compounds *poē*-, grass, a grassy place. *Ex:* Po-aceae*; po-ad (Ecol.); Poa*; Poe-bro-therium (Mam.); Poe-phagus (Mam.); po-ium (Ecol.); poo-cola; Poo-ecctes (Av.); poo-phyta (Ins.).

**poc**—Gr. *pokos*, fleece. *Ex:* Poco-cera (Ins.); Poco-desmus (Myr.).

**pocad**—Gr. *pokas*, genit. *pokados*, hair, wool. *Ex:* Pocadi-cnemis (Arach.); Pocadi-opsis (Ins.); Pocadius (Ins.).

**pocas**—See **pocad.**

**pocill**—See **pocul.**

**pocillat**—L. *pocillator*, the bearer of a cup.

**pocul**—L. *poculum*, dim. *pocillum*, cup, bowl. *Ex:* Pocillo-pora (Coel.); Pocillon (Coel.); poculiform.

**pod**—Gr. *pous*, genit. *podos*, a foot; *podion*, a small foot; *podotēs*, footed; *podōkēs*, swift of foot; LL. *podicus*, belonging to a foot. *Ex:* pod-al; pod-al-ic version; Pod-axon (Prot.); pod-ite; Pod-ura (Ins.); podi-leg-ous; Podistera*; Podica (Av.); Podo-cnemys (Rept.); Podo-sphaera*; Podoces (Av.); A-pus (Crust.); Brady-pus (Mam.); Di-podo-mys (Mam.); Eu-podotis (Av.);mero-pod-ite; Poly-podium*; Ptero-poda (Moll.); Scapho-poda.

**podabr**—Gr. *podabros*, tender-footed. *Ex:* Podabrus (Mam.).

**podagr**—Gr. *podagra*, gout in the feet. *Ex:* Podagr-aria*.

**podagric**—L. *podagricus*, gouty.

**podalyr**—Gr. *Podaleirios*, son of Asclepias. *Ex:* Podalyri-eae*; podalyri-ae-folia; Podalyria*.

**podapion**—Gr. *podapion*, a small foot. *Ex:* Poda pion (Ins.).

**podarg**—Gr. *podargos*, swift-footed. *Ex:* Poda rgus (Av.).

**podeti**—NL. *podetium*<Gr. *pous*, genit. *podos*, foot. *Ex:* podetii-form; podetium.

**podex**—See **podic.**

**podic**—L. *podex*, genit. *podicis*, the fundament, anus, rump. *Ex:* podex; podic-al; Podice-ps (Av.), *ps*, a contraction of pes. See also pod.

**podilymbus**—NL. *podilymbus*, barbarous combination of Podiceps+Columbus. *Ex:* Podilymbus (Av.).

**podism**—Gr. *podismos*, measuring by feet. *Ex:* Podism-opsis (Ins.).

**podistr**—Gr. *podistra*, a foot-trap. *Ex:* Podistrella (Ins.); Podistra (Ins.).

**podium**—See **pod.**

**poe**—Gr. *poieō*, to make, produce; *poiēsis*, a production, creation; *poiētikos*, creative. *Ex:* Poeo-bius (Gephyr.); chylo-poiesis; haemapoietic; Poro-poea (Ins.); Sphaero-poeus (Myr.). See also po<Gr. *poa*, grass.

**poecil**—Gr. *poikilos*, variegated, various, many-colored>*poikilias*, name of a spotted fish. *Ex:* Poecil-ichthys (Pisc.); Poecil-odus, (Pisc.); Poecili-um (Ins.); Poecilo-chroa (Arach.); poecilo-cyttar ous; Poecilo-tettix (Ins.); Poecilo-toma (Ins.); Poecilo-zon-ites (Ins.); poikilo-thermal; Lepto-poecile (Av.).

**poeetes**—Gr. *poiētēs*, a maker. *Ex:* Sceno-poeetes (Av.).

**poemen**—Gr. *poimēn*, a shepherd; *poimēnion* =*poimnion*, a single head of cattle. *Ex:* Poemen-orthus (Ins.); Poemenia (Ins.).

**poen**—L. *poena*, punishment, satisfaction. *Ex:* Di-poena (Arach.).

**poescop**—S. African Dutch *poeskop*, local name for a hump-back whale. *Ex:* Poescopia (Mam.).

**pog**—Gr. *pōgōn*, genit. *pōgōnos*, dim. *pōgōnion*, beard; *pōgōnias*, bearded. *Ex:* Pogo-gnathus (Pisc.); Pogon-ephydra (Ins.); Pogonia*; pogonium; Pogono-myrmex (Ins.); A-pogon*; Mero-pogon (Av.).

**pogon**—See **pog.**

**poiesis**—See **poe.**

**poietic**—Gr. *poiētikos*, making, forming.

**poikil**—See **poecil.**

**pol**—Gr. *polos*, a pivot, an axis>L. *polus*, the end of an axis, pole; NL. *polaris*, of or pertaining to a pole *Ex:* pol-ar body; pol-ar-form; polo-cyte; Para-polia (Rot.). See also poli and poly.

**polanisi**—NL. *polanisia*, a plant name<*poly*, many+*aniso*, unequal, dissimilar+*ia*. *Ex:* Polanisia*.

**polar**—See **pol.**

**polem**—Gr. *polemos*, war, fight; *polemikos*, of or for war, war-like; *polemistēs*, a fighter. *Ex:* Polem-aetus (Av.); Polemistes (Ins.); Polemocoris (Ins.); Polemon (Ins.).

**polemarch**—Gr. *polemarchos*, a leader. *Ex:* Polemarchus (Ins.).

**polemon**—Gr. *polemōnion*, a kind of plant. *Ex:* Polemoni-aceae*; Polemonium*.

**poli**—Gr. *polios*, hoary, gray>NL. *polius*, gray; *poliotēs*, grayness of hair. *Ex:* Poli-alurus (Mam.); Poli-anthes*; poli-encephal-itis (Med.); poli-folium; Polio-aetus (Av.); Poliocoris (Ins.); Polo-thyrsis*. See also polis and poly.

**polimen**—See **polimin.**

**poliment**—L. *polimenta*, the testicles.

**polimin**—L. *polimen*, genit. *poliminis*, brightness.

**polinices**—Gr. *Polyneikēs*, son of Oedipus. *Ex:* Polinices (Moll.).

**polis**—Gr. *polis*, a city; *polistēs*, founder of a city; *polismos*, the building of a city. *Ex:* poli-clinic (Med.); Polisma (Ins.); Polistes (Ins.).

**polistes**—See **polis.**

**polisto**—See **poly.**

**polit**—L. *politus*, smooth, polished, cultivated <*polio*, to smooth, polish, adorn. *Ex:* Polita (Moll.); Politi-tapes (Moll.); Polito-spiculum (Nemat.); Acro-politus (Ins.).

**politor**—L. *politor*, a polisher.

**pollac**—Gr. *pollakis*, often, frequent. *Ex:* pollaci-uria. See pollach.

**pollach**—Gr. *pollachē*, often, many times, in diverse manners; *pollakis*, often. *Ex:* pollach-anthic; pollachi-genus; Pollachi-us (Pisc.).

**pollen.**—L. *pollens*, genit. *pollentis*, powerful, vigorous<*polleo*, to be strong. See also pollin.

**pollex**—See **pollic.**

**pollic**—L. *pollex*, genit. *pollicis*, the thumb; *pollicaris*, of or belonging to a thumb; NL. *pollicatus*, having thumbs. *Ex:* pollex; Polliceps (Arth.); pollic-al; Pollicata (Mam.).

**pollin**—L. *pollen*, genit. *pollinis*, fine flour<Gr. *palē*, fine meal. *Ex:* Pollenia* (Ins.); pollin-ate; pollin-od-ium; pollini-ger-ous; pollinia.

**polonic**—NL. *polonicus*<L. *Polonia*, Poland.

**polph**—Gr. *polphos*, macaroni. *Ex:* Polph-opsis (Ins.); Polpho-peza (Ins.).

**polul**—L. *polulus*, little.

**poly**—Gr. *poly*, neut. sing. of *polys*, much, many; *polistos*, superlative, most. *Ex:* Pol-acanthus (Rept.); Pol-anisia* (Gr. *anisos*, unequal); Poli-carpa (Tun.); Poli-clynium (Tun.); Polisto-trema (Pisc.); Poly-chaeta (Ann.); Poly-gala*; Poly-gonum*; Poly-gordius (Ann.); Poly-my-aria (Nemat.); Poly-podi-aceae*.

**polybor**—Gr. *polyboros*, greedy. *Ex:* Polyborus (Av.).

**polymn**—Gr. *Polymnia* (lit. many hymns), muse of oratory and sacred lyric. *Ex:* Polymnia* = Polyhymnia*.

**polynices**—L. *Polynices*, a mythological name. *Ex:* Polynices (Moll.).

**polynoe**—Gr. *Polynoē*, the daughter of Nereus and Doris. *Ex:* Polynoe (Ann.).

**polyp**—L. *polypus*=Gr. *polypous* (lit., many-footed), a polyp, an aquatic animal. *Ex:* polyp-ide, see -ite; polypi-form; Bathy-polypus (Moll.).

**polyxene**—Gr. *Polyxenē*, mythological daughter of Priam and Hecuba, betrothed to Achilles.

**pom**—1. Gr. *pōma*, genit. *pōmatos*, a cover. *Ex:* Pom-acanthus*; Pom-aulax (Moll.); Poma-derris*; poma-rinus=pomato-rhinus; Pomatomus (Pisc.); Pomato-ceros (Ann.); Pomatonota (Amph.); Lyo-pomata (Brach.); Macropoma (Pisc.): 2. L. *pomus*, a fruit-tree of any kind; *pomum*, fruit of any kind, an apple. *Ex:* pom-aceous; pomi-granate (lit. many-seeded apple); pomi-fera.

**pomast**—Gr. *poma*, a lid, cover; dim. *pomasterion*. *Ex:* Rhino-pomastus (Av.).

**pomeridian**—L. *pomeridianus*, in the afternoon.

**pomp**—Gr. *pompos*, a conductor, escort. *Ex:* Pompo-nema (Nemat.); Pompo-stola (Ins.).

**pompat**—L. *pompatus*, magnificent.

**pomph**—Gr. *pomphos*, genit. *pomphygos*, a blister; *pompholyx*, a bubble. *Ex:* Pompho-rhynchus (Nemat.); Pompholyx (Rot.); Pompholyxia (Prot.).

**pompholyx**—See **pomph.**

**pompil**—Gr. *pompilos*, a kind of fish that follows ships. *Ex:* Pompili-oides (Ins.); Pompilus (Pisc.).

**pomponi**—NL. *pomponius*, with a tuft or top-knot<Fr. *ponpon*, a pompon or tuft.

**pon**—Gr. *poneō*, to toil; *ponos*, work, toil; *ponikos*, hard-working. *Ex:* Amblyo-pon-ides (Ins.)=Ambly-opo-pone (Ins.); Meli-pona (Ins.); Ornitho-ponus (Ins.); Boo-ponos (Ins.).

**poncir**—Fr. *poncire*, a kind of citron. *Ex:* Pon cirus*.

**ponderos**—L. *ponderosus*, weighty, ponderous. *Ex:* Ponderosi-ana (Moll.).

**poner**—Gr. *ponēros*, wicked, vicious, useless, difficult. *Ex:* Poner-amoeba (Prot.); poner-ine; Poner-odon (Pisc.); Ponera (Ins.); Poneroplax (Moll.); Ponero-psar (Av.).

**pong**—West African *mpongwe*, name of a tribe and the region they inhabit. *Ex:* Pongo (Mam.).

**pons**—See **pont.**

**pont**—1. Gr. *pontos*, sea; *pontios*, of the sea; *pontistēs*, one who casts into the sea. *Ex:* Pont-acarus; pont-ine; Ponti-vaga (Mam.); Pontio-ceramus (Echin.); Pontis (Ins.); Pontistes (Mam.); Ponto-bdella (Ann.); Pontodrilus (Ann.); Ponto-leo (Mam.); ponto-phyta;

Ponto-poria (Mam.): **2.** L. *ponto*, a small boat, a pontoon. *Ex:* Pont-ella (Crust.): **3.** L. *pons*, genit. *pontis*, a bridge. *Ex:* pont-al; ponto-cerebellar.

**popan**—Gr. *popanon*, a round cake. *Ex:* Popano-cerus (Moll.).

**poples**—See poplit.

**poplit**—L. *poples*, genit. *poplitis*, joint of the knee, the knee. *Ex:* poplite-al.

**popul**—L. *populus*, people, a great number; also the poplar tree, so-called from the number and continual motion of its leaves. *Ex:* populi-folia; Populus*.

**populne**—L. *populneus*, of poplars < *populus*, the poplar.

**por**—**1.** *porus*, a pore, passage. *Ex:* Por-ambon-ites (Brach.); Por-ites (Coel.); Pori-fera; Porono-tus (Pisc.); A-poro-blatt-ina (Ins.); blasto-pore; neuro-pore; nulli-pore; Tubuli-pora (Polyzoa): **2.** Gr. *poros*, a soft stone. *Ex:* madre-por-ite; madre-pore: **3.** Gr. *poros*, blind: **4.** Gr. *poros*, callus. *Ex:* Poro-chilus (Pisc.); Poro-temnus (Mam.); Oxy-porus (Av.).

**porc**—L. *porcus*, dim. *porculus = porcellus =* NL. *porcinola*, pig. *Ex:* Porca (Mam.); Porcinolus (Ins.); Porco-rhinus (Ins.); Porcula (Mam.); Porcus (Pisc.).

**porcat**—NL. *porcatus*, ridged < L. *porca*, the ridge between two furrows.

**porcell**—L. *porcellio*, genit. *porcellionis*, a wood-louse, sowbug. *Ex:* Porcell-idium (Crust.); Porcellio (Crust.); Porcellion-ides (Crust.).

**porcellan**—Ital. *porcellana =* the Venus shell or the nacre of the shell—"The curved shape of the upper surface of the shell was thought to resemble the raised back of a hog" (Ency. Dict.) < L. *porcus*, pig. *Ex:* Porcellan-aster-idae (Echin.); Porcellana (Crust.); Porcellano-pagurus (Crust.).

**porcin**—L. *porcinus*, of a hog < *porcus*, a pig.

**pord**—Gr. *pordē*, breaking wind. *Ex:* Ono-pordon*.

**pore**—**1.** Gr. *poreuō*, to ferry, to cross a river. *Ex:* Porcuo-mena (Ins.); Ponto-poria (Mam.): **2.** Gr. *poreia*, journey, course, gait.

**poreuo**—See pore.

**poriz**—Gr. *porizō*, to invent, come upon; also to bring about. *Ex:* Porizon (Ins.).

**porn**—Gr. *pornē*, a prostitute; *pornotrips*, a fornicator. *Ex:* porn-erast-ic; Pornotrips (Ins.).

**porom**—Gr. *pōrōma*, a callus.

**poros**—Gr. *pōrōsis*, a callosity. *Ex:* porosis.

**porp**—Gr. *porpē*, a buckle, pin, a brooch; *porp-ema = porpama*, a garment fastened with a brooch. *Ex:* Porp-alia (Coel.); Porp-ita (Coel.); Porpe-phyllum (Coel.); Porpema (Coel.); oto-porpae.

**porpac**—Gr. *porpax*, genit. *porpakos*, a ring, loop. *Ex:* Porpac-ella (Ins.); Porpax (Ins.).

**porpax**—See porpac.

**porpem**—See porp.

**porphyr**—**1.** Gr. *porphyros*, red brown, russet, also purple; *porphyra = porphyreos*, the purple fish; NL. *porphyreticus*, purplish. *Ex:* Porphyr-aspis (Ins.); Porphyra*; Porphyreo-cephalus (Av.); porphyro-leucus; Porphyro-stoma (Moll.) **2.** Gr. *porphyriōn*, the water hen < *porphyreos*, purple. *Ex:* Porphyrio (Av.); Porphyrion (Av.).

**porphyretic**—L. *porphyreticus*, purple-colored.

**porr**—L. *porrus*, the leek. *Ex:* porr-aceous; porri-folius.

**porrect**—L. *porrectus*, projected, extended forward horizontally, long < *porrigo*, to stretch out, to put forth. *Ex:* porrect.

**porrh**—Gr. *porrhō = prosō*, forward, distant = L. *porro*, forward. *Ex:* Porrh-omma (Arach.); Porrho-pus (Ins.); Porro-stoma (Ins.).

**porrig**—L. *porrigo*, genit. *porriginis*, dandruff. See also porrect.

**porro**—See porrh.

**port**—**1.** L. *porta*, a gate > ML. *portalis*, pertaining to a gate. *Ex:* porta; portal: **2.** L. *porto*, to carry; *portabilis*, portable. *Ex:* port-able: **3.** Gr. *portis*, a calf = *portax*, see portac: **4.** L. *portus*, a haven, a retreat.

**portab**—See port 2.

**portac**—Gr. *portax*, genit. *portakos*, a calf. *Ex:* Portax (Mam.).

**portari**—L. *portarius*, a door-keeper.

**portax**—See portac.

**portentos**—L. *portentosus*, also *portentuosus*, monstrous, hideous, unnatural.

**porthe**—Gr. *portheō*, to destroy; *porthēsis*, the sack of a town; *porthētēs*, also *porthētōr*, a destroyer. *Ex:* Porthe-odon (Mam.); Porthe-us (Pisc.); Portho-cyon (Mam.); Porthesia (Ins.); Chasma-porthetes (Mam.).

**porthm**—Gr. *porthmos*, a narrow passage, a tube; *porthmeus*, a ferry-man. *Ex:* Porthm-idius (Ins.); Porthm-ornis (Av.); Porthmeus (Pisc.).

**portho**—See porthe.

**portulac**—L. *portulaca*, a name for purslane. *Ex:* Portulac-aceae*; Portulac-aria*; Portulaca*.

**portun**—L. *Portunus*, a Roman deity, god of the port or harbor < *portus*, a harbor. *Ex:* Portuni-cepon (Crust.); Portunion (Crust.); Portunus (Crust.).

**porzan**—It. *porzana*, the crake. *Ex:* Porzan-ula (Av.); Porzana (Av.).

**pos**—**1.** Gr. *posis*, drinking, drink: **2.** Gr. *posos*, how much?, how many? *Ex:* poso-logy: **3.** Gr. *posis*, a husband. *Ex:* Di-posis*.

**posidon**—Gr. *Poseidōn*, Neptune, god of the sea. *Ex:* Posidonia (Moll.); Posidono-mya (Moll.).

**positor**—L. *positor*, a builder, founder, placer< *pono*, to place, put. *Ex:* ovi-positor.

**post**—L. *post*, after, behind; *posterior*, coming after, comp. of *posterus*, following, behind; *prosteritas*, futurity. *Ex:* post-cava; post-mortem; Post-eu-tatus (Mam.); Post-pithecus (Mam.); post-zyg-apophsis; posterior; pos terity.

**poster-** —NL. *poster-*, combining form<L. *posterior*, back, posterior. *Ex:* posteri-ad; Postero-branchus (Moll.); postero-lateral.

**posterior**—See post.

**posth**—Gr. *posthē*, membrum virile, the phallus; also the prepuce; *posthōn*, one with a large phallus. *Ex:* Posthon (Ins.); Artio-posthia (Verm.); Cyclo-posthium (Prot.).

**postic**—L. *posticus*, that which is behind. *Ex:* posticous.

**postrem**—L. *postremus*, the hindmost, the last.

**postum**—L. *postumus*, that which is at the end, final; superlative of *posterus*, next, following.

**pot**—Gr. *potos*, a drinking; *potēs*, genit. *potētos*, a drinking, a drink; *potēs*, a drinker. *Ex:* poteto-meter; Potosia (Ins.); Hydro-potes (Mam.).

**potam**—Gr. *potamos*, river. *Ex:* Potam-acmaea (Moll.); Potam-archus (Mam.); Potam-ides (Moll.); Potamo-gale (Mam.); Potamo-geton*; hippo-potamus; Pseudo-potam-illa (Ann.).

**potator**—L. *potator*, genit. *potatoris*, a drinker; *potatorius*, drinking.

**poten**—L. *potentia*, power, force, might< *potens*, genit. *potentis*, ppr. of *posse*, to be able>Sp. *potencia*, power. *Ex:* potency; Potent-illa*; potential.

**poteri**—L. *poterium*, a cup<Gr. *potērion*, a drinking cup; also a kind of plant. *Ex:* Poterio-crinus (Echin.) Poterio-dendron (Prot.); Poterion (Por.); Poterium*.

**poterion**—Gr. *potērion*, a plant name, perh. the tragacanth. See poteri.

**potet**—Gr. *potētos*, flying, winged. See pot.

**pothin**—Gr. *pothinos*, desired, longed for. *Ex:* Pothinus (Av.).

**potho**—1. NL. *pothos*<native Celonese name of some climbing plant. *Ex:* Potho-ideae*; Pothos*: 2. Gr. *Pothos*, winged youth who was strong with desire.

**pothos**—See potho.

**potoro**—Native Australian *potoroo*, the rat kangaroo. *Ex:* Potoro-ine (Mam.); Potorous (Mam.).

**potorous**—NL. *potorous*<native Australian name (*potoroo*) for a rat-kangaroo. *Ex:* Potorous (Mam.).

**potos**—See pot.

**pra**—Gr. *praos*, mild, gentle, delightful. *Ex:* Pra-obdella (Ann.); Praia (Ins.); Prao-pus (Mam.); Prao-therium (Mam.); Praon (Ins.).

**prae-** —L. *prae-*, prefix meaning before, in front. *Ex:* Prae-coces (Av.); prae-cocial; prae-cox; Prae-heter-odonta (Moll.); prae-morsus; prae-natica (Moll.); Prae-sorex (Mam.). S.

**prae-altus**—L. *prae-altus*, very high, very deep.

**praeceps**—See praecipit.

**praecipit**—L. *praeceps*, genit. *praecipitis*, a steep place, declivity; as adj. headlong, hasty.

**praemors**—L. *praemorsus*, bitten off, pp. of *praemordeo*, to bite off.

**praetect**—L. *praetectus*, covered over, protected; pp. of *praetego*, to cover.

**praetios**—See pretios.

**praetor**—L. *praetor*, one who goes before, a leader. *Ex:* Praetores (Ins.).

**praeust**—L. *praeustus*, scorched, brown at the summit.

**praevar**—L. *praevarus*, irregular, unsteady.

**pragm**—Gr. *pragma*, genit. *pragmatos*, an act, a thing done. *Ex:* Pragmat-odes (Ins.); Pragmo-pholas (Moll.).

**prao**—See pra.

**pras**—Gr. *prason*, leek; *prasinos*=*prasios*, leek-green; *prasokouris*, a larva that feeds on leek. *Ex:* Prasino-xena (Ins.); Prasi-ola*; Praso-pora (Bry.); Prasocuris (Ins.).

**prasin**—See pras.

**prasinat**—L. *prasinatus*, with leek-green dress< *prasinus*, leek-green. See pras.

**prasocur**—See pras.

**prason**—See pras.

**prat**—L. *pratum*, a meadow, *pratensis*, pertaining to or growing in a meadow; related to *pratens*, genit. *pratentis*, green. *Ex:* Prat-incola (Av.); prat-incol-ous; Prati-cola (Mam.); Prato-bombus (Ins.).

**prav**—L. *pravus*, deformed, distorted; *pravissimus*, very crooked.

**prax**—Gr. *praxis*, action, business. *Ex:* Praxi-thea (Ins.); Praxis (Ins.).

**praxill**—Gr. *Praxilla*, poetess of Sicyon. *Ex:* Praxill-ula (Ann.); Praxilla (Ann.).

**pre-** —Eng. *pre-* (<L. *prae*), prefix meaning before, in front, priority in time, place, rank. *Ex:* Pre-Cambrian; pre-caval; pre-dental; pre-natal. See also prae.

**precat**—L. *precatus*, requested; *precativis*, obtained through prayer, entreaty.

**precatori**—L. *precatorius*, relating to division or partitioning.

**precipit**—L. *praecipito*, to cast down, precipitate; pp. *praecipitatus*, thrown down. *Ex:* precipitation; precipit-in. See -in 1.

**predator**—L. *praedator*, a plunderer< *praeda*, booty. *Ex:* predator-y.

**pregnant**—L. *praegnatus*, pregnancy; *praegnans*, genit. *praegnantis*, with child, ppr. of a supposed *pregno*, to be pregnant< *prae*, before +*gno*, to bear, an obsolete verb with pp.

*gnatus*=*natus*. *Ex:* pregnant; im-pregnat-ion.

**pregnat**—See **pregnant**.

**prehens**—L. *prehensus*, seized, pp. of *prehendo*, to take, to seize>Fr. prehensile. *Ex:* prehen sile.

**premn**—Gr. *premnon*, the trunk or stump of a tree, the stem. *Ex:* Premn-ornis (Av.); Premna*; Premno-copus (Av.); Premno-plex (Av.); Poly-premnum*.

**pren**—Gr. *prēnēs*, drooping. *Ex:* Pren-anthes* Preneo-pogon (Ins.); Prenes (Pisc.); Preno-lepis (Ins.).

**prenes**—See **pren**.

**preon**—Gr. *preōn*, a headland, peninsula; top of a mountain. *Ex:* Preon-anthus*.

**prep**—Gr. *prepō*, to be visible, conspicuous, to resemble; *prepōdēs*, fit, suitable. *Ex:* Prepo-therium (Mam.); Charito-prepes (Ins.): **2.** Gr. *prepō*, to excell, to be very much like. *Ex:* Prepo-pharus (Ins.). See phar. 1.

**prept**—Gr. *preptos*, distinguished, renowned. *Ex:* Prepto-ceras (Mam.); Preptos (Ins.).

**prepuc**—Fr. *prepuce*<L. *praeputium*, the fore-skin<*prae*+Gr. *posthion*, the penis. *Ex:* prepuce; preputi-al.

**prepus**—Gr. *prepousa*, conspicuous. *Ex:* Prepusa*.

**preput**—See **prepuc**.

**presby**—Gr. *presbytēs*, fem. *presbytis*, an old person; *presbys*, an elder. *Ex:* Presby-ornis (Av.); Presby-pithecus (Mam.); Presbyt-iscus (Mam.); Presbytis (Mam.).

**press**—L. *pressus*, pressed, pp. of *premo*, to press; *impressus*, marked with slight impressions. *Ex:* Presso-odonta (Moll.); Pressi-bombus (Ins.); ad-press-ed; ap-press-orium.

**prest**—NL. *prestis*<Gr. *prethrō*, to blow up. *Ex:* Bu-prestis (Ins.); Bu-prest-idae (Ins.).

**pretios**—L. *pretiosus*, of great value.

**pretor**—NL. *pretor*<L. *praetor*, a commander< *prae*, before+*ire*, to go. *Ex:* Pretori-ana (Ins.); Pretoria (Ins.).

**pri**—See **prion**.

**pria**—See **prion**.

**priap**—Gr. *Priapos*, son of Aphrodite and Diony-sus, god of gardens and reproduction, repre-sented by a red painted figure with club or garden knife and with large phallus as a sym-bol of procreative powers; also the male gen-erative organ itself. *Ex:* Priap-ismus (Ins.); Priap-odes (Ins.); Priap-ulus (Gephyr.); Priapus (Gephyr.).

**prim**—L. *primus*, first; *primaevus*, young, youth-ful; *primordium*, the beginning; *primordialis*, original; *primitivus*, the first or earliest of its kind; *primitus*, at first; LL. *primula*, the prim-rose, because of its early flowering. *Ex:* prim-ordial; Primates (Mam.); primitive; primo-geniture; Primo-spinus (Pisc.); Primoevus (Mam.); Primul-aceae*; Primula*.

**primaev**—See **prim**.

**primar**—L. *primarius*, chief.

**primine**—Fr. *primine*<L. *primus*, first. *Ex:* primine.

**primno**—NL. *primnoa*, Lamarck's name for some gorgonian polyp. *Ex:* Primno-dendron (Coel.); Primno-ella (Coel.); Primnoa (Coel.).

**primord**—See **prim**.

**primul**—L. *primulus*, dim. of *primus*, first. *Ex:* Primula*. See prim.

**prin**—Gr. *prinos*, an evergreen oak, the holm oak=L. *prinus*. *Ex:* Prino-bius (Ins.).

**princ**—L. *princeps*, first, in front, chief. *Ex:* Princ-idium (Ins.); Princeps (Ins.).

**princeps**—See **princ**.

**prinia**—Javanese *prinya*, the name of a certain bird. *Ex:* Prinia (Av.).

**prion**—Gr. *priōn*, a saw; *priō*, to saw; *prionōdēs*, like a saw; *prionōtos*, made like a saw, jagged. *Ex:* Pri-acanthus (Pisc.); Pri-odon (Mam.); Pri-onychus (Ins.); Prion (Av.); Prion-ace (Elasm.); Prion-ops (Av.); Priono-cidaris (Echin.); Priono-myrmex (Ins.); Prionodes (Pisc.); Prionoto-lytta (Ins.); Prionotus (Ins.); Prionus (Ins.); Myrmeco-pria (Ins.).

**prisc**—L. *priscus*, of or belonging to former times, primitive. *Ex:* Prisca (Bry.); Prisco-delphinus (Mam.); Prisco-physeter (Mam.).

**prism**—Gr. *prisma*, genit. *prismatos*, a prism. *Ex:* Prisma (Moll.); Prismat-idium (Prot.); Prismato-monas (Prot.); Prismo-zoon (Prot.).

**prist**—**1.** Gr. *pristēs*, a sawyer; also a file, saw. *Ex:* Prist-urus (Rept.); Pristi-cercus (Rept.); Pristi-phor-idae (Elasm.); Pristio-nema (Nemat.); Pristo-ceutho-philus (Ins.): **2.** Gr. *pristis*, a large fish of the whale kind. Prist-idae (Elasm.); Pristis (Elasm.).

**pristin**—L. *pristinus*, primitive. *Ex:* Pristino-cetus (Mam.).

**priv**—L. *privus*, individual, peculiar; *privatus*, individual.

**pro-**—Gr. *pro-*, a prefix with wide usage making up many puzzling compounds and meaning before, in front of, forth, forward. *Ex:* Pro-avis (Ins.); pro-boskis=pro-boscis; Pro-cyon (Mam.); pro-gnathus; pro-phase; pro-polis; Pro-rhynchus (Platy.); pro-stomium; Pro-xylo-bius (Ins.); Pro-xyrodes (Ins.); etc. **S.**

**probat**—**1.** Gr. *probaton*, anything that walks forward, sheep and goats; *probatikos*, of sheep or goats. *Ex:* Probaticus (Ins.): **2.** L. *probator*, a tester, examiner<*probatus*, tested: **3.** Gr. *probateus*, a grazier, shepherd. *Ex:* Pro-bateus (Av.).

**probl**—Gr. *problēs*, genit. *problētos*, jutting out, projecting; *probolē*, a projection, a putting forward; *problētos*, thrown away, rejected. *Ex:* Proble-tomus (Ins.); Probles (Ins.); Probolo-ptera (Ins.); Probolus (Ins.).

**probol**—See **probl**.

**probosc**—L. *proboscis*, genit. *proboscidis*, a pro-boscis<Gr. *proboskis*, an elephant's trunk< *pro*+*boskō*, to feed. *Ex:* Probosc-idae (Mam.); Probosci-myia (Ins.); proboscidi-an; Pro-boscido-phora (Ins.).

**proboscid**—See **probosc.**

**proc**—Gr. *prox*, genit. *prokos*, a kind of deer. *Ex:* Proco-bus (Mam.); Prox (Mam.).

**procac**—L. *procax*, genit. *procacis*, shameless, in-solent.

**procax**—See **procac.**

**procella**—L. *procella*, a tempest. *Ex:* Procell-aria (Av.); Procell-arus (Av.).

**procer**—L. *procerus*, stretched out, long, high. *Ex:* Procerus (Ins.).

**process**—L. *processus*, process, progression. *Ex:* process.

**prochny**—Gr. *prochny*, kneeling. *Ex:* Prochny-anthes\*.

**procne**—See **progn.**

**proct**—Gr. *prōktos*, the anus, the hinder parts, tail. *Ex:* proct-odeum; Procto-porus (Rept.); Endo-procta (Brach.); Mega-proctus (Ins.); peri-proct.

**procumbens**—L. *procumbens*, falling forwards, prostrate.

**procurrens**—L. *procurrens*, genit. *procurrentis*, extending, advancing.

**prod**—L. *prodo*, to disclose, to show, bring forth, pp. *proditus*, brought forth, revealed. *Ex:* Prodi-domus (Arach.).

**prodic**—Gr. *prodikos*, judged first. *Ex:* Prodicus (Myr.).

**prodigios**—L. *prodigiosus*, full of wonder, marvelous.

**prodit**—See **prod.**

**prodot**—Gr. *prodotos*, betrayed, abandoned< *prodotēs*, a betrayer; *prodosia*, betrayal, aban-donment; *prodotikos*, traitorous. *Ex:* Prodotes (Ins.); Prodotis (Ins.).

**prodox**—Gr. *prodoxos*, judging of a thing prior to experience. *Ex:* Prodoxus (Ins.).

**prodrom**—**1.** Gr. *prodromos*, going in advance, going before. *Ex:* Prodromo-crinus (Echin.); Prodromo-teuthis (Moll.); Prodromus (Ins.): **2.** L. *prodromus*, a kind of early fig.

**product**—L. *productus*, lengthened, led forth, pp. of *produco*, to bring or lead forth. *Ex:* Productus (Brach.).

**proe-**—Gr. prefix *proe-* or *proei-*<*proeis*, before. *Ex:* Proe-palpus (Ins.); Proe-scutella (Echin.).

**proedri**—Gr. *proedria*, the seat of honor, *proedros*, one who sits in the first place. *Ex:* Proedrium (Mam.); Proedrus (Ins.).

**profund**—L. *profundus*, genit. pl. *profundorum*, of the depths, deep. *Ex:* profunda femoris.

**progan**—NL. *progano*<Gr. *pro*, before+*ganos*, brightness, brilliance. *Ex:* Progano-sauria (Rept.).

**progn**—Gr. *Proknē*, daughter of Pandiōn, trans-formed by the gods into a swallow. *Ex:* Progn-aster (Echin.); Progne=Procne (Av.); Hydro-progne (Av.)=Hydro-procne (Av.).

**progrediens**—L. *progrediens*, advancing.

**proi**—Gr. *prōios*, early in the day or the year. *Ex:* Proi-cene; proio-gony.

**projicien**—L. *projiciens*, genit. *projicientis*, pro-jecting, ppr. of *projicio*, to project. *Ex:* projicient.

**prol**—L. *proles*, offspring. *Ex:* proles; proli-fer-ation; proli-ficum; proli-ger-ous.

**prolat**—L. *prolatus*, bringing forward.

**prolix**—L. *prolixus*, long, extended, drawn out.

**promach**—Gr. *promachos*, a challenger. *Ex:* Promacho-crinus (Echin.).

**promenaea**—Gr. *Promeneia*, prophetess of Do-dona. *Ex:* Promenaea\*.

**prometh**—**1.** Gr. *promēthēs*, provident, wary. *Ex:* Prometh-ichthys (Pisc.): **2.** Gr. *Promē-theus*, son of the Titan, Iapetus. *Ex:* Prometheus (Ins.).

**prominen**—L. *prominens*, genit. *prominentis*, jutting out.

**prominul**—L. *prominulus*, projecting, prominent.

**promiscu**—L. *promiscuus*, mixed, common, with-out discrimination.

**prompt**—L. *promptus*, manifest, easy to see.

**pron**—**1.** Gr. *prōn*, genit. *prōnos*, a promontory, headland. *Ex:* Prono-cephalus (Verm.); Prono-phlebia (Ins.): **2.** L. *pronus*, leaning for-ward, bending down; *pronatus*, bowed. *Ex:* pronat-ion; pronat-or; prone.

**pronoe**—Gr. *Pronoē*, daughter of Nereis. *Ex:* Pronoe (Moll.).

**pronopi**—Gr. *pronōpius*, in front of the door. *Ex:* Pronopius (Arach.).

**pronub**—L. *Pronuba*, goddess of marriage. *Ex:* Pronuba (Ins.).

**propatul**—See **patul.**

**propendens**—L. *propendens*, suspended, hanging down, dangling; *propensus*, hanging down.

**propens**—See **propendens.**

**propinquus**—L. *propinquus*, near, neighboring; related to some other.

**propodeum**—NL. *propodeum*<Gr. *pro-*, before +*podex*, the anus. See -deum.

**propodi**—Gr. *propodios*, in front of or before the feet. *Ex:* propodium, pl. propodia; propodi-al; propodi-alia.

**propol**—Gr. *propolis*, a waxy material gathered by bees to be used as a cement. *Ex:* propolis.

**propri**—L. *proprius*, one's own, special. *Ex:* proprio-ceptor; proprio-genic.

**proptos**—Gr. *proptōsis*, a falling forward, an inclination. *Ex:* proptosis.

**propus**—Gr. *propous* genit. *propodos*, one with large feet. *Ex:* Propus (Ins.).

**propylae**—Gr. *propylaia*, an entrance. *Ex:* Propylaea (Ins.).

**pror**—Gr. *prōra*, a prow, front; *prōratēs*, the officer in charge of a ship. *Ex:* Pror-odon (Prot.); Prora-stomus (Mam.); Prorates (Ins.); proridens; Proro-phora (Ins.); Aetho-prora (Pisc.); Calli-prora*; Za-prora (Pisc.).

**prorat**—See pror.

**prors**—L. *prorsus*, straight on; also truly, absolutely. *Ex:* Prorsi-ceras (Moll.).

**pros**—See proso.

**proserp**—L. *proserpo*, to creep>*proserpinaca*, name of a kind of plant. *Ex:* Proserpinaca*.

**proserpin**—Fr. *Proserpine*<L. *Proserpina*, wife of Pluto. *Ex:* Proserpin-ella (Moll.); Proserpinidae (Moll.); Proserpina (Moll.). See also proserp.

**proso**—Gr. *pros*, forward, towards, in advance of, near. *Ex:* pros-enchyma; pros-thermo-taxis; Proso-branchiata (Moll.); proso-pyle; Prosostomata (Platy.).

**prosop**—1. Gr. *prosōpon*, the face, a mask; also a kind of wild herb. *Ex:* Prosopi-gastra (Ins.); Prosopium (Pisc.); Prosopo-theca (Arach.); Prosopon (Crust.); Platy-prosopos (Mam.): 2. Gr. *prosōpis*, an unidentified plant, the name probably from *prosōpon*, face, mask. *Ex:* Prosopis*.

**prospalt**—Gr. *Prospalta*, name of a deme. *Ex:* Prospalta (Ins.).

**prosphy**—Gr. *prosphyō*, to cause to grow; also to hang upon, cling to. *Ex.* Prosphy-odontes (Mam.).

**prosphor**—Gr. *prosphoros*, like, similar. *Ex:* Prosphoro-cichla (Av.).

**prospoiet**—Gr. *prospoiētos*, assumed, adopted. *Ex:* Prospoietus (Av.).

**prost**—See prosthec.

**prosth**—Gr. *prosthen*=*prosthe*, before, in front of. *Ex:* Prostho-gonimus (Platy.); not prosthenic, see pro and stheno.

**prosthec**—Gr. *prosthēkē*=*prosthēma*, an appendage, an addition; also an aid, assistance. *Ex:* Prost-anthera*; Prosthe-cotyle (Platy.); Prosthec-arthron (Ins.); Prosthema-dera (Av.).

**prosthem**—See prothec.

**prosthesis**—Gr. *prosthesis*, application, an addition. *Ex:* prosthesis.

**prosthio**—Gr. *prosthion*, neut. of *prosthios*, foremost. *Ex:* Prosthio-stomum (Platy.); prosthion.

**prostyp**—Gr. *prostypos*, embossed. *Ex:* prostypus.

**prot**—Gr. *prōtos*, first, primary. *Ex:* Prot-agrion (Ins.); Prot-amoeba (Prot.); Prot-echidna (Mam.); Prot-echinus (Echin.); Prot-elotherium (Mam.); Prot-ura (Ins.); Protobranchia (Moll.); proto-nema; Proto-phyta*; proto-plasm; proto-pod-ite; Proto-zoa.

**protact**—Gr. *protaktos*, placed before. *Ex:* Pro tactis (Echin.); Protacto-clymenia (Moll.).

**protasis**—Gr. *protasis*, a stretching forward; also a problem, question. *Ex:* Protasis (Ins.); Pachy-protasis (Ins.).

**protax**—Gr. *protaxis*, a placing before or in front. *Ex:* Protaxo-crinus (Echin.).

**prote**—Gr. *Prōteus*, a sea god, Neptune's herdsman who changed his shape at will. *Ex:* Prote-idae (Amph.); protei-form; Protea*; Proteo-saurus (Rept.); Proteus (Amph.).

**protein**—Ger. *protein*<Gr. *prōtos*, first. *Ex:* protein; prote-id=prote(in)+-*id*<Gr. *eidos*, like.

**protemn**—Gr. *protemnō*, to cut short. *Ex:* Protemn-odon (Mam.).

**protens**—L. *protensus*, extended, stretched out.

**protero**—Gr. *proteros* (comp. of *pro*), before either in time or space. *Ex:* Protero-cetus (Mam.); protero-phragma; protero-zoic.

**proterv**—L. *protervus*, violent.

**protict**—Gr. *protiktō*, to bring forth before. *Ex:* Protict-ops (Mam.).

**protimes**—Gr. *protimēsis*, preference. *Ex:* Pro times-ius (Arach.).

**protin**—NL. *protinus*, extended<Gr. *proteinō*, to extend, to show, to stretch forth.

**protist**—Gr. *prōtistos*, first of all. *Ex:* Protista.

**protium**—NL. *protium*, a plant name of unknown origin. *Ex:* Protium*.

**proto**—See prot.

**protom**—Gr. *protomē*, the front, the foremost part.

**protrit**—L. *protritus*, worn out, stale, vulgar<*protero*, to crush, wear down.

**protuber**—L. *protubero*, to swell out, to grow forth. *Ex:* protuber-ance.

**provect**—L. *provectus*, advanced, increased.

**provid**—L. *providus*, cautious.

**prox**—See proc.

**proxenetes**—Gr. *proxenētēs*, a negotiator, agent. *Ex:* Proxenetes (Platy.).

**proxim**—L. *proximus*, the nearest, next; superl. of *prope*, near. *Ex:* proxim-al.

**proxy**—NL. *proxys* (origin uncertain), name of certain bugs given by Spinola. *Ex:* Proxys (Ins.). See pro.

**pruin**—L. *pruinosus*, frosted, rimy<*pruina*, frost, winter>NL. *pruinatus*, frosted.

**prun**—L. *prunum*, dim. *prunulum*, a plum; *prunus*, a plum-tree. *Ex:* pruni-form; Prunocarpus (Prot.); Prunul-etta (Prot.); Prunulum (Prot.); Prunum (Moll.); Prunus*; not Prunella* which comes from a German word meaning quinsy or croup which certain plants of the genus Prunus were supposed to cure.

**prunul**—See prun.

**prurien**—L. *pruriens*, genit. *prurientis*, itching, ppr. of *prurio*, to itch. *Ex:* prurient.

**prymn**—Gr. *prymnos*, the hind-most, undermost, end-most; *prymnon*, the lower part; *prymnē*, the stern of a ship; also the bottom. *Ex:* Prymn-echinus (Echin.); Prymno-pteryx (Ins.); Belo-prymnus (Mam.); Crypto-prymna (Ins.); Eu-prymna (Moll.).

**prymnetes**—Gr. *prymnētēs*, the steersman. *Ex:* Prymnetes (Pisc.).

**psac**—Gr. *psakas*, genit. *psakados*=*psekas*, any small piece broken off, a grain. *Ex:* Psacaphora (Ins.); Psacadia (Ins.); Psacado-notus (Ins.); Psacadium*; Psacus (Ins.).

**psacad**—See **psac.**

**psadar**—NL. *psadara*<Gr. *psathyros*, fragile. *Ex:* Psadara (Moll.).

**psaenyth**—Gr. *psainythios*, false, vain. *Ex:* Psaenythia (Ins.).

**psaer**—Gr. *psairō*, to barely scrape, to touch lightly, to flutter. *Ex:* Psaero-ptera (Ins.); Psairo-neura (Ins.).

**psair**—See **psaer.**

**psal**—Gr. *psalis*, genit. *psalidos*, a pair of shears; also a razor. *Ex:* Psal-idium (Ins.); Psali-odus (Pisc.); psalido-dect; Psalido-myrmex (Ins.); Psalido-procne (Av.); Macro-psalis (Av.); Proto-psalis (Mam.).

**psalid**—See **psal.**

**psalist**—Gr. *psalistos*, chipped. *Ex:* Psalist-ops (Arach.); Psalistus (Ins.).

**psalm**—Gr. *psalmos*, a pulling, twitching; also a psalm, hymn. *Ex:* Psalmo-charias (Ins.).

**psalter**—Gr. *psaltērion*, the psalter, a book of many leaves; also a harp. *Ex:* Psalter (Av.); psalterium.

**psaltr**—Gr. *psaltria*, fem. of *psaltēs*, a harper. *Ex:* Psaltri-parus (Av.); Psaltri-tes (Av.); Psaltria (Av.).

**psamath**—Gr. *psamathos*, sand; *psammathōdēs*, sandy. *Ex:* Psamathio-myia (Ins.); Psamathocrita (Ins.).

**psamm**—Gr. *psammos*, sand. *Ex:* Psamm-echinus (Echin.); Psammo-bia (Moll.); Psammo-donax (Moll.); psammo-phil-ous.

**psammath**—See **psamath.**

**psaphar**—Gr. *psapharos*=Ionic *psapheros*, friable, crumbling. *Ex:* Psaphar-acis (Ins.); Psaphara (Ins.); Psapharo-mys (Ins.).

**psar**—Gr. *psar*, genit. *psaros*, the starling. *Ex:* Psar (Av.); Psar-alector (Av.); Psari-somus (Av.); Psaro-colius (Av.); Platy-psaris (Av.).

**psaron**—Modern Gr. *psaron*, a little fish. *Ex:* Ptero-psaron (Pisc.).

**psaroni**—NL. *psaronius*, speckled like a starling < Gr. *psar*, a starling. Cf. Ger. *starstein*, "starling stone," because of its speckled surface. *Ex:* Psaronius*.

**psathur**—See **psathyr.**

**psathyr**—Gr. *psathyros*, brittle, friable; *psathyrotēs*, brittleness, looseness of consistency. *Ex:*

Psathura (Moll.); Psathyro-metra (Echin.); Psathyrotes*; Psathyrus (Ins.).

**psechr**—Gr. *psēchros*, thin. *Ex:* Psechrus (Arach.).

**psectr**—Gr. *psēktra*, a scraper. *Ex:* Psectra-pus (Ins.); Psectro-gaster (Pisc.).

**psedn**—Gr. *psednos*, thin, scanty, naked. *Ex:* Psedno-blennis (Pisc.); Psedno-serica (Ins.); Psednos (Pisc.).

**psegm**—Gr. *psēgma*, genit. *psēgmatos*, shavings, scrapings. *Ex:* Psegma-phora (Ins.); Psegmato-pterus (Ins.); Psegmo-ptera (Ins.).

**psel**—See **psell.**

**pselact**—Gr. *pselaktos*, to be touched. *Ex:* Pselactus (Ins.).

**pselaph**—Gr. *psēlaphaō*, to feel about. *Ex:* Pselaphe-phila (Ins.); pselapho-theca; Pselaphon (Mam.); Pselaphus (Ins.).

**psell**—Gr. *psellion*=*pselion*, an armlet or anklet. *Ex:* Psellio-phorus (Av.); Psellio-pus (Ins.); Pselium*.

**psen**—1. Gr. *psēn*, an insect which lives in the plant of the Capri fig. *Ex:* Psen (Ins.); Psenulus (Ins.); Psenia (Ins.): 2. Gr. *psēnos*, smooth, bald. *Ex:* Pseno-bolus (Ins.); Psenocerus (Ins.).

**psene**—NL. *psenes*, a bird name, appar. < Gr. *psēnē*, the osprey. *Ex:* Psene (Av.); Psenes (Pisc.).

**pseph**—1. Gr. *psēphos*, also *psēphis*, genit. *psēphidos*, dim. *psēphidion*, a pebble. *Ex:* Pseph-idae (Moll.); Pseph-odus (Elasm.); Pseph-otus (Av.); Psepho-derma (Rept.); Psepho-phorus (Rept.): 2. Gr. *psephos*= *psephas*, obscurity, darkness, smoke; *psepharos*, gloomy, cloudy. *Ex:* Pseph-urus (Pisc.); Psephio-cera (Ins.): 3. Gr. *psephō*, to be afraid.

**psephen**—Gr. *psephēnos*, dark, obscure. *Ex:* Psepheno-saurus (Rept.); Psephenus (Ins.).

**psett**—Gr. *psētta*, a kind of flatfish. *Ex:* Psett-ichthys (Pisc.); Psett-odes (Pisc.); Psetta (Pisc.); Para-psettus (Pisc.).

**pseud**—Gr. *pseudēs*, false, deceptive. *Ex:* pseudapo-spory; pseud-axis; Pseud-echis (Rept.); Pseud-eu-antha (Ins.); Pseudi-conus (Ins.); pseudo-coel; pseudo-podium.

**pseust**—Gr. *pseustēs*, a liar, cheat. *Ex:* Pseustoplaca (Ins.); Phyllo-pseustes (Av.).

**psiad**—Gr. *psias*, genit. *psiados*, a drop. *Ex:* Psiadia*; Psiado-sporus (Ins.).

**psiath**—Gr. *psiathos*, a rush mat; also a bedfellow. *Ex:* Psiatho-lasius (Ins.).

**psidium**—NL. *psidium*, generic name of the guavas<a supposed Gr. *psidion*, name of some plant. *Ex:* Psidium*.

**psil**—Gr. *psilos*, naked, smooth. *Ex:* Psil-actis*; Psil-otum*; psil-ium (Ecol.); psilo-paedic; Psilo-soma (Rept.); Psilo-strophe*; A-psil-ops

(Ins.); Amphi-psila (Echin.); ?Lam-psilis (Moll.).

psithyr—Gr. *psithyros*, whispering, twittering; *psithyrismos*, a whispering. *Ex:* Psithyr-oedus (Av.); Psithyrus (Ins.).

psitt—Gr. *psittakē=psittakos*, a parrot; L. *psittacina*, parrot-like, of a parrot. *Ex:* psittac-osis (Med.); Psittacus (Av.); Psitto-spiza (Av.); Rhyncho-psitta (Av.).

psoa—Gr. *psoa*, the muscles of the loins. *Ex:* Psoa (Ins.); psoas major.

psoc—Gr. *psōchō*, to rub away, grind. *Ex:* Psocidae (Ins.); Psoc-idium (Ins.); Psoco-ptera (Ins.); Psocho-desmus (Myr.); Psochus (Ins.); Psocus (Ins.).

psoch—See psoc.

psol—1. Gr. *psōlos=*L. *psoleos*, the penis, also one circumcised or with prepuce retracted. *Ex:* Psol-idium (Echin.); Psolo-cystis (Echin.); Psolus (Echin.): 2. Gr. *psolos*, soot, smoke. *Ex:* Psolo-cnemis (Ins.); Psolo-desmus (Ins.).

psom—Gr. *psōmos*, a morsel, bit. *Ex:* Psomophilus (Av.); Psomus (Ins.).

psoph—Gr. *psophos*, noise; *psophētikos*, able to make a noise. *Ex:* Psopha (Av.); Psopheticus (Crust.); Psophia (Av.); Psopho-carpus*; Temno-psophus (Ins.).

psophetic—See psoph.

psor—Gr. *psōra*, the itch, scurvy; *psōriasis*, a being itchy; *psōraleos*, itchy, scabby. *Ex:* Psoroptes (Arth.) (*<psora+Sarcoptes*); Psoralea*; psoriasis (Med.); Psoro-phora (Ins.).

psorale—See psor.

psoth—Gr. *psothos*, a tumult, noise; also dirt, filth. *Ex:* Psothus (Ins.).

psych—1. Gr. *psychē*, mind, soul, understanding, breath. *Ex:* psych-iatry; Psych-ichthys (Pisc.); psycho-logy: 2. Gr. *Psychē*, a Greek nymph sometimes represented as a butterfly. *Ex:* Psych-ine*, because of the butterfly-like appendage of the pods; Psyche (Ins.); Psychomy-idae (Ins.); Hydro-psych-idae (Ins.): 3. Gr. *psychos*, genit. *psycheos*, frosts, cold weather, winter. *Ex:* Psycheo-trephes (Echin.); Psycho-bius (Ins.).

psychod—NL. *psychod<*Gr. *psychē*, a butterfly*+od<eidos*, form. *Ex:* Psychod-idae (Ins.).

psychotri—Gr. *psychōtria*, vivifying. *Ex:* Psychotria*.

psychr—Gr. *psychros*, cold, frosty; also mean, miserly; *psychroloutēs*, a bather in cold water. *Ex:* psychro-phylic; psychro-phytes; Psychropotes (Echin.); Psychrolutes (Pisc.).

psydr—Gr. *psydros*, untrue. *Ex:* Psydrus (Ins.).

psydrac—Gr. *psydrax*, genit. *psydrakos*, a pustule, blister. *Ex:* Psydrax*.

psydrax—See psydrac.

psygm—Gr. *psygma*, genit. *psygmatos*, anything that cools, a fan; *psygmos*, chilliness, damp-

ness. *Ex:* Psygmato-cera (Ins.); Psygmatolepis (Pisc.); Psygmo-phyllum*.

psyll—Gr. *psylla*, a flea*>psyllion*, a kind of plant, fleawort. *Ex:* Psyll-idae (Ins.); Psyllopsis (Ins.); Psylla (Ins.); Psylli-odes (Ins.); Psyllium*; Psyllo-sphex (Ins.); Cini-psyllum (Ins.); Sarco-psylla (Ins.). See also psil and psyllax.

psyllax—Gr. *psyllax=psylla=psyllos*, a flea; also, a kind of spider.

ptaer—Gr. *ptairō*, to sneeze. *Ex:* Ptaero-xylon*.

ptaesm—Gr. *ptaisma*, genit. *ptaismatos*, a false step, a mistake.

ptarmic—Gr. *ptarmikē*, yarrow*<ptarmikos*, causing to sneeze. *Ex:* ptarmic-al; Ptarmica*.

ptarmigan—L. *ptarmigan*, name for a kind of grouse. *Ex:* Ptarmigania (Tri.).

ptele—Gr. *ptelea*, the elm. *Ex:* Ptelea*; Pteleobius (Ins.); Eu-ptelea*.

pten—Gr. *ptēnos*, feathered, winged. *Ex:* Ptenidium (Ins.); Pten-ura (Av.); Pteno-glossa (Moll.); A-pteno-dytes (Av.); Lipo-ptena (Ins.).

ptenophyll—Late Gr. *ptenophyllos*, with deciduous leaves. *Ex:* ptenophyllo-phyta.

pter—Gr. *pteron*, dim. *pteridion*, wing, fin; *pteridios*, feathered. *Ex:* Pter-an-odon (Rept.); Pter-aspis (Pisc.); Pter-engraulis (Pisc.); Pteri-aphis (Ins.); pteridium; ptero-pegum; ptero-saur; Pteron-ura (Mam.); Pteron-ella (Platy.); Pterono-tropis (Pisc.); A-ptera (Ins.); Coleo-ptera (Ins.); Di-ptera (Ins.); Sci-uro-pterus (Mam.). See pterid.

pterelas—Gr. *Pterelas*, one of Actaeon's hounds. *Ex:* Pterelas (Crust.).

pterid—1. Gr. *pteris*, genit. *pteridos*, a kind of fern*<pteron*, wing. *Ex:* Pterido-phyta*; Pteris*; Pteriso-podus (Crust.); Dryo-pteria*: 2. Gr. *pteridion*, a fine small feather. *Ex:* Pterido-monas (Prot.); Pterido-phora (Av.).

pterigyn—Gr. *pteryginos*, winged.

pterin—Gr. *pterinos*, made of feathers, feathered. *Ex:* Pterino-crinus (Echin.).

ptern—1. Gr. *pterna*, heel; *pternistēs*, one who strikes with the heel. *Ex:* pterna; Pternistes (Av.); Pterno-pterus (Mam.); Eri-pterna (Mam.); Lito-pterna (Mam.): 2. Gr. *pternis*, a kind of hawk. *Ex:* Pternes (Av.); Leuco-pternis (Av.): 3. Gr. *pternis* genit. *pternidos*, the bottom of a dish.

pternes—See ptern.

pternist—See ptern.

pterocl—NL. *pterocles<*Gr. *pteron*, feather*+ kleis*, the tongue of a clasp. *Ex:* Pterocl-urus (Av.); Pterocles (Av.); pteroclo-morphic.

pteron—See pter.

pterot—Gr. *pterōtos*, winged with handles. *Ex:* Pteroto-blastus (Echin.); Pteroto-ceras (Moll.).

pteryg—Gr. *pteryx*, genit. *pterygos*, the wing, fin,

feather; *pterygion*, a little wing; in Anatomy the combining form *pterygo-* indicates connection with the pterygoid bone or process. *Ex:* pterg-oid; Pterigi-fer (Nemat.); Pterygioteuthis (Moll.); pterygium; pterygo-spinous; Pteryx (Ins.); A-pteryx (Av.); Tri-pterygium*.

**pterygist**—NL. *pterogistes*, a flutterer < Gr. *pterygizō*, to flutter. *Ex:* Pterygistes (Mam.)

**pterygot**—Gr. *pterygōtos*, winged. *Ex:* Pterygota; pterygote.

**pteryl**—NL. *pteryla* < Gr. *pteron*, feather, wing + *hylē*, wood, forest. *Ex:* pteryl-osis; pterylae, pl. of *pteryla*; pterylo-graphy.

**pteryx**—See pteryg.

**ptes**—Gr. *ptēsis*, a flying, flight. *Ex:* Ptesio-mya (Ins.); A-ptesis (Ins.).

**ptesim**—Gr. *ptēsimos*, winged, able to fly.

**ptich**—See ptych.

**ptil**—Gr. *ptilon*, a wing, feather-down, anything like a feather or wing. *Ex:* Ptil-ella (Coel.); ptil-inum, pl. ptil-ina; ptil-osis; Ptilia (Ins.);

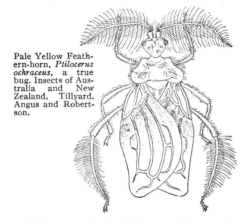

Pale Yellow Feathern-horn, *Ptilocerus ochraceus*, a true bug. Insects of Australia and New Zealand, Tillyard. Angus and Robertson.

Ptilio-gonys (Av.); Ptilo-gonys (Av.); ptilo-paedic; Ptilo-sarcus (Coel.); Ptilon-ella*; Ptilono-rhynchus (Av.); Polio-ptila (Av.).

**ptilinum**—NL. *ptilinum* < Gr. *ptilon*, a wing, a wing-like membrane. *Ex:* ptilinum.

**ptilon**—See ptil.

**ptilot**—Gr. *ptilōtos*, winged, stuffed with feathers. *Ex:* Ptilota*; Ptiloto-pus (Ins.); Ptilotus (Mam.).

**ptin**—NL. *ptinus* < Gr. *ptēnos*, feathered. *Ex:* Ptin-idae (Ins.); Ptino-bius (Ins.); Ptinus (Ins.); A-ptino-thrips (Ins.); Xyle-(p)tinus (Ins.).

**ptistes**—Gr. *ptistēs*, a winnower. *Ex:* Ptistes (Av.).

**ptoch**—Gr. *ptōchos*, one who crouches, a beggar. *Ex:* Ptoch-ella (Ins.); Ptocho-ptera (Av.); Ptocho-stola (Ins.); Ptochus (Ins.); Para-ptochus (Ins.).

**ptolemais**—Gr. *Ptolemais*, name of a tribe in Attica. *Ex:* Ptolemais (Mam.).

**ptom**—Gr. *ptōma*, genit. *ptōmatos*, a fallen body, a corpse; also a fall, misfortune. *Ex:* Ptoma-phagus (Ins.); Ptoma-phila (Ins.); Ptomato-phila (Ins.).

**ptor**—See phthor.

**ptortho**—Gr. *ptorthos*, a young branch, shoot, sapling. *Ex:* Ptorth-odius (Ins.); Ptortho-cera (Ins.).

**ptos**—Gr. *ptōsis*, a falling, an inflexion; *ptōsimos*, fallen, *ptōtos*, apt to fall. *Ex:* Ptosima (Ins.); ptosis (Med.); ptot-ic.

**ptosim**—See ptos.

**ptot**—See ptos.

**pty**—1. Gr. *ptyō*, spit out; *ptysis*, a spitting. *Ex:* Ptyo-iulus (Myr.); Ptyo-nodus (Pisc.); Ptyo-phora (Ins.); plasmo-ptysis: 2. Gr. *ptyon*, a fan < *ptyō*, to winnow. *Ex:* ptyo phyllum.

**ptyad**—Gr. *ptyas*, genit. *ptyados*, the spitter, a supposed serpent which spit poison into the eyes of those who molested it. *Ex:* Ptyas (Rept.).

A deep-sea crustacean *Ptychogaster defensa*, with prominent limbs and well-marked "stomach folds." Redrawn from Volume 26, Proceedings of United States National Museum.

**ptyal**—Gr. *ptyalon*, salvia; *ptyalismos*, a spitting, expectoration. *Ex:* ptyal-ose.

**ptyas**—See ptyad.

**ptych**—Gr. *ptychē* = *ptyx*, genit. *ptychos*, a fold, leaf, layer; *ptychōdēs*, in folds or layers. *Ex:* Ptych-emys (Rept.); Ptych-otis*; Ptycho-bothrium (Platy.); Ptycho-cetus (Mam.); Ptychodes (Ins.); Asio-ptych-aspis (Tri.).

**ptyct**—Gr. *ptyktos*, folded, doubled up. *Ex:* Ptyct-odus (Pisc.); Ptycto-notus (Ins.); Ana-ptycta (Ins.).

**ptyg**—NL. *ptygo*, irreg. < Gr. *ptyx* = *ptychē*, a fold, leaf, layer. *Ex:* Ptyg-ura (Rot.); Ptygo-derus (Rept.); ptyxis.

**ptygm**—Gr. *ptygma*, genit. *ptygmatos*, anything folded. *Ex:* Ptygmatis (Moll.); Ptygamto-phora (Ins.); Mono-ptygma (Moll.).

ptygmat—See ptygm.

ptylo—See ptil.

ptyng—Gr. *ptynx*, genit. *ptyngos*, the eagle-owl. *Ex:* Ptynx (Av.); Pholeo-ptynx (Av.).

ptynx—See ptyng.

ptyon—Gr. *ptyon*, a shovel or fan used in winnowing. *Ex:* Ptyon-odus (Pisc.); Ptyon-ornis (Av.); Ptyono-cera (Ins.).

ptyrtic—Gr. *ptyrtikos*, timorous. *Ex:* Ptyrticus (Av.).

ptysis—See pty.

ptysm—Gr. *ptysma*, genit. *ptysmatos*, spittle. *Ex:* ptysm-agoge (Med.); Ptysma-phora (Ins.).

ptyss—Gr. *ptyssō*, to fold. *Ex:* Ptysso-phorus (Mam.); Ptysso-stoma (Prot.).

ptyx—See ptyg, also ptych.

pub—1. L. *pubes*, any of the hair of adolescence but particularly the hair of the genitals; *puber* = *pubes* also *pubis*, adult, downy, that which has arrived at puberty, i.e. with hairiness: 2. L. *pubis*, the region of the pubes or the pubic bone. *Ex:* puber-al; pubes; pubi-gerous; pubis; pubo-femoral.

puber—See pub.

pubert—L. *pubertas*, the grown up age, manhood, the marriagable age; also the beard. *Ex:* pubert-y.

puberulen—ML. *puberulens*, genit. *puberulentis*, dim.<L. *pubens*, arrived at puberty. *Ex:* puberlulent.

pubescen—L. *pubescens*, genit. *pubescentis*, with hairs of puberty, downy; ppr. of *pubesco*, to reach maturity. *Ex:* pubescence; pubescent.

pubic—See pub.

pucras—Nepalese *pokras*, a pheasant. *Ex:* Pucrasia (Av.).

pudend—L. *pudendum*, pl. *pudenda*, the external genitals, especially of the female. *Ex:* pudendum muliebre.

pudibund—L. *pudibundus*, full of shame, modest <*pudor*, shame.

pudic—L. *pudicus*, modest<*pudeo*, to be ashamed. *Ex:* pudic vein.

pudu—S. A. Indian *pudu*, name of a small deer. Pudu (Mam.); Pudua (Mam.).

puell—L *puella*, a girl>*puellaris*, pertaining to young women, i.e. pretty. *Ex:* Puell-ina (Bry.); Puell-ula (Rept.); Puella (Moll.).

puer—L. *puer*, a male child. *Ex:* Puer (Crust.); Puer-ulus (Crust.).

pueril—L. *puerilis*, childish.

puffin—NL. *puffinus* from Eng. *puff*, a puff+*in*, a dim. ending. *Ex:* Puffin-aria (Av.); Puffin-uria (Av.); Puffinus (Av.).

pugil—L. *pugilis*, a pugilist. *Ex:* Pugil (Moll.); Pugil-arca (Moll.).

pugillat—L. *pugillator*, a message bearer.

pugio—L. *pugio*, genit. *pugionis*, a dagger. *Ex:* Pugio (Moll.); Pugio-droso-phila (Ins.); pugioni-formis; Pugionum*.

pugion—See pugio.

pugm—See pygm.

pugn—L. *pugnus*, a fist. *Ex:* Pugn-ellus (Moll.); Pugno-oides (Brach.); Pugnus (Moll.).

pugnac—L. *pugnax*, genit. *pugnacis*, fond of fighting, quarrelsome; *pugnacitas*, readiness in fight. *Ex:* pugnaci-ous; pugnacit-y; Pugnax (Brach.).

pugnax—See pugnac.

pukn—See pucn.

pulchell—L. *pulchellus*, beautiful. *Ex:* Pulchelli-scala (Moll.); Pulchellia (Moll.).

pulchr—L. m. *pulcher*, fem. *pulchra*, beautiful, fair. *Ex:* Pulchri-pitta (Av.); Pulchro-soma (Nemat.).

pulegi—L. *pulegium*=*puleium*, pennyroyal.

pulex—See pulic.

pulic—L. *pulex*, genit. *pulicis*, a flea; *pulicosus*, full of fleas. *Ex:* Pulex (Ins.); Pulic-aria*; Pulic-idae (Ins.); Pulici-phora (Ins.); pulicose.

pullar—L. *pullarius*, of or belonging to young animals.

pullat—L. *pullatus*, clothed in a black robe.

pullig—L. *pulligo*, genit. *pulligonis*, a dark color < *pullus*, dark-colored.

pullorum—L. *pullus*, genit. pl. *pullorum*, a young fowl, chicken, also the young of animals. *Ex:* Pullorum, a disease of chickens.

pullul—L. *pullulus*, a young animal; as an adj., blackish, dusky, gray. See pullus.

pullus—L. *pullus*, dusky, dark-colored. *Ex:* pull-ous; Pullus (Ins.).

pulmo—L. *pulmo*, genit. *pulmonis*, a lung; *pulmonarius*, pertaining to the lungs, beneficial to the lungs; NL. *pulmonatus*, having lungs. *Ex:* Pulmo-branchia (Moll.); Pulmonaria*; pulmonary; Pulmonata (Moll.).

pulmon—See pulmo.

pulp—L. *pulpa*, solid flesh. *Ex:* pulp; pulpi-fy.

puls—L. *pulsus*, a push, blow; also the pulse; *pulsatus*, beaten, struck; *pulsator*, a fighter; *pulsatrix*, a female fighter. *Ex:* puls-ellum; pulsat-ile; ?Pulsat-illa*; Pulsator (Prot.); Pulsatoria (Prot.); Pulsatrix (Av.); pulse.

pulsat—See puls.

pultiphagonides—L. *Pultiphagonides*, the pap-eater<*puls*, genit. *pultis*, pap+Gr. *phagō*, to eat. *Ex:* Pultiphagonides (Mam.).

pulv—L. *pulvis*, dust, powder.

pulver—L. *pulverulentus*, dusty, akin to *pulvereus*, full of dust; *pulver*, genit. *pulveris*, dust, powder.

pulvi—*pulvinus*, a little cushion, pillow, dim. *pulvinulus*, contr. to *pulvillus*; *pulvinaris*, of or belonging to a cushion; *pulvinatus*, cushion-shaped, elevated; *pulvinulus*, a little bank of earth. *Ex:* pulvi-plume; Pulvilli-gera (Ins.); pulvillus; Pulvin-ella (Ins.); Pulvinaria (Ins.); Pulvinulus (Prot.).

**pulviscul**—L. *pulvisculus*, fine dust<*pulvis*, dust+dim. *-culus.*; *cum pulvisculo* "dust and all," i.e. completely.

**puma**—Peruvian *puma*, name of a native cat. *Ex:* Puma (Mam.).

**pumex**—See **pumic**.

**pumic**—L. *pumex*, genit. *pumicis*, a pumice-stone, a porous stone. *Ex:* Pumex (Por.); Pumicia (Por.).

**pumil**—L. *pumilus*, diminutive, dwarfish; *pumilo* also *pumilio*, a pygmy. *Ex:* Pumilio (Ins.); Pumilo-myia (Ins.).

**punct**—L. *punctus*, a stinging, a puncture; *punctura*, a prick, puncture<*pungo*, to prick, puncture; *punctatus*, spotted as with punctures;

Spotted Gilia, *Gilia punctata*. Redrawn from Desert Wild Flowers—Jaeger. Stanford University Press.

*punctulatus*, dotted. *Ex:* Puncta-ptychus (Moll.); punctate; Punctati-ana (Moll.); Puncti-scala (Moll.); punctulate; punctum; Punctur-ella (Moll.).

**punctat**—See **punct**.

**punctul**—See **punct**.

**pung**—L. *pungo*, to prick, to puncture, ppr. *pungens*, genit. *pungentis*, stinging.

**pungiti**—NL. *pungitius*, pungent, sharp<L. *pungo*, to puncture.

**punic**—1. L. *punicans*, reddish, blushing; *puniceus*, reddish, purple-colored: 2. *Punicus*, ancient name for Carthage; *punica*, the Punic apple or pomegranate. *Ex:* Punica*.

**pup**—L. *pupa*, baby, child, doll; also a NL. name given by Linnaeus to the chrysalis of Lepidoptera. *Ex:* Pup-idae (Moll.); Pup-illa (Moll.); Pup-oides (Moll.); Pup-opsis; Pupa (Moll.); pup-ate; Pupi-para (Ins.); Bothrio-pupa (Moll.).

**pur**—L. *purus*, clean, pure.

**purg**—See **pyrg**.

**purgan**—L. *purgans*, genit. *purgantis*, cleansing, ppr. of *purgo*, to cleanse.

**purgat**—L. *purgatus*, cleansed<*purgo*, to make clean.

**purpur**—L. *purpuratus*, clad in purple; *purpurascens*, purplish, grown purple; *purpureus*, reddish, violet, purple; *purpura*, the purple-snail. *Ex:* Purpur-ella (Ins.); Purpura (Moll.); Purpurei-pitta (Av.); purpuri-parous.

**purulent**—L. *purulentus*, festering<*pus*, genit. *puris*, pus+*-lentus*, full of. *Ex:* purulent.

**pus**—See **pod**, also **pusill**.

**pusa**—See **pusiola**.

**pusill**—L. *pusillus*, very small, weak. *Ex:* Pusill-aster (Echin.); Pusill-ina (Moll.).

**pusio**—L. *pusio*, genit. *pusionis*, a little child.

**pusiola**—1. L. *pusa*, dim. *pusiola*, a girl. *Ex:* Pusa (Mam.); Pusiol-ina (Moll.); Pusiola (Moll.): 2. L. *pusio*, genit. *pusionis*, a lad.

**pusion**—See **pusiola**.

**pustul**—L. *pustula*, a pimple; *pustulosus*, full of pimples; *pustulatus*, blistered. *Ex:* Pustul-aria (Moll.); Pustula (Brach.); Pustuli-fer (Moll.); Pustulo-pora (Bry.); Choneti-pustula (Brach.).

**pusul**—L. *pusula*, a pimple, blister; *pusulatus*, blistered, made pure.

**put**—L. *putus*, pure, clean.

**putamen**—See **putamin**.

**putamin**—L. *putamen*, genit. *putaminis*, a shell, husk, pod.

**putat**—L. *putator*, a pruner of trees.

**putativ**—L. *putativus*, imaginary.

**putid**—L. *putidus*, rotten, fetid.

**putill**—L. *putillus*, a little boy. *Ex:* Putilla (Moll.).

**putor**—L. *putor*, genit. *putoris*, a foul odor, stench. *Ex:* Putori-odus (Mam.); Putoria*; Putorius (Mam.).

**putrescen**—L. *putrescens*, genit. *putrescentis*, becoming rotten, ppr. of *putresco*, to grow rotten.

**putrid**—L. *putridus*, putrid, decayed. See putid.

**py**—Gr. *pyon*, pus. *Ex:* py-oid; pyo-cyte; pyo-rhaea.

**pychn**—See **pycn**.

**pycn**—Gr. *pyknos*=*pychnos*, compact, dense, solid, strong. *Ex:* Pycn-anthemum*; pycn-ic; pycn-idium; Pycn-ophion (Ins.); pycnium; Pycno-gonum (Arach.); Pycno-notus; Pycno-podia (Echin.); Pychno-stachys*.

**pycnos**—Gr. *pyknōsis*, condensation. *Ex:* Pycnosiphorus (Ins.).

**pyct**—Gr. *pyktēs*, a boxer. *Ex:* Pyct-ornis (Av.); Pyctes (Av.).

**pydn**—Gr. *Pydna*, Macedonian city. *Ex:* Pydna (Ins.).

**pyel**—Gr. *pyelos*, an oblong trough, any vat-shaped vessel; also the pelvis of the kidney; *pyelōdēs*, like a trough, hollow. *Ex:* pyel-itis (Med.); pyelo-nephritis (Med.); Pyelo-somum (Platy.).

**pyg**—Gr. *pygē*, the rump, buttocks. *Ex:* Pyg-acanthus (Pisc.); Pyg-aera (Ins.), Gr. *airō*, to lift up; pyg-al; Pyg-eretmus (Mam.); pyg-idium; Pyga-thrix (Mam.); Pygo-podes (Av.); Pygo-sceles (Av.); cyto-pyge; Doro-pygus (Arth.); Pigo-styla (Prot.); steato-pyg-ous.

**pygarg**—L. *pygargus*, a kind of eagle; also a kind of antelope < Gr. *pygargos*, white rump. *Ex:* Pygargus (Av.).

**pygm**—Gr. *pygmē*, the fist; L. *pygmaeus*, pygmy-like, dwarfish, like a tom-thumb < Gr. *pygmaios*, dwarfish; *Pygmaioi*, the Pygmies. *Ex:* Pygm-ura (Mam.); Pigm-ornis (Av.); Pygmae-phorus (Arach.); Pygmaeo-drilus (Ann.); Pygme-odon (Mam.); Pigmeo-phorus (Arach.); Pygmo-crates (Ins.).

**pyl**—Gr. *pylē*, a gate, entrance > Gr. *pylōros*, a gate-keeper > NL. *pylorus*, the opening from the stomach. *Ex:* pyl-angium; pyla; Pylo-capsa (Prot.); pylor-ic; pylorus; apo-pyle; Di-pyl-idium (Platy.); micro-pyle.

**pylad**—L. *pylades*, faithful friend < *Pylades*, friend of Orestes.

**pylaemen**—Gr. *Pylaimenēs*, king of the Paphlagonians. *Ex:* Pylaemenes (Ins.).

**pylai**—Gr. *pylaios*, at the gate < *pylē*, a door, entrance. *Ex:* Pylai-ella*.

**pylon**—Gr. *pylōn*, genit. *pylōnos*, a gateway. *Ex:* Pylon-issa (Prot.); Pylon-ium (Prot.); Amphi-pylon-ium (Prot.).

**pyr**—1. ML. *pyrum* = L. *pirum*, dim. *pirula*, a pear. *Ex:* Pyr-ola*; pyri-form; Pyro-plasma (Prot.); Pyrul-aria*; Pyrula (Moll.); Pyrum (Moll.); Pyrus*: 2. Gr. *pyros*, wheat. *Ex:* Agro-pyron*; Dios-pyros*; Fago-pyrum*: 3. Gr. *pyr*, genit. *pyros*, fire. *Ex:* Pyr-acantha*; Pyr-alauda (Av.); pyr-ium (Ecol.); Pyri-glena (Av.); Pyro-felis (Mam.); Pyro-stegia*; Pyro-teuthis (Moll.).

**pyracmon**—Gr. *Pyrakmōn*, servant of Vulcan. *Ex:* Pyracmon (Ins.).

**pyral**—Gr. *pyralis*, a kind of insect supposed to live on fire < *pyr*, fire. *Ex:* Pyral-idae (Ins.); Pyralis (Ins.); Pyralo-morpha (Ins.).

**pyram**—Gr. *pyramis*, genit. *pyramidos*, a pyramid. *Ex:* Pyrami-don (Mam.); pyramid-al; Pyramid-ella (Moll.); Pyramid-ula (Moll.); Pyramido-phorus (Ins.); Acro-pyramis (Prot.).

**pyramid**—See **pyram**.

**pyramin**—Gr. *pyraminos*, of wheat, wheaten.

**pyraust**—Gr. *pyraustēs*, a fire dwelling insect of fable. *Ex:* Pyrausta (Ins.).

**pyren**—Gr. *pyrēn*, kernel, the pit of a fruit. *Ex:* pyren-oid; Pyrene (Moll.); pyren-ium; pyreno-carp; Pyreno-mycetes*.

**pyret**—Gr. *pyretos*, a burning heat, fever. *Ex:* pyreto-logy (Med.).

**pyrethr**—Gr. *pyrethron*, a hot spicy plant of the pellitory kind, *Anthemis pyrethrum*. *Ex:* Pyrethrum*.

**pyrg**—Gr. *pyrgos*, a tower; *pyrgitēs*, a house sparrow, lit., bird of a tower; *pyrginos*, tower-like. *Ex:* Purgosia*; Pyrg-elix (Moll.); Pyrg-idium (Moll.); Pyrgi-soma (Av.); Pyrgit-ina (Av.); Pyrgit-opsis (Av.); Pyrgita (Av.); Pyrgo-cystis (Echin.); Pyrgo-phylax (Ins.); Pyrg-ula (Moll.); Pyrgus (Moll.).

**pyrgit**—See **pyrg**.

**pyrgom**—See **pyrgot**.

**pyrgot**—Gr. *pyrgōma*, a tower; *pyrgōtos*, fem. *pyrgōtis*, made like a tower. *Ex:* Pyrgotes (Ins.).

**pyrin**—Gr. *pyrinos*, of fire. 2. Gr. *pyrinos*, made of wheat.

**pyros**—Gr. *pyrōsis*, a burning, inflammation. *Ex:* pyrosis. See also pyr.

**pyrr**—See **pyrrh**.

**pyrrh**—Gr. *pyrrhos*, flame-colored, reddish < *pyr*, fire. *Ex:* Pyrrh-aspis (Ins.); Pyrrho-coris (Ins.); pyrrho-melas; Pyrrho-pappus*.

**pyrrhul**—L. *pyrrhula*, a bullfinch < *pyrrhos*, red, purplish. *Ex:* Pyrrhul-oxia (Av.), see lox; Pyrrhula (Av.); Pyrrhulo-rhynchus (Av.).

**pyrul**—NL. *pyrula* < L. *pyrum*, *pyrus*, a pear. *Ex:* Pyrula (Moll.); Pyrulo-rhynchus (Moll.).

**pystis**—Gr. *pystis*, a listening to, an inquiring. *Ex:* Poly-pystis (Ins.).

**pyth**—1. Gr. *pythō*, to cause rot, to decay. *Ex:* Pithium*; pytho-genic. See also pith: 2. Gr. *Pythō*, old name for Delphi. *Ex:* Pytho (Ins.); Pytho-dora (Ins.); Pytho-plesius (Ins.).

**pythi**—Gr. *Pythia*, a priestess. *Ex:* Pythia (Moll.).

**python**—Gr. *Pythōn*, a serpent said to destroy men and cattle about Delphi. *Ex:* Python (Rept.); Python-aster (Echin.); Python-iscus (Rept.); Pythono-morpha (Rept.).

**pyx**—1. Gr. *pyxis*, dim. *pyxidion* = L. *pyxis*, genit. *pixidis*, a box. *Ex:* Pyx-in-ae*; Pyxi-cephalus (Amph.); Pyxi-poma (Moll.); Pyxid-anthera*; pyxid-ate; pyxidium; Centro-pyxis (Prot.): 2. Gr. *pyx*, later form of *pygē*, the rump, buttock. *Ex:* Sphaero-pyx (Ins.).

**pyxid**—L. *pyxidatus*, box-like, cubical < *pyxis*, a box.

**pzamm**—See **psamm**.

# Q

quadr—L. *quadrus*, fourfold. *Ex:* quadri-capsular; Quadri-lateral (Crust.); Quadri-sulcata (Mam.); Quadru-mana (Mam.); corpora quadri-gemmina.

quadrat—L. *quadratus*, squared. *Ex:* quadrate.

quadrul—LL. *quadrula*, a little square. *Ex:* Quadrula (Moll.).

qual—L. *qualis*, of what kind or sort. *Ex:* qualimeter.

quamasia—See **camass.**

quamoclit—Gr. *quamoclit* < Gr. *kuamos*, a bean + *klitus*, a slope or hillside, or perhaps from some native Mahratta name. *Ex:* Quamo-clit*.

quasi- —L. *quasi*-, as if; often used as a prefix to Eng. words to denote resemblance. *Ex:* quasi-radiate.

quass—L. *quassus*, shaken, pp. of *quatio*, to shake; *quassatus*, shaken violently, shattered, pp. of *quasso*, to shake. *Ex:* Quassi-labia (Pisc.).

quatern—L. *quaterni*, four each > *quaternarius*, consisting of four, arranged in fours. *Ex:* Quaternary.

quebracho—Pg. *quebracho*, lit. axe-breaker; the name of several hard-wooded South American trees. *Ex:* Quebracho*.

quelea—NL. *quelea*, probably from native African word for the crimson-beaked weaver-bird. *Ex:* Quelea (Av.).

quercin—L. *quercinus*, of oak leaves.

quercus—L. *quercus*, an oak. *Ex:* Querc-ineae*; Quercus*.

quernal—L. *quernus*, pl. *quernales*, oaken. *Ex:* Quernal Alliance*; Quernales*.

querquedula—L. *querquedula* < Gr. *kerkouris*, a kind of duck. *Ex:* Querquedula (Av.).

querul—L. *querulus*, complaining. *Ex:* Querula (Av.).

quillai—NL. *quillaia* = *quillaja* < Chilian *quillai*, name for some rosaceous tree. *Ex:* Quillaiae*; Quillaja*.

quinari—L. *quinarius*, containing five.

quinat—NL. *quinatus*, in fives < L. *quini*, five each.

quinqu—L. *quinque*, five, consisting of or pertaining to the number five. *Ex:* quinque-angular; quinque-costate.

quiscalus—*quiscalus* < ML. *quiscula, quisquila,* etc., a quail. *Ex:* Quiscal-inae (Av.); Quiscalus (Av.).

quisqualis—NL. *quisqualis* < L. *quis,* who + *qualis,* of what kind. *Ex:* Quis-qualis*.

# R

rab—L. *rabus*, dark-colored. *Ex:* Rabo-cerus (Ins.).

rabd—See **rhabd.**

rabduch—NL. *rabduchus* < Gr. *rhabdouchos*, a judge, a staff-bearer. *Ex:* Rabducho-petalus (Myr.); Rabduchus (Ins.).

rabid—L. *rabidus*, fierce, furious. *Ex:* Rabida (Pisc.).

rabul—L. *rabula*, a wrangler. *Ex:* Rabula (Pisc.).

racem—L. *racemus*, dim. *racemulus*, the stalk of a cluster, a bunch of berries, a cluster of grapes. *Ex:* racem-ose; raceme; racemi-ferous; Racemul-ina (Prot.).

rach—See **rhach.**

rachi—Gr. *rhachia*, a rocky shore, a crag, cliff, a ridge; also a dashing of waves, uproar. *Ex:*

Rachi-callis*; Rachi-discus (Ins.) Rachianectes (Mam.). See also **rhach.**

rachist—See **rhachist.**

rachit—See **rhach** 1.

rachy—See **rhach.**

radi—L. *radius*, dim. *radiolus*, a ray, spoke of a wheel; *radiatus*, rayed. *Ex:* radi-al-is; Radiaster (Echin.); Radiat-odonta (Moll.); radiospermic; Radio-sphaera (Prot.); Radiol-aria (Prot.); Radiolus (Moll.); Radius (Moll.).

radian—L. *radians*, genit. *radiantis*, beaming with light, brightly shining

radiat—See **radi.**

radic—L. *radix*, genit. *radicis*, a root; dim. *radicula*; *radicatus*, rooted; *radicans*, striking root < *radico*, to take root. *Ex:* Radic-ula*;

Radici-pes (Coel.); Radici-spongia (Por.); radicle; Radix (Moll.).

**radin**—See **rhadin.**

**radiol**—See **radi.**

**radix**—See **radic.**

**radul**—L. *radula,* a scraper <*rado,* to scrape. *Ex:* radul-ate; radul-inus; Radula (Moll.); Radulo-pecten (Moll.).

**raeb**—See **rhaeb.**

**raet**—L. *Raetus* < *Raeti,* a people living north of the Po. *Ex:* Raet-ella (Moll.); Raeta (Moll.); Raeto-lucina (Moll.); Raeto-mya (Moll.).

**rai**—L. *raia,* a flatfish, skate. *Ex:* Rai-idae = Raj-idae (Elasm.); Raia = Raja (Elasm.).

**raj**—See **rai.**

**rale**—Fr. *râle,* a sound (usually of morbid origin) which may accompany the sounds normally heard on auscultation of the chest <*raler,* to rattle in the throat. *Ex:* rale (Med.). See also rall.

**rall**—**1.** NL. *rallus* <Fr. *râle,* a kind of bird, the rail. *Ex:* Rall-idae (Av.); Ralli-cola (Ins.); Rallus (Av.): **2.** L. *rallus,* thin.

**ram**—L. *ramus,* dim. *ramulus = ramusculus,* a branch; *ramealus =* NL. *ramulis,* pertaining to a branch; *rameus,* belonging to branches; *ramosus,* dim. *ramulosus,* full of branches. *Ex:* ram-al; ram-astrum; rami-ger-ous; Rami-spongia (Por.); ramose; Ramul-aria (Pisc.)· ramuli-ferous; Ramulus (Ins.); ramus; bi-ram-ous.

**ramal**—L. *ramale,* pl. *ramalia,* shoots, twigs. *Ex:* Ramal-ina*. See also ram.

**rame**—L. *rameus,* belonging to tree trunks and branches <*ramus,* a branch.

**rament**—L. *ramentum,* pl. *ramenta,* a scale, chip; *ramentosus,* full of chips, scaly. *Ex:* rament-aceous; ramenti-ferous; ramentum = rament.

**rameus**—See **ram.**

**ramex**—See **ramic.**

**ramic**—L. *ramex,* genit. *ramicis,* a rupture, hernia. *Ex:* Ramici-forma (Prot.).

**ramph**—See **rhamph.**

**ramul**—See **ram.**

**ramuscul**—L. *ramusculus,* a little branch; a dim. of *ramus,* a branch.

**ran**—L. *rana,* dim. *ranunculus,* a frog > NL. *raninus,* pertaining to frogs. *Ex:* Ran-etta (Amph.); Ran-idae (Amph.); Ran-odon (Amph.); Ran-unculus*; Rana (Amph.); Rani-ceps (Pisc.); Ranina (Crust.); ranine artery; Rano-soma (Amph.).

**ranatra**—NL. *ranatra,* etym. unknown. *Ex:* Ranatra (Ins.).

**rancen**—L. *rancens,* genit. *rancentis,* stinking, putrid.

**rancid**—L. *rancidus,* disgusting, offensive.

**rangifer**—NL. *rangifer,* name applied to a genus of reindeer <OSw. *ren,* a reindeer+L. *fera,* a wild beast. *Ex:* Rangifer (Mam.); rangifer-ine.

**ranin**—See **ran.**

**ranuncul**—See **ran.**

**rap**—L. *rapa,* also *rapum,* a turnip. *Ex:* rap-aceous; Rapa (Moll.); rapi-formis.

**rapac**—L. *rapax,* genit. *rapacis,* greedy. *Ex:* rapac-ious; Rapaces (Mam.).

**rapanea**—NL. *rapanea* <Tropical Amer. native name for some plant. *Ex:* Rapanea*.

**rapate**—NL. *rapatea,* from native name in Guiana. *Ex:* Rapatea*.

**rapax**—See **rapac.**

**raph**—Gr. *rhaphē,* a seam, suture. *Ex:* raphe; Ortho-rrhaphe (Ins.); not Raphia*, which is derived from *raffia,* a native name.

**raphan**—Gr. *rhaphanos,* a cabbage, akin to *rhaphanis = rhaphanē,* a radish < Gr. *ra,* quickly +*phainomai,* to appear. *Ex:* Raphan-istrum*, evidently a combination of Gr. *rhaphanos,* cabbage and L. *rapistrum,* the wild turnip; Raphano-crinus (Echin.); Raphanus*.

**raphid**—Gr. *rhaphis,* genit. *rhaphidos,* a needle, pin. *Ex:* Raphi-cerus (Mam.); Raphid-echinus (Echin.); Raphidi-idae (Ins.); Rhaphidi-ophrys (Prot.); Raphidia (Ins.); Rhaphido-phor-inae (Ins.); Raphio-lepis*; raphis.

**raphis**—See **raphid.**

**rapid**  L. *rapidus,* tearing away, seizing. See also raphid.

**rapinator**—L. *rapinator,* genit. *rapinatoris,* a robber.

**rapistr**—L. *rapistrum,* the wild turnip. *Ex:* Rapistrum (Prot.).

**rapt**—Gr. *rhaptos,* stitched, sewed; *rhaptēs,* a mender, patcher <*rhaptō,* to sew, patch. *Ex:* Rapt-omphalus (Moll.); Rapti-formica (Ins.).

**raptator**—See **raptor.**

**raptor**—L. *raptor = raptator,* a robber; NL. pl. *raptores = raptatores. Ex:* Raptores (Av.) = Raptatores (Av.); raptatori-al.

**rar**—L. *rarus,* infrequent, rare. *Ex:* Rari-squamosa (Ins.).

**rasbor**—NL. *rasbora* <native name of a fish. *Ex:* Rasbor-ella (Pisc.); Rasbor-ichthys (Pisc.); Rasbora (Pisc.).

**rasil**—L. *rasilis,* neut. *rasile,* shaved, polished.

**rasor**—L. *rasor,* a scraper, NL. pl. *rasores* (applied to a fiddler) <*rado,* pp. *rasus,* to scrape, to scratch. *Ex:* Rasores (Av.); rasori-al.

**rastr**—L. *rastrum,* a rake, dim. *rastrella* also *rastellus* > NL. *rastratus,* covered as if with longitudinal scratches. *Ex:* rastr-ite; Rastr-ites (Coel.); Rastrelli-ger (Pisc.); rastrellus; Ras tro-graptus (Coel.); Rastrum (Pisc.).

**ratel**—South Afr. Dutch, *rateld,* lit. honey-comb badger. *Ex:* ratel; Ratelus (Mam.).

rathym—See rhathym.

ratis—See ratit.

ratit—L. *ratis*, a raft or flat-bottomed boat> *ratitus*, marked with the figure of a raft. *Ex:* Ratis (Coel.); Ratit-ae (Av.); ratite.

ratt—L. *rattus*, rat. *Ex:* Rattus (Mam.).

rav—L. *ravus*, tawny, gray-yellow. *Ex:* ravi-venter.

ravid—L. *ravidus*, gray, dark-colored. *Ex:* ravidous.

ravidul—NL. *ravidulus*, somewhat grayish <L. *ravidus*, dark-colored, grayish +-*ulus*, a dim.

re-—L. *re*, back, again. *Ex:* re-generation; re-plicatile; re-production.

reagen—NL. *reagens*, genit. *reagentis*, acting again <L. *re-*, again+*agens*, ppr. of *ago*, to perform, to move. *Ex:* reagent (re-+Eng. *agent*).

rebell—L. *rebellis*, rebellious, revolting.

recapitul—L. *recapitulo*, to go over the principal points again. *Ex:* recapitul-ation.

recept—L. *receptor*, a receiver. *Ex:* receptor.

receptacul—L. *receptaculum*, a reservoir. *Ex:* receptacle; Receptacul-ites (Por.).

receptiv—NL. *receptivus*, receiving, restoring, recovering <*receptus*, falling back+-*ivus*, tendency toward. *Ex:* receptive.

recess—L. *recessus*, drawn back, receding.

reciproc—L. *reciprocus*, going backward and forward. *Ex:* reciproc-al.

recis—L. *recisus*, cut back, pp. of *recido*, to cut short.

reclinat—L. *reclinatus*, bent back.

recliv—L. *reclivis*=*reclivus*, inclined, leaning backwards.

recolligens—L. *recolligens*, gathering up; ppr. of *recolligo*, to collect.

recondit—L. *reconditus*, concealed. *Ex:* recondite.

recrudesc—L. *recrudesco*, to make raw or primitive again, to become raw; ppr. *recrudescens*, genit. *recrudescentis*. *Ex:* recrudescence.

rect—1. L. *rectus*, straight. *Ex:* Rect-axis (Moll.); Recti-dens (Moll.); recti-serial; Recto-gloma (Moll.); rectus: 2. L. *rectum*, the rectum <*rectus*, straight. *Ex:* recto-vaginal; rectum.

recter—Gr. *rhektēr*, genit. *rhektēros*, strenuous, active, nimble. *Ex:* Recter (Av.).

rectric—L. *rectrix*, pl. *rectrices*, a directress< *rector*, a ruler. *Ex:* rectrici-al.

rectrix—See rectric.

recul—L. *recula*, a small thing, trifle, a matter of small importance<*res*, a thing.

recumben—L. *recumbens*, genit. *recumbentis*, reclining, ppr. of *recumbo*, to lie down. *Ex:* recumbent.

recurs—L. *recursus*, a going back, a retreat.

recurv—L. *recurvo*, to bend backward. *Ex:* Recurv-aria (Ins.); Recurvi-rostra (Av.); re-curvo-rostrate.

recutit—L. *recutitus*, having a fresh or new skin; also skinned, circumcised.

redact—L. *redactus*, restored, returned, pp. of *redigo*, to give back.

redia—NL. *redia*, name applied to one of the numerous larvae produced within sporocysts of certain digenetic trematodes<Redi, an Italian naturalist. *Ex:* redia.

rediviv—L. *redivivus*, revived (sometimes applied to rediscovered or resurrected species). *Ex:* Rediviva (Ins.).

redolen—L. *redolens*, genit. *redolentis*, emitting an odor. *Ex:* redolent.

reduct—L. *reductus*, brought back, returned.

redunc—L. *reduncus*, curved backwards. *Ex:* Redunca (Mam.).

reduvi—L. *reduvia*, a hangnail. *Ex:* Reduvi-idae (Ins.); Reduvi-olus (Ins.); Reduvia (Ins.).

reflex—L. *reflexus*, reflected, turned back, pp. of *reflecto*, to turn back. *Ex:* reflex-or.

refract—L. *refractus*, broken. *Ex:* refracted.

reg—L. *rex*, genit. *regis*, dim. *regulus*, a king; fem. *regina*; *regius*, royal; *regalis*, royal. *Ex:* Reg-alecus (Pisc.), (*rex*+*allec*=*alec*, a herring); Regina (Rept.); Regul-oides (Av.); Regulus (Av.); Rex (Av.).

regal—See rex.

regill—L. *regillus*, regal, splendid.

regin—See reg.

regius—See reg.

regma—See rhegm.

regn—L. *regno*, to be lord, to rule. *Ex:* Regno-saurus (Rept.).

regul—See reg.

regular—L. *regularis*, regular. *Ex:* Regularia (Echin.); Regularis (Por.).

regurgitat—ML. *regurgito*, to be thrown back, pp. *reguritatus*, thrown back. *Ex:* regurgitate.

reicul—L. *reiculus*, useless, worthless.

reism—See erism.

reithr—See rheithr.

relat—L. *relatus*, returned, carried back.

relict—L. *relictus*, forsaken, abandoned, pp. of *relinquo*, to abandon. *Ex:* relict; relictus.

rem—1. Gr. *rhēma*, genit. *rhēmatos*, what has been said, a word, a song. *Ex:* ?Rem-aster (Echin.); Rema (Ins.): 2. L. *remus*, an oar. *Ex:* remi-ped; Remi-ped-ella (Ins.); Remi-pes (Crust.); Remo-pleura (Tril.); Remus (Ins.); Pent-rem-ites (Blast.): 3. *Rheims*, city of France. *Ex:* Remi-ornis (Av.).

remig—L. *remes*, pl. *remiges*, dim. *remigulus*, a rower; *remigium*, a rowing. *Ex:* Remigia (Ins.); Remigo-lepis (Pisc.); Remigulus (Crust.).

**remiss**—L. *remissus*, bent back, bent up.

**remmius**—L. *Remmius*, name of a Roman gens. *Ex:* Remmius (Arach.).

**remor**—L. *remora*, one who holds back. *Ex:* Remor-opsis (Pisc.); Remora (Pisc.); remora (Med.).

**remorat**—L. *remoratus*, lingering, remaining; pp. of *remoror*, to linger, stay, remain.

**remot**—L. *remotus*, distant, removed, moved back. *Ex:* remoti-florus.

**remulc**—L. *remulceo*, to droop; also to soothe; *remulcus*, drooping.

**ren**—1. Gr. *rhēn*, a sheep or lamb. *Ex:* Reno-cera (Ins.): 2. L. *ren*, pl. *renes*, a kidney; LL. dim. *reniculus*. *Ex:* Ren-illa (Coel.); reni-cardiac; Reni-ceps (Elasm.); Reni-cola (Platy.); Reni-fer (Platy.); reni-form; reno-parietal.

**renal**—NL. *renalis* < L. *renes*, the kidneys.

**renat**—L. *renatus*, arisen, pp. of *renascor*, to spring up, to be born again. *Ex:* Renatus (Moll.).

**renn**—Mid. Eng. *renne*, to run; OD. *rinnen*, to press, curdle. *Ex:* rennin; rennet.

**renod**—L. *renodis*, loose, untied.

**reo**—See **rheo**.

**rep**—Gr. *rhepō*, to sink, incline downwards > *rhepsis*, an inclination. *Ex:* rep-ium; Repo-rhamphus (Pisc.); repsis. See also repen.

**repagul**—L. *repagula*, bolts, limits.

**repand**—L. *repandus*, bent backwards, turned up. *Ex:* repand.

**repen**—1. *repens*, genit. *repentis*, creeping, crawling, ppr. of *repo*, to creep. *Ex:* Repentia (Rept.); Repo-trudis (Pisc.): 2. L. *repens*, genit. *repentis*, sudden, new, unlooked for.

**repentin**—L. *repentinus*, unexpected, hasty, giving surprise.

**repert**—L. *repertus*, found again, rediscovered; also discovered

**repertici**—L. *reperticius*, met with by accident.

**replet**—L. *repletus*, filled to the full, complete.

**replum**—L. *replum*, a bolt. *Ex:* replum.

**reps**—See **rep**.

**rept**—L. *repto*, to crawl; ppr. *reptans*, genit. *reptantis*, crawling, pp. *reptatus* > *reptilis*, creeping > LL. *reptile*, a crawling animal, a reptile. *Ex:* Reptat-ores (Av.); reptant; Reptilia; reptili-ferous.

**reptan**—See **rept**.

**reptat**—See **rept**.

**reptil**—See **rept**.

**resart**—L. *resartus*, restored, repaired; pp. of *resarcio*, to mend.

**resect**—L. *resectus*, cut off, pared, *Ex:* resect-ion.

**resed**—L. *reseda*, name of a kind of plant < *resedo*, to assuage, calm. *Ex:* Resed-aceae*; Reseda*.

**resid**—L. *reses*, genit. *residis*, motionless, inactive.

**resil**—L. *resilio*, to leap or dart back, recoil. *Ex:* resili-fer; resilium.

**resim**—L. *resimus*, turned or bent back.

**resin**—L. *resina*, resin. *Ex:* resini-ferous; resino-cysts.

**resonans**—L. *resonans*, resounding; ppr. of *resono*, to resound.

**respicien**—L. *respiciens*, genit. *respicientis*, ppr. of *respicio*, to consider.

**resplenden**—L. *resplendens*, genit. *resplendentis*, glittering, shining, ppr. of *resplendeo*, to shine, be resplendent.

**resso**—See **rhess**.

**rest**—L. *restis*, dim. *resticula*, a rope. *Ex:* resti-brachium; resti-form; Resticula (Rot.); Resticuli-scala (Moll.); Restia*; Resti-aceae*.

**restan**—L. *restans*, genit. *restantis*, standing still, ppr. *resto*, to stand still, to remain behind. *Ex:* restant.

**restibil**—L. *restibilis*, restored, tilled every year, new < *re*, back, again + *stabilis*, standing firm, stable, enduring. *Ex:* restible.

**resticul**—See **rest**.

**restrict**—L. *restrictus*, made fast, bound tight, pp. of *restringo*, to tighten, make fast.

**restru**—L. *restruo*, to restore, make new; ppr. *restruans*, genit. *restruantis*, restoring; pp. *restructus*, restored.

**restruct**—See **restru**.

**resupin**—L. *resupinus*, bent back. *Ex:* Resupin-ata (Moll.).

**ret**—L. *rete*, dim. *reticulum*, a net > *reticularis*, netted; *reticulatus*, made like a net; *retiarius*, one who fights with a net; ML. *retina*, a fine net. *Ex:* Ret-aster (Echin.); Ret illa (Ins.); Rete-pora (Bry.); rete testis; Reteo-crinus (Echin.); Reti-spongia (Por.); Retia (Por.); Retiaria (Arach.); Reticularia*, (Prot.); Reticuli-termes (Ins.); Reticulo-ceras (Moll.); reticulo-cyte; Retin-ella (Moll.); retin-ule; retina; retino-phora; Retio-graptus (Coel.).

**retent**—L. *retentus*, held back < *retineo*, to retain.

**retiar**—See **ret**.

**reticul**—See **ret**.

**retin**—Gr. *rhētinē*, resin of the pine. *Ex:* Retini-phyllum*; Retino-spora*. For retina, see ret.

**retinacul**—L. *retinaculum*, a band, holdfast. *Ex:* retinaculum.

**retinen**—L. *retinens*, genit. *retinentis*, retained, ppr. of *retineo*, to keep back. *Ex:* retinent.

**retort**—Fr. *retorte*, a retort < L. *retorqueo*, to twist back. *Ex:* Retorta-monas (Prot.).

**retro**—L. *retro*, backward > *retroversus* = *retrorsus*, bent or turned backward. *Ex:* Retr-oculus (Pisc.); retro-mingent; Retro-pluma (Crust.); Retro-teuthis (Moll.).

retrors—See retro.

retus—L. *retusus*, dulled, made blunt, pp. of *retundo*, to dull. *Ex:* Retus-ites (Tri.); Retusa (Moll.); Retusum (Moll.).

revect—L. *revectus*, carried back.

revent—L. *reventus*, a return.

revert—L. *revertus*, turned back, returned.

revols—L. *revolsus*, torn off, plucked <*revello*, to pull out, pluck.

revolut—L. *revolutus*, rolled back, pp. of *revolvo*, to turn back. *Ex:* revolute.

revuls—L. *revulsus*, torn off <*revello*, to tearoff.

rex—See reg.

rhab—See rhabd.

rhabd—Gr. *rhabdos*, a rod; *rhabdion*, a little rod; *rhabdōma*, a bundle of rods; *rhabdōtos*, striped, streaked. *Ex:* Rhab-osteus (Mam.); Rabdiodon (Mam.); Rhabd-ites (Moll.); Rabdion (Rept.); Rhabdo-coela (Platy.); Rhabdomonas*; rhabdom; Rhabdoto-cephalus (Rept.); stato-rhab; Tri-rhabda (Ins.).

rhac—Gr. *rhakos*, rags, tatters. *Ex:* Rhaco-chilus (Pisc.); Rhaco-disc-ula (Por.); Rhaco-notus (Ins.); Pleo-rhacus (Myr.).

rhach—1. Gr. *rhachis*, dim. *rhachion*, a spine, *rhachitēs*, of the spine. *Ex:* Racheo-pora (Bry.); Rachi-centron (Pisc.); Rachi-ptera (Ins.); Rachi-trema (Rept.); Rachio-cephalus (Rept.); Rachio-pogon (Ins.); Rachis (Myr.); Rachisellus (Ins.); Rachis-poda (Ins.); Rachites (Rept.); Racho-gaster (Ins.):   2. Gr. *rhachos*, a thorn, briar. *Ex:* Rhacho-cnemis (Ins.); Tricho-rhachus (Ins.).

rhachia—Gr. *rhachia*, surf. *Ex:* Rhachia-nectes (Mam.).

rhachist—Gr. *rhachistos*, cut up, divided. *Ex:* Rhachistus (Ins.).

rhadin—Gr. *rhadinos*, slender, slim, delicate, slight; also graceful. *Ex:* Radin-acantha (Ins.); Rhadin-ichthys (Pisc.); Rhadina (Ins.); Rhadino-ceras (Moll.). See also bradin.

rhaeb—Gr. *rhaibos*, crooked, bent. *Ex:* Rhaeba (Amph.); Rhaebo-ceras (Moll.); Rhaibo-sceles (Ins.).

rhaest—Gr. *rhaistēr*, genit. *rhaistēros*, a destroyer, also a hammer. *Ex:* Rhaestes (Ins.).

rhaetic—L. *Rhaeticus*, mountain province of Rome.

rhag—1. Gr. *rhagas*, genit. *rhagados*, a chink, break; *rhagos*, ruptured. *Ex:* Rhag-onycha (Ins.); Rhaga-therium (Mam.); rhagades (Med.); hemo-rrhage (Med.); Homolo-rhagae (Rot.):   2. Gr. *rhax*, genit. *rhagos*, a berry, a kernel, grape; *rhagōdēs*, like grapes. *Ex:* Rhag-odon (Mam.); Rhagodia*; rhagon; Rhax-ella (Por.):   3. Gr. *rhaga*, vigor, violence:   4. *rhagion*, a kind of spider. *Ex:* Rhagio-morpha (Ins.); Rhagium (Ins.).

rhagio—Gr. *rhagion*, a kind of poisonous spider—Rhagio (Ins.); Rhagion-idae (Ins.).

rhagion—See rhagio.

rhaib—See rhaeb.

rhammat—Gr. *rhamma*, genit. *rhammatos*, a seam, a thread, also a patch. *Ex:* Rhammato-cerus (Ins.); Rhammato-pora (Bry.).

rhamn—Gr. *rhamnos*, name of a kind of prickly plant, the buckthorn. *Ex:* Rhamn-idium*; rhamno-xanthin; Rhamnus*.

rhamph—Gr. *rhamphis*, genit. *rhamphidos*, a hook; *rhamphos*, a beak, crooked beak. *Ex:* Ramph-alcyon (Av.); Rhamph-odon (Av.); Ramphi-stoma (Moll.); Rhamphido-phyllum*; Ramphis (Ins.); Rhampho-rhynchus (Rept.).

rhamphast—NL. *rhamphastus* <Gr. *rhampazomai*, to have a break or<*rhamphos*, a beak. *Ex:* Rhamphastus (Av.).

rhanid—Gr. *rhanis*, genit. *rhanidos*, a drop, a spot, rain. *Ex:* Rhanido-phora (Ins.); Rhanis (Ins.).

rhanis—See rhanid.

rhant—Gr. *rhantos*, sprinkled, spotted with dew. *Ex:* Rhant-istes (Av.); Rhantus (Ins.).

rhap—Gr. *rhapis*, genit. *rhapidos*, a rod. *Ex:* Rhapis*.

rhaph—See raph, also raphid.

rhaphan—See raphan.

rhapid—See raphid.

rhapontic—LL. *rhaponticum*, name of the Pontic plant called "rha" <Gr. *rha*, rhubarb. *Ex:* rhapontic; Rhaponticum*; see rheum.

rhapt—Gr. *rhaptō*, to sew>*rhaptēs*, a sewer mender. *Ex:* Syr-rhaptes (Av.).

rhathym—Gr. *rhathymos*, indifferent, inactive thoughtless. *Ex:* Rhathymo-scelis (Ins.) Rhathymus (Ins.).

rhax—See rhag 2.

rhe—Gr. *Rhea*, daughter of Uranus and Gaea, mother of Zeus. *Ex:* Rhe-oideae (Av.); Rhea (Av.).

rhect—Gr. *rhektēs*, a breaker; *rhektēr*, active, strenuous. *Ex:* Rhectes (Av.); Rhecto-psammia (Coel.).

rheg—Gr. *rhegos*, a blanket. *Ex:* Rheg-aster (Echin.); rhego-lith = rego-lith.

rhegm—Gr. *rhēgma*, genit. *rhēgmatos*, a break, tear. *Ex:* regma; regma-carp; Rhegma (Pisc.); Rhegmo-clema (Ins.); Rhegmato-phila (Ins.).

rhegn—Gr. *rhēgnymi* = *rhēgnyō*, to break off, fracture. *Ex:* Rhegn-opsis (Mam.); Rhegno-pteri (Pisc.). See rhegm.

rheithr—Gr. *rheithron*, a stream, a channel. *Ex:* Reithr-odon(Mam.); Rheithro-sciurus (Mam.).

rhem—Gr. *rhēma*, word, report, speech. *Ex:* A-rhemon (Av.).

rhemb—Gr. *rhembō*, to wander, to turn, to spin, reel about. *Ex:* Rhembo-bius (Ins.); Rhembus (Ins.); Poly-rhembia (Ins.).

**rhen**—L. *Rhenus*, the river Rhine. *Ex:* Rhen-aster (Echin.); Rheno-crinus (Echin.); Rheno-ptera (Pisc.); Rheno-squama (Echin.).

**rheo**—Gr. *rheō*, to flow; *rheos*, a stream. *Ex:* Rheo-chara (Ins.); rheo-stat; rheo-taxis; not Rhe-oideae (Av.); see rhe.

**rhep**—Gr. *rhepō*, to incline, lean; to be favorable or kind. *Ex:* Rhepo-coris (Ins.).

**rhes**—Gr. *Rhēsos*, king of Thrace, who came to the assistance of Priam at Troy. *Ex:* Rhesus (Mam.); Audebert who applied the name to the species stated that it had no meaning.

**rhess**—Gr. *rhēsso*, to break, to make ragged. *Ex:* Rhesso-pygus (Echin.); Rhesso-cephalis (Pisc.).

**rhetin**—Gr. *rhētinē*, resin of the pine. *Ex:* Rhetin-angium*.

**rheum**—1. ML. *rheum*, rhubarb, perhaps an accommodation form of Gr. *rha*, rhubarb, or <Gr. *rheō*, to flow (referring to its purgative properties). *Ex:* Rheum*: **2.** L. *rheuma*, genit. *rheumatis*, a flooding, flowing, a cold <Gr. *rheō*, to flow; *rheumatismos*, liability to flux, flow. *Ex:* rheumatism (Med.).

**rhex**—Gr. *rhēxis*, a rupture, breaking. *Ex:* rhexi-genetic; Rhexi-neura (Ins.); Rhexi-stoma (Moll.); Rhexia*, (Ins.); rhexo-lytic; karyo-rhexis.

**rhicn**—Gr. *rhiknos*, shrivelled, crooked, stiff with cold, bent, old. *Ex:* Rhicno-pelte (Ins.); Ricn-odon (Rept.); Acro-rhicnus (Ins.).

**rhig**—Gr. *rhigoō*, to be cold, to be frosty, to shiver; *rhigos*, frost, cold. *Ex:* Rhig-opsis (Ins.); Rhigio-glossa (Ins.); Rhigo-nema (Nemat.); Rhigoon (Mam.); Rhig-ops-idius (Ins.); Rhigus (Ins.).

**rhigel**—Gr. *rhigēlos*, chilling, hence rigid. *Ex:* Rhigel-ura (Av.); Rhigelus (Ins.).

**rhimph**—Gr. *rhimpha*, nimbly, with leaps and bounds, swiftly; *rhimphaleos*, swift. *Ex:* Rhimphalea (Ins.); Rhimpho-ctona (Ins.).

**rhimphal**—See rhimph.

**rhin**—1. Gr. *rhinē*, a file, rasp; also, name of a kind of shark with rough skin. *Ex:* Rhin-acanthus*; Rhin-anthus*; Rhina (Elasm.); Rhine-odon (Elasm.); Rhine-somus (Pisc.); Rhino-batus (Elasm.); Rhino-ptera (Elasm.); Rin-odus (Pisc.); Rin-osteus (Pisc.); Scyllio-rhinus (Elasm.): **2.** Gr. *rhis*, genit. *rhinos*, a nose. *Ex:* rhin-enceph-alon; Rhina (Ins.); Rhino-ceros (Mam.); Rhino-cerot-idae (Mam.); rhino-phore; rhino-theca; Rhino-thorax (Ins.); Platy-rrhina (Mam.).

**rhinch**—Gr. *rhynchos*, snout. *Ex:* Rhincho-nycteris (Mam.). See also rynch.

**rhio**—Gr. *rhion*, peak, promontory. *Ex:* Rhio-bia (Ins.); Rhio-stoma (Moll.); Rhion (Arach.).

**rhip**—1. Gr. *rhipis*, genit. *rhipidos*, dim. *rhiphidion*, a fan. *Ex:* Rhipi-pallus (Ins.); Rhipid-istia (Pisc.); rhiphidium; Rhipido-ptera (Ins.); not Rhipsalis*, see rhips; Delo-rrhipis (Arach.)

Schizo-rhipis (Av.): **2.** Gr. *rhipē*, flight, sweep, swing. *Ex:* Rhip-ornis (Av.).

**rhiphid**—See rhip.

**rhips**—Gr. *rhips*, a mat, a plaited work of rushes >NL. irreg. *rhipsalis*, name applied to a genus of cacti because of the plaited woody framework. *Ex:* Rhipo-auchenia (Moll.); Rhipsalis*.

Roxell's Ape, *Rhinopithecus roxellanae*, with peculiar upturned nose. Redrawn from Tiergrotesken—Ramme.

**rhipt**—Gr. *rhiptos*, thrown out <*rhiptō*, to throw, fling. *Ex:* Rhipto-glossa (Rept.).

**rhis**—See rhin 2.

**rhisc**—Gr. *rhiskos*, a box, chest. *Ex:* Rhiscosoma (Myr.).

**rhithr**—See rheithr.

**rhiz**—Gr. *rhiza* also *rhizōma*, a root >Fr. rhizome. *Ex:* rhiz-anthus; Rhiz-oecus (Ins.); rhiz-oid; Rhiz-ina*; Rhizo-geton (Por.); Rhizo mys (Mam.); rhizome; Rhizota (Rot.); Bio-rrhiza (Ins.).

**rhochm**—Gr. *rhōchmos*, a cleft, crack; also a snoring, croaking. *Ex:* Rochmo-gaster (Ins.).

**rhod**—Gr. *rhodon*, a rose, hence red; *rhodeos*, of roses; *rhoditēs*, pertaining to a rose, flavored with a rose (said of wine). *Ex:* Rhod-(h)ymenia*; Rhod-oeca (Ins.); Rhode-ina (Pisc.); Rhodeus (Pisc.); Rhodi-ola*, the last element a dim.; Rhodites (Ins.); rhodo-chrous; Rhododendron*; Rhodo-leia*; rhodo-plast; Rhodosphaera (Prot.); Rhodora*; Chamae-rhodos*.

**rhodan**—1. L. *Rhodanus*, the river Rhone <Gr. *rhodanos*, very rapid, as a subs., the River Rhone. *Ex:* Rhodano-mys (Mam.): **2.** Gr. *rhodanē*, thread.

**rhodin**—Gr. *rhodinos*, prepared from roses. *Ex:* Rhodino-cichla (Av.); Rhodino-soma (Myr.).

**rhodop**—Gr. *Rhodopē*, a mythological name. *Ex:* Rhodope (Moll.).

**rhodymenia**—See rhod.

rhoead—L. *rhoeas*, genit. *rhoeadis*, name of the common red poppy. *Ex:* Rhoeades*.

rhoec—1. Gr. *rhoikos*, crooked. *Ex:* Rhoecus (Ins.): 2. Gr. *rhoikos*, he who suffers from a flux, failing, weak. *Ex:* ?Rhoeco-ptera (Ins.).

rhog—See rhogm.

rhogad—Gr. *rhōgas*, genit. *rhōgados*, rent, ragged. *Ex:* Rhogad-opsis (Ins.).

rhogm—Gr. *rhōgē*=*rhōgmē*, a cleft, chink, fracture; *rhogas*, a rent in a wall. *Ex:* Rhogas (Ins.); Rhogo-gastera (Ins.); Rhogmus (Ins.); Brechmo-rhoga (Ins.); Tri-rhogma (Ins.).

rhomal—See romal.

rhomb—Gr. *rhombos*, whirling, turning; also a rhomb, lozenge and in this sense usually used in generic names. *Ex:* rhomb-oid; Rhomb-oplites (Pisc.); Rhomb-ura (Av.); Rhombi-chiton (Moll.); Rhombi-fera (Echin.); rhombi-form; Rhombo-sepion (Moll.); Rhombus (Moll.).

rhop—Gr. *rhōps*, genit. *rhōpos*, underwood, brush. *Ex:* Rhop-ornis (Av.); Rhopo-cichla (Av.).

rhopal—Gr. *rhopalon*, a club, stick; the phallus; *rhopalōtos*, shaped like a club; *rhopalizō*, to brandish a club. *Ex:* Rhophal-apion (Ins.); rhopalia; Rhopalizus (Ins.); Rhopalo-cera (Ins.); Rhopalot-ella (Ins.); Rhopalotus (Ins.); Ropal-actis (Coel.); Amphi-rhopalum (Prot.); Za-rhopalus (Ins.).

rhope—Gr. *rhopē*, a leaning, bending. *Ex:* Di-rrhope (Ins.).

rhoph—Gr. *ropheō*, to swallow, gulp down. *Ex:* Rhoph-ites (Ins.).

rhophet—Gr. *rhophetikos*, absorbing, drawing in. *Ex:* Meli-rhophetes (Av.), lit. a honey sucker.

rhopi—Gr. *rhōpēia*, bushes. *Ex:* Rhopias (Av.); Micro-rhopias (Av.).

rhopo- —Gr. *rhōpo-*, petty, weak. *Ex:* Rhopo-cichla (Av.); Rhopo-spiza (Av.).

rhops—See rhop.

rhopt—Gr. *rhoptos*=*rhophētos*, something absorbed. *Ex:* rhopto-meter; Rhopto-tricha (Ins.).

rhoptr—Gr. *rhoptron*, a staff, club. *Ex:* Rhoptr-urus (Arach.); Rhoptro-cerus (Ins.); Rhoptro-merus (Ins.); Rhoptrum (Por.); Aphano-rrhoptrum (Ins.).

rhorus—Gr. *rhōros*, strong. *Ex:* Rhorus (Ins.).

rhus—L. *rhus*<Gr. *rhous*, sumac. *Ex:* Rhus*.

rhyac—Gr. *rhyax*, genit. *rhyakos*, a brook. *Ex:* Rhyac-ornis (Av.); Rhyacia (Ins.); Rhyaco-phila (Ins.); rhyaco-phyte.

rhych—NL. *rhychus*<Gr. *rhynchos*, a beak, snout. *Ex:* Achor-rhychus (Ins.).

rhygch—See rhynch.

rhym—Gr. *rhyma*, genit. *rhymatos*, a rope, cable; also a protection, defence. *Ex:* Rhym-odus (Pisc.).

rhymb—Gr. *rhymbos*=*rhombos*, a whirling; also a top. *Ex:* Rhymbo-cochlias (Moll.).

rhync—See rhynch.

rhynch—Gr. *rhynchos*, a beak, snout. *Ex:* Rhynch-actis (Pisc.); Rhyncha-ceros (Av.); Rhyncheo-lithus (Moll.); Rhyncho-cephalia (Rept.); Rhygcho-psalis (Av.); Rhynchosia*; Ornitho-rhynchus (Mam.).

rhyno—See rhin 2.

rhyp—See rhypar.

rhypar—Gr. *rhyparos*, filthy, dirty<*rhypos*, dirt, filth. *Ex:* Rhyparo-bia (Ins.); Rhyparus (Ins.).

rhyph—1. Gr. *rhyphos*, bent, crooked. *Ex:* Rhypho-pteryx (Ins.); Rhyphus (Ins.): 2. Gr. *rhypheō*, to gulp down. *Ex:* Rhyph-odon (Mam.).

rhyptic—Gr. *rhyptikos*, fit for cleaning. *Ex:* Rhypticus (Pisc.).

rhys—See rhyss.

rhysi—Gr. *rhysios*, defending, delivering. *Ex:* Rhysi-polis (Ins.).

rhysimon—Gr. *rhysimon*=*erhysimon*, hedge-mustard.

rhyss—Gr. *rhyssos*=*rhysos*, wrinkled, shrivelled; *rhysōdēs*=*rhyssōdēs*, wrinkled looking; *rhysēma*, less correctly *rhyssēma*, a wrinkle. *Ex:* Rhyso-desmus (Myr.); Rhyss-alus (Ins.); Rhyssa (Ins.); Rhyssemus (Ins.); Rhysso-labus (Ins.); Rhyssod-idae=Rysod-idae (Ins.);

rhyssem—See rhyss.

rhyssod—See rhyss.

rhyt—Gr. *rhytis*, genit. *rhytidos*, a wrinkle. *Ex:* Rut-idia*; Ruti-therium (Mam.); Ruti-deres (Ins.); Rhyt-elminthus (Platy.); Rhyt-ina (Mam.); Rhyti-glossa*; Rhitid-acris (Ins.); Rhytid-ura (Arach.); Rhytido-concha (Moll.); Rhytis-odon (Mam.); Ryti-odon (Mam.); Calli-rhytis (Ins.).

rhythm—Gr. *rhythmos*, rhythm, symmetry, fitness, rule. *Ex:* Rhythmo-notus (Ins.).

rhytid—See rhyt.

rhytis—See rhyt.

rhytism—Gr. *rhytisma*, a darn or patch. *Ex:* Rhytisma*.

rhyz—Gr. *rhyzō*, to growl, snarl; inf. *rhyzein*. *Ex:* Rhyzaena. (Mam.), a name coined by Illiger.

ribes—Ar. *rībās*, a plant with sour sap. *Ex:* Ribes*.

ribodon—NL. *ribodon*, contr. of Gr. *rhybdēn*, with a noise+*odōn*, tooth. *Ex:* Ribodon (Mam.).

ric—L. *rica*, dim. *ricula*, a veil. *Ex:* rici-ferous.

ricin—L. *ricinus*, the castor oil plant, the castor oil bean; also name of a kind of tick. *Ex:* Ricin-ella*, (Moll.); Ricin-ula (Moll.); Ricino-carpos*; Ricinulus (Arach.); Ricinus*.

ricn—See rhicn.

**rict**—See **ring.**

**rid**—L. *rideo*, to laugh. *Ex:* ridi-bundus.

**rigens**—See **rigio.**

**rigesc**—L. *rigescens*, genit. *rigescentis*; ppr. of *rigesco*, to grow stiff, to stand up. *Ex:* rigescent.

**rigid**—See **rigio.**

**rigio**—L. *rigeo*, to be stiff or numb, ppr. *rigens*, genit. *rigentis*, stiffening, becoming rigid > *rigidus*, stiff. *Ex:* Rigid-antenna (Ins.); Rigido-mastix (Prot.); Rigio-pappus*.

**rigo**—See **rhig.**

**rigu**—L. *riguus*, watered.

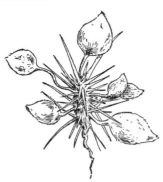

Rigid Spiny-Herb, *Chorizanthe rigida*, a woody dwarf flourishing on black pebble beds of the desert. The name *Chorizanthe*, meaning divided flower, refers to the parted involucre. Redrawn from Desert Wild Flower —Jaeger. Stanford University Press.

**rim**—L. *rima*, dim. *rimula*, a fissure. *Ex:* rim-osus; Rima-cephalus (Platy.); rimi-colus; Rimul-opsis (Moll.); Rimula (Moll.); Rimulus (Moll.).

**rimat**—L. *rimator*, an inquirer < *rimor*, to investigate. *Ex:* Rimator (Av.).

**rimph**—See **rhimph.**

**rimul**—See **rim.**

**-rimus**—L. *-rimus*, superlative ending of Latin adjectives ending in *-er*. *Ex:* acer-rimus < *acer*, keen, sharp.

**rin**—See **rhin.**

**ring**—L. *ringor*, to open wide the mouth, gape, ppr. *ringens*, genit. *ringentis*, gaping, pp. *rictus*, gaped, opened. *Ex:* rict-al; Rict-axis (Moll.); Ricto-cyma (Moll.); rictus; ringent; ringenti-florus; Ringi-cardium; Ringi-culo-spongia (Por.), see -culus; Ringo-ceras (Moll.).

**ringent**—See **ring.**

**ringicul**—See **ring.**

**ripa**—L. *ripa*, pl. *ripae*, the bank of a stream; *riparius*, frequenting or belonging to a stream bank. *Ex:* Ripae-cola (Av.); Riparia (Av.); ripari-an.

**ripar**—See **ripa.**

**ripers**—NL. *ripersia*, anagram of Perissia. *Ex:* Ripersia (Ins.).

**ripid**—See **rhip.**

**ris**—L. *risor*, a laugher, one who mocks; *risus*, laughter < *rideo*, to laugh. *Ex:* risori-us, risori-al.

**risc**—L. *riscus*, dim. *risculus*, a chest. *Ex:* Risculus (Crust.).

**risor**—See **ris.**

**rissa**—Icelandic *rissa*, the kittiwake.

**riv**—L. *rivus*, dim. *rivulus*, a channel, groove, stream. *Ex:* riv-ose; Rivo-bates (Arach.); Rivul-aria*; rivul-aris.

**rival**—L. *rivalis*, belonging to a brook; also a rival.

**rivul**—See **riv.**

**rix**—L. *rixosus*, quarrelsome < *rixo* = *rixor*, to quarrel; *rixator*, a wrangler. *Ex:* Rixator (Pisc.).

**rixator**—See **rix.**

**robor**—L. *robor*, genit. *roboris*, the oak, the hard wood of the oak > *roboreus*, made of oak, strong. *Ex:* robor-inus; roboreous.

**roborat**—L. *roboratus*, strengthened, made vigorous < *roboro*, to make strong.

**roboror**—L. *rober* = *robur*, oak-wood; by transfer, strength; *roborosus*, stiff, rigid like oak.

**robust**—L. *robustus*, oaken, strong, robust. *Ex:* Robusta (Moll.); Robusto-bombus (Ins.).

**robuste**—L. *robusteus*, of oak, oaken, of oaken strength. See robor.

**rocc**—Port. *rocca*, a rock. *Ex:* Roccus (Pisc.); not Roccella*, which is probably derived from It, *rorcella*, the name of a lichen < *rocca*, a rock.

**rochm**—Gr. *rhōchmos*, a cleft, opening. *Ex:* Rochmo-soma (Ins.).

**roden**—L. *rodens*, genit. *rodentis*, gnawing, ppr. of *rodo*, to gnaw, eat away. *Ex:* Rodent-ia (Mam.); Rodento-caulus (Nemat.); Corrodent-ia (Ins.).

**rodiotherium**—anagram of Diorotherium (Mam.).

**roll**—NL. *rollus*, a roll < Fr. *rôle*, a roll. *Ex:* Strapa-rollus (Moll.).

**romal**—Gr. *rhōmaleos*, strong of body. *Ex:* Romal-orina (Ins.); Romalea (Ins.); Romaleo-syrphus (Ins.); Romaleon (Crust.); Romalo-cera (Ins.).

**romul**—L. *Romulus*, one of the mythical founders of Rome. *Ex:* Romulea*.

**roncador**—Sp. *roncador*, a snorer < L. *rhoncho*, to snore. *Ex:* Roncador (Pisc.).

**ropal**—See **rhopal.**

**ropt**—See **rhopt.**

**ror**—L. *ros*, genit. *roris*, dew; *roridus*, dewy; *rorulentus*, full of dew. *Ex:* Ros-marinus*; Rorid-ula*; rorulent.

**rorid**—See **ror.**

**rorip**—NL. *roripa* = *rorippa*, a name of unexplained origin. *Ex:* Roripa*, also Rorippa*.

**rorulent**—See **ror.**

**ros**—L. *rosa*, dim. *rosella*, a rose<Gr. *rhodon*, a rose. *Ex:* Ros-aster (Echin.); Rosa*. See also ror.

**roscid**—L. *roscidus*, full of dew, wetted<*ros*, genit. *roris*, dew.

**rose**—L. *roseus*, red like the rose, ruddy.

**rosmar**—Dan. *rosmar*, a walrus. *Ex:* Rosmarus (Mam.).

**rosor**—L. *rosor*, a gnawer. *Ex:* Rosor-es (Mam.); rosori-al.

**ross**—It. *rosso*, red. *Ex:* Rosso-crinus (Echin.); Rosso-phyllum (Coel.).

**rostell**—See **rostr.**

**rostr**—L. *rostrum*, a bill, snout, beak; dim. *rostellum*; *rostratus*, beaked, hooked. *Ex:* Rostella (Moll.); Rostrata (Moll.); rostrum; rostri-form; Rostri-palpus (Ins.); Rostro-spirifer (Brach.); rostrum.

**rosulat**—NL. *rosulatus*, like a small rose<L. *rosula*, a little rose.

**rot**—L. *rota*, dim. *rotula*, a wheel=NL. *rotella*; *rotalis*, having wheels; *rotalarius*, round, circular. *Ex:* Rotal-ina (Prot.); Rotell-omphalus (Moll.); Roti-fera; Rotul-echinus (Echin.); bi-rotul-ate.

**rotal**—See **rot.**

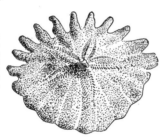

The Wheel urchin, *Rotula*. Redrawn from New Natural History—Thomson. George Newnes & Co. Ltd., London.

**roth**—Gr. *rothos*, a torrent. *Ex:* roth-oecus (Ecol.); Rothus (Arach.).

**rotul**—See **rot.**

**rotund**—L. *rotundus*, round. *Ex:* Rotund-aster (Echin.); rotundi-folius; Rotundi-termes (Ins.); Rotund-ula (Coel.).

**rrhin**—See **rhin.**

**rrhophet**—See **rhoph.**

**ru**—See **rusa.**

**rub**—1. L. *rubeo*, to be red, ppr. *rubens*, genit. *rubentis*, reddening; *rubesco*, to become red, ppr. *rubescens*, genit. *rubescentis*, becoming red; *ruber* (fem. *rubra*, neut. *rubrum*), red. *Ex:* rubescent; Rubia*; Rubo-crinus (Echin.); Rubri-capella (Av.); rubri-cauda:     2. L. *rubus*, a bramble<*rubeo*, to be red. *Ex:* Rubicola (Av.); Rubus*.

**rubell**—L. *rubellus*, reddish; *rubellianus*, reddish.

**rubesc**—See **rub 1.**

**rubicund**—L. *rubicundus*, red.

**rubid**—L. *rubidus*, reddish, deep red.

**rubig**—L. *rubigo*, genit. *rubiginis*, rust; *rubiginosus*, rusty, rust-colored<*rubeo*, to be red. *Ex:* rubigin-ous; Rubiginos-ana (Moll.); rubigo.

**rubr**—See **rub 1.**

**rubric**—L. *rubrica*, red ochre. *Ex:* rubric-ous; Rubrica (Ins.).

**rud**—1. L. *rudis*, rough, unused, raw, wild> *rudimentum*, a beginning. *Ex:* rudiment-al:     2. L. *rudis*, a rod, spatula:     3. L. *rudus*, rubble. *Ex:* rud-aceous.

**ruden**—1. L. *rudens*, genit. *rudentis*, crying out, bellowing, ppr. of *rudo*, to bray, to roar:     2. L. *rudens*, genit. *rudentis*, a stay, prop. *Ex:* Rudens (Moll.).

**ruder**—L. *rudus*, genit. *ruderis*, old rubbish, waste>NL. *ruderalus*, growing in waste places. *Ex:* ruderal.

**ruderal**—See **ruder.**

**rudit**—1. L. *ruditas*, ignorance:     2. L. *ruditus*, a roaring, bellowing, pp. of *rudo*, to bray, bellow.

**ruf**—L. *rufus*, dim. *rufulus*, reddish; *rufesco*, to become red or reddish, ppr. *rufescens*, genit. *rufescentis*, reddening, becoming red. *Ex:* rufescent; Rufi-brenta (Av.); rufi-pileus.

**rufescen**—See **ruf.**

**rufid**—NL. *rufidulus*, somewhat red<L. *rufus*, red, reddish.

**rug**—L. *ruga*, dim. *rugula*, a wrinkle, a crease of the face; *rugosus*, NL. dim. *rugulosus* wrinkled, full of wrinkles; *rugositas*, the state of being wrinkled; *rugatus*, creased, wrinkled. *Ex:* Rug-anodont-ites (Moll.); Rugat-ula (Moll.); Rugati-scala (Moll.); Rugi-fera (Moll.); Rugo-pharynx (Nemat.); rugosit-y; Rugosa (Coel.); rugose.

**rugat**—L. *rugator*, a wrinkler<*ruga*, a wrinkle.

**rum**—1. L. *rumen*, the throat<*rumino*, to chew the cud, ppr. *ruminans*, genit. *ruminantis*, chewing again. *Ex:* Rum-ella (Moll.); rumen; ruminant; Ruminant-ia (Mam.):     2. L. *ruma*; a dart. *Ex:* Ruma (Moll.); rumi-form.

**rumen**—See **rum.**

**rumex**—See **rumic.**

**rumia**—L. *Rumia*, goddess of infants. *Ex:* Rumia*.

**rumic**—L. *rumex*, genit. *rumicis*, sorrel. *Ex:* Rumex*.

**rump**—L. *rumpo*, to burst, break, ppr. *rumpens*, genit. *rumpentis*, breaking, bursting; pp. *ruptus*, ruptured, broken; *ruptor*, a breaker; *ruptura*, a break. *Ex:* rupti-nervis; rupture; e-rumpent.

**runcin**—1. L. *runcina*, a kind of tool, a plane> NL. *runcinatus*, with divisions pointing toward the base, much as in the blade of a plane-leaf. *Ex:* runcinate:     2. L. *Runcina*, rural goddess presiding over weeding<*runco*, to weed. *Ex:* Runcin-ella (Moll.); Runcina (Moll.).

rup—L. *rupes*, genit. *rupis*, a rock; *rupina*, a rock-cleft or rocky chasm; NL. *rupestris*, growing among rocks. *Ex:* Rup-ornis (Av.); Rupestr-ella (Moll.); rupestr-ine; Rupestri-ana (Moll.); Rupi-cola (Av.); Rupi-capra (Mam.); Rupi-sitta (Av.).

rupestr—See rup.

rupex—See rupic.

rupic—L. *rupex*, genit. *rupicis*, a rustic. *Ex:* Rupic-ula (Av.).

rupin—See rup.

rupt—See rump.

rural—L. *ruralis*, belonging to the country. *Ex:* Ruralis (Ins.).

rurs—L. *rursus*, turned back, backwards. *Ex:* Rursi-ceras (Moll.).

rusa—Malay, *rusa*, deer. *Ex:* Rusa (Mam.); Ru-cervus (Mam.).

rusc—L. *ruscum*, butcher's broom; *ruscarius*, of or for butcher's broom. *Ex:* Ruscarius (Pisc.); Ruscus*.

russ—L. *russus*, dim. *russulus*, reddish; *russeus*, reddish, dim. *russeolus*, somewhat red. *Ex:* Russula*.

russat—L. *russatus*, clothed in red.

ussul—L. *russulus*, reddish.

rustic—L. *rusticus*, rural, rustic; *rusticola*, a country dweller. *Ex:* Rusticola (Av.); Rusticus (Ins.).

rut—Fr. *rut*<L. *rugitus*, roaring, in allusion to the sounds made by deer in the mating season. *Ex:* the rut; rutting season. See rhyta, also ruta.

ruta—L. *ruta*, rue. *Ex:* Rut-aceae*; Ruta*.

rutabul—L. *rutabulum*, a stirring spoon, a shovel.

rutel—See rutil.

ruticill—NL. *ruticilla*, name applied to a genus of birds<L. *rutilus*, red+dim. suffix *-cilla*, a small tail. *Ex:* Ruticilla (Av.).

rutid—See rhyt.

rutil—L. *rutilo*, to be red, pp. *rutilatus*, reddened; *rutilus*, ruddy. *Ex:* Rutila (Ins.); Rutilo-dexia (Ins.). See ruticill.

rutilans—L. *rutilans*, glowing, red.

rutr—L. *rutrum*, a shovel, spade. *Ex:* Rutri-palpus (Arach.); Rutro-phora (Ins.).

ryncho—See rhynch.

rypar—Gr. *rhyparos*, dirty. *Ex:* Ryparo-bius*.

ryt—See rhyt.

ryz —See rhyze.

# S

sa—Gr. *saos*=*sōs*, healthy, sound. *Ex:* Sa-pyga (Ins.); Sa-pyg-ites (Ins.); Sa-therium (Mam.).

sabal—Etym. doubtful, perhaps Mex. or S. Amer. *sabal*, name for a certain plant. *Ex:* Sabal-ites*.

sabatin—L. *Sabatinus*, belonging to the town or lake of Sabate. *Ex:* Sabatina (Moll.).

sabell—NL. dim. of L. *sabulum*, sand, gravel. *Ex:* Sabell-aria (Ann.); Sabell-astarte (Ann.); Sabella (Ann.); Sabelli-philus (Crust.); Sabello-chares (Crust.).

sabi—Pers. *sabza*, greenness, verdure>Hind. *sabjā*>NL. *sabia*. *Ex:* Sabi-aceae*; Sabia*.

sabin—L. *sabinus*, of Sabine; adj. applied to a kind of juniper, *Juniperus sabina*<*Sabinus*, the Sabine River or territory where it grows. *Ex:* Sabin-ella (Moll.); Sabina*; sabini-ana.

sabr—L. *Sabrina*, the modern river Severn; also a river nymph. *Ex:* Sabrina (Bry.).

sabul—L. *sabulum*=*sabulo*, sand; *sabulosus*, sandy. *Ex:* Sabul-arius (Ins.); Sabul-odes (Ins.); Sabuli-cola (Ins.); sabulose.

saburr—L. *saburra*, sand. *Ex:* Saburra (Arach.).

sac—Gr. *sakos*, a shield; *sakesphorus*, shield-bearering. *Ex:* Sac-odes (Ins;) Sacesphorus (Arach.); Andro-sace*. See also sacc.

sacali—Mod. Gr. *sakalia*, a name for the jackals <an oriental name. *Ex:* Sacalius (Mam.).

Flagellated Sac-throat, *Saccopharynx flagellum*, a deep sea fish allied to the eels. Re drawn from Oceanic Ichthyology—Goode and Bean.

**sacc**—L. *saccus*, dim. *sacculus* < Gr. *sakkos*, a sac, strainer > NL. *saccatus*, of the form of a sack. *Ex:* sac; Sacc-amoeba (Prot.); Saccata (Cten.); sacci-form; Sacco-labium*; Saccomys (Mam.); saccul-ar; Sacculo-cochlear.

**sacchar**—Gr. *sakchar* = *sakcharon*, sugar. *Ex:* Sacchar-issa (Ins.); sacchari-ferous; Saccharomyces*; saccharo-phylly; Saccharum*.

**saccin**—Gr. *sakkinos*, made of sack cloth; also a cloth for mourning, black. *Ex:* Saccino-baculus (Prot.).

**sacer**—L. *sacer*, fem. *sacra*, neut. *sacrum*, sacred.

**saces**—See sac.

**sacr**—NL. (os)*sacrum*, the sacral bone, formerly offered in sacrifices < L. masc. *sacer*, neuter *sacrum*, sacred. *Ex:* sacr-al; sacri-lumbar; sacro-coccygeus.

**sacrestinus**—"(*Sacré Chien*), a local name of *Aphareus furcatus* on the Ile de France." *Ex:* Sacrestinus (Pisc.).

**sact**—Gr. *saktos*, crammed, stuffed. *Ex:* Sact-ortho-ceras (Moll.); Sacto-gaster (Ins.).

**sadina**—NL. *sadina*, a corruption of Gr. *sardēnē* the sardine.

**sadypus**—anagram of *Dasypus*. *Ex:* Sadypus (Mam.).

**saen**—Gr. *sainō*, to move. *Ex:* Saen-ura (Ann.); Saeno-lophus (Prot.).

**saep**—See sep.

**saev**—L. *saevus*, cruel, savage.

**saevitat**—L. *saevitas*, genit. *saevitatis*, rage, fury.

**sag**—1. Gr. *sagos*, a covering, cloak; *sagē*, armour, harness. *Ex:* Sage-phora (Ins.); Sago-sphaera (Prot.); Sago-plegma (Prot.): Ec-copto-sage (Ins.); schisti-sagus:    2. Gr. *sagis*, genit. *sagidos*, a pouch. *Ex:* A-sagis (Mam.):    3. Malay *sāgu*, the starch of the sago palm. *Ex:* Sagus*; sago.

**sagac**—L. *sagax*, genit. *sagacis*, keen-scented, quick.

**sagar**—Gr. *sagaris*, a weapon, a sword. *Ex:* Sagar-ites (Por.); Sagar-opsis (Ins.); Sagari-phora (Ins.); Sagaris (Ins.).

**sagart**—NL. *sagartia*, a name of uncertain origin. *Ex:* Sagartia (Coel.).

**sagax**—See segac.

**sagen**—Gr. *sagēnē* = L. *sagena*, a seine. *Ex:* Sagen-aria*; Sagen-ichthys (Pisc.); Sagena (Prot.); Sageno-crinus (Echin.).

**sagin**—L. *sagina*, a stuffing, feeding < *sagino*, to stuff full. *Ex:* Sagina*; Sagina-fusus (Moll.).

**sagis**—See sag 2.

**sagitt**—L. *sagitta*, an arrow; *sagittarius*, an archer; also belonging to an arrow. *Ex:* sagitt-al; Sagitt-ula (Platy.); Sagitta (Ins.), (Ann.); Sagittaria*; sagitti-formis; Sagitto-ceras (Moll.)

**sagm**—Gr. *sagma*, genit. *sagmatos*, a saddle; L. *sagmarius*, pertaining to a saddle; Gr. *sagma-*

*rion*, a pack-horse. *Ex:* Sagm-idium (Prot.); Sagma-plaxus (Moll.); Sagmarium (Prot.); Sagmat-ias (Mam.); Sagmato-rrhina (Av.).

**sagoin**—Fr. *sagouin* < Braz. *sahui*, native name of a mammal. *Ex:* Sagoin-us (Mam.).

**saguer**—NL. *saguerus* < Pg. *sagueiro*, the sago palm. *Ex:* Saguerus*.

**saig**—Russ. *saiga*, antelope. *Ex:* Saiga (Mam.).

**sainour**—Gr. *sainouros*, wagging the tail < *sainō*, to wag + *oura*, tail. *Ex:* Sainourus (Prot.).

**sair**—Gr. *sairō*, to clean. *Ex:* Sair-anthus*.

**sal**—1. L. *sal*, genit. *salis*, salt; *salinae*, salt works, salt pits; *salinator*; a dealer in salt; ML. *salinus*, salty. *Ex:* Sali-cornia*; Salinator (Moll.); saline; salini-form:    2. Gr. *salos*, an open roadstead, roads. *Ex:* Sale-bius (Ins.): lit. living by the road-side:    3. Gr. *salos*, silly.    4. Gr. *salos*, any unsteady motion.

**salac**—1. Gr. *salax*, genit. *salakos*, a miner's sieve. *Ex:* Salax (Ins.):    2. L. *salax*, genit. *salacis*, fond of leaping, lustful, unchaste. *Ex:* salaci-ous; See also salacia.

**salacia**—L. *Salacia*, sea-goddess, wife of Neptune. *Ex:* Salacia (Moll.).

**salamandr**—Gr. *salamandra*, a salamander. *Ex:* Salamandr-ops (Amph.); Salamandra (Amph.); salamandri-form.

**salang**—Gr. *salanx*, genit. *salangos*, a kind of fish. *Ex:* Salang-idae (Pisc.); Salanx (Pisc.).

**salanx**—See salang.

**salari**—L. *salarius*, of or belonging to salt. *Ex:* Salar-ichthys (Pisc.); Salaria (Pisc.); Salari-gobius (Pisc.).

**salax**—See salac.

**sald**—NL. *Salda*, a proper name. *Ex:* Sald-arius (Ins.); Sald-idae (Ins.); Sald-ula (Ins.); Salda (Ins.).

**salebr**—L. *salebra*, the roughness of an uneven road, harshness; *salebrosus*, rough, uneven.

**salen**—NL. *salenia*, name for a genus of fossil sea urchins (orig. unknown, perhaps from a proper name). *Ex:* Salen-ida (Echin.); Salenia (Echin.); Saleno-cidarus (Echin.).

**sali**—L. *salio*, to leap > *Salii*, *Salius*, the Leapers, a group of dancing priests at Rome. *Ex:* Salio-stethus (Ins.); Salio-coccus (Ins.); Salius (Ins.); for Salicornia* see sal.

**salic**—L. *salix*, genit. *salicis*, the willow. *Ex:* Salic-aceae*; Salic-aria (Av.); salic-etum; Salici-passer (Av.); salico-log-ist; not Salicornia, see sal; Salix*.

**salien**—L. *saliens*, genit. *salientis*, ppr. of *salio*, to leap, spring, jump. *Ex:* Salient-ia (Amph.).

**salifa**—NL. *salifa*, an arbitrary combination of letters used to designate a genus of leeches. *Ex:* Salifa (Ann.).

**salign**—L. *salignus*, of willow, used in the sense of willow-like.

**salin**—See sal.

salit—L. *salito*, to dance, ppr. *salitans*, dancing.

salius—L. *Salius*, of or belonging to the Salii or leapers, see sali. *Ex:* Salius (Pisc.), etc.

saliva—L. *saliva*, spittle, slime. *Ex:* saliv-arium; saliva.

salix—See salic.

salm—L. *salmo*, genit. *salmonis*, the salmon. *Ex:* Salmo (Pisc.); Salmo-perca (Pisc.); salmoniform.

salmac—Gr. *Salmakis*, fountain said to make weak those who drank its waters. *Ex:* Salmacopsis (Echin.); Salmacis (Mam.), (Echin.).

salmon—See salm.

salp—L. *salpa*, a kind of stockfish < Gr. *salpē*. *Ex:* Salpa (Tun.); Salpi-cola (Prot.); for Salpiglossis* and Salpornis (Av.); see salping.

salpinct—See salping.

salping—Gr. *salpinx*, genit. *salpingos*, a war trumpet; *salpinktēs*, a trumpeter. *Ex:* Salpornis (Av.); Salpi-glossis*; Salpinctes (Av.); Salpingo-capsa (Prot.); salpingo-nasal; Salpinx (Ins.); Uro-salpinx (Moll.).

salpinx—See salping.

salpiz—Gr. *salpizō*, to proclaim, to announce with the sound of the trumpet. *Ex:* Salpizo (Av.).

sals—L. *salsus*, salted, pp. of *salio*, to salt, to sprinkle with salt. See also sal and sali.

salsol—ML. *salsola*, the saltwort < L. *sal*, salt; *salsus*, salted. *Ex:* salsol-aceus; Salsola*; Salsolo-myia (Ins.).

salsugin—ML. *salsuginosus*, growing in places overflowed by salt or brackish water; L. *salsugo*, saltness < *salsus*, salted. *Ex:* salsugin-ous.

salt—1. L. *saltus*, a leap < *salio*, to spring, leap. *Ex:* salti-grade: 2. L. *saltus*, a woodland; *saltuarius*, a forester; *saltuensis*, belonging to a *forest*; *saltuosus*, well-wooded: 3. L. *salto*, to dance, *saltator*, a dancer; *saltatrix*, a dancing girl > LL. *salticus*, a dancer. *Ex:* saltatori-al; Saltatoria (Ins.); Saltatric-ula (Av.); Saltatrix (Ins.); Salticus (Arach.); Salto-pus (Rept.); Salto-po-suchus (Rept.).

saltan—L. *saltans*, genit. *saltantis*, dancing; ppr. of *salto*, to dance.

saltator—See salt 3.

saltatrix—See salt 3.

saltu—See salt 2.

salubr—L. *salubris*, healthful, wholesome

salutar—L. *salutaris*, beneficial < *salus*, health, a being sound.

salv—L. *salvus*, whole, sound, uninjured; well preserved, *salvia*, sage. *Ex:* Salvia*.

salvelin—NL. *salvelinus* < Ger. *Salbling*, a small salmon. *Ex:* Salvelinus (Pisc.).

sam—Gr. *Samos*, an island in the Turkish Archipelago. *Ex:* Samo-therium (Mam.).

samandura—Singhalese *samandarā*, an East Indian tree. *Ex:* Samandura*.

samanea—NL. *samanea*, a corrupt. of native Spanish *zaman*, a plant name. *Ex:* Samanea*.

samar—L. *samara* = *samera*, the seed of the elm. *Ex:* samar-oid; samara; samari-form; Samaroblatta (Ins.).

samaris—NL. *samaris*, name of a fish of Chinese waters, etym. not apparent. *Ex:* Samar-iscus (Pisc.); Samaris (Pisc.).

sambuc—L. *sambucus*, the elder tree < Gr. *sambukē*, a stringed musical instrument, parts of which were sometimes made from the elder. *Ex:* sambuc-eus; Sambuci-fex (Ins.); Sambucus*.

samia—L. *Samius*, fem. *Samia* < Gr. *Samos*, island of the Aegean Sea. *Ex:* Samia (Ins.); Callo-samia (Ins.).

samol—L. *samolus*, the anemone. *Ex:* samoli-folia; Samolus*.

samyd—Gr. *samyda*, a plant, perh. the birch tree. *Ex:* Samyd-aceae; Samyda*.

sanct—L. *sanctus*, sacred, saintly.

sandal—Gr. *sandalon*, a wooden shoe, sandal; also a flat-fish; *sandalōdēs*, sandal-like. *Ex:* Sandal-ops (Moll.); Sandalium (Moll.); Sandalo-litha (Coel.); Sandalodes (Arach.); Eu-sandalum (Ins.).

sangui—L. *sanguis*, blood; *sanguinolentus*, bloody, full of blood. *Ex:* sangui-col-ous; sangui-renal; Sangui-sorba*; Sangui-suga (Ann.); Sanguinaria*; sanguini-vorous.

sani—*Sani*, Indian deity. *Ex:* Sani-therium (Mam.).

sanicula—NL. *sanicula*, a plant name, a dim. < *sano*, to heal. *Ex:* Sanicula*.

sanid—Gr. *sanis*, genit. *sanidos*, a board, tablet; *sanidōma*, genit. *sanidōmatos*, a planking, a framework. *Ex:* sanid-aster; Sanid-astr-ella (Por.); Sanido-phyllum (Coel.).

sanios—L. *saniosus*, full of bloody pus or of purple fluid.

sanit—L. *sanitas*, genit. *sanitatis*, soundness, health; *sanitarius*, promoting health. *Ex:* sanit-ation; sanitary.

sannio—L. *sannio*, genit. *sannionis*, one who mimics, a clown. *Ex:* Sannion-ite (Moll.).

santal—Pers. *chandal* < Sanskrit *chandama*, name of the sandalwood tree. *Ex:* santal-aceous; Santal-ales*; Santalum*.

santolin—It. *santolina*, the sacred flax < L. *sanctum*, holy + *linum*, flax. *Ex:* Santolina*.

santonic—L. *Santonicus*, of or belonging to the Santoni, a people of Aquitania.

saot—Gr. *saōtēs* = *saōtēr* = *sōtēr*, a deliverer, a preserver < *saoō* = *sōzō*, to save. *Ex:* Saotis (Ins.); Saotus (Ins.).

sap—See sapon.

saperd—Gr. *saperdēs*, name for a salted fish. *Ex:* Saperd-opsis (Ins.); Saperda (Ins.).

saph—Gr. *saphēs*, distinct; *saphōs* = Ion. *sapheōs*, clearly, plainly; *saphēnēs*, the plain truth. *Ex:* saphen-ous vein; Sapheo-pipo (Av.); A-saphes (Ins.); A-saphin-ella (Moll.); A-sapho-morpha (Ins.); Lepido-saphes (Ins.).

saphen—See saph.

sapid—L. *sapidus*, savory, well-tasted.

sapien—L. *sapiens*, genit. *sapientis*, of good taste; also mighty, wise, sensible < *sapio*, to taste, to have a taste of something.

sapindus—See sap.

sapine—L. *sapineus*, belonging to the fir tree, smelling like the resin of the pine or fir.

sapium—NL. *sapium* < L. *sapinus*, a kind of pine or fir. *Ex:* Sapium*.

sapon—L. *sapo*, genit. *saponis*, soap. *Ex:* Sap-indus*; sapon-aceous; Sapon-aria*; saponi-fer-ous; saponi-fy.

sapor—L. *sapor*, savor; *saporus*, savory; *sapo-ratus*, seasoned, savory. *Ex:* sapori-fic.

sapot—NL. *sapota* < Mex. *tzapotl*, Sp. *zapote*, name of a plant. *Ex:* Sapot-aceae*; Sapota*.

sapphirin—L. *sapphirinus*, of sapphire < *sap-phirus*, a sapphire; *sapphiratus*, adorned with sapphire. *Ex:* Sapphirina (Crust.).

sappho—Gr. *Sapphō*, poetess of Lesbos. *Ex:* Sappho (Av.).

sapr—Gr. *sapros*, putrid. *Ex:* Sapro-legnia*; sapro-phytic; sapro-plankton; Sapro-via (Ins.); oligo-sapro-bia.

sapyg—NL. *sapyga*, name for a genus of digger wasps. *Ex:* Sapyg-idae (Ins.); Sapyg-ina (Ins.); Sapyga (Ins.).

sar—Gr. *saron*, also *sarōtron*, a broom; *sarōtēs*, a sweeper; *sarapous*, one who sweeps with the feet; i.e. one who turns out the feet when walk-ing. *Ex:* Saro-phorus (Arach.); Saro-scelis (Ins.); Sarotes (Arach.); Acmo-sara (Ins.); Hedy-sarum*.

sarc—Gr. *sarx*, genit. *sarkos*, flesh; *sarkion*, a bit of flesh; *sarkōdēs*, fleshy. *Ex:* Sarc-anthus*; Sarcio-phorus (Av.); sarco-lemma; sarco-plasm; Sarco-rhamphus (Av.); Sarcod-ina (Prot.); sarcode; Sarcodes*; Sarkodi-ornis (Av.); ecto-sarc.

sarcin—L. *sarcina*, bundle. *Ex:* Sarcina*; sarcini-form; Plano-sarcina*.

sarcod—See sarc.

sarcolip—Gr. *sarcolipēs*, lean, poor in flesh.

sarcotac—Gr. *sarkotakēs*, wasting the flesh < *sarx*, genit. *sarkos*, flesh. *Ex:* Sarcotaces (Crust.).

sard—Gr. *sarda*, a kind of tunny caught near Sardinia; *sardē* = *sardēnē*, the sardine. *Ex:* Sarda (Pisc.).

sarg—Gr. *sargos*, a kind of sea-fish. *Ex:* Sargo-somus (Pisc.); Sargus (Ins.).

sargan—Gr. *sarganē*, a plait, band. *Ex:* Sargan-ura (Av.).

sargass—Sp. *sargaza*, seaweed. *Ex:* Sargasso Sea; Sargassum*.

sark—See sarc.

sarm—1. Gr. *sarmos*, a heap of earth or sand: 2. Gr. *sarma*, genit. *sarmatos*, a chasm in the earth.

sarmatic—L. *Sarmaticus* < Gr. *Sarmatēs*, a name for a region west of the River Volga. *Ex:* Sarmaticus (Moll.).

sarment—L. *sarmentum*, twigs; *sarmentosus*, twiggy. *Ex:* sarment; sarmenti-fer-ous; sarmentose; sarmentum.

sarom—Gr. *sarōma*, genit. *sarōmatos*, sweepings.

saron—See sar.

sarotes—See sar.

sarothr—NL. *sarothrum* < Gr. *sarōthron*, broom. *Ex:* Sarothr-ura (Av.); Sarothro-myia (Ins.); sarothrum.

sarotro—See sar.

sarpedon—Gr. *Sarpēdōn*, king of Lycia. *Ex:* Sarpedon (Arach.).

sarritor—L. *sarritor*, one that scrapes, a hoer. *Ex:* Sarritor (Pisc.).

sartor—L. *sartor*, genit. *sartoris*, a tailor, a mender. *Ex:* sartorius.

sat—1. L. *satus*, a planting; *sativus*, that which is sown < *sero*, to sow. *Ex:* sati-form: 2. L. *satis*, sufficiently, moderately.

satan—Gr. *Satan* = *Satanas*, the Devil, Satan *Ex:* Satan-ellus (Mam.); Satano-perca (Pisc.).

sathr—Gr. *sathros*, decayed, broken; *sathrotēs*, rottenness, weakness; *sathrōma*, a flaw. *Ex:* Sathra (Ins.); sathro-phyta; Sathro-pterus (Ins.).

sativ—See sat.

satrap—Gr. *satrapēs*, a ruler, one wearing a golden crown. *Ex:* Satrap-archis (Ins.); Satrapa (Av.); Satrapes (Ins.); Satrapo-doxa (Ins.).

satur—L. *saturo*, to fill, to saturate; pp. *saturatus*, filled; ppr. *saturans*, genit. *saturantis*, filling. *Ex:* saturate; saturant.

saturat—L. *saturatus*, of full rich color < *saturo*, to fill full, saturate.

satureia—L. *satureia*, the plant known as savory. *Ex:* Satureia*.

saturn—L. *Saturnius*, pertaining to Saturn; *Saturnalis*, of or belonging to Saturn; *Satur-ninus*, a Roman surname. *Ex:* Saturn-ulus (Prot.); Saturnalis (Prot.); Saturni-idae (Ins.); Saturnia (Ins.); Saturninus (Prot.).

saturnus—NL. *saturnus*, lead, leaden (a derived meaning) < L. *Saturn*, a planet name.

satyr—Gr. *Satyros*, sylvan deity given to merri-ment and insatiable lasciviousness > L. *satyrus*, a satyr. *Ex:* Satyr (Ins.); satyri-asis; Satryium*.

sauci—L. *saucius*, wounded, weakened.

saucr—Gr. *saukros*, beautiful, graceful. *Ex:* Saucro-pus (Ins.).

saul—Gr. *saulos*, waddling. *Ex:* Saulo-procta (Av.).

saur—Gr. *sauros*, lizard; *saurōtos*, spotted like a lizard. *Ex:* Saur-an-odon (Rept.); Saur-opsida; Saur-urus*; sauri-an; Sauro-cetes (Mam.); Saur-omalus (Rept.); dino-saur; Plesio-saur (Rept.); Ptilo-saurus (Coel.); Tyranno-saurus (Rept.).

saurit—NL. *sauritus*, lizard-like < Gr. *sauros*, a lizard.

saurot—Gr. *saurōtēr*, genit. *saurōtēros*, a ferrule or spike; *saurōtos*, furnished with a spike. See saur.

savanna—Sp. *sabana* = *zavana*, a meadow. *Ex:* savanna.

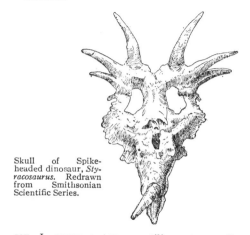

Skull of Spike-headed dinosaur, *Styracosaurus*. Redrawn from Smithsonian Scientific Series.

sax—L. *saxum*, a stone; *saxatilis*, neut. *saxatile*, dwelling or growing among rocks; *saxialis*, of rock or stone. *Ex:* Saxi-cava (Moll.); saxifrag-ous; Saxi-fraga*; saxatile.

saxital—See sax.

scab—L. *scaber*, rough; *scabiosus*, rough, scurfy. *Ex:* Scabiosa*.

scabell—L. *scabellum*, a footstool, dim. of *scamnum*, a bench, stool. *Ex:* Scabellia (Mam.); scabellum.

scabios—See scab.

scabr—L. *scaber*, fem. *scabra*, rough; *scabridus*, rough, rugged; *scabratus*, roughened. *Ex:* scabrate; Scrabri-cola (Moll.); Scabri-fer (Mam.); scabrid-ul-ous; scabr-ose.

scabrat—See scabr.

scabrid—See scabr.

scae—See scai.

scaer—See scair.

scaev—L. *scaevus*, left. *Ex:* Scaevo-gyra (Moll.)

scaevit—L. *scaevitas*, left-handed, awkward.

scaevol—L. *Scaevola*, dim. of *Scaeva*, the Left-handed One. *Ex:* Scaevola (Moll.).

scai—Gr. *skaios*, clumsy, crooked. *Ex:* Scaeopus (Mam.); Scaeo-rhynchus (Av.).

scair—Gr. *skairō*, to skip, dance. *Ex:* Scaerophyla (Ins.).

scal—1. L. *scala*, a ladder, a staircase. *Ex:* Scala (Moll.); scala vestibuli; Scali-bregma (Ann.):     2. Gr. *skalops*, a mole < *skallō*, to dig. *Ex:* Scalo-pus (Mam.); Scalopo-saurus (Rept.):     3. Gr. *skaleys*, a digger. *Ex:* Scales (Ins.):     4. Gr. *skalis*, genit. *scalidos*, a hoe, a mattock; also a bowl or cup. *Ex:* Scal-idia (Ins.); Scalido-gnathus (Arach.).

scalar—L. *scalaris*, pertaining to a ladder; NL. *scalator*, a ladder climber. *Ex:* scalari-form; Scalari-pora (Bry.); Scalaria (Moll.); Scalator (Moll.).

scalat—See scalar.

scald—L. *Scaldis*, the river Scheldt. *Ex:* Scaldicetus (Mam.).

scalen—Gr. *skalēnos*, uneven, crooked; also a kind of triangle. *Ex:* Scalen-aria (Moll.); Scalen-arthrus (Ins.); Scaleno-stoma (Moll.); scalenus anticus.

scalid—See scal 4.

scall—Gr. *skallō*, to hoe, to stir up the earth. *Ex:* Scallo-dera (Ins.).

scalm—Gr. *skalmos*, the pin with which the Greek oar was fastened. *Ex:* Scalmi-cauda (Ins.); Scalmo-phorus (Ins.); Scalmus (Ins.).

scalop—See scal 2.

scalpel—NL. *scalpel* < L. *scalprum*, dim. *scalpellum*, a small surgical knife, a lancet. *Ex:* scalpeli-form; Scalpell-opsis (Crust.); Scalpellum (Crust.).

scalpell—See scalpel.

scalpen—L. *scalpens*, genit. *scalpentis*, cutting, scraping; ppr. of *scalpo*, to carve, cut.

scalpr—L. *scalprum*, a chisel, knife. *Ex:* scalpriform; scalprum.

scalprat—L. *scalpratus*, with a sharp edge, made for cutting.

scalpt—L. *scalptus*, scratched, carved < *scalpo*, to engrave.

scalpturat—NL. *scalpturatus*, engraved < *scalpturio*, to scratch, engrave, incise.

scamb—Gr. *skambos*, curved, bent. *Ex:* Scambula (Moll.); Scambo-neura (Ins.); Scambus (Ins.).

scammat—Gr. *skamma*, genit. *skammatos*, a trench. *Ex:* Scammato-cera (Ins.).

scamn—Gr. *skamnos*, a couch. *Ex:* Scamno-ceras (Moll.).

scan—1. Gr. *skanos* = *skēnos*, a dead body, corpse. *Ex:* Scani-ornis (Av.):     2. Gr. *skana*, a tent.

**scand**—L. *scando*, to climb, to ascend, ppr. *scandens*, genit. *scandentis*, climbing; *scansor*, a climber. *Ex:* scandent; Scandentes (Av.); scansor-ial; Scansor-es (Av.).

**scandalon**—Gr. *skandalon*, a trap for an enemy. *Ex:* Scandalon (Ins.).

**scandic**—L. *scandix*, genit. *scandicis* < Gr. *skandix*, the herb called chervil. *Ex:* Scandic-idae*; Scandix*.

**scandix**—See **scandic**.

**scansor**—See **scand**.

**scap**—L. *scapus*, the stalk of a plant, a shaft < Gr. *skapos*, a staff. *Ex:* scap-oid; scapi-ger-ous; scap-oid-eus; scape.

**scapan**—Gr. *skapanē*, a digging tool; *skapanētēs*, a digger. *Ex:* Scapan-ulus (Mam.); Scapanetes (Ins.); Scapania (Ins.); Scapano-rhynchus (Pisc.); Scapanus (Mam.).

**scaph**—Gr. *skaphē*, anything dug out, a bowl, ship; *skaphidion*, a small ship; *skaphion*, a bowl > L. *scaphium*, a hollow vessel; Gr. *skapheutēs* = *skapheus*, a digger; *skaphis*, genit. *skaphidos*, a skiff; also a shovel. *Ex:* Scaph-ander (Moll.); Scaph-ella (Moll.); Scaph-erpeton (Amph.); scaph-oid; Scaphi-rynchus (Pisc.); Scaphid-urus (Av.); Scaphidium (Ins.); Scaphido-morphus (Ins.); Scaphio-dactylus (Ins.); Scapho-gnathus (Rept.); Eu-scaphis*.

**scaphid**—See **scaph**.

**scaphis**—See **scaph**.

**scaphit**—Gr. *skaphitēs*, a steersman. *Ex:* Scaphita (Moll.).

**scapt**—Gr. *skaptō*, to dig; *skaptēr*, a digger; *skaptos*, dug. *Ex:* Scapt-onyx (Mam.); Scapter-us (Ins.); Scaptero-mys (Mam.); Scapto-chirus (Mam.); Para-scaptor (Mam.); Syco-scapter (Ins.).

**scapter**—See **scapt**.

**scapul**—L. *scapula*, the shoulder blade; NL. dim. *scapulet*. *Ex:* Scapulo-clavicular; scapulet.

**scar**—1. Gr. *skaros*, the parrot-fish, Scarus cretensis. *Ex:* Scar-ichthys (Pisc.); Scaro-soma (Pisc.); Scarus (Pisc.); Pseudo-scarus (Pisc.): 2. Gr. *skaris*, genit. *skaridos*, a little worm. *Ex:* Scar-idium (Rot.); Scaris (Ins.).

**scarabae**—L. *scarabaeus*, a beetle, a scarab. *Ex:* Scarabae-idae (Ins.); scarabae-id-oid; Scarabaeus (Ins.).

**scariol**—NL. *scariola*, wild lettuce; Fr. *scariole*, endive.

**scariosus**—ML. *scariosus*, thin, dry, membranous < ML. *scaria*, a spinous shrub; or < Prov. Eng. *scare*, lean, scaly. *Ex:* scariose.

**scaris**—See **ascar**.

**scarit**—Gr. *skaritis*, genit. *skaritidos*, a stone colored like the fish, Scarus. *Ex:* Scarites (Ins.); Scarit-ides (Ins.); Scarito-derus (Ins.).

**scart**—Gr. *skartēs*, springing, quick, nimble < *skairō*, to skip, dance. *Ex:* Scart-urus (Mam.);

Scartes (Mam.); Scarto-myzon (Mam.); Cardio-scarta (Ins.).

**scarthm**—Gr. *skarthmos*, a leaping, running. *Ex:* Eu-scarthmus (Av.).

**scat**—Gr. *skōr*, genit. *skatos*, dung. *Ex:* Scat-ops (Ins.); scato-logy; Scato-phag-idae (Ins.); Ornitho-scat-oides (Arach.).

**scaturig**—L. *scaturigex*, genit. *scaturiginis*, gushing water < *scaturio*, to spring. *Ex:* scaturiginous (Ecol.).

**scaur**—Gr. *skauros*, with projecting ankles. *Ex:* Scaurus (Ins.).

**scaveng**—ME. *scavager*, an inspector, one attending to scavage, inspection > Eng. *scavenger*, one who collects filth. *Ex:* scavenger.

**scedas**—Gr. *skedasis* = *skedasmos*, a scattering. *Ex:* Scedasus (Ins.).

**scel**—1. Gr. *skelis*, genit. *skelidos*, ribs of beef; also a leg of ham, the leg; akin to *skelos*, a leg. *Ex:* Sceli-mena (Ins.); scelides (NL. pl. of *skelis*); Scelido-saurus (Rept.); Scelido-therium (Mam.); Eu-scelis (Ins.): 2. Gr. *skelos*, the leg. *Ex:* Scel-odonta (Ins.); Scelo-dromus (Ins.); Scelo-physa (Ins.); Scelo-tes (Rept.); Scelo-tretus (Rept.); Scelos-odis (Ins.). See also scelio.

**sceler**—L. *scelero*, to pollute; *scelerus*, abominable; *sceleratus*, defiling.

**scelerat**—See **sceler**.

**scelest**—L. *scelestus*, wicked, knavish.

**scelet**—Gr. *skeletos*, dried up, dried like a mummy; *skeletōdēs*, like a mummy. *Ex:* Sceleto-merus (Ins.); skeleton.

**scelid**—See **scel**.

**scelio**—L. *scelio*, an impious man, a scoundrel. *Ex:* Scelio (Ins.); Scelio-trachelus (Ins.).

**sceliphron**—Gr. *skeliphros*, dry, lean. *Ex:* Sceliphron (Ins.).

**scen**—Gr. *skēnē*, a sheltered place, a tent, a scene; L. *scenicus* = Gr. *skēnikos*, pertaining to a tent, scene. *Ex:* Scen-odon (Rept.); Sceni pinus (Ins.) apparently from Gr. *skēnopois*, tent-making; Sceno-poeetes (Av.); Scenopoeus (Av.); Archi-scenium (Prot.); Sago-scena (Prot.). See also scan.

**scenop**—See **scen**.

**scep**—Gr. *skepē*, covering, shelter < *skepō*, to cover. *Ex:* Hedy-scepe*; Macro-scepis*.

**sceparn**—Gr. *skeparnon*, an adze. *Ex:* Sceparn-odon (Mam.).

**scepasm**—Gr. *skepasma*, genit. *skepasmatos*, a covering. *Ex:* A-scepasma (Ins.).

**sceptic**—Gr. *skeptikos*, observant. *Ex:* Scepticus (Ins.).

**sceptr**—L. *sceptrum* = *scaeptrum* < Gr. *skēptron*, a royal staff, sceptre. *Ex:* Sceptr-ella (Por.); Sceptro-neis (Prot.); Sceptrum (Moll.).

**sched**—Gr. *schedē*, a tablet. *Ex:* Schedo-philus (Pisc.); Schedo-trigona (Myr.).

schedon—Gr. *schedon*, close, almost, all but, perhaps. *Ex:* ?Schedo-cardia (Moll.).

schem—Gr. *schēma*, genit. *schēmatos*, form, shape; *schēmatizō*, to form. *Ex:* Schemato-rrhages (Ins.); Schematiza (Ins.); Echino-schema (Ins.).

schematic—Gr. *schēmatikos*, resembling, of similar form: *Ex:* schematic.

schematiz—See schem.

schesis—Gr. *schesis*, condition, nature. *Ex:* Amaro-schesis (Ins.).

schid—Gr. *schidion*, something split off, a chip, a splinter of wood, dim. of *schiza*, a thin plank. *Ex:* schidi-gera; Schidium (Ins.); Schido-nychus (Ins.). See also shiz.

schidac—Gr. *shidax*, genit. *schidakos*, a piece of wood split off, a lath splinter. *Ex:* Schidax (Ins.).

schidax—See schidac.

schin—NL. *schinus* < Gr. *schinos*, the mastic tree. *Ex:* Schinus*.

schindyles—Gr. *schindylēsis*, a splitting into fragments. *Ex:* schindylesis.

schism—Gr. *schisma*, genit. *schismatos*, a splitting; *schismos*, a cleaving. *Ex:* Schismato-glottis*; Schismo-thele (Arach.); Schismus*; Macro-schisma (Moll.).

schist—1 Gr. *schistos*, divided, cleft. *Ex:* Schist-ura (Pisc.); Schisto-cera (Ins.); Schisto-stoma: 2. L. *schist*, slate which is in layers; *schistaceus*, slaty, i.e. cleft < Gr. *schistos*, see 1. above.

schiz—Gr. *schizō*, to split, cleave, ppr. *schizōn*, genit. *schizontos*, cleaving. *Ex:* Schiz-anthus*; Schiz-ymenia*; Schiz-aea*; Schizia (Ins.); schizo-gamy; Schizo-metopa (Ins.); Schizo-phora (Ins.); schizont.

schoen—Gr. *schoinos*, a rush, reed; *schoinis*, genit. *schoinidos*, a rope, cord; *schoininos*, made of rushes. *Ex:* Schoenia (Av.); Schoenis (Ins.); Schoenis-cera (Ins.); Schoeno-lirion*; Schoenus*.

schoin—See schoen.

scholast—Gr. *scholastēs*, one who lives at ease; *scholastikos*, enjoying leisure. *Ex:* Scholastes (Ins.); Scholasticus (Ins.).

sci—Gr. *skia*, a shadow, a phantom. *Ex:* Sci-ur-odon (Mam.); Sci-uri-pter-us (Mam.); Sci-urus (Mam.); Scia-mys (Mam.); Scia-pteryx (Ins.); Scio-phila (Ins.); scio-phil-ous; Epi-scia*; Poly-scias*.

sciad—Gr. *skias*, genit. *skiados*, a canopy, an arbor; an umbel, and in this sense is often used in naming genera of umbelliferous plants; *skiadēphoros*, carrying an umbrella; *skiadeion*, umbrella, sunshade. *Ex:* Sciade-oides (Pisc.); Sciadio-phora (Prot.); Sciado-pitys*; Sciado-stoma (Prot.); Helo-sciadium*, Gr. *helos*, a marsh.

sciaen—Gr. *skiaina*, a sea-fish. *Ex:* Sciaen-urus (Pisc.); Sciaena (Pisc.).

sciar—Gr. *skiaros*, shady, dark-colored. *Ex:* Sciara (Ins.).

sciasm—Gr. *skiasma*, a shelter, a shadow. *Ex:* Sciasma (Ins.); Sciasmo-myia (Ins.).

sciather—Gr. *skiathēras*, a sundial, a shadow-catcher. *Ex:* Sciather-odes (Ins.); Sciatheras (Ins.).

sciatic—LL. *sciaticus* < L. *ischiadicus*, relating to gout in the hip. *Ex:* sciatic.

sciatroph—Gr. *skiatrophia*, sheltered, brought up in the shade. *Ex:* Sciatrophes (Ins.); Sciatrophus (Ins.).

scier—Gr. *skieros*, shady, dark-colored. *Ex:* Sciero-pepla (Ins.); Scierus (Ins.).

scill—L. *scilla*, squill. *Ex:* Scilla*. For Scillium (Pisc.) see scyll.

scinac—Gr. *skinax*, genit. *skinakos*, quick, nimble. *Ex:* Scinaco-pus (Ins.); Scinax (Rept.).

scinax—See scinac.

scinc—Gr. *skingkos* = L. *scincus*, a kind of lizard. *Ex:* scinci-cauda; Scinco-saurus (Rept.); Scincus (Rept.).

scintill—L. *scintilla*, a spark, glimmer; *scintilla-tus*, pp. of *scintillo*, to throw sparks. *Ex:* Scintill-orbis (Moll.); Scintilla (Moll.); scintillate.

scintillan—L. *scintillans*, genit. *scintillantis* twinkling, glowing < *scintillo*, to sparkle.

scio—See sci.

scion—Fr. *scion*, a twig, shoot. *Ex:* scion; not Cyno-scion (Pisc.), the last element is from the Gr. *skiaina*, a sea-fish.

sciot—Gr. *skiōtos*, shaded, striped with colors shading into one another. *Ex:* Sciota (Ins.).

scipon—Gr. *skipōn*, genit. *skipōnos*, a staff. *Ex:* Scipono-ceras (Moll.).

scir—See scirr.

scirrh—See scirr.

scirp—L. *scirpus*, a rush, bulrush; *scirpeus*, of rushes; *scirpiculus*, a basket made of rushes. *Ex:* Scirpe-aria (Coel.); Scirpo-phaga (Ins.); Scirpus*.

scirr—Gr. *skirrhos* = *skirhos*, gypsum, stucco, any hard coat or covering; ML. *scirrhos*, a tumor or hardened swelling. *Ex:* schirrh-osis (Med.); scirros-ity = scirrhosity; Scirrus (Arach.).

scirrh—See scirr.

scirt—Gr. *skirtaō*, to leap, bound; *skirtēsis*, a leaping; *skirtētēs*, a leaper. *Ex:* Scirtes (Ins.); Scirtet-ellus (Ins.); Scirtetes (Mam.); Scirto-mys (Mam.); Scirto-thrips (Ins.).

scirtes—See scirt.

scirtet—See scirt.

sciss—L. *scissus*, rent, slit; *scissio*, a cleaving < *scindo*, to cut, to rend. *Ex:* Sciss-ur-ella (Moll.); Scissi-labra (Moll.); scissi-par-ous; circum-sciss-ile.

**scit**—L. *scitus*, wise, skillful.

**scitul**—L. *scitulus*, handsome, neat, slender.

**sciur**—Gr. *skiouros* = L. *sciurus*, a squirrel. *Ex:* sciur-oid, bushy like a squirrel's tail; Sciur-avus (Mam.); Sciuro-ptera (Mam.); Sciurus (Mam.). See also sci.

**scler**—Gr. *sklēros*, hard; *sklēria*, hardness; *sklērotēs*, hardness, stiffness. *Ex:* scler-ite; scler-enchyma; scler-oma (Med.); Scleria*; Sclero-stomum (Nem.); sclerot-ic.

**sclerit**—NL. *sclerite*, a hard plate or spicule < Gr. *sklēros*, hard. *Ex:* Sclerito-derma (Por.).

**scob**—L. *scobis*, sawdust, scrapings. *Ex:* scobi-form.

**scobin**—L. *scobina*, a rasp. *Ex:* scobin-ate; Scobin-ichthys (Pisc.); Scobina (Ins.); Scobino-pholas (Moll.).

**scol**—Gr. *skōlos*, a thorn, prickle. *Ex:* Scol-oplos (Ann.); Scol-ops (Ins.); Scolo-derus (Arach.); scolo-pale; Scolo-thrips (Ins.); Scolos-anthus*; Cnido-scolus*.

**scole**—See **scolec.**

**scolec**—Gr. *skōlēx*, genit. *skōlēkos*, a worm; *skōlēkiasis*, being worm eaten. *Ex:* Scole-toma (Ann.); scoleci-asis (Med.); Scoleco-campa (Ins.); Scoleco-phagus (Mam.); scolex; Rhyncho-scolex (Platy.).

**scolex**—See **scolec.**

**scoli**—Gr. *skolios*, bent, crooked; *skoliōsis*, crookedness. *Ex:* Scoli-odon (Elasm.); ?Scolia (Ins.); Scolio-pus*; scoliosis (Med.).

**scolia**—NL. *skolia*, name of a genus of wasps, perh. < Gr. *skolios*, bent. *Ex:* Scolia (Ins.).

**scolic**—See **scolec.**

**scolop**—Gr. *skolops*, genit. *skolopos*, anything pointed. *Ex:* Scolopo-crypt-ops (Myr.); Scolops-is (Pisc.); Scolopso-morpha (Ins.).

**scolopac**—Gr. *skolopax*, genit. *skolopakos*, the woodcock. *Ex:* Scolopac-idae (Av.); Scolopac-inus (Av.); Scolopax (Av.).

**scolopax**—See **scolopac.**

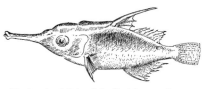

Much-spined Snipe-fish, *Centriscus scolopax*, a small fish of the Mediterranean. After Brehm.

**scolopendr**—Gr. *skolopendra*, a centipede > *skolopendrion*, a kind of fern, hart's tongue. *Ex:* Scolopendra (Myr.); Scolopendr-ella (Myr.); Scolopendr-opsis (Myr.); Scolopendrium*.

**scolops**—See **scolop.**

**scolym**—Gr. *skolymos*, an artichoke. *Ex:* Scolym-astra (Por.); Scolymus (Moll.); Scolymus*.

**scolypt**—See **scolyt.**

**scolyt**—NL. *scolytus* < Gr. *scolyptō*, to shorten, mutilate. *Ex:* Scolyt-idae (Ins.); Scolyto-proctus (Ins.); Scolypto-caulus (Ins.); Scolytus (Ins.).

**scolythr**—Gr. *skolythros*, low, shabby. *Ex:* Scolythrus (Ins.).

**scomber**—See **scombr.**

**scombr**—Gr. *skombros* = L. *scomber*, mackerel. *Ex:* Scomber-esox (Pisc.); Scombero-morus (Pisc.); Scombr-ops (Pisc); scombr-oid; Scombro-clupea (Pisc.); Chloro-scombrus (Pisc.).

**scop**—1. Gr. *skopos*, a watchman, one that looks about; *skopeō*, to see. *Ex:* Scopo-therium (Mam.); micro-scope; Phyllo-scopus (Av.): 2. L. *scopa*, dim. *scopula*, thin branches, twigs, a broom. *Ex:* Scop-aria*; scop-arius; scopi-ped; Scopo-phorus (Mam.); scopula; not Scopola*, which was named for Scopoli, 18th century Austrian naturalist; scopuli-form.

**scopae**—Gr. *skōpaios*, a dwarf. *Ex:* Scopaeo-thrips (Ins.); Scopaeus (Ins.).

**scoparches**—Gr. *skoparchēs*, the leader of a scouting party. *Ex:* Scoparches (Ins.).

**scopel**—Gr. *skopelos*, a cliff, high rock or peak. *Ex:* Scopel-archus (Pisc.); Scopelo-genys (Pisc.); Scopelus (Pisc.).

**scopes**—Gr. *skopēsis*, an excuse. *Ex:* Scopesis (Ins.).

**scopim**—Gr. *skopimos*, suitable to a purpose.

**scopior**—Gr. *scopiōros*, a watcher. *Ex:* Scopiorus (Ins.).

**scops**—1. Gr. *skōps*, genit. *skōpos*, nom. pl. *skōpes*, a little horned owl. *Ex:* Scops (Av.): 2. Gr. *skōpsis*, mockery.

**scopt**—Gr. *skōptēs*, a mimic, a mocker; *skōptikos*, given to jesting. *Ex:* Scoptes (Ins.); Scopticus (Arach.); Oro-scoptes (Av.).

**scopul**—L. *scopulus*, a rock, cliff, crag; *scopulosus*, full of rocks, craggy. *Ex:* Scopuli-fer (Ins.); Scopuli-pedes (Ins.). See also scop.

**scord**—Gr. *skordon* = *skorodon*, garlic. *Ex:* Scordonia (Ins.); Notho-scordum*.

**scorp**—Gr. *skorpiōn* also *skorpios*, a scorpion; *skorpaina*, name for a fish with a poisonous sting. *Ex:* Scorpaen-ichthys (Pisc.); Scorpaena (Pisc.); Scorpaeno-pterus (Pisc.); scorpi-oides; Scorpio (Arach.); Scorpio-teleia (Ins.); Scorpion-ura (Crust.).

**scort**—L. *scortum*, hide, leather; *scorteus*, made of leather. *Ex:* Scort-ima (Prot.); Scorteus (Prot.).

**scorzonera**—Sp. *scorzonera*, "snake-weed" used for snake-bite. *Ex:* Scorzonera*.

**scot**—Gr. *skotos*, darkness; *skotasmos*, becoming dark; *skoteinos*, dark; *skotaios*, obscure; *skotios*, dark, in secret; *skoteyō*, to hide in darkness. *Ex:* Scot-oecus (Mam.); Scota-eu-mys (Mam.);

Scotaeus (Av.); Scoteinus (Mam.); Scot-oecus (Mam.); Scoti-myza (Ins.); Scotio-neurus (Ins.); Scoto-glaux (Av.); Scoto-philus (Mam.). See also scops.

**scotasm**—See scot.

**scotein**—See scot.

**scoteu**—See scot.

**scotic**—L. *scoticus*, Scotch, Scottish.

**scotin**—NL. *scotinus* < Gr. *skoteinos*, obscure hidden.

**scrin**—L. *scrinium*, a case, box. *Ex:* scrin-aceus.

**script**—L. *scriptus*, written, pp. of *scribo*, to write.

**scrob**—L. *scrobis*, a trench, ditch; *scrobiculus*, a little trench. *Ex:* Scrob-odus (Pisc.); scrobe; Scrobi-gera (Ins.); scrobicul-ar.

**scrobicul**—See scrob.

**scrof**—L. *scrofa*, a sow. *Ex:* Scrofa (Pisc.), (Mam.).

**scrophularia**—NL. *scrophularia* < L. *scrofulae* (dim. derived < *scrofa*, a sow), a swelling of the glands of the neck > *scrofula*, a disease. *Ex:* Scrophularia*.

**scrot**—L. *scrotum*, a pouch, the scrotum. *Ex:* scroti-form; scroto-cele; scrotum.

**scrup**—L. *scrupus*, a small sharp stone; *scrupulosus*, rough, jagged; *scrupens*, genit. *scrupentis*, sharp, rough. *Ex:* Scrupo-cell-aria (Polyzoa).

**scrupe**—L. *scrupeus*, difficult to climb, full of stones, sharp.

**scrupen**—See scrup.

**scurr**—NL. *scurrus*, ludicrous < L. *scurra*, a jester.

**scut**—L. *scutum*, dim. *scutulum*, an oblong shield; *scutatus*, armed with a shield; *scuta* = *scutra*, dim. *scutula*, a flat dish; *scutella*, a salver, dish; *scutarius*, of a shield; *scutellatus*, covered with small plates. *Ex:* Scut-emys (Rept.); scutate; Scutell-aria*; Scutell-era (Ins.); Scutella (Echin.); Scuti-sorex (Mam.); Scuto-lampas (Echin.); Scutul-ina (Moll.); Scutulum (Echin.). See also scyt.

**scutic**—L. *scutica*, a whip. *Ex:* Scutic-aria*; Scutica (Pisc.).

**scutulat**—L. *scutulatus*, diamond or lozenge-shaped, chequered.

**scybal**—Gr. *skybalon*, dung, refuse; *skybaliktos*, dirty, mean. *Ex:* Scybal-icus (Ins.); Scybal-ista (Ins.).

**scydmaen**—Gr. *skydmainos*, angry, gloomy, sad. *Ex:* Scydmaen-ichnus (Ins.); Scydmaen-illa (Ins.); Scydmaenus (Ins.).

**scyl**—1. Gr. *skylos*, a hide, the skin, the husk of a nut. *Ex:* Anthribo-scyla (Ins.): 2. Gr. *skylon*, spoils, booty.

**scylac**—Gr. *skylax*, genit. *skylakos*, a dog, also a young dog, puppy. *Ex:* Scylac-ops (Rept.); Scylaco-gnathus (Rept.); Scylaco-saurus (Rept.); Scylax (Arach.). See also scyll.

**scylax**—See scylac.

**scyli**—See scyll.

**scyll**—1. Gr. *skyllos* = *skylax*, genit. *skylakos*, a dog, also a young pup, puppy, see skylac: 2. Gr. *Skylla*, a monster with twelve arms and six necks, inhabiting a cavern in the Straits of Sicily and fabled to be girt about with barking dogs; L. *scyllaeus*, pertaining to *Scylla* < Gr. *skyllō*, to mangle, tear in pieces. *Ex:* Scylla (Crust.); Scylla-ea (Moll.): 3. Gr. *skylion*, a dogfish, probably related to *skylax*, a dog and to *skyllō*, to mangle. *Ex:* Scylio-rhinus (Elasm.); Scyllium (Elasm.); Chilo-scyllium (Elasm.); Para-scyllium (Elasm.).

**scyllar**—ML. *scyllarus* < Gr. *skyllaros*, a kind of crab, see cyllar. *Ex:* Scyllar-idae (Crust.); Scyllar-ella (Crust.); Scyllar-ides (Crust.); Scyllarus (Crust.).

**scyllium**—See scyll 3.

**scylm**—Gr. *skylmos*, a tearing, laceration < *skyllō*, to mangle, tear.

**scymn**—Gr. *skymnos*, a cub, whelp; also an ancient name for some shark. *Ex:* Scymn-odes (Ins.); Scymn-odon (Rept.); Scymnia (Moll.); Scymno-gnathus (Rept.); Scymnus (Elasm.); Centro-scymnus (Elasm.).

**scyph**—Gr. *skyphos* = *skythos*, a cup; *skypheios*, cup-like. *Ex:* Scyph-anthus*; scyphi-form; Scyphi-stoma (Coel.); Scyphius (Pisc.); Scyphozoa (Coel.); Chen-endro-scyphia (Por.).

**scyr**—1. Gr. *skyros*, chippings of stone, gypsum, stucco; also hard; *Skyros*, island of the Aegean, so called from its ruggedness. *Ex:* Scyr-ornis (Av.); Scyra (Crust.): 2. Gr. *skyron* = *askyron*, a kind of plant, St. John's wort. *Ex:* Ascyrum*.

**scyt**—Gr. *skytos*, leather; *skytinos*, leathern; *skytōdēs*, like leather. *Ex:* Scuto-pterus (Ins.); Scytina (Rept.); scytinum; Scyto-siphon*; Scytodes (Arach.).

**scytal**—Gr. *skytalē*, dim. *skytalina*, a staff; a serpent of uniform roundness and thickness; a cylinder. *Ex:* Scytal-ichthys (Pisc.); Scytal-idae (Rept.); Scytal-ina (Pisc.); Scytale-crinus (Echin.); Scytalo-crinus (Echin.); Scytalus (Rept.).

**scyth**—Gr. *skythos* = *skyphos*, a cup. *Ex:* Scyth-odonta (Ins.); Scytho-centro-pus (Ins.).

**scythr**—Gr. *skythros*, sullen, angry; *skythrōpos*, gloomy or angry looking. *Ex:* Scythr-odes (Ins.); Scythr-ops (Av.); Scythropus (Ins.); Scythropia (Ins.).

**scythrop**—See scythr.

**scytin**—See scyt.

**se-** —L. *se-*, prefix, denoting a going or separating. *Ex:* se-cretion.

**seb**—L. *sebum*, grease, tallow, wax; Low L. *sebaceous*, like lumps of tallow. *Ex:* sebaceous; sebi-fic; sebum.

sebasm—Gr. *sebasmos*, reverence; *sebasma*, genit. *sebasmatos*, that which is held in reverence. *Ex:* Sebasmia (Ins.).

sebast—Gr. *sebastos*, magnificent, venerable> NL. *sebastes*, a fish name. *Ex:* Sebast-avus (Pisc.); Sebast-icus (Pisc.); Sebastes (Pisc.); Sebasto-nyma (Ins.).

sec—Gr. *sēkos*, an enclosure, a nest, a chapel; *sēkōdēs*, chapel-like. *Ex:* Secod-ella (Ins.); Secodes (Ins.).

secal—L. *secale*, an old name for rye. *Ex:* Secale*.

secamon—NL. *secamone*<Ar. *squamouna*, a plant name. *Ex:* Secamone*.

sechi—Fr. *sechion*, perhaps<Gr. *sikyos*, a cucumber. *Ex:* Sechium*.

seclus—L. *seclusus*, hidden, secluded, separated.

secod—See sec.

secret—L. *secretus*, severed, placed apart< *secerno*, to part, separate<*se*+*cerno*, to sift, separate. *Ex:* secret-ion; secret-ory; Secreti-pes (Ins.).

sect—L. *sectus*, cut, cleft<*seco*, to cut; *sectilis*, cut or cleft; *sector*, one who cuts off; *secula*, a sickle. *Ex:* Secti-lumen (Moll.); Secti-pecten (Moll.); sectori-al.

sectator—L. *sectator*, a follower or attendant. *Ex:* Sectator (Pisc.).

sector—See sect.

secul—See sect.

secund—L. *secundus*, following or second< *sequor*, to follow, to succeed. *Ex:* secund; secundi-florus; secundo-phore.

secundat—1. L. *secundatus*, adjusted, pp. of *secundo*, to adapt, to make prosperous: 2. L. *secundatus*, a noun meaning the second place or rank.

secur—L. *securis*, axe, hatchet; *securiclatus*, shaped like a hatchet. *Ex:* Secur-illa (Moll.); securi-form; Securi-nega*; Securi-palpus (Ins.); Securidaca*, an error for *securiclata*, name of a weed<*securiclatus*.

securiclat—See secur.

secutor—L. *secutor*, genit. *secutoris*, a follower; *secutorius*, following<*sequor*, to follow. *Ex:* Secutor (Pisc.).

sedell—NL. *sedella*, a dim.<*sedum*, a plant name. *Ex:* Sedella*.

sedentar—L. *sedentarius*, sitting. *Ex:* sedentar-y; Sedentari-ae (Arach.).

sedul—L. *sedulus*, diligent, careful. *Ex:* Sedulo-thrips (Ins.).

sedum—L. *sedum*, the houseleek<*sedeō*, to sit. *Ex:* Sedum*.

segest—L. *Segestes*, German prince, friend of the Romans. *Ex:* Segest-idae (Ins.); Segestes (Ins.).

segestr—L. *segestria*=*segestre*, genit. *segestris*, a covering, wrapper, mantle. *Ex:* Segestri-ella

(Arach.); Segestri-oides (Arach.); Segestria (Arach.).

seget—L. *seges*, genit. *segetis*, a corn-field; *segetalis*, of or belonging to standing crops (sometimes taken to mean growing in grain-fields). *Ex:* Segetia (Ins.).

segment—L. *segmentum*, a slice, zone<*seco*, to cut; *segmentatus*, trimmed, ornamented. *Ex:* segment-al; Segment-aria (Moll.); Segment-inus (Moll.); segmentat-ion.

segn—L. *segnis*, slow, sluggish; *segnitas*, slow-ness, tardiness, slothfulness.

segnit—See segn.

segreg—L. *segregus*, separated.

segund—Sp. *segundar*<NL. *segundus*, to be second<L. *secundus*, second.

segutil—L. *segutilum*, a kind of earth found with gold. *Ex:* Segutilum (Pisc.).

sei—Gr. *seios*, divine. *Ex:* Seius (Arach.). See also seis.

seir—1. Gr. *seira*, a chain, a rope. *Ex:* Seir-anota (Amph.); Seir-odonta (Ins.); Seiri-ola (Por.); seiro-spore: 2. Gr. *Seirēn*, a Siren. *Ex:* Seiren (Av.).

seis—Gr. *seiō*, to shake, to move to and fro, to heave; *seismos*, an earthquake, a shaking. *Ex:* Sei-urus (Av.); Seio-ptera (Ins.); Seis-ura (Av.); seismo-tropism; Sej-urus (Av.).

seison—Gr. *seisōn*, an earthen vessel for shaking beans in while being roasted. *Ex:* Seison (Rot.).

sej—See seis.

sejug—1. L. *sejugis*, from *sex*, six+*jugum*, a yoke, i.e. yoked in sixes. *Ex:* sejug-ous: 2. L. *sejugis*, disjoined, separate; *sejugo*, to dis-join, pp. *sejugatus*, disjoined. *Ex:* sejugate.

sejunct—L. *sejunctus*, separated, severed, pp. of *sejungo*, to part, sever.

sel—Gr. *selis*, genit. *selidos*, a plank, a leaf or sheet of papyrus. *Ex:* Seli-odus (Ins.); Seli-trichus*; Selid-acantha (Ins.); Selido-sema (Ins.).

sela—Gr. *selas*, genit. *selatos*, brightness, light; *selaō*, to shine, brighten; *selasma*, genit. *selasmatos*, a shining. *Ex:* Sela-derma (Ins.); Selao-phora (Arach.); Selas-phorus (Av.); Selato-somus (Ins.); A-selasma (Ins.).

selag—L. *selago*, genit. *selaginis*, a kind of club moss. *Ex:* Selagin-ella*; Selago*.

selagi—Gr. *selageō*, to enlighten, illumine. *Ex:* Selagia (Ins.); Selagis (Ins.).

selagin—See selag.

selas—See sela.

selasm—See sela.

selat—See sela.

selen—Gr. *selēnē*, the moon; *selēnis*, a small moon. *Ex:* Selen-idium (Prot.); Selen-ichnus (Amph.); Seleni-cereus*; selen-odont; Selen-ops (Arach.); Seleni-dera (Av.); Selenia*;

Selenis (Moll.); Selen-ites (Moll.); Seleno-stomum (Prot.).

**seleucia**—Gr. *Seleukeia*, celebrated city in Babylonia. *Ex:* Seleucia (Ins.).

**seleucid**—*seleucis*, genit. *seleucidis*, a kind of bird. *Ex:* Seleucidis (Av.).

**selid**—See **sel**.

**selin**—Gr. *selinon*, parsley>L. *selinas*, a kind of cabbage resembling parsley. *Ex:* Selino-carpus*; Selinum*; Petro-selinum*.

**sell**—L. *sella*, a seat, saddle; *sellaris*, of or belonging to a seat. *Ex:* Sella-cotyle (Platy.); sella-turcica; selli-fer; Sello-saurus (Rept.).

**sellul**—L. *sellula*, a small seat or stool, dim. of *sella*, a seat.

**selm**—1. Gr. *selmis*, genit. *selmidos*, a noose. *Ex:* Zygo-selmis (Prot.): **2.** Gr. *selma*, genit. *selmatos*, a seat, throne; also logs or building timber. *Ex:* ?Selma (Moll.).

**sem**—Gr. *sēma*, genit. *sēmatos*, a sign, mark, standard, *sēmeion*, a mark; *sēmeiōtos*, marked. *Ex:* semat-ic; Semato-neura (Ins.); semato-phore; Seme-carpus; Semeio-phorus (Av.);

Skirmishing Sail-bearer, *Semiophorus velitans*, a Miocene fish from Europe. Redrawn from Story of Fishes—Gunther.

Semio-notus (Pisc.); Semo-stoma (Coel.); Dis-sem-ur-opsis (Av.); Ptero-sema (Ins.).

**semae**—Gr. *sēmaia*, a military standard. *Ex:* Semaeo-stomata (Coel.); Semaio-phyllum (Coel.).

**semaio**—See **semae**.

**semant**—Gr. *sēmantos*, marked, emphatic. *Ex:* Semant-idium (Prot.); Semant-iscus (Prot.).

**semantic**—Gr. *sēmantikos*, significant, designating, pointing out.

**semantor**—Gr. *sēmantōr*, a leader, one who gives a signal. *Ex:* Semantor (Mam.).

**semantr**—Gr. *sēmantron*, a seal. *Ex:* Semantrum (Prot.).

**semasi**—Gr. *sēmasia*, a mark, a marking. *Ex:* Semasia (Ins.).

**semat**—See **sem**.

**semeio**—See **sem**.

**semel**—Gr. *Semelé*, earth goddess, daughter of Cadmus and Hermione. *Ex:* Semel-artemis (Moll.); Semel-ina (Moll.); Semele*; Semelo-seris (Coel.).

**semen**—L. *semen*, genit. *seminis*, seed; *seminalis*, pertaining to seed; *semino*. to sow, pp. *seminatus*, sown. *Ex:* semen; semin-al; semini-fer-ous; Semin-ula (Moll.); in-seminate.

**sement**—L. *sementivus*, belonging to seed, or to seed time; *sementis*, a sowing; *semento*, to bear seed. See -ivus. *Ex:* sementi-fer.

**semi**—L. prefix *semi-*, half. *Ex:* semi-amplectus; Semi-donta (Ins.); Semi-limax (Moll.); semi-lunar.

**semin**—See **semen**.

**seminator**—L. *seminator*, one who produces, originates, plants seed.

**semn**—Gr. *semnos*, sacred, solemn, placed high, majestic; *semnotēs*, dignity. *Ex:* Semn-ophrys (Ins.); Semn-ornis (Av.); Semno-pithecus (Mam.); Semnotes (Ins.).

**semon**—L. *Semonia*, Roman goddess who watched over crops. *Ex:* Semonia (Platy.).

**semot**—L. *semotus*, removed, separated, distant < *semoveo*, to place apart.

**semper**—L. *semper*, forever, ever, always. *Ex:* semper-virens; Semper-vivum*; not Semperia (Moll.) nor Semper-ella (Moll.) which are names made in honor of Karl Semper, German malacologist.

**sen**—1. L. *seni*, six each. *Ex:* Sen-oculus (Arach.); Sen-odon (Mam.); seni-ped. Neither Seno-gaster (Ins.), Seno-basis (Ins.), Seno-metopia (Ins.), Seno-propopis (Ins.); Seno-pternia (Ins.), nor Seno-stoma (Ins.), genera described by Macquart who badly mutilated the Gr. *stenos*, narrow. **2.** NL. *seno-*, anagram of *neso*, in the following. *Ex:* Seno-don (Mam.); Seno-nycteres (Mam.).

**senar**—L. *senarius*, belonging to or containing six. *Ex:* Senario-crinus (Echin.).

**senecio**—See **senic**.

**senect**—L. *senectus*, aged, very old. *Ex:* Senecti-dens (Moll.); Senectus (Moll.).

**senesc**—See **senic**.

**senex**—See **senic**.

**senic**—L. *senex*, dim. *seniculus*, an old man < *senesco*, to grow old; *senilis*, of or belonging to old people; *senium*, the feebleness of age, an old man. *Ex:* Senecio*; Senecio-bius (Ins.); senescence; Senex (Av.); Senexi-bombus (Ins.); Seniculus (Moll.); senile; Senilia (Moll.); Senio-cebus (Mam.).

**senil**—See **senic**.

**senio**—See **sen**, also **senic**.

**sens**—L. *sensus*, feeling, sense; Low L. *sensitivus*, having sense or feeling; LL. *sensualis*, endowed with feeling. *Ex:* sensitiv-ity; sensual. See also sensor.

**sensibil**—L. *sensibilis*, sensible < *sensus*, sense + *-ilis*, capable of.

**sensil**—L. *sensilis*, sensitive. *Ex:* sensile.

**sensim**—L. *sensim*, gradually, slowly.

**sensor**—NL. *sensorius*, pertaining to sense < L. *sensus*, sense; LL. *sensorium*, an organ of sense, see -ium 1. *Ex:* sensori-al; sensori-motor; sensorium; sensory.

**sent**—L. *sentis*, a thorn, bramble; *sentus*, thorny, rough; *sentosus* also *senticosus*, full of thorns, thorny. *Ex:* sentic-etum; senticose.

**senticos**—See **sent**.

**sentinel**—Fr. *sentinelle*, a watcher. *Ex:* Sentinelia (Por.).

**seors**—L. *seorsus*, separate, apart.

**sep**—L. *sepes*, genit. *saepis* = *saepes*, a hedge, fence. *Ex:* sepi-cola; sepi-col-ous.

**sepal**—NL. *sepalum*, formed as a rhyming analogue of "petal," *cf.* L. *separ*, different, separate, disjoined. *Ex:* sepal; sepali-fer-ous Sepalo-spyris (Moll.).

**separat**—L. *separatus*, separated. *Ex:* Separato-bombus (Ins.).

**sepedo**—Gr. *sēpedōn*, rottenness, decay. *Ex:* Sepedo-morphus (Ins.); Sepedon (Rept.), (Ins.); Sepedono-philus (Myr.).

**sepia**—Gr. *sēpia* = *sēpias*, genit. *sēpiados*, a cuttle-fish = L. *sepia*; Gr. *sepion*, the bone of the cuttle-fish. *Ex:* sepi-aceus; Sepia (Moll.); Sepia-lites (Moll.); Sepiad-arium (Moll.); Sepio-phora (Moll.); Sepiola (Moll.); Acantho-sepion (Moll.); Belemno-sepia (Moll.).

**sepidi**—Gr. *sēpidion*, dim. of *sēpia*, a cuttle-fish. *Ex:* Sepidi-acis (Ins.); Sepidio-coris (Ins.); Sepidium (Ins.), (Moll.).

**sepiment**—See **sept**.

**sepio**—See **sepia**.

**sepium**—L. *sepium*, genit. pl. of *sepes*, a hedge, enclosure.

**seps**—Gr. *sēps*, genit. *sēpos*, a putrefying sore; also a kind of lizard; *sēpsis*, putrefaction. *Ex:* Seps (Rept.); Seps-idae (Rept.); Seps-ophis (Rept.); Sepsi-soma (Ins.); Sepsis (Ins.).

**sept**—1. L. *septum*, pl. *septa*, a hedge or enclosure, a partition; *septuosus*, obscure < *sepio*, to hedge in; *sepimentum*, a partition, a division. *Ex:* sept-al; Sept-oria*, see -torius; septa; septa-sternum; septi-fer-ous; Septo-branchium (Ins.); dis-sepiment:    2. Gr. *sēptos*, putrefying; *sēptikos*, septic. *Ex:* Septis (Ins.).

**septentrional**—L. *septentrionalis*, belonging to the north, northern; *septentriones*, the constellation of the dipper.

**septic**—L. *septicus*, causing decay.

**sepult**—L. *sepultus*, buried, sleeping < *sepelio*, to bury, bury in sleep.

**sequax**—L. *sequax*, as adj., following, pursuing; as noun, one who follows, an attendant.

**sequestr**—L. *sequestro*, to surrender, to segregate. *Ex:* Sequestria (Arach.).

**ser**—1. L. *sero*, to plant, put in a row; also to establish, to produce. *Ex:* sere (Ecol.); xero-sere (Ecol.):    2. L. *sero*, to fasten, bind together, join, interweave:    3. L. *serus*, late, late in bearing; *serotinus*, late ripe, backward. *Ex:* serotinus:    4. L. *serum*, whey, the watery parts of things. *Ex:* ser-osa; ser-osi-ty; ser-ous: 5. Gr. *seris*, endive. *Ex:* Hali-ser-ites*. See also seri.

**serang**—Gr. *sēranx*, genit. *sērangos*, a hollow in a rock, cave; *sērangōdēs*, full of caverns, porous. *Ex:* Serangi-um (Ins.); Serangodes (Myr.).

**seranx**—See **serang**.

**serap**—1. Gr. *Serapis*, deity derived from the Egyptian Apis, an ox worshiped as a god. *Ex:* Serap-ista (Ins.); Serapis (Ins.):    2. Gr. *serapias*, genit. *serapiados*, an orchidaceous plant. *Ex:* Serapias*.

**serenan**—NL. *serenanus*, a name made to honor Sereno Watson, American botanist.

**serescen**—1. L. *serescens*, genit. *serescentis*, growing dry < *seresco*, to grow dry:    2. L. *serescens*, turning to whey < *seresco*, to turn to whey.

**serg**—Fr. *serge*, silken stuff < L. *serica*, silken. *Ex:* Sergi-olus (Arach.).

**sergest**—L. *Sergestus* < Gr. *Sergestēs*, companion of Aeneas. *Ex:* Sergestes (Crust.).

**seri**—Gr. *Sēr*, the Seres, an Indian people from whom the first silk came > L. *ser*, genit. *seris*, silk; Gr. *sērikon*, silk; *sērikos*, silken; L. *sericatus*, dressed in silk; *sericeus*, pertaining to silk. *Ex:* seri-culture; seri-fic; seric-in-ous; Seric-ulus (Av.); Serico-spilus (Ins.); Serico-stoma (Ins.); Gastro-sericus (Ins.); Lopho-serinae (Coel.). See also series.

**seriat**—ML. *seriatus*, pp. of *serio*, to arrange in series; *seriatim*, in regular order; L. *series*, a row. *Ex:* seriat-ions; seriati-ly; seriatim; Seriato-pora (Coel.).

**seric**—See **seri**.

**serid**—Gr. *seris*, genit. *seridos*, endive or lettuce. *Ex:* Micro-seris*.

**series**—L. *series*, a row of things, a series. *Ex:* seri-al; series.

**serin**—NL. *serinus*, of or pertaining to the small fringillid bird known as the *serin*, the Zeesig of the Germans; Fr. *serin*. Kock (1816) used the specific name *serin* as the basis of the generic name *Serinus*.

**seriol**—NL. *seriola* < It. native name for a fish. *Ex:* Seriol-ella (Pisc.); Seriola (Pisc.).

seriph—Gr. *seriphos*, a kind of wormwood; also a kind of locust. *Ex:* Seripha (Ins.).

seris—See serid.

serosa—NL. *serosa*, a serous membrane < L. *serum*, serum. *Ex:* serosa.

serotin—See ser 3.

serp—NL. *serpis* < Gr. *herpēs*, a serpent. *Ex:* serpis; not Serpo-phaga (Av.), see serph.

serpent—L. *serpens*, genit. *serpentis*, a serpent. *Ex:* Serpent-arius (Av.); Serpenti-cola (Arach.).

serph—Gr. *serphos*, a gnat, mosquito. *Ex:* Serpho-phaga (Av.) = Serpo-phaga (Av.).

serpul—L. *serpula*, a little snake. *Ex:* Serpulopsis (Ann.); Serpul-orbis (Moll.); Serpula (Ann.); Serpulo-spira (Moll.).

serpyll—L. *serpyllum* = *serpillum*, thyme. *Ex:* serpylli-folia.

serr—L. *serra*, dim. *serrula*, a saw; *serratus*, saw-shaped, serrated; NL. *serrator*, one who saws. *Ex:* Serra-salmus (Pisc.); serrat-ulus; serrate; serrati-folius; Serrator (Av.); Serri-cornia (Ins.); Serrula (Moll.).

serrag—L. *serrago*, genit. *serraginis*, sawdust.

serran—LL. *serranus*, native name of a fish along many coasts of the Mediterranean < L. *serra*, "on account of the dentition of their preopercle." *Ex:* Serran-ichthys (Pisc.); Serranidae (Pisc.); Serranus (Pisc.).

serrat—See serr.

sert—L. *serta*, dim. *sertula*, a garland < *sero*, to interweave. *Ex:* Sert-aria (Coel.); Sertul-aria (Coel.); sertum.

sertul—See sert.

serumal—NL. *serumal*, pertaining to serum < L. *sera*, whey + -al.

serv—L. *servus*, a slave.

serval—L. *cervus*, a stag > Fr. *cerval* > NL. *serval. Ex:* Serval (Mam.); Serval-ina (Mam.).

ses—ML. *sesia* < Gr. *sēs*, genit. *seos*, a moth = the later form, *sētos. Ex:* Ses (Ins.); Ses-aspis (Ins.); Sesi-ura (Ins.); Sesia (Ins.), (Av.); Scsio-phaga (Ins.); Seto-gyps (Av.); Seto-phaga (Av.).

sesam—Gr. *sēsamē*, sesame; *sēsamon*, the seed or fruit of the sesame plant. *Ex:* sesam-oid; Sesam-odon (Rept.); sesame*; Sesamum*.

sesban—NL. *sesbania* < Arabic, *seiseban*; Pers. *sisaban*, a kind of tree. *Ex:* Sesbania*.

sesel—Gr. and L. *seselis*, the meadow saxifrage.

sesqui- —L. *sesqui-*, prefix meaning one and one-half. *Ex:* sesqui-alter; sesqui-ocell-us; Sesquipes (Rept.); sesqui-terti-ous fascia.

sensibil—L. *sensibilis*, having feeling, able to feel.

sessil—L. *sessilis*, of or belonging to sitting; having a broad foot; of plants, low, dwarf. *Ex:* sessile; Sessili-ocles (Crust.).

sessor—L. *sessor*, genit. *sessoris*, one that is seated  a resident.

sestro—Gr. *sēstron*, a sieve. *Ex:* Sestro-dictyon (Por.); Sestro-podium (Prot.); Astro-sestron (Prot.).

sesuvi—L. *Sesuvium*, land of the Sesuvii, a Gallic tribe. *Ex:* Sesuvium*.

set—L. *seta* = *saeta*, a bristle. *Ex:* Set-aria*; Set-onix (Mam.); seta; Seta-cera (Ins.); seti-gera; seti-parous; seto-branch; Seto-gyps (Av.); Cata-setum*; Equi-setum*; Tri-setum*. See also ses.

seth—Gr. *sēthos*, a sieve, a sifter < *sēthō*, to sift. *Ex:* Setho-discus (Prot.); Setho-perdium (Prot.); not Sethia*, which is from a personal name.

seto—See ses.

setodoc—Gr. *sētodokis*, a butterfly. *Ex:* Setodocis (Ins.).

setos—L. *saetosus* = *setosus*, full of hairs, hairy. *Ex:* Setos-ella (Bry.); Setos-ura (Av.).

setul—L. *saetula*, a small bristle. *Ex:* setul-ose.

sev—L. *sevum* = *sebum*, suet, grease.

sever—L. *severus*, stern, serious.

sex—1. L. *sex-*, prefix meaning six or six-fold. *Ex:* Sex-arthrus (Ins.); sex-locular; sex-valent: 2. Fr. *sexe* < L. *sexum*, acc. of *sexus*, sex; prob. lit. a division < *seco*, to cut; *sexualis*, pertaining to sex. *Ex:* sexual.

si—Gr. *seiō*, to wave, move to and fro. *Ex:* Si-urus (Av.); sio-trop-ism; A-sio-gonum (Ins.).

siagon—Gr. *siagōn*, genit. *siagonos*, the jaw, cheek-bone; *saigonitēs*, the jaw musclc. *Ex:* Siagon-odon (Pisc.); Siagona (Ins.); Siagon-ium (Ins.); Siagono-phorus (Prot.).

sial—1. Gr. *sialis*, a kind of bird. *Ex:* Sialia (Av.); Sialis (Ins.): 2. Gr. *sialos*, fat, grease: 3. Gr. *sialon*, saliva, spittle. *Ex:* sial-oid; sialo-gogue (Med.); ?Sialo-scarta (Ins.).

sibi—L. *sibi*, self. *Ex:* sibi-conjugate.

sibil—L. *sibilo*, to hiss, whistle, ppr. *sibilans*, genit. *sibilantis*, whistling; pp. *sibilatus*, whistling; *sibilator*, fem. *sibilatrix*, a whistler. *Ex:* sibil-ous; sibilan-ce; sibilant; sibilate; Sibilatrix (Av.).

sibilant—See sibil.

sibin—L. *sybina* = *sibina*, a kind of spear < Gr. *sibynē*, a spear.

sibyll—Gr. *Sibylla*, a sibyl, a prophetess. *Ex:* Sibyll-ina (Ins.); Sibylla (Ins.); Sibyllo-nautilus (Moll.).

sibyn—Gr. *sibynē* = *sibynēs*, a spear. *Ex:* Sibinia (Ins.); Sibyn-ophis (Rept.); Sibynes (Ins.); Sibyno-morphus (Rept.); Sibynon (Rept.).

sic—L. *sica*, dim. *sicula*, a dagger. *Ex:* Sicul-odes (Ins.); sicula; Siculi-fer (Moll.); Nebulo-sicus (Echin.).

sicari—Gr. *sikarios*, an assassin. *Ex:* Sicaria (Arach.); Sicario-ides (Arach.).

sicc—L. *siccus*, dry; *siccitas*, dryness, drought; *siccificus*, drying; *siccaneus*, dry; NL. *sicca-*

*tus*, dried <L. *sicco*, to dry. *Ex:* siccaneous; siccit-y; ex-siccata.

**sicul**—1. L. *Siculus*, of Sicily: 2. L. *sicula*, a small dagger.

**sicy**—1. Gr. *sikyos*, dim. *sikydion*, the wild cucumber or gourd. *Ex:* Sicydium (Pisc.); Sicyocrinus (Achin.); Sicyos*: 2. L. *Sicyon*, a mythological name. *Ex:* Sicyon (Crust.).

**sicyas**—Gr. *sikyasis*, a cupping. *Ex:* Sicyasus (Pisc.).

**sid**—1. Gr. *Sida*, a nymph. *Ex:* Sida (Crust.): 2. Gr. *sidē*, a pomegranate tree; also a water-plant, perhaps the water lily. *Ex:* Sidalcea*; Sida*.

**sider**—1. Gr. *sidēros*, iron or things made of it; *sidēritēs*, name of various plants including the ironwort. *Ex:* Sideri-aster (Echin.); Sideritis*; Sidero-therium (Mam.); Sidero-xylon*; Metrosideros*: 2. L. *sidus*, genit. *sideris*, a star. *Ex:* Sider-actis (Coel.); Sider-ina (Prot.).

**sig**—1. Gr. *sigē*, silence; *sigaō*, to keep silence, to be silent; *Sigalion*, Egyptian God of Silence; *sigalos*, disposed to silence; *sigēros*, silent. *Ex:* Sig-erpes (Ins.); Siga (Ins.); Sigalion (Ann.); Sigara (Ins.); Sigelus (Av.); Sigo-desmus (Mys.); sigo-lutes; Codo-siga (Av.); Diplo-siga (Prot.): 2. Gr. *sigaloeis*, fem. *sigaloessa*, glossy, glittering, splendid. *Ex:* Sigalo-cera (Moll.); Sigaloessa (Arach.).

**sigal**—See **sig**.

**sigan**—NL. *siganus* <Ar. *sijān*, a kind of fish. *Ex:* Siganus (Pisc.).

**sigaret**—NL. *sigaret* <*sagaret*, a native mollusk name latinized by Adanson. *Ex:* Sigaretotrema (Moll.); Sigaretus (Moll.).

**sigel**—See **sig**.

**siger**—See **sig**.

**sigill**—L. *sigilla*, a seal, mark, sign; *sigillatus*, adorned with images or figures; *sigillaria*, little images presented on the last days of the feast of Saturnalia. *Ex:* Sigill-ina (Tun.); Sigilla (Ins.); Sigillaria*; sigillate.

**sigl**—Gr. *siglos*, a sheckel; also an earring. *Ex:* Siglo-phora (Ins.).

**sigm**—1. Gr. Σ, the letter *sigma*, genit. *sigmatos*, equivalent to the English S, hence meaning curved like the letter sigma. *Ex:* Sigm-istes (Pisc.); Sigm-odon (Mam.); sigm-oid flexure; Sigmato-gyne*; Sigmato-salix*; Sigmogomphius (Mam.); Sigmo-mys (Mam.): 2. Gr. *sigmos*, a hissing such as is made by tortoises.

**sigmat**—See **sigm 1**.

**sign**—L. *signo*, to mark, designate; *signum*, a mark, sign; *signatus*, sealed, guarded, preserved, marked; *signatura*, fem. sing. of fut. part. of *signo*. *Ex:* Signi-phora (Ins.); Signosoma (Ins.); Doctrine of Signatures.

**signat**—See **sign**.

**signet**—See **cygn**.

**sigr**—Gr. *sigrai*, a kind of wild swine. *Ex:* Echino-sigra (Echin.).

**sik**—Jap. *sika*, dim. *sikaillus*, a kind of Japanese deer. *Ex:* Sik-elaphus (Mam.); Sika (Mam.); Sikaillus (Mam.).

**sil**—L. *silus* = Gr. *silos*, snub-nosed. *Ex:* Sil-opa (Ins.); Silis (Ins.); Silo (Ins.); Silo-trachelus (Ins.); Silusa (Ins.).

**sila**—L. *silaus*, a kind of parsley. *Ex:* silaifolium. See silaus.

**silaus**—L. *silaus*, the pepper saxifrage. *Ex:* Silaus*.

**silen**—1. L. *silenus* <Gr. *sialon*, spittle, foam. *Ex:* Silene*: 2. Gr. *Seilēnos* = L. *Silenus*, leader of the satyrs, constant attendant of Bacchus, represented as drunken, bald-headed, flat-nosed and foam-covered. *Ex:* Silenus (Mam.).

**silent**—L. *silens*, genit. *silentis*, still, calm <*sileo*, to be silent.

**siler**—L. *siler*, genit. *sileris*, a kind of willow. *Ex:* Siler*.

**silib**—See **silyb**.

**silic**—L. *silex*, genit. *silicis*, flint; *siliceus*, of flint or lime-stone. *Ex:* silici-ous; Silici-fibris (Por.); Silici-spongiae (Por.); not silicle, see siliqu.

**silicul**—See **siliqu**.

**siliqu**—L. *siliqua*, dim. *silicula*, a pod or husk. *Ex:* silicle; Silicul-aria (Coel.); silicul-ose; Siliqu-aria (Moll.); Siliqua (Moll.); Siliqu-ose*; silique.

**silph**—1. Gr. *silphē*, a grub; an insect emitting an unpleasant odor. *Ex:* Silph-idium (Ins.); Silphites (Ins.); Silpha (Ins.); Silpho-telus (Ins.): 2. Gr. *silphion*, a plant, the juice of which was used in medicine or for food. *Ex:* Silphium*.

**silub**—Gr. *sillybos* = *silybon*, a kind of thistle. *Ex:* Silubo-lepis (Rept.); Silubo-ura (Rept.).

**silur**—1. L. *silurus*, a kind of river fish <Gr. *silouros*. *Ex:* Silur-an-odon (Pisc.); Silur-ina (Pisc.); Silur-odes (Pisc.); Silurus (Pisc.): 2. L. *Silures*, an old people of Britain who inhabited Wales; the combining form *silur*- when used in forming generic names refers to Silurian Age. *Ex:* Silur-aster (Echin.); Silurian; Siluri-spongia (Por.).

**silus**—See **sil**.

**silv**—L. *silva*, a wood, forest; *Silvanus*, deity of the woods; *silvaticus*, belonging to woods or trees; *silvestris*, of or belonging to a wood. *Ex:* Silvano-phloeus (Ins.); Silvestr-ella (Ins.); silvi-culture; Silvi-tettix (Ins.); Silvius (Ins.); silv-ose.

**silyb**—L. *silybum* <Gr. *sillybos* = *silybon*, a kind of thistle with edible stems. *Ex:* Silibum*.

**sim**—1. Gr. *simos*, flat-nosed, snub-nosed; also concave; *simotēs*, snubbiness. *Ex:* Sim-orhynchus (Av.); Sima (Ins.); Simo-cyon (Mam.); Simo-bison (Mam.); Simotes (Mam.); not

Simi-meryx which in part involves the place-name Simi, of California; Ryncho-simus (Av.): **2.** L. *simia*, dim. *simiolus*, ape. *Ex:* Simi-idae (Mam.); Simia (Mam.); simi-an; Simo-lestes (Rept.); **3.** L. *simus*, with flat nose.

**simaba**—Guianese, *simaba*, the cedron tree. *Ex:* Simaba*.

**simarub**—NL. *simaruba* < supposed native Caribbean name for the bitter-wood. *Ex:* Simaruba*; Simarub-aceae*.

**simbl**—Gr. *simblos*, a bee-hive, a store or hoard. *Ex:* Simble-philus (Ins.); Simblum*.

**simethis**—L. *Symaethis*, Sicilian nymph. *Ex:* Simethis*.

**simil**—L. *similis*, like. *Ex:* simil-ize; Simil-onyx (Av.); simili-florus; Simili-pecten (Moll.); as-simil-ate.

**simot**—See sim.

**simoter**—NL. *simoterus*, snub-nosed < Gr. *simos*, snub-nosed.

**simpl**—L. *simplus*, simple.

**simplex**—See simplic.

**simplic**—L. *simplex*, genit. *simplicis*, simple. *Ex:* Simplic-ella (Por.); Simplici-dentata (Mam.); Simplico-dolium (Moll.).

**simpul**—L. *simpulum*, a small ladle. *Ex:* Simpul-inus (Moll.); Simpulum (Moll.).

**simul**—L. *simulo*, to make like, ppr. *simulans*, genit. *simulantis*, imitative. *Ex:* Simul-idium (Ins.); Simul-opsis (Moll.); Simuli um (Ins.).

**sin**—**1.** Gr. *sinas*, genit. *sinados*, destructive; *sinos*, damage, hurt; *sinis*, genit. *sinidos*, a destructive one, a plunderer; *sinaros*, hurt, damaged. *Ex:* Sin-anther-ina (Rot.); Sino-dendron (Ins.); Sino-xylon (Ins.); Carpo-sina (Ins.); Hyel-sinus (Ins.): **2.** L. *Sinae*, an Oriental people mentioned by Ptolemy, now the Chinese > NL. *sinensis*, of China. *Ex:* Sin-anthropus (Mam.); Sin-ictis (Mam.); Sini-gobio (Pisc.); Sino-blatta (Ins.); Sino-castor (Mam.); Sino-menium*: **3.** L. *sinus*, a bay of the sea, a curve; *sinuose*, intricately < *sinuo*, to bend, pp. *sinuatus*, bent, curved. *Ex:* Sin-ella (Ins.); sinu-palliate; sinus; sinus-oid; Epi-sinus (Arach.). See also syn.

**sinad**—See sin 1.

**sinap**—L. *sinapis*, mustard; Gr. *sinapismos*, the use of a mustard plaster. *Ex:* Sinapis*; sinapism (Med.).

**sinas**—See sin 1.

**sincer**—L. *sincerus*, sound, clean, natural, without mutilation.

**sincipit**—L. *sinciput*, genit. *sincipitis*, the fore part of the head. *Ex:* sincipit-al; sinciput.

**sinciput**—See sincipit.

**sindron**—Gr. *sindrōn*, genit. *sindrōnos*, mischievous.

**singul**—L. *singulus*, separate; *singularis*, singular, solitary. *Ex:* singuli-form.

**sinic**—NL. *sinicus*, Chinese < L. *Sinae*, the Chinese.

**sinid**—See sin 1.

**sinis**—Gr. *Sinis*, mythical robber of Corinth. *Ex:* Sinus (Arach.).

**sinistr**—L. *sinister*, -*tra*, -*trum*, left, on the left. *Ex:* sinistr-al; Sinistr-aspis (Ins.); sinistro-caudate; Sinistro-porus (Platy.).

**sinotes**—See sinotic.

**sinotic**—Gr. *sinōtikos*, mischievous; *sinotēs*, faultiness.

**sint**—Gr. *sintōr*, genit. *sintoros* = *sintēs*, tearing, ravenous; also a thief. *Ex:* Sinto-cephalus (Rept.); Sintor (Ins.); Sintor-ops (Ins.).

**sintoc**—East Ind. *sintoq*, a kind of cinnamon.

**sinu**—See sin 3.

**sinuat**—See sin 3.

**sio**—See si.

**sion**—**1.** Heb. Mount Zion < *tsion*, exposed on a sunny place. *Ex:* Siona (Ins.) "on account of the barrenness of its markings": **2.** Gr. *sion*, genit. *siontos*, a marsh plant.

**siont**—See sion 2.

**siopel**—Gr. *siōpēlos*, silent, quiet.

**sipal**—Gr. *sipalos*, deformed; also a defect. *Ex:* Sipalo-cyon (Mam.); Sipalo-lasma (Arach.); Sipalus (Mam.).

**siph**—See sipho.

**siphl**—**1.** Gr. *siphlos* = *siphnos*, crippled, maimed. *Ex:* Siphl-urus (Ins.): **2.** Gr. *siphneus*, a mole < *siphnos*, maimed, purblind (because of its supposed blindness). *Ex:* Siphneus (Mam.).

**siphn**—See siphl.

**sipho**—L. *sipho*, a siphon < Gr. *siphōn*, genit. *siphōnos*, a tube, siphon; *siphōnizō*, to draw off with a pipe. *Ex:* Siph a-gonus (Pisc.); Siph-opsis (Coel.); Sipho-cypraea (Moll.); Sipho-stoma (Pisc.); Siphon-a-ptera (Ins.); Siphonia (Coel.); Siphono-cetus (Mam.); siphono-glyph; Siphono-phora (Coel.); Siphoniz-antia (Myr.).

**siphon**—See sipho.

**sipuncul**—L. *sipunculus* = *siphunculus*, a little siphon < *siphon*, a tube. *Ex:* sipuncle; Sipunculus (Ann.).

**sir**—Gr. *siros*, a pit for keeping corn in, a pit-fall, a trench. *Ex:* Sir-odes (Ins.); Siro (Arach.); not siro-gonium nor Siro-siphon*, see seir.

**sirden**—*syrdēn*, dragging, in a long line. *Ex:* Sirdenus (Ins.).

**sired**—Gr. *seirēdōn*, a siren. *Ex:* Siredon (Amph).

**siren**—L. *siren*, a Siren, one of the mythical birds with faces of virgins who enticed ashore those who were sailing by and then killed them; a mermaid < Gr. *seirēn*, a siren; also a kind of wasp or bee; Late Gr. *seirēdōn*, a siren > LL. *sirex*, genit. *siricis*, a mermaid, siren. *Ex:* Siredon (Amph.); Sirenia (Mam.); Sireno-pyga (Ins.); Sirex (Ins.); Siric-idae (Ins.); Lepido-siren (Pisc.).

sirex—See **siren**.

siric—See **siren**.

sirtal—NL. *sirtalis*, like a garter.

siryct—Gr. *syrictēr*, the male crane. *Ex:* Syricter (Av.).

-sis—Gr. *-sis*, suffix denoting act of. *Ex:* diagno-sis; hypothe-sis; synthe-sis.

sisenn—L. *Sisenna*, Roman surname. *Ex:* Sisenna (Arach.); Sisennes (Ins.).

sison—Gr. *sisōn*, the honewort. *Ex:* Sison\*. Per-haps < Celtic *sisun*, a running stream since "some of the plants formerly contained in this genus were inhabitants of running streams."— Don.

sistr—L. *sistrum* < Gr. *seistron*, a rattle < *seiō*, to shake. *Ex:* Sistr-urus (Rept.); Sistrum (Moll.).

sisymbr—L. *sisymbrium*, a sacred herb < Gr. *sisymbron* = *sisymbrion*, name of a sweet-smell-ing plant, perhaps < *sisybos* = *thysanos*, a fringe. *Ex:* Sisymbrium\*.

sisyphus—Gr. *Sisiphos*, cunning robber of Cor-inth. *Ex:* Sisyphus (Arach.).

sisyr—Gr. *sisyra* = *sisyrna*, a garment of skin. *Ex:* Sisyr-opa (Ins.); Sisyra (Ins.); Sisyra-cera (Ins.); Sisyro-stolus (Ins.); Sisyrno-phorus (Ins.).

sisyrinch—Gr. *sisyrinchion*, a bulbous plant of the Iris kind. *Ex:* Sisyrinchium\*.

sisyrn—See **sisyr**.

sit—Gr. *sitos* = *sition*, grain for food, grain; *sitarion*, a little corn or food. *Ex:* Sitarea (Ins.); Sitarion (Ins.); Sitaris (Ins.); Sitaro-morpha (Ins.); Sito-mys (Mam.); Sito-philus (Ins.); Sito-troga (Ins.); Oligo-sita (Ins.); para-site.

sita—*Sita*, wife of the Hindu hero-god, Rama. *Ex:* Sit-ana (Rept.).

sitar—See **sit**.

sitarion—See **sit**.

siteut—Gr. *siteutēs*, one who feeds cattle; *siteutos*, fed up, fatted. *Ex:* Siteutes (Ins.); Siteytes (Ins.).

sitiens—L. *sitiens*, thirsty, dry.

siton—Gr. *sitōn*, a corn-field; also a granary; *sitōnēs*, a corn merchant < *sitos*, corn, grain. *Ex:* Sitona (Ins.); Sitones (Ins.).

sitt—Gr. *sittē*, the nuthatch. *Ex:* Sitt-idae (Av.); Sitta (Av.); Sitti-parus (Av.); Sitto-somus (Av.) = Sitta-somus (Av.).

situl—L. *situla*, an urn. *Ex:* Situl-aspis (Ins.); Situl-ites (Echin.).

situs—L. *situs*, situation; Low L. *situatus*, pp. of *situo*, to place.

sium—Gr. *sion*, a kind of marsh or meadow plant. *Ex:* Sium\*.

siva—Skr. *Civa*, Hindu god of destruction and restoration. *Ex:* Siva-meles (Mam.); Siva-meryx (Mam.); Siva-therium (Mam.).

sival—NL. *Sival* < *Siwalik* Hills of India. *Ex:* Sival-hippus (Mam.).

skaph—Gr. *skaphē*, a skiff, anything hollow, a digging. *Ex:* Scaph-arca (Moll.); scapha; Scapho-poda (Moll.); skapho-plankton.

skelet—Gr. *skeleton*, a dried body, mummy < *skeletos*, dried up < *skellō*, to dry up, to parch. *Ex:* skelet-al; skeleton; endo-skeleton.

skemmat—Gr. *skemma*, genit. *skemmatos*, a sub-ject for speculation. *Ex:* Skemmato-pyge (Tri.).

sken—Gr. *skēnē*, dim. *skēnidion*, a tent, a cov-ered place; *skēnōtēs*, a comrade in a tent. *Ex:* Sken-ella (Moll.); Skenidium (Moll.); Skeno-toka (Coel.); not Skenea (Moll.), named after Dr. Skene, of Aberdeen, contemporary of Lin-naeus.

skeo—NL. *skeo* < Gr. *skaios*, to the left, on the left. *Ex:* skeo-cyt-osis.

skimmia—Jap. *skimmi*, a hurtful fruit. *Ex:* Skimmia\*.

skua—Faeroëse *skua*, a name for some bird.

smaragd—Gr. *smaragdos*, a precious stone of a light green color, probably the emerald. *Ex:* Smaragd-esthes (Ins.); Smaragd-itis (Av.); Smaragdo-chroa (Av.).

smaris—Gr. *smaris*, a small sea fish. *Ex:* Smaris (Pisc.); Smaris (Arach.).

smect—Gr. *smēktēs*, one who rubs, cleanses, washes. *Ex:* Hydro-smecta (Ins.).

smegm—Gr. *smēgma*, genit. *smēgmatos*, soap. *Ex:* smegma, smegmat-ic.

smerd—L. *Smerdis*, son of Cyrus. *Ex:* Smerdis (Pisc.).

smerdale—Gr. *smerdaleos*, terrible.

smering—See **mering**.

smerinth—Gr. *smērinthos* = *mērinthos*, a cord, string. *Ex:* Smerinth-ulus (Ins.); Smerintho-thrips (Ins.); Smerinthus (Ins.).

smerinx—See **mering**.

smet—Gr. *smaō*, 3rd. sing. pass. *smētai*, to be-smear, spread over. *Ex:* Mela-smetus (Ins.).

smicr—Gr. *smikros*, Ionic for *mikros*, small. *Ex:* Smicr-ornis (Av.); Smicra (Ins.); Smicro-lusus (Ins.).

smil—1. Gr. *smilē*, dim. *smilion*, a carving knive, chisel. *Ex:* Smil-erpeton (Amph.); Smil-odon (Mam.); Smile-ceras (Ins.); Smili-ornis (Av.); Smilio-pus (Ins.); Smilium (Crust.); Smilo-dectes (Mam.): 2. Gr. *smilos* = *milos*, the yew.

smilac—Gr. *smilax*, genit. *smilakos*, the yew; also a bind-weed. *Ex:* Smilac-ina\*; Smilax\*.

smilax—See **smilac**.

sminth—Gr. *sminthos*, old Cretan word for a field-mouse. *Ex:* Sminth-opsis (Mam.); Sminth-urus (Ins.); Smintho-coris (Ins.); Sminthus (Mam.); Aco-sminthus (Mam.).

**smodicum**—See **smoding.**

**smoding**—Gr. *smōdix*, genit. *smōdingos*, a swollen bruise. *Ex:* Smodicum (Ins.); Smoding-ium*; Smodingo-ceramus (Moll.).

**smodix**—See **smoding.**

**smyrn**—Gr. *smyrnion*, a kind of plant, horse-parsley<*smyrna*, myrrh. *Ex:* Smyrnium*.

**sob**—Gr. *sobeō*, to drive away. *Ex:* Myio-sobus (Av.).

**sobar**—Gr. *sobaros*, arrogant, rapid, violent. *Ex:* Sobaro-cephala (Ins.); Sobarus (Ins.).

**sobol**—L. *soboles* = *suboles*, a sprout, offshoot. *Ex:* sobol; soboli-fer-ous.

**sobri**—L. *sobrius*, moderate, sober, clever.

**sobrin**—L. *sobrinus*, masc.; *sobrina*, fem., a cousin. *Ex:* ab-sobrinus; con-sobrinus.

**soccat**—L. *soccatus*, wearing a sock or slipper.

**soci**—L. *socius*, a companion; *socialis*, of or belonging to companionship; *societas*, fellowship, union<*socio*, to join. *Ex:* social; societ-y; socio-logy; con-socies.

**sodal**—L. *sodalis*, a comrade; *sodalitas*, companionship; Sodali-scala (Moll.).

**sol**—L. *sol*, genit. *solis*, the sun; *solaris*, of or belonging to the sun; *solatus*, sun-burned; *solarium*, a sun dial. *Ex:* Sol-aster (Echin.); Sol-puga (Arth.); Solar-ites (Moll.); solar plexus; Solari-cllum (Moll.); Solario-conulus (Moll.); Solarium (Moll.); not Solandra* named after Charles Solander of Sweden; Solo-mys (Mam.).

**solan**—1. L. *solor*, to comfort, to soothe>L. *solanum*, a plant, the nightshade. *Ex:* Solan-aceae*; Solano-phagus (Ins.); Solano-phila (Ins.); Solanum*: 2. L. *solanus*, a hot easterly wind<*sol*, sun: 3. *Solano*, Suisune Indian Chief. *Ex:* Solanoa*.

**solar**—See **sol.**

**soldan**—NL. *soldana*<It. *soldo*, a coin. *Ex:* Soldan-ella*.

**sole**—L. *solea*, a sandal. *Ex:* Solea (Pisc.); soleae-form; solei-form; soleus. See also solen.

**solen**—Gr. *sōlēn*, genit. *sōlēnos*, channel, pipe, a grooved tile; also the penis, the cavity of the spine. *Ex:* Sole-mya = Soleno-mya (Moll.); Sole-gnathus (Pisc.); Solen (Moll.); Solen-aria (Moll.); Solen-iscus (Por.); Solen-odon (Pisc.); Soleni-curtus (Moll.); soleno-cyte; Dis-solena*; Leuco-solen-ia (Por.); Strepto-solen*; typhlo-sole.

**solid**—L. *solidus*, sound, firm, genuine; *solidum*, soundly, thoroughly; *solidatus*, made firm. *Ex:* Solid-ulus (Moll.); Solido-pus (Rept.).

**solidag**—ML. *solidago*, golden-rod<L. *solido*, to put together, to make firm. *Ex:* Solidago*.

**solium**—L. *solium*, a seat, throne.

**solivag**—L. *solivagus*, wandering alone. *Ex:* Solivaga (Moll.).

**sollicit**—L. *sollicito*, to vex, disturb; ppr. *sollicitans*, genit. *sollicitantis*, disturbed, agitated.

**solpug**—See **sol.**

**solstitial**—L. *solstitialis*, belonging to mid-summer<*solsticium*, summer-time.

**solus**—L. *solus*, single.

**solut**—L. *solutus*, free, loose; *solutum*, thin, diffused. *Ex:* solut-ion; Soluti-scala (Moll.).

**som**—Gr. *sōma*, genit. *sōmatos*, the body; *sōmatikos*, of or for the body. *Ex:* som-ite; Somat-eria (Av.); somatic; Somat-ium (Ins.); somato-pleure; Somi-leptus (Pisc.); Somo-platus (Ins.); Tamio-soma (Echin.).

**somat**—See **som.**

**-some**—Eng. *-some*, adj. suffix meaning like, same, and serving to intensify the quality of the first element of the compound. *Ex:* blithe-some; tooth-some.

**somn**—L. *somnus*, sleep; *somnulentus* = *somnolentus*, sleepy, drowsy; *somnolentia*, sleepiness; *somniosus*, sleepy, overcome by sleep; *somnialis*, dream-bringing<*somnio*, to dream. *Ex:* Somniosus (Elasm.); somnol-ence.

**somph**—Gr. *somphos*, spongy, porous; *somphōdēs*, of spongy porous nature. *Ex:* Sompho-spongia (Por.).

**sonans**—L. *sonans*, resounding, making a noise.

**sonch**—Gr. *sonchos*, the sow thistle. *Ex:* Son chus*.

**sonit**—L. *sonitus*, noise, sound.

**sonivi**—L. *sonivius*, noisy<*sonus*, sound.

**soph**—Gr. *sophia*, cleverness, wisdom; *sophos*, clever, skillful; *sophistēs*, a wise man, an expert. *Ex:* Sophia*; Sophista (Ins.).

**sophod**—Gr. *sophōdēs*, of wise character.

**sophora**—Ar. *sophera*, a papilionaceous tree. *Ex:* Sophora*.

**sophro**—Gr. *sōphrōn*, sensible; *sōphroneō*, to be sound of mind, to show self-control. *Ex:* Sophr-ops (Ins.); Sophro-rhinus (Ins.); Sophron (Ins.); Sophron-ica (Ins.); Sophrono-merus (Ins.).

**sopor**—L. *sopor*, genit. *soporis*, sleep; *soporator*, a sleeper<*sopio*, to sleep. *Ex:* sopori-fer-ous; sopori-fic.

**sor**—1. Gr. *sōros*, dim. *sōridion*, a heap, mound; *sōrēdon*, in heaps; *sōreitēs* = *sōritēs*, heaped up; *sōreutos*, heaped up; *sōreuma*, a heap, pile; *sōreusis*, an accumulation. *Ex:* Sor-ictis (Mam.); sorema; Soreuma (Prot.); Soreuto-neura (Ins.); Sorites (Prot.); Soro-sphaera (Prot.); sorus: 2. Gr. *soros*, a vessel for holding anything, a coffin.

**sorb**—1. L. *sorbeo*, to absorb, ppr. *sorbens*, genit. *sorbentis*, absorbing. *Ex:* sorb-ile; sorbe-facient; sorbent; ab-sorb; ab-sorp-tion; Sangui-sorba*: 2. L. *sorbus*, the mountain ash. *Ex:* Sorbus*.

**sorbillan**—L. *sorbillans*, genit. *sorbillantis*, sipping, taking sparingly, ppr. of *sorbillo*, to take drop by drop, to sip.

**sordens**—L. *sordens*, becoming dirty, of no account <*sordeo*, to be dirty.

**sordescen**—L. *sordescens*, genit. *sordescentis*, becoming dirty, vile, mean, ppr. of *sordesco*, to soil.

**sordid**—L. *sordidus*, dirty, foul.

**sored**—ML. dim. *soredium*<Gr. *sōros*, dim. *sōridion*, a heap. *Ex:* soredi-ferous.

**sorem**—See sor.

**soreum**—See sor.

**soreus**—See sor.

**sorex**—See soric.

**sorghum**—Low L. *surgum*, great millet. *Ex:* Sorghum*.

**soric**—L. *sorex*, genit. *soricis*, the shrew-mouse< Gr. *hyrax*, a shrew-mouse; L. *soricinus*, of a shrew. *Ex:* Sorex (Mam.); Soric-iscus (Mam.); Soric-ulus (Mam.); Sorici-dens (Pisc.); Neosorex (Mam.).

**sorindeia**—NL. *sorindeia*, a plant name. *Ex:* Sorindeia*.

**sorites**—See sor 1.

**-sorius**—L. *-sorius*, *-a*, *-um*, adj. suffix meaning belonging to. See also -arius and -torius.

**sorori**—L. *sororius*, of or belonging to a sister; *soror*, a sister.

**sos**—Gr. *sōs*, neut. *sōn*, sure, safe, sound. *Ex:* Sos-ylus (Ins.), see hyl; Sos-yl-opsis (Ins.).

**-sote**—NL. *sote*<Gr. *sōzō*, to preserve, keep. *Ex:* creo-sote (Gr. *kreas*, flesh). See soz.

**soter**—Gr. *sōtēr*, genit. *sōtēros*, a watcher, also a deliverer, preserver; *sōtēria*, safety. *Ex:* Soter (Ins.); Dendro-soter (Ins.); Hypo-soter (Ins.).

**soz**—Gr. *sōzō*, to save, preserve, keep. *Ex:* soz-in; soz-albumin.

**spadic**—Gr. *spadix*, genit. *spadikos*=L. *spadix*, genit. *spadicis*, a palm branch or frond>NL. *spadiceus*, date-brown, nut-brown. *Ex:* spadici-form; spadix; Phyllo-spadix*.

**spadix**—See spadic.

**spadon**—1. L. *spado*, genit. *spadonis*<Gr. *spadōn*, genit. *spadōnos*, one who has no generative powers, an eunuch; an unfruitful or seedless plant. *Ex:* spadon-ism: 2. Gr. *spadōn*, genit. *spadonos*, a convulsion, cramp, a tear, rent. *Ex:* ?Spadono-tettix (Ins.).

**spalac**—Gr. *spalax*, genit. *spalakos*=*aspalax*, a mole. *Ex:* Spalac-odon (Mam.); Spalaco-mys (Mam.): Spalax (Mam.); Aspalo-mys (Mam.).

**spalax**—See spalac.

**span**—Gr. *spanios*, scarce, rare; *spanis*, scarcity; *spanistos*, most scarce. *Ex:* span-anthus; Span-a-gon-icus (Ins.); Span-odontus (Brach.); spani-pelagic; Spani-aster (Echin.); Spaniodon (Pisc.); Spania (Ins.); Spanio-therium (Mam.); Spanista (Ins.); Spano-tecnus (Ins.).

**spar**—Gr. *sparos*, name of some sparoid fish< *spairō*, to gasp. *Ex:* Spar-opsis (Pisc.); Spari-soma (Pisc.); Sparo-soma (Pisc.); Sparo-ides (Pisc.); Sparus (Pisc.).

**sparact**—See sparagm.

**sparagm**—Gr. *sparagmos*=*sparaxis*, a tearing, mangling; *sparagma*, genit. *sparagmatos*, a piece torn off; *sparaktēs*, one who tears to pieces. *Ex:* Sparagm-ites (Rept.); Sparaxis*; Sparact-odon (Pisc.); Sparactus (Ins.); Catosparactes (Av.).

**sparass**—Gr. *sparassō*=Att. *sparattō*, to tear, rend in pieces. *Ex:* Sparassion (Ins.); Sparasso-cynus (Mam.); Sparatto (Ins.); Sparatto-sperma*.

**sparatt**—See sparass.

**sparax**—See sparagm.

**sparg**—Gr. *spargaō*, to be full>*spargōsis*, a swelling. *Ex:* spargosis (Med.).

**spargan**—Gr. *sparganon*, a swaddling band, a ribbon. *Ex:* Spargan-ium*; Spargano-philus (Ann.); Sparganum (Platy.).

**sparn**—Gr. *sparnos*, poet. for *spanios*, rare. *Ex:* Sparn-odus (Pisc.); Sparnus (Ins.).

**spars**—L. *sparsus*, few, scattered. *Ex:* Sparsi-cavea (Bry.); Sparsi-spongia (Por.).

**spart**—1. Gr. *spartos*, sown, scattered; also a shrub known as broom. *Ex:* Sparto-centrum (Moll.); Spartio-myia (Ins.); spiro-spart: 2. Gr. *spartē*, a rope, cord; *sparton*=*spartinē*, dim. *spartion*, a rope, a cable made of broom. *Ex:* Sparte-cerus (Ins.); Spartina*; Sparto-cera (Ins.); Sparty-cerus (Ins.): 3. L. *spartum*= Gr. *spartos*, a kind of plant, Spanish broom (Spartum)>L. *sparteus*, consisting of broom. *Ex:* Spartium*; Sparto-phila (Ins.); Lepido-spartum*.

**sparver**—L. *sparverius*, pertaining to a sparrow. *Ex:* Sparverius (Av.).

**spasm**—1. Gr. *spasmos*, a convulsion. *Ex:* A-spasmo-gaster (Pisc.): 2. Gr. *spasma*, genit. *spasmatos*, a piece torn off; also a spasm. *Ex:* Spasma (Mam.); Spasmo-stoma (Prot.).

**spastic**—Gr. *spastikos*, absorbing; also relating to spasms. *Ex:* spastic; Spastica (Ins.).

**spatag**—See spatang.

**spatal**—Gr. *spatalē*, luxury, lewdness. *Ex:* Spatal-ura (Rept.).

**spatalistes**—Gr. *spatalistēs*, a profligate. *Ex:* Spatalistes (Ins.).

**spatang**—Gr. *spatangēs*, a kind of sea urchin. *Ex:* Spatagus (Echin.), evidently an error as is also Spatago-brissus (Echin.); Spatango-idea (Echin.); Spatango-morpho (Echin.)

**spath**—L. *spatha*, a spatula<Gr. *spathē*, a blade, spatula; a staff. *Ex:* spath-aceous; spath-illa; Spathe-cera (Ins.); Spathe-philus (Ins.); Spatho-pterus (Av.); Spathyema*; Spathius

(Ins.); Spatho-centrus (Ins.); A-spatha (Av.); Erio-spatha*.

**spathali**—Gr. *spathalion*, a kind of bracelet.

**spathul**—L. *spathula*, dim. of *spatha*, a spatula, a flat broad knife>NL. *spathulatus*, shaped like a spatula. See spatula.

**spathyema**—See spath.

**spati**—L. *spatium*, pl. *spatia*, space, room.

**spatiat**—L. *spatiatus*, roving<*spatior*, to walk about.

**spatil**—Gr. *spatilē*, parings; thin excrement.

Spiny Abrojo, *Condalia spathulata*, with small spatulate leaves. Redrawn from Desert Wild Flowers—Jaeger. Stanford University Press.

**spatios**—L. *spatiosus*, full of room, large, long.

**spatul**—L. *spatula*, spoon, a broad piece. *Ex:* spatul-ate; Spatula (Av.).

**speci**—L. *species*, a shape, kind or sort, a particular kind. *Ex:* speci-fic; speci-ation; species.

**specimen**—L. *specimen*, an example, evidence, sign.

**specios**—L. *speciosus*, showy, brilliant.

**spectabil**—L. *spectabilis*, visible, remarkable.

**spectan**—L. *spectans*, genit. *spectantis*, looking forward, watching, ppr. of *specto*, to look at, to watch.

**spectr**—L. *spectrum*, an image, appearance. *Ex:* Spectr-ellum (Mam.); Spectro-bates (Ins.); Spectrum (Mam.).

**specu**—L. *specus*, a hole, cave, ditch. *Ex:* specui-cola.

**specul**—L. *speculum*, a mirror<*specto*, to look at, to gaze.

**speir**— See spir.

**spel**—Gr. *spēlaion*=L. *spelaeum*, a cave. *Ex:* Spel-erpes (Amph.); Spelaei-acris (Ins.); Spelaeo-bates (Ins.); Spelaeus (Mam.); spele-an; Spele-arctos (Mam.); Spelo-bia (Ins.); speleus.

**speo**—Gr. *speos*, a cave, cavern, opening. *Ex:* Speo(ri)-fera (Mam.), the *ri* for the sake of euphony; Speo-thos (Mam.); Speo-tyto (Av.).

**sperat**—L. *speratus*, desired, hoped for, promised.

**sperch**—Gr. *sperchō*, to be in haste. *Ex:* See next entry.

**sperchius**—L. *Sperchius*, a river in Thessaly< *sperchō*, to be rapid. *Ex:* Spercheus (Ins.); Sperchius (Crust.).

**spergul**—NL. *spergulinus*, scattering<*spargo*, to scatter. *Ex:* Spergula*; sperguli-folia, with leaves like Spergula.

**sperm**—Gr. *sperma*, genit. *spermatos*, seed, semen. *Ex:* sperm; spermat-ic; Spermatophyta*; spermato-zoon; spermo-phile; Angiosperm*; Gymno-sperm*.

**spermolog**—Gr. *spermologos*, picking up seeds. *Ex:* Spermologa (Av.).

**sphacel**—1. Gr. *sphakelos*, gangrene>NL. *sphacelo*, to mortify, pp. *sphacelatus*, poisoning, killing. *Ex:* Sphacel-aria*; Sphacel-oma*; Sphacel-odes (Ins.); Sphacel-ura (Nemat.): 2. Gr. *sphakos*, sage. *Ex:* Sphacele.* S.

**sphact**—Gr. *sphaktēs*, a slayer, murderer.

**sphadasm**—Gr. *sphadasmos*, a convulsion. *Ex:* Sphadasmus (Ins.).

**sphaen**—See sphen.

**sphaer**—Gr. *sphaira*, dim. *sphairion*, a ball, sphere; *sphairitis*, a kind of cypress with globular fruit; *sphairikos*, globular; *sphairōtos*, rounded. *Ex:* Sphaer-alcea*; Sphaer-id-ops (Ins.); Sphaer-ul-aria (Nemat.); Sphaeriodiscus (Echin.); Sphaerion (Ins.); Sphaerium (Moll.); Sphaero-stigma*; Sphaero-zoum (Prot.); Sphaira (Prot.); Mela-sphaer-ula*.

**sphag**—1. Gr. *sphax*, genit. *sphagos*=*sphagē*, the throat, the spot where the victim is struck; also slaughter, butchery. *Ex:* Sphag-odus (Pisc.); Sphage-branchus (Pisc.): 2. Gr. *sphageus*, a slayer, cut-throat. *Ex:* Sphageus (Ins.).

**sphagi**—Gr. *sphagios*, slaying, deadly. *Ex:* Sphagio-crates (Ins.).

**sphagn**—Gr. *sphagnos*, a kind of moss. *Ex:* Sphagn-ales*; Sphagnum*.

**sphair**—See sphaer.

**sphaler**—See sphall 2.

**sphall**—1. Gr. *sphallos*, a round leaden plate, a round block of wood with two holes for the feet, a pair of stocks. *Ex:* Sphallo-morpha (Ins.): 2. Gr. *sphallō*, to deceive, to trip up>*sphaleros*, deceiving, treacherous. *Ex:* Sphallero-carpus*; sphalero-carpum.

**sphalm**—Gr. *sphalma*, genit. *sphalmatos*, a trip, fault, error. *Ex:* Sphalma (Ins.); Sphalmato-blattina (Ins.); A-sphalmus (Ins.).

**spharag**—Gr. *spharagos*, a bursting with noise; *spharageomai*, to crackle, to sputter. *Ex:* Spharagemon (Ins.).

sphax—See sphag.

sphec—Gr. *sphēx*, genit. *sphēkos*, a wasp. *Ex:* Sphec-idae (Ins.); Sphec-odes (Ins.); Spheci-gaster (Ins.); Spheco-theres (Av.); Sphex (Ins.).

sphedan—Gr. *sphedanos*, violent. *Ex:* Sphedano-lestes (Ins.); Sphedanus (Arach.).

sphel—Gr. *sphelas*, genit. *sphelatos*, a pedestal, a footstool. *Ex:* Sphelatus (Echin.).

sphen—Gr. *sphēn*, genit. *sphēnos*, dim. *sphē-narion*, a wedge; *sphēniskos*, a small wedge, a wedge-shaped bandage, a solid of three unequal dimensions. *Ex:* Sphen-odon (Rept.); sphen-oid; Sphenarium (Ins.); Spheniscus (Ins.); Spheno-clea*; Spheno-pholis*; Spheno-phorus (Ins.).

spher—See sphaer.

sphex—See sphec.

sphigg—Gr. *sphingō*, to bind >NL. *sphiggo*, to bind; Gr. *sphingion*, a bracelet, necklace. *Ex:* Sphigg-urus (Mam.); Sphiggo-mys (Mam.).

sphigm—Gr. *sphygmos*, the pulse. *Ex:* sphigmo-mano-meter; Sphigmo-cephalus (Ins.).

sphinct—Gr. *sphinktēr*, that which binds tight; a circular muscle closing an opening; *sphinktos*, tightly bound. *Ex:* sphincter; Sphincto-ceras (Moll.); Sphinctus (Ins.); Sphincter-ella (Por.).

sphing—1. Gr. *sphingō*, to bind tight. *Ex:* Sphing-urus (Mam.); Sphingo-cladia (Ins.): 2. L. *sphinx*, genit. *sphingis*, a sphinx<Gr. *sphingō*, to bind. *Ex:* Sphing-idae (Ins.); Sphing-ulus (Ins.); Sphingi-campus (Ins.); Sphinx (Ins.).

sphinx—See sphing.

sphodr—Gr. *sphodros*, active, strong; *sphodrotēs*, violence. *Ex:* Sphodr-istus (Ins.); Sphrodro-mys (Mam.); Sphodros (Arach.); Sphodrotes (Ins.); Rhyso-sphodrus (Ins.).

sphondyl—1. Gr. *sphondylē*, a stinking insect>L. *sphondylion*, a plant name, the odorous cowparsnip: 2. Gr. *sphondylios = sphondylos*, a vertebra. See also spondyl.

sphrag—Gr. *sphragis =* Ionic *sphrēgis*, genit. *sphrēgidos*, a seal; *sphragistikos*, pertaining to seals. *Ex:* Sphragi-fera (Ins.); Sphragio-pora (Bry.); Sphragido-phorus (Ins.); sphragistic.

sphragid—See sphrag.

sphrig—Gr. *sphrigaō*, to be full, to be bursting; *sphriganos*, to be strong, plump. *Ex:* sphrig-osis.

sphynx—See sphing 2.

sphyr—1. Gr. *sphyra*, a hammer, mallet; *sphyra = sphyraina*, the hammer-fish. *Ex:* Sphyr-an-ura (Platy.); Sphyra-picus (Av.); Sphyraena (Pisc.); Sphyrion (Crust.); Sphyrna (Pisc.), one of Rafinesque's efforts at word-butchery; Sphyro-cephalus (Mam.); Calli-sphyrum*: 2. Gr. *sphyron*, the ankle. *Ex:* Syn-tomo-sphyrum (Ins.).

spic—L. *spica*, dim. *spiculum*, a point; also a dart, spike, top, tuft; *spico*, to furnish with spikes; pp. *spicatus*, spiked; *spiceus*, consisting of ears of corn; *spiculus*, pointed. *Ex:* Spica (Moll.); spicate; Spicat-or (Moll.); Spici-pora (Prot.); spicule; spiculi-form; Spiculo-caulus (Nemat.).

spican—L. *spicans*, genit. *spicantis;* ppr. of *spico*, to furnish with points, spikes or ears.

spicul—See spic.

spil—Gr. *spilos*, genit. *spilados*, a spot, blemish; *spilōtos*, stained, soiled. *Ex:* Spil-anthes*; Spilo-gale; Spilo-soma (Ins.); Spilotes (Rept.); A-spil-ota (Ins.); A-spila (Ins.); Cirro-spilus (Ins.). See also spilad.

spilad—Gr. *spilas*, genit. *spilados = spilos*, a rock wet with sea spray, a slab, cave; also stony, chalky. *Ex:* spilado-philus.

spilot—See spil.

spin—1. L. *spina*, dim, *spinula*, thorn, spine; *spinatus*, spined, with spines. *Ex:* spina-sternum; Spinacea*, cf. Low L. *spinacia*, spinach; spinate; Spini-fex*; spini-form; Spini-gera (Mam.): 2. Gr. *spinos = L. spinus*, a linnet or some related bird. *Ex:* Spinus (Av.).

Broad-leaved Stillingia, *Stillingia spinulosa*, with leaves edged with many small spines. The generic name honors Dr. Benj. Stillingfleet, English botanist. Redrawn from Desert Wild Flowers —Jaeger. Stanford University Press.

spindle—A.S. *spinl*, a spindle<*spinan*, to span, to spin. The *d* of spindle is excrescent as it is also in such words as soun*d* and thun*d*er. *Ex:* achromatic spindle.

sping—Gr. *spingos = spinos*, a finch. *Ex:* Chloro-spingus (Av.).

spinthar—Gr. *spintharis = spinthēr*, a spark. *Ex:* Spintharis (Ins.); Spintharus (Arach.); Spin tharo-bolus (Pisc.).

spinther—See spinthar.

spinul—See spin.

spio—L. *Spio*, a sea-nymph. *Ex:* Spio (Ann.); Spio-chaeto-pteras (Ann.); Spio-phagnes (Ann.).

spir—1. Gr. *speira*, anything wound or wrapped round > *speiraia*, the meadow-sweet, used in making wreaths < L. *spira*, a coil. *Ex:* Speiroceras (Ins.); Spir-al-aria (Bry.); spir-aster; spir-al; Spir-anthes*; spir-icle; Spir-illum*; Spir-orbis (Moll.); Spir-ula (Moll.); Spiraea*; Spiri-fera (Brach.); Spiro-dela*; Spiro-loculinus (Coel.); Spiro-phytum; Crypto-spira (Moll.):     2. L. *spiro*, to breathe, to blow. *Ex:* in-spir-ation.

spiracle—See spiracul.

spiracul—L. *spiraculum*, a pore, air-hole. *Ex:* spiracle; Spiraculum (Moll.).

spiramen—L. *spiramen*, genit. *spiraminis*, an air-hole, vent.

spirem—Gr. *speirēma*, genit. *speirēmatos*, a coil, a twisted thread. *Ex:* Spirem-arium (Prot.); spireme.

spirill—See spir.

spiss—L. *spissus*, compact, crowded; *spissatus*, thickened. *Ex:* spissat-ed; Spissi-pedes (Ins.).

spitham—L. *spithama*, a span < Gr. *spithamē*, a span. *Ex:* spithama-eus.

spiz—Gr. *spiza*, dim. *spizion*, a finch < *spizō*, to chirp; *spizilēs*, a titmouse. *Ex:* Spiz-aetus (Av.); Spiza-circus (Av.); Spizi-a-pteryx (Av.); Spizites (Av.); Spizo-corys (Av.); Amphi-spiza (Av.); Melo-spiza (Av.).

splachn—NL. *splachnum* < a supposed Gr. *splachnon*, moss. *Ex:* Splachnum*.

splanchn—Gr. *splanchnon*, an entrail, viscera. *Ex:* splanchn-ic; splanchno-coele; A-splanchna (Rot.).

splen—1. Gr. *splēn*, genit. *splynos*, the spleen; *splēnikos*, of the spleen; *splēnion*, a kind of fern, spleenwort > L. *asplenum*, a spleenwort. *Ex:* splenic; A-splenium, the *a* euphonic; Chryso-splenium*:     2. Gr. *splēnion*, a bandage, compress. *Ex:* spleni-al; splenium, an anatomical term; splenius muscle.

splendid—L. *splendidus*, splendid, bright, glittering.

spleniat—L. *spleniatus*, plastered, having a patch on.

spod—Gr. *spodos*, ashes; *spodios*, ash-gray. *Ex:* Spodio-psar (Av.); spodo-chrous; Spodo-lepis (Ins.).

spoggod—Gr. *spongōdēs*, spongy < *spongos*, dim. *spongion*, a sponge (In transliteration the double gamma (γγ) is ordinarily rendered by ng). *Ex:* Spoggodes (Por.).

spoliat—L. *spoliatus*, plundered, made poor.

spondias—Gr. *spondias*, a kind of plum-tree. *Ex:* Spondias*.

spondyl—Gr. *spondylos* = *spondylē*, vertebra. *Ex:* Spondyl-aspis (Ins.); Spondylio-soma (Pisc.); Spondylus (Moll.); Palaeo-spondylus (Pisc.); stereo-spondyl-ous.

spong—Gr. *spongos* = L. *spongia*, a sponge. *Ex:*

spong-oid; Spong-aster (Prot.); Spong-echinus (Prot.); Spong-olena (Por.); spongi-form; spongio-plasma; spongo-coel; Spongo-trochus (Prot.); Hippo-spongia (Por.); Ver-ongia (Por.).

spons—L. *sponsus*, promised, bound, pp. of *spondeo*, to bind, promise.

sponsa—L. *sponsa*, a bride < *spondeo*, to bind, pledge one's self. *Ex:* Sponsa (Moll.).

spont—L. *sponte*, of one's free will, freely > *spontaneus*, freely. *Ex:* spontaneous.

spor—Gr. *spora*, a seed. *Ex:* spor-idium; spor-ont; spore; spori-desm; sporo-cyst; Sporozoa (Prot.); mega-spore; Pitto-sporum*; Ryncho-spora*.

sporad—Gr. *sporas*, genit. *sporados*, scattered; *sporadikos*, scattered. *Ex:* Sporadi-pus (Echin.); Sporado-cyphus (Echin.); sporado-phyt-ium (Ecol.).

sporidiol—NL. *sporidiolum*, dim. of *sporidium*, dim. of Gr. *spora*, a spore. *Ex:* sporidiolum.

sport—L. *sporta*, a basket. *Ex:* Sport-ella (Moll.).

spretus—L. *spretus*, despised, pp. of *sperno*, to sever, reject, scorn.

spuda—Gr. *spoudē*, haste, zeal; *spoudaios*, quick, active; *spoudastikos*, zealous, earnest. *Ex:* Spudaea (Ins.); Spudastica (Ins.).

spudastic—See spuda.

spum—L. *spuma*, foam; *spumescens*, genit. *spumescentis*, foaming, ppr. of *spumesco*, to grow foamy; *spumidus*, foamy, spongy. *Ex:* Spum-aria (Prot.); spume; spumi-gena; Spumi-spongia (Por.); spumid.

spumid—See spum.

spurc—L. *spurcus*, unclean; *spurcatus*, most foul. *Ex:* Spurco (Pisc.).

spuri—L. *spurius*, false. *Ex:* spurious; Spurio-stylo-ptera (Ins.).

sputat—L. *sputator*, a spitter.

spyr—See spir.

spyrid—Gr. *spyris*, genit. *spyridos*, a basket. *Ex:* Spyridia*; Spiridio-crinus (Echin.); Spyrido-botrys (Prot.); Antho-spyris (Prot.); Tholo-spyr-idae (Prot.).

spyris—See spyrid.

squal—1. L. *squalus*, a kind of sea fish. *Ex:* Squal-odon (Mam.); Squali-raja (Elasm.); Squaliobarbus (Elasm.); Squalo-delphis (Mam.); Squalus (Elasm.):     2. L. *squalus*, dirty, filthy.

squalid—L. *squalidus*, squalid, filthy, neglected; also dry.

squam—L. *squama*, dim. *squamula*, a scale; *squamatus*, scaly; *squamosus*, scaly, covered with scales. *Ex:* Squam-apion (Ins.); Squam-aria*; Squam-aster (Echin.); squam-ous; squami-ger-ous; Squamo-discus (Platy.); squamos-al; Squamato-ornis (Av.); Squamul-ina (Prot.).

squamat—See squam.

squarros—L. *squarrosus*, rough, scurfy. *Ex:* squarrose; squarroso-dentate.

squatarol—Venetian *squatarola*, the black-bellied plover. *Ex:* Squatarola (Av.).

squatin—L. *squatina*, a kind of shark, skate. *Ex:* Squatin-ella (Rot.); Squatina (Elasm.); Squa tini-raja (Elasm.).

squill—L. *squilla* = *scilla*, a sea-onion, leek, squill; also a prawn, shrimp. *Ex:* Squill-erichthus (Crust.), see erict; Squilla (Crust.).

stabil—L. *stabilis*, firm, steady, stable < *sto*, to stand; *stabilimentum*, a prop, stay.

stabul—L. *stabulo*, to have a resting place, abode; ppr. *stabulans*, genit. *stabulantis*, abiding, resting.

stachy—Gr. *stachys*, an ear of grain, spike. *Ex:* Stachy-colobus (Mam.); Stachy-spongia (Por.); Stachynia (Ins.), the *n* supplied by Micquart who was much given to such practices in coining words; Stachyo-crinus (Echin.); peri-stachy-um; Stachys*.

stachyer—NL. *stachyerus* < Gr. *stachyēros*, bearing ears of grain, spike-like.

stact—Gr. *staktos*, oozing out by drops, dropping. *Ex:* Stacto-bia (Ins.); Stacto-cichla (Av.).

stadi—NL. *stadium*, pl. *stadia*, a stage, station, a site or position temporarily occupied < Gr. *stadion*, a measure of length. *Ex:* stadium.

stag—Gr. *stagōn*, a drop; *stagetos*, a drop. *Ex:* Stag-odon (Mam.); Stageto-morphus (Ins.); Stagono-lepis (Rept.).

staget—See stag.

stagm—Gr. *stagma*, genit. *stagmatos*, a drop, that which is dropped. *Ex:* Stagmato-ptera (Ins.); Stagmo-pimpla (Ins.); Tri-stagma*.

stagn—L. *stagnum*, a pool, pond, swamp. *Ex:* Stagni-cola (Moll.); Stagni-gradi (Ins.).

stagnat—L. *stagnatus*, formed into pools of still water; also made fast, strong. *Ex:* stagnation.

stagon—See stag.

stalact—Gr. *stalaktikos* = *stalaktos*, a dripping or dropping. *Ex:* Stalact-ella (Moll.); stalact-ite.

stalagm—Gr. *stalagmos*, a dropping, a dripping. *Ex:* stalagm-ite; Stalagmo-pygus (Ins.).

stalic—Gr. *stalix*, genit. *stalikos*, a stake to which nets are fastened. *Ex:* Stalix (Pisc.); Stigmato-stalix.

stalix—See stalic.

stalsis—Gr. *stalsis*, genit. *stalseōs*, a compression, restriction. *Ex:* peri-stalsis.

stamen—See stamin.

stamin—L. *stamen*, genit. *staminis*, a thread, fiber. *Ex:* stamen; stamini-ferous.

stamn—Gr. *stamnos*, a jar. *Ex:* Stamn-odes (Ins.); Stamno-cnemis (Por.).

stamnar—Gr. *stamnarion*, an urn, dim. of *stamnos*, a wine jar. *Ex:* Stamnaria*.

stan—See stans.

stans—L. *stans*, standing, ppr. of *sto*, to stand. *Ex:* Tri-stania*.

stap—Low L. *stapes*, stirrup. *Ex:* stapes; stapi-form.

staphyl—Gr. *staphylē*, a cluster of grapes; also the uvula, when swollen. *Ex:* Staphylea*; staphylo-coccus; Staphylo-cystis (Platy.).

staphylin—Gr. *staphylinos*, a kind of insect; also a kind of carrot. *Ex:* Staphilino-chrous (Ins.); Staphylinus (Ins.).

stas—Gr. *stasis*, a placing, standing, posture. *Ex:* stasis (Med.); staso-philus (Ecol.); homoeo-stasis; for dia-stase, see diastas.

stasiastic—Gr. *stasiastikos*, seditious. *Ex:* Stasi asticus (Av.).

stasiotes—Gr. *stasiōtēs*, a body-guard. *Ex:* Stasiotes (Rept.).

stat—Gr. *statos*, standing, placed; *statikos*, causing to stand. *Ex:* static; stato-blast; stato-cyst; stato-rhab; Hebe-statis (Arach.).

stathm—Gr. *stathmē*, a carpenter's rule. *Ex:* Stathme-pora (Bry.); Stathmo-notus (Pisc.).

static—Gr. *statikē*, an astringent herb. *Ex:* Statice*; Statico-bium (Ins.).

stativ—L. *stativus*, standing still.

statumin—L. *statumen*, genit. *statuminis*, a support; *statuminatus*, supported. *Ex:* Statuminat-ae*.

staur—Gr. *stauros*, a cross. *Ex:* Staur-opsis*; Stauro-nereis (Ann.); Stauro-stigma*; Stauro-teuthis (Moll.).

stax—Gr. *staxis*, a dropping.

stear—See steat.

steat—Gr. *stear*, genit. *steatos*, fat, tallow. *Ex:* Stearo-ceras (Moll.); Steat-oda (Arach.); Steat-ornis (Av.); Steato-mys (Mam.); steato-pygy.

steg—Gr. *stegē*, also *stegos*, a covering, roof. *Ex:* Stego-cephalia (Amph.); Stego-saurus (Rept.); branchio-steg-al; Loxo-stege (Ins.).

stegan—Gr. *steganos*, water-tight, close, covered. *Ex:* Stegan-apsis (Arach.); stegan-ophthalm-ate; Stegana (Ins.); Stegano-podes (Av.).

stegnos—Gr. *stegnos* = *steganos*, something sheathed or closed; also water-tight, made tight; *stegnōtikos*, astringent, causing stoppage; *stegnōsis*, a stoppage of pores, a making close. *Ex:* stegnosis; stegnotic.

stegnot—See stegnos.

stein—Gr. *steinos*, a narrow confined space. *Ex:* Stein-ella (Prot.); Steino-myia (Ins.).

steir—1. Gr. *steiros*, sterile. *Ex:* Steira (Moll.); Steiro-nema*; Ano-stirus (Ins.); 2. Gr. *steira*, a keel. *Ex:* Steir-axis (Moll.); Steir-odon (Ins.); Steiro-mys (Mam.).

**stel**—Gr. *stēlē*, a prop, stay, a boundary post of stone; *stēlidion*, a small pillar, monument. *Ex:* stel-ar; stele; Stelidio-crinus (Echin.); stelidium; stelo-lemma; Stelo-spongia (Por.); meristele.

**stele**—Gr. *steleon*, a handle. *Ex:* Steleo-pyga (Ins.); Steleo-xiphus (Ins.).

**stelech**—Gr. *stelechos*, the crown of the root from which the stem springs; *stelechōdēs*, with stem or trunk. *Ex:* stelech-ite; Stelecho-pus (Ann.).

**stelgid**—Gr. *stelgis*, genit. *stelgidos*, a scraper. *Ex:* Stelgido-pteryx (Av.).

**stelidi**—See **stel.**

**stell**—**1.** L. *stella*, a star; *stellaris*, starry, speckled; *stellio*, the starry one; *stellatus*, starred, i.e., spotted. *Ex:* Stell-acantha (Prot.); Stelleria*; not Stelleria (Av.), named after G. W. Steller; stellato-pilosus; stelli-form; Stellis (Rept.). See also **stol:** **2.** Gr. *stellō*, to set in order, to bring together.

**stellat**—See **stell 1.**

**stellerid**—Fr. *stellerides*, starry, star-like<L. *stellaris*, starry. *Ex:* Stellerida (Echin.).

**stellio**—L. *stellio*, a lizard with star-like spots on its back (cf. L. *stella*, a star); also a crafty, deceitful person. *Ex:* Stellio (Rept.).

**stellulat**—NL. *stellulatus*, with small star-like markings<L. *stellula*, a little star.

**stelm**—Gr. *stelma*, a crown. *Ex:* Brachy-stelma*; Phylo-stelma*.

**stem**—Gr. *stēma*, genit. *stēmatos*, the exterior sheath of the phallus; the stamen of a flower; *stemōn*, a thread, a stamen. *Ex:* stema-poda; Stemo-dia*, Gr. *diakris*, two tips or points; Stemon-idium (Pisc.); Stemona*; Stemonocera (Ins.); hexa-stemon-ous; Laci-stema*; Tricho-stema*.

**stemm**—Gr. *stemma*, genit. *stemmatos*, a crown, garland. *Ex:* stemma; stemmata; Stemmato-steres (Ins.); Agro-stemma*; Dichelo-stemma*; Tetra-stemma (Nemert.); Tri-stemma*.

**stemod**—NL. *stemodia*, abbreviation of P. Brown's plant genus, Stemodiacra—Stemodia*.

**stemon**—See **stem.**

**sten**—Gr. *stenos*, narrow, straight; *stenos*, genit. *steneos*, a narrow confined space; *stenōdēs*, somewhat narrow; *stenotēs*, narrowness. *Ex:* Sten-andrium*; Sten-otus*; Steneosaurus (Rept.); Stenodes (Nemat.); Apo-stenus (Arach.). See stenos.

**stenia**—NL. *stenia*, name for certain beetles. *Ex:* Stenia (Ins.).

**stenist**—Gr. *stenistos*, very narrow, most narrow; supl. of *stenos*, narrow, straight. *Ex:* Stenisto-mera (Ins.). See sten.

**stenos**—Gr. *stenōsis*, a being narrowed. *Ex:* stenosis; Stenosi-pora (Bry.).

**stenot**—See **sten.**

**stentor**—Gr. *Stentōr*, a Grecian herald with strong voice who yelled as loud as fifty men together. *Ex:* Stentor (Prot.), (Mam.).

**stenygr**—Gr. *stenygros*, Ionic for *stenos*, narrow. *Ex:* Stenygro-cerus (Arach.).

**steph**—See **stephan.**

**stephan**—Gr. *stephanos*=poet. *stephos*, a crown <*stephō*, to crown. *Ex:* Stephano-meria*;

Coronated Garland-horn Ammonite, *Stephanoceras coronatus*. Redrawn from Textbook of Palaeontology—Zittel. The Macmillan Co.

Stepho-myia (Ins.); Stephus (Crust.); Andro-stephium*; Eu-stephanus; not Stephania*, named after S. Stephan, botanist of Moscow.

**ster**—**1.** Gr. *stereos*, solid, hard, firm. *Ex:* Ster-echinus (Echin.); stere-id; Stere-ornithes (Av.); Stereo-spermum*; stereo-spondylous; Podi-stera*; Stemmato-steres (Ins.). **2.** Gr. *stēr*, fat, tallow; a contraction of *stear*, see steat. *Ex:* Sepio-stera (Moll.).

**sterc**—L. *stercus*, genit. *stercoris*, dung; *stercorarius*, having to do with dung; *Sterculius*, Latin god who presided over manuring. *Ex:* stercoraceous; Stercorarius (Av.); Sterculia*.

**steres**—Gr. *sterēsis*, a plundering, privation, a deprivation. *Ex:* Bio-steres (Ins.); Tricho-steresis (Ins.); tylo-steresis (Med.).

**sterigm**—Gr. *sterigma*, genit. *stērigmatos*, a prop, support; also a fork. *Ex:* Sterigma*; sterig mata; Acro-sterigma (Moll.).

**steril**—L. *sterilis*, unfruitful; *sterilitas*, barrenness. *Ex:* steril-iz-ation; sterile; sterilit-y.

Big-headed Tortoise, *Platysternum megacephalum.*

**stering**—Gr. *stērinx*, genit. *stēringos*, a support, prop. *Ex:* Steringo-trema (Platy.).

**steripho**—Gr. *steriphos*, firm, solid, hard; *steriphōma*, genit. *steriphōmatos*, a firm foundation; *Ex:* Steriph-otis (Ins.); Steripho-pus (Arach.); Steriphoma*.

**stern**—**1.** Gr. *sternon*, breast, chest. *Ex:* stern-al; stern-ebra, see vertebr; stern-ite; Stern-odes (Ins.); Stern-oxus (Ins.); Sterno-therus (Rept.);

sternum: **2.** NL. *sterna*, from English, *stern, starn* or *tern*, a tern. *Ex:* Stern-idae (Av.); Sterna (Av.): **3.** L. *sterno*, to extend; pp. *stratus*, stretched out, extended. *Ex:* humistratus.

**sternut**—L. *sternuto*, to sneeze; *sternutatio*, genit. *sternutationis*, a sneezing; *sternumentum*, a sneezing. *Ex:* sternutation; sternutat-ory.

**sterop**—Gr. *Steropēs*, one of three Cyclopes < *steropē*, a flash of lightning, a gleam. *Ex:* Sterope (Moll.); Steropes (Ins.).

**sterquilin**—L. *sterquilinum*, a dung-pit.

**sterr**—Gr. *sterros*, rigid, strong, rugged as pertaining to countries. *Ex:* Sterri-chrotes (Rept.); Sterro-lophus = Sterrho-lophus; sterro-philus (Ecol.); Sterrho-ptilus (Av.).

**sterrh**—See sterr.

**stesichor**—Gr. *Stēsichoros*, name of a Grecian lyric poet. *Ex:* Stesichorus (Ins.).

**steth**—Gr. *stēthos*, the breast. *Ex:* steth-idium; Steth-orus (Ins.); stetho-scope; Meco-stethus (Ins.); meta-steth-ium; Pedano-stethus(Arach.); Pro-stethes (Echin.); Rhodo-stethia (Av.); Sphincto-stethus (Ins.); Tomo-stethus (Ins.).

**sthen**—Gr. *sthenos*, strength. *Ex:* Stheno-meris (Mam.); a-sthenia; a-stheno-biosis; Callisthenes (Ins.); Mega-sthena (Mam.); prosthen-ic; cali-sthen-ics.

**sthenar**—Gr. *sthenaros*, strong, mighty. *Ex:* Sthenaro-saurus (Rept.).

**stib**—Gr. *stibos*, a track, footstep, path. *Ex:* Stibo-scopus (Ins.).

**stibad**—Gr. *stibas*, genit. *stibados*, a bed of straw or leaves, a nest. *Ex:* Stibado-derus (Ins.).

**stibar**—Gr. *stibaros*, strong. *Ex:* Stibaro-bdella (Ann.); Stibaro-stoma (Ins.); Stibarus (Mam.).

**stibeut**—Gr. *stibeutēs = stibeus*, a walker, tracker. *Ex:* Stibeutes (Ins.).

**stica**—See stich.

**stich**—Gr. *stichos*, dim. *stichidion*, a row of soldiers, a line or row of things; as a combining form, *sticho-* often denotes a row of rod-like processes. *Ex:* Stich-aster (Echin.); Stich-aeus (Pisc.); Stichidium*; Sticho-tricha; Di-stichis*; para-stichy; Poly-stichium*; Rhecho-stica (Arach.); rhipido-stichous; tetra-stichous.

**stict**—Gr. *stiktos*, punctured, dotted, dappled. *Ex:* Stict-ichneumon (Ins.); Sticta*; Stictigramma (Ins.); Sticto-cephala (Ins.); Aposticto-pterus (Ins.); Leuco-sticte (Av.); Rhinostictus (Mam.).

**stigeo**—Gr. *stigeys*, genit. *stigeōs*, an awl or needle used in puncturing or tattooing, a pointed instrument. *Ex:* Stigeo-clonium*.

**stigm**—Gr. *stigma*, genit. *stigmatos*, a point, a pricked mark; *stigmatizō*, to prick, to puncture. *Ex:* Stigm-aria*; stigma; stigmat-ic; Stigmat-omma (Ins.); stigmati-ferous; Stig

mato-teuthis (Moll.); Stigmo-sphaera (Prot.); Stigmus (Ins.); a-stigmat-ism (Med.); Metastigmata (Arth.); Pro-stigmata (Arth.).

**stigmae**—NL. *stigmaeus*, spotted < Gr. *stigma*, a spot.

**stigmat**—See stigm.

**stigmosus**—L. *stigmosus*, full of points, marks, branded. *Ex:* stigmosa.

**stigon**—Gr. *stigōn*, genit. *stigōnos*, one who marks.

**stil**—**1.** L. *stilus*, a stake < Gr. *stylis*, a pillar, a post; *stylitēs*, standing on a pillar. *Ex:* Stilpontia (Ins.); Stili-fer (Moll.); Stilo-phora*; Stilo-therium (Mam.): **2.** Gr. *stilē*, a drop.

**stilb**—Gr. *stilbōn*, the shining, glittering one; the planet Mercury; *stilbē*, a lamp; *stilbōma*, genit. *stilbōmatos*, a glittering ornament; *stilbōtēs*, a polisher; L. *stilbius*, shining. *Ex:* Stilb-ella*; Stilbe*; Stilbia (Ins.); Stilboma (Ins.); Stilbotes (Ins.); Stilbum*; Stilbum (Ins.); A-stilbe*; Chloro-stilbon (Av.).

**still**—L. *stilla*, a drop; *stillatus*, dropping, pp. of *stillo*, to drop, trickle. *Ex:* stillati-ous; stilliform; stillatim.

**stillatici**—L. *stillaticius*, dripping, with drops < *stilla*, a drop.

**stilpn**—Gr. *stilpnos*, a shining. *Ex:* Stilpno-soma (Ins.); Stilpnus (Ins.).

**stimul**—L. *stimulus*, a prick, goad; *stimulans*, genit. *stimulantis*, stimulating, ppr. of *stimulo*, to stimulate. *Ex:* stimulant; stimulus.

**stimule**—L. *stimuleus*, made of prickles.

**stip**—**1.** L. *stipes*, genit. *stipitis*, pl. *stipes*, a log, stem, the branch of a tree; dim. *stipula*, a stalk, blade, stipule; ML. *stipitatus*, having or borne on a stipe. *Ex:* stipe; stipel; stipes; stipi-form; Stipit-urus (Av.); stipiti-form; stipito-cardinal; stipo-dema; stipule; Stipulicida*: **2.** L. *stipa*, tow, the coarse part of flax. *Ex:* Stipa*.

**stipat**—L. *stipatus*, compressed, surrounded. *Ex:* stipate.

**stiphr**—Gr. *stiphros*, firm, stout, sturdy. *Ex:* Stiphr-ornis (Av.).

**stipit**—See stip 1.

**stipt**—Gr. *stiptos*, trodden down.

**stipul**—See stip.

**stir**—See steir 1.

**stiri**—L. *stiria*, an icicle. *Ex:* stiri-ated; stiri-ous; Stiria (Ins.).

**stirp**—L. *stirps* also *stirpis*, the stem of plants; also stock, race, lineage. *Ex:* stirpi-culture.

**stix**—Gr. *stixis*, puncture. *Ex:* Stixis (Ins.); osteo-stixis.

**stiz**—Gr. *stizō*, to prick, puncture. *Ex:* Stizolobium*; Stizo-stedion (Pisc.); Stizus (Ins.); Crypto-helco-stizus (Ins.).

**stoa**—Gr. *stoa*, a roofed colonnade, a storehouse. *Ex:* Stoa-stoma (Moll.).

stoch—See stochasm.

stochasm—Gr. *stochasma*, genit. *stochasmatos*, a thing aimed, an arrow, spear; *stochasmos*, a guess, inference < *stochos*, a guess. *Ex:* Stochasmus (Crust.); Stocho-mys (Mam.).

stoeb—Gr. *stoibē*, a padding, heap. *Ex:* Stoeberhinus (Ins.).

stoech—1. Gr. *stoichos*, a row. *Ex:* Orthostoechus (Pisc.): 2. Gr. *stoichas*, an aromatic plant. *Ex:* Lavandula stoechas*.

stoechio—Gr. *stoicheion*, a part, element, one of a series, unit. *Ex:* stoechio-logy.

stoich—Gr. *stoichos*, a row. *Ex:* Stoich-actis (Echin.).

stol—Gr. *stolē* = *stolos* = L. *stola*, a garment, an armor; also a white band worn by priests. *Ex:* Stol-asterias (Echin.); Stola (Moll.); Stolephorus (Pisc.); Stolo-teuthis (Moll.); Agathostola (Arth.).

stolid—Gr. *stolis*, genit. *stolidos*, a garment, robe. *Ex:* Stolid-ophides (Rept.); Stolida (Av.); Stolido-soma (Ins.).

stolis—See stolid.

stolon—L. *stolo*, genit. *stolonis*, a shoot, a branch. *Ex:* stolon; Stoloni-fera (Coel.); stoloni-form; Stolono-clypus (Echin.).

stom—Gr. *stoma*, genit. *stomatos*, mouth; *anastomoō*, to furnish with a mouth. *Ex:* Stomaster (Echin.); Stom-ina (Ins.); Stom-oisia*; Stom-oxys (Ins.); stoma-podi form; Stomatella (Moll.); stomati-ferous; Stomato-poda (Crust.); stomo-deum; Stomo-lophus (Coel.); Stomo-nema (Ins.); anastom-osis; Callio-stoma (Moll.); Cyclo-stomata (Pisc.); Di-stomum (Platy.).

stomach—Gr. *stomachos*, a mouth, the gullet, stomach < *stoma*, a mouth. *Ex:* stomach; stomach-ic; Stomachi-cola (Platy.); Stomacho-myia (Ins.).

stomais—Gr. *stomais*, hard-mouthed, "mouthy" < *stoma*, mouth. *Ex:* Stomais (Pisc.).

stomat—See stom.

stomb—Gr. *stombos*, deep-sounding, noisy. *Ex:* Stombus (Amph.).

stomic—Gr. *stōmix*, genit. *stōmikos*, a wooden beam.

stomix—See stomic.

stomph—Gr. *stomphos*, loud, bombastic. *Ex:* Stompho-sphinctes (Moll.).

stomphac—Gr. *stomphax*, genit. *stomphakos*, a ranter; *stomphastēs*, a great talker. *Ex:* Stomphastes (Ins.) Stomphax (Ins.).

stomphax—See stomphac.

stomyl—Gr. *stōmylos*, wordy, talkative. *Ex:* Stomylo-myia (Ins.); Stomylus (Platy.).

stonych—Gr. *stonyx*, genit. *stonychos*, any sharp point. *Ex:* Stonycho-phora (Ins.).

stonyx—See stonych.

storth—Gr. *storthē*, a point, spike, tyne of an antler. *Ex:* ?Storthia (Ins.); Stortho-sphaera (Prot.).

storthyng—Gr. *storthynx*, genit. *storthyngos*, a point, the tyne of an antler. *Ex:* Storthing-ura (Crust.); Storthingo-crinus (Echin.).

strab—L. *strabo*, genit. *strabonis*, a squinter, *strabus*, squinting < Gr. *strabos*, squinting, distorted; *strabismos*, a squinting. *Ex:* Strabops (Ins.); strabismus (Med.); Strabo-mantis (Amph.); Strabos-odon (Mam.); Strabus (Ins.).

strabos—See strab.

stragul—L. *stragulus*, a covering; *stragulum*, a blanket, carpet.

stramin—L. *stramen*, genit. *straminis*, straw; *stramineus*, made of straw.

stramon—NL. *stramonium*, name applied to the genus of thorn apples < Gr. *strychnos*, nightshade + *manikos*, mad. *Ex:* Stramon-ita (Moll.) Stramonium*

strang—Gr. *strangos*, twisted, crooked. *Ex:* Strangia (Por.); A-strangia (Coel.).

strangal—Gr. *strangalē*, a halter. *Ex:* Strangalina (Ins.); Strangalo-stoma (Moll.).

strangulat—L. *strangulatus*, choked, constricted; pp. of *strangulo*, to choke.

strapar—NL. *strapar* < Gr. *streptos*, twisted. *Ex:* Strapar-ollus (Moll.), see roll.

strat—1. L. *strata*, a paved road, a layer; *stratum*, neut. sing. of *strata*; *stratus*, spread out, layered, pp. of *sterno*, to spread out. *Ex:* Stratodus (Pisc.); Strata (Mam.); strati-fy; Stratipora (Coel.); Strato-phyllum (Coel.); stratum; humi-stratum: 2. Gr. *strataō*, to be encamped. *Ex:* Amphi-strate (Rept.).

strati—Gr. *stratios*, warlike, *stratiōtēs*, a soldier; also a water plant, the millfoil with sword-like leaves. *Ex:* Stratio-mys (Ins.); Stratiotes*.

strebl—Gr. *streblos*, twisted. *Ex:* Strebl-odus (Pisc.); Strebla (Ins.); Streblo-ceras (Moll.); Streblo-nema*; Streblus (Prot.).

stremmat—Gr. *stremma*, genit. *stremmatos*, a thread, anything twisted. *Ex:* Stremmat-opsis (Moll.); Stremmato-gnathus (Rept.).

stren—Gr. *strēnēs*, strong, hard, harsh; *strēnos*, haughtiness. *Ex:* Streno-ceras (Moll.); Strenoloma (Ins.); A-strenis (Ins.).

strenu—L. *strenuus*, restless, quick, nimble. *Ex:* Strenu-ella (Tri.).

streper—LL. *streperus*, noisy.

streph—Gr. *strephō*, fut. *strepsō*, to twist, turn *strepsis*, a twisting. *Ex:* Streph-uris (Ann.); Strepho-basis (Moll.); Streps-axis (Moll.); Strepsi-las (Av.); Strepsi-ceros (Mam.); Strepsiptera (Ins.); cata-strepsis; Omma-strephes (Ins.). See also stroph.

strepitan—L. *strepitans*, genit. *strepitantis*, making a great noise; ppr. of *strepito*, to make a loud noise.

**streps**—See **streph**.

**strept**—Gr. *streptos*, twisted, bent. *Ex:* Strept-axis (Moll.); Strepto-ceryl (Av.); strepto-coccus; Strepto-neura (Moll.).

**stri**—L. *stria*, pl. *striae*, a furrow, channel, NL. dim. *striola*; *striatus*, striped <*strio*, to furnish with furrows; NL. *striolatus*, finely grooved. *Ex:* Stri-acanthus (Pisc.); stri-al; Stri-aria*; Stri-luna (Moll.); Stria-ptera (Ins.); Striato-pora (Coel.); striola; Strio-lucina (Moll.); strio-punctate; striolate; Striolatus (Moll.); Pyro-stria*.

**striat**—See **stri**.

**strict**—L. *strictus*, drawn tight; pp. of *stringo,* to bind, press together; *strictura*, a contraction. *Ex:* Strict-echinus (Echin.); Stricto-gonia (Ins.); strictus (as used by botanists, meaning straight, erect); stricture.

**striden**—See **stridul**.

**stridul**—L. *stridulus*, harsh, creaking <*strideo,* to creak; ppr. *stridens*, genit. *stridentis*, creaking, noisy. *Ex:* strident, stridul-ate; stridul-ous; Striduli-velia (Ins.).

**strig**—1. Gr. *strix*, genit. *strigos* (>NL. plural *striges*), an owl, a night-bird >NL. *string-*, an owl. *Ex:* Strig-idae (Av.); Striges (Av.); Strigi-philus (Ins.); String-ops (Av.); Stringo-cephalus (Brach.); Strix (Av.): **2.** L. *striga*, furrow, streak; *strigatus*, streaked, striped. *Ex:* striga; Strigat-ella (Moll.); Strigi-chiton (Moll.); Strigo-cuscus (Mam.): **3.** NL. *stria*, appressed stiff hairs, bristles or scales, often of unequal length >NL. strigose, beset with striae. *Ex:* strigose.

**strigat**—See **strig**.

**strigil**—L. *strigilis* = *strigula*, a scraper, a flesh-brush; NL. *strigilatus*, furnished with a scraper. *Ex:* strigilis; Strigil-ina (Ins.); strigilat-or; Strigilla (Moll.); Strigilo-delima (Moll.); Strigula (Moll.).

**strigos**—L. *strigosus*, thin, lean, destitute, barren. See also strig 3.

**string**—See **strig**.

**striol**—See **stri**.

**striphn**—Gr. *striphnos*, firm, solid. *Ex:* Striphno-pteryx (Ins.).

**strix**—See **strig**.

**strob**—Gr. *strobos*, a whirling round. *Ex:* Strobo-ceras (Moll.).

**strobil**—Gr. *strobilos*, anything twisted, a pine cone. *Ex:* strobil-ation; Strobil-ops (Moll.); Strobil-urus (Rept.); strobila; strobili-ferous; Strobili-phaga (Av.); Strobilo-cephalus (Platy.)

**strom**—Gr. *strōma*, genit. *strōmatos*, a mattress, bed. *Ex:* stroma; Stromat-actis (Coel.); Stromateus (Pisc.); Stromato-pora (Coel.); A-stroma (Ins.); cono-stroma.

**stromat**—See **strom**.

**stromb**—L. *strombus* = Gr. *strombos*, a turban, a top; also a kind of spiral snail; Gr. *strombōdēs,* like a top. *Ex:* Stromb-ella (Moll.); Stromb-idium (Moll.); Strombi-formis (Moll.); Strombodes (Prot.); Strombus (Moll.).

**strongul**—See **strongyl**.

**strongyl**—Gr. *strongylos*, round, compact. *Ex:* Strongyl-iscus (Pisc.); Strongylo-centr-otus (Echin.); Strongulo-gnathus (Mam.); Strongylus (Nemat.).

**strongylot**—Gr. *strongylotēs*, roundness. *Ex:* Strongylotus (Moll.).

**stroph**—Gr. *strophos*, twisted, *strophē*, a turning; *stropheus*, one of the vertebrae <*strephō*, to turn about, to twist; *strophalos*, a top or whirling instrument; *strophios*, a twisting, slippery fellow. *Ex:* Stroph-anthus*; Stroph-ella (Moll.); Stroph-esia (Moll.); Stroph-urus (Rept.); Strophalosia (Moll.); Strophi-odonta (Brach.); Strophi-ona (Ins.); Strophia (Moll.); stropho-genesis; Stropho-lirion*; Stropho-poda (Ins.); epi-stropheus; Sys-trophia (Moll.).

**strophad**—Gr. *strophas*, genit. *strophados*, winding, circling, revolving.

**strophal**—See **stroph**.

**strot**—Gr. *strōtos*, spread, laid. *Ex:* Stroto-crinus (Echin.); carpo-strotes (Ecol.).

**struic**—L. *struix*, genit. *struicis,* a heap, a mass of things.

**struix**—See **struic**.

**strum**—L. *struma*, a scrofulous tumour; *stru-mosus*, swollen. *Ex:* Strum-ella*; struma; strumi-fer-ous; Strumi-genys (Ins.); Strumi-ger (Ins.); Strumosa (Moll.).

**struth**—L. *struthio*, genit. *struthionis*, an ostrich; *strouthos*, any small bird. *Ex:* Struth-ida (Av.); struthi-form; Struthio (Av.); Struthio-laria (Moll.); Struthio-mimus (Rept.); Struthionis (Av.); Struthio-scelis (Ins.).

**strychn**—Gr. *strychnos*, a kind of nightshade. *Ex:* Strychnos*.

**strygo**—See **strig**.

**stryphn**—Gr. *stryphnos*, astringent. *Ex:* striphn-ic; Stryphno-dendron*.

**stult**—L. *stultus*, foolish, simple. *Ex:* Stulta (Av.).

**stup**—L. *stupa*, tow, the coarse parts of flax; *stupeus*, made or consisting of tow; ML. *stuposus*, bearded. *Ex:* stupeous; stupose; stup-ul-ose; Larno-stupa (Prot.).

**stupr**—L. *stupro*, to corrupt, defile; *stupratus*, de-filed, pp. of *stupro; stuprator*, a defiler. *Ex:* stuprat-ion. See masturb.

**stur**—LL. *sturio*, genit. *sturionis*, the sturgeon. *Ex:* Sturio (Pisc.); Sturion-idae (Pisc.).

**sturn**—L. *sturnus*, a starling. *Ex:* Sturn-ella (Av.); Sturn-ornis (Av.); Sturna (Av.); Sturno-paster (Av.); not Sturn-ira (Mam.), named after the ship, *Starling*, consort of the H.M.S. Sulphur in 1836, when the type was collected.

stych—See stich.

styg—Gr. *Styx*, genit. *Stygos*, the lower world; *Stygios*, pertaining to the Styx < *stygeō*, to hate; *stygos*, hatred; *styganos*, odious; *stygnos*, abhorred, hated, sullen. *Ex:* Styg-ides (Ins.); Stygan-odon (Moll.); Stygi-cola (Pisc.); Stygia (Ins.); Stygio-chelifer (Arach.); Stygni-cranus (Arach.); Stygo-genes (Pisc.); Stygno-hydrus (Ins.); Stygnus (Arach.).

stygan—See styg.

stygen—See styg.

styger—*stygeros*, wretched, hateful. *Ex:* Stygeromyia (Ins.).

styget—Gr. *stygētos*, hated, despised. *Ex:* Stygeto-blatta (Ins.).

styl—Gr. *stylos*, a style, stake, pillar; *stylōtos*, having pillars; *styloō*, to prop with pillars. *Ex:* Styl-ac-odon (Mam.); Styl-act-ella (Coel.); Styl-actis (Coel.); Styl-idium*; Styl-in-odon (Mam.), see ino 1.; Styl-ochus (Platy.); Stylommato-phora (Moll.); Stylat-ula (Coel.); Styl-phorus (Pisc.); styli-form; Stylo-trochus (Coel.); Stylon-urus (Crust.); Stylos-anthes*; Fimbri-stylis*; pygo-style.

stylon—See styl.

stylot—See styl.

stym—Gr. *styma*, priapism > *styō*, fut. *stysō*, to make stiff or erect. *Ex:* stym-ose.

stymphal—Gr. *Stymphalos*, a mountain of Arcadia. *Ex:* Stymphalus (Crust.).

styp—1. Gr. *stypos*, a stump, stem. *Ex:* Styp-odon (Pisc.); Stypo-lophus (Mam.):    2. Gr. *styppē* = *stypē*, the coarse fiber of hemp or flax. *Ex:* Stypo-larcus (Prot.).

styphel—Gr. *styphelos*, solid, hard, rough. *Ex:* Styphelia*.

styphl—Gr. *styphlos*, rough, harsh; also sour, astringent *Ex:* Styphlo-trema (Platy.); Styphlos (Por.).

styptic—Gr. *styptikos*, astringent. *Ex:* styptic.

styrac—1. Gr. *styrax*, genit. *styrakos*, ancient name for a tree producing a fragrant gummy resin called storax by Pliny and Vergilius Maro. *Ex:* Styrac-aceae*; Styrax*:    2. Gr. *styrax*, genit. *styrakos*, the spike at the lower end of the shaft of a spear. *Ex:* Styrac-aster (Echin.); Styraco-teuthis (Moll.); Styrax (Ins.); Styraxo-desmus (Myr.).

styrax—See styrac.

stys—Gr. *stysis*, an erection, a setting up. *Ex:* Stysanus*. See also stym.

stytic—Gr. *stytikos*, causing erection, priapism. See also sty.

su—L. *sus*, genit. *suis*, the pig; *suillus*, pertaining to swine. *Ex:* Sui-dae (Mam.); Suill-ine; suillus; Sus (Mam.); Sy-arctos (Mam.); Sy-ornis (Av.); Syo-therium (Mam.); Syspotamus (Mam.).

suav—L. *suavis*, sweet, agreeable; *suavitas*, sweetness, pleasantness; *suaveolens*, sweet smelling < *suave*, sweetly, + *olens*, smelling. *Ex:* Suavi-psitta (Av.); Suavo-trochus (Moll.).

sub-  —L. *sub-*, prefix meaning under, below, almost, somewhat, near. The *b* is sometimes changed into the consonant with which the next syllable begins. *Ex:* sub-auratus, slightly gilt; sub-maxillary; Sub-ursus (Mam.); succint; suc-cavus; suf-farcinate; suf-fuse.

California Styrax, *Styrax californica*, a white flowered shrub of the mountains.

subbullien—L. *subbulliens*, genit. *subbullientis*, gently bubbling or foaming; ppr. of *subbullio*, to foam.

subcub—L. *subcubo* = *succubo*, to be under. *Ex:* subcub-ous.

suber—L. *suber*, cork, the cork oak; *subereus* = *suberinus*, of the cork tree; *suberosus*, corky in texture. *Ex:* Suber-ites (Por.); suberi-fic-ation; suberin; Subero-corona (Por.).

subis—L. *subis*, name of some unknown bird that breaks eagle's eggs.

subit—L. *subitus*, sudden, unexpected; *subitarius*, done suddenly, hastily < *subeo*, to approach, to spring upon.

sublat—L. *sublatus*, elated, proud < *tollo*, to lift up, elevate.

subol—L. *suboles*, an offspring, a sprout.

subsessor—L. *subsessor*, genit. *subsessoris*, one who lies in wait.

subsolan—L. *subsolanus*, eastern, oriental, lit. lying under the sun.

subter-  —L. *subter-*, prefix meaning below, beneath, underneath; *subternus*, that is underneath, lower.

subtil—L. *subtilis*, slender, minute, delicate.

subul—1. L. *subula*, an awl. *Ex:* subul-ate; Subul-aria*; Subul-ina (Moll.); subuli-fer-ous.

**2.** L. *subulo*, one who plays the flute; also a kind of deer with pointed horns. *Ex:* Subulo (Mam.).

**subvex**—L. *subvexus*, inclining upwards.

**suc**—See sub.

**succ**—L. *succus* = *sucus*, juice, sap; *succosus* = *sucosus*, juicy, sappy; *succulentus*, juicy, succulent < *succus*, juice + *-lentus*, full of. *Ex:* Succulent-ae*; succus entericus.

**succedan**—L. *succedaneus*, that follows after, succeeds to; also as a noun, a substitute.

**succin**—L. *succinum*, amber; *succineus*, of amber color; *succinaceus*, amber-colored, prob. < *succus*, juice, sap. *Ex:* Succinea (Moll.).

**succinct**—L. *succinctus*, short, small, contracted, pp. of *succingo*, to gird, tuck up. *Ex:* Succincta (Moll.).

**succiss**—L. *succissus*, cut down, trimmed off below < *succido*, to cut down.

**succos**—See succ.

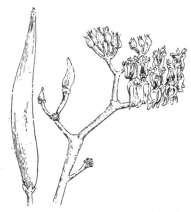

*Asclepias subulata*, a milk weed with awl-shaped leaves. Redrawn from Desert Wild Flowers—Jaeger. Stanford University Press.

**succulent**—See succ.

**such**—Gr. *souchos*, an Egyptian name for the crocodile. *Ex:* Sucho-saurus (Rept.); Eu-suchia (Rept.); Para-suchia (Rept.); Rampho-suchus (Rept.).

**sucr**—Fr. *sucre*, sugar. *Ex:* sucr-ase; sucr-ore.

**suct**—L. *suctus*, sucking, pp. of *sugo*, to suck; *suctus*, a sucking, ML. *suctori(al)*, adapted for sucking; OFr. *suction*, a sucking. *Ex:* Melli-suga = Meli-suga (Av.); suction; suctori-al; Suctoria (Prot.).

**sudis**—L. *sudis*, a stake, pile, pillar, thorn, arrow; also a fish called pike. *Ex:* Sudis (Pisc.); Omo-sudis (Pisc.).

**sudor**—L. *sudor*, sweat. *Ex:* sudori-fer-ous; sudori-fic.

**suecic**—L. *Suecica*, Swedish. *Ex:* suecicus.

**sued**—Ar. *Suaed*, a soda yielding plant. *Ex:* Sueda*.

**suf**—See sub-.

**sufflamen**—L. *sufflamen*, genit. *sufflaminis*, impediment.

**suffocat**—L. *suffocatus*, choked, stifled.

**suffult**—L. *suffultus*, propped up, supported; pp. of *suffulcio*, to support with props.

**sug**—See suct.

**sugillat**—L. *sugillatus*, beaten black and blue < *sugillo*, to bruise.

**suill**—See su.

**sula**—Icelandic *sula* = *sule*, said to mean an awkward fellow and applied to the solan-goose or the gannet. *Ex:* Sula (Av.).

**sulc**—L. *sulcus*, a furrow, NL. dim. *sulculus*; *sulcatus*, furrowed. *Ex:* Sulc-astrum (Moll.); Sulci-basis (Moll.); sulci-form; Sulcul-aria (Moll.); Sulco-bombus (Ins.); sulcus; bi-sulcate.

**summat**—**1.** ML. *summatus*, summed up, pp. of *summo*, to sum up. *Ex:* summat-ion: **2.** L. *summas*, genit. *summatis*, noble, distinguished: **3.** L. *summatim*, on the surface, with briefness.

**sundar**—Sansk. *sundara*, elegant. *Ex:* Sundar-us (Ins.).

**super**—L. *super*, over, above, on top; *superus*, upper, higher, exceeding in size and length;

Supernumerary teats of male.

*supero*, to rise above, surmount; ppr. *superans*, genit. *superantis*, rising above, surmounting. *Ex:* Super-lestes (Ins.); super-numer-ary; Superi-cornes (Ins.).

**superbiens**—L. *superbiens*, being proud, haughty.

**superflu**—L. *superfluus*, running over, overfull, redundant.

**supern**—L. *supernus*, higher, celestial; *superne*, upwards, above. *Ex:* Supern-ola (Ins.).

supernat—1. L. *supernas*, genit. *supernatis*, of or from the upper country: **2.** L. *supernatus*, growing above.

superst—L. *superstes*, genit. *superstitis*, a witness, that which remains to testify even after death.

suphalasca—Anagram of *Ascalaphus. Ex:* Suphalasca (Ins.).

supin—L. *supinus*, lying with face upwards, i.e. on the back, prostrate, bent backwards. *Ex:* supin-ation; supine; Re-supinata (Moll.).

supplex—See supplic.

supplic—L. *supplex*, genit. *supplicis*, humble, earnestly begging.

supra- —L. *supra-*, prefix meaning above, over, beyond, usually in a positional sense. Used much in the same way as super. *Ex:* supra-renal; supra-scapular.

sur- —1. Fr. *sur-*, prefix, contracted from L. *super. Ex:* sur-anal; sur-angulare: **2.** L. *sur-* prefix, assumed form of *sub-*, before syllables beginning with r. *Ex:* sur-reptiti-ous; sur-ruf-ous.

surcat—NL. *surcatus*, having shoots or suckers< *surculus*, a young branch or shoot.

surcul—L. *surculus*, a sprout, young branch; *surculosus*, woody. *Ex:* Surcul-ina (Moll.); Surcul-ites (Moll.) Surcula (Moll.); Surculo-fusus.

surd—L. *surdus*, deaf. *Ex:* Surdi-sorex (Mam.).

surg—L. *surgo*, to rise > *adsurgens* = *assurgens*, rising upward, ascending.

suricat—Javanese *suracatje* < *sura*, a chief + *catje*, a kitten. *Ex:* Suricat-oecus (Ins.); Suricata (Mam.).

surni—NL. *surnia*, a bird name given by Dumeril in 1806, etymology unknown. *Ex:* Surnia (Av.).

surrect—L. *surrectus*, erect, nearly straight.

surs—L. *sursum*, upwards, high up, above.

sus—See sub.

suscept—L. *susceptio*, genit. *susceptionis*, a taking in hand; *susceptus*, undertaken, pp. of *suscipio*, to support, to take up. *Ex:* intus-suscept-ion.

suspect—L. *suspectus*, distrusted, suspicioned.

suspicac—L. *suspicax*, genit. *suspicacis*, suspicious.

suspicax—See suspicac.

susurr—L. *susurro*, to buzz, ppr. *susurrans*, genit. *susurrantis*, buzzing, humming.

sutil—L. *sutilis*, sown together.

sutor—L. *sutor*, genit. *sutoris*, a shoe-maker.

sutur—L. *sutura*, a seam < *suo*, to sew. *Ex:* Sutur-aspis (Ins.); Sutur-odes (Myr.); suture; suturi form.

swimmer—Eng. *swimmer*, one that swims < AS. *swimman*, to swim. *Ex:* swimmer-et (-*et*, a dim.).

sy—See su.

syaci—Gr. *syakion*, a name for some small fish; dim. of *syax*, genit. *syakos*, a kind of fish; also a kind of legume. *Ex:* Syacium (Pisc.).

syad—Gr. *syas*, genit. *syados*, a sow.

syas—See syad.

sybax—Gr. *sybax*, genit. *sybakos*, hoglike, swinish. *Ex:* Sybax (Ins.).

sybin—Gr. *sybinē* = *sybēnē*, a flute-case.

syc—Gr. *sykon* dim. *sykidion*, a fig: *sykinos*, of the fig tree; *sykitēs*, fig-like. *Ex:* Syc-andra (Por.); Syc-arium (Por.); Syc-etta (Por.); Syc-opsis*; Syc-ur-ella (Por.); syca-more; Sycinula (Por.); Syco-carpus (Por.); Syco-nycteris (Mam.); Sycon (Por.); Bu-sycon (Moll.).

sychneon—Gr. *sychneōn*, genit. *sychneōnos*, a thicket.

sycophant—L. *sycophanta*, a deceiver < Gr. *sykophantēs*, a false accuser.

sylleg—Gr. *syllegō*, to collect, gather. *Ex:* Syl lego-mydas (Ins.); Syllego-ptera (Ins.).

syllexis—Gr. *syllexis*, a contribution. *Ex:* Syl lexis (Ins.).

sylv—L. *sylva*, a wood; *sylvaticus*, growing among trees; *Sylvanus*, god of the woods. *Ex:* Syl vania (Av.); Sylvano-cochlis (Moll.); Sylvanus (Mam.); Sylvi-lagus (Mam.); Sylvio-cantor (Av.).

sym—See syn.

symbio—Gr. *symbiōsis*, also *symbios*, a living together. *Ex:* Symbio-chara (Ins.); symbiosis.

symmach—Gr. *symmachis*, genit. *symmachidos*, allied, near; *symmachia*, an alliance. *Ex:* Symmachis (Ins.); Symmachia (Ins.).

symmctr—Gr. *symmetrōs*, in accord with, suitable; *symmetria*, symmetry. *Ex:* Symmetro-pleura (Ins.); symmetr-y.

symp—Gr. *sympiezō*, to squeeze together; *sympyknos*, pressed together. *Ex:* Symp-etrum (Ins.); Sympieza*; Sympycna (Ins.).

sympaestr—Gr. *sympaistria*, a playmate. *Ex:* Sympaestria (Ins.).

sympect—See sympex.

sympex—Gr. *sympexis*, a concretion; *sympēktos*, joined together, constructed.

sympher—Gr. *sympheron*, useful. *Ex:* Symphero-bius (Ins.).

symphy—Gr. *symphyō*, to glue together. *Ex:* Symphyo-loma*.

symphyt—Gr. *symphyton*, name of a kind of plant with healing properties, comfrey. *Ex:* Symphytum*.

sympiez—See symp.

symploc—Gr. *symplokē*, a twisting together. *Ex:* Symplo-carpus*.

sympycn—See symp.

**syn-** —Gr. *syn-*=*sym-*, (sometimes *syr-* or *sys-*), together; prefixes entering into to many compounds. *Ex:* sym-biosis<Gr. *symbiōsis*, a living together; Sym-bor-odon (Mam.); Symphemia (Av.); sym-phile; Sym-phori-carpos*; Sym-phytum*; syn-apsis; Syn-apta (Echin.); Syn-apto-mys (Mam.); syn-cyt-ium; Syn-ed

Skull of Radcliff's East African Buffalo, *Syncerus coffer radclifferi.*

ella*; syn-oeketes; Syn-osma*; syn-otic tectum; Syn-therisma*; Syr-rhaptes (Av.); syssarcis; Sys-saura (Ins.); Sys-sphinx (Ins.); etc.

**synairema**—Gr. *synairema*, a union, a drawing together. *Ex:* Synairema (Ins.).

**synallax**—Gr. *synallaxis*, exchange, interchange. *Ex:* Synallaxis (Av.).

**synanceia**—Gr. *synankeia*, a narrow valley in which streams meet. *Ex:* Synanceia (Pisc.).

**synaph**—Gr. *synaphē*, connection, union; *synaphēs*, united, connected. *Ex:* Synapha (Ins.); Synapho-branchus (Pisc.); Erio-synaphe*.

**synaps**—Gr. *synapsis*, a union, joining together. *Ex:* synapse; synapsis. See hapt.

**synarmog**—Gr. *synarmogē*, a combination. *Ex:* Synarmog-oidea (Ins.).

**synars**—Gr. *synarsis*, union. *Ex:* Synarsis (Ins.).

**synchn**—Gr. *synchnos*, long in point of time, thick, many. *Ex:* Sychno-portus (Ins.); A-sychna (Ins.).

**syncras**—Gr. *synkrasis*; a mixing, mingling, blending. *Ex:* idio-syncras-y.

**syndes**—Gr. *syndesis*, a binding together; *syndetos*, well knit together. *Ex:* syndesis; Syn-deto-crinus (Echin.); Syndetus (Ins.).

**syndet**—See **syndes.**

**syndy**—Gr. *syndyo*, two together, double. *Ex:* Syndyo-ceras (Mam.); Syndyo-graptus (Coel.).

**synech**—Gr. *synechō*, to hold together, unite. *Ex:* Syneches (Ins.); Synecho-cryptus (Ins.).

**synemosyn**—Gr. *synēmosynē*, ties of relationship. *Ex:* Synemosyna (Arach.).

**synetaer**—Gr. *synetairos*, a comrade, mate. *Ex:* Synetaeris (Ins.).

**synod**—Gr. *synodos*, an assembly, a coming together. *Ex:* Synodus (Crust.).

**synophrus**—Gr. *synophrys*, lit. brow brought together, sad, frowning, cruel, proud. *Ex:* Synophrus (Ins.).

**synovi**—NL. *synovial*, pertaining to the viscous fluid of the joint cavity, perh.<Gr. *syn*, with+*ovum*, an egg (referring here to the egg white)+*-al*. *Ex:* synovi-al.

**syntherism**—Gr. *syntherismos*, reaping together, crop-making. *Ex:* Syntherisma*.

**syphar**—Gr. *syphar*, a wrinkled skin, also a decrepit old person. *Ex:* Sypharo-chiton (Moll.); Micro-syphar*.

**syphe**—Gr. *sypheos*, a hogsty. *Ex:* Syphe-otis (Av.).

**syphon**—See **siphon.**

**syr-** —See **syn-.**

**syr**—Gr. *syra*, skin, a coat of goat's skin. *Ex:* Syro-mastes (Ins.).

**syring**—Gr. *syrinx*, genit. *syringos*, a pipe, a tube. *Ex:* Syringa*; Syringo-pora (Coel.); Syringo-dea*; syrinx; Cymato-syrinx (Moll.).

**syrinx**—See **syring.**

**syristes**—Gr. *syristēs*, a piper. *Ex:* Syristes (Av.).

**syrm**—1. Gr. *syrma*, genit. *syrmatos*, something trailed or dragged along; also sweepings, refuse. *Ex:* Syrma (Arach.); Syrmatia (Ins.); Syrmato-phora (Moll.); Syrmo-cerus (Ins.); A-syrma (Av.); Bathmi-syrma (Av.): 2. Gr. *syrmas*, genit. *syrmados*, a drift (of dust or snow).

**syrmatic**—Gr. *syrmatikos*, trailing<*syrma*, a robe that trails; refuse; also in medicine, a place where the skin has been torn off. *Ex:* Syrmaticus (Av.).

**syrni**—Gr. *syrnion*, a bird of evil omen. *Ex.* Syrnium (Av.).

**syrph**—Gr. *syrphos*, a small flying insect. *Ex.* Syrph-idae (Ins.); Syrpho-phagus (Ins.); Syrphus (Av.).

**syrraxis**—Gr. *syrraxis*, a clashing together. *Ex:* Syrraxis (Pisc.).

**syrrhapt**—Gr. *syrrhaptos*, sewn together, see syn and rhapt. *Ex:* Syrrhapt-idae (Av.); Syr rhapt-oecus (Ins.); Syrrhaptes (Av.).

**syrrhiz**—Gr. *syrrizos*, with roots united. *Ex:* Syrrhiz-odes (Ins.); Syrrhizus (Ins.).

**syrt**—Gr. *syrtos*, washed down by a stream, trailing. *Ex:* Syrt-odes (Ins.).

**syrtid**—Gr. *syrtis*, Ionic genit. *syrtidos*, a quicksand. *Ex:* syrtido-phyta (Ecol.); syrtis (Ins.).

**syrtis**—See **syrtid.**

**sys**—See **syn-** also **su.**

**syschid**—Gr. *syschidēs*, rent asunder.

**sysci**—Gr. *syskios*, thickly shaded; also a closely shaded place; *syskiazō*, to throw a shade over. Syscia (Ins.); Syscio-blatta (Ins.).

**sysphing**—Gr. *sysphingō*, to bind close together. *Ex:* Kalio-sysphinga (Ins.).

**syssarcos**—Gr. *syssarkōsis*, overgrown with flesh. *Ex:* syssarcosis.

**systalt**—Gr. *systaltikos*, contractile. *Ex:* Sys talto-cerus (Ins.).

**systell**—Gr. *systellō*, to draw together. *Ex:* Sys tello-cerus (Ins.).

**system**—Gr. *systēma*, genit. *systēmatos*, a complex whole put together; *systēmatikos*, pertaining to a system, according to system. *Ex:* system-ic; System-odon (Mam.); systemat-ist; systematic.

**systen**—Gr. *systenos*, running to a fine point. *Ex:* Systeno-gnathus (Ins.); Systenus (Ins.).

**ystol**—Gr. *systolē*, a contraction, drawing to-gether. *Ex:* Systol-ides (Rot.); systole; Systole-derus (Ins.); Systolo-tettix (Ins.).

**systom**—Gr. *systomos*, having a narrow mouth, i.e., one drawn together. *Ex:* Systoma (Amph.); Systomus (Pisc.).

**systroph**—Gr. *systrophē*, a twisting together < *systrephō*, to twist up, roll up. *Ex:* Systropha (Ins.); Systrophia (Moll.); Systropho-ceras (Moll.).

**syzyg**—Gr. *syzygos*, a joining or yoking together < *syn*, together + *zygon*, a yoke. *Ex:* Syzy gium*; Syzygo-phyllia (Coel.); Syzygoni (Ins.); syzygy.

# T

**a**—Gosiute Indian, *tai*, to emerge, project. *Ex:* Ta-iulus (Arach.).

**tabacari**—NL. *tabacarius*, pertaining to or of tobacco, of the odor or color of tobacco < NL. *tabacum*, tobacco. *Ex:* Tabacaria (Pisc.).

**tabal**—Gr. *tabala*, a drum. The Moorish word is *atabal*, later adopted by the Spanish.

**taban**—L. *tabanus*, the horse fly. *Ex:* Taban-idae (Ins.); Tabano-cella (Ins.); Tabanus (Ins.).

**tabebuia**—Tupi *tabebuya*, ant-wood. *Ex:* Tabebuia*.

**tabell**—L. *tabella*, a little tablet; also a fan. *Ex:* Tabell-aria (Prot.).

**tabes**—See **tabesc**.

**tabesc**—L. *tabescens*, genit. *tabescentis*, wasting away, ppr. of *tabesco*, to waste away; *tabes*, a consuming, wasting away. *Ex:* tabescent.

**tabid**—L. *tabidus*, shrinking or wasting away, putrefying < *tabeo*, to languish, waste away. *Ex:* Tabidia (Ins.).

**tabul**—L. *tabulatus*, boarded, floored, layered; *tabularis*, flat-surfaced, relating to boards < *tabula*, a floor. *Ex:* Tabula (Ins.); tabulare; Tabulata (Coel.); Tabulo-phyllum (Coel.); not Tabulo-digitus (Por.) nor Tabularia (Coel.), which are errors for Tubulo-digitus and Tubul-aria.

**tabulat**—See **tabul**.

**tacc**—Malay *tacca*, arrowroot. *Ex:* Tacc-aceae*; Tacca.*

**tacer**—Gr. *takeros*, tender. *Ex:* Tacerus*.

**tach**—Gr. *tachys*, swift = *tacheos*; *tachos*, swift-ness, speed = poet. *tachinos*; *tachytēs*, swift of foot; *tachypetēs*, flying fast; *tachykinētos*, moving rapidly. *Ex:* Tach-uris (Av.); Tachea (Moll.); Tacheo-campylaea (Moll.); Tachina (Ins.); Tachin-aria (Ins.); Tachino-ptera (Ins.); Tachinus (Av.); Tachipetes (Av.); Tachus (Arach.); Tachy-cineta (Av.); tachy-genesis; Tachy-sphex (Ins.); Tachypetes (Av.); Tachytes (Ins.); not Tachardia (Ins.), named after G. Tachard, French Jesuit missionary in the East Indies.

**tacheo**—See **tach**.

**tachin**—See **tach**.

**tachy**—See **tach**.

**tachyeres**—Gr. *tachyērēs*, rowing rapidly, rapid. *Ex:* Tachyeres (Av.).

**tacso**—Peruvian *tacsa*, small. *Ex:* Tacsonia*.

**tact**—**1**. L. *tactus*, a handling, touch < *tango*, to touch. *Ex:* tact-ile; Tacto-derus (Ins.); tactu-al: **2**. Gr. *taktos*, prescribed; *taktikos*, fit for ordering. *Ex:* Tacto-comus (Ins.); a-tacto-desmic.

**tactil**—L. *tactilis*, pertaining to touch.

**tadarid**—NL. *tadarida*, a word created by Rafinesque to designate a genus of bats. Concerning its origin and meaning, he left us no word. *Ex:* Tadarida (Mam.).

**tadorn**—NL. *tadorna* < Fr. *tadorne*, a sheldrake, a kind of duck. *Ex:* Tadorna (Av.).

**taed**—L. *taeda*, a kind of pine.

**taedios**—L. *taediosus*, disgusting, loathsome.

**taen**—Gr. *tainia*, band, ribbon > L. *taenia*; NL. *taeniatus*, stripped. *Ex:* Taen-arthrus (Ins.); taen-idium; Taenia (Platy.); Taenio-poda

(Ins.); Taenio-pteryx (Ins.); Taeno-soma Tainis*; Taino-ceras (Moll.); Tenia (Platy.); Tenia-notus (Pisc.); A-taenia*; Lino-taenia (Arthr.); Peri-taenius (Ins.).

**tag**—Gr. *tagos*, a commander, leader *Ex:* Tagona (Ins.).

**tagal**—Malayan *tagala*, Phillipine branch of the Malayan language. *Ex:* Tagalo-psocus (Ins.); Tagalus (Ins.).

**tagassu**—NL. *tagassu*<Tupi *tayacu*, pig. *Ex:* Tagassu (Mam.).

**tagetes**—NL. *tagetes*, a plant name said to be <L. *Tages*, Etrurian god, grandson of Jupiter, who sprang from the earth as a boy and taught the art of ploughing to the Etrurians. *Ex:* Tagetes*.

**tagi**—Gr. *tageia*, stewardship, management. *Ex:* Tagi-ades (Ins.).

**tagma**—Gr. *tagma*, genit. *tagmatos*, that which has been ordered or arranged; also a division; *tagmatikos*, of or for a division. *Ex:* tagmatic; Mono-tagma*.

**tagmat**—See **tagma**.

**tain**—See **taen**.

**tal**—1. L. *talus*, the ankle, also the heel>Fr. *talon*, the heel. *Ex:* tali-form; talo-tibital; Talona (Moll.): 2. Gr. *talas*, genit. *talanos*, poor, wretched. *Ex:* Tal-orchestria (Crust.); Talanes (Ins.); A-talo-phlebia (Ins.).

**talaepor**—Gr. *talaipōria*, hard work; *talaipōros*, suffering, miserable. *Ex:* Talaeporia (Ins.); Taleporia (Ins.).

**talan**—See **tal 2**.

**talant**—Gr. *talanton*, a balance, pair of scales; *talantōsis*, a weighing, a swinging to and fro. *Ex:* Talanto-discus (Moll.).

**talar**—Gr. *talaros*, a basket. *Ex:* Talaro-crinus (Echin.).

**talasi**—Gr. *talasia*, wool-spinning. *Ex:* Talasius (Ins.).

**talaum**—Abor. *talauma*, native name of a tree. *Ex:* Talauma*.

**tale**—1. New Guinean *tale*, native name for a water-hen. *Ex:* Tale-gallus (Av.): 2.L. *talea*, a rod.

**talin**—NL. *talinum*, a plant name, etym. unknown, perh. <Gr. *thaleia*, full of bloom, luxuriant. *Ex:* Talin-opsis*; Talinum*.

**talitr**—L. *talitrum*, a rap with the finger. *Ex:* Talitr-ella (Crust.); Talitrus (Crust.).

**talp**—L. *talpa*, a mole. *Ex:* Talp-avus (Mam.); Talpa (Mam.); Talpa-sorex (Mam.); Talp-oides (Mam.); Gryllo-talpa (Ins.).

**tamandua**—Brazilian *tamandua*, an ant-trap. *Ex:* Tamandua (Mam.).

**tamaric**—L. *tamarix*, genit. *tamaricis*, a tamarisk <Tamaris River on the border of the Pyrenees where it grows; called also *tamariscus*. *Ex:* Tamarisca (Coel.); Tamaric-aceae*; Tamarix*.

**tamarind**—ML. *tamarindus*, tamarind<Arabic *tamr*, a dried date+ *Hindi*, Indian, the Indian date. *Ex:* Tamarindi-formis (Moll.); Tamarindus*.

**tamarisc**—See **tamaric**.

**tamarix**—See **tamaric**.

**tamia**—Gr. *tamias*, a storer, a distributor. *Ex:* Tamia-sciurus (Mam.); Tamias (Mam.): Tamio-soma (Crust.); Eu-tamias (Mam ).

**tamio**—See **tamia**.

**tamn**—See **temn**.

**tamonea**—NL. *tamonea*, a plant name, etym unknown. *Ex:* Tamonea*.

**tamus**—NL. *tamus*<L. *tamnus*, a kind of wild climbing plant. *Ex:* Tamus*.

**tan**—Gr. *tanaos*, stretched, tall, long<*tanyō*, to stretch, to spread; *tanysipteros*, with spreading wings; *tanystys*, a stretching. *Ex:* Tan-arthr-ipsis (Ins.); Tanao-rhamphus (Nem).; Tany-ops (Mam.); Tany-pus (Arach.); Tany-stylum (Ins.); Tanysiptera (Av.): Tanystus (Rept.).

**tanacet**—See **tenacet**.

**tanagr**—See **tangar**.

**tanais**—Gr. *Tanais*, the river now known as the Don. *Ex:* Tanais (Crust.).

**tanao**—See **tan**.

**tandan**—NL. *tandanus*<native Australian name for a fresh water fish.

**tang**—L. *tangens*, genit. *tangentis*, touching, ppr. of *tango*, to touch. *Ex:* tangenti-al; tango-receptor.

**tangar**—NL. *tangar* prob.<Tupi *tangara*, name for some bright-colored bird, a martin. *Ex:* Tangar-idae (Av.); Tanagara (Av.); ?Tangaro-saurus (Rept.); Tanagr-idae (Av.); Tanagra (Av.).

**tangent**—See **tang**.

**tantal**—Gr. *Tantalos*, mythical king of Phrygia sent for punishment to the infernal regions. *Ex:* Tantal-ops (Av.); Tantalus (Av.).

**tantill**—NL. *tantilla*, an unexplained name. *Ex:* Tantilla (Rept.).

**tany**—See **tan**.

**tanystys**—See **tan**.

**tao**—See **taon**.

**taon**—Gr. *taōs*, genit. *taō*, also *taōn*, acc. *taōn*, the peacock. *Ex:* Tao-perdix (Av.); Taon-iscus (Av.); Taon-urus*; Taonus (Av.); Pseudo-taon (Av.).

**tapes**—See **tapet**.

**tapet**—L. *tapete*=Gr. *tapēs*, a carpet, tapestry, hangings>NL. *tapesium* and *tapetum*. *Ex:* Tapes (Moll.); tapesium; tapet-al; tapetum nigrum.

**taph**—1. Gr. *taphos*, a grave; also a funeral. *Ex:* Taph-aetus (Av.): Tapho-nycteris (Mam.);

Tapho-zous (Mam.); **2.** Gr. *taphos*, astonishment.

**taphr**—Gr. *taphros*, a ditch, trench. *Ex:* taphr-ad (Ecol.); Taphr-aster (Echin.); taphr-enchyma; Taphr-orychus (Ins.); Taphria*; Taphr-ina*; Taphro-campa (Rot.); Steno-taphrum*.

**tapin**—Gr. *tapeinos*, mean, base, humble; *tapenōma*, humility. *Ex:* Tapin-auchenius (Arach.); Tapin-opa (Arach.); Tapina*; Tapino-therium (Mam.); Tapinoma (Ins.).

**tapir**—Braz. *tapyra, tapir*, "probably from *tapy*, thick, in ref. to the thickness of the hide." *Ex:* Tapir-avus (Mam.); Tapiro-porcus (Mam.)

**tapiscia**—Anagram of Pistacia. *Ex:* Tapiscia*.

**tarach**—Gr. *tarachē*, trouble, disorder. *Ex:* Tarache (Ins.); Tarachia (Ins.); Tarachomantis (Ins.).

**taract**—Gr. *taraktēs*, a disturber <*taraktos*, disturbed; *taraktikos*, disturbing. *Ex:* Taractes (Pisc.); Taracticus (Ins.); Taracto-pora (Coel.); Tarakto-genos*.

**taragm**—Gr. *taragma*, uneasiness. *Ex:* Taragma (Ins.).

**tarakt**—See **taract**.

**tarand**—Gr. *tarandos* = L. *tarandrus*, the reindeer or the elk. *Ex:* Tarand-ichthys (Pisc.); Tarandus (Mam.).

**tarandr**—See **tarand**.

**tarant**—It. *tarantola* <L. *Taranto*, town in the south of Italy. *Ex:* tarant-iam; Tarantula (Arach.); Tarantulo-ides (Arach.).

**tarass**—Gr. *tarassō* = Attic *tarattō*, to stir up trouble, to confuse. *Ex:* Taratto-stichus (Ins.).

**taratt**—See **tarass**.

**tarax**—Gr. *taraxis*, disorder, confusion. *Ex:* Taraxi-neura (Ins.); Taraxis (Ins.).

**taraxac**—NL. *taraxacum*, perh. < Pers. *tarashqun*, dandelion. *Ex:* Taraxacum*.

**tarb**—Gr. *tarbos*, terror <*tarbeō*, to be frightened. *Ex:* Tarb-ophis (Rept.); Bary-tarbes (Ins.).

**tarbale**—Gr. *tarbaleos*, fearful, terrible. *Ex:* Tarbale-opsis (Ins.); Tarbaleus (Ins.).

**tard**—L. *tardus*, slow, sluggish. *Ex:* Tardi-grada (Mam.); Tardi-vola (Av.).

**tardiv**—NL. *tardivus*, slow growing <L. *tardus*, slow.

**tarentol**—It. *tarantola* (< *Tarantum*, a town in southern Italy = L. *Tarentum*) a salamander; also the gecko lizard, *Tarentola mauri-tanicus*. *Ex:* Tarentola (Rept.).

**tarich**—Gr. *tarichos*, a mummy, something pickled in salt. *Ex:* Taricha (Amph.).

**tarph**—Gr. *tarphos*, a thicket. *Ex:* Stachy-tarpha* = Stachy-tarph-eta*.

**tarphio**—See **tarphy**.

**tarphy**—Gr. *tarphys*, thick, close = *tarphetos*, a thickening. *Ex:* Tarphio-mimus (Ins.), i.e., mimicking Tarphius; Tarphius (Ins.); Tarphyceras (Moll.); Tarphy-pygus (Echin.).

**tarr**—See **tars**.

**tars**—Gr. *tarsos* = Attic *tarros*, a flat basket, any broad flat surface, such as the flat of the foot; the tarsus; *tarsōdēs*, woven, like basket work. *Ex:* Tarr-asius (Pisc.); Tarro-grantia (Por.); Tarrus (Por.); Tars-aster (Echin.); Tarsius (Mam.); tarso-meta-tarsus.

**tartar**—NL. *tartar* <L.L. *tatarum* <Late Gr. *tataron*, a crust deposited on the sides of wine barrels. *Ex:* tartar-eus, tartar.

**tartaro**—Gr. *Tartaros*, dark prison of the Titans. *Ex:* Tartaro-thyas (Arach.).

**tas**—Gr. *tasis*, genit. *taseōs*, stretching. *Ex:* Taseo-conia (Por.); taseo-meter = tasi-meter.

**taseo**—See **tas**.

**tass**—Gr. *tassō*, to dispose. *Ex:* Di-tassa*.

**tath**—See **tein**.

**-tatos**—Gr. *-tatos*, adj. superlative ending. *Ex:* anchio-tatos, most worthy. See also istos.

**tatu**—Port. *tatu* = Fr. *tatusie*, native name for the armadillo in Paraguay. *Ex:* Tatu (Mam.); Tatusia (Mam.); Eu-tatus (Mam.).

**tatus**—See **tatu**.

**tau**—Gr. *tau*, the Greek letter Τ.

**taum**—See **thaum**.

**taur**—L. *taurus*, bull, ox, steer. *Ex:* Taur-ichthys (Pisc.); tauri-cornis; Tauro-tettix (Ins.); Tauro-tragus (Mam.); Taurus (Mam.).

**taure**—L. *taureus*, bull-like, of or pertaining to cattle.

**taurin**—L. *taurinus*, steerlike, tough-hided.

**taut**—Gr. *tauta*, so far, to that extent. *Ex:* Tauto-zelus (Ins.).

**tautog**—Indian *tautog*, a fish name. *Ex:* Tautoga (Pisc.); Tautogo-labrus (Pisc.).

**tax**—**1.** Gr. *taxō* = fut. of *tassō*, to put in order, arrange. *Ex:* taxeo-pod-ous; taxi-dermy; taxonomy; Taxo-crinus (Echin.); A-taxio-ceras (Moll.); Haplo-taxis (Ann.); **2.** NL. *taxus*, a badger. *Ex:* Tax-odon (Mam.); Taxo-therium (Mam.); **3.** L. *taxus*, a yew-tree, also, a javelin made of yew-wood <Gr. *toxon*, a bow, the yew-wood being used for bows. *Ex:* Tax-aceae*; Taxo-odium*; Taxo-xylon*; Taxus*.

**teano**—NL. *teano* <NL. *Teonoma*, a rodent genus. *Ex:* Teano-pus (Mam.).

**tebenn**—Gr. *tēbenna*, a toga, robe of state. *Ex:* Tebenna (Ins.); Tebenno-toma (Ins.).

**techn**—Gr. *technē*, art, craft, skill, handiness. *Ex:* Techno-crinus (Echin.); Techno-myrmex (Ins.); Hetero-tecno-mera (Ins.).

**technit**—Gr. *technitēs*, one skilled in work. *Ex:* Technit-ella (Ins.).

**tecn**—Gr. *teknon*, a child, the young. *Ex:* Tecnophilus (Ins.); Spano-tecnus (Ins.). See also techn.

**tecoma**—Aztec *Tecomaxochitl*, name for a plant, Tecoma. *Ex:* Tecoma*.

**tect—1.** L. *tectum*, dim. *tectulum*, roof, covering < *tego*, to cover; *tectus*, covered; *detectus*, naked, without cover. *Ex:* tect-al; Tect-aria*; Tectiscuti (Ins.); Tecto-spondyli (Elasm.); Tectulipora (Bry.); ob-tect: **2.** Gr. *tektōn*, a carpenter, a builder. *Ex:* Tecton (Arach.), not Tectona*, which is from Malay *tekku*, teakwood; tecto-logy: **3.** Gr. *tēktos*, melted; also soluble. *Ex:* eu-tect-ic.

**tecton**—See tect.

**tectrices**—NL. *tectrix*, pl. *tectrices*, the wing or tail coverts < *tecto*, to cover. *Ex:* tectrices.

**tecturat**—NL. *tecturatus*, covered < L. *tectura*, a cover.

**teg**—Gr. *tegos*, roof. *Ex:* Tego-capsis (Ins.); Tego-notus (Arach.); diplo-tegia.

**tege**—Gr. *Tegea*, a town in Arcadia > NL. *tegenaria*, name for a spider genus. *Ex:* Tegea (Arach.); Tegenar-idae (Arach.); Tegenaria (Arach.); Tegeo-cranus (Arach.).

**tegenar**—See tege.

**teges**—See teget.

**teget**—L. *teges*, genit. *tegetis*, a covering, mat. *Ex:* Teget-icula (Ins.).

**tegetari**—L. *tegetarius*, a mat maker.

**tegmen**—See tegmin.

**tegmin**—L. *tegmen*, genit. *tegminis*, pl. *tegmina*, also *tegimen*, a cover; *tegmentum = tegumentum*, a covering. *Ex:* tegmen; tegmentum; tegminal; tegmin-alia (neut. pl. of NL. *tegminalis*); tegmina; tegument.

**tegul**—L. *tegula*, tile; *tegulum*, a covering, roof. *Ex:* Tegula (Moll.); Teguli-fera (Brach.); Tegulo-rhynchia (Brach.); pro-tegulum.

**tegumen**—See tegmen.

**tei**—See tej.

**teich**—Gr. *teichos*, a wall. *Ex:* Teicho-bia (Ins.); teicho-some; Tich-odon (Mam.); Ticho-leptus (Mam.); Diocho-tichus (Mam.).

**teiche**—Gr. *teicheō*, to build walls. *Ex:* Dysteicheus (Ins.).

**tein**—Gr. *teinō*, to stretch > *tathē*. *Ex:* Teinodactyla (Ins.); Teino-palpus (Ins.); Tino-ceras (Rept.); Tino-spora*; neo-tein-ic; Tathi-carpus (Pisc.); Tatho-delta (Ins.).

**tej**—*teju*, native S. A. name for a lizard. *Ex:* Tej-idae (Rept.) = Tii-idae (Rept.); Tejo-varanus (Rept.); Tijus (Rept.).

**tel**—**1.** Gr. *tēle*, far, far off, at a distance. *Ex:* tele-dendron; Tele-gonus (Ins.); Telea (Ins.): **2.** L. *tela*, a web. *Ex:* tel-arius; Terro-tel-aria (Arach.); Orbi-telae (Arach.): **3.** Gr. *telos*, the end, the end of life. *Ex:* Tel-acodon (Mam.); Telo-spor-idea*; Telo-tremata (Brach.); Raphi-telus (Ins.).

**telamon**—Gr. *telamōn*, a supporting band. *Ex:* Telamon (Av.); Telamon-anthe (Ins.).

**telchin**—Gr. *telchin*, a mischievous person. *Ex:* Telchinia (Ins.).

**tele**—Gr. *teleios* and *teleos*, perfect, entire, without blemish. *Ex:* Tele-basis (Ins.); Telei-dosaurus (Rept.), -do- from *eidos*, form; Teleia (Ins.); Teleio-crinus (Echin.); Teleo-saurus (Rept.); Ambly-teles (Ins.); Hoplo-teleia (Ins.); for Brachy-teles, see brachy; Pro-teles (Mam.); Xylo-teles (Ins.).

**teleo**—See tele.

**telephi**—NL. *telephium*, < Gr. *tēlephion*, a plant called sedum or orpine. *Ex:* telephii-folium; Telephium*.

**teles**—See tele, also atel.

**telest**—Gr. *telestēs*, poet. *telestōr*, an official, one who accomplishes; *telestikos*, fit for finishing, accomplishing. *Ex:* Telestes (Pisc.); Telesto (Ins.).

**telet**—Gr. *teletos*, perfect.

**teletha**—Gr. *tēlethaō*, blooming, flourishing.

**teleut**—Gr. *teleutē*, a finishing, end. *Ex:* Teleut-aea (Ins.), the last element of uncertain meaning; teleuto-spore; A-teleute (Ins.).

**telio**—Gr. *telos*, genit. *teleos*, the end, completion > NL. *telium*. *Ex:* telio-spore, telio-stage; telium.

**tell**—L. *tellus*, the earth; *Tellus*, goddess of the Earth. *Ex:* Tellea (Pisc.).

**tellima**—Anagram of Mitella. *Ex:* Tellima*.

**tellin**—Gr. *tellinē*, a kind of shell fish. *Ex:* Tellin-ites (Moll.); Tellina (Moll.); Tellino-mya (Mam.).

**telm**—Gr. *telma*, genit. *telmatos*, a swamp, pond. *Ex:* Telma-tettix (Ins.); Telmat-ornis (Av.); Telmato-saurus (Rept.); Telmia (Ins.).

**telmat**—See telm.

**telop**—Gr. *tēlōpos*, seen from afar. *Ex:* Telopea*; Telopes (Ins.).

**telson**—Gr. *telson*, limit, boundary. *Ex:* telson.

**telur**—Gr. *tēlouros*, distant, far away. *Ex:* Telura (Ins.).

**tem**—Sp. *temo*, name of a shrub in Chile. *Ex:* Temus*.

**temach**—Gr. *temachos*, a slice of salt meat. *Ex:* Temachia (Bry.).

**temen**—Gr. *temenos*, a piece of land withheld and marked off from common uses; *temenouchos*, holding a piece of land. *Ex:* Temenuchus (Av.); Temenus (Ins.).

**temenuch**—See temen.

**temn**—Gr. *temnō* = Dor. *tamnō*, to cut. *Ex:* Temn-odon (Pisc.); Temno-chila (Ins.); temno-spondylus; A-temnus (Arach.).

**temnibil**—L. *temnibilis*, neut. *temnibile*, contemptible, beneath notice.

**temperat**—L. *temperatus*, moderate, calm, with limits.

**temul**—L. *temulus*, also *temulentus*, drunken, nodding, top-heavy.

ten—Gr. *tenōn*, genit. *tenontos*, a tendon, a tightly stretched bandage. *Ex:* teno-tomy (Med.); Tenonto-myia (Ins.). See also taen.

tenac—L. *tenax*, genit. *tenacis*, gripping, holding <*teneo*, to hold; *tenaculum*, a holder. *Ex:* Tenaci-ella (Por.); Tenaci-pes (Moll.); Tenacia (Por.); tenaculum.

tenacet—LL. *tenacetum*, a name for tansy. *Ex:* tenaceti-folia; Tenacetum*.

tenag—Gr. *tenagos*, shoal-water, a shallow pond; *tenagōdēs*, standing or living in pools. *Ex:* Tenago-gonus (Ins.); Tenago-mysis (Arth.); Tenagodes (Moll.).

tenax—See tenac.

tend—1. Gr. *teinō*, to stretch; L. *tendo*, to stretch, to extend, to cause to expand>NL. *tendon*, a tendon. *Ex:* tendin-ous (Fr. *tendineux*, sinewy); ex-tend-ed; neo-teny:    2. Gr. *tendō*, to gnaw, gnaw at something.

tendril—Fr. *tendrillons*; OFr. *tendron*, a tender fellow; also a tendril <*tendre* = *tender* <L. *tener*, tender, thin. *Ex:* tendril.

-tene—NL. *-tene* <Gr. *tainia*, a band. *Ex:* pachytene.

tenebri—L. *tenebrio*, genit. *tenebrionis*, a lover of darkness. *Ex:* Tenebrio (Ins.); Tenebrion-idae (Ins.).

tenebros—L. *tenebrosus*, dark, full of gloom.

tenell—L. *tenellus*, quite delicate. *Ex:* Tenellia (Moll.).

tener—L. *tener*, neuter *tenerum*, soft, tender, delicate. *Ex:* tener-al; ?Tenero-mimus (Ins.); ?Tenerus (Ins.).

tenont—See ten.

tenrec—Fr. *tenrec*, the tenrec. *Ex:* Tenrec (Mam.); Tenrec-idae (Mam.).

tent—L. *tentus*, stretched, extended; pp. of *tendo*, to extend.

tentacul—LL. *tentaculum*, a feeler <L. *tento*, to touch, feel. *Ex:* tentacle; tent-illum< *tenaculum* + *-illum*;    Tentacul-ata    (Cten.); Tentaculi-fera (Prot.).

tenthes—Gr. *tenthēs*, a dainty feeder. *Ex:* Tenthes (Ins.).

tenthred—Gr. *tenthrēdōn* = *tenthrēn* = NL. *tenthredo*, genit. *tenthredinis*, a kind of wasp; Gr. *tenthrēnion*, the nest of a certain wasp. *Ex:* Tenthredin-idae (Ins.); Tenthredo (Ins.); Tenthren-odes (Por.).

tenthren—See tenthred.

tentigin—L. *tentigo*, genit. *tentiginis*, a stretching; also lewdness. *Ex:* tentigin-ous.

tentori—L. *tentorium*, a tent. *Ex:* Tentorium (Por.).

tenu—L. *tenuis*, thin, narrow, slender. *Ex:* Tenui-aspis (Ins.); tenu-ous; tenui-rostral; Tenui-rostres (Av.); Tenuia (Ins.); Tenuo-pus (Ins.).

teny—See tend.

teo—poorly made anagram of *neo*. *Ex:* Teonoma (Mam.); not Teo-disca (Ann.) which in part involves Gr. *theō*, to run.

tephr—Gr. *tephra*, ashes; *tephraios* = *tephros*, ash-colored, gray; L. *tephritis*, an ash-colored stone; Gr. *tephroō*, to burn to ashes; *tephrōsis*, a burning to ashes; *tephrōdēs* = *tephrinos*, ash-colored. *Ex:* Tephr-aphis (Ins.); Tephritis (Ins.); Tephrae-ops (Pisc.); Tephraea (Ins.); Tephrina (Ins.); Tephrites (Moll.); Tephritus (Ins.); Tephrod-ornis (Av.); Tephronia (Ins.); Tephros-anthus*; Tephrosia*; tephrosius.

tephrit—See tephr.

tephros—See tephr.

tepid—L. *tepidus*, lukewarm.

tepidari—L. *tepidarius*, belonging to warm water.

tepor—L. *teporatus*, warmed; *teporus*, warm.

ter—1. Gr. *teras*, genit. *teratos*, a monster, wonder; *terateia*, jugglery. *Ex:* Teras (Ins.); Teratodes (Arach.); terato-logy; Terato-lytta (Ins.); idio-ter-y:    2. Gr. *tēros*, a warden, guard. *Ex:* Tero-bia (Ins.).

teram—Gr. *teramōn*, soft>NL. *teramus*. *Ex:* Teramo-cerus (Ins.).

teramn—Gr. *teramnos*, soft. *Ex:* Teramnos*.

terasti—Gr. *terastios*, monstrous. *Ex:* Terastiozoon (Ins.).

terat—See ter 1.

tere—Gr. *terēn*, genit. *terenos*, smooth, delicate. *Ex:* Tere-myia (Ins.); Terenia (Ins.); Tereniotriccus (Av.).

tereb—See terebr.

terebinth—Gr. *terebinthos*, the terebinth tree. *Ex:* Terebinthus*.

terebr—L. *terebra*, a boring <*terebro*, to bore, perforate, ppr. *terebrans*, genit. *terebrantis*, boring; *terebratus*, a boring, perforation. *Ex:* Tereb-ella (Ann.); Terebra (Moll.); Terebra-spira (Moll.); Terebrantia (Ins.); terebrat-ing; Terabrat-ula (Brach.).

tered—Gr. *terēdōn*, the wood worm = ML. *teredina*; Gr. *terēdonizomai*, to be worm-eaten, full of holes. *Ex:* Tered-ina (Moll.); Teredinidae (Moll.); Teredo (Moll.); Teredo-soma (Ins.); Teredon (Ins.).

terek—NL. *terekia* <*Terek*, a river in the Caucasus. *Ex:* Terekia (Av.).

teren—Gr. *terēn*, genit. *terenos*, smooth, soft, delicate.

teren—See tere.

teres—See teret.

teret—L. *teres*, genit. *teretis*, smooth, rounded off, well turned; also elegant. *Ex:* terete; teretiformis.

teretr—Gr. *teretron*, a borer, *teretrion*, a small borer. *Ex:* Teretrius (Ins.); Teretro-poma.

terg—L. *tergum*, the back. *Ex:* terg-ite; tergi-ferous; Tergi-pes (Moll.); tergo-pleural.

tergemin—L. *tergeminus*, poetic form of *trigeminus*, born three at one time, triple.

tergin—L. *terginum*, leather, raw-hide.

teri—Gr. *tereō*, to pierce. *Ex:* io-terium (Gr. *ios*, poison).

-terium—Gr. *-terion*, a formative suffix meaning a place for. *Ex:* osme-terium; not io-terium, see teri.

term—1. Gr. *terma*=L. *termen*, genit. *terminis*, the terminis, boundary, end; L. *terminalis*, terminal, belonging to a boundary. *Ex:* termen; terminal; Terminal-inus (Ins.); phloe-terma; Platy-terma (Ins.): 2. L. *termes* =*tarmes*, a wood-worm. *Ex:* term-ite; termit-arium; Termit-aphis (Ins.); Termito-pulex (Ins.); Hami-termes=Ami-termes (Ins.); Leuco-termes (Ins.).

termen—See term.

termin—See term.

termit—See term 2.

tern—L. *terni*, three each=NL. *ternatus*. *Ex:* ternate; terni-pes; Terni-dens (Nem.).

ternari—L. *ternarius*, made of three parts, ternate.

ternat—See tern.

terph—Gr. *terphos*, a skin, shell, covering. *Ex:* Terphis (Ins.); Lio-terphus (Ins.).

terpn—Gr. *terpnos*, delightful, agreeable. *Ex:* Terpna (Ins.); Terpno-myia (Ins.); Terpnosia (Ins.).

terps—Gr. *terpsis*, genit. *terpseōs*, enjoyment, delight; as a combining form *terpsi-* means gladdening, delighting, *terpsinoos*, heart-gladdening; *noos*, the heart, mind, senses, reasoning. *Ex:* Terpsi-phone (Av.); Terpsis (Av.).

terr—L. *terra*, earth; *terrestris*, of or belonging to the earth. *Ex:* terrestri-al; Terri-cola (Platy.); Terri-tel-aria (Arach.).

terrapen—NL. *terrapene*<a name of American Indian origin. *Ex:* Terrapene (Rept.).

terren—L. *terrenus*, earthy, earthen, of clay.

terrest—See terr.

territan—L. territans, genit. *territantis*, alarmed <*territo*, to frighten.

terrulent—L. *terrulentus*, belonging to the earth.

ters—L. *tersus*, wiped off clean, neat; as a noun, a cleansing. *Ex:* tersi-formis.

tert—L. *tertius*, the third; *tertiarius*, containing a third part, now often taken to mean belonging to the third part. *Ex:* tertiari-es; Tertiary.

tertiar—See tert.

-tes—Gr. *-tes*, suffix denoting agent or having to do with. *Ex:* Brach-alle-tes (Mam.); Chiro-pe-tes (Mam.), see petes; Tryngi-tes (Av.).

tesot—NL. *tesotus*<Sp. *tieso*, stiff, hard, firm.

tessar—Gr. *tessares*, four. *Ex:* Tessar-odon (Ins.); Tessar-omma (Ins.); Tessar-ops (Arach.); Tessara-toma (Ins.); Tessaro-merus (Ins.).

tessel—L. *tessella*, a little cube; *tesselatus*, made of small square stones, checkered. *Ex:* tessellat-ed; Tessellata (Echin.).

tesser—Gr. *tesseres*, four. *Ex:* Tesser-omma (Ins.); Tessero-pora (Crust.).

test—L. *testa*, a shell, brick, tile. *Ex:* test, Test-amoeba (Prot.); test-arum; Test-ed-ium (Ins.), see hed; Testa-spongia (Por.).

testace—L. *testaceus*, consisting of brick or tile, covered with a shell; *testaceum*, a shell-covered animal<*testa*, a tile, brick. *Ex:* Testacea (Moll.); testaceo-logy.

testes—See testis.

testis—L. *testis*, a witness<L. *testis*, dim. *testiculus*, pl. *testes*, a testicle. *Ex:* testi-cond; testicle; testicul-ar.

testud—L. *testudo*, genit. *testudinis*, a tortoise, ML. *testudinarius*, tortoise-like; *testudineus*, resembling tortoise shell. *Ex:* Testudinaria*; Testudini-tes (Rept.); Testudo (Rept.).

tetan—Gr. *tetanos*, stiffness or spasm of the neck; as adj. stiff, stretched<*tetanoō*, to stretch. *Ex:* tetan-ize; Tetan-ura (Ins.); Tetana (Ins.); tetani-form; Tetano-cera (Ins.)

tetart—Gr. *tetartos*, fourth, the fourth time. *Ex:* Tetarto-peus (Ins.).

teter—L. *teter*, *-ra*, *-trum*=*taeter*, hideous, foul, abominable.

teth—1. Gr. *tēthē*, grandmother; also a nurse. *Ex:* Tethe-odon (Pisc.); Tethe-opsis (Mam.); Tethea (Ins.): 2. Gr. *tēthis*, genit. *tēthidos*, an aunt. *Ex:* Teth-ina (Ins.).

tethel—Gr. *tethēlōs*, luxuriant, flourishing; *tethēla*, to be flourishing<*thallō*, to abound. *Ex:* tethel-in.

tethid—See teth 2.

tethm—Gr. *tethmos*, law, custom. *Ex:* Sema-tethmos (Coel.).

tethy—Gr. *Tēthys*, a sea goddess, sister of Saturn and wife of Oceanus. *Ex:* Tethy-aster (Echin.); Tethy-idea (Moll.); Tethya (Por.); Tethyo-phaena (Por.); Tethys (Moll.); Tethyum (Tun.).

tetill—Sp. *tetilla*, dim. of *teta*, a teat. *Ex:* Tetilla* (Por.); Tetill-opsis (Por.).

tetr—Gr. *tetra*, four; *tetras*, genit. *tetradis*, the number four; *tetracha*, in four parts; *tetradion*, quaternary. *Ex:* Tetr-actin-ell-idae (Por.); Tetr-aechma (Ins.); tetr-axon; tetra-cotyl; Tetra-gono-lobus*; Tetracha (Ins.); tetrad; Tetradium (Moll.); Tetro-bothrium (Platy.).

tetrach—See tetr.

tetrad—See tetr.

tetrao—Gr. *tetraōn*, genit. *tetraōnos*, a pheasant. *Ex:* Tetrao (Av.); Tetrao-phasis (Av.); Tetraon-idae (Av.).

tetraplo—Gr. *tetraploos*, fourfold. *Ex:* Tetraplo-odon (Mam.).

tetrax—Gr. *tetrax*, a pheasant. *Ex:* Hetero-tetrax (Av.).

**tetric**—L. *tetricus*, harsh, fierce. *Ex:* Tetrica (Ins.); Tetrico-dina (Ins.).

**tetrix**—Gr. *tetrix*, a bird, supposed to be a grouse. *Ex:* Tetrix (Av.).

**tettig**—Gr. *tettix*, genit. *tettigos*, a kind of grasshopper; *tettigonion*, a small cricket or grasshopper. *Ex:* Tettig-ades (Ins.); Tettigo-myia (Ins.); Tettigon-idae (Ins.); Eu-tettix; Para-tettix (Ins.).

**tettix**—See **tettig**.

**teuch**—Gr. *teuchos*, a tool, armour, arms; *teuchēstēs*, a warrior. *Ex:* Teuchestes (Ins.); Teucho-pora (Bry.); A-teuchestus (Ins.); A-teucho-pus (Ins.).

**teucr**—Gr. *teukrion*, a plant of the germander kind < *Teucer*, king of Troy. *Ex:* teucri-folium; Teucrium*.

**teuth**—Gr. *teuthis*, genit. *teuthidos*, a squid. *Ex:* Teuth-ophrys (Prot.); Teuthi-discus (Moll.); Teuthid-idae (Moll.); Teuthis (Moll.); Architeuthus (Moll.); Onycho-teuthis (Moll.).

**text**—L. *textus*, a tissue, structure; *textor*, a weaver; *textilis*, woven, braided. *Ex:* Texti-spongia (Por.); Texti-venus (Moll.); Textil-aria = Textul-aria (Prot.); Textilli-forma (Por.).

**textil**—See **text**.

**thaer**—See **thair**.

**thair**—Gr. *thairos*, the hinge of a door or gate. *Ex:* Thaira (Moll.); Thairo-pora (Bry.); Rexi-thaerus (Moll.). See ther 3.

**thais**—Gr. *Thais*, celebrated courtesan of Athens. *Ex:* Thais (Ins.), (Moll.).

**thal**—Gr. *thalos*, a shoot, a young twig; also a young person. *Ex:* Thal-otia (Rot.); not Thalia*, named after Johann Thalius, German naturalist. See also thalassa

**thalam**—Gr. *thalamos*, an inner room, chamber. *Ex:* Thalam-archis (Ins.); thalam-encephalon; Thalamo-crinus (Echin.); Thalamus (Moll.); Conio-thalam-eae*; mono-thalam-ous.

**thalass**—Gr. *thalassa* = Attic. *thalatta*, the sea; *thallassios*, marine. *Ex:* Thal-arctos (Mam.); Thalass-ema (Gephyrea); Thalass-ictes (Mam.); Thallasina (Crust.); thalassinus; Thalasso-geron (Av.); Thalassio-mastix (Prot.); Thalatto-saurus (Rept.).

**thalatt**—See **thalass**.

**thalaz**—Gr. *thēlazō*, to suck. *Ex:* Thalazia = Thelazia (Nemat.).

**thale**—See **thalei**.

**thalei**—Gr. *thaleia*, blooming. *Ex:* Thale-ichthys (Pisc.).

**thaler**—Gr. *thaleros*, green, blooming. *Ex:* Thaler-astria (Ins.); Thalera (Ins.); Thalero-phaga (Ins.).

**thalesia**—Gr. *Thalēs*, Greek philosopher. *Ex:* Thalesia*.

**thalestr**—Gr. *Thalēstris*, a mythical personage. *Ex:* Thalestr-ella (Crust.); Thalestris (Crust.).

**thalia**—1. Gr. *thalia*, luxuriance, abundance; *thaleia*, luxuriant. *Ex:* Odon-thalia*: 2. Gr. *Thalia*, a muse, patroness of comedy who made life seem glorious, abundant. *Ex:* Thal-ida (Tun.); Thali-acea (Tun.); Thalia* (Ins.).

**thalictr**—Gr. *thaliktron*, meadow-rue. *Ex:* Thalic trum*.

**thalid**—See **thalia**.

**thall**—Gr. *thallos*, a branch, a young shoot; *thallinos*, pertaining to a green shoot; *thallō*, to grow green. *Ex:* thall-oid; Thalli-cera (Rot.); thalli-form; thalline; thallium; thallo-gen; Thallo-graptus (Coel.); Thallo-phyta*; thallus.

**thalp**—Gr. *thalpos*, heat. *Ex:* Thalpo-chares (Ins.); Thalpo-mys (Mam.); en-thalpy.

**thalpor**—Gr. *thalpōros*, warm.

**thalusi**—Gr. *thalysia*, first of the harvest. *Ex:* Thalusia (Ins.).

**thalycr**—Gr. *thalykros*, hot, glowing, red, brilliant, white with heat. *Ex:* Thalycr-odes (Ins.); Thalycra (Ins.).

**tham**—NL. *euthamia* < Gr. *euthēmōn*, neat, well-ordered, pretty. *Ex:* Euthamia*.

**thamb**—Gr. *thambos*, astonishment. *Ex:* Thambo-ceras (Moll.); Thambo-tricha (Ins.); Thambus (Ins.).

**thamn**—Gr. *thamnos*, a shrub; *thamnion*, a small shrub. *Ex:* Thamn-idium*; Thamn-ophis (Rept.); Thamn-osma*; Thamnea*; thamnium; thamno-blastus; Thamno-cephalus (Annel.); Thamno-tettix (Ins.); Calli-thamnion*; Chryso-thamnus*.

**thamyris**—Gr. *Thamyris*, boastful singer of Thracian mythology. *Ex:* Thamyris (Crust.).

**thanas**—See **thanat**.

**thanat**—Gr. *thanatos*, death; *thanasimos*, murderous; *thanatikos*, deadly. *Ex:* Thanasimus (Ins.); thanat-osis; Thanata (Arach.); Thanatophilus (Ins.); Thanatus (Arach.); A-thanas (Crust.).

**thaps**—Gr. *thapsia*, a name for the deadly carrot brought from the Isle of Thapsus. *Ex:* Thapsia*.

**thapsin**—Gr. *thapsinos*, yellow. *Ex:* Thapsin-illas (Av.). See also thaps.

**thapt**—Gr. *thaptō*, to honor with burial rites. *Ex:* Thapto-mys (Mam.).

**tharr**—Gr. *tharros*, quick, alert, bold. *Ex:* Tharr-ops (Ins.).

**tharrhale**—Gr. *tharrhaleos*, bold. *Ex:* Tharrhalea (Arach.); Tharrhaleus (Av.).

**thars**—Gr. *tharsos*, courage, confidence; *tharsaleos*, daring. *Ex:* Tharsalea (Ins.); Tharsus (Ins.).

**thaspi**—NL. *thaspium*, a transposition of *thapsia*, see thaps. *Ex:* Thaspium*.

**thass**—Gr. *thassō*, to sit. *Ex:* Coni-thassa (Ins.).

**thaum**—Gr. *thauma*, genit. *thaumatos*, a wonder; *thaumasmos*, a marvelling; *thaumasteos*, to be admired; *thaumastos*, wonderful, extraordinary; *thaumaleos=thaumasios*, wondrous, marvelous. *Ex:* Thaumalea (Ins.); Thaumast-ura (Av.); Thaumasto-myrex (Ins.); Thaumat-ibis (Av.); Thaumato-lampas (Moll.); Batho-thauma (Moll.).

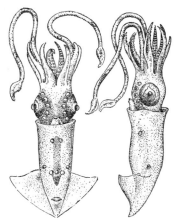

Jewelled Wonder-torch, *Thaumato-lampas diadema*, a squid. Redrawn from The Seas—Russell and Yonge.

**thaumal**—See **thaum.**

**thaumant**—Gr. *Thaumas*, genit. *Thaumantos*, name of god known for his miracles. *Ex:* Thaumant-idae (Coel.); Thaumas (Prot.). See **thaum.**

**thaumas**—See **thaumant.**

**thaumast**—See **thaum.**

**the**—1. Gr. *theō=theiō*, to run. *Ex:* Ane-thum*, Gr. *ano*, up; Croce-thia (Av.): 2. Gr. *theion*, brimstone. *Ex:* Theio-picus (Av.): 3. Gr. *theos*, also *thios*, God, the Gods. *Ex:* Theo-broma*, (Moll.); Theo-capsa (Prot.).

**thea**—1. Latinized *thea* from Chinese *tcha*, name of the tea plant. *Ex:* Thea*: 2. Gr. *thea*, a looking at, an aspect. *Ex:* Liby-thea (Ins.); Praxi-thea (Ins.).

**theat**—Gr. *theatēs*, a spectator. *Ex:* Theat-ops (Myr.).

**thec**—Gr. *thēkē*, dim. *thēkion*, a case for something, a box=L. *theca*; L. *thecatus*, encased. *Ex:* Thec-idium (Moll.); Thec-oidea (Echin.); Thec-odonto-saurus (Rept.); Theca-cera (Moll.); Thecata (Crust.); thecium; Theco-diplosis (Ins.); Theco-somata (Moll.); Theco-stegites (Coel.); apo-thecium; phaino-theca; Prosopo-theca (Arach.); sperma-theca.

**thecl**—NL. *thecla*<Gr. *Thekla*, a personal name. *Ex:* Thecla (Ins.).

**thect**—Gr. *thēktos*, sharpened. *Ex:* Thect-ura (Ins.); Thecto-phila (Ins.).

**theg**—Gr. *thēgō*, to sharpen. *Ex:* Theg-ornis (Av.).

**theka**—Tamil *thekku*, the teak tree or Indian oak. *Ex:* Theka*, see tect.

**thel**—1. Gr. *thēlē*, nipple. *Ex:* Thel-avia (Ins.); Thel-ida (Ins.); Thel-odus (Pisc.); Thela-stoma (Nemat.); Thele-phora*; Thele-pus (Ann.); Thele-sperma*; Thelo-cactus*; Aego-theles (Av.); epi-theli-um; Meso-thelae (Arach.); Opistho-thelae (Arach.): 2. Gr. *thēlys*, tender, delicate; also female. *Ex:* Theli-derma (Moll.); Theli-domus (Moll.). See also thelys.

**thelaz**—Gr. *thēlazō*, to suckle; also to suck. *Ex:* Thelazia (Nemat.); Thelazo (Nemat.); Thelazo-menus (Av.).

**thelg**—Gr. *thelgō*, to charm, enchant, trap, cheat; *thelgētron*, a charm or spell. *Ex:* Thelgetrum (Ins.).

**thelgetr**—See **thelg.**

**thelic**—Gr. *thēlykos*, feminine. *Ex:* Thelic-onus (Moll.), see con; thelycum.

**thelphus**—Gr. *Thelphousa*, Greek nymph who gave her name to a town in Arcadia. *Ex:* Thelphus-idae (Crust.); Thelphusa (Crust.); Geo-thelphusa (Crust.).

**thelx**—Gr. *thelxis*, a bewitching, an enchanting. *Ex:* Thelxi-novum (Moll.); Thelxi-ope (Crust.).

**thely**—Gr. *thēlys*, female. *Ex:* Thely-gonum*; Thely-mitra*; Thely-phonus (Arach.); Thely-podium*; thelyo-toky; Micro-thely-phon-ida (Arach.).

**thelyc**—See **thelic.**

**thelydri**—Gr. *thēlydrias*, an effeminate person. *Ex:* Thelydrias (Ins.).

**-them**—NL. *-them*, a suffix<Gr. *tithēme*, to put. *Ex:* epi-them.

**then**—Gr. *thenar*, genit. *thenaros*, the palm of the hand. *Ex:* then-al; thenar; Thenaro-crinus (Echin.); opis-thenar. See also -then.

**-then**—Gr. *-then*, from. *Ex:* Eo-theno-mys (Mam.).

**thenar**—See **then.**

**theon**—L. *Theon*, satirical Latin poet. *Ex:* Theon-ella (Por.).

**theor**—1. Gr. *theōros*, a spectator. *Ex:* Theorus (Rot.): 2. *theōria*, a viewing, a sight, spectacle; *theōrēma*, a spectacle, hence a subject for contemplation. *Ex:* theorem; theory.

**theos**—Gr. *theos*, god; also fortune. *Ex:* Theos-odon (Mam.), here used in the sense of fortune.

**ther**—1. Gr. *thēr*, a wild beast; *thērion*, a wild animal; *thēridion*, a small animal, a beast. *Ex:* Ther-ina (Ins.); Therid-idae (Arach.); Theridion (Arach.); Therido-mys (Mam.); Therio-morpha=Thero-morpha (Rept.); Therion (Ins.); Aco-ther-ulum (Mam.); Pinno-theres (Arthr.); Titano-therium (Mam.): 2. Gr. *theros*, summer. *Ex:* thero-phyll-ous: 3. NL. *therus*<Gr. *thairos*, hinge. *Ex:* Sterno-therus (Rept.): 4. Gr. *thēraō*, to hunt after, to find. *Ex:* Thero-saurus (Rept.); Thero-zoaria (Moll.); Sauro-thera (Av.). See also **there**

theragr—Gr. *thēragros*, a snare for catching wild beasts. *Ex:* Theragra (Pisc.); Theragr-etes (Arach.).

therap—Gr. *therapis*, obedient; *therapōn*, an attendant. *Ex:* Therapis (Ins.); Therapon (Pisc.); Theraps (Pisc.) < Gr. *theraps*, a slave.

therapeutic—Gr. *therapeutikos* < *therapeutēs*, a servant; *therapeuō*, to attend, to take care of medicinally. *Ex:* therapeutic; therapeutics.

therat—See there.

there—1. Gr. *thēreuō*, also *thēraō*, to hunt after, to chase; *thēratēs*, a hunter. *Ex:* Therates (Ins.); Therato-scirtus (Arach.); Thereu-therium (Mam.); Thereua (Ins.); Thereuo-nema (Myr.); Thereva (Ins.); Copro-theres (Av.); Helmi-therus (Av.); Sauro-thera (Av.): 2. Gr. *thēreios*, of wild beasts. *Ex:* Therei-ceryx (Av.); Thereo-myia (Ins.).

thereu—See there 1.

therev—See there 1.

theri—Gr. *thērion*, a wild animal. *Ex:* Therio-suchus (Rept.).

theriac—L. *theriacus* < Gr. *thēriakos*, serving as an antidote against the poison of animals.

therid—See ther.

therism—Gr. *therismos*, harvest, reaping time. *Ex:* Syn-therisma*.

therist—Gr. *theristēs*, a reaper, fem. *theristria*; *theristikos*, of or for reaping. *Ex:* Theristes (Ins.); Theristicus (Av.); Theristria (Ins.).

therm—1. Gr. *thermos* = *thermē*, heat; also hasty, rash, *thermotēs*, heat. *Ex:* Thermesia (Ins.); Thermo-bia (Ins.): 2. Gr. *thermos*, lupine. *Ex:* Therm-opsis*.

thermal—NL. *thermalis*, warm, of warm springs < Gr. *thermē*, heat. *Ex:* thermal.

thermant—Gr. *thermantos*, promoting heat. *Ex:* A-thermantus (Ins.).

thermastr—Gr. *thermastris*, a pair of tongs, pliers. *Ex:* Thermastris (Ins.).

thersit—Gr. *Thersitēs*, the audacious, reviling one. *Ex:* Thersitia (Ins.); Thersites (Moll.); Thersito-myia (Ins.).

therus—See ther 3.

thes—Gr. *thesis*, a putting or setting in order, arranging; also a deposit. *Ex:* theso-cytes; photo-syn-thesis. See also thet.

thescel—Gr. *theskelos*, marvelous, divine. *Ex:* Thescelo-cichla (Av.); Thescelo-saurus (Rept.); Thescelus (Rept.).

thesium—L. *thesium*, a kind of plant < Gr. *thēseion*, "the plant Theseus crowned Ariadne with." *Ex:* Thesium*.

thesm—Gr. *thesmos*, that which is made sure, a law.

thespesi—Gr. *thespesios*, divine, wonderful. *Ex:* Thespesia*; Thespesio-psyllus (Crust.).

thet—1. Gr. *thēs*, genit. *thētos*, a serf, a hired servant. *Ex:* Thes (Ins.); Theto-mys (Mam.):

2. Gr. *thetos*, placed, adopted. *Ex:* dys-thetic (Med.): 3. Gr. *Thetis*, = L. *Thetis*, genit. *Thetidus*, one of the sea-deities, mother of Achilles. *Ex:* Theti-opsis (Moll.); Thetis (Moll.); Thetis-ella (Ann.).

thetes—Gr. *thetēs*, one who pledges, places. *Ex:* Ortho-thetes (Moll.).

thetid—See thet 3.

theut—NL. *theutes*, anagram of *teuthis*, a squid. *Ex:* Sepio-theutes (Moll.).

thez—Gr. *thēzas*, a point. *Ex:* Thezera*.

thi—Gr. *theion*, sulphur. *Ex:* Thi-ornis (Av.); Thi-osmus (Mam.); Thio-scelis (Ins.); Thio-tricha (Ins.); Pros-thio-stomum (Platy.).

thia—Gr. *theiō*, to run. *Ex:* Croce-thia (Av.).

thialot—Gr. *thialōtēs*, one of a great throng of revellers. *Ex:* Thialotes (Ins.).

thias—Gr. *thiasos*, a company of singers marching together; *thiasōdēs*, festive. *Ex:* Thiaso-phila (Ins.).

thiasot—Gr. *thiasōtes*, a member of a dancing, singing company. See thialot.

thigm—Gr. *thigma*, genit. *thigmatos*, touch. *Ex:* Thigmo-phyra (Prot.); thigmo-taxis.

thin—Gr. *this*, genit. *thinos*, the beach, the shore; also a sand heap. *Ex:* thin-ium (Ecol.); Thino-batis (Ins.); Thino-bius (Ins.); Thino-dromus (Ins.); Thino-lestes (Mam.); thino-phyta (Ecol.); Apo-thinus (Ins.); Epi-thinia*.

this—See thin.

thisanot—See thysan.

thix—Gr. *thixis*, a touching. *Ex:* thixo-tropic; Noto-thixos*.

thla—Gr. *thlaō*, to crush. *Ex:* Thlae-odon (Mam.); Thlao-soma (Arach.).

thladi—Gr. *thladias*, an eunuch. *Ex:* Thladi-antha*.

thlasp—Gr. *thlaspi*, a kind of cress, the ground seed of which was used like mustard < *thlaō*, to crush. *Ex:* Thlaspi*.

thlast—Gr. *thlastos*, crushed, broken in pieces; *thlastēs*, a crusher. *Ex:* Thlasto-coris (Ins.); Eu-thlasto-blatta (Ins.); Eu-thlastus (Mam.).

thlatt—Gr. *thlattō*, to crush. *Ex:* Thlatt-odus (Pisc.).

thlib—Gr. *thlibō*, to press, squeeze > *thlipsis*, pressure; *thlyptikos*, oppressive. *Ex:* Thlibo-ptera (Ins.); Thlips-ura (Crust.); Thlipt-odon (Moll.); Thlypso-gaster (Ins.); Syn-thlibo-rhamphus (Av.); Syn-thlipsus*.

thlips—See thlib.

thlipt—See thlib.

thlyp—Gr. *thlypis*, a kind of finch, also an alleged proper name. *Ex:* Thlyp-opsis (Av.); Compso-thlypis (Av.).

thlyps—See thlib.

thnet—Gr. *thnētos*, mortal. *Ex:* Thneto-schistus (Ins.); Thnetus (Ins.).

**tho**—1. Gr. *thoos*, fem. *thoē*, quick, nimble; *thoazō*, to move quickly. *Ex:* Thoa-therium (Mam.); Thoe (Crust.); Amphi-thoe (Arth.); Cymo-thoa (Crust.); Harmo-thoe (Ann.); Hippo-thoa (Bry.): 2. Gr. *thōs*, genit. *thōos*, a jackal. *Ex:* tho-oid; Thoo-pterus (Mam.); Thos (Mam.); Speo-thos (Mam.).

**-thode**—NL. *-thode* < Gr. *hodos*, a way. *Ex:* hydra-thode, modelled after the word cathode < Gr. *kathodos*, a going down.

**thoe**—See **tho**.

**thol**—1. Gr. *tholos*, a dome; *tholōtos*, built like a dome. *Ex:* Thol-artus (Prot.); Thol-ichthys (Pisc.); Tholo-spir-idae (Prot.); Amphi-tholos (Prot.): 2. Gr. *tholos*, mud, dirt; *tholoō*, to make muddy; *tholōdēs*, like mud, turbid; *tholōma*, muddiness.

**tholer**—Gr. *tholeros*, muddy, foul. *Ex:* Tholer-astis (Ins.); Tholera (Ins.).

**tholod**—See **thol 2**.

**tholom**—See **thol 2**.

**tholot**—See **thol 1**.

**thom**—Gr. *thōmos*, a heap. *Ex:* Thomo-mys (Mam.).

**thominx**—See **thomis**.

**thomis**—Gr. *thōminx* = *thomix*, a string; *thomissō*, to bind. *Ex:* Thominx (Nemat.); Thomis-idae (Arach.); Thomis-ops (Arach.); Thomisus (Arach.); Amphi-thomisus (Arach.).

**thoo**—See **tho 2**.

**thor**—1. Gr. *thouros*, rushing, furious. *Ex:* Thoro-phos (Pisc.); Thoro-sphaera (Prot.); Cisto-thorus (Av.); Thryo-thorus (Av.): 2. Gr. *thorē* = *thoros*, the semen. 3. Gr. *thoros*, limit, rule. *Ex:* Pen-thorum.

**thorac**—Gr. *thōrax*, genit. *thōrakos*, a breast-plate, the chest. *Ex:* Thoraci-poda (Arth.); Thoraco-saurus (Rept.); Thoraxo-phorus (Ins.).

**thorax**—See **thorac**.

**thorect**—Gr. *thōrēktēs*, armed with a breast-plate. *Ex:* Thorect-andra (Por.); Thorecta (Por.); Thorecto-psamma (Por.); Thorectus (Ins.); Thorictis (Rept.); Thoricto-soma (Ins.); Thorictus (Ins.).

**thoric**—Gr. *thorikos*, of or for the seed, the seminal duct.

**thorict**—See **thorect**.

**thorus**—NL. *thorus* < L. *torus*, a protuberance, a cushioned bed. *Ex:* Thryo-thorus (Av.).

**thoryb**—Gr. *thorybos*, noise. *Ex:* Thorybes (Ins.); Thorybo-thrips (Ins.).

**thos**—See **tho 2**.

**thracia**—L. *Thracia*, Thrace; *Thracius*, belonging to Thrace. *Ex:* Thraci-ella (Moll.); Thraci-opsis (Moll.); Thracia (Av.), (Moll.).

**thran**—1. Gr. *thranos*, dim. *thranion*, a bench, seat. *Ex:* Thranium (Prot.): 2. Gr. *thranis*, the sword-fish.

**thras**—See **thrasy**.

**thrass**—Gr. *thrassō*, to trouble, to disturb. *Ex:* Thrassis (Ins.).

**thrasy**—Gr. *thrasys*, bold. *Ex:* Thras-aetus = Thrasy-aetus (Av.); Thrasy-chirus (Arach.); Thrasy-doxa (Ins.); Thrasys (Av.).

**thraup**—Gr. *thraupis*, a small bird. *Ex:* Hemi-thraupis (Av.).

**thraust**—Gr. *thraustos*, brittle; *thraustēs*, torn to pieces. *Ex:* Thrausto-colus (Ins.); Caryo-thraustes (Av.).

**threctic**—Gr. *threktikos*, swift, able to run. *Ex:* Threcticus (Ins.).

**thremm**—Gr. *thremma*, genit. *thremmatos*, a nursling. *Ex:* Thremma (Ins.); Thremmato-philus (Av.).

**thren**—Gr. *thrēnos*, a lament; *threnōdēs*, like a dirge. *Ex:* Threno-dyta (Av.); Threno-pipo (Av.); Threnodes (Ins.); Threnus (Ins.).

**threnetic**—Gr. *thrēnētikos*, lamenting, wailing. *Ex:* Threnetica (Coel.).

**threps**—Gr. *threpsis*, nutrition. *Ex:* a-threpsia.

**thrept**—Gr. *threptos*, a servant, one nourished. *Ex:* Campo-threptus (Ins.); Heleo-threptus (Av.); Meli-threptes (Av.).

**thresc**—See **thresk**.

**thresk**—Gr. *thrēskeia*, religious worship < *thrēs-keuō*, to worship. *Ex:* Thresci-ornis (Av.); Thresci-ornith-idae (Av.).

**thriambeut**—Gr. *thriambeutēs*, one who revels in triumph. *Ex:* Thriambeutes (Ins.).

**thridac**—Gr. *thridax*, genit. *thridakos*, lettuce.

**thridax**—See **thridac**.

**thrinac**—Gr. *thrinax*, genit. *thrinakos*, a three-pronged fork. *Ex:* Thrinac-odus (Mam.); Thrinax*.

**thrinax**—See **thrinac**.

**thrinc**—Gr. *thrinkos*, a battlement, eaves, cornice; also a wall, fence. *Ex:* Thrinc-ia*.

**thring**—Gr. *thringos*, cornice, eaves. *Ex:* Thringo-rhina (Av.).

**thrip**—See **thrips**.

**thrips**—Gr. *thrips*, genit. *thripos*, a wood-worm. *Ex:* Thrip-idae (Ins.); Thrip-onax (Av.); Thrips (Ins.); Thrips-aphis (Ins.); Thripso-phagus (Ins.).

**thriss**—Gr. *thrissos* = *thrissa*, a kind of fish. *Ex:* Thriss-acanthias (Echin.); Thriss-ops (Pisc.); Thrissa (Pisc.); Thrisso-charax (Pisc.); Thrisso-pterus (Pisc.).

**thrix**—See **trich**.

**thromb**—Gr. *thrombos*, a clot, a lump; *thrombōsis*, a becoming clotted. *Ex:* thromb-in; thrombosis (Med.); thrombus; Thrombus (Por.).

**thron**—Gr. *thronos*, a seat; *thronistēs*, one who en-thrones. *Ex:* Thronistes (Ins.).

**throsc**—Gr. *thrōskō*, to leap. *Ex:* Throsc-idae (Ins.); Throsc-idium (Ins.); Throscus (Ins.); Throskus (Ins.); Aulono-throscus (Ins.).

**thry**—Gr. *thryon*, a reed. *Ex:* Thryo-manes (Av.); Thryo-spiza (Av.); Thryono-mys (Mam.).

**thryon**—See **thry.**

**thrypt**—Gr. *thryptō*, to break, to enfeeble; *thryptikos*, made feeble, delicate. *Ex:* Thrypt-ac-odon (Mam.); Thryptico-myia (Ins.); Thrypticus (Ins.); Thrypto-cera (Ins.).

**thu**—Gr. *thya*=*thyia*, an African tree with fragrant, durable wood; ML. *thuia, thuja* and *thuya*, the arbor vitae. *Ex:* Thu-idium*; Thui-aria (Coel.); Thuites*=Thuytes*; Thuj-opsis*; Thuja*.

**thuj**—See **thu.**

**thula**—Chilean *thula*, name of the snowy egret> *Egretta thula*, scientific name of the snowy egret. See also thylac.

**thunn**—L. *thunnus*=Gr. *thynnos*, the tunny-fish. *Ex:* Thunn-idae (Pisc.); Thunnus (Pisc.).

**thur**—Gr. *thyra*=sometimes transliterated *thura*, a door; *thyrion*, a little door; *thyris*, a window. *Ex:* Mega-thura (Moll.); Pelago-thuria (Echin.). See also thyr and thus.

**thurs**—See **thyrs.**

**thus**—L. *thus*, genit. *thuris*, incense, frankincense. *Ex:* thuriferous, yielding frankincense.

**thyad**—Gr. *thyas*, genit. *thyados*, a mad woman, mad for love. *Ex:* Thyas (Arach.); Thyas-ides (Arach.).

**thyas**—See **thyad**

**thyatir**—Gr. *Thyatira*, ancient city of Asia-minor. *Ex:* Thyatir-odes (Ins.); Thyatira (Ins.).

**thyell**—Gr. *thyella*, a hurricane, a storm of the worst kind. *Ex:* Thyello-droma (Av.).

**thyen**—L. *Thyene*, nurse of Jupiter and Bacchus. *Ex:* Thyen-illus (Arach.); Thyen-ula (Arach.); Thyene (Arach.).

**thyest**—Gr. *Thyestēs*, wicked son of Pelops. *Ex:* Thyestes (Pisc.).

**thyla**—See **thylac.**

**thylac**—Gr. *thylax*, genit. *thylakos*, a sack or pouch. *Ex:* Thyla-cynus (Mam.); Thylac-ella (Ins.); Thylach-ium*; Thylaci-phorus (Ann.); Thylaco-crinus (Echin.); Thylax (Mam.); Thylo-gale (Mam.); Lio-thula (Ins.).

**thylax**—See **thylac.**

**thylo**—See **thylac.**

**thym**—1. Gr. *thymos*=*thymon*, thyme<*thyō*, to perfume, to sacrifice, perhaps because it was burned on altars; *thymelē*, a place of sacrifice. *Ex:* Thymelae-aceae*; Thymele (Ins.); Thymo-phylla*; Thymus*: 2. Gr. *thymos*, a warty lump, also the thymus gland "so called because of its resemblance to a bunch of thyme." *Ex:* thymus gland: 3. Gr. *thymos*, the soul, mind, will.

**thymall**—Gr. *thymallos*, name of some fish having the odor of thyme. *Ex:* Thymall-idae (Pisc.); Thymallus (Pisc.). See thym 1.

**thymel**—See **thym 1.**

**thymelic**—Gr. *thymelikos*, scenic, muscial. *Ex:* Thymelicus (Ins.).

**thynn**—Gr. *thynnos*, the tunny, so called because of its quick motions<*thynō*, to rush, to dart along. *Ex:* Thynn-ascaris (Nemat.); Thynn-ichthys (Pisc.); Thynni-cola (Crust.); Thynnus (Pisc.), (Ins.).

Thyme Penny-royal, *Hedeoma thymoides*, an aromatic herb from limestone mountains of the S. W. deserts of the United States. Redrawn from Desert Wild Flowers—Jaeger. Stanford University Press.

**thyone**—Gr. *Thyōnē*, the raging one, mother of the fifth Bacchus. *Ex:* Thyon-ella (Echin.); Thyon-idium (Echin.); Thyone (Crust.), (Echin.).

**thyr**—1. Gr. *thyra*, a door; *thyris*, genit. *thyridos*, a window. *Ex:* thyr-idium; Thyr-ina (Pisc.); Thyrido-pteryx (Ins.); Thyris (Ins.); Thyro-ptera (Mam.): 2. Gr. *thyreos*, a door-stone, an oblong shield shaped like a door. *Ex:* thyre-oid=thyr-oid; Thyreo-cera (Ins.); Thyreus (Moll.); thyrio-thecum; Thyro-ptera (Mam.); Aco-thyreus (Ins.); Acro-thyrea (Ins.).

**thyreo**—See **thyr.**

**thyrid**—See **thyr.**

**thyrs**—Gr. *thyrsos*, a stalk, wand. *Ex:* Thyrs-ites (Pisc.); thyrs-oid; Thyrsito-cephalus (Pisc.); Thyrso-pteris*; Thyrso-phorus (Prot.); thyrsus; Polio-thyrsis*.

**thysan**—Gr. *thysanos*, a fringe, tassel; *thysanōtos*, tasseled. *Ex:* Thisanotia (Ins.); Thysan-otus*; Thysan-ura (Ins.); Thysano-ptera (Ins.); Thysano-soma (Platy.); Thysanote (Crust.); A-thysanus*.

**thysio**—Gr. *thysia*, a rite, an offering. *Ex:* Thysio-torus (Ins.).

**tiar**—Gr. *tiara*=*tiaras*, a Persian head-dress worn on great occasions; L. *tiaratus*, turban-like. *Ex:* Tiar-echinus (Echin.); Tiar-ella*; Tiar-opsis (Coel.); Tiara-dendron (Coel.); Tiari-concha (Moll.); Tiaro-cera (Ins.); Cosmo-tiara (Av.).

**tibi**—L. *tibia*, the shin bone; also a pipe or flute originally made from this bone. *Ex:* tibi-al;

Tibi-ella (Por.); Tibio-nema (Ins.); tibio-tarsal.

**tibicen**—L. *tibicen*, one who plays on a flute made from the tibia, a flute-player. *Ex:* Tibicen (Ins.).

**tibicin**—L. L. *tibicen*, genit. *tibicinis*, a piper, flutist < *tibia*, a pipe + *cano*, to sing, play > NL. *tibicinus*, flute-like, pipe-like.

**tibur**—1. L. *Tibur*, genit. *Tiburis*, ancient town of Latium on the River Anio: 2. Sp. *tiburon*, also *tiburo*, a shark. The *-on* ending in Spanish often refers to largeness in size.

**-tic**—Gr. *-tikos*, suffix entering into the formation of Greek verbal adjectives denoting relation, fitness, inclination or ability; source of the Eng. ending *-ic*, to which is often added the ending *-al* < L. *-alis*, as in phys-ic-al, method-ic-al.

**tich**—See **teich.**

**tict**—Gr. *tiktō*, to give birth. *Ex:* ticto-logy.

**-ticus**—L. *-ticus*, suffix, denoting made of or belonging to, as in *silvaticus*, belonging to the forest.

**tig**—Fr. *tige*, dim. *tigelle*, a stem, stalk. *Ex:* tigell-ate; tigella.

**tigill**—L. *tigillum*, a small piece of wool > NL. *tigillaris*, pertaining to a bit or tuft of wool.

**tigli**—NL. *tiglium*, croton; perh. < Gr. *tilos*, a liquid stool or manure.

**tigr**—L. *tigris*, a tiger; *tigrinus*, tiger-like, striped like a tiger. *Ex:* Tigr-idia*; Tigr-ina (Mam.); Tigri-soma (Av.); Tigris (Mam.).

**tigrin**—See **tigr.**

**til**—1. Gr. *tilos*, flock, down. *Ex:* Tilo-pteris*: 2. Gr. *tileis*, fut. of *tillō*, to pull, pluck. *Ex:* Tili-ornis (Av.).

**tilia**—L. *tilia*, the linden tree. *Ex:* Tilia*.

**till**—Gr. *tillō*, to pluck, nip, tear. *Ex:* Till-agra (Ins.); Till-ides (Ins.); Till-odonta (Mam.); Tilli-cera (Ins.); Tillo-mys (Mam.); Tillo-therium (Mam.); Tillus (Ins.).

**tilm**—Gr. *tilma*, genit. *tilmatos*, anything torn. *Ex:* Tilmat-ura (Av.).

**tilt**—Gr. *tiltos*, shredded, plucked < *tillō*, to pluck. *Ex:* Mnio-tilta (Av.).

**tim**—Gr. *timē*, esteem, honor; *timoros*, conferring honor, aiding; *timētos*, honorable; *timios*, held in honor < *timaō*, to honor, adorn. *Ex:* Tim-andra (Ins.); Tim-aspis (Ins.); Timeta (Av.); Timorus (Ins.); Iso-tima (Ins.). See also timor.

**timai**—Gr. *timaios*, highly prized, held in esteem.

**timali**—NL. *timalia*, a bird name, etym. uncertain. *Ex:* Tamali-idae (Av.); Tamalia (Av.). Timalia is an earlier generic name.

**timarch**—Gr. *timarchia*, a state in which civil and political honors are disbursed according to value of property. *Ex:* Timarcha (Ins.).

**timbal**—Fr. *timbale*, from Arab. *thabal*, a timbal. *Ex:* timbal.

**timet**—See **tim.**

**timi**—Gr. *timios*, worthy, costly. *Ex:* Timia (Ins.).

**timid**—L. *timidus*, fearful, timid. *Ex:* timid-ul-ous.

**timor**—*Timor*, island of the Dutch East Indies. *Ex:* Timor-ites (Moll.); Timoria (Moll.); Timoro-blastus (Echin.); Timoro-phyllum (Coel.). See also tim.

**tin**—1. Gr. *tinō*, to avenge, to punish. *Ex:* Tino-ceras (Mam.); Tino-saurus (Rept.): 2. L. *tinea*, a worm, a bookworm, moth. *Ex:* Tin-odes (Ins.); Tine-idae (Ins.); Tinea (Ins.); Tineo-mima (Ins.); Tineo-myza (Ins.); Tineo-phaga (Ins.): 3. Gr. *teinō*, to stretch. *Ex:* Tin-otis (Moll.); Tino-soma (Moll.); Tino-tarsus (Ins.): 4. L. *linus*, a name for Viburnum, the wayfaring tree.

**tinact**—Gr. *tinaktōr*, a shaker. *Ex:* Tinactor (Av.).

**tinagm**—Gr. *tinagma*, a shaking, an agitation. *Ex:* Tinagma (Ins.).

**tinam**—NL. *tinamus* < *tinamou*, a native name for certain S.A. birds. *Ex:* Tinam-otis (Av.); Tinami-formes (Av.); Tinamou (Av.); Tinamus (Av.).

**tinc**—L. *tinca*, some small fish, the tench. *Ex:* Tinca (Pisc.).

**tinct**—L. *tinctus*, dyed, pp. of *tingo*, to dye; *tinctorius*, of or belonging to dyeing.

**tine**—L. *tinea*, a moth, worm. *Ex:* Tine-idae (Ins.); Tine-ina (Ins.); Tinea (Ins.).

**ting**—1. Tupi Indian, *tinga*, white, bright, ornamented. *Ex:* Urubi-tinga (Av.): 2. NL. *tingis*, name for certain bugs. *Ex:* Ting-idae (Ins.); Tingis (Ins.).

**tingen**—L. *tingens*, genit. *tingentis*, a dyer < *tingo*, to dye.

**tinn**—L. *tinnio*, to ring, ppr. *tinniens*, genit. *tinnientis*, ringing; *tinnitus*, a ringing; *tinnulus*, ringing, tinkling, shrill-sounding; *tinnunculus*; a kind of hawk or kestrel with "querulous bell-like note." *Ex:* tinnient; tinnitus (Med.); Tinnunculus (Av.).

**tinnuncul**—See **tinn.**

**tintinn**—NL. *tintinnus*, a bell < L. *tintinno*, to ring, jingle. *Ex:* Tintin-idium (Prot.); Tintinnus (Prot.).

**tintinnabul**—L. *tintinnabulum*, a bell. *Ex:* Tintinnabulum (Coel.).

**-tion**—Eng. *-tion*, suffix denoting result of an action, state of. *Ex:* absorp-tion.

**tiph**—Gr. *tiphē*, a kind of insect—Tiph-ia (Ins.).

**tipul**—L. *tipula*, a water spider. *Ex:* Tipul-aria*; Tipula (Ins.); Tipula-toma (Ins.); Tipuli-forma (Ins.); Tipulo-gaster (Ins.).

**tiranites**—See **tyrann.**

**tisiphon**—Gr. *Tisiphonē*, one of the Furies,

avenger of murder. *Ex:* Tisiphone (Rept.); Tisiphonia (Coel.).

**tissue**—Fr. *tissu*, tissue. *Ex:* tissue.

**titan**—Gr. *Titan*, Titan, the giant. *Ex:* Titan-ichthys (Pisc.); Titan-ornis (Av.); Titano-mys (Mam.); Titano-therium (Mam.); Titanus (Mam.).

**tithen**—Gr. *tithēnos*, a nurse. *Ex:* Tithene (Ins.).

**tithon**—Gr. *Tithōnos*, consort of Aurora. *Ex:* Tithonia*; Tithonus (Ins.).

**tithymal**—Gr. *tithymalos*, name of a spurge. *Ex:* Tithymalus*.

**titill**—L. *titillatio*, a tickling <*titillo*, to tickle> Fr. *titillation*. *Ex:* titillation.

**titth**—Gr. *titthē*, udder. *Ex:* Eu-tittha (Arach.).

**titub**—L. *titubo*, to stagger, ppr. *titubans*, genit. *titubantis*, tottering, wavering, restless.

**tityr**—Gr. *tityros*, a pheasant. *Ex:* Tityra (Av.).

**tityrus**—L. *Tityrus*, a shepherd mentioned by Virgil.

**tityus**—L. *Tityus*, river of Illyria. *Ex:* Tityo-buthus (Arach.); Tityo-lepreus (Arach.); Tityus (Arach.).

**tla**—Gr. *tlaō*, to bear, suffer; part. *tlas*, suffering, undergoing hardship. *Ex:* Tlasia (Ins.).

**tlam**—Gr. *tlamon*, genit. *tlamonos*=*tlēmon*, genit. *tlēmonos*, wretched, suffering, miserable, stout-hearted. *Ex:* Tlemon (Ins.).

**tlasia**—See **tla**.

**tlemon**—See **tlam**.

**tmar**—L. *Tmarus*, a mountain in Epirus. *Ex:* Tmarus (Arthr.).

**tmeg**—Gr. *tmēgō*, to cut, to divide. *Ex:* Tmego-ceras (Moll.).

**tmem**—Gr. *tmēma*, genit. *tmēmatos*, a section, a portion, something cut off.

**tmes**—Gr. *tmēsis*, a cutting, dividing, making in pieces; *tmētikos*, capable of cutting. *Ex:* Tmesi-pterus*; Tmesi-sternus (Ins.); Tmeso-rhina (Ins.); Tmetic-ides (Arach.); Tmeticus (Arach.); Tmeto-cera (Ins.); dia-tmesis; Plagio-tmesis (Ins.).

**tmetic**—See **tmes**.

**tmol**—Gr. *Tmōlos*, a mountain of Lydia. *Ex:* Tmolus (Ins.).

**toc**—Gr. *tokos*, offspring; *toketos*, bearing, bringing forth young. *Ex:* a-tok-ous; Em-bio-toca (Pisc.); epi-tok-ous; Mono-toca*; Skeno-toka (Coel.); Taenio-toca (Pisc.); Zoo-toca (Mam.).

**tocet**—See **toc**.

**tod**—L. *todus*, name for some small bird, a tody. *Ex:* Tod-idae (Av.); Todus (Av.).

**toen**—See **taen**.

**togat**—L. *togatus*, clad in a toga.

**toich**—Gr. *toichos*, a wall. *Ex:* Toicho-poma (Coel.).

**tok**—See **toc**.

**tolm**—Gr. *tolma*, daring, boldness; *tolmēros*, hardy, audacious. *Ex:* Tolm-archus (Av.); Tolm-odus (Av.); Tolmero-lestes (Ins.); Tol merus (Ins.); Tolmo-lestes (Av.).

**tolmer**—See **tolm**.

**tolu**—NL. *tolu*, tolu balsam first brought from Santiago de Tolu, a seaport of Colombia. *Ex:* Tolui-fera*.

**tolype**—Gr. *tolypē*, a ball of wool; *tolypeuō*, to wind off wool; to accomplish; *tolypeutikos*, of or for accomplishing. *Ex:* Tolype (Ins.); Tolype-ceras (Moll.); Tolypeu-tes (Mam.).

**tom**—1. Gr. *tomos*, a cut, slice; as verb. adj., cutting, sharp; *tomia*, a cutting; *tomikos*, of or for cutting. *Ex:* Tom-arctus (Mam.); Tomi-stoma (Rept.); Tomi-therium (Mam.); Tomic-odon (Pisc.); Tomico-merus (Arach.); Tomo-pteris (Ann.); ana-tomy; Dio-tome-odon (Mam.); en-tom-ion; en-tomo-logy; mero-tomy; Neo-toma (Mam.); Phloeo-tomus (Av.); sclero-tome:    2. Gr. *tomē*, a section, stump. *Ex:* Tom-opeas (Mam.).

**toment**—L. *tomentum*, a stuffing of wool or hair for cushions, cushioning; *tomentosus*, densely covered with matted wool or short hairs, i.e. full of stuffings.

**tomi**—NL. *tomium*, pl. *tomia*, the cutting edge of a bird's beak <Gr. *temnein*, inf. of *temnō*, to cut. *Ex:* tomium.

**tomic**—See **tom** 1.

**-tomy**—Eng. *-tomy* <Gr. *tomō*, to cut, sever+-*y*, a noun suffix. *Ex:* ana-tomy.

**tomyr**—Gr. *Tomyris*, Scythian Queen who slew the elder Cyrus. *Ex:* Tomyris (Rept.), (Moll.).

**ton**—Gr. *tonos*, something stretched, a brace, a strain; *tonikos*, capable of extension. *Ex:* ton-esis; Tono-desmus (Moll.), (Myr.); tono-plast; tono-tropism; Chiro-ton-ctes, the last element of unknown meaning; chordo-tonal; eco-tone; Epi-tonium (Moll.).

**tonic**—See **ton**.

**tono**—Anagram of *noto-* in Notorhinus (Mam.). *Ex:* Tono-rhinus (Mam.). See also ton.

**tonsill**—L. *tonsilla*, tonsil. *Ex:* tonsill-itis.

**tonsurans**—NL. *tonsurans*, shaving <L. *tonsura*, a shearing, shaving.

**top**—Gr. *topos*, a place. *Ex:* topo-taxis; Topo-trita (Ins.); topo-type; bio-tope; hetero-top-ous; ec-top-ic.

**topaz**—Gr. *topazos*, a precious stone, usually yellowish. *Ex:* Topaza (Av.).

**toph**—L. *tophus*=*tofus*, a porous stone, tufa. *Ex:* tophus.

**tor**—1. L. *torus*, a bulge, swelling, knot; *torosus*, full of muscle, fleshy; NL. *toric*, pertaining to a torus. *Ex:* Toric-elli-ceras (Moll.); Toro-dinium (Prot.); Toroso-myia (Ins.):    2. Gr. *toros*, piercing; also a borer. *Ex:* Cocco-torus (Ins.).

**-tor**—L. *-tor*, noun suffix (masculine) denoting agent or doer of an action, akin to Eng. *er*. *Ex:* peti-tor, from L. *peto*, to seek. See -or.

**tora**—Abyss. *tora*, name of the hartebeest of E. Africa.

**torda**—Swedish *torda*, a name for the razor-billed auk.

**tordyl**—Gr. *tordylion*, a plant name<*tornos*, a lathe+*illō*, to turn. *Ex:* Tordyl-opsis*; Tordy lium*.

**toret**—See **toreu.**

**toreu**—Gr. *toreuō*, to bore through, to shape; *toreutēs*, a turner; *torētos*, bored, pierced; *toreutos;* turned on a lathe; *toreuma*, carved work; also a whirling motion. *Ex:* Toreto-cnemus (Rept.); Toreum-atica (Echin.); A-toreuteus (Ins.); Eu-toreuma (Ins.).

**torg**—Gr. *torgos*, a vulture. *Ex:* Torgos (Av.).

**toric**—See **tor 1.**

**-torius**—L. *-torius*, *-a*, *-um*, adj. suffix meaning, belonging to, possessing. *Ex:* Sept-(t)oria*; Call-(t)oria*. See also -arius, and -sorius.

**torix**—NL. *torix*, an arbitrary combination of letters used to designate a genus of leeches. *Ex:* Torix (Ann.).

**torm**—Gr. *tormos*, a hole, a socket. *Ex:* torm-odont; Tormo-crinus (Echin.); tormo-gen; epi-torma; Odonto-tormae (Pisc.).

**tormin**—L. *torminosus*, causing colic, subject to colic<*tormina*, colic. *Ex:* tormin-alis.

**torn**—L. *torno*, to turn; Gr. *tornos*, a compass; also that which is turned, a circle, a wheel; ML. *tornatus*, turned, well wrought, finished; *tornatella* ML. dim. of L. *tornus*, a turner's wheel. *Ex:* torn-aria; Torn-axis (Moll.); torn-ote; Tornat-ella (Moll.); Tornat-ina (Moll.); Torno-sinus (Ins.); tornus.

**tornat**—See **torn.**

**torneu**—Gr. *torneuō*, to bore through, to work in relief, to round off; *torneuma*, shavings, embossed work; *torneutēs*, one who works in relief. *Ex:* Torneutes (Ins.); Torneuto-uras (Moll.).

**torp**—L. *torpens*, genit. *torpentis*, torpid; *torpidus*, inactive.

**torped**—L. *torpedo*, genit. *torpedinis*, a benumbing, a stiffness. *Ex:* Torpedin-idae (Elasm.); Torpedo (Elasm.).

**torque**—L. *torques* and *torquis*, a twisted neck-chain<*torqueo*, to twist; *torquatus*, adorned with a necklace. *Ex:* Torquat-ella (Prot.); torquate; torque-ate; Torque-nympha (Prot.); Torquati-scala (Moll.); Torquis (Moll.).

**torr**—L. *torrus*, a fire-brand. *Ex:* Torr-ell-isca (Moll.); Torr-ella (Moll.).

**torren**—L. *torrens*, genit. *torrentis*, a torrent; also inflamed, hot. *Ex:* Torrent-aria (Av.); Tor renti-cola (Arach.).

**torrens**—See **torrent.**

**torrent**—L. *torrens*, genit. *torrentis*, a torrent. *Ex:* torrential. See also torren.

**torror**—L. *torror*, genit. *torroris*, a drying up, scorching.

**tort**—L. *tortus*, a twisting, winding<*torqueo*, to twist, to twist awry, torment. *Ex:* torti-folius: Torti-sternum (Rept.); Torto-pus (Ins.); Nas-turti-um*, here used in sense of tormenting since the acrid taste affects the nose muscles.

**tortil**—L. *tortilis*, twisting, winding.

**tortiv**—L. *tortivus*, squeezed dry.

**tortric**—NL. *tortrix*, genit. *tortrices*<L. *tortus*, twisted. *Ex:* Tortric-idae (Ins.); Tortric-odes (Ins.); Tortrici-forma (Ins.); Tortrico-morpha (Ins.); Tortrix (Ins.).

**tortrix**—See **tortric.**

**tortul**—L. *tortula*, a small twist. *Ex:* Tortul-osa (Moll.); Tortula (Moll.).

**torul**—L. *torulus*, a pencil, tuft of hair; also a muscular part of the body; *torus*, a bulge or swelling, cushion, an elevation, the bank of a stream. *Ex:* Torula (Moll.); Toruli-pora (Bry.); torulose.

**torv**—L. *torvus*, wild, cruel, savage; *torvidus*, savage, wild.

**torym**—NL. *torymus*, said to be from <Gr. *toreō*, to bore, pierce, to engrave. *Ex:* Torymus (Ins.).

**toryn**—Gr. *torynē*, a stirrer, a spoon; *torynētos*, stirred about. *Ex:* Toryni-fer (Brach.); Toryno-crinus (Echin.).

**tos**—1. Gr. *tosos*, in compounds tosa-, etc., so much, so very. *Ex:* Tosa-stroma (Coel.): 2. *Tosa*, a province of Japan. *Ex:* Tosi-aster (Echin.).

**tot**—L. *totus*. the whole, entire. *Ex:* toti-palmate.

**totanus**—It. *totano*, the moor hen. *Ex:* Totanus (Av.).

**tourac**—Fr. *touraco*, an imitation of the touraco's cry. *Ex:* turac-in; Turaco (Av.).

**tox**—Gr. *toxon*, dim. *toxarion*, a bow; *toxotēs*, a bow-man; *toxikos*, belonging to arrows or archery; *toxeutēs*, a bowman. *Ex:* Tox-odontia (Mam.); Toxares (Ins.); Toxarium (Prot.); Toxeutēs (Ins.); toxius; Toxo-stoma (Av.); toxon; Toxotes (Pisc.); To(xo)-xylon*; See also toxic.

**toxar**—See **tox.**

**toxeres**—Gr. *toxērēs*, furnished with a bow. *Ex:* Toxeres (Ins.).

**toxeum**—Gr. *toxeuma*, an arrow. *Ex:* Toxeu-morpha (Ins.); Toxeum-ella (Ins.); Toxeuma (Ins.).

**toxeut**—See **tox.**

**toxic**—Gr. *toxikos* (poison) for smearing on arrows<*toxon*, a bow. *Ex:* tox-one; toxic-ity; Toxic-ophis (Rept.); Toxico-dendron*; Toxicoa (Rept.); toxico-logy; toxin; toxo-phil.

**toxot**—See **tox.**

trab—L. *trabs*, genit. *trabis*, a beam, club; *trabalis*, of or belonging to beams. *Ex:* Trabala (Ins.); Trabalia (Mam.); Hypo-trabala (Ins.).

trabal—See trab.

trabea—L. *trabea*, a toga trimmed with purple stripes >*trabeatus*, wearing a robe of state. *Ex:* Trabea (Arach.).

trabeat—See trabea.

trabecul—L. *trabecula*, a small beam; NL. *trabeculatus*, marked with cross bars. *Ex:* Trabecula (Moll.); Trabeculate; Trabeculus (Ins.)

trach—L. *trachia*, the wind-pipe, the "rough artery" <*trachys*, rough. *Ex:* trach-enchyma; Trache-aria (Arach.); trache-ate; trache-id; trachea; Tracheo-philus (Nemat.); Tracheophonae (Av.). See also trachy.

trachel—Gr. *trachēlos*, the neck, throat. *Ex:* Trachel-acanthus (Pisc.); Trachel-aeum (Ins.); the last element of unknown meaning; Tracheli-pus (Arth.); Trachel-ium*; Trachelocampus (Arach.); Trachelus (Ins.); Silotrachelus (Ins.).

trachi—See trachy.

trachin—M.L. *trachinus*, the horse mackerel. *Ex:* Trachin-ops (Pisc.); Trachino-cephalus (Pisc.). See Trachinus under trachy.

trachy—Gr. *trachys*, rough; *trachōdēs*, of rough nature; *trachytēs*, roughness. *Ex:* Trach-odon (Rept.); Trachi-dermus (Pisc.); Trachinus (Pisc.), ML. <Gr. *trachouros*, rough tail; Trachodes (Ins.); Trachusa (Ins.); Trachyarus (Ins.), the last element of unknown meaning; Trachy-cnemus (Av.); Trachy-medusae (Coel.); Trachyt-ella*; not Trachyopus (Ins.) which is an anagram of Tachyporus.

trachyt—See trachy.

tract—L. *traho*, to draw, haul; pp. *tractus*, drawn. *Ex:* Tracto-lira (Moll.); con-tract-ile; protract-or; re-tract-or.

trag—Gr. *tragos*, dim. *tragulus*, a he-goat <*tragō* to nibble. *Ex:* Trag-elaphus (Mam.); Tragopogon*; Tragul-ichthys (Pisc.); Tragulotherium (Mam.); Tragulus (Mam.); Tragus (Mam.); Oreo-tragus (Mam.).

tragisc—Gr. *tragiskos*, a young he-goat. *Ex:* Tragisco-coris (Ins.); Tragiscus (Ins.).

trajectil—NL. *trajectilis*, passing over <*trajectus*, a passing over.

tralatiti—L. *tralatitius*, usual, common.

tram—L. *trama*, the woof; also something thin. *Ex:* Tram-etes*; Trama (Ins.); tramo-sericeous.

tran—Gr. *tranēs*, clear, distinct; *tranoō*, to make clear; *tranotēs*, clearness. *Ex:* Tranes (Ins.); Trano-cera (Ins.); Ari-tranis (Ins.); Epitranus (Ins.).

tranot—See tran.

trans-—L. *trans*-, prefix meaning across. *Ex:* Trans-pithecus (Mam.); trans-position.

transfus—L. *transfusio*, a pouring out; Fr. *transfusion*. *Ex:* transfusion.

trapel—Gr. *trapelos*, easily turned, changeable. *Ex:* Trapelo-cera (Tri.); Trapelus (Ins.).

trapez—Gr. *trapēza*, a four-legged table, also a grinding surface. *Ex:* Trapezi-cepon (Crust.); Trapezo-dera (Ins.).

traphe—See traphec.

traphec—Gr. *traphēx*, genit. *traphēkos*, a spear, plank. *Ex:* Traphe-corynus (Ins.); Traphecocorynus (Ins.).

trapher—Gr. *trapheros*, fattening, well fed. *Ex:* Traphera (Ins.).

traphex—See traphec.

traum—Gr. *trauma*, genit. *traumatos*, a wound. *Ex:* Traum-oecia (Ins.); Traumato-crinus (Echin.).

trebac—L. *trebax*, genit. *trebacis*, crafty, cunning, smooth <Gr. *tribakos*, rubbed, worn.

trebax—See trebac.

trebia—1. L. *Trebia*, a river in upper Italy: 2. L. *Trebius*, name of a Roman gens. *Ex:* Trebius (Crust.).

trech—1. Gr. *trechō*, to hasten, to run. *Ex:* Trechi-gnathus (Ins.); Trecho-corys (Ins.); trecho-meter; Trecho-mys (Mam.); Trechus (Ins.); Ammo-trecha (Arach.); Hemero-trecha (Arach.); Xylo-trechus (Ins.): 2. *trechō*, rough; also rough, rocky country.

trechale—Gr. *trechaleos*, poet. for *trechys*, rough, rugged. *Ex:* Trechalea (Arach.).

trechn—Gr. *trechnos*, a branch. *Ex:* Colotrechnus (Ins.).

trem—1. Gr. *trēma*, genit. *trēmatos*, a hole; also the female pudendum; *trēmatōdēs*, perforated. *Ex:* Trem-andra*; Trema*; Trema-dictyon (Por.); Tremato-notus (Moll.); Trematoda (Playt.); Tremex (Ins.), irregularly formed; Tremo-pora (Bry.); Haplo-trema (Moll.); Mono-tremata (Mam.). 2. L. *tremo*, to shake. *Ex:* Trem-andra*: Trem-ell-ina*; Trem-ella*.

tremul—L. *tremulus*, a quaking, trembling. *Ex:* tremul-oides.

trep—1. Gr. *trepō*, to turn. *Ex:* Trep-ophrys (Ins.); Trepo-carpa*; Trepo-monas (Prot.); Trepo-nema (Prot.); Trepo-stomata (Bry.): 2. Gr. *trepōn*, genit. *trepōnos*, timorous, shy.

trepano—See tryp.

treph—Gr. *trephō*, to thicken; also to nurture, support, feed. *Ex:* Trephi-onus (Ins.); Campotrephus (Ins.); Zoo-trephes (Ins.).

trepid—L. *trepidus*, restless, alarmed; *trepidulus*, anxious, shy. *Ex:* Trepidulus (Ins.).

trepon—See trep.

treps—Gr. *trepsis*, a turning. *Ex:* Trepsi-chrois (Ins.).

trept—Gr. *treptos*, to be turned about or changed.

*treptikos*, changeable. *Ex:* Trept-ichnus (Ins.); Trepto-gon (Ins.); Spiro-treptus (Myr.).

trer—Gr. *trērōn*, fearful, shy, generally in reference to a dove. *Ex:* Trero-laema (Av.); Treron (Av.); Osmo-treron (Av.); Phapsi-treron (Av.).

tres—Gr. *trēsis*, a boring, a perforation. *Ex:* Tresus (Arach.).

trest—Gr. *trestēs*, one who shakes, a coward. *Ex:* Trestis (Ins.).

tret—Gr. *trētos*, pierced, perforated. *Ex:* Tret-aspis (Tri.); Tret-echinus (Echin.); Treto-calyx (Por.); Amphi-tretus (Moll.); Hyper-treti (Cycl.); Phyllo-treta (Ins.).

tri—Gr. *treis*, three=L. *tri*, three; *triens*, genit. *trientis*, a third part; *trientalis*, containing the

Three-caruncled Bell-bird, *Chasmorhynchus tricarunculatus.*

third part of a foot; Gr. *tris*, thrice. *Ex:* Tri-adenum*; Tri-cerat-ops (Rept.); Tri-clad-ida (Playt.); Tri-lob-ita (Arth.); Tri-odia*; Tri-prot-homo (Mam.); Trientalis*; Trio-spyridium (Prot.); Tris-toma (Moll.); Tristus (Moll.); Calli-tris*.

tria—Gr. *trias*, genit. *triados*, the number three, a triad. *Ex:* Tria-crinus (Echin.); Tria-toma (Ins.); triad; Triado-cidaris (Echin.); Trias (Geol.); Triass-ic (Geol.), s doubled as in Jurassic; Philo-tria*.

triad—See tria.

triaen—Gr. *triaina*, a trident. *Ex:* Triaen-ops (Mam.); Triaena (Ins.); Triaeno-phorus (Platy.).

trias—See tria.

trib—Gr. *tribō*, to rub; *tribē*, a wearing away. *Ex:* Trib-odon (Mam.); trib-ium (Ecol.); Tribo-stethus (Ins.); Tribon (Arach.).

tribac—Gr. *tribax*, genit. *tribakos*, worn down, rubbed. *Ex:* Tribax (Ins.).

tribax—See tribac.

tribel—Gr. *tribelēs*, three pointed. *Ex:* Tribelo-cephala (Ins.).

tribol—Gr. *tribolos*, three-pointed. *Ex:* Tribol-odon (Pisc.); Tribol-ium (Ins.) or perhaps as Agassiz suggests it may be derived from *tris*, thrice and *bolē*, a throw; Tribolo-ceras (Moll.).

tribon—Gr. *tribōn*, a thread-bare cloak. *Ex:* Tribon-ium (Ins.); Tribono-phorus (Mam.); Tribono-sphaera (Prot.).

tribul—1. L. *tribulus* < Gr. *tribolos*, three-pointed, a caltrop; also a kind of thorn. *Ex:* Tribul-ina (Prot.); Tribulus*:　2. L. *tribulis*, one of the same tribe.

tricc—Gr. *trikkos*, a small bird. *Ex:* A-phano-triccus (Av.).

trich 1.—Gr. *thrix*, genit. *trichos*, dim. *trichion*, the hair; a single hair; *trichinos*, of hair; *trichōdēs*, hairy. *Ex:* Trich-echus (Mam.), see trich; trich-ite; Trichia*; Trichin-ella (Nemat.); Trichina (Nemat.); Trichini-um*; Trichio-cerus (Ins.); tricho-cyst; Tricho-dectes (Ins.); Thricho-proctos (Ins.); Trichoda (Prot.); Ulo-thrix*; Zono-trichia (Av.):　2. Gr. *tricha*, in three parts: *Ex:* Trich-ilia*, the last element of unknown origin or meaning.

trichad—Gr. *trichas*, genit. *trichados*, a thrush *Ex:* Trichas (Av.).

trichas—See trichad.

trichil—Gr. *tricheilos*, three-lipped. *Ex:* Trichilia*; Trichilo-gaster (Ins.).

trichin—Gr. *trichinos*, hairy, of hair. *Ex:* Trichina (Nemat.); Trichino-chaeta (Ins.); Trichinus (Nemat.).

trichod—Gr. *trichōdēs*, hairy, hirsute < *thrix* genit. *trichos*, a hair. *Ex:* Trichodo-cerus (Ins.).

trichot—Gr. *trichōtos*, furnished with hair. *Ex:* Trichot-ichnus (Ins.).

trichth—Gr. *trichtha*, into three parts. *Ex:* Trichtha-ceras (Crust.).

trichthadi—Gr. *trichthadios*, three-fold.

tridacn—Gr. *tridaknos*, eaten at three bites. *Ex:* Tridacna (Moll.); Tridacno-caris (Crust.).

tridu—L. *triduum*, the space of three days.

trienni—L. *triennium*, the space of three years.

trient—See tri.

trigl—Gr. *triglē* the red mullet. *Ex:* Lepido-trigla (Pisc.); Trigle (Pisc.); Trigla (Pisc.).

trigloch—Gr. *triglōchis*, genit. *triglōchinos*, three-barbed < *tri-*, three + *glōchin*, a projecting point or barb. *Ex:* Triglochin-ura (Arach.); Triglochinos*.

trigon—Gr. *trigōnos*, triangular, three-cornered. *Ex:* Trigon-ella*; Trigon-istis (Ins.); Trigono-bunus (Arach.).

trilisa—anagram of Liatris*. *Ex:* Trilisa*.

trillium—NL. *trillium*, an herb with leaves in whorls of three, perh. < L. *trilix*, woven with three sets of leashes, triple < *tri*, three + *licium*, the ends of a weaver's thread. *Ex:* Trilli-aceae*; Trillium*.

trimm—Gr. *trimma*, genit. *trimmatos*, that which is rubbed, fine meal. *Ex:* Trimmato-thelia*.

trin—L. *trinus*, also *trini*, three each, triple. *Ex:* Trin-ella (Rot.).

tring—Gr. *tryngas* > ML. *tringa* and *trynga*, name for a sandpiper. *Ex:* Tryng-ites (Av.); Tringa (Av.).

trinit—L. *trinitas*, genit. *trinitatis*, a trinity, a triad.

triodit—Gr. *triodītēs*, a worthless fellow, a lounger, of the streets; also common, vulgar. *Ex:* Triodites (Ins.).

trion—1. NL. *trionum* < Gr. *trionon*, name of a malvaceous plant: 2. L. *Triones*, genit. *Trionum*, the constellation of the Great Dipper, hence something northern.

tripan—See tryp.

triphas—Gr. *triphasios*, triple. *Ex:* Triphasia*.

tripl—L. *triplus* = Gr. *triploos*, triple. *Ex:* triplo-blastic; Triplo-pus (Mam.).

tript—Gr. *triptēr*, a pestle; *triptos*, rubbed or pounded. *Ex:* Tripto-rhinus (Av.).

tripudi—L. *tripudians*, dancing, ppr. of *tripudio*, to dance, leap, caper.

triquetr—L. *triquetrus*, having three corners, triangular. *Ex:* Triquetra (Moll.); Triquetri-cornis (Ins.).

tris—See tri.

triss—Gr. *trissōs*, threefold, triple. *Ex:* Triss-acantha (Ins.); Trisso-phaes (Ins.).

trist—L. *tristis*, neut. *triste*, sad, melancholy; often now used to imply dullness of color.

trit—Gr. *tritos*, the third. *Ex:* Trit-aria (Moll.); Trito-micrus (Ins.); Topo-trita (Ins.). See also triton.

tritic—L. *triticum*, an old name for wheat; *triticeus*, of wheat, wheaten. *Ex:* Triticum*.

triton—Gr. *Triton*, demigod of the sea. *Ex:* Triturus (Amph.); Triton (Amph.); Triton-aster (Echin.); Tritoni-doxa (Moll.); Tritono-harpa (Moll.); Typhlo-triton (Amph.).

trivi—1. L. *trivium*, a place where three roads meet, a cross-road. *Ex:* trivium: 2. *Trivia*, a name for Diana. *Ex:* Trivia (Moll.).

-trix—1. L. *-trix*, suffix denoting agent or doer of an action. *Ex:* can-trix < *cano*, to sing; Sibila-trix (Av.) > *sibilo*, to hiss; Gyra-trix (Platy.), < *gyro*, to move or turn around: 2. L. *trix*, fem. ending corresponding to masc. *-tor*.

trix—Gr. *trixos*, threefold. *Ex:* Trix-agus (Ins.); Trixis*; Trixo-stomus (Ins.); not Trixa (Ins.) which is poorly derived from *thrix*, a hair; Buccula-trix (Ins.).

trixag—L. *trixago*, genit. *trixaginis*, name of a plant called germander.

troch—1. Gr. *trochos* = L. *trochus*, dim. *trochatella* = *trochillus*, a wheel, hoop; Gr. *trochia*, the track left by wheels, *trochiskos*, a small wheel, a small ball, a pellet; *trochalos*, rotund, round. *Ex:* Trochalo-nota (Ins.); Trochatella (Av.);

Trochilla (Moll.); Trochiscus (Moll.); trocho-phore; Trocho-toma (Moll.); Trochus (Moll.); Colpo-trochia (Ins.): 2. Gr. *trochos*, a badger. *Ex:* Troch-ictis (Mam.); Trochotherium (Mam.).

trochal—See troch 1.

trochant—Gr. *trochantēr*, a runner; also the ball on which the hip-bone turns. *Ex:* trochant-in; Trochant-odon (Ins.); trochanter; Trochantero-cerus (Arach.).

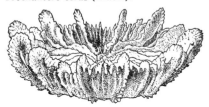

Diademed Horn-wheel Coral, *Ceratotrochus diadema*. Redrawn from Voyage of The Challenger—Thomson. Courtesy of Joseph McDonough Co.

trochat—See troch 1.

trochil—1. Gr. *trochilos*, a bird of the sand-piper kind; *trochos*, a running. *Ex:* Trochilus (Av.): 2. Gr. *trochil-*, a pulley, wheel. *Ex:* Trochil-ina (Moll.); Trochilo-glossa (Ins.); Trochilo-pora (Bry.).

trochisc—See troch 1.

trochlea—L. *trochlea*, a pulley < Gr. *trochalia*, the roller of a windlass. *Ex:* trochle-arts; trochlea.

troct—Gr. *trōktēs*, a nibbler, gnawer; *trōctos*, eatable. *Ex:* Troctes (Ins.); Trocto-cerus (Ins.); Platy-troktes (Pisc.); Proto-troctes (Pisc.).

troctic—Gr. *trōktikos*, greedy. *Ex:* Trocticus (Ins.).

trog—Gr. *trōgō*, to gnaw; ppr. *trōgōn*, gnawing; *trōx*, genit. *trōgos*, a gnawer, weevil. *Ex:* Trogosus (Mam.); Trogon (Av.); Trogon-therium (Mam.); Trogus (Ins.); Trox-ites (Ins.); Troximon*, the application not obvious; Di-glosso-trox (Ins.); Rhizo-trogus (Ins.); Sito-troga (Ins.).

trogl—Gr. *trōglē*, a hole made by gnawing. *Ex:* Trogl-ichthys (Pisc.); Troglo-dytes (Av.) "lit., one who creeps into holes"; Troglo-hyphantes (Arach.).

trogon—See trog.

troil—Gr. *Troilos*, son of Priam.

trokt—See troct.

trollius—NL. *trollius* < Hung. *torolya*, name of an herb, perh. < Ger. *trollen*, to stroll. *Ex:* Trollius*.

trom—Gr. *tromos*, a quivering. *Ex:* Tromo-sternus (Ins.); Peri-tromus (Prot.).

tromb—NL. *trombidium*, a little timid one < Gr. *tromeō*, to tremble; Agassiz derives it from a Gr. *trombōdēs*, timid, but no such word occurs in Scott and Lidell's Lexicon. *Ex:* Tromb-ella

(Arach.); Trombide-idae (Arth.) Trombidium (Arach.).

**tromer**—See **tromic.**

**tromic**—Gr. *tromikos* = *tromeros*, trembling. *Ex:* Tromiko-soma (Echin.).

**tromik**—See **tromic.**

**trop**—1. Gr. *tropos*, a turn, change in manner; *tropē*, a turning <*trepō*, to turn. *Ex:* trop-ism; trop-ic-al; Homo-tropus (Ins.); Loxo-tropa (Ins.): **2.** Gr. *tropis*, genit. *tropēos*, late genit. *tropidos*, a keel. *Ex:* Trop-odon (Mam.); trope-ic; Tropeo-notus (Ins.); Tropi-sternus (Ins.); Tropid-emys (Rept.); Tropid-ischia (Ins.); Tropido-clonion (Rept.); Tropido-notus (Rept.); Tri-mero-tropis (Ins.).

**tropae**—1. Gr. *tropaion*, a monument of the enemy's defeat made of shields and weapons, a trophy = L. *tropaeum*, a sign of victory. *Ex:* Tropaeolum*, a dim. <*tropaion*; Tropaeum (Moll.): 2. Gr. *tropaia*, a change in heart or mind. *Ex:* Tropaea (Ins.).

**tropeo**—See **trop 2.**

**troph**—Gr. *trophos*, one who feeds; *trophis*, big, well fed; *trophon*, food, that which feeds. *Ex:* Trophis*; Tropho-discus (Echin.); Tropho-myia (Ins.); tropho-some; tropho-zo-ite; Trophon (Moll.); Trophoni-um (Moll.); a-trophy; auto-troph-ic.

**trophoni**—L. *Trophonius*, Roman deity. *Ex:* Trophonia (Ann.).

**tropic**—Gr. *tropikos*, of the solstice, tropical. *Ex:* Tropic-orbis (Moll.); Tropico-perdix (Av.).

**tropid**—See **trop 2.**

**tropio**—See **trop 2.**

**tropis**—See **trop 2.**

**tros**—Gr. *trōsis*, an injury to a tree. *Ex:* Trosia (Ins.).

**trot**—Gr. *trōtos*, vulnerable. *Ex:* Troto-sema (Ins.).

**trox**—See **trog.**

**truc**—L. *trux*, genit. *trucis*, fierce, savage; *truci-ter*, fiercely. *Ex:* Truci-felis (Mam.); Truxo-mantis (Ins.).

**trucher**—Gr. *trycheros*, ragged, worn. *Ex:* Trycher-odon (Pisc.); Truchero-gnathus (Ann.)

**truculent**—L. *truculentus*, fierce, cruel, savage.

**trud**—L. *trudis*, a pointed pole, a spike. *Ex:* Repo-trudis (Pisc.).

**trull**—1. L. *trulla*, a bricklayer's trowel; *trullisatus*, plastered. trowelled. *Ex:* trulli formis. 2. L. *trullium* = *trulleum*, a basin.

**trull**—L. *trulla*, a dipper, small cup, basin. *Ex:* Trull-ula*; trulli-fer.

**-trum**—L. *-trum* (<Gr. *-tron*), suffix denoting means or instrument, e.g. Gr. *loutron*, a bath <*louō*, to wash> NL. loutrum.

**trunc**—1. L. *truncus*, dim. *trunculus*, a trunk, stem. *Ex:* Trunc-illa (Moll.); Truncul-ites

(Moll.); truncus arteriosus: **2.** L. *trunco*, to maim, disfigure; NL. *truncatus*, mutilated, cut off. *Ex:* Truncat-ella (Moll.); truncate.

**truncat**—See **trunc.**

**trup**—See **tryp.**

**trus**—L. *trudo*, to push, shove, thrust; pp. *trusus* pushed. *Ex:* re-trus-ion.

**trutin**—L. *trutina*, a pair of scales; *trutinator*, one who examines, passes judgment. *Ex:* Trutina (Moll.).

**trutt**—Low L. *trutta*, trout. *Ex:* Trutta (Pisc.).

**trux**—See **truc.**

**try**—1. Gr. *tryō*, to rub, wear-out. *Ex:* Xyro-trya (Moll.): **2.** NL. *trya*, an opening, hole, evidently <Gr. *tryma*, a hole. *Ex:* Hydro-trya*, and in several other genera of fungi.

**trybl**—Gr. *tryblion*, a cup, bowl. *Ex:* Trybl-idium (Moll.); Tryblio-crinus (Echin.).

**trych**—Gr. *trychō*, to consume, eat up; *trychōsis*, a consuming. *Ex:* Trycho-proctus (Echin.); Trychosis (Ins.).

**trycher**—See **trucher.**

**trychin**—Gr. *trychinos*, ragged. *Ex:* Trychine (Ins.).

**trychn**—1. Gr. *trychnos*, nightshade, solanum: **2.** Gr. *trychnoō*, to waste, to wear out. *Ex:* Trychno-mera (Ins.); Trychno-phylla (Ins.).

**tryg**—Gr. *trygē*, ripe fruit. *Ex:* Tryge-nycteris (Mam.).

**trygon**—Gr. *trygōn*, a dove; also a sting-ray. *Ex:* Trygon (Av.); Trygon-idae (Elasm.); Trygono-batus (Elasm.).

**trym**—Gr. *tryma*, genit. *trymatos* = *trymē*, a hole. *Ex:* Trym-ochthe-bius (Ins.); Trymato-derus (Ins.); Trymo-sternus (Ins.); Dia-tryma (Av.).

**trymn**—Gr. *trymnos*, the hindmost, the farthest under.

**tryng**—See **tring.**

**tryos**—Gr. *tryos*, labor, work. *Ex:* tryos-in-ase, tryos-in.

**tryp**—Gr. *trypa*, a hole; *trypanon*, a borer < *trypaō*, to bore through. *Ex:* Tryp-auchen (Pisc.); Trypano-corax (Av.); Trypano-soma (Prot.); Trypanus (Ins.); Trypo-dendron (Ins.); Trypo-pitys (Av.); Coryno-trypa (Bry.); Hylo-trupes (Ins.); Plagio-trypes (Ins.).

**trypan**—See **tyrp.**

**trypet**—Gr. *trypētos*, bored; *trypētēs*, a borer. *Ex:* Trypet-idae (Ins.); Trypetes (Ins.).

**tryph**—Gr. *tryphē*, delicacy, softness. *Ex:* Trypho-mys (Mam.).

**trypher**—Gr. *trypheros*, delicate, dainty. *Ex:* Trypher-opsis (Amph.); Tryphera (Ins.); Tryphero-mera (Ins.); Trypherus (Ins.).

**tryphos**—Gr. *Tryphōsa*, a fem. name, the dainty one. *Ex:* Tryphos-ites (Crust.); Tryphosa (Crust.).

**trypsin**—Gr. *tribō*, fut. *tripsō*, to wear down; *trypsis*, a rubbing, a wearing out (hence to digest) <*tryō*, to wear out > Ger. *Tryptone*, a peptone which is the product of tryptic digestion. *Ex:* trypsin; trypsino-gen; trypto-phane; tryptone.

**trypt**—See **trypsin**.

**trysi**—Gr. *trysis*, a wearing away, a coming to exhaustion > *tryō*, to wear out. *Ex:* Trysicampe (Ins.).

**tryss**—Gr. *tryssos*, dainty. *Ex:* Trysso-thele (Arach.).

**tryz**—Gr. *tryzō*, to coo like a dove. *Ex:* Tryzusa (Av.).

**tsuga**—Jap. *tsuga*, the larch. *Ex:* Tsuga*; Pseudo-tsuga*.

**tub**—L. *tuba*, a trumpet; *tubicen*, a trumpeter; *tubus*, dim. *tubulus*, a water-pipe, a tube. *Ex:* tubi-col-ous; Tubi-fera (Coel.); tubo-abdominal; Tubul-aria (Coel.); tubuli-florous; tubulus; hydra-tuba.

**tuber**—L. *tuber*, dim. *tuberculum*, a tumor, a knob, a hump; *tuberosus*, full of humps. *Ex:*

Lower Jaw of Tubercled-tooth Iguana, *Iguana tuberculata*. Redrawn from Textbook of Palaeontology—Zittel. The Macmillan Co., Publishers.

tuber; Tuber*; Tuber-aster (Echin.); tuberculate; Tubero-cephalus (Ins.).

**tubercul**—See **tuber**.

**tubicen**—See **tub**.

**tubul**—See **tub**.

**tucan**—1. Mex. *tucan*, a pocket gopher. *Ex:* Tucanus (Mam.): 2. Tupi Indian *tucan*, a bird name. *Ex:* Tucanus (Av.).

**tud**—L. *tudes*, genit. *tudis*, a hammer. *Ex:* Tudes (Moll.).

**-tude**—Eng. *-tude* < L. *-tudo*, genit. *-tudinis*, suffix denoting condition or quality. *Ex:* magni-tude < *magnus*, great.

**tudicul**—L. *tudiculo*, to stir about, pp. *tudiculatus*, moved about. *Ex:* Tudicula (Moll.).

**tuditan**—1. L. *Tuditanus*, a family name. *Ex:* Tuditanus (Rept.): 2. L. *tuditans*, pushing or driving on < *tundo*, to strike, beat.

**tudor**—Eng. *Tudor* < Welsh *Tewdyr*, Theodore. *Ex:* Tudora (Moll.).

**tuf**—See **toph**.

**tuitans**—L. *tuitans*, defending; ppr. of *tueor*, to defend, to watch.

**tulip**—Fr. *tulipe* < Turk. *tulbend*, a turban. *Ex:* Tulipa*.

**tulo**—See **tyl**.

**tumid**—L. *tumidus*, swollen. *Ex:* Tumidi-coxa (Ins.); Tumido-nautilus (Moll.).

**tumul**—L. *tumulus*, a hill, *Ex:* tumuli-cola.

**tunic**—L. *tunica*, a garment; *tunicatus*, clothed with a tunic. *Ex:* Tunica*; tunicat-ed; Tunicata (Urochorda).

**tupai**—Latinized Malay *tupai*, a name "applied to various small animals which have the external form and agility of the squirrel". *Ex:* Tapaia (Mam.).

**tupid**—Gr. *tupis*, genit. *tupidos*, mallet. hammer. *Ex:* Tupid-anthus*; Tupistra*, the last portion of the word of undetermined origin.

**tupis**—See **tupid**.

**turac**—See **tourac**.

**turb**—1. L. *turbo*, genit. *turbinis*, anything that whirls around, such as a whirl-wind, a top; *turbinatus*, cone- or top-shaped; *turbineus*, cone-shaped. *Ex:* Turbin-aria (Coel.); Turbinidae (Moll.); Turbinata (Moll.); Turbo (Moll.); Turbon-illa (Moll.); 2. L. *turba*, dim. *turbellae*, a disturbance; *turbatus*, troubled, disturbed. *Ex:* Turbell-aria (Platy.).

**turbon**—See **turb**.

**turci**—L. *turcicus*, of Turkey. *Ex:* sella turcica, "Turkish saddle."

**turd**—L. *turdus*, a thrush > *Turdus*, generic name of the American robin. *Ex:* Turd-idae (Av.); Turdus (Av.).

**turg**—L. *turgeo*, to swell, ppr. *turgescens*, genit. *turgescentis*, swelling; *turgidus*, swollen. *Ex:* Turgenia*; turgescent; turgid; turgor.

**turio**—L. *turio*, genit. *turionis*, a shoot, tendril. *Ex:* turio = turion; turioni fer-ous.

**turion**—See **turio**.

**turm**—L. *turma*, a troop; *turmalis*, of or belonging to a troop.

**turnic**—NL. *turnix*, genit. *turnicis*, a shortened form of L. *coturnix*, genit. *coturnicis*, a quail, "to show that it was a mutilated form, wanting the hind toe." *Ex:* Turnico-morphae (Av.); Turnix (Av.).

**turnix**—See **turnic**.

**turp**—L. *turpis*, neut. *turpe*, ugly, foul.

**turpili**—L. *Turpilia*, a Roman family name. *Ex:* Turpilia (Ins.); Turpilio-ides (Ins.).

**turr**—L. *turris*, dim. *turritella*, a tower, *turritus*, furnished with towers. *Ex:* Turri-capsia (Moll.); Turri-culum (Moll.); Turri-lepas (Crust.); turri-form; Turrit-opsis (Coel.); Turritella (Moll.); Turrito-spira (Moll.); Turritis*.

**turrit**—See **turr**.

**tursi**—L. *tursio*, a porpoise. *Ex:* Tursi-ops (Mam.); Tursio (Mam.).

**turt**—See **tort**.

**turtur**—L. *turtur*, a turtle-dove. *Ex:* Turtur (Av.); Turtur-oena (Av.).

**-tus**—1. L. *-tus*, adj. suffix denoting possession or fullness. *Ex:* ala-tus, winged < *ala*, a wing; cornu-tus, horned < *cornu*, a horn: 2. L. *-tus, -ta, -tum*, perfect participial suffix often changed to *-sus, -sa, -sum.* The perfect participle is formed (1) from a theme consisting of a root (most verbs in *-ere* and *ēre*). *Ex:* gestus<*gero,* to carry; sparsus<*spargo,* to sprinkle: (2) from a theme in long *a* or in long *i* (from denominatives in *-āre* or *-ire*). *Ex:* laudatus<*laudo,* to praise; castratus< *castro,* to emasculate; auditus<*audio,* to hear: **3.** *-tus (-sus),* L. suffix added to verb stems, or roots to form most Fourth Declension Latin nouns. *Ex:* cantus, a song<*cano,* to sing; exultatus, an exile<*exsulo,* to be an exile.

**tuscar**—NL. *tuscar*<*Tuscarora Deep* in the Pacific Ocean. *Ex:* Tuscar-antha (Prot.); Tuscar-ella (Prot.).

**tuss**—L. *tussis,* a cough. *Ex:* per-tussis (Med.).

**tussilago**—L. *tussilago,* an herb called the colt's foot. *Ex:* Tussilago*.

**tutacul**—See **tutat.**

**tutat**—L. *tutor,* to protect; *tutatus,* protected; *tutator,* a defender; *tutatrix,* a female protector; *tutaculum,* a defence.

**tutel**—L. *tutela,* a safeguard.

**tutican**—L. *Tuticanus,* a personal name. *Ex:* Tuticanus (Arach.).

**-ty**—Eng. *-ty,* suffix used to form nouns of quality or condition from words of French or Latin origin. *Ex:* beau-ty.

**tych**—Gr. *tychē,* good luck, accident; *tychaios,* by accident; also common; *tycheros,* lucky. *Ex:* Tychaeus (Ins.); Tyche-psephenus (Ins.); Tycherus (Ins.); Tycho-styl-ops (Mam.); Tychus (Ins.); Eu-stych-ides (Arach.).

**tycher**—See **tych.**

**tycn**—Gr. *tyknos,* dense. *Ex:* Tycno-saccus (Echin.).

**tydeus**—Gr. *Tydeus,* son of Oeneus, king of Calydon. *Ex:* Tydeus (Arach.).

**tyl**—Gr. *tylos,* a knot, a callus, a knob on a club; also the phallus; *tylē,* a swelling, a cushion; *tylōtos,* knobbed. *Ex:* Tul-odon (Mam.); Tylenchus (Nemat.); Tylo-poda (Mam.); Tylophora*; Tylos-urus (Pisc.); tylot-oxea; Tylote; Eremo-tylus (Ins.); Peri-tyle*.

**tylar**—NL. *tylarus* cushioned<Gr. *tylē,* a cushion. *Ex:* tylarus

**tylot**—See **tyl.**

**tymb**—Gr. *tymbos,* a tomb. *Ex:* Tymb-archa (Ins.); Tymbo-desmus (Myr.).

**tymm**—Gr. *tymma,* a blow<*typtō,* to strike. *Ex:* Tymmo-phorus (Ins.); Chaere-tymma (Ins.); Philo-tymma (Ins.).

**tympan**—Gr. *tympanon*=L. *tympanum,* a drum; also a panel in a door. *Ex:* tympan-ic membrane; Tympan-uchus (Av.); tympanohyal; Tympano-palpus (Ins.).

**tyntlast**—Gr. *tyntlastēs,* a mud-dauber<*tyntlazō,* to work in mud. *Ex:* Tyntlastes (Pisc.).

**typ**—Gr. *typē,* a wound; *typos,* a blow, product of a blow, an impression, image, type; *typikos,* typical, figurative. *Ex:* A-typena (Arth.); a-typic; A-typus (Arth.); Hol-ec-typus (Echin.); Idio-typa (Ins.).

**typh**—1. Gr. *typhos,* smoke, cloud. *Ex:* typh-oid; Typhis (Moll.):    2. Gr. *typhē,* a plant used for stuffing beds, such as the cat's tail, hence sometimes used in the sense of hairy, fluffy. *Ex:* Typha*; Typh-aceae*; Typhis (Crust.).

**typhl**—Gr. *typhlos,* blind; *typhlotēs,* blindness; *typhlinēs,* a kind of snake like the blind-worm. *Ex:* Typhl-ops (Rept.); Typhlines (Rept.); Typhlo-plana (Platy.).

**typhoeus**—Gr. *Typhōeus,* a fierce-eyed hundred-headed monster smitten by Jupiter and buried under Mt. Aetna. *Ex:* Typhoeus (Ins.); ?Typhaeus (Ann.).

**typhon**—Gr. *typhōn,* stormy wind, a whirlwind. *Ex:* Typhonia (Ins.); Typhono-dorsum*.

**typot**—Gr. *typōtos,* moulded; *typōtēs,* one who forms, molds. *Ex:* Za-typota (Ins.).

**tyr**—1. Gr. *tyros,* cheese. *Ex:* Tyro-glyphus (Arth.); Tyro-phaga (Ins.):    2. Gr. *Tyrō,* daughter of Salmoneus. *Ex:* Tyro (Crust.).

**tyrann**—L. *tyrannus,* a tyrant. *Ex:* Tyrann-ina (Av.); Tyrann-ites (Moll.); Tyrann-ula (Av.); Tyranno-chelifer (Arach.); Tyrannus (Ins.).

**tyrianthin**—Gr. *tyrianthinos,* having a purple color.

**tyto**—Gr. *tytō,* a kind of night owl. *Ex:* Tyto (Av.); Tyton-idae (here the Gr. *tyto* is treated as a third declension Latin noun with a genitive *tytonis*); Speo-tyto (Av.).

**tytth**—Gr. *tytthos,* small, young. *Ex:* Tytth-onyx (Ins.); Tyttho-conus (Mam.).

**tyx**—Gr. *tyxis*=*teyxis,* a creating, making, an attainment. *Ex:* Katae-tyx (Pisc.).

# U

uber—L. *uber*, genit. *uberis*, fruitful, plump; also by transference, of the earth. *Ex:* uberi-color.

ubiquit—L. *ubique*, everywhere>Fr. *ubiquité*. *Ex:* ubiquit-ous.

uca—Tupi Indian *uca*, a fiddler crab. *Ex:* Uca (Crust.).

ucalegon—Gr. *Oukalegōn*, the name of a Trojan. *Ex:* Ucalegon (Ins.).

ud—1. L. *udus*, moist, damp. *Ex:* udo-meter: 2. Gr. *oudas*, the ground, earth: 3. Gr. *oudos*, path, way, step. *Ex:* Udo-baenus (Mam.), but perh. from 2: 4. Gr. *oudos*, an entrance, threshold.

udam—Gr. *oudamos*, not even one. *Ex:* Udam-acantha (Ins.); Udamo-chiras (Ins.).

udamin—Gr. *oudaminos*, good for nothing. *Ex:* Udamina (Ins.).

udder—A.S. *ūder*, Skr. *ūdhar*, the udder.

uden—Gr. *ouden*, none, no one. *Ex:* Uden-odon (Rept.); Udenia (Ins.).

udeo—Gr. *oudas*, genit. *oudeos*, the ground, earth. *Ex:* Udeo-dromus (Ins.).

udeter—Gr. *oudeteros*, neuter, neither of the two. *Ex:* Udeterus (Ins.).

-ugo—NL. *-ugo*, suffix with meaning unknown. *Ex:* Hyps-ugo (Mam.); Nann-ugo (Mam.); Vesper-ugo (Mam.).

uint—*Uinta*, the Uinta Mts. of Utah. *Ex:* Uinta-cyon (Mam.); Uinta-therium (Mam.).

uji—Jap. *uji*, a maggot. *Ex:* Uji-myia (Ins.).

ul—1. Gr. *oulon* pl. *oula*, the gums. *Ex:* Ul-ias (Mam.); Haem-ulon (Pisc.): 2. Gr. *oulos*, woolly, curly, twisted. *Ex:* Ulio cnemis (Ins.); Ulo-borus (Arach.); Ulo-chaetes (Ins.); Ulo-thrix*: 3. Gr. *oulos=ulios*, destructive. *Ex:* Uli-odon (Arach.): 4. Gr. *oulē*, a scar. *Ex:* ul-erythema (Med.): 5. Gr. *oulos*, Ion. for *holos*, whole, entire.

-ula—See -ule.

ulcer—L. *ulcus*, genit. *ulceris*, an ulcer. *Ex:* ulcer-ous; ulcero-membran-ous (Med.).

-ule—Eng. *-ule*, suffix<L. *ulus-a-um*, diminutive suffix. *Ex:* blast-ula; animac-ule; Pogoni-ulus (Av.).

ulex—See ulic.

uli—Gr. *oulios*, baneful, destructive. *Ex:* Ulio-cnemis (Ins.); Ulio-soma (Ins.).

ulic—L. *ulex*, genit. *ulicis*, a shrub resembling rosemary. *Ex:* Ulex*.

ulig—L. *uligo*, genit. *uliginis*, moisture, *uliginosus*, swampy, marshy, moist. *Ex:* uliginose.

ullucus—NL. *ullucus*<some native name for a certain chenopodiaceous plant. *Ex:* Ullucus*.

-ullus—NL. *-ullus*, *-a*, *-um*, diminutive<L. *ulus*, a diminutive. *Ex:* Citr-ullus*.

ulm—L. *ulmus*, the elm; *ulmeus*, of or belonging to an elm tree. *Ex:* Ulm-aceae*; Ulmus*.

ulmar—NL. *ulmaria*<mediaeval name for the goats-beard, a plant with elm-like leaves<L. *ulmus*, the elm; *ulmarium*, a nursery of elms. *Ex:* Ulmaria*; Ulmaris (Coel.).

ulna—L. *ulna*, the elbow, also the lower arm; *ulnaris*, pertaining to the ulna; NL. *ulnare*, the cuneiform bone of the carpus.

ulobor—Gr. *ouloboros*, lethal, with deadly bite. *Ex:* Ulobor-idae (Arach.); Uloborus (Arach.).

ulon—See ul 1.

ultim—L. *ultimus*, fartherest, the last, extreme. Ultimo-stomias (Pisc.).

ultra—L. *ultra*, beyond, in excess. *Ex:* ultra-montane (Ecol.); Ultra-pithecus (Mam.).

ultrone—L. *ultroneus*, voluntary.

ulul—L. *ulula*, a screech owl; *ululatus*, a wailing, a mournful cry<*ululō*, to howl as if in pain, ppr. *ululans*, genit. *ululantis*, wailing, crying. *Ex:* Ulula (Av.); ululate; ululant.

-ulus—L. *-ulus*, adj. ending denoting tendency, often toward evil. See -ax. *Ex:* garr-ulus<L. *garrio*, to chatter. See also -ule and garr.

ulv—NL. *ulva*, sea-lettuce<L. *ulva*, sedge. *Ex:* Ulva*; Ulvi-cola (Pisc.).

-um—L. *-um*, regular first and second declension adjectival ending for the nom. sing.; also the ending of second declension nouns. For other uses consult a Latin grammar.

uma—NL. *uma*< *Yuma*, an Indian tribe and place in Arizona. *Ex:* Uma (Rept.).

umbell—L. *umbella*, dim. *umbellula*, a sunshade. *Ex:* umbel; Umbella (Moll.); umbelli-fer-ous; Umbellul-aria*; Umbelluli-fera (Coel.).

umbilic—L. *umbilicus*, the navel; L. *umbilicatus*, navel-shaped; LL. *umbilicaris*, pertaining to the navel. *Ex:* umbili-fer-ous; umbilic-al; Umbilicaria*; Umbilico-sphaera (Prot.); umbilicus.

umbo—See umbon.

umbon—L. *umbo*, genit. *umbonis*, a shield; NL. *umbonatus*, shielded. *Ex:* umbo.

umbrat—L. *umbratus*, shading, spreading over; pp. of *umbro*, to shade; *umbraticus*, belonging to shade, belonging to seclusion.

umbros—L. *umbrosus*, full of shade, shady.

umect—L. *umectus*, moist, damp.

un—L. *unus*, one.

unc—L. *uncus*, a hook; *uncinus*, dim. *uncinulus*, a hook; *uncinatus*, barbed. *Ex:* unci; unci-form; Unci-rostrum (Av.); Uncinulus (Brach.); uncinus.

uncat—L. *uncatus*, bent inwards, hooked.

uncial—L. *uncialis*, the twelfth part of anything, especially of a pound or foot<*uncia*, a twelfth; *unciola*, a little ounce. *Ex:* Unciola (Crust.).

uncin—See unc.

unciol—See uncial.

-uncle—See -uncul.

-uncul—L. -unculus, -a, -um, suffix meaning little, generally used with L. 3rd. decl. nouns ending in -on. Ex: carb-uncle; Did-unculus (Av.); Gladi-unculus (Pisc.); Lollig-uncula (Moll.).

und—L. unda, dim. undula, a wave. Ex: Unda (Prot.); Undul-aria (Moll.); Unduli-fer (Ins.).

undat—L. undatus, in a wavy or wave-like form <undo, to rise in waves; unda, a wave, water.

undin—NL. undina, a water spirit, a spirit of the waves<unda, a wave. Ex: Undin-ula (Crust.); Undina (Pisc.).

undul—See und.

undulat—L. undulatus, diversified as if with waves, undulated <undo, to rise in waves. Ex: undulat-ion.

uned—L. unedo, name for the fruit of the strawberry tree, Arbutus unedo; also the name of the tree itself. According to Pliny, the fruit is so bad that un, one (only) do I eat (edo).

ungui—L. unguis, dim. unguiculus, nail, claw; also a talon or hoof. Ex: Ungui-termes (Ins.); Unguicul-ata; unguis.

unguicul—See ungui.

ungul—L. ungula, hoof, claw; ungulina, like a claw or hoof. Ex: Ungul-aspis (Ins.); Ungul-ata (Mam.); unguli-folia; unguli-grade; Ungulina (Moll.).

uni—L. unus, one. Ex: Uni-cornus (Mam.); uni-cursal; Uni-dens (Moll.); uni-lateral; Uni-locul-ina (Prot.).

unic—L. unicus, one only, single.

unio—L. unio, genit. unionis, unity, union; also a single pearl. Ex: Unio (Moll.); Union-idae (Moll.); Unioni-cola (Arach.); Unioni-tes (Moll.).

uniol—L. uniola, an ancient plant name. Ex: Uniola*.

upen—Gr. upēnē, the hair on the upper-lip, the moustache; also the upper lip. Ex: Upenus (Pisc.).

uper—Gr. uper, over, above. Ex: Uper-odon (Mam.); Uper-odonta (Rept.).

uperan—Gr. uperanō, over, above. Ex: Uperan-odon (Rept.).

upo-  —NL. upo-<Gr. hypō-, prefix, meaning under, below. Ex: Upo-corpus (Ins.). See hyp.

upsil—Gr. upsilon, the Greek letter Υ. Ex: Upsil-odon (Moll.); Upsilo-porpa (Ins.).

upti—Gr. uptios, bent backwards, supine; uptiotēs, flatness, the position of a body when lying backwards. Ex: Uptiotes (Arach.); Gaster-uption (Ins.).

upup—L. upupa, a hoopoe. Ex: Upupa (Av.); Upupi-certhia (Av.).

ur—1. Gr. oura, the tail. Ex: Ur-aria*; Ur-auges (Av.); ur-ite; Ur-odela (Amph.); Uro-cyon

(Mam.); Uro-plectes (Arach.); uro-stege; Our-amoeba (Prot.); Oura-pteryx (Ins.); Ouro-teuthis (Moll.); An-ura (Amph.); Brachy-ura (Crust.); Centr-urus (Arach.); leuc-urus; Loph-yrus (Ins.); sci-ur-oid:    2. Gr. ouros =oros, genit. oreos, a mountain. Ex: Uri-threptus (Ins.); not Uronemus (Pisc.); which is an anagram of numerous. This same curious Gr. word ouros may mean a fair wind, a guardian, a boundary, a trench for hauling up ships; also a buffalo. Unless the describer of a genus concisely states his origins and applications the student is left almost wholly at sea when he attempts to decipher the literal meaning of words beginning with ouro or ur:    3. Gr. ouron, urine>L. urina, urine. Ex: ur-agoge; ur-ic; urea; urin-ary; uro-genital=urino-genital; uro-logy=ouro-logy; glycos-uria; Lys-urus*. See also urena.

urach—Gr. ourachos, the foetal urinary canal. Ex: urachus.

urae—Gr. ouraios, hindmost, of the hinder part, of the tail. Ex: Uraeo-typhlus (Amph.); Uraeus (Rept.).

urag—Gr. ouragia, the rear; ouragos, leader of the rear-guard. Ex: Uragis (Av.).

uran—Gr. ouranos, the sky, a vaulted roof; also the roof of the mouth or palate; uranios, heavenly. Ex: Uran-ichthys (Pisc.); Uran-odon (Mam.); Ouranion (Ins.); Urano-kyrtus (Mam.); Urano-scopus (Pisc.); Urano-tes (Ins.).

urax—See ourax.

urbic—L. urbicus, of or pertaining to a city.

urce—L. urceus, dim. urceolus, a pitcher. Ex: Urceo-labrum (Moll.); Urceol-ina*; Urceola*; Urceoli-pora (Bry.); Urceus (Moll.).

urceol—See urce.

urea—See ur 3, also urena.

urech—NL. urechis<Gr. urichos, a wicker basket. Ex: Urech-ites*; Urechis (Geph.).

ured—L. uredo, genit. uredinis, a blight, a burning itch<uro, to burn. Ex: Uredin-ales*; uredini-um;   uredinio-spore;   uredo-gonium; uredo-spore.

uren—L. urens, genit. urentis, stinging, scorching, ppr. of uro, to burn, scorch.

urena—Malabar, uren, name of a kind of mallow. Ex: Urea*.

urens—See uren.

urera—NL. urera, a plant name<L. urere, to burn. Ex: Urera*.

ureter—Gr. ourētēr, the ureter. Ex: ureter; uretero-tomy (Med.).

urethr—Gr. ourēthra, the urethra. Ex: urethra urethro-scope.

urgin—Algerian Ben Urgin, name of an Arab tribe. Ex: Urginea*.

uria—Gr. ouria=L. uria, a kind of diving bird <L. urinor, to dive. Ex: Uria (Av.).

urich—Gr. *urichos*, a basket of wicker-work.

urigin—L. *urigo*, genit. *uriginis*, lust, desire.

urin—1. L. *urinor*, to dive. *Ex:* Urino-phila (Pisc.). See also ur 3: 2. L. *urinus*, full of wind: 3. L. *urina*, urine. *Ex:* urin-ate; urine; urino-genital.

urinator—L. *urinator*, a diver < *urinor*, to dive. *Ex:* Urinator (Av.).

urinatrix—L. *urinatrix*, a female diver < *urinor*, to dive.

urinor—See urin.

-urn—L. *-urnus*, suffix, meaning belonging to > *di-urnus*, daily. *Ex:* di-urn-al; noct-urn-al.

urs—L. *ursus*, bear; fem. *ursa*, dim. *ursula*, a little she-bear; *ursinus*, resembling a bear. *Ex:* Urs-avus (Mam.); Ursi-taxus (Mam.); Ursino-pereio-pus (Crust.); Ursinus (Mam.); Urso-taxus (Mam.); Ursula (Ins.); Ursus (Mam.); Uva-ursi*.

ursin—See urs.

urtic—L. *urtica*, a nettle < *uro*, to burn. *Ex:* Urtic-aceae*; Urtic-astrum*; Urtica*.

urub—Tupi Indian *urubu*, a vulture. *Ex:* Urubi-tinga (Av.); Urubu (Av.).

urus—L. *urus*, a kind of wild ox. *Ex:* Urus (Mam.).

-us—1. L. *-us, -a, -um*, ending often added to geographical names to make "geographical adjectives." *Ex:* Riab-us < Pulo Riaba, island of the South China Sea; Kans-us < Kansu Province in China. See also -ensis: 2. L. *-us*, nom. ending in certain 2nd declension Latin masculine singular nouns and masc. nom. sing. ending of many Latin adjectives.

us—Gr. *ousia*, one's substance, property, condition. *Ex:* Chaen-usa (Ins.); Rhamn-usium*.

usitat—L. *usitatus*, customary, common, familiar < *usitor*, to use often, to be in the habit of using. *Ex:* usitat-issimum.

usne—Ar. *oshnah*, moss. *Ex:* Usnea*.

ustici—L. *usticius*, brown, produced by burning.

ustilag—LL. *ustilago*, a prickly, thistle-like plant < L. *ustulatus*, burned, scorched—Ustilago*.

ustulat—L. *ustulatus*, scorched, singed, of a warm russet color < *ustulo*, to burn, scorch.

uta—NL. *uta* < *Utah*, name originally given to a Territory of the U. S. < *Ute*, name of an Indian tribe. *Ex:* Uta (Rept.).

uter—L. *uterus*, the womb < *uter*, a bag, bottle made of skin. *Ex:* uter-algia (Med.); uter-ine; Uteri-porus (Platy.); utero-cervical; Utero-somus (Ins.).

utetheis—NL. *utetheisia* an insect name, etym. unknown. *Ex:* Utetheisia (Ins.).

utibil—L. *utibilis*, fit, useable.

util—L. *utilis*, beneficial, advantageous; *utilitas*, advantage, profit. *Ex:* Utilit-aria (Ins.).

utilit—See util.

utricul—L. *utriculus*, a little bag, small skin, dim. of *uterus*, a bag. *Ex:* Utricul-aria*; utricul-ate; utricle; utriculi-form; Utriculo-fera (Ins.); utriculus.

utricl—See utricul.

-utus—L. *-utus -a -um*, adjectival ending meaning provided with, having, as in corn-utus, provided with horns < *cornu*, a horn.

-uus—L. *-uus*, termination denoting quality expressed by the verb. Adjectives ending in *-uus*, derived from active verbs, take a passive meaning, as *congruus*, agreeing < *congruo*, to agree; *innocuus*, harmless < *innoceo*, to harm not; *irriguus*, well watered < *irrigo*, to water.

uv—L. *uva*, a grape, dim. *uvula*, which also refers to the pendant portion of the soft palate. *Ex:* Uv-aria*; Uv-ella (Prot.); Uva-ursi*; uve-ous; uvi-ferous; uvul-ar; Uvul-aria*; Uvuli-fer (Platy.).

uvid—L. *uvidus*, damp.

uvul—See uv.

uxor—L. *uxor*, a wife.

# V

vacc—L. *vacca*, a cow; *vaccinus*, of or from cows. *Ex:* Vacc-aria*; vaccin-ate.

vaccini—L. *vaccinium*, the whortleberry. *Ex:* Vaccinium*.

vacill—L. *vacillans*, genit. *vacillantis*, swayed, ppr. of *vacillo*, to sway to and fro, to stagger.

vacillans—See vacill.

vaciv—L. *vacivus*, empty.

vacu—L. *vacuus*, empty; *vacuum*, neut. sing. of *vacuus* < *vacuo*, pp. *vacuatus*, to empty; *vacuolum*, ML. dim. of *vacuum*. *Ex:* vacuole; e-vacuate.

vad—L. *vadum*, a shallow in water; *vadosus*, full of shallows.

vafer—L. *vafer*, fem. *vafra*, crafty, sly.

vafr—See vafer.

vag—L. *vago*, to wander, ppr. *vagans*, genit. *vagantis*, wandering, unsettled; in anatomical terms the combining form *vago-* generally refers to the vagus or so-called wandering nerve. *Ex:* vag-al; vagus; Areni-vaga (Ins.).

vagabund—L. *vagabundus*, strolling about. *Ex:* Vagabundus (Av.).

vagat—L. *vagatus*, strolling about < *vagor*, to roam. *Ex:* vagate.

vagin—L. *vagina*, a sheath, a scabbard; ML. *vaginans*, genit. *vaginantis*, sheathing. *Ex:* vagin-ate; vagina; Vagini-cola (Prot.); vaginifer-ous; viginant; vagino-vulvar; in-vagination.

valen—L. *valens*, genit. *valentis*, made strong < *valeo*, to be strong; *valentulus*, stout, strong.

valentul—See valen.

valerian—L. *valeriana*, valerian < *valeo*, to be strong, because of the powerful medicinal quality of some species; by some said to be named after Valerius. *Ex:* Valeriana*.

valg—L. *valgus*, bow-legged, awry; *valgatus*, provided with bow-legs, but now mostly used in sense of enlarged at the bottom, club-footed. *Ex:* Valgi-pes (Mam.); Acantho-valgus (Ins.).

valid—L. *validus*, strong, robust. *Ex:* valid.

vall—1. L. *vallis* = *valles*, a valley, dim. *vallicula* = *vallecula*:   2. L. *vallum*, a rampart < *vallo*, to enclose; *vallatus*, pertaining to walls; *vallaris*, of or belonging to a wall or rampart. *Ex:* Vallaris*; circum-vallate.

vallicul—See vall 1.

valv—L. *valva*, a leaf of a folding door; *valvatus*, having folding doors. *Ex:* valvate; valviform; uni-valve.

vampyr—Serb. *vampira*, a nocturnal demon supposed to eat out the heart and soul and suck the blood of its victim. *Ex:* Vampyr-ina (Prot.); Vampyr-ops (Mam.); Vampyroteuth-idae (Moll.); Vampyrus (Mam.).

van—L. *vanus*, empty, vacant.

vand—Skr. *vandā*, a parasitic plant. *Ex:* Vanda*.

vandois—Fr. *vandoise*, name for the dace. *Ex:* vandois-ulus.

vanell—It. *vanello*, the lapwing. *Ex:* Vanellus (Av.).

vanescen—L. *vanescens*, genit. *vanescentis*, vanishing, ppr. of *vanesco*, to vanish. *Ex:* e-vanescent.

vanessa—a character in Swift's poem, Cadenus and Vanessa < Gr. *Phanes*, a mystic divinity of the Orphic rites. *Ex:* Vaness-ula (Ins.); Vanessa (Ins.).

vanilla—Sp. *vaynilla* (dim. of *vayna*, a knife), a sheath or scissor's case. *Ex:* Vanilla*.

vann—L. *vannus*, dim. *vannellus*, a fan for winnowing grain. *Ex:* vann-al fold.

vannius—L. *Vannius*, a king of the Quadi in the time of Tiberius. *Ex:* Vannius (Ins.).

var—L. *varus*, grown inwards, bent. *Ex:* talipes varus.

varan—Fr. *varan* < Ar. *waran*, the name for a monitor lizard. *Ex:* Varan-idae (Rept.); Varanus (Rept.).

vari—L. *vario*, to variegate, change, pp. *variatus*, modified, changed; *varius*, different, changing, varying; *variabilis*, variable. *Ex:* vari-colored; variabil-ity; Vario-lepis (Moll.); vario-tinted.

varian—L. *varians*, genit. *variantis*, changing; ppr. of *vario*, to change.

variat—See vari.

varic—L. *varix*, genit. *varicis*, pl. *varices*, a dilated vein; as applied in conchology, a ridge; *varicosus*, full of dilated veins, ridges. *Ex:* varic-ell-ate; varici-form; varix.

variolari—NL. *variolarius* ( < *variola*, smallpox + *-arius*, adj. suffix), variegated, marked with small dots or indentations.

variolat—ML. *variolatus*, variegated < L. *vario*, to diversify. *Ex:* variolate.

varix—See varic.

vas—L. *vas*, genit. *vasis*, dim. *vasculum*, a vessel. *Ex:* vas deferens; vascul-ar; vasculi-fer-ous; vasi-form; vaso-dentin.

vasc—L. *vascus*, a kind of flute. *Ex:* Vasco-ceras (Moll.).

vascul—See vas.

vast—L. *vasto*, to lay waste; *vastator*, a destroyer; *vastus*, empty, unoccupied; applied to a waste desert, also to any vast space. *Ex:* vastus externus muscle; eu-vast-al (Ecol.).

vastatrix—L. *vastatrix*, a female waster.

vastit—L. *vastitas*, genit. *vastitatis*, an empty place, a desert.

vates—L. *vates*, a prophet, a poet.

**vati**—L. *vatius*, bow-legged, with legs bent outwards.

**vaticin**—L. *vaticinius*, prophetic<*vaticinor*, to foretell. *Ex:* Vaticin-odus (Pisc.).

**vect**—1. L. *veho*, to carry, pp. *vectus*, carried; *vectitatus*, carried about. *Ex:* vect-or; Vectura (Ins.): 2. L. *vectis*, a bar, lever.

Various-spined Urchin, *Salenia varispina*. Salenia is an euphonius combination of letters without meaning. Redrawn from Voyage of the Challenger—Thomson. Courtesy of Joseph McDonough Co.

**vectabil**—L. *vectabilis*, portable.

**vectius**—L. *Vectius*, name of a Roman gens. *Ex:* Vectius (Arach.).

**vedalia**—NL. *vedalia*, a coined name. *Ex:* Vedalia (Ins.).

**veget**—L. *vegetus*, lively, vigorous; *vegetabilis*, animating; *vegetatus*, animated<*vegeto*, to enliven, quicken. *Ex:* veget(able)-arian; veget-al; vegetate.

**vehicul**—L. *vehiculum*, a conveyance, a means of transport.

**vejovis**—Etruscan *Vejovis*, god of the underworld. *Ex:* Vejovis (Arach.).

**vel**—L. *velum*, a veil, covering. *Ex:* vel-ate; Velella (Coel.); veli-ger; velum.

**velamen**—See velamin.

**velament**—L. *velamentum*, a cover, screen. *Ex:* velament, velament-ous.

**velamin**—L. *velamen*, genit. *velaminis*, a cover, covering. *Ex:* velamen.

**veled**—L. *Veleda*, a prophetic virgin among the Germans. *Ex:* Veled-ella (Ins.); Veleda (Moll.).

**velit**—L. *velitor*, to fight, skirmish; ppr. *velitans*, genit. *velitantis*, skirmishing.

**vell**—1. L. *vellus*, wool, down; *vellosus*, fleecy, full of hair: 2. NL. *vella*, the cress-rocket. *Ex:* Vell-idae*; Vella*.

**veloc**—L. *velox*, genit. *velocis*, swift, speedy.

**velox**—See veloc.

**velutin**—M.L. *velutinus*, velvety<*vellus*, a fleece. *Ex:* Velutina (Moll.); velutinous.

**vemet**—NL. *vemetus*<L. *vermes*, a worm. *Ex:* Vemetus (Moll.).

**ven**—L. *vena*, dim. *venula*, a vein. *Ex:* ven-ation; ven-ose; ven-ous; venul-osus.

**venan**—L. *venans*, genit. *venantis*, hunting, ppr. of *venor*, to hunt. *Ex:* Venantes (Arach.).

**venatic**—L. *venaticus*, pertaining to the hunt; *canis venaticus*, the hunting dog.

**venator**—L. *venator*, a hunter; *venatorius*, belonging to the chase or hunt. *Ex:* Venator (Arach.).

**venefic**—L. *venefic* a poisoner, a sorceress; *veneficus*, poisc s, magical. *Ex:* Venefica (Pisc.).

**venen**—L. *veneno*, to poison; *venenata*, venomous animals; *venenarius*, belonging to poison; *venenosus*, full of poison, very poisonous; Eng. *venomous*<Old French *venimeux*. *Ex:* venenifer-ous; veneno-salivary = venomo-salivary; venom-ous.

**vener**—L. *Venus*, genit. *Veneris*, goddess of love. *Ex:* Vene-rupis (Moll.); Vener-idae (Moll.); venere-al; venere-us; veneris; Venus (Moll.).

**venerup**—NL. *venerupis*<L. *Venus*+*rupes*, a rock. *Ex:* Venerupis (Moll.).

**venet**—L. *venetus*, sea-colored, bluish.

**venill**—L. *Venilia* (1), mother of Turnus; (2), wife of Jason. *Ex:* Venilia (Ins.).

**venom**—See venen.

**vent**—L. *ventus*, wind; *ventosus*, full of wind, swift, puffed up.

**venter**—See ventr.

**ventil**—L. *ventilo*, to fan<*ventus*, the wind; *ventilatus*, ventilated, fanned. *Ex:* Ventil-ago* (L. *ago*, to drive away); ventilate.

**ventr**—L. *venter*, genit. *ventris*, the belly, dim. *ventriculus*; *ventralis*, of or belonging to the belly. *Ex:* venter; ventral; ventri-dorsal; ventricle.

**venul**—See ven.

**venus**—See vener.

**venust**—L. *venustus*, charming, elegant.

**veprecul**—L. *veprecula*, a little briar bush, dim. of *vepres*, a thorn-bush. *Ex:* Vepreculae*.

**ver**—L. *verus*, true, real. *Ex:* Ver-aphis (Ins.); Ver-atrum*; Ver-ongia (Por.) S.

**veratr**—L. *veratrum*, name for the hellebore. *Ex:* Veratrum*.

**verbasc**—L. *verbascum*, mullein. *Ex:* Verbascum*, according to some authorities it should have been Barbascum, because of the bearded filaments.

**verben**—L. *verbenae*, pl. of *verbena*, sacred boughs; also a class of plants used medicinally. *Ex:* Verbena*.

**verbesina**—NL. *verbesina*, a plant name altered from *verbena*. *Ex:* Verbesina*.

**verecund**—L. *verecundus*, shy, unassuming.

**vered**—L. *veredus*, a horse for pursuit.

**veretill**—L. *veretillum*, dim. of *veretrum*, the private parts. *Ex:* Veretill-ium (Coel.).

**veretr**—See veretill.

**veridic**—L. *veridicus*, genuine.

**verm**—L. *vermis*, pl. *vermes*, dim. *vermiculus*, a worm; *vermiculatus*, wormy, worm-shaped. *Ex:* Verm-etus (Moll.); Verm-illa (Moll.); Vermes; vermi-form; Vermi-leo (Ins.); Vermi-linguia (Rept.); vermicul-ar.

**vern**—1. L. *verno*, to seem like spring, to be verdant, to bloom, to grow young again; *vernalis*, belonging to spring; *vernatus*, renewed; ML. *vernicosus*, full of spring; also new, varnished, shiny: 2. L. *vernus*, native. S.

**vernan**—L. *vernans*, becoming green; ppr. of *verno*, to become green.

**vernat**—See vern 1.

**vernic**—NL. *vernix*, genit. *vernicis*, varnish.

**vernicos**—See vern 1.

**vernix**—See vernic.

**veronic**—NL. *Veronica*, traditional name of a woman who was cured of an issue of blood (Mark v. 24—34); also the name of a plant known as speed-well. *Ex:* Veronica,* flower of St. Veronica; Veronico-bius (Ins.).

**verp**—L. *verpa*, the penis; *verpus*, a circumcised man. *Ex:* verpa-myia; Verpa*.

**verruc**—L. *verruca*, a wart, a height, a steep place; *verrucosus*, full of warts; *Verucaria herba*, a plant able to remove warts; Fr. *verrue*, wart. *Ex:* Verru-sus (Mam.); Verruca*; Verrucaria*; verruci-form; Verrucosa (Arach.).

**vers**—1. L. *versus*, turned<*verto*, to turn, to change. *Ex:* versi-color; trans-verse: 2. *versus*, a furrow. *Ex:* versi-form.

**versabil**—L. *versabilis*, changeable.

**versipell**—L. *versipellis*, that which changes its appearance or form.

**versut**—L. *versutus*, shrewd, versatile.

**vert**—L. *verto*, to turn, turn about; ppr. *vertens*, genit. *vertentis*, twining, whirling. *Ex:* di-verticul-um.

**vertebr**—L. *vertebra*, a joint, a vertebra, something turned<*verto*, to turn; *vertebratus*, jointed, vertebrated. *Ex:* vertebr-al; Vertebrata.

**vertens**—See vert.

**vertex**—See vertic.

**vertic**—L. *vertex*, genit. *verticis*, a wheel, top; also a summit; *verticalis*, pertaining to the vertex or top of the head.

**verticill**—L. *verticillus*, a whorl, the whorl of a spindle<*verto*, to turn; NL. *verticillatus*, whorled. *Ex:* Verticill-aria*; verticill-aster; Verticill-ium*; verticillus.

**vertumn**—L. *Vertumnus*, god of the changing year. *Ex:* Vertumnus (Crust.).

**veru**—L. *veru*, a dart, javelin; sometimes taken to mean a ridge or elevation, as in verumontanum, an anatomical term.

**veruculat**—L. *veruculatus*, furnished with a small pike or javelin.

**verut**—L. *verutus*, armed with a dart.

**vesc**—L. *vescus*, small, thin, feeble.

**vesic**—L. *vesica*, dim. *vescicula*, blister, a bladder; *vesicarius*, of or belonging to a bladder. *Ex:* Vesicaria*; vesicle; Vesico-mya (Moll.); vesico-uteral; vesicul-ar; vescicula seminalis.

**vesicat**—See vesic.

**vesp**—L. *vespa*, a wasp. *Ex:* Vesp-idae (Ins.); Vespa (Ins.).

**vesper**—L. *vesper*, the evening; *vesperus=vespertinus*, of or belonging to the evening; *vespertilio*, lit. animal of the evening, a bat; *vesperugo*, the evening star; also a bat. *Ex:* vesper-ine; Vesperi-mus (Mam.); Vespertili-avus (Mam.); Vespertilio (Mam.); Vesperugo (Mam.); Vesperus (Mam.).

**vespertili**—See vesper.

**vesti**—L. *vestis*, clothes, a covering. *Ex:* Vesti-pedes (Av.).

**vestibul**—L. *vestibulum*, a place of entrance. *Ex:* vestibul-ar; vestibule.

**vestigi**—L. *vestigium*, a foot-print, a track, a trace. *Ex:* vestige; vestigi-al; Vestigi-fera (Ins.).

**vestiment**—L. *vestimentum*, clothing.

**vestit**—L. *vestitus*, dress, attire.

**vet**—L. *vetus*, old; *vetula*, a little old woman; *vetensis*, pertaining to age.

**vetensis**—See vet.

**veterin**—L. *veterinus*, drawing burdens, burdened.

**vetiver**—Tamil *vettivēru*, a root dug up. *Ex:* Vetiveria*.

**vetula**—See vet.

**vetust**—L. *vetustus*, old age, old.

**vexan**—L. *vexans*, genit. *vexantis*, aggitated, pp. of *vexo*, to harass, damage.

**vexativ**—L. *vexativus*, causing trouble, quarrelsome.

**vexill**—L. *vexillum*, a banner, flag; *vexillarius*, a standard bearer. *Ex:* vexilla; Vexillaria (Echin.).

**via**—L. *via*, a way. *Ex:* bi-vium, see bivius; tri-vium.

**viable**—Fr. *viable*, likely to live.

**vialis**—L. *vialis*, of or belonging to the highways or road.

**viari**—L. *viarius*, of the roadside, along by-ways.

**viatic**—L. *viaticus*, pertaining to a road or journey. *Ex:* viatic-al.

**viator**—L. *viator*, genit. *viatoris*, a traveller.

vibex—See vibic.

vibic—L. *vibex*, genit. *vibicis*, the mark of a whip, a weal. *Ex:* Vibex (Ins.).

vibracul—NL. *vibraculum*, long specialized hair-like processes of cheilostome polyzoans<*vibro*, to quiver, to move rapidly to and fro. *Ex:* vibracul-arium; Vibracul-ina (Bry.).

vibrio—NL. *vibrio*, name given to a genus of bacteria<*vibro*, to vibrate; Fr. *vibrion*, a motile bacterium. *Ex:* Vibrion-idae.*

vibriss—L. *vibrissa*, a hair of the nostrils, so called, it is said, because its removal causes a person to shake the head<*vibro*, to agitate, shake. *Ex:* vibrissa.

viburn—L. *viburnum*, the wayfaring-tree. *Ex:* Viburnum*.

vicari—L. *vicarius*, a substituting, a substitute.

vicia—L. *vicia*, vetch; *viciarius*, belonging to vetches, said to be<*vinciō*, to bind together. *Ex:* Vicia*.

vicilin—L. *Vicilinus*, the Watchful One.

vicin—L. *vicinus*, near, neighboring; *vicinalis*, neighboring; *vicinor*, near. *Ex:* vicin-al; vicin-ism.

vicugna—Peruv. *vicuna*, vicugna, a mammal. *Ex:* Vicugna (Mam.); Vicuna (Mam.).

vicuna—See vicugna.

videns—See vident.

vident—L. *videns*, genit. *videntis*, a prophet.

vidu—L. *viduus*, mateless, widowed. *Ex:* viduous; not *Vidua* (Av.) which is derived from the name *Whidah*, a territory in East Africa.

viduat—L. *viduo*, to deprive of a husband, to make in want of something; pp. *viduatus*, bereft; *viduus*, mateless, bereft of a lover. *Ex:* viduat-orius.

viduus—See viduat.

viet—L. *vietus*, shrunken, shrivelled. *Ex:* Vieto-morpha (Ins.).

vigescen—L. *vigescens*, genit. *vigescentis*, lively, vigorous, ppr. of *vigesco*, to thrive, to flourish.

vigeth—NL. *vigethia* (*vig*<Viguiera+*ethia*< Wyethia, related genera of Composites). *Ex:* Vigethia*.

vigil—L. *vigil*, genit. *vigilis*, alert, awake; *vigilax*, genit. *vigilacis*, watchful; *vigilabilis*, watchful<*vigilo*, to watch, be wakeful.

vigilax—L. *vigilax*, watchful.

vil—L. *vilis*, mean, worthless, repulsive, cheap. *Ex:* vile.

vill—L. *villus*, pl. *villi*, shaggy hair>*villosus*, hairy, shaggy, rough. *Ex:* villi; villi-form; villose.

vimen—See vimin.

vimin—L. *vimen*, genit. *viminis*, a switch, a twig; *vimeneus*, made of wicker-work; *viminalis*, pertaining to twigs. *Ex:* vimen; vimin-al; Vimin-aria*; Viminia (Ins.).

vin—L. *vinum*, wine; *vineus*, made of or belonging to wine, sometimes used in sense of wine-colored.

vinace—L. *vinaceus*, of or concerning wine of the grape, wine-colored.

vinca—NL. *vinca*<L. *vincapervinca*, a plant, the periwinkle. *Ex:* Vinca*.

vince—NL. *vince*<L. *vinco*, to conquer. *Ex:* Vince-toxicum*.

vinctus—L. *vinctus*, bound, banded, pp. of *vincio*, to bind, fetter.

vincul—L. *vinculum*, a bond, a cord; *vinculatus*, chained. *Ex:* Vincul-aria (Bry.); vinculate.

vineal—L. *vinealis*, belonging to vines.

vineus—See vin.

vinitor—L. *vinitor*, a dresser of vines, a pruner.

vinnul—L. *vinnulus*, delightful.

vinolent—L. *vinolentus*, drunk on wine, now sometimes taken to mean wine-colored.

viol—L. *viola*, the violet. *Ex:* viol-escent; Viola*.

viper—L. *vipera*, the viper<*vivus* alive+*pario*, to produce. *Ex:* Viper (Rept.); Viper-idae (Rept.).

vipio—L. *vipio*, genit. *vipionis*, name of a kind of small crane. *Ex:* Vipio (Ins.).

vir—See virus.

virect—L. *virectum*, a green place. *Ex:* Virecta*.

virens—See vireo.

vireo—1. L. *vireo*, genit. *vireonis*, a kind of bird, acc. to some the greenfinch<*vireo*, to be green. *Ex:* Vireo (Av.); Vireo-sylva (Av.): 2. L. *vireo*, to be green or verdant; ppr. *virens*, genit. *virentis*, becoming green, green.

virescens—L. *virescens*, genit. *virescentis*, becoming or growing green, flourishing, prospering, ppr. of *viresco*, to grow green.

virg—L. *virga*, dim. *virgula*, a twig, sprout; *virgatus*, twiggy, made of twigs; also striped; *virgultus*, full of bushes, shrubby. *Ex:* Virgul-aria (Coel.); Virgil-ia*.

virgat—See virg.

virgine—L. *virgineus*, maidenly, belonging to a virgin.

virgulat—L. *virgulatus*, striped.

virgult—See virg.

virid—L. *viridis*, green; *viridulus*, greenish. *Ex:* Virido-bucco (Av.).

viridican—L. *viridicans*, genit. *viridicantis*, becoming green, greenish, ppr. of *virido*, to make green, to cause to grow green.

viridicans—NL. *viridicans*, becoming green< *viridis*, green.

viridicat—L. *viridicatus*, made green, green.

viriditas—L. *viriditas*, greenness, verdure.

virios—L. *viriosus*, strong, robust.

viror—L. *viror*, genit. *viroris*, greenness, verdure.

**viros**—1. L. *virosus*, muddy, covered with slime, poison, fetid. *Ex:* virose: 2. L. *virosus*, lustful, longing after men.

**virt**—L. *virtus*, genit. *virtutis*, vigor, strength, courage.

**virulent**—L. *virulentus*, full of poison, poisonous.

**virus**—L. *virus*, slime, poison. *Ex:* virus; Betae-vir; Citro-vir.

**vis**—L. *visus*, seen, viewed, pp. of *video*, to see. *Ex:* bi-visus.

**visc**—1. L. *viscum*, bird-lime, made from the berries of the mistletoe; *viscosus*, sticky; *viscatus*, smeared with bird-lime; *viscidus*, clammy, sticky like bird-lime. *Ex:* Visc-aria*; Viscum*; 2. *viscus*, an entrail, pl. *viscera*. *Ex:* viscer-al; viscero-motor.

**viscat**—See **visc 1**.

**viscer**—See **visc 2**.

**viscid**—See **visc 1**.

**vishnu**—*Vishnu*, Hindu god, the Preserver. *Ex:* Vishnu-therium (Mam.).

**visnaga**—Sp. Amer. *visnaga*, tooth-pick; *bisnaga* =*visnaga*, name of a barrel cactus<Nahuatl *huitzli* and *nahuac*, around, i.e., covered with spines. From the spines of the visnaga the people of New Spain made toothpicks.

**vison**—?Icel. or Sw. *vison*, a kind of marten or weasel<Dan. and Sw. *vissen*, withered, shrivelled.

**visor**—L. *visor*, genit. *visoris*, a scout.

**vit**—L. *vita*, life; *vitalis*, of or pertaining to life; *vitalitas*, vital force, vitality. *Ex:* vit-amine, later vitamin. The final *e* of *vitamine* was dropped to prevent an original mistaken chemical connection with *amines* derived from ammonia. See -amine; vital; vital-ity.

**vitabilis**—L. *vitabilis*, neut. *vitabile*, that which ought to be shunned, avoided, declined.

**vital**—See **vit**.

**vitell**—L. *vitellus*, yolk. *Ex:* vitell-aria; vitell-ine; Vitell-inus (Ins.).

**vitex**—See **vitic**.

**viti**—L. *vitis*, dim. *viticula*, a vine. *Ex:* viti-culture; Vitis*.

**vitic**—L. *vitex*, genit. *viticis*, the chaste tree. *Ex:* Vitex*.

**vitil**—L. *vitilis*, interwoven, made of wickerwork.

**vitios**—L. *vitiosus*, defective.

**vitis**—See **viti**.

**vitr**—L. *vitrum*, glass; *vitreus*, like glass, brittle, brilliant. *Ex:* Vitr-ina (Moll.); Vitrea (Moll.); vitreous.

**vitt**—L. *vitta*, a chaplet, a ribbon; *vittatus*, bound with a ribbon or chaplet, striped. *Ex:* Vitt-aria*; e-vittatus.

**vittat**—See **vitt**.

**vitul**—L. *vitulus*, a calf; also a sea-calf, seal. *Ex:* Vitula (Ins.); vituli-fera. S.

**vitulin**—L. *vitulinus*, of a calf; as subst., the flesh of a calf.

**vium**—See **via**.

**viv**—L. *vivus*, alive. *Ex:* Vivi-par-idae (Moll.); vivi-par-ous; vivi-section.

**vivac**—L. *vivax*, genit. *vivacis*, vivacious, long-lived.

**vivat**—L. *vivatus*, animated, vigorous, spry.

**vivax**—See **vivac**.

**viverr**—L. *viverra*, a ferret. *Ex:* Viverr-avus (Mam.); Viverra (Mam.); Viverri-ceps (Mam.).

**vivescen**—L. *vivescens*, genit. *vivescentis*, becoming active, springing forth, ppr. of *vivesco*, to get life, to spring forth, grow strong. *Ex:* vivescent.

**vivid**—L. *vividus*, living.

**vix**—L. *vix*, barely, with much difficulty. *Ex:* vix-gregari-ous.

**vol**—1. L. *vola*, the palm of the hand. *Ex:* vol-ar: 2. L. *volo*, to will. *Ex:* vol-ition. See volans.

**volador**—Sp. *volador*, a flyer.

**volans**—L. *volans*, genit. *volantis*, flying, ppr. of *volo*, to fly.

**volat**—L. *volatus*, a flight; *volatilis*, flying, with wings; *volaticus*, fleeting, winged.

**volitan**—L. *volitans*, genit. *volitantis*, flying, ppr. of *volito*, to fly, to flutter. *Ex:* volitant.

**volitor**—NL. *volitores*, birds which are able to fly. *Ex:* Volitores; volitori-al.

**vologes**—L. *Vologesus*, name of several kings of Parthis. *Ex:* Vologesia (Echin.).

**volsell**—L. *volsella*=*vulsella*, forceps, pincers.

**volubil**—L. *volubilis*, twining, rolling, turning.

**volucell**—NL. *volucella*, name for a genus of flies, also of a genus of mammals<L. *volucris*, flying, fitted for flight. *Ex:* Volucella (Mam.), (Ins.).

**volucr**—L. *volucer*, genit. *volucris*, winged, swift, *volucritas*, swiftness of flight. *Ex:* volucr-ine; Volucres (Av.); Volucris (Ins.).

**voluntar**—L. *voluntarius*, of one's free will. *Ex:* voluntary.

**volut**—L. *voluta*, a spiral scroll<*volvo*, to roll; *volutatio*, genit. *volutationis*, a rolling about. *Ex:* Voluta (Moll.).

**volv**—L. *volva*=*vulva*, dim. *volvula*=*vulvula*, a wrapper, covering; also the womb<*volvo*, to roll. *Ex:* volv-aceus; volv-ate; vulvi-form; vulvo-vaginal; Volvul-ina (Por.); Volvuli-fex (Arth.).

**volvocin**—NL. *volvox*, genit. *volvocinis*, name for a genus for protozoans<L. *volvo*, to roll. *Ex:* volvocin-aceous; Volvox (Prot.).

**volvox**—See **volvocin**.

**volvul**—See **volv**.

**vomer**—L. *vomer*, a plowshare; also membrum virile, the penis. *Ex:* vomer; vomer-ine.

**vomic**—L. *vomicus*, ulcerous, filthy.

**vor**—L. *voro*, to devour; *vorax*, genit. *voracis*, voracious; *voratus*, the devouring one; also a chasm. *Ex:* vorac-ious; Vorates (Ins.); herbi-vor-ous; Musci-vora (Av.).

**vorac**—See vor.

**voragin**—L. *voraginosus*, full of pits or whirl-pools<*vorago*, genit. *voraginis*, a depth, whirl-pool.

**vorat**—See vor.

**vorax**—See vor.

**vortex**—See vortic.

**vortic**—L. *vortex*, genit. *vorticis*, a whirlpool, eddy <*verto*, to turn. *Ex:* Vortic-ella (Prot.).

**vorticos**—L. *vorticosus*, full of whirlpools.

**vot**—L. *votus*, dedicated<*voveo*, to consecrate. *Ex:* pre-votus.

**vulcan**—L. *Vulcan*, the fire god. *Ex:* Vulcan-ella (Por.); Vulcano-myia (Moll.).

**vulgari**—L. *vulgaris*, genera., common, usual< *vulgus*, the multitude.

**vulgat**—L. *vulgatus*, made common, wide-spread, general<*vulgo*, to spread among, to publish.

**vulgivag**—L. *vulgivagus*, roving, changeable, un-certain, variable, capricious.

**vulner**—L. *vulnerator*, one who wounds or mutilates<*vulnus*, a wound; *vulnero*, to injure; pp. *vulneratus*, wounded, injured.

**vulp**—L. *vulpes*=*vulpis*, a fox; also cunning, craftiness; *vulpinus*, of or belonging to a fox. *Ex:* Vulp-avus (Mam.); Vulpes (Mam.); Vulpi-canis (Mam.); vulpine.

**vuls**—L. *vulsus*, shorn, smooth; pp. of *vello*, to pluck hair or feathers. *Ex:* Vuls-iculus (Pisc.); Aviculo-vulsa (Moll.).

**vultur**—L. *vultur*, a vulture; *vulturinus*, vulture-like. *Ex:* Vultur (Av.).

**vulv**—See volv.

**vulvari**—L. *vulvarius*, of foul smell, of evil odor< *vulva*, the womb, also the external organs of generation of the female.

# W

**whorl**—ME. *wharwyl*, *whorwyl*, the whorl of a spindle. *Ex:* whorl; whorl-ed.

# X

**xalapense**—NL. *xalapense*<*ex*+*halapense*, of or belonging to Aleppo=Apelo.

**xani**—Gr. *xanion*, a comb, a card for combing wool. *Ex:* Xanio-pelma (Ins.); Zanio-elpis (Pisc.), errorously for Xanio-lepis (Pisc.).

**xanth**—Gr. *xanthos*, the various shades of yellow; *xanthisma*, anything dyed yellow; *xanthion*, a kind of plant used to dye hair yellow. *Ex:* xanth-ism; Xanth-ispa (Ins.), see hisp; xanth-in; xanth-odont; xanthi-uria (Med.); Xanthisma*; Xanthium*; Xantho-cephalus

(Av.); xantho-derma; Xantho-ura (Av.); Antho-xanthum*; Zoo-xanth-ella*.

**xem**—NL. *xema*, a bird name of uncertain origin, perh. < some native name. *Ex:* Xema (Av.).

**xen**—Gr. *xenos*, a stranger, a foreigner; *xenikos*, of a stranger, foreign; *xenios*, hospitable. *Ex:* Xen-altica (Ins.), see haltic; Xen-apates (Ins.); xen-arthral; Xen-omi (Pisc.); Xen-otis (Pisc.); Xenicus (Av.); Xeno-mys (Mam.); Xeno-picus (Av.); lipo-xen-ous (here taken in the sense of a host).

**xenic**—See **xen.**

**xenism**—Gr. *xenisma*, amazement < *xenizō*, to astonish. *Ex:* Xenisma (Pisc.); Xenismus (Ins.).

**xer**—Gr. *xēros*, dry; *xērasia*, dryness; *xērōdēs*, looking dry; *xēransis*, a drying; *xērotēs*, dryness, thirst. *Ex:* Xer-ano-bium (Ins.); xeransis; xeras-id; xeras-ium; xerio-boles; Xeris (Ins.); xero-phytic; Phyllo-xera (Ins.); Xerus (Mam.).

**xerans**—See **xer.**

**xerod**—See **xer.**

**xerot**—See **xer.**

**xes**—Gr. *xesis*, a polishing, scraping; *xesma*, that which is scraped; scrapings. *Ex:* Xes-urus (Pisc.); Xesm-odon (Mam.).

**xesm**—See **xes.**

**xest**—Gr. *xestos*, polished by scraping or filing. *Ex:* Xestia (Ins.); Xesto-lept-ura (Ins.); Xesto-phya (Ins.); Xesto-spongia (Por.); Eu-xesta (Ins.).

**xiph**—Gr. *xiphos*, dim. *xiphidion*, sword; *xiphydrion*, a little sword; *xiphistēr* = *xiphistēs*,

Heller's Sword-bearer, *Xiphophorus helleri*. Redrawn from Fishes of North and Middle America—Jordan.

a sword belt; *xiphias*, anything sword-shaped, a swordfish. *Ex:* Xiph-agrostis*; Xiphister (Pisc.); Xiphistes (Pisc.); Xiphos-ura (Arth.); Xiphydria (Ins.); Brachy-xiphus (Ins.).

**xiphos**—See **xiph.**

**xiron**—NL. *xiron* < Gr. *xyron*, a razor. *Ex:* Xirono-drilus (Ann.).

**xorid**—NL. *xorides*, an insect generic name created by Latreille. Origin unknown. *Ex:* Xoridi-formis; Xorides (Ins.).

**xot**—Anagram of *tox*. *Ex:* Xot-odon (Mam.).

**xuth**—Gr. *xouthos*, yellowish, tawny. *Ex:* Xutho-trichis (Ins.); Xuthus (Ins.).

**xy**—Gr. *xyō*, to scrape. *Ex:* Xya (Ins.).

**xyel**—Gr. *xyēlē*, a sharp cutting instrument, a tool for scraping wood. *Ex:* Xyela (Ins.).

**xyl**—Gr. *xylē* = *xylon*, wood; *xylinos*, wooden; *xyleus*, a wood-cutter. *Ex:* Xyl-ia*; Xyle-tinus (Ins.), see ptin; Xylina*; Xylo-bi-um*; Xylo-copa (Ins.); Xylo-crinus (Ins.); Xylo-pinus (Ins.), see pin 4.; Xylo-pia*, see picr; Xylo-teles (Ins.); Xylo-terus (Ins.); Lyme-xylon (Ins.); Trypo-xylon (Ins.).

**xylebor**—Gr. *xylēboros*, eating wood. *Ex:* Zyleborus (Ins.).

**xylem**—Ger. *xylem* < Gr. *xylon*, wood. *Ex:* xylem.

**xylic**—Gr. *xylikos*, of wood, like wood. *Ex:* Xylica (Ins.).

**xylin**—See **xyl.**

**xylit**—Gr. *xylitēs*, wooden, like wood. *Ex:* Xylita (Ins.); Xylites (Ins.).

**xyloch**—Gr. *xylochos*, a thicket, forest. *Ex:* Xylochus (Ins.).

**xylod**—Gr. *xylōdēs*, woody, hard as wood. *Ex:* Xylodes (Coel.).

**xylon**—See **xyl.**

**xylopia**—NL. *xylopia* < Gr. *xylopikron*, bitter wood < *xylon* + *pikros*, bitter. *Ex:* Xylopia*.

**xyn**—Gr. *xynos*, common, general. *Ex:* Xyno-bius (Ins.).

**xyr**—1. Gr. *xyron*, razor. *Ex:* Xyr-auchen (Pisc.); Xyr-ichthys (Pisc.): 2. Gr. *xyris* genit. *xyridos*, a kind of iris. *Ex:* Xyrid-aceae* Xyris*.

**xyrid**—See **xyr** 2.

**xyris**—See **xyr** 2.

**xysil**—Gr. *xysilos*, shaven, smooth

**xysm**—Gr. *xysma*, genit. *xysmatos*, shavings, scrapings, lint, threads. *Ex:* Xysmia-bolium*; Xysmato-doma (Ins.).

**xyst**—1. Gr. *xystos*, a covered smooth-floored porch: 2. *xystos*, made smooth, scraped, grated. *Ex:* Xysta (Ins.).

**xyster**—Gr. *xystēr* = *xystēs*, one that scrapes, a raker; *xystos*, scraped, polished; *xystra* = *xystris*, genit. *xystridos*, a scraper; *xystikos*, of or for scraping; *xystrōtos*, scraped, fluted; *xyō*, to rub, scrape. *Ex:* Xysticus (Arachn.); Xystr-opiites (Pisc.); Xystro-cera (Ins.).

**xystic**—Gr. *xystikos*, of or for scraping. *Ex:* Xysticus (Av.).

**xystr**—See **xyster.**

# Y

-y—**1.** -y. Eng. suffix used to form adjectives from nouns, and meaning, partaking of the nature of (chalk-y), abounding in, full of (flower-y), having the quality of (silk-y), having (speckl-y): **2.** -y, Eng. suffix of abstract nouns or of nouns once abstract but now concrete. *Ex:* ana-tom-y; histor-y, theor-y: **3.** -y, dim. of certain Eng. nouns, both common and proper. *Ex:* kitt-y, dogg-y.

yal—Gr. *yalos*, glass. *Ex:* Delo-yala (Ins.).

yaleo—Gr. *yaleos*, glassy, bright. *Ex:* Yaleosaurus (Rept.).

yla or ylae—See yle.

yle—Gr. *hylē*, a wood, forest; also matter; *hylaios* belonging to a forest, savage. *Ex:* prot-yle.

ymenia—See hymen.

yolk—ME. *yolke*; A.S. *geoleca=gioleca*, the yolk, lit., the yellow part < *geolu*, yellow.

yper—Gr. *yper* also *ypeir*, over, above, across, beyond, instead of; much used in Greek compounds; rendered *hyper-* in English compounds.

yph—Gr. *yphos*, *yphē*, a web. *Ex:* Ypho-myrmex (Ins.).

yphant—Gr. *yphantos*, woven > NL. *yphantes.* a weaver. *Ex:* Yphantes (Av.).

ypn—See hypn.

ypo-—Gr. *ypo-*, prefix meaning under, somewhat > *ypophaios*, somewhat gray. *Ex:* Ypophae-myia (Ins.). See hyp.

yponom—Gr. *yponomos*, going underground; also an underground passage.

yponomeut—Gr. *yponomeuō*, to go underground. to mine. *Ex:* Yponomeuta (Ins.).

ypophae—See ypo-.

yps—Gr. *ypsi-*, prefix meaning high. *Ex:* Ypsi-stoma (Prot.); Ypso-lophus (Ins.). See hyps.

ypsil—Gr. *ypsilon*, the Gr. letter Υ. *Ex:* ypsili-form; Ypsilo-neura (Ins.).

yrus—See ur 1.

yss—Gr. *yssos*, a javelin.

yucc—West Indian *yucca*, a name for manihot = Sp. *yuca*. *Ex:* Yucc-aceae*; Yucca*; Yucca-borus (Ins.).

yung—NL. *yunx* < Gr. *iunx*, genit. *iungos*, a bird, the wryneck. *Ex:* Yung-idae (Av.) = *Iung*-dae (Av.); Yungi-picus (Av.); Yunx (Av.).

yunx—See yung.

# Z

za—Gr. *za*, an intensive meaning very. *Ex:* Za-glyptus (Ins.); Za-lamb-odont; Za-lophus (Mam.); Za-phrentis (Coel.), see phren; Za-pod-idae (Mam.); Za-prora (Pisc.); Za-pterus (Av.); Za-pus (Mam.); Za-rhinchus (Av.).

zabr—Gr. *zabros*, gluttonous. *Ex:* Zabro-morphus (Ins.); Zabrus (Ins.).

zachol—Gr. *zacholos*, wrathful. *Ex:* Zacholus (Rept.).

zachresta—NL. *zachrestra* < Gr. *zachrēēs*, attacking with violence, raging. *Ex:* Zachresta (Ins.).

zal—Gr. *zalē*, a storm, the surging of the sea, sea spray, surf; *zalos*, muddy foam. *Ex:* Zal-embius (Pisc.); Zal-ieutes (Pisc.), see alieut; Zal-ocys (Pisc.); Zal-ypnus (Pisc.); Zale (Ins.); Zale-scopus (Pisc.); Zalo-bius (Ins.); Zalo-chelidon (Av.).

zaloi—Gr. *zaloeis*, stormy. *Ex:* A-zalois (Pisc.)

zamen—Gr. *zamenēs*, forceful, mighty, raging. *Ex:* Zamen-ophis (Rept.); Zamenis (Rept.).

zami—L. *zamia*, loss, damage > NL. *zamia*, name applied by Linnaeus to a cycad with sterile-appearing male cone < L. *zamia* "a pine cone, which when suffered to decay on the tree, injured the succeeding crop" (Pliny). *Ex:* Zam-ites*; Zamia*; Zamio-strobus*; Macro-zamia*.

zamric—NL. *zamricus*, name for an extinct genus of small-toothed mammals < Gr. *za*, very + *mikros*, small. *Ex:* Zamicrus (Mam.).

zancl—Gr. *zanklon*, sickle. *Ex:* Zancl-odon (Rept.); Zanclo-stomous (Av.); Zanclus (Pisc.).

ze—Gr. *zeō*, to boil. *Ex:* a-zeo-tropic. See also zeus.

**zea**—L. *zea*, a kind of grain < Gr. *zea=zeia*, a kind of grain. *Ex:* Zea*; zea-col-ella.

**zebr**—Abyssinian *zibra* > Fr. *zebre* and NL. *zebra*, a zebra; *zebrinus*, zebra-marked, striped. *Ex:* zebr-ass; zebr-oid; zebr-ule < *zebr(a)+(m)ule*; Zebra (Moll.); Zebra-picus (Av.); Zebrina*, Zebro-myia (Ins.).

**zel**—Gr. *zēlos*, emulation, zeal. *Ex:* Zelo-typa (Ins.); Zelus (Ins.).

**zele**—Gr. *zēlē*, a female rival. *Ex:* Zele (Ins.).

**zelotes**—Gr. *zēlōtēs*, a zealot, one who is jealous; also an emulator. *Ex:* Zelotes (Arach.).

**zem**—Gr. *zema*, a drink. *Ex:* Chori-zema*, Gr. *choros*, a joyful dance, the juice of the plant having been a welcome refreshment to thirsty travellers.

**zemi**—Gr. *zēmia*, loss, hurt, damage. *Ex:* Zemi-stephanus (Moll.); Zemia (Av.); Zemi-ophron (Ins.); Macro-semia*.

**zemiotes**—Gr. *zēmiōtēs*, one who punishes or causes damage. *Ex:* Zemiotes (Ins.).

**zen**—Fr. *zenith*, from Arabic, the zenith. *Ex:* Zenithi-cola (Ins.); Zenitho-ptera (Ins.); zeno-tropism. See also xen and zeus.

**zenith**—See zen.

**zephyr**—Gr. *Zephyros*, the west wind. *Ex:* Zephyr-anthes*; Zephyrus (Ins.).

**zeren**—NL. *zerena* and *zeritis*, etc. < Gr. *xērainō*, to parch, dry up. *Ex:* Zeren-idae (Ins.); Zerena (Ins.); Zeritis (Ins.).

**zeritis**—See zeren.

**zest**—Gr. *zestos*, seethed, boiled (until soft), also hot. *Ex:* Zest-ichthys (Pisc.); Zest-idium (Pisc.); Zestis (Pisc.); Zesto-carus.

**zet**—1. Gr. *zēteō*, to search for, to seek. *Ex:* Zet-odon (Mam.); Zeteo-lyga (Ins.); Basso-zetus (Pisc.).      2. L. *zeta*, a chamber.

**zetesim**—Gr. *zētēsimos*, to be searched. *Ex:* Zetesima (Ins.).

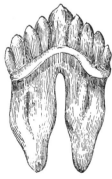

Tooth of Zeuglodon. "The name was chosen because the first section of a molar examined was taken from the base of the crown where it was beginning to divide into roots, and thus it looked like two single teeth yoked or linked together."

**zetet**—Gr. *zētētēs*, a seeker, one who inquires. *Ex:* Zetetes (Ins.); Zetet-icus (Ins.); Myio-zetetes (Av.).

**zeth**—Gr. *Zēthus*, one of Jupiter's sons. *Ex:* Zeth-oides (Ins.); Zethus (Ins.); Zethus-culus (Ins.).

**zeuct**—Gr. *zeuktos*, joined. *Ex:* zeucto-coelomata; Uro-zeuctes (Crust.).

**zeug**—Gr. *zeugos*, things paired, a yoke of beasts of burden. *Ex:* Zeug-onyx (Ins.); Zeug-orchis (Platy.); Zeugo-ptera (Pisc.).

**zeugl**—Gr. *zeuglē*, the strap of a yoke, the bar joining the halves of a double rudder. *Ex:* Zeugl-odon (Mam.).

**zeugm**—Gr. *zeugma*, genit. *zeugmatos*, a bond, band. *Ex:* Zeugma (Ins.); Zeugmato-thrips (Ins.).

**zeus**—1. L. *Zeus*, father of gods and men; *Zen* (a poet. form of *Zeus*), dim. *Zenion*. *Ex:* Zen-opsis (Pisc.); Zenion (Pisc.):      2. L. *zeus*, a kind of fish. *Ex:* Ze-idae (Pisc.); Zeus (Pisc.).

**zeux**—Gr. *zeuxis*, a joining, yoking. *Ex:* Zeux (Ins.); Zeux-idia (Ins.); Zeuxi-diplosis (Ins.); Zeuxis (Moll.).

**zeuzera**—NL. *zeuzera*, a change by misprint or otherwise from Zenzera. *Ex:* Zeuzera (Ins.).

**zexmenia**—Anagram of Ximenezia. *Ex:* Zex menia*.

**zibeth**—Fr. *zibet*=Ital. *zibetto*=Gr. *zibeth*, the civet > NL. *zibethicus*, civet-odored, musty-odored. *Ex:* Zibeth-ailurus (Mam.); Zibetha (Mam.).

**zigzag**—Fr. *zigzag* < Ger. *zickzack*, zigzag. *Ex:* Zigzag-ites (Moll.); Zigzagi-ceras (Moll.).

**zill**—NL. *zilla* < Ar. *sillah*, a plant name. *Ex:* Zilla*.

**zin**—Gosiute Indian *dsina=zina*, needle or spine. *Ex:* Zin-iulus (Arth.); Zin-aria (Arth.)

**zingiber**—Gr. *zingiberis*=L. *zingiberi*, ginger. *Ex:* Zingiber*.

**ziph**—NL. *ziph-* < Gr. *xiphos*, a sword; *xiphios*, a sword-fish. *Ex:* Ziph-ac-odon (Mam.); Ziphi-opsis (Mam.); Ziphius (Mam.); most names of mammals beginning with *ziphi-* and *ziphio-* refer to Ziphius, the beaked whale; Zipho-theca (Pisc.).

**zizani**—Gr. *zizanion*, darnel, the tares of Scriptural parable. *Ex:* Zizani-opsis*; Zizania*.

**ziziph**—Gr. *zizyphon*, name of the Mediterranean jujube < Ar. *Zizouf*, name of the lotus. *Ex:* ?Ziziphinus (Moll.); Zizipho-myia (Ins.); Zizyphus*.

**zo**—Gr. *zōon*, an animal, a living being; Attic. *zōē*, life; *zōikos*, pertaining to life; *zōos*, alive, living. *Ex:* zo-id (*zo*+dim. *-idion*)=zoo-id; Zoe; zoea; zoic; zoo-geo-graphy; zoo-logy; zoo-spore; Zoo-toca (Mam.); zoo-xanth-ella; Antho-zoa (Coel.); Chelido-zoum (Bry.); di-zoic; Scoto-zous (Mam.); spermato-zoa.

**zoarc**—Gr. *zōarkēs*, life-supporting, refreshing. *Ex:* Zoarc-idae (Pisc.); Zoarc-ites (Pisc.); Zoarces (Pisc.); Zoarcus (Pisc.).

**zodio**—Gr. *zōdion*, a small animal, dim. of *zōon*, animal—Zodio-myces*.

**zoic**—See **zo.**

**zoid**—NL. *zoid*, a diminutive animal, a sperm cell. *Ex:* zoid; zoidio-phil-ous; zoido-gamous. See also zo.

**zom**—Gr. *zōmos*, soup, sauce; also a corpulent greasy fellow. *Ex:* Zomo-therapy (Med.).

**zon**—Gr. *zonē*, a girdle, belt=L. *zona*; L. *zonatus*, banded; *zonarius*, pertaining to a girdle. *Ex:* Zon-iscus (Prot.); Zon-ites (Moll.); Zon-urus (Rept.); zonar-y; Zonaria*; zono-placental; Zono-trichia (Av.).

**zonal**—L. *zonalis*, belted, belonging to a zone.

**zonat**—See **zon.**

**zoph**—See **zopher.**

**zopher**—Gr. *zopheros*, dusky, gloomy<*zophos*, a darkness. *Ex:* Zopherus (Ins.); Soleno-zopheria (Ins.).

**zor**—1. Gr. *zōros*, pure, strong, sheer. *Ex:* Zor-aptera (Ins.); Zora (Arach.); Zoro-typus (Ins.): 2. Sp. *zorra, zorro*, dim. *zorilla*, a fox. *Ex:* Zorilla (Mam.): 3. L. *Zoroaster*, founder of Zoroastrianism. *Ex:* Zoro-aster (Echin.), see aster.

**zorill**—See **zor.**

**zoster**—Gr. *zōstēr*, a girdle or band. *Ex:* Zoster-ops (Av.); Zoster-ornis (Av.); Zostera*; Zosteri-cola (Moll.); Zostero-thrix (Ins.).

**zothec**—Gr. *zōthēkē*, a closet, chamber. *Ex:* Zotheca (Ins.).

**zoum**—See **zo.**

**zoyph**—Gr. *zōyphion*, a little animal, dim. of *zōon*. *Ex:* Zoyphium (Ins.).

**zyg**—Gr. *zygos*=*zygon*, yoke; *zygosis*, a joining. *Ex:* Zyg-adenus*; zyg-antra; zyg-apophysis; zyg-oma; Zygo-mys (Mam.); Zygo-nectes (Pisc.); zygo-sphere; A-zygo-phleps (Ins.); Meta-zygia (Arach.).

**zygaen**—Gr. *zygaina*, ancient name for the hammer-headed shark<*zygon*, yoke. *Ex:* zygaena.

**zylebor**—See **xylebor.**

**zym**—Gr. *zymē*, leaven, yeast; *zymōma*, a fermented mixture; *zymōsis*, fermentation; *zymōtikos*, pertaining to or causing fermentation. *Ex:* zym-ase; zymo-gen; zymotic; en-zyme.

**zyx**—NL. *zyxis*=Gr. *zeuxis*, a yoking, joining. *Ex:* Zyx-omma (Ins.).

**zyzz**—NL. *zyzza* perh.<Sp. *ziszas*, zigzag. *Ex:* Zyzza (Ins.); Zyzzo-geton (Ins.).

# Addendum

## A

aages—Gr. *aagēs*, hard, unbroken.

ab- —L. *ab-*, prefix meaning from, away, without. Placed before Latin stems beginning with a vowel<L. *ab*, from, away. *Ex:* ab-err-ation; ab-or-al; also placed before stems beginning with a consonant in making English words. *Ex:* ab-norm-al; ab-neur-al.

abat- —Gr. *abatos*, pure, inviolate.

abdicat- —L. *abdicatus*, disowned, renounced, abandoned.

abebae- —Gr. *abebaios*, uncertain, wavering, changeable.

abhorr- —L. *abhorreo*, to dislike, shrink from.

abige- —L. *abigeus*, a cattle stealer.

ablabes—Gr. *ablabēs*, harmless, innocent.

ablechr- —Gr. *ablēchros*, weak, feeble.

ablepsia—Gr. *ablepsia*, blind, blindness.

aboethet- —Gr. *aboēthētos*, hopeless, incurable.

abol- —Gr. *abolos*, uncast, unshed, held on to.

abolit- —L. *abolitus*, terminated, destroyed, cancelled.

abrict- —Gr. *abriktos*, wakeful.

aborigin—L. *aborigines*, the original Romans. *Ex:* aborigin-us; aborigin-al.

abroch- —Gr. *abrochos*, dry, waterless.

abscons—L. *absconsus*, -a, -um, concealed, hidden away.

abson- —L. *absonus*, inharmonious, disagreeing with.

-ac—Adj. suffix (<L. -acus, <Gr. -akos) meaning *concerning*. *Ex:* ile-ac; mani-ac; obeli-ac.

acanthod—Gr. *akanthōdēs*, thorny. *Ex:* acanthod-us.

acanthyll- —L. *acanthyllis*, a little bird, probably a titmouse.

accept- —L. *acceptus*, welcome, agreeable.

aceros—L. *acerosus*, -a, -um, needle-shaped, full of needles; also, chaffy.

achlis—L. *achlis*, some northern wild animal, probably the elk.

acnis—Gr. *aknisos*, without fat, lean, spare.

acti—L. *actius*, -a, -um, pertaining to Actium, a promontory and city of Greece.

-aculum—Suffix of verb stems, forming nouns. It conveys the idea of *place for, a thing for. Ex:* hibern-aculum, a place for spending the winter<L. *hiberno*, to spend the winter; NL. tent-aculum<L. *tento*, to touch, to feel, to probe.

-acy—Suffix (<L. -acia, -atia or Fr. -acie) denoting *state* or *quality. Ex:* accur-acy; leg-acy; priv-acy.

ad- —ad-, prefix<L. *ad*, meaning *to, toward, near*, or *upon, against;* also, *increase*, and often changed to *ac-, ag-, af-*, etc. following the Latin rule that the final consonant of a prefix may be changed to correspond to the letter which follows it. *Ex:* ad-oral (*ad-*, to); ad-renal (near or upon); ad-maxill-ary (to); aggregation (toward); an-nectens (upon); assimilation (to); etc.

adaet- —Gr. *adaētos*, unknown, unknowing.

adait- —Gr. *adaitos*, inedible.

adapanet- —Gr. *adapanētos*, inexhaustible.

adibilis—L. *adibilis*, accessible.

adim- —Gr. *adeimos*, fearless, bold.

adin- —Gr. *adinos*, close, thick, crowded.

admirabil—L. *admirabilis*, -e, worthy of admiration, wonderful.

adnix—L. *adnixtus*, -a, -um, leaning upon, pressing against.

ado—<*Aden* in Arabia. *Ex:* ado-ensis.

adocim- —Gr. *adokimos*, false, base, spurious.

adpress—L. *adpressus*, -a, -um, pressed toward.

adscenscion—L. *adscenscio*, genit. *adscensionis*, an ascending.

adsit—L. *adsitus*, -a, -um, sown, planted, set near something.

adulterin—L. *adulterinus*, -a, -um, not genuine.

adventit—NL. *adventitius*<L. *adventicius*, that is added from abroad, outside; NL. *adventitia* (neut. pl. of *adventitius*), the external coat of a blood vessel. *Ex:* adventiti-ous; adventitia.

aegae—<*Aegean* archipelago. *Ex:* aegae-us.

aegypt—<L. *Aegyptus*, Egypt. *Ex:* aegypt-acus, aegypt-ius.

aenet- —Gr. *ainetos*, praiseworthy.

aeruginos—L. *aeruginosus*, -a, -um, full of copper, rusty.

aestival—L. *aestivalis*, -e, of summer-flowering, pertaining to summer.

aethere—L. *aethereus*, -a, -um, heavenly.

aethiop—L. *Aethiopia*, Ethiopia. *Ex:* aethiopicus, aethiop-um.

aezeo- —Gr. *aizeos*, strong, active, vigorous.

-age—Fr. *-age*<LL. *-aticum*. Suffix used in formation of nouns to indicate: (1) function, an act or process; as in breakage, clottage, cleavage, marriage, shrinkage; (2) a state or condition, as in blockage; (3) a place of abode, as in orphanage.

aget- —Gr. *agētos*, admirable.

-ago—Old L. suffix denoting disease or malfunction. *Ex:* lumb-ago.

agonos—Gr. *agonos*, unfruitful, sterile.

aigrette—See egrett.

ainu—Etym. doubtful; supposed to be a corruption of Japanese *inu*, a dog, a word applied contemptuously by certain aboriginal tribes of Japan.

aiz—NL. *aizoides*, like *aizoon* or house-leek.

ajaja—South American name of the spoonbill.

alb—L. *Albis*, the river Elbe. *Ex:* alb-ensis.

albidul—L. *albidulus, -a, -um*, whitish.

-ale—1. NL. noun suffix (<NL. sing. of NL. *-alia*, a group), *meaning one of a group.* See -alia 2. *Ex:* carp-ale; dist-ale; orbit-ale; tars-ale.    2. L. *-ale*, neut. adj. ending. *Ex:* norm-ale.

alema—Gr. *alēma*, flour, meal.

alepp—<*Aleppo*, Syria. *Ex:* alepp-icus; alepp-ensis.

-ales—L. masc. and fem. plural adj. ending. *Ex:* intercost-ales.

aleut—<*Aleutian* Islands in Alaska. *Ex:* aleut-icus.

-alia—1. NL. *-alia*, suffix denoting *condition.* *Ex:* algo-psych-alia; psych-alia.    2. NL. *-alia* (pl. of *-ale*), noun suffix, meaning *one of a group*<Gr. *halia*, an assembly. *Ex:* carp-alia; dist-alia; im-placent-alia; tars-alia. Compare with Mamm-alia, Anim-alia.

alkal—Ar. *al qaliy*, the ashes of saltwort or glasswort, which abound in soda. *Ex:* alkal-ine.

aloides—NL. resembling *Aloe*, a genus of liliaceous plants. *Ex:* Aloides*.

alp—L. *alpis*, also *alpes*, the alps. *Ex:* alp-estris, alp-igenus, alp-iganus, alp-inus.

alpaca—<Fr. *alpaca*, transferred from Sp. Peruvian *paco*, native name of this mammal.

alsa—<*Alsace*, France. *Ex:* alsa-ticus.

altai—<The Altai Mountains. *Ex:* altai-cus, altai-ensis.

altiuscul—L. *altiusculus, -a, -um*, rather high, a little too high.

amazili—NL.<*amazili*, probably of South American origin. Name applied by Lesson to a species of humming-birds.

ambagios-—L. *ambagiosus*, full of digressions.

ambian—<*Ambiani*, now Amiens, France. *Ex:* ambian-ensis.

amenen-—Gr. *amenēnos*, feeble, weak.

amerind—*Amerind*, a coined name<*American*+*Indian.*

amethyste—L. *amethysteus, -a, -um*, bluish, amythyst-colored.

amianthium—NL.<Gr. *umiantos lithos*, an unspotted stone. *Ex:* Amianthium*.

amnes—Gr. *amnēsia*, forgetfulness. *Ex:* amnesia; amnes-ic.

amorg—<the island of *Amorgos*, Greece. *Ex:* amorg-inus.

amot-—Gr. *amotos*, furious, savage.

amphrysus—L.<Gr. *Amphrysos*, a river in Thessaly.

ampullat—L. *ampullatus, -a, -um*, boxed, bottled.

ampyx-—Gr. *ampyx*, a headress, fillet.

ana-—Gr. *ana-*, prefix meaning: (a) *up, upon, up against;* (b) *back again, backward;* (c) *similar to;* (d) *excessive;* (e) *apart*, as in analysis and anatomy; (f) *throughout*, as in anasarca. *Ex:* ana-bol-ism; ana-bios-is; ana-dips-ia *ana-*, an intensive); ana-genesis (*ana-*, again);

ana-kat-estes-ia (*ana-*, up); ana-kineto-mer-e (*ana-*, up); ana-rrhea (*ana-*, up).

anaid—NL. *anaidus, -a, -um*, reckless,<Gr. *anaidēs*, reckless, shameless.

anam—<*Anam*, India. *Ex:* anam-iticus.

anaphal—NL. *anaphalis*<a Greek name for a plant called "everlasting".

anaspid—L. *anaspidea*<Gr. *an*, not+*aspis*, shield. *Ex:* anaspides.

anatol—<*Anatolia*, Turkey. *Ex:* anatol-icus.

anaton—L. *anatonus, -a, -um*, extending upward.

anax—Gr. *anax*, pl. *anakes*, a lord.

anaxi-—Gr. *anaxios*, unworthy, worthless.

-ance—Fr. *-ance, -ence* (<L. *-antia, -entia*), noun suffix added to: (1) verb stems to denote *act of* or *fact of doing, a being*, as in toler-ance and continu-ance; (2) to adj. bases to denote *condition* or *quality of being*, as in defi-ance (<defiant) and signific-ance (<significant).

anchistin-—Gr. *anchistinos*, close, crowded, thick.

-ancy—Compound suffix (<-*ant* and *-cy*), denoting a *state* or *fact of performance*, as in continuancy. Similar forms are *-ance, -ence*, and *-ency.*

and—<*Andes* Mountains, South America. *Ex:* and-icolus, and-inus.

andalusi—<*Andalusia*, a large area in southern Spain. *Ex:* andalusi-an.

andira—NL. *andira*, <Tupi Indian name.

andr—In modern botany, a stamen<Gr. *andros* male. *Ex:* pelt-andra*.

andromeda—L. *Andromeda*, daughter of Cepheus. *Ex:* Andromeda*.

aneclipt-—Gr. *anekleiptos*, endless, incessant.

anetik-—Gr. *anetikos*, denying, negative.

anglic—NL. *anglicus, -a, -um*, English.

angol—<*Angola*, West Coast of Africa. *Ex:* angol-ensis.

angora—<*Angora*, a city in Turkey.

ani—Brazilian *ani*, name of some bird.

annal—L. *annalis, -e*, continuing a year.

annu—L. *annuus, -a, -um*, yearly, annual.

anomal—L. *anomalus, -a, -um*, irregular, deviating from rules.

anostraca—NL. *anostraca*, <Gr. *an-*, not+*ostrakon*, a shell.

anoz-—Gr. *anozos*, with few or no branches, bare.

antaug-—Gr. *antaugēs*, reflecting light, sparkling.

-anth—NL. *-anthes;* also *-anthus, -a, -um*, flowered, <Gr. *anthos*, flower.

-anthem—NL. *-anthemus, -a, -um*, flowered, <Gr. *anthos*, flower.

anthracin—NL. *anthracinus, -a, -um*, coal-colored<L. *anthrax*, genit. *anthracis*, coal.

-anus—L. *-anus, -a, -um*, adj. endings, sometimes appearing as noun endings. *Ex:* membrana<L. *membrum*, limb.

aor-—Gr. *aoros* 1. pendulous, something hanging or waving.    2. untimely, unripe, unfit.

apag-—Gr. *apagēs*, not firm, soft, flabby.

apennin—<the *Apennine* Mountains. *Ex:* apennin-us.

apeor- —Gr. *apēŏros*, hanging on high, soaring.

aphaur- —Gr. *aphauros*, feeble, weak, powerless.

aphelin—NL. *Aphelinidae*, <Gr. *apo*, from+ *hēlios*, sun+*n*+*idae*.

apical—NL. *apicalis*, concerning, or of, the top<L. *apex*, genit. *apicis*, the top.

aplet- —Gr. *aplētos*, immense, large.

apolegm- —Gr. *apolēgma*, a hem, border.

aposit- —Gr. *apositos*, hunger.

apoxyr- —Gr. *apoxyros*, sharp, sheer.

appropinquat—L. *appropinquatus, -a, -um*, approaching (in appearance).

apul—<*Apulia*, in southern Italy. *Ex:* apulicus, apul-us.

arab- —Gr. *arabos*, a chattering, rattling or gnashing of teeth.

aral—<*Aral* Sea, in Asiatic Russia. *Ex:* aralensis.

arane—L. *araneus, -a, -um*, pertaining to a spider.

arbore—L. *arboreus, -a, -um*, tree-like.

archilochus—L. *Archilochus*, name of a 7th century B.C. Greek poet.

arctic—L. *articus, -a, -um*, northern<Gr. *arktos*, a bear.

arcuat—L. *arcuatus, -a, -um*, in the form of a bow.

arculari—NL. *arcularius, -a, -um*, pertaining to or of a box, <L. *arca*, a box.

argal—Mongolian *argali*, sheep.

argyre—NL. *argyreus, -a, -um*, silvery<Gr. *argyreios*, silvery.

aristotel—*Aristotle* (384–322 B.C.)<Gr. *Aristotelēs*, Father of natural history. Among first to use the scientific method. *Ex:* aristotel-ian.

-arium—L. *-arius, -a, -um*, suffix added to noun stems to form adjectives denoting the *place of a thing, belonging to* or *connected with, a thing like*. *Ex:* cerc-aria (-*aria*, connected with); coel-arium (-*arium*, belonging to); sanit-arium; sol-arium; utric-aria.

arizona—Arizona, a state name. *Ex:* Arizona (Rept.).

armeni—L. *armenius, -a, -um*, of Armenia.

aromatic—L. *aromaticus, -a, -um*, spicy, fragrant.

arrhem- —Gr. *arrhemōn*, silent, speechless.

artemi—NL. *artemia*<Gr. *artēma*, that which hangs, <*artanō* to hang upon, to fasten to.

arundinace—L. *arundinaceus, -a, -um*, of or like a reed or cane.

arvon—<Carn*arvon*, Wales. *Ex:* arvon-icus.

-ary—1. Suffix<L. *-arium*, denoting *the place of a thing, place where*. *Ex;* mortu-ary; natu-ary. 2. Adj. suffix meaning *related to*. *Ex:* capill-ary; evolution-ary; unit-ary.

aryball- —Gr. *aryballos*, a bag, purse.

ascit- —L. *ascitus*, alien, foreign.

asinin—L. *asininus, -a, -um*, of or produced by an ass; also, foolish.

-asm—Suffix (<Gr. *-asmos*) occurring after *-i-*, as in E. enthusiasm<Gr. *enthusiasmos*. *Ex:* ecdysi-asm.

asodes—Gr. *asōdēs*, muddy, slimy.

asomat- —L. *asomatus* <Gr. *asomatos*, without a body, incorporeal.

asperat—L. *asperatus, -a, -um*, made rough, uneven; also exasperating.

aspernat—NL. *aspernatus, -a, -um*, despised, rejected.

aspers—L. *aspersus, -a, -um*, scattered, sprinkled over with, be-spattered.

asser- —L. *asser*, a beam, stake, pole.

assimil—L. *assimilis, -e*, similar, like.

assit- —L. *assitus*, near.

asthm—Gr. *asthma*, genit. *asthmatos*, a panting, breathing hard. *Ex:* asthma, asthmat-ic.

asynteles—Gr. *asyntelēs*, useless, worthless.

-ate—1. *-ate* (<L. suffix *-atum*, neut. of *-atus*), used to end names of salts and esters if the acids from which they are formed have names ending in *-ic*. *Ex:* carbon-ate. 2. E. *-ate* (<L. *-atus, -a, -um*), a suffix meaning *provided with*, added to noun stems to form adjectives. *Ex:* cune-ate; ligul-ate.

ateleut- —Gr. *ateleutos*, endless, eternal, always.

atmet- —Gr. *atmētos*, uncut, undivided.

atoll—probably<Malayan *adal*, closing, uniting.

-atory—Adj. suffix<L. *-atorius*<*ator*+*-ius*. Termination of adjectives derived from L. or NL. nouns ending in *-ator*. *Ex:* respir-atory; vibr-atory.

atres—Gr. *atrētos*, imperformation>in modern medicine, in the sense of an absence of, or closure of, a natural opening. *Ex:* atres-ia (in the sense of disintegration or death); atret-ic.

atret—See atres.

atyph- —Gr. *atyphos*, frightful, fearful.

aurochs—Ger. *auerochse*, the wild ox.

austriac—NL. *austriacus, -a, -um*, belonging to the south<L. *auster*, the south wind.

austrin—L. *austrinus, -a, -um*, southern.

avenace—NL. *avenaceus, -a, -um*, of oats, <L. *avena*, oats.

aver—<*Auvergue*, France. *Ex:* aver-n-ensis.

avicular—NL. *avicularis, -e*, pertaining to little birds.

avit—L. *avitus, -a, -um*, ancestral.

avocet—Fr. *avocet*, a name for a bird.

axillar—NL. *axillaris, -e*, growing in the axils, <L. *axilla*, the armpit.

axitios- —L. *axitiosus*, acting together, in harmony, in combination.

# B

babax—G. *babax*, a chatterer, loud talker.

babul- —L. *babulus*, babbler, fool.

baccan—NL. *baccans*, genit. *baccantis*, with berries, berry-like, pulpy<L. *bacca*, a berry.

baccat—NL. *baccatus, -a, -um*, with berry-like fruit, <L. *bacca*, a berry.

bahia—<*Bahia*, Brazil.

baical—<Lake *Baikal*. *Ex:* baical-ensis.

bald—<Mount *Baldo*, in Lombardy. *Ex:* bald-ensis.

balearic—<the *Balearic* Isles in the Mediter-ranean. *Ex:* balearic-us.

ballator—L. *ballator*, a dancer.

bambus—Malay *bambusa*, the bamboo. *Ex:* bambus-oides.

barbad—<the Isle of *Barbados* in West Indes. *Ex:* barbad-ensis.

barbar—<*Berber*, in Sudan. *Ex:* barbar-us.

basal—NL. *basalis*, basal<L. *basis*, a base.

batal-—Gr. *batalōs*, the anus, hinder parts, rump.

batav—L. *Batavia*, the peninsula of, in the Netherlands. *Ex:* batav-us.

becc-—L. *beccus*, a beak, bill.

belamcanda—NL. *belamcanda*, <Malabar name of the plant.

bellicos—NL. *bellicosus, -a, -um*, full of fight, <L. *bellum*, war, fight+-*osis*, full of.

bembr-—Gr. *bembros*, dull.

benguel—<*Benguela*, Angola. *Ex:* benguel-ensis.

berothidae—Said to be derived<L. *Beroe*, name of a nymph+-*idae*.

betarmon—Gr. *bētarmōn*, a dancer.

-bility—Noun suffix corresponding to -*ble*+-*ity*.

binari—L. *binarius, -a, -um*, containing two, consisting of two.

binoculat—NL. *binoculatus, -a, -um*, two-eyed, <L. *bini*, two+*oculus*, eye+-*atus*.

bipes—L. *bipes*, two-footed.

bisnaga—Sp. *biznaga, viznaga*, <Ar. *bisnāj, bashuāga*, <L. *pastinaca*, a parsnip.

bisulc—L. *bisulcus, -a, -um*, divided into two parts.

bithyn—<*Bithynia*, in Northwestern Asia Minor. *Ex:* bithyn-icus.

blastes—Gr. *blastēsis*, a budding. *Ex:* blastesis.

bles—<*Bois*, on the Loire in France. *Ex:* bles-ensis.

bobolink—name derived from bird's call; earlier used name is *boblincoln*.

boeot—<*Boeotia*, Greece. *Ex:* boeot-icus, boeot-ius, boeot-us.

bogot—<*Bogota*, Columbia. *Ex:* bogot-ensis.

bom-—Gr. *bōmos*, base, stand, altar.

bonari—<*Bonaria*, in Buenos Aires. *Ex:* bonari-ensis.

bonasus—L. *bonasus*, a buffalo.

borag—ML. *borago*, per. <L. *burra*, a shaggy garment, referring to the rough foliage. *Ex:* borag-e; borago; Borag-in-aceae*.

borne—<the Island of *Borneo*. *Ex:* borne-ensis.

botanodes—Gr. *botanōdēs*, herbaceous, <*botanē*, grass, fodder+-*ōdēs*, like.

bovista—NL. *bovista*, <Ger. *bovista*, name for a certain fungus.

bracat—L. *bracatus, -a, -um*, with breeches, <*braca*, trousers.

brachium—L. *brachium*, the arm.

branta—Eng. *branta*, the snow-goose.

brephal—NL. *brephalos*, the first larval stage of a crustacean after emerging from the egg<Gr. *brephos*, a young animal. *Ex:* brephalos.

brochet-—Gr. *brochetos*, rain, a wetting.

bromi—NL. *bromius, -a, -um*, full of sound, <Gr. *bromios*, noisy.

-brum—Neuter suffix of Latin nouns of means or instrument. Such nouns are formed from verb-stems (rarely noun-stems). *Ex:* labrum<the root, *la*, to lick.

brutt—L. *Bruttium*, Calabria, in southern Italy. *Ex:* brutt-ius.

bubal—L. *bubalus, -a, -um*, of the wild ox.

buccat—NL. *buccatus, -a, -um*, big jawed, with big cheeks, <L. *bucca*, the cheek.

buchloe—NL. *buchloe*, <Gr. *bous*, buffalo+*chloē*, verdure, grass. *Ex:* Buchloe*.

bufoni—NL. *bufonius, -a, -um*, having to do with toads<L. *bufo*, a toad.

-bulum—Neuter suffix denoting means or instrument. It is added to L. verb-stems (rarely noun-stems) to form nouns. Other similar suffixes are: -*culum*, -*brum*, and -*trum*. *Ex:* man-dibulum<L. *mando*, to chew; vehiculum< *veho*, to carry.

bunting—ME. *buntynge* or *bountying*, a bird name, apparently from MLGer. *bunt*, speckled.

burro—Sp. *burro*, an ass.

buthus—NL. *buthus*, perhaps a fabricated name. *Ex:* Buthus (Ins.).

butom—NL. *butomus, -a, -um*, <Gr. *butomomos*, a kind of water plant, <*bous*, an ox+*temnō*, to cut. *Ex:* Butom-acea*, butom-opsis.

bycan-—Gr. *bykanē*, a trumpet.

bysma—Gr. *bysma*, a plug.

byzant—L. *Byzantium*, Istanbul. *Ex:* byzant-inus, byzant-ius.

# C

cabul—<*Kabal*, capital of Afghanistan. *Ex:* cabul-icus.

cachlec-—Gr. *kachlex*, genit. *kachlekos*, pebble, gravel, stone in stream bed.

caerulescens—NL. *caerulescens, -entis*, becoming blue, <L. *caeruleus*, dark colored, dark blue.

caf—Fr. *cafe*, coffee. *Ex:* cafe-ine=caffe-ine.

calcare—NL. *calcareus, -a, -um*, limy, pertaining to lime, <L. *calcarius*, limy.

calendul—NL. *calendulus, -a, -um*, of the first of

the month, <L. *kalendae*, first day of the month.

calyculat—NL. *calyculatus, -a, -um*, provided with a calyx, <Gr. *kalyx*, genit. *kalykos*, a calyx, a cup.

calypso—L. *calypso*, <Gr. name of several female personages in Mythology, particularly the nymph who held Ulysses on her island.

calyptrae—NL. *calyptraeus, -a, -um*, hooded, hel-meted, <Gr. *kalyptēr*, a sheath.

cambr—<*Cambria*, Wales. *Ex:* cambr-icus.

campani—NL. *campanius, -a, -um,* of the field, <L. *campus,* a plain, a field.

canace—NL. *Canace,* <Gr. *Kanakē,* daughter of Aeolus.

canari—<the *Canary* Islands. *Ex:* canari-ensis.

candidul—L. *candidulus, -a, -um,* shining white <*candidus,* white.

cantabr—<the *Cantabrian* Mountains in northern Spain. *Ex:* cantabr-icus.

cantaloupe—Fr. *cantaloup*<*Cantalupo,* a castle in Italy where the melon was first grown in Europe.

canutus—NL. *canutus,* to king *Canute,* specific name of the American Knot (*Calidris canutus rufus*). See Manual Of The Vertebrate Animals, David Starr Jordan, p. 293.

capibara—Tupi, *capibara,* a grass-eater.

capitat—L. *capitatus, -a, -um,* having a head, growing in a head.

capuchin—<Fr. *capuche,* a hood.

carapace—Eng. *carapace*<Sp. *carapacho,* a gourd.

Cardon—OFr. *cardon,* a thistle, <L. *carduus,* a thistle

carduace.—NL. *carduaceus, -a, -um,* thistle like, <L. *carduus,* a thistle.

cardui—L. *cardui,* of the heart, <Gr. *kardia,* heart.

caribou—Fr. *caribou,* one who paws, a scratcher; the name now applied to the American reindeer.

carit-—L. *caritus,* lacking, devoid of, freedom.

carlina—NL. *carlina,* <L. *carduus,* a thistle.

carunculat—NL. *carunculatus, -a, -um,* like a little piece of flesh<L. *caruncula,* a small piece of flesh.

cassandra—L. *Cassandra,* <Gr. daughter of Priam, King of Troy.

cassicul-—L. *cassiculus,* small net, cobweb.

castell—<*Castile,* Spain. *Ex:* castell-anus.

cata-—*cata-,* a prefix<Gr. *kata,* meaning *down, downwards, lower, under;* also, *away,* **Ex:** cata-basi-al (*basion*<Gr. *basis,* a base); cata-basis (*basis*<Gr. *bainō,* to go); cata-bol-ic; etc. The meanings of the many other words beginning with *cata-* may be inferred from the definitions of their combined elements.

catalpa—NL. *catalpa,* an aboriginal name, or perhaps<Gr. *kutuhlpa,* a head with wings.

catamon-—Gr. *katamonos,* permanent.

catarrh—Fr. *catarrhe*<L. *catarrhus*<Gr. *katarrhos,* a running down, a flow of humors from the head. The ancients thought the phlegm which they called *pituita,* was manufactured in the brain (pituitary portion) and reached the nose through the cribiform plate.

cathartic—NL. *catharticus, -a, -um,* cleansing, purifying. <Gr. *kathartikos,* cleansing.

catherp—NL. *catherpes,* <Gr. *katherpein,* to creep down. *Ex:* Catherpes (Av.).

caunac-—Gr. *kaunakēs,* thick cloth.

cayenn—<*Cayenne,* in French Guiana. *Ex:* cayenn-ensis.

-ce—The terminus of many words derived through Fr. from L. It is seen in the suffixes *-ace, -ice,* and *-ance. Ex:* pallescen(t)-ce.

celandine—<Gr. *chelidonion,* swallowwart.

cembroides—NL. *cembroides,* like the *Cembra* pine = Swiss Stone Pine<*cembra*+-oides.

centaur—Gr. *kentauros,* a mythical monster half-man and half-horse.

centrum—L. *centrum,* a sharp point, the point around which a circle is described.

centur—L. *centuria,* a company of a hundred, >Fr. *centurion. Ex:* Centurio (Mam.).

ceod-—Gr. *keōdēs,* fragrant, like incense.

ceph—See cephen.

cephalot—NL. *cephalotus, -a, -um,* <Gr. *kephalōtos,* with a head.

-ceptor—NL. *-ceptor*<L. *ceptus* (pp. of *capio,* to take)+-or, one who or one which, lit., "one who takes back" (*re-,* back, being understood). *Ex:* ambo-ceptor; chemo-ceptor; photo-ceptor; proprio-ceptor; re-ceptor.

cerambycidae—NL. *cerambycidae,* <Gr. *kerambyx,* a kind of horned beetle, +-idae.

cercalis—NL. *cercalis, -e,* tailed, <Gr. *kerkos,* a tail.

cerchneis—Gr. *kerchneis,* the kestrel.

cerciat—NL. *cerciatus, -a, -um,* tailed, with a tail<Gr. *kerkos,* a tail.

cessator—L. *cessator,* an idler, loiterer.

cestr—<*Chester,* Pennsylvania. *Ex:* cestr-icus.

cetur-—L. *ceturus,* the other, the remainder.

chacalaca—Sp. *chacalaca,* <Nahuatl, the twittering of a bird.

chamois—Fr. *chamois,* the fallow-deer.

chaunos—Gr. *chaunos,* spongy, porous.

chelisoch—<Gr. *chela,* claw+*isos,* equal. *Ex:* Chelisoch-idae (Ins.), the final *s* of *isos,* equal, has been changed for the sake of euphony into ch.

chickadee—a name given in reference to the bird's unusual call-note.

chimpanzee—Fr. Port. *chimpanze,* <native Angola name.

chincha—Sp. *quecha,* lit. strong+little.

chirot-—Gr. *cheirōtos,* tamable, able to be tamed.

chleuastes—Gr. *chleuastēs,* a mocker, a scoffer.

chloe—See chlo in main body of text.

chnaur-—Gr. *chnauros,* dainty.

cholla—Sp. *cholla,* head. A name now applied to a cane-cactus.

chondrilla—NL. *chondrilla,* a name given by Dioscorides for some gum-excreting plant.

chord—L. *chordus, -a, -um,* produced late.

chordeiles—NL. *chordeiles,* late-ringing<Gr. *chordē,* the chord of a lyre+*deilē,* evening.

cicutari—NL. *cicutarius, -a, -um,* like *Cicuta,* a genus of Umbelliferae.

-cide—Eng. *-cide*<L. *cid,* root of *caedo,* to cut, to cut the throat with intent to kill+Eng. silent *-e. Ex:* fratri-cide; scabi-cide; sui-cide.

ciliaris—NL. *ciliaris, -e,* fringed as with eyelashes<L. *cilium,* an eye-lid.

cingulat—NL. *cingulatus, -a, -um,* girdled, encircled, zoned<L. *cingulum,* a girdle.

cinnamomin—L. *cinnamoninus, -a, -um,* of or from cinnamon.

circellat—NL. *circellatus, -a, -um,* with small ring or rings, <L. *circellus,* a small ring.

circumcis—L. *circumcisus, -a, -um,* cut off around, cut off.

cirrhat—NL. *cirrhatus,* <L. *cirratus, -a, -um,* curled, having curls.

-cision—NL. *-cision* (<*-cision,* in circumcision), a cutting. See circumcis. *Ex:* electro-cision.

claus—L. *clausus, -a, -um,* enclosed, shut.

claviculat—NL. *claviculatus, -a, -um,* provided with bars, provided with tendrils, <L *clavicula,* a key, a bar.

cleronomy—Eng. *cleronomy,* <Gr. *klōronomia,* an inheritance.

closmat- —Gr. *klosma,* genit. *klosmatos,* a spindle.

clypeat—L. *clypeatus, -a, -um,* shielded, with shields, <*clypeus,* a shield.

cmelethro—Gr. *kmelethron,* a beam.

co- —a prefix (<L. *co-* = *con-*) meaning *with, together with, in connection with, jointly. Ex:* co-ad-un-ation (*un* <L. *una,* one) = co-ad-un-ition (bringing of dissimilar material into a single mass); co-ossi-fic-ation; co-ordinat-e. See col-, com-, and con-.

cohibil- —L. *cohibilis,* shortened.

col- —Assimilated form of *con-,* meaning *together with.* See co-, com-, con-. *Ex:* col-lect (*lect-* <*lectus,* pp. of L. *lego,* to gather).

colch—L. *Colchis,* province in Asia, east of the Black Sea. *Ex:* colch-icus.

colloid—Gr. *kollōdēs,* like glue. *Ex:* colloid-al.

colubris—L. *colubris,* of a serpent, genit. of *coluber,* a serpent.

columbin—L. *columbinus, -a, -um,* pertaining to a dove<*columba,* a dove.

columen—L. *columen,* the top, crown, summit.

colytic- —Gr. *kolytikos,* checking, hindering, preventing.

com- —Prefix from L. *cum,* meaning *with, in conjunction.* This form is used before word stems beginning with *b, p, m* and sometimes *f.* See co-, col-, and con-.

comatos—L. *comatosus, -a, -um,* morbidly drowsy, <*coma,* sleep+-*osus,* full of. *Ex:* comatos-e.

commod- —L. *commodus,* fit, proper, suitable.

comos—L. *comosus, -a, -um,* with long hair, hairy; also, leafy; lit. full of hair.

compress—L. *compressus, -a, -um,* squeezed together, straight, narrow.

comyth- —Gr. *komys,* genit. *komythos,* bundle, sheath.

con- —Prefix meaning *together, with*<L. *cum,* with. See co-, col-, com-. Used before stems beginning with consonants other than *b, h, l, m, p, r,* and *w.* For the origin of most words beginning with *con-* consult the meaning of the stems following *con-.* Thus, for congestion, see the entry gest.

conar- —Gr. *konaros,* fat, well fed.

condit—L. *conditus, -a, -um,* seasoned well.

confert—L. *confertus, -a, -um,* pressed together, crowded, dense.

confoss- —L. *confossus,* pierced through, full of holes.

confract—L. *confractus, -a, -um,* uneven, broken.

confus—L. *confusus, -a, -um,* confused, perplexed.

conglomerat—L. *conglomeratus, -a, -um,* rolled together, crowded, heaped.

congregat—L. *congregatus, -a, -um,* collected.

conic—NL. *conicus, -a, -um,* cone-like<L. *conus,* a cone.

conjunctiv—L. *conjunctivus, -a, -um,* joining, connecting<*con,* with+*junctus,* joined+-*ivus,* a suffix added to verb stems to express quality or tendency; NL. *conjunctiva,* the membrane "joining" the eye-ball to the lids (an instance of the fem. form of an adjective being used as a noun). *Ex:* conjunctiv-al, conjunctiv-itis.

consanguine—L. *consanguineus, -a, -um,* related by blood.

continu—L. *continuus, -a, -um,* joining, continuous.

convalleri—NL. *convallarius, -a, -um,* pertaining to or of a valley, <L. *convallis,* a valley, a basin enclosed on all sides.

cootie—slang for body-louse, perhaps a dim. <*cute,* clever, shrewd.

cordov—<*Cordova,* Spanish province. *Ex:* cordov-ensis.

coriace—L. *coriaceus, -a, -um,* made of leather, leathery.

coriz—Gr. *koris,* a bed-bug. *Ex:* Coriz-idae (Ins.)

cornicul—L. *corniculus, -a, -um,* a small horn.

cornub—<*Cornwall,* England. *Ex:* cornub-i-ensis.

corthy- —Gr. *korthys,* genit. *korthyos,* a heap.

cott—<the *Cottian* Alps, between France and Italy. *Ex:* cott-icus.

counter- —Combining form or prefix<the Eng. adverb *counter,* meaning, when combined with nouns or adjectives, *against, in opposition to, negating, reversing;* also, *in addition to. Ex:* counter-irritant; counter-opening (a second opening); counter-poison; counter-stain (a second stain of different color).

coyote—<Nahuatl, a wild dog.

cracov—<*Cracow,* Poland. *Ex:* cracov-i-ensis.

craepn- —Gr. *kraipnos,* swift, rushing.

craget- —Gr. *kragetēs,* a screamer.

crappie—Name of obscure origin given to a sunfish.

cret—<the Island of *Crete. Ex:* cret-ensis, cret-icus, cret-is, cret-aeus.

crinit—L. *crinitus, -a, -um,* covered with hair.

criont- —Gr. *kreion,* genit. *kreiontos,* a ruler, master.

crisia—NL. *crisia,* probably a coined, meaningless name given to a genus of polyzoans. *Ex: Crisia.*

ctedono- —Gr. *ktēdōn,* genit. *ktedonos,* fiber, thread.

ctisto—Gr. *ktistos,* founded, established.

cultellus—L. *cultellus*, a little knife<*culter*, a knife.

cunctat- —L. *cunctatus*, delayed.

cupidonia—NL. *cupidonia*, <L. *cupido*, desire, wish; also lust.

cuprit—L. *cupritus*, *-a*, *-um*, desired.

curlew—OFr. *corlieu*, probably in reference to the bird's cry.

curussav—<a Latinization of *Curacao*, West India. *Ex:* crussav-icus.

cuscuta—NL. *cuscuta*, <Ar. name for a kind of twining, leafless, parasitic vine.

cuticl—L. *cuticulus*, a little skin. *Ex:* cuticle.

cyaro- —Gr. *kyar*, genit. *kyaros*, a hole, orifice.

cybelido—Gr. *kybēlis*, genit. *kybēlidos*, ax, cleaver.

cyclosis—NL. *cyclosis*, a whirling, <Gr. *kyklos*, a whirling, a circulation.

cymatil- —L. *cymatilis*, sea colored, blue.

cymos—L. *cymosus*, *-a*, *-um*, full of shoots, <Gr. *kyma*, a shoot, sprout. *Ex:* cymose.

# D

dactylis—NL. *dactylis*, <Gr. *daktylos*, finger. *Ex:* Dactylis*.

daemoni—NL. *daemonius*, *-a*, *-um*, elfine, strange, marvelous, <Gr. *daimonios*, belonging to a god; also, grand, marvelous, <*daimōn*, a ghost, an elf.

dalmat—<*Dalmatia*, Yugoslavia. *Ex:* dalmat-icus.

dan—<*Danmark*, Denmark. *Ex:* dan-icus.

dat—L. *datum*, pl. *data*, a giving, a presenting, pp. of *do*, to give.

daunus—L. *Daunus*, <Gr. *Daunos*, fabled king of part of Apulia.

deci- —Prefix<L. *decimus*, ten. *Ex:* deci-bel (-*bel*<A. G. Bell, who invented the telephone); deci-para. See deca-.

decollat—L. *decollatus*, *-a*, *-um*, beheaded.

decoloran—L. *decolorans*, *-antis*, without color, <*decoloro*, to discolor, to deprive of color.

decumar—NL. *decumaria*, pertaining to the 10th part, <L. *decuma*, the 10th part+-*aria*. *Ex:* Decumaria*.

defass- —L. *defassus*, wearied, fatigued.

delemon—Gr. *dēlēmōn*, baneful, noxious.

delicatul—NL. *delicatulus*, *-a*, *-um*, somewhat charming, delightful, <L. *delicatus*, alluring, delightful.

delicul- —L. *deliculus*, blemished, defective.

demi- —Fr. *demi*, half<L. *dimidius*, half (<Gr. *dia-*, apart+*medius*, middle). *Ex:* demi-lune (Fr. *lune*, moon<L. *luna*, moon); demi-monstrosity; demi-penni-form.

dempt- —L. *demptus*, taken away, removed.

denaeo- —Gr. *dēnaios*, long lived, old.

denotat- —L. *denotataus*, marked out, notable, conspicuous.

denticulat—L. *denticulatus*, *-a*, *-um*, having small teeth.

derobio—Gr. *dērobios*, long lived.

des- —Fr. *des*, from, away from, asunder<L. *dis*, neg. See dis. *Ex:* des-anim-(m)ania; des-amid-ize; des-hydr-em-ia; de(s)-squam-ation.

deseret—L. *deseretus*, *-a*, *-um*, deserted, solitary.

desiccant—L. *desiccans*, genit. *desiccantis*, drying up; pp. of *desicco*, to dry up.

desipiens—L. *desipiens*, silly, foolish.

desis—Gr. *desis*, a binding together.

desmod—Gr. *desmōdēs*, like a chain. *Ex:* desmod-ium.

desuet- —L. *desuetus*, discontinued, disused.

det—Gr. *detos*, bound<*deō*, to fasten, to tie one's self. *Ex:* Auto-det-us (Ann.).

deus- —Gr. *deusō*, to dye, stain.

deversor—L. *deversor*, guest, inmate.

di- —1. *di-*<Gr. *dia-*, through, as in NL. *diuresis*<Gr. *dioureō*, to pass urine. 2. *di-*, shortened form of L. *dis-*, apart, as in Eng. *di-vergent* (L. *vergo*, to bend).

dickcissel—name derived from the bird's call— "dick-dick—cissel-cissel-cissel."

diedr- —Gr. *diedros*, sitting apart. *Ex:* Diedrocephala (Ins.).

dienec—Gr. *dienekēs*, continuous, unbroken.

diesthio—Gr. *diesthiō*, to eat through, consume.

dignabil—L. *dignabilis*, *-e*, worthy.

dilucul- —L. *diluculum*, daybreak, dawn.

dimer—NL. *dimerus*, *-a*, *-um*, <Gr. *di-*, two+ *meros*, part.

dioic—NL. *dioicus*, *-a*, *-um*, <Gr. *di-*, two+ *oikos*, household, a house) in two households, dioecious.

diomede—NL. *diomedea*, <Gr. *Diomēdēs*, hero at the siege of Troy. *Ex:* Diomedea (Av.).

diphetor—Gr. *diphētōr*, a searcher.

dis- —L. *dis*, asunder, apart, in two; also used as a negative. *Ex:* dis-chron-ation (neg.); dis-harmoni-ous (neg.); dis-impact-ion (neg.); dis-integrat-ion (apart); dis-joint (apart); dis-locat-ion (apart); etc.

disiderat—NL. *disideratus*, *-a*, *-um*, <Gr. *di-*, (two) twice+L. *sideratus*, to be sunstruck.

disjunct—L. *disjunctus*, *-a*, *-um*, separated, distant, remote, disjoined.

displicat- —L. *displicatus*, spread out, scattered.

dissect—L. *dissectus*, *-a*, *-um*, cut up, cut asunder; pp. of *disseco*, to cut in pieces.

dissidens—L. *dissidens*, differing, disagreeing.

distich—L. *distichus*, *-a*, *-um*, <Gr. *distichos*, in two rows.

ditissim- —L. *ditissimus*, very rich, abundant.

diurnal—L. *diurnalis*, daily<*diurnus*, of or belonging to the days<*dies*, day.

div—L. *dives*, *-itis*, rich, splendid, precious.

dmeter—Gr. *dmētēr*, a tamer, one who tames.

domicil—L. *domicilium*, an habitation, an abode. *Ex:* domicil-e.

doming—<San *Domingo* of the Bahama Islands. *Ex:* doming-ensis.

dorea—Gr. *dōrea*, a gift, present. *Ex:* Chamae-dorea.*

dovekie—apparently<Eng. *dove*+dim. *-kie.*

dowitcher—perhaps of American Indian origin.

dumetorum—L. *dumetorum*, of thickets; genit. pl. of *dumetum*, a thicket.

dyac—NL. *dyacus, -a, -um*<Gr. *dyakis*, twice.

dyssodia—See dysodia.

# E

e- —Prefix (<L. *ex*) meaning *out, out of, from, completely, beyond, without, away from;* etc. See **ex-**. *Ex:* e-caudate (without); e-masculation (from); e-margin-ate (without); etc.

-e—Eng. silent *-e*, added to classical verb, adjective, and noun stems to form nouns denoting *state, condition, being,* or *action. Ex:* delet-e, etc.

-eal—Adjectival suffix consisting of euphonic -e-+-al, added to the stem (found in the genitive) of Greek nouns ending in *-nx.* Thus: laryng-eal<Gr. *larynx*, genit. *laryngos;* mening-eal<Gr. *meninx*, genit. *meningos;* phalangeal<Gr. *phalanx*, genit. *phalangos;* pharyngeal<Gr. *pharynx*, genit. *pharyngos.* There are a few other Gr. nouns to which *-eal* may be added. *Ex:* paraphys-eal<*paraphysis.*

ecdysis—See dysis.

-ed—AS. adjectival suffix, meaning *having characteristics of, provided with. Ex:* crook-ed; pointed-ed; wretch-ed.

edentul—L. *edentulus, -a, -um*, toothless.

eel—<AS. *oel*=Swedish *ål*, eel.

efferat- —L. *efferatus*, fierce, savage, wild.

egen—L. *egenus, -a, -um*, needy, in want of, poor, worthless.

egg—<Old Nordic, an egg.

egret—<Fr. *aigrette*, a heron.

-el—Dim. suffix<OFr. *-el, -ele*<L. *-ellus, -a, -um. Ex:* kern-el; mors-el; tunn-el.

ellogimo- —Gr. *ellogimos*, esteemed, famous.

elongat—L. *elongatus, -a, -um*, removed, kept aloof.

em- —Assimilated form of *en-* before labials (*b, m, p*) conveying the idea of *in, into, within. Ex:* em-bol-ism; em-bryo; em-physis; em-pyema.

embia—NL. *embia*, said to be from Gr. *embios*, living.

en- —Fr. *en-*, meaning *in, upon. Ex:* en-amel.

-en—Suffix *-en* (<Gr. *en*, in), meaning *in. Ex:* then-en, lit., "in the palm."

encaust—L. *encaustus, -a, -um*, burned in.

-ency—Noun suffix (<L. *-entia*) denoting *condition, state,* or *quality.* See **-ance** and **-cy.** *Ex:* emerg-ency; pot-ency.

enerter—NL. *enerterus, -a, -um*, lower, <Gr. *enerteros*, lower.

enneaphyll—L. *enneaphyllus, -a, -um*, nine-leaved.

-ent—Fr. *-ent* (<L. *-entem*), suffix used in Eng. words to form adjectives or nouns showing agency. *Ex:* absterg-ent; deterg-ent.

-eon—<Fr. *-eon*, (*-ion*), an ending added to verb stems to form nouns (<L. *-io*, genit.

*-ionis*). *Ex:* accord-ion=accord-eon; accou-lat-eon; accoustic-on, the *e* omitted.

epacmo- Gr. *epakmos*, in full bloom, mature.

epactio- —Gr. *epaktios*, on the shore.

epicteto- —Gr. *epiktētos*, acquired, gained.

epir—L. *Epirus*, part of Balkan peninsula, now Albania. *Ex:* epir-o-ticus.

-er—Suffix (<AS. *-ere*<L. *-arius*), denoting *agent, actor, one who engages in, one who has to do with. Ex:* mak-er; malinger-er; practition-er.

erigenia—NL. *erigenia*, <Gr. *ērigeneia*, born in the spring.

eritmo- —Gr. *eritimos*, precious.

-ery—ME. *erie*<L. *-eria*, formative noun suffix signifying: a: Qualities collectively, character, behavior, conduct, or the like; as in green-ery, etc.    b: Act, art, trade, occupation, as in tann-ery, surg-ery, etc.    c: Place (where some trade is practiced, or where something is kept, or grows, or is done, or gathered), as in tann-ery, win-ery, etc.    d: Collection or aggregation.

erysimum—See rhysimon.

erythrae—L. *erythraeus, -a, -um*, reddish, <Gr. *erythros*, red. *Ex:* Erythraea*.

erythre—<Erythrea, in Abyssinia. *Ex:* erythre-us.

es- —NL. *es-*, <Gr. *eis*, a prefix meaning *in, into. Ex:* es-od-ic (*-od-*<Gr. *hodos*, way); es-pno-ic.

-esis—A suffix of Gr. origin denoting *a state* or *condition* = *-iasis*, which see. *Ex:* lagn-esis, full of lust = lagn-osis.

-et—Dim. ending (<OFr. *-et, -ete*) used as a noun suffix. *Ex:* quadrupl-et.

-etic—Suffix meaning *of* or *pertaining to*<Gr. *-etikos*<*-ikos. Ex:* aig-etic; gen-etic; ul-etic.

-etorum—Genit. pl. of suffix *-etum*, denoting in Ecology a plant community. *Ex:* querc-etorum.

-ety—Eng. *-ety* (<L. *-etus, -a, -um*), an ending of certain nouns derived from L. adjectives in *-ius.* Thus, L. *varius*>*varietas* and Eng. vari-ety.

euclea—Gr. *eukleia*, glory. *Ex:* Euclea*.

eunice—NL. *Eunice*<Gr. *Eunikē*, one of the Nereids.

-eur—Fr. *-eur*, masc. suffix. *Ex:* accouch-eur; mass-eur. See **-euse.**

europhil—NL. *europhilus, -a, -um*, <Gr. *eyros*, the south wind+*philō*, I love.

-eus—L. *-eus, -a, -um*, adjective ending with sense of *belonging to. Ex:* Musculus cephalo-pharyng-eus; Musculus cerato-cric-oid-eus; lact-eus.

**-euse**—Fr. *-euse*, fem. suffix. *Ex:* accouch-euse; mass-euse. See **-eur.**

**euthenics**—Eng. *euthenics*, the science of human living, condition and betterment, <Gr. *euthēnĕo*, to thrive, to be well off.

**evax**—a classical personal name. *Ex:* Evax*.

**excelsior**—L. *excelsior*, still higher.

**exciple**—NL. *exciple = excipulum*, the outer covering of the apothecium of lichens, <L. *excipula*, receptacles.

**excult-** —L. *excultus*, adorned, polished, refined.

**excurrent**—L. *excurrens*, genit. *excurrentis*, ppr. of *excurro*, to run out <*ex*, out+*curro*, to run. *Ex:* excurrent.

**exeches**—Gr. *exechēs*, prominent.

**exhomilo-** —Gr. *exhomilos*, strange, unfamiliar.

**exon**—<*Exeter*, in Devonshire, England. *Ex:* 'exon-i-ensis.

**extens**—L. *extensus, -a, -um*, stretched out, extended, spread out, pp. of *extendo*, to stretch.

**extero-**—Prefix<L. *exterior*, outside. *Ex:* extero-ceptor; extero-fect-ive.

**extra-** —Prefix<L. *extra*, beyond, outside, in addition to, independent of. *Ex:* extra-bronchi-al; extra-bucc-al; extra-cell-ul-ar; extra-vas-ation; etc.

**extro-** —Prefix<L. *extra*, outside. *Ex:* extro-spect-ion; extro-version; extro-vert.

# F

**fabricat-** —L. *fabricatus*, made or wrought.

**factios-** —L. *factiosus*, powerful, strong.

**factiti-** —L. *factitius*, artificial, unnatural.

**fang**—AS. *feng, fang*, a tusk, a tooth.

**fascies**—L. *fascies*, face, figure, shape.

**fascinum**—L. *fascinum*, the penis.

**fasiolar**—NL. *fasiolar*<L. *fasciola*, a strip of cloth+*-ar*, pertaining to.

**fastigiat**—NL. *fastigiatus, -a, -um*, narrowing toward the top, with erect parallel branching, united in a conical bundle, <L. *fastigium*, the extreme part, that which is at the top, high.

**fat**—<AS. *foet*, fat.

**favulos**—NL. *favulosus, -a, -um*, full of small cells<L. *favus*, a honey comb.

**fer-de-lance**—Fr. *fer*<L. *ferrum*, iron+*-de-*<L. *de*, of+*lance*<L. *lancea*, a light spear.

**ferret**—MEng. *feret, ferette*, <OFr. *furet*, perhaps<L. *fur*, a thief.

**ferrugine**—L. *ferrugineus, -a, -um*, dark red, rust-colored, dusky.

**-fiant**—L. *fiant*, "let there be made," 3rd person plu. (Active Subjunctive) present tense of *facio*, to make. *Ex:* ole-fiant, lit., "let there be made oil."

**-ficans**—Suffix, *-ficans*, genit. *-ficantis*, making <*fic*, weak root of *facio*, to make. *Ex:* acidi-ficans.

**-fication**—Fr. *-fication*, a making<L. *-ficatio*, genit. *-ficationis* (in composition)<*fico*, to make. *Ex:* clari-fication.

**figwort**—AS. *figwort*, <*fic-*, name of a certain disease+*wort*, a plant being used as a remedy for *ficus*, a disease.

**filicula**—L. *filicula*, the rock fern.

**finch**—AS. *finc*, a chaffinch.

**flexuos**—L. *flexuosus, -a, -um*, full of turns or windings, tortuous, crooked.

**foetid**—L. *foetidus, -a, -um*, ill smelling, foul, stinking.

**foetidissim**—NL. *foetidissimus, -a, -um*, most fetid, foul-odored<L. *foetidus*, stinking+ *-issimus*, an intensive.

**forastic-** —L. *forasticus*, out of doors, public.

**fore-** —Prefix of AS. origin, meaning *before* (in *position* or *time*). *Ex:* fore-arm; fore-brain; fore-finger; fore-gut; fore-head; fore-kidney; fore-skin; fore-pleasure; etc.

**fortax**—L. *fortax*, carrier, bearer. <Gr. *phorax.*

**fox**—AS. *fox*, a fox.

**fracon**—<*Fraconia*, a name once given to a major part of western Europe. *Ex:* fracon-icus.

**fragros**—L. *fragrosus, -a, -um*, fragile.

**fraterculus**—L. *fraterculus*, a little brother; also, as an adj., of unknown parents. *Ex:* fratercul-a.

**freget**—NL. *fregata*, <Fr. a frigate. *Ex:* Frigata (Av.).

**frequen**—L. *frequens, -entis*, often, repeadly.

**frustulent**—L. *frustulentus, -a, -um*, filled with small pieces; dim. of *frustrum*, a bit, a piece.

**fruticos**—L. *fruticosus, -a, -um*, shrubby, bushy, full of bushes.

**fruticulos**—NL. *fruticulosus, -a, -um*, putting forth many small shoots<L. *frutico*, to put forth shoots+*ul*, a dim.+*-osus*, full of, many.

**fucos**—L. *fucosus, -a, -um*, painted, colored, beautified; also, spurious.

**fug**—NL. *fuga*, a fleeing, a taking flight, a falling off or fading early. *Ex:* fug-acious.

**fulgid**—L. *fulgidus, -a, -um*, glittering, flashing.

**fulgurans**—L. *fulgurans*, genit. *fulgurantis*, flashing, glittering, pp. of *fulguro*, to glisten, to flash.

**fume**—L. *fumeus, -a, -um*, smoky, full of smoke.

**funebral**—NL. *funebralis, -e*, pertaining to the dead, <L. *funebris*, funereal+*alis.*

**furv**—L. *furvus, -a, -um*, dark, dusky, black.

# G

galact—NL. *galactus, -a, -um,* milky, milky white, <Gr. *gala,* genit. *galaktos,* milk.

galago—NL. *galago,* <African native name for a monkey. *Ex:* Galago (Mam.).

galericulat—NL. *galericulatus, -a, -um,* hooded as with a skullcap, <L. *galerus,* a skull cap.

gambus—NL. *gambus,* <American Sp. *gambusino,* nothing. *Ex:* Gambus-ia, a genus of fish effective in destroying mosquito larvae.

gamphel- —Gr. *gamphēlai,* the jaws of animals.

gange—<*Ganges* River in India. *Ex:* gange-ticus.

gangrenos—NL. *gangrenosus, -a, -um,* full of eating sores<Gr. *gangraina,* an eating, festering ulcer+L. *-osus,* an intensive meaning full of.

gape—Scandinavian *gape,* to yawn.

gargan—<*Gargano,* the ancient *Garganus* in Apulia, Italy. *Ex:* gargan-icus.

garzetta—It. *garzetta,* <Sp. *garza,* an egret. *Ex:* Garzetta (Av.).

gastrula—NL. *gastrula,* <Gr. *gastros,* stomach+ L. dim. *-ula.*

gazell—OFr. *gazella,* <Ar. *ghazēl,* <Pers. *ghazāl,* a gazelle. *Ex:* Gazella (Mam.).

gegono- —Gr. *gegōnos,* loud-sounding, sonorous.

gelasinat—NL. *gelasinatus, -a, -um,* with dimples<L. *gelasinus,* a dimple.

generos—L. *generosus, -a, -um,* of noble birth, eminent, superior, excellent.

gentil—L. *gentilis, -e,* belonging to the same stock.

georg—<*Georgia* or *Grusia* in Transcaucasia, Russia. *Ex:* georg-icus.

gerae—Gr. *geraios,* old. *Ex:* Geraea*.

geraro- —Gr. *geraros,* dignified, honored.

gigante—L. *giganteus, -a, -um,* belonging to the giants, very large.

gill—<Dan. *gjalle,* gill, jaw, lip.

glagero- —Gr. *glageros,* full of milk.

glaucin—NL. *glaucinus, -a, -um,* blue-gray, <Gr. *glaukos,* silvery, gray.

glischr—NL. *glischrus, -a, -um,* <Gr. *glischros,* sticky, clammy.

gnu—< *Kafir* name for this animal.

gob—<*Gobi,* a desert in Eastern Asia. *Ex:* gob-icus.

goby—<L. *gobio, gobius,* a gudgeon<Gr. *kōbios.*

godwit—<early modern Eng., possibly<AS. *gōd,* good+*wiht,* creature, in reference to the eating quality of the flesh.

gogo—<*Gogo* or *Ghaogo* in the western Sudan. *Ex:* gogo-ensis.

gondol—NL. *gondolus, -a, -um,* shaped like a boat, <It. *gondola,* a kind of boat.

gopher—Fr. *gaufre,* a honey-comb. *Ex:* gopher-us, because of its burrowing habits.

goshawk—AS. *gōshafoc,* goose-hawk.

gracilent—NL. *gracilentus, -a, -um,* very slender, <L. *gracilis,* slender, thin.

grackle—<L. *graculus,* a jackdaw, after its note, "gra gra."

graec—L. *Graecia,* Greece. *Ex:* graec-us.

gramp—A corruption of It. *gran pes,* great fish, or Sp. *grand pez*<L. *grandis piscis,* a great fish. *Ex:* Gramp-us (Mam.).

granat—<*Granada,* in southern Spain. *Ex:* gran-at-ensis.

grand—L. *grandis, -e,* large, great, full, abundant.

grebe—Fr.<Ger. *grebe,* a grebe, name perhaps given in reference to the crested species; perhaps<Welsh *cribyn,* a crest, the tuft of feathers on a bird's head.

grisescen—ML. *grisescens,* genit. *grisescentis,* tending toward or becoming gray.

grosbeak—Fr. *grosbec,* <*gros,* large, stout+*bec,* beak.

grouse—Origin uncertain, perhaps<ML. *gruta,* a gallinaceous bird, or from the singular form of OFr. *grice,* gray. Compare *mouse* and *mice* and *grouse* and *grice.*

guian—<*Guiana,* in South America. *Ex:* guian-ensis.

guillemot <Fr.<Eng. *gull.*

guine—<the coast of *Guinea,* West Africa. *Ex:* guine-ensis.

gullinule—L. *gallina,* hen+dim. *-ule.*

gynase—Gr. *gynē,* female+*-ase,* a ferment.

gyrfalcon—<an old name for a greedy vulture.

# H

haereto- —Gr. *hairetos,* chosen, desirable, taken.

hafni—L. *Hafnia,* Copenhagen =*Köbenhavn. Ex:* hafni-ensis, havni-ensis.

hake—AS. *hake,* a hook.

haliaeet—NL. *haliaeetus,* <Gr. *haliaetos,* name of some sea bird, perhaps the sea eagle<*hals,* the sea+*aetos,* an eagle. *Ex:* Haliaeetus (Av.).

halibut—ME. *hali,* holy+*but,* flounder.

haliphthoro- —Gr. *haliphthoros,* a pirate.

hamamel—NL. *hamamelis,* <Gr. *hamamēlis,* a kind of medlar tree<*hama,* at the same time+ *mēlon,* an apple. *Ex:* Hamamelis*.

hare—<AS. *hara,* a hare.

harlequin—OFr. *harlequin,* a demon, Satan.

harmosto- —Gr. *harmostos,* adapted, suitable, fit.

hass, hess—< *Hesse* in Germany. *Ex:* hassi-acus, hess-ian.

helicoid—NL. coiled like a snail shell, <Gr. *helikē,* a coil, a winding+NL. *-oid,* like.

hercyn—L. *Hercynia silvia,* Hercynian forest in ancient Germany. *Ex:* hercyn-icus, hercyn-ianus.

heron—OFr. *heron,* a heron.

herpestic—L. *herpesticus, -a, -um*, spreading, eating (as of a sore) <Gr. *herpestikos*, spreading.
heterosis—Gr. *heteros*, other+NL. *-osis*, full of.
hiantul—NL. *hiantulus*, with a small opening <L. *hiantus*, an opening+-*ulus*, a diminutive.
histri—L. *Istria*, an area on the eastern shore of the Adriatic Sea. *Ex:* histri-cus, histri-onicus.
hoof—<AS. *hof*, a hoof.
horn—AS. *horn*, a horn.
humect- —L. *humectus*, moist, damp.

hybridus—NL. *hybridus, -a, -um*, <L. *hybrida*, a mongrel.
hylister—Gr. *hylister*, a collander, filter, strainer.
hymeni—NL. *hymenium*, a membrane, <Gr. *hymen*, a membrane. *Ex:* hymenium.
hypogae—NL. *hypogaeus, -a, -um*, underground, <Gr. *hypogaios*, beneath the earth, *ge*.
hyrax—Gr. *hyrax*, genit. *hyrakos*, a shrew mouse. *Ex:* Hyrac-odon (Mam.); Hyrac-oid-ea (Mam.); hyrac-o-theri-um; hyrax.

# I

-iac—NL. *-iac* <Gr. *-iakos*, affected with, suffering from. *Ex:* man-iac.
-ialis—LL. *-ialis*, a suffix of adjectives. *Ex:* LL. *partialis* <L. *pars*, genit. *partis*, a part, a share.
-ian—Eng. *-ian*, adjectival and noun suffix (<*i*, +-*an*. See -an), denoting *one who works in, one who works as. Ex:* clinic-ian; physic-ian; laborator-ian.
-iasis—A suffix <Gr. *-iasis*, denoting (1) a course of action, as in odontiasis, the process of having teeth <*odontos*, a tooth. (2) a diseased condition, as in amoebiasis or (3) a state resulting from the action of the verb, as in psoriasis <*psoriao*, to have the mange or itch; pthir-iasis; etc. *-iasis* may occasionally be used instead of *-iosus* or *-osis;* thus, amebiasis =amebiosus, trichiniasis=trichinosis.
iberi—<L. *Hiberia*, the country of Iberia, near the Caucasus (now Georgia). *Ex:* iberi-cus.
-ible—Adjectival suffix<L. *-ibilis, -abilis*, capable of, susceptible of, fit to. *Ex:* im-misc-ible; in-compat-ible; prehens-ible.
-ic—Suffix (<L. *-icus*, prob. Gr. *-ikos*) signifying *of* or *pertaining to, of the nature of, after the manner of, resembling, having, consisting of, causing. Ex:* a-clin-ic (having); aquat-ic (pertaining to) cata-lyt-ic (causing); cephal-ic (pertaining to); tox-ic (causing or pertaining to).
-ical—1. LL. -icalis (<*-icus*+-*alis*, an adj. ending). Adj. suffix of nouns ending in *-ic;* thus, clinical<*clinic;* medical<*medic;* physical< *physic.* 2. *-ical*, suffix meaning *pertaining to, relieving,* etc. *Ex:* din-ical, relieving dizziness.
-icle—1. Dim. suffix of noun stems<L. *-iculus, -a, -um. Ex:* part-icle<L. *particula*, dim. of *pars*, genit. *partis*, a part; article<L. *articulus*, dim. of *artus*, a joint. 2. Suffix of verb stem conveying the idea of *a place for, a thing that. Ex:* cub-icle<L. *cubiculum*, a bedchamber<*cubo*, to lie down; veh-icle<L. *vehicula*<*veho*, to carry; L. *perpendiculum*, a plummet<*per-pendo*, to hand down.
-ics—Eng. pl. suffix (in imitation of L. *-ica* and Gr. *-ika*), meaning, *the science or art of. Ex:* biot-ics; evect-ics; mathemat-ics; statist-ics. Such names are now regarded as singular nouns although their endings are properly plural.
ida—L. *Ida*, a high mountain in Crete. *Ex:* ida-eus.

idalimo- —Gr. *eidalimos*, comely, shapely.
idano- —Gr. *idanos*, fair, comely.
idechth- —Gr. *eidechthes*, ugly, of repulsive appearance.
-ify—Eng. *-ify*, a suffix meaning *making* (<Latin *i*, from+L. or NL. stem of *facio*, to make, to do+-y)=-*fy. Ex:* clar-ify; magn-ify; oss-ify.
-igenus—Suffix (<*-i*, combining letter+*gen*, to originate+-*us, -a, -um*, L. adjectival endings) meaning, *coming from* or *growing in. Ex:* alp-igenus, of or from the alps.
-igo—Old L. suffix denoting *disease* or *malfunction. Ex:* impet-igo; vert-igo<L. *vertigo*, a turning.
-ile—*-ile*, adjectival suffix<Fr. *-il, -ile*<L.*-ilis*, signifying *of* or *pertaining to;* also, *capable of* or *suited to. Ex:* juven-ile; mob-ile; pen-ile; pont-ile; prehens-ile; sept-ile.
illabefact- —L. *illabefactus*, unbroken, unimpaired.
illecebr—NL. *illecebrum*, <L. *illecebra*, an allurement, a charm.
imant—Gr. *imas*, genit. *imantos*, a strap, a thong. *Ex:* Imant-odes (Rept.)
immutabil—L. *immutabilis, -e*, changed, altered.
-imus—L. suffix meaning *most. Ex:* Bul-imus (Moll.).
in- —1. Prefix denoting *intensification of an action. Ex:* in-ebriet-y; in-spissat-ed; in-toxication. 2. Prefix<L. *in-*, inwards. *Ex:* in-cyclo-phor-ia; in-cyclo-troph-ia. 3. *in-*, a prefix meaning *again*, as in in-ost-osis, the re-formation of bone to replace bone tissue previously destroyed. 4. a combining form denoting fibrin or fibrous tissue (Gr. *is*, gen. *inos*, fiber, muscle, nerve). *Ex:* in-axon; in-em-ia (presence of fibrin in the blood); in-itis (inflammation of a muscle) ino-gen-ous. See also ino.
-in—NL. *-in*, suffix used to form names of organic substances. *Ex:* chromat-in; fibr-in; melan-in; prote-in.
incarnat—NL. *incarnatus, -a, -um*, made flesh, made red, reddish, <L. *incarnatus*, made flesh, pp. of *incarno*, to make flesh.
incas—L. *incasus, -a, -um*, to meet unexpectedly, pp. of *incido*, to meet, to fall upon.
incis—L. *incisus, -a, -um*, cut into, pp. of *incido* to cut into.

increpit—L. *increpitus, -a, -um,* making a noise, rattling, rebuking, pp. of *increpo,* to make a noise, to reprove.

incubus—LL. *incubus,* nightmare<*incubo,* genit. *incubonis,* an incubus, one that lies upon<*in-,* in, upon+*cubo,* to lie. In ML., *incubus* refers to a mythical being of the male sex who supposedly clandestinely consorted with women while asleep, deformed children being the off-spring. Female demons who consorted with men while asleep were called succubi (<*suc*= *sub,* under+*cubo,* to lie).

incurv—L. *incurvus, -a, -um,* bent, curved.

indecoris—L. *indecoris,* unbecoming, inglorious.

-ine— -*ine,* Suffix used in Chemistry in the formation of names of (1) certain elements and compounds, such as brom*ine,* fluor*ine;* (2) of organic bases, esp., of an alkaloid nature; (3) of compounds of a basic nature.

infest—L. *infestus, -a, -um,* unsafe, infested, troublesome.

ingenit—L. *ingenitus, -a, -um,* innate, instilled by birth, pp. of *ingigno,* to implant, to engender.

iniqu— —L. *iniquus,* uneven, unjust, hurtful.

inodor—L. *inodorus, -a, -um,* without smell.

insubid- —L. *insubidus,* stupid, foolish.

insuls—L. *insulsus, -a, -um,* unsalted, insipid, tasteless.

intermedi—L. *intermedius, -a, -um,* intermediated.

interrupt—L. *interruptus, -a, -um,* broken, parted, interrupted.

interstice—See interstit.

interstit—L. *interstitio,* genit. *interstitionis,* a pause, a space between, an interval>Fr. *interstice,* an opening, a space. *Ex:* interstice; interstiti-al.

intine—<L. *intus,* within.

intro- —Prefix denoting *in* or *into*<L. *intro,* to the inside, into. *Ex:* Intro-flex-ion; intro-gastr-Ic; Intro-Itus (-*itus*<*iter,* a way), intro-miss-ion; intro-spect-ion; intro-vers-ion; intro-vert; intro-vision.

-inus—L. *-inus,* diminutive. See **-imus.** *Ex:* Bulinus (Moll.).=Bulimus (Moll.).

inusitat- —L. *inusitatus,* rare, unusual.

invict—L. *invictus, -a, -um,* unconquered, invincible.

involut—L. *involutus, -a, -um,* intricate, obscure, involved.

-ious—An adj. suffix<thematic -*i*+*ous.* Found in such words as infect-ious, praecoc-ious; various.

ir- —1. L. *ir-* (<*in-,* in, into), prefix before stems beginning with *r. Ex:* ir-radi-ation; ir-rigat-ion.    2. L. *ir-* (<*in-,* not), prefix before stems beginning with *r. Ex:* ir-reduc-ible; ir-regular; ir-re-suscit-able.

iridoprocne—NL. *iridoprocne,* <Gr. *iris,* genit. *iridis,* a rainbow+*Proknē,* who was changed into a swallow.

-is—Genitive singular ending of Third Declension L. nouns but sometimes erroneously used as a genitive ending of nouns of other declensions in forming NL. names, as in the terms tort*i*-coll-is and pyr*i*-form-is. L. *forma* is a First Declension noun and the ending should be *-ae.*

-ite—Suffix denoting: (a) a salt or ester formed from an acid with name ending in *-ous. Ex:* nitr-ite<nitr-ous. (b) a derived sugar. *Ex:* abiet-ite<L. *abies,* genit. *abietis,* the fir tree.

-iticus—NL. *-iticus,* suffix<Gr. *-itikos,* meaning *pertaining to, of the nature of. Ex:* anam-iticus, pertaining to Anam, in India.

-itious—Adj. suffix<L. *-icius,* meaning *of the nature of, characterized by, affected by. Ex:* advent-itious; excrement-itious; icter-itious, affected by jaundice.

ium— -*ium,* genit. pl. ending of L. Third Declension nouns. *Ex:* compressor nar-ium, lit., "compressor of the nostrils."

iva—NL. *iva*<ML. so named by Linnaeus because of its odor being similar to that of *Ajuga Iva,* the ground pine.

-ive—Eng. *-ive,* suffix (<L. *-ivus, -a, -um*) added to verb stems to form verbal adjectives denoting *quality, tendency to, fitness, pertaining to. Ex:* abort-ive; act-ive; audit-ive; convuls-ive; dis-junct-ive; irritat-ive; recess-ive; sedat-ive; tard-ive; etc.

-ize—Verb-forming suffix (<Gr. *-izo*), added to noun and adj. stems to form compound verbs. Such verbs are often accompanied by nouns of action ending in *-ation:* thus, *civilize*>*civilization.* The suffix *-ize* (*-ise*) generally denotes *the doing of a thing,* expressed or implied, in the noun or adj. base to which it is added. Added to proper nouns it may denote the doing of a thing in the way the person did it. Thus *Pasteur*>*pasteur-ize. -izo* is said to be the only ending of a Gr. verb which, in modified form (*-ize*), comes over into English.

# J

jacana— <Tupi Indian.

jaguar—Tupi-Guarani *jaguara,* name for tigers and dogs.

jalapa—NL. *jalapa,* <Sp.<Nahuatl, sand+ water.

jerboa—NL. *jerboa,* <Barbary, the jerboa, also the flesh of the back and loins.

juda—<*Judea,* Palestine. *Ex:* juda-icus.

# K

kahir—<*Cairo*, Egypt. *Ex:* kahir-icus, kahir-inus.

kalahari—<the desert of *Kalahari*, in South Africa. *Ex:* kalahari-ensis.

keto- —Prefix<NL. *ketone* (<NL. *acetone*), a chemical compound containing the carboxyl group CO. *Ex:* keto-lytic; keto-em-ia; keto-gen-ic; keto-plast-ic; etc.

kew—<the Botanical Gardens at *Kew*. *Ex:* kew-ensis.

kidney—ME. *kidney*, origin uncertain, kidney.

kiwi—<Maori, imitation of the cry of this bird.

kob—NL. *kob*, <African native name, *koba*, for an antelope. *Ex:* Kob-us (Mam.).

korai—< *Korea*. *Ex:* korai-anus, korai-ensis.

kurd—< *Kurdistan*, western Asia. *Ex:* kurd-icus.

# L

laciniat—L. *laciniatus*, *-a*, *-um*, with jagged edges.

lagenari—NL. *lagenarius*, *-a*, *-um*, of or pertaining to a bottle or flask<L. *lagena*, a flask+ -*arius*. See -arium.

laph—See elaph.

lapide—L. *lapideus*, *-a*, *-um*, of stone, stony <*lapis*, a stone.

laro- —Gr. *laros*, agreeable, pleasant.

laut- —L. *lautus*, washed, neat, elegant.

lechri- —Gr. *lechrios*, slanting, crosswise.

lechuguilla—Sp. dim. of *lechuga*, lettuce.

lecyth- —L. *lecythus*, flask, bottle, vase, jug.

legat—L. *legatus*, *-a*, *-um*, appointed, chosen.

lemph- — Gr. *lemphos*, mucus from the nose.

leon—1. <Sierra *Leone*. *Ex:* leon-ensis. 2. L. *leo*, genit. *leonis*, lion. *Ex:* leon-ine.

leporin—L. *leporinus*, *-a*, *-um*, of a hare.

-let—Dim. suffix<Fr. -*el* and -*et*. *Ex:* plate-let; ring-let.

leucelen—Etym. unknown. *Ex:* Leucelene*.

leuro- —Gr. *leuros*, smooth, even.

lever—<L. *levator*, a lifter.

libethr- —Gr. *libēthron*, a meadow, field.

libit- —L. *libitus*, pleasing, agreeable.

lich—See leich.

licno- —Gr. *liknon*, a basket or cradle used in winnowing.

lict- —L. *lictus*, abandoned, forsaken.

liger—L. *Liger*, Loire River in France. *Ex:* liger-icus.

limbal—NL. *limbalis*, -*e*, bordered, with a hem, <L. *limbus*, an edge.

limero- —Gr. *limēros*, hungry.

limpet—<AS. *lempedu*, a lamprey.

-ling—1. Suffix<AS. (-*l*+-*ing*) denoting origin, but often used as a diminutive, as in duckling, starling, sanderling, nestling, gosling. 2. Suffix of adjectives<AS. -*ling*, as in darkling (beetle).

linurgo- —Gr. *linourgos*, a weaver.

liparo- —Gr. *liparos*, greasy, sleek, shiny.

lipaug- —Gr. *lipaugēs*, dark, sunless, shady.

-lith—Gr. *lithos*, a stone>-*lith*, a suffix, denoting the mineral nature of a thing. See also -ite. *Ex:* gastro-lith; nephro-lith.

litigator—L. *litigator*, a disputant.

liver—<AS. *lifer*, liver.

locell—L. *locellus*, a small compartment. *Ex:* locell-ate.

longaev- —L. *longaevus*, aged, ancient, old.

longinqu—L. *longinquus*, *-a*, *-um*, long, of length.

lorice—NL. *loriceus*, *-a*, *-um*, clothed in armor <L. *lorica*, a corslet made of thong.

louse—<AS. *lūs*, perhaps, lit. a destroyer, a louse.

luctat- —L. *luctator*, a wrestler.

luculent—L. *luculentus*, *-a*, *-um*, clear, bright, splendid, <*lux*, light+-*lentus*, full of.

luin—Anagram of *Inula*, a plant name. *Ex:* Luina*.

lund—Icelandic, *lund*, the puffin. *Ex:* Lunda (Av.).

lung—<AS. *lungen*, lung.

luziol—NL. *luziola*, a name modified from *Luzula*, a genus of Juncaceae. *Ex:* Luziola*.

lygod—See lyg.

lytta—NL. *lytta*, < Gr. *lytta*, frenzy, rage, canine madness.

# M

macellarius—L. *macellarius*, of or belonging to a meat-seller.

macraeo- —Gr. *makraiōn*, long lasting.

madaro- —Gr. *madaros*, bald, bare.

mader—<the island of *Madeira*. *Ex:* mader-ensis.

magdalen—<the *Magdalen* River, South America. *Ex:* magdalen-icus.

malai—<The *Malay* Archipelago. *Ex:* malai-anus.

maler- —Gr. *maleros*, fierce, fiery, mighty.

mandalo- —Gr. *mandalos*, a bolt, bar.

mandrill—said to be from a native West African name, <*man*, early+*drill*, baboon.

mansuet- —L. *mansuetus*, tame, gentle.

marmoset—ME. *marmeset*, a beast>*marmoset*, a little ape or monkey.

martes—L. *martes*, a marten. *Ex:* Martes (Mam.).

maur—L. *Mauri*, northwest Africa. *Ex:* maurorum.

mauritan—L. *Mauritania*, northwestern Africa; *Mauritanicus*, of or belonging to *Mauritania*. *Ex:* mauritan-icus.

means—L. *means*, going, passing; sometimes used in sense of quick-moving <*meo*, to go.

meleagrid—L. *meleagris*, genit. *meleagridis*, of the guinea-fowl.

meligeryo- —Gr. *meligērys*, sweet voiced.

memnoni- —L. *memnonius*, black, dark. <Memnōn, fabled king of Egyptians.

-ment— *-ment* (<Fr. *-ment* <L. *-mentum*). Suffix added to verbs, often of French origin, to form nouns. *Ex:* derange-ment; labio-placement; liga-ment. See **-mentum**.

mentit- —L. *mentitus*, counterfeit, feigned.

-mentum—L. suffix added to verb stems or roots to denote *acts*, *results* or *means of acts*. Thus: L. *regimentum*, rule <*regi*, stem of *rego*, to direct; *ligamentum*, a band, a ligature <*ligo*, to bind.

merac- —L. *meracus*, pure, unmixed.

merinth- —Gr. *mērinthos*, a cord, line, string.

merops—Gr. *merops*, a bee-eater.

mescal—Sp. *mescal*, a drink distilled from agave (*mescal*).

mesembri- —Gr. *mesembria*, midday, noon.

mesic— <Gr. *mesos*, middle+*-ikos*, pertaining to.

messan— <*Messina*, in Sicily. *Ex:* messan-ensis, messan-ius.

miarch—NL. *miarchus*, <Gr. *myia*, a fly, + *archos*, a ruler. *Ex:* Mi-archus (Av.).

mic—NL. *mic*, small, <Gr. *mickros*, small. *Ex:* Mic-urus (Rept.).

micid- —L. *micidus*, thin, poor.

micron—NL. *micron*, a measure <Gr. *mikros*, small.

mid- —As. *mid*, a preposition meaning *together with*. *Ex:* mid-wife.

milli- —A prefix denoting 1/1000 <L. *mille*, a thousand. *Ex:* milli-gram; milli-meter; milli-normal.

mirific—L. *mirificus, -a, -um*, wonderful, strange.

mirounga—NL. *mirounga*, from an Australian native name.

molann—Said to come from L. *mola*, a millstone. *Ex:* Molann-idae (Ins.).

mongol—Mongolian *Mongol*, said to be from *mong*, brave. *Ex:* mongol-ian; mongol-oid.

mono- —Combining form or prefix (<Gr. *monos*, single). It denotes *one* or *a single thing, confined or limited to one part or thing. Ex:* monocephal-us; mono-choic; mono-clinic; monocyesis; mono-cyt-e; mono-dipl-op-ia; monogenesis; mono-infection; mono-mel-ic; monomoria; mono-neur-ic; mono-phob-ic; monopleg-ic; etc.

morrhua—L. *morrhua*, cod.

morul- —L. *morulus*, dark, dark-colored, black.

mouflon—Fr. *mouflon*, the Corsican wild sheep.

mulatto—Sp. *mulatto*, <*mule*, an animal of mixed breed.

muntiac—NL. *muntiacus*, <Malay, *muntjac*, name of some mammal. *Ex:* Muntiacus (Mam.).

murorum—L. *murus*, genit. pl. *murorum*, wall.

murre—a bird name, origin obscure.

musang—Malay, *musang*, name for a viverroid mammal.

musimon—LL. *musimon*, the mouflon, <L. *musimo*, a Sardinian animal supposed to be the mouflon.

must- —L. *mustus*, new, young, fresh.

myrrhis—NL. *myrrhis*, a plant name, <L. *myrrha*, the balsamic juice of the Arabian myrtle. *Ex:* Myrrhis.*

mytilin—NL. *mytilinus, -a, um*, shell-like in form, <Gr. *mytilos*, a sea mussel.

# N

naevi—NL. *naevius, -a, -um*, spotted with moles, with blemishes, <L. *naevus*, a mole, a wart.

napell—NL. *napellus, -a, -um*, a little turnip, <L. *napus*, a turnip+*ellus*, a dim.

nauc—NL. *naucinus, -a, -um*, a trifle, <L. *naucum*, a trifle, something lightly esteemed +*-inus*.

naucler—L. *nauclerus*, a ship-owner, <Gr. *nauklēros*, a ship-owner.

naval—L. *navalis, -e*, belonging to ships.

neanderthal—Ger. *Neanderthal*, the Valley of Neander, in the Rhine Province. *Ex:* neanderthal-ensis.

nechaleo- —Gr. *nēchaleos*, swimming.

nefrens—L. *nefrens*, toothless, that cannot bite.

nemoros—L. *nemorosus, -a, -um*, full of foliage, bushy; also, woody, shady.

nemorum—L. *nemus*, pl. genit. *nemorum*, of woods, of a grove.

neomexican— <*New Mexico. Ex:* neomexican-us.

nepet—L. *nepeta*, catmint, catnip. *Ex:* Nepeta.*

nerithmo- —Gr. *nērithmos*, countless, immense.

-ness—Suffix (<AS. *-ness*) used to form abstract nouns denoting *quality* or *state of being. Ex:* amativ-ness; dim(n)-ness; ill-ness; sick-ness; etc.

netro- —Gr. *nētron*, a spindle.

neyraudi—Anagram of *Reynaudia. Ex:* Ney raudia*.

nicobar— <the islands of Nicobar, in the Indian Ocean. *Ex:* nicobar-icus.

nigritell—NL. *nigritellus, -a, -um*, dark, nearly black, <L. *nigritia*, blackness+*ellus*, a dimutive.

nipon— <*Nipon*, a name for Japan. *Ex:* nipon-icus.

nival—L. *nivalis, -e*, snowy, belonging to snow, <*nix*, genit. *nivis*, snow.

nochele- —Gr. *nŏchelēs*, moving slowly, sluggish.

non- —Prefix<L. *non*, not. *Ex:* non-con-ductor; non-par-ous; non-resistant; non-sept-ate; non-vi-able.

notidano- —Gr. *nŏtidanos*, with pointed dorsal fin.

nubeculat—NL. *nubeculatus, -a, -um*, cloudy, with dark spots, <L. *nubecula*, something cloudy+*-atus.*

nuthatch—ME. *nuthake*, nut+to hack.

# O

obstip- —L. *obstipus*, inclined to one side, oblique.

ocior- —L. *ocior*, swifter.

ocule—L. *oculeus, -a, -um*, full of eyes, <L. *oculus*, eye.

odax—Gr. *odax*, by biting with the teeth.

odocoileus—See odont.

okape—<an African native name for some giraffe-like animal.

-ol—Chemical suffix denoting that the substance is an alcohol or phenol. *Ex:* chole-ster-ol; xyl-ol.

oleic—<L. *oleum*, oil+*-ic.*

olethreut—<Gr. *olethros*, destruction; the *-eut-* is of unknown origin. *Ex:* Olethreut-idea*.

-on—1. Nom. sing. ending of certain 2nd (o-) declension Gr. neut. nouns such as *axon* and *neuron;* also the neut. ending of certain Gr. adjectives and participles; thus, *argon* is neut. of *argos*, inactive, idle.  2. Arbitrarily meaningless ending of certain NL. scientific names such as arch-nephron. See also -eon.

-one—Suffix<Gr. *-ōnē*, signifying a female offspring. It is often used in chemistry in the sense of something derived from, a descendent. *Ex:* acet-one; andro-ster-one; keyt-one.

oneros- —L. *onerosus*, burdensome, heavy.

onocrotal- —L. *onocrotalus*, a pelican.

opthalm—Gr. *opthalmos*, the eye. *Ex:* opthalm-ic; an-opthalm-ia.

orbicular—L. *orbicularis, -e*, circular, in the shape of an orb.

orchesti—NL. *orchestia*, <Gr. *orcheisthai*, to leap. *Ex:* Orchestia (Arth.).

-orial—Adj. suffix corresponding to nouns in *-or* and *-ory. Ex:* equat-orial; purgat-orial; suct-orial.

oriental—L. *orientalis, -e*, belonging to *oriens*, the East. *Ex:* oriental.

original—L. *originalis, -e*, primitive, original. *Ex:* original.

oriund—L. *oriundus, -a, -um*, descended, sprung from.

ornatul—L. *ornatulus, -a, -um*, fine, smart.

-ory—*-ory*<L. *-orius, -a, -um*, adj. suffix meaning *pertaining to, serving for, marked by. Ex:* audit-ory; claudicat-ory; secret-ory.

orymagdo- —Gr. *orymagdos*, a loud noise, din, roar.

orythmo- —Gr. *ōrythmos*, a howling.

-ose—*1.* Suffix (<L. *-osus*) forming adjectives meaning *full of, of the nature of. Ex:* bulb-ose; comat-ose; glob-ose; glomerul-ose; siliqu-ose. *2.* Suffix<Fr. *-ose*, indicating that the substance is (a) *a carbohydrate* (fruct-ose; galat-ose), or (b) *a protein product* (prote-ose).

osmyl—Gr. *osmylē*, a strong smelling sea-polypus. *Ex:* Osmyl-idae (Ins.).

osprey—<OFr. *osprey*<L. *ossifragus*, a bone-breaker.

-otic—*1.* suffix<Gr. *otikos* forming adjectives from nouns ending in *-osis;* as, cyanotic<*cyanosis;* osmotic<*osmosis;* narcotic<*narcosis;* thalp-otic<*thalposis.*  *2.* Combining form <the adj. *otic*, meaning, *pertaining to the ear Ex:* peri-otic.

otrero- —Gr. *otrēros*, nimble, quick, busy.

ouzel—<ME, *osel*, <AS. *osel*, an ouzel.

owl—<AS. *ūle;* probably originally based on an imitation of bird's strange cry.

oxyop—Gr. *oxyopēs*, sharp-eyed. *Ex:* Oxyopes (Arach.); Oxyop-idae (Arach.).

oxyri—NL. *oxyria*, <Gr. *oxys*, sharp. *Ex:* Oxyria*.

oxyrop—Gr. *oxyrropos*, turning quickly>NL. *oxyropus. Ex:* Oxyropus (Rept.).

oxyten—Gr. *oxytenēs*, pointed. *Ex:* Oxyten-ia*.

# P

paepalimo- —Gr. *paepalimo*, sly, subtle.

palabund- —L. *palabundus*, wandering about.

palitans—L. *palitans*, wandering about.

palyn—Gr. *palynō*, to scatter, to strew. *Ex:* paly no-logy, the science of pollen and spore dispersal; melito-palynology.

panda—Nepalese name of this animal.

paniculat—NL. *paniculatus, -a, -um*, having pannicles or tufts of flowers, dim. of *panus*, an ear of millet.

panorp—NL. *panorpus*, <Gr. *pan*, all+*harpē*, sickle. *Ex:* Panor-p-idae (Ins.).

para- —Prefix<Gr. prep. *para*, used in sense of amiss, incorrect, apart from, aside, abnormal, faulty, a going aside, contrary to. *Ex:* para-bleps-ia; para-bul-ia; par-acanth-oma; para-cephal-us; para-chol-ia; para-chromat-ism; para-cines-ia; para-crisis; para-cyesis; para-eccrisis; par-enter-al; para-function-al; para-geus-ia; para-glossa; para-hypn-osis; para-kerat-osis; par-alges-ia; para-men-ia; para-mnes-ia; para-morph-ia; par-onom-ia; para-pedesis; para-phora; para-phren-ia; para-pycno-morph-ous; par-arthr-ia; para-sigmat-ism; para-steat-osis; para-thym-ia; para-trichosis; para-tripsis; para-vitamin-osis; para-encephal-ia; par-encephal-us.

parabilis—L. *parabilis*, easily procured.

parakeet—<Sp. Port. *periquito*, dim. of Sp. (not Port.) *perico*, a parrot.

parit- —L. *paritus*, present, visible.

paroro- —Gr. *parōros*, untimely, unseasonable.

patasso- —Gr. *patassō*, to beat, throb.

pecan—<Algonkian, *pakin*, nut.

pecari—Sp. *pecari*, *peccary*, a word of Cariban origin. *Ex:* Pecari (Mam.).

pedano- —Gr. *pedanos*, short.

pejor—L. *pejor*, worse.

pelecin—Gr. *pelekinos*, a water-bird of the pelican kind. *Ex:* Pelecin-idae (Ins.).

peli—NL. *pelius*, -a, -um, <Gr. *pelios*, black, livid.

pelican—LL. *pelicanus*, *pelecanus*, a pelican <Gr. *pelekinos*, a pelican.

penelope—Gr. *Pēnelopē*, wife of Ulysses.

penn—L. *pennus*, -a, -um, pointed, sharp, <*pinna*, a wing.

perpingu—L. *perpinguis*, -e, very rich.

petrel—ML. *petrel*, "Little Peter."

petunia—NL. *potunia*, <Fr. *petun*, tobacco, <Port. *petume*, <Tupi Indian *petuma*.

phanes—See phaner.

phoebe—named after one of the phoebe's call notes.

phrenitic—NL. *phreniticus*, -a, -um, <Gr. *phrenitikos*, mad. delirious.

pigeon—ME. *pigeon*, *pijon*, <L. *pipio(n)*, a young piping or chirping bird <*pipo*, to chirp.

pika—<Tunguse, *piika*, a small rodent.

pipit—named after bird's note.

pipixican—<an Aztec name.

piscin—NL. *piscinus*, <L. *piscis*, fish. *Ex:* piscin-e.

pisinn- —L. *pisinnus*, small, little.

pletur- —L. *pletura*, fullness.

plover—OFr. *plover*, the rain-bird.

podeon—Gr. *podeōn*, genit. *podeōnos*, the neck of a wine-skin bottle <*pous*, foot.

pollosto- —Gr. *pollostos*, smallest, least.

portland—<the isle of *Portland*, English Channel. *Ex:* portland-icus.

prae- —Prefix meaning *very*. *Ex:* prae-longus, very long; prae-potent.

praecoc—See praecox.

praecox—L. *precox*, genit. *praecocis*, before time, immature. *Ex:* praecoci-al.

praev—L. *praevius*, -a, -um, gone before, in front of, leading the way. *Ex:* praevia.

prepes—L. *prepes*, flying swiftly, fleet, swift.

primiti—L. *primitiae*, first fruits of their kind. *Ex:* Primiti-a (Crust.).

primiveris—NL. *primiveris*, first of the spring, <L. *primus*, first+*ver*, genit. *veris*, spring.

pro- —Prefix signifying *before* or *forward*, *in front of*, *premature*. It may even mean *in favor of*, as in *pro-gest-in* (favoring gestation) and *pro-lact-in* (favoring lactation). Most often combined with words of Greek or Latin origin. *Ex:* pro-al; pro-amnion; pro-chondr-al; pro-ger-ia; pro-gnosis; pro-labium; pro-na-us.

prolat- —L. *prolatus*, extended, elongated.

prolifer—NL. *proliferus*, -a, -um, bearing progeny, reproducing freely <L. *pro*, before+*fero*, to bear.

prosator—L. *prosator*, an ancestor.

pterygi—NL. *pterygius*, -a, -um, winged, with wing-like spot, <Gr. *pterygion*, a little wing. *Ex:* pterygi-um.

pulcherrim—L. *pulcherrimus*, -a, -um, most beautiful, superlative of *pulcher*, beautiful.

pullulat—L. *pullulatus*, -a, -um, <*pullulo*, to sprout, to bud, to germinate. *Ex:* pullulat-ion.

pulsatill—NL. *pulsatillus*, -a, -um, somewhat beaten (by the wind) <L. *pulsatus*, beaten +*-illus*, a diminutive.

pupil—L. *pupilla*, dim. of *pupa*, a doll, an image of a little girl. Later this word was transferred to the pupil or central spot of the eye in which small images (*pupilla*) may be seen. *Ex:* pupillary; pupil-a-ton-ia.

pupill—See pupil.

putori—NL. *putorius*, -a, -um, with foul odor, <L. *putor*, genit. *putoris*, a foul smell, rottenness.

# Q

quadrangul—L. *quadrangulus*, -a, -um, four-cornered.

quadrupl—L. *quadruplus*, fourfold. *Ex:* quadruplet

quail—ML. *quaquila* <MDutch *quakele*, in reference to bird's note.

quarantine—ML. *quarantena*, a period of forty days, or a measure of forty rods.

quartile—ML. *quartilus*, <L. *quartus*, fourth.

quintupl—ML. *quintuplus*, fivefold, <L. *quintus*, fifth, +*-plus*.

# R

rail—<Ger. *ralle*, so called from bird's note.

ratio—L. *ratio*, a reckoning, an account, a relation, a reference.

raven—<AS. *hraefn*, in reference to bird's coarse note.

recidiv- —L. *recidivus*, recurring, returning.

redimicul—L. *redimiculum*, a head band, a girdle, a necklace, <*redimio*, to encircle, to crown.

redpole—See redpoll.

redpoll—a name given to a bird because of the red color on the head (MDutch *pol*, the head).

reservoir—Fr. *reservoir*, a tank, a cistern, <L. *reservo*, to keep back, to save up.

retrus—L. *retrusus, -a, -um,* distant, hidden, pp. of *retrudo,* to thrust back.

rhaet—L. *Rhaeti,* the Rhaedian Alps. *Ex:* rhaeticus, rhaet-us.

rhamnus—NL. *rhamnus,* <Gr. *rhamnos,* a thorn or prickly shrub, the buck thorn.

rigidul—NL. *rigidulus, -a, -um,* dim. of L. *rigidus,* rigid, stiff, hard.

rimos—L. *rimosus, -a, -um,* full of cracks or fissures, <*rima,* a cleft, crack.

robiginos—L. *robiginosus, -a, -um,* rusty.

rops—Gr. *rōps,* a bush. *Ex:* Chamae-rops*.

ros-—L. *rosus,* gnawed.

ross—<*Russia. Ex:* ross-icus.

rostral—NL. *rostralis, -e,* of or concerning a beak or snout, <L. *rostrum,* a beak.

ruginos—L. *ruginosus, -a, -um,* wrinkled, <*ruga,* a facial crease, a wrinkle.

ruid—L. *ruidus, -a, -um,* rough.

-ry—Suffix<*-ery*<OFr. *-erie* (<L. *arius+ie* <L. *-ia*), signifying: (1) *an art or occupation,* as in dentist-ry; (2) *place where,* as in the word baptist-ry.

# S

sanderling—<AS. *sand,* sand, gravel+*-er-,* of uncertain meaning+*-ling,* a diminutive.

sanguinal—L. *sanguinalis, -e,* bloody, of blood, bloodthirsty.

sanguine—L. *sanguineus, -a, -um,* bloody, of blood, blood-red.

saxos—L. *saxosus, -a, -um,* stony, full of rocks, <*saxum,* a rock.

scallop—<OFr. *escalope,* a shell.

scaup—said to be from Eng. *scalp,* the top of the head.

scelion—NL. *scelion,* an insect name said to be derived from Gr. *skelos,* a leg, rib or side, in reference to the keeled sides of the abdomen. *Ex:* Scelion-idae (Ins.).

scorzonella—NL. *scorzonella,* like *Scorzonera,* a genus of European herbs.

scoter—Etym. doubtful, possibly from Icel. *skoti,* shoot, in reference to its dives or darts.

screat-—L. *screatus,* a hawking, coughing.

scurvy—ME. *scurvy, scurfy,* scurvy.

secalin—NL. *secalinus, -a, -um,* resembling rye, <L. *secale,* rye.

sectatrix—L. *sectatrix,* a female follower.

sediment—NL. *sedimentarius, -a, -um,* of the nature of settlings<L. *sedimentum,* a settling, a sediment+*-arius.*

segetum—NL. *segetum,* of cornfields, <L. *seges,* genit. *segetis,* a cornfield.

seleucides—NL. *Seleucides,* <Gr. a descendent of Seleucus, a general of Alexander the Great.

sempervirens—NL. *sempervirens,* evergreen, <L. *semper,* always, ever+*virens,* becoming green.

sensual-—L. *sensus,* feeling sense. >L.L. *sensualis,* endowed with feeling: *Ex:* con-sensual; sensuality.

setace—NL. *setaceus, -a, -um,* bristly, with bristles, <L. *seta,* a stiff, thick hair+*-aceus.*

simultaneous—ML. *simultaneous, -a, -um,* at the same time, extended, <L. *simul,* together, at the same time.

sinew—<ME. <AS. a cord, a tendon.

siskin—<LGer. *sieske,* probably<Slovenian, *chizhek.*

skua—Faeroese name of a bird.

skull—Eng. *skull,* <Swedish dialect *skulle,* <Swedish *skollt.*

sloth—<AS. *slāw,* slow.

smyrne—Gr. *smyrnion*<*smyrna,* myrrh>NL. *Smyrnium,* a genus of Umbelliferae, so named by Dioscorides because of its myrrh-like odor.

snake—Eng. *snake,* <AS. *snaca,* a serpent.

snipe—ME.<Icel. *snīpa,* a snipe, probably from some word meaning to clip or to snap.

sonor—<the river or state of *Sonora* in Mexico. *Ex:* sonor-ae; sonori-ensis.

sotol—Sp.<Nahuatl name.

spadonius—L. *spadonius,* barren, seedless, <*spado,* an unfruitful tree, an unfruitful person.

spelae—NL. *spelaeus, -a, -um,* pertaining to or of caves, <L. *spelaeum,* a cavern.

speltus—NL. *speltus,* <LL. *spelta,* a kind of wheat.

sphacel—NL. *sphacelus,* dark, withered looking<Gr. *sphakelos,* gangrene.

spider—<AS. *spither,* <*spinther,* <*spinnan,* to spin+*ther,* an agent.

spinos—L. *spinosus, -a, -um,* full of spines or thorns.

spinosior—L. *spinosior,* more spiny. See **-ior.**

spizella—NL. *spizella,* dim.<Gr. *spiza,* a finch.

splenetic—LL. *spleneticus,* with spleen<L. *splen,* spleen.

sponge—Eng. *sponge,* <L. *spongia,* a sponge.

sponsal—L. *sponsalis, -e,* of or belonging to betrothal, <*sponsus,* an engagement.

stagnin—L. *stagninus, -a, -um,* growing in swampy places, <*stagnum,* a swamp, a pool.

starling—<AS. *staer,* to stare+*-ling,* a dim. See **sanderling.**

stilt—in reference to the bird's long stilt-like legs.

stipular—NL. *stipularis, -e,* having stipules, pertaining to stipules, <L. *stipula,* a blade, stem +*-aris.*

stolid—L. *stolidus, -a, -um,* unmovable, dull, stupid.

stomach—L. *stomachus,* the throat, gullet, the stomach.

stupulos—L. *stupulosus, -a, -um,* covered with fine, short hairs.

stygn—Gr. *stygnos,* a pillar, a post, a mast of a ship. *Ex:* Bune-stygnus (Arach.).

suberect—L. *suberectus, -a, -um,* raised up, pp. of *suberigo,* to raise up.

**subid-** —L. *subidus*, knowing, sensible.

**sublest-** —L. *sublestus*, slight, trifling, trivial.

**submers**—NL. *submersus, -a, -um*, submerged, plunged under water, <L. *sub-* under+*mersus*, immersed, pp. of *mergo*, to dip in.

**sumac**—<Ar. *summaq*, sumac.

**surculace**—L. *surculaceus, -a, -um*, woody, like wood, <*surculus*, a spout, a shoot+*-aceus*.

**surculat**—L. *surculatus, -a, -um*, trimmed, pruned, pp. of *surculo*, to trim.

**sustentacul**—L. *sustentaculum*, a prop, a support<*sustento*, to hold up<*sus* (for *sub*, under)+*teneo*, to hold. *Ex:* sustentacul-ar; sustentacul-um.

**syconium**—NL. *syconium*, a fleshy collective fruit as such the fig<Gr. *sykon*, a fig.

**syletes**—Gr. *sylētōr*, a robber.

**sylvestris**—See silv.

**systyl**—NL. *systylus, -a, -um*, with styles together, <Gr. *syn-*, together+*stylos*, a style, a pillar.

# T

**-t**—*1.* Suffix used to make nouns from verbs. *Ex:* fligh-t; weigh-t.     2. Suffix sometimes added to verb stems to form past participles (pp.) and participial adjectives. *Ex:* buil(d)t; burn-t.

**tanager**—<NL. *tanagra*, <Tupi Indian *tangara*, a bird name.

**tectorum**—L. *tectorum*, of roofs, genit. pl. of *tectum*, a roof, a cover.

**tenuissim**—NL. *tenuissimus, -a, -um*, most weak or thin.

**teretiuscul**—NL. *teretiusculus, -a, -um*, most smooth, well-rounded, cylindrical, <L. *teres*, genit. *teretis*, smooth | NL. dim., *-usculus*.

**tern**—<Danish *terne*, a name for this bird.

**-tery**—Suffix<Gr. *-tērion*, place of, means of. *Ex:* ceme-tery (*ceme-*<Gr. *koimaō*, to put to sleep); cau tery (*can-*<Gr. *kaiō*, to burn). The Greeks had a word, *kautērion*, for the branding iron.

**teter**—L. *teter, -tra, -trum*, offensive, foul, loathsome.

**tetragonum**—L. *tetrogonum*, a quadrangle<Gr. *tetragonon*.

**texas**—<the state of *Texas*, genit. *texanus*, Texas. *Ex:* texan-us, of Texas

**-th**—*1.* Suffix forming nouns of *action* from verbs, and nouns of *state* or *quality* from adjectives. *Ex:* bir-th (<the root *bher*, to bear); grow-th (<AS. *grōwan*, to grow); heal-th (<AS. *hāēlon*, to make whole<*hal*, whole). *2.* Suffix forming ordinal numbers. *Ex:* six-th; ten-th.

**thalasse**—NL. *thalasseus, -a, -um*, of the sea, <Gr. *thalassa*, the sea.

**thalissin**—L. *thalissinus, -a, -um*, sea-green, <Gr. *thalassinos*

**themist**—Gr. *themistos*, lawful, sanctioned by law. *Ex:* Themisto-cles-ia*.

**thrasher**—Perhaps a variant of *thrusher, thrushel*, a kind of thrush, or<*thresher*, <AS. *thrysce*.

**thuya**—NL. *thuya*, <Gr. *thyia*, an African tree with sweet smelling wood.

**thygater**—Gr. *thygatēr*, a daughter .

**tiburon**—Sp. *tiburon*, a shark.

**tingitan**—L. *Tingitanus, -a, -um*, pertaining to Tangiers, <*Tingi*, Tangiers. *Ex:* tingitan-us.

**tonitr-** —L. *tonitrus*, thunder.

**tonsil**—L. *tonsilla*, a tonsil. *Ex:* tonsil-itis.

**tornatil-** L. *tornatilis, -e*, finished, rounded, beautifully wrought<*torno*, to turn in a lathe.

**tost-** —L. *tostus*, burned, parched.

**towhee**—said to be named from bird's note.

**toxicum**—NL. *toxicum*, poison, <Gr. *toxikos*, <*toxon*, a bow. *Ex:* Vince-toxicum*.

**-trics**—NL. *-trics*<L. *-trix*, a fem. ending. *Ex:* ob-ste-trics (*-ste-*<L. *sto*, to stand).

**trifid**—L. *trifidus, -a, -um*, cut into three parts.

**tripartit**—L. *tripartitus, -a, -um*, divided into three parts.

**triplasis**—L. *triplusis*, <Gr. *triplasios*, triple.

**tripsacum**—NL. *tripsacum*, word of unknown origin, but according to some<Gr. *tripsis*, a rubbing, friction, alluding to the smooth joints.

**trisulc**—L. *trisulcus, -a, -um*, three pointed, triple.

**trivial**—L. *trivialis, -e*, common, ordinary, found everywhere.

**tunga**—Port. (Brazilian) *tunga*, a chigger. *Ex:* Tunga (Ins.).

# U

**umbellat**—NL. *umbellatus, -a, -um*, umbelled, with umbels, <L. *umbra*, shade.

**umbrin**—NL, *umbrinus, -a, -um*, darkened, shady, <L. *umbra*, shade+*-inus*.

**un-** —Inseparable prefix meaning *not*=AS. *an-*, <L. *in-*, <Gr. *an-*. Joined to a verb, it implies a reversal of the meaning of the simple verb. *Ex:* un-balance; un-rest; un-sex; un-strip-ed.

**undos**—L. *undosus, -a, -um*, full of waves<*unda*, a wave+*-osus*.

**ungava**—New Quebec territory of Quebec Province (Canada), east of Hudson Bay.

**urban**—L. *urbanus, -a, -um*, belonging to the city, refined, elegant, nice, <*urbs*, genit. *urbis*, a city.

**-ure**—Eng. *-ure* (<L. *-ura*), suffix used in the formation of abstract nouns. It means *act of, agent or result of the act of*. *Ex:* dent-ure; enclosure; expos-ure; fract-ure; ligat-ure; post-ure; seiz-ure; unct-ure.

# V

**vaginal**—NL. *vaginalis, -e*, sheathed <L. *vagina*, a sheath+*-alis*.

**vaginat**—NL. *vaginatus, -a, -um*, sheathed <L. *vagina*, a sheath+*-atus*.

**variegat**—L. *variegatus, -a, -um*, of various colors, various, manifold, pp. of *variego*, to make various sorts of colors.

**veery**—probably named after one of its notes.

**vellerosus**—NL. *vellerosus*, woolly, fleecy, <L. *vellus*, a fleece+*-osus*, full of.

**ver**—L. *ver*. genit. *veris*, spring. *Ex:* primi-veris, lit. "first of spring."

**verdin**—<Fr. the yellowhammer, probably from *verd*, green; applied to *Auriparus flaviceps* with yellowish head.

**vern**—L. *vernus, -a, -um*, of or belonging to spring, <*ver*. spring.

**versatil**—L. *versatilis, -e*, able to be turned around <*verso*, to whirl about.

**versicolor**—L. *versicolor*, color changing, of various colors.

**vertagus**—L. *vertagus*, a grey-hound.

**vestiari**—L. *vestiarius, -a, -um*, belonging to clothes.

**vesticeps**—L. *vesticeps*, genit. *vesticipis*, bearded.

**vilfa**—ML. *vilfa*, <an old name for some grass. *Ex:* Vilfa*.

**vireo**—L. *vireo*, a kind of bird, a greenfinch, prob. <*vireo*, to be green.

**viscari**—NL. *viscarius, -a, -um*, <L. *viscum* bird-lime, slimy+*-arius*.

**visnaga**—See **bisnaga**.

**vison**—NL. *vison*, a mink, particularly the American mink.

**vituliferus**—NL. *vituliferus*, <L. *vitulus*, a bull calf+*fero*, to bear+adj. ending *-us*. Sometimes used in sense of bearing fruits, resembling the scrotal sac of a calf.

**volubl**—<L. *volubilis, -e*, turning, able to climb <*volvo*, to turn.

# W

**warbler**—<Middle High German *werbelen*, to whirl, to be busy about.

**-ward**—Eng. *-ward*, a suffix denoting *direction toward which* <AS. *-weard*, as in *tóweard*=Eng. toward; cf. Old High German *-wert, -wart* with similar meaning. This suffix may be added to adverbs or prepositions. *Ex:* back-ward; forward; to-ward; up-ward.

**willet**—named after bird's note.

**wolg**—<the river *Wolga* or *Volga*, in Russia. *Ex:* wolg-aricus; wolg-ensis.

# Y

**-y**—1. Many Greek nouns which end in *-ia* appear in English with the *-ia* changed to *-y*. Thus, Gr. *arteria*=Eng. artery. 2. *-y*, a substitute for the Latin ending *-ius*. Thus we have *armator-y* <L. *armator-ius*. 3. *-y*, noun suffix denoting *result of an action*. *Ex:* perjur-y; surger-y.

# Z

**zambes**—<the *Zambesi* River, South Africa. *Ex:* zambes-inus.

**zuni**—<the *Zuni* Mountains or *Zuni* Indians in New Mexico. *Ex:* zuni-ensis.

**zygote**—NL. *zygote*, <Gr. *zygōtos*, yoked.

# Brief Biographies of Men in whose Honor
# Commemorative Names Have Been Given

## A

**Abert,** James William (1820–1897). Lt., later, Major U. S. Army. The type specimen of A-bert's Towhee (*Pipilo aberti*) was collected in New Mexico. The name of the Tassel-eared Squirrel (*Sciurus aberti*) recognizes Abert's field work in the southwestern U. S.

**Adanson,** Michael (1727–1806). Eminent and arduous French naturalist after whom the plant genus *Adansonia* was named by Linnaeus. *Adansonia digitata* is the boabab tree of Africa, and *Adansonia gregorii*, the cream-of-tartar tree of Australia. Adanson did much to open the way for the final establishment of the natural method of classification of plants. His explorations in Senegal and collections and descriptions of its plants and animals were noteworthy.

**Adlum,** Major John (1759–1836). In his honor the eccentric Rafinesque named the plant genus *Adlumia*. Adlum published the first book on grapes in America (1823 and 1836).

**Agassiz,** Louis (1807–1873). Swiss zoologist and student of Cuvier. Founder of the Museum of Comparative Zoology at Harvard University. Famous as the teacher of many of America's foremost zoologists. The desert tortoise (*Gopherus agassizii*) honors his name.

**Albizzia.** An Italian family who introduced the silk-tree, now known as *Albizzia*, into Italian gardens.

**Alexander,** Miss Annie Montague (1867–1950). Founder of the Museum of Vertebrate Zoology of the University of California at Berkeley. Long active patron of Zoology, traveller, for many years in field work in Alaska, California and Nevada. A subspecies of a Ptarmigan (*Lagopus leucurus alexandrae*) and a form of the white-breasted Nuthatch (*Sitta canadensis alexandrae*) was named in her honor.

**Alexandre,** Dr., of Mexico. Says Dr. T. S. Palmer: "Practically nothing is known of Dr. Alexandre except that he discovered the Black-chinned Hummingbird, now known as *Archilochus alexandri*, which was described and dedicated to him by Bourcier and Mulsant in 1846."

**Allen,** Charles Andrew (1841–1930). Collector of Marin County, California. Henry W. Henshaw named the hummingbird, *Selasphorus alleni* in his honor.

**Amman,** Paul (1634–1691). German physician-botanist after whom the plant genus *Ammania* was named.

**Amson,** Charles. 18th century Virginia Colonial physician and traveler after whom the plant genus *Amsonia* was named by Walter.

**Anna.** Wife of Prince Victor Massena. In honor of the Duchess the first described hummingbird from California was named *Calypte anna*; the type specimen is in the Philadelphia Academy of Sciences.

**Anthony,** Alfred Webster (1865–1939). Known for his ornithological field work in southern California and Lower California. Birds named in his honor include Anthony's Green Heron (*Butorides virescens anthonyi*), Anthony Vireo, and Anthony Towhee.

**Audubon,** John James (1785–1851). American ornithologist celebrated for his drawings of birds. Author *Birds of America*, sold by subscription at $1000 per copy. Name perpetuated in Audubon's Warbler (*Dendroica auduboni*) and Audubon's Caracara (*Polyborus cheriway auduboni*).

## B

**Babes,** Victor (1854–1926). Roumanian bacteriologist after whom the protozoan genus *Babesia* was named. He worked out a successful anti-rabic serum.

**Bachman,** Dr. John (1790–1874). Close friend of Audubon. Name perpetuated in *Quadrupeds of N. A.* by Audubon and Bachman and the Black Oystercatcher (*Haematopus bachmani*).

**Baer,** Karl Ernst von (1792–1876). Russian zoologist; "Father of modern embryology." He established the germ-layer theory. For him the plant genus *Baeria* was named.

**Bailey,** Vernon (1864–1942). For many years chief field naturalist of the U. S. Biological Survey; author of many publications on birds and mammals, among them *Mammals and Life Zones of Oregon*, *Mammals of New Mexico*, *Wild Animals of Glacier National Park* (birds by Mrs. Bailey, Florence Merriam, author of *Hand Book of Birds of the Western United States*. In her honor Dr. Joseph Grinnell named Bailey's Mountain Chickadee, *Penthestes gambeli baileyae*).

**Baird,** Spencer Fullerton (1823–1887). Founder of the U. S. National Museum and organizer of the zoological work of the Pacific Railroad Surveys. Says T. S. Palmer: "Baird did more than any other man of his time to advance the study of ornithology and zoology." Among birds commemorating his name we have Baird's Sandpiper (*Pisobia bairdi*), Baird's Cormorant (*Phalacrocorax pelagicus resplen-*

*dens*), Baird's Wren (*Thryothorus bewicki bairdi*), and Baird's Woodpecker (*Balanosphyra formicivora bairdi*).

**Banks,** Sir Joseph (1743–1820). English naturalist after whom the molluscan genus *Banksia* was named. He was a member of Captain Cook's first Pacific expedition and president of the Royal Society from 1778 until 1820. To him is given credit for introducing botanical lectures into Oxford University.

**Barbour,** Thomas (1884–1946). Curator of Reptiles and Amphibians of the Museum of Comparative Zoology, Harvard University. *Graptemys barbouri*, *Micrurus fulvius barbouri* and *Sistrurus miliarius barbouri* are reptile names recognizing his contributions of the science of herpetology.

**Barton,** Benjamin Smith (1766–1815). American naturalist in whose honor the plant genus *Bartonia* was named by Sims. He was one of the earliest professors of botany at the College of Philadelphia, now the University of Pennsylvania.

**Bartram,** John (1699–1777). American botanist in whose honor the moss genus *Bartramia* was named. Bartram established the first botanical garden in America; probably the first American to perform successful experiments in plant hybridization.

**Bartsch,** Johann (?–1738). German physician after whom the plant genus *Bartsia* was named.

**Batchelder,** Charles Foster (1856–1954). Ornithologist of Cambridge, Massachusetts and one of the founders of the American Ornithologist's Union. A Downy Woodpecker from New Mexico is called Batchelder's Woodpecker.

**Beckman,** Johann (1739–1811). Founder of scientific technology; professor of philosophy at Goettingen. After him the plant genus *Beckmannia* was named.

**Beechey,** Frederick William (1776–1856). Captain, and later, Rear Admiral of the British Navy. In command of the ship "H.M.S. Blossom" he visited the west coast of Mexico and California in 1828. The Mexican Crested Jay (*Cissolopha beecheyi*) and the California Ground Squirrel (*Citellus beecheyi*) commemorate his name.

**Begon,** Michel (1638–1710). French patron of science, once governor of Canada. The well known plant genus *Begonia* honors his name.

**Belding,** Lyman (1829–1917). "Nestor of California ornithologists." Author: "Land Birds of the Pacific District" (1890). Worked on reptiles of Baja California. *Rallus beldingi, Goethlypis beldingi, Passerculus beldingi*, are names given to do him honor. He also did field work in Baja California where he collected Belding's Rail and Belding's Yellow Throat. The Marsh Sparrow (*Passerculus beldingi*) was named by Dr. Ridgway.

**Bell,** John Graham (1812–1889) of New York who went as taxidermist with Audubon on his Missouri River journey in 1843. He visited California in 1849. Bell's Vireo (*Virio belli*) was discovered and described by Audubon. John Cassin named the Bell Sparrow (*Amphispiza belli*) in his honor.

**Bendire,** Charles Emil (1836–1897). Author *Life Histories of North American Birds.* Ardent and conscientious collector. While Major, U. S. Army, collected in Arizona and Oregon. The California Screech Owl (*Otus asio bendirei*) and Bendire's Thrasher (*Toxostoma bendirei*) are among interesting birds bearing his name.

**Berlandier,** Jean Louis, a Belgian who for some years was in the service of the Mexican government. He was probably the first botanist to visit San Antonio, Texas: there he collected about 1827. A genus of plants *Berlandiera* was named for him.

**Berry,** Dr. Samuel Stillman (1887–    ). Enthusiastic and able student of mollusks, particularly western land snails and mollusca of the Pacific. World authority on Cephalopoda and Chitons of western North America and Pacific waters. Author of many species.

**Bewick,** Thomas (1753–1828). Famous English wood engraver and bird artist, naturalist and friend of Audubon. Among famous books he illustrated were: A General History of Quadrupeds (1790) and The History of British Birds. Audubon honored him in the name Bewick's Wren (*Thyromanes bewicki*).

**Bichat,** Marie Francois Xavier (1771–1802). French anatomist and physiologist. Founder of histology and pathogical anatomy.

**Bigelow,** Dr. J. M. (1804–1878). American naturalist connected with various Pacific Coast Railroad Surveys. Collected in the Western U. S.; for him not only *Opuntia bigelovii* but many other California plants were named.

**Bignon,** Abbe (1662–1743). Librarian of Louis XV, after whom the plant genus *Bignonia* was named by Linnaeus.

**Blainville,** Henri Marie Ducrotay (1777–1850). French naturalist. In 1825 admitted as member to the Academy of Sciences. He succeeded J. B. Lamarck in the chair of natural history at the Museum. On the death of Cuvier he was appointed to the chair of comparative anatomy, a position he filled for 18 years. Worked and wrote on the classification of reptiles and amphibians. *Phyrnosoma blainvilli* was named for him.

**Blanchard,** Dr. Frank Nelson (1888–1937). Zoologist at the University of Michigan, Ann Arbor, specializing in life histories and distribution of amphibians and snakes. Especially known for his taxonomic studies of North American snakes. Author of: Revision of the King Snakes of the Genus Lampropeltis. Several snakes, *Sonora semiannulata blanchardi, Arizona elegans blanchardi, Diadophis regalis blanchardi, Ophiodrys vernalis blanchardi*, and a frog, *Acris gryllus blanchardi* carry his name.

**Bloomer,** Hiram G. (1821–1874). Early California botanist in whose honor the plant genus *Bloomeria* was named by Dr. John Allen Veatch.

**Boehmer,** G. R. (1723–1803). German botanist after whom the plant genus *Boehmeria* was named by Jacquin.

**Borrich,** Olaf. Danish botanist after whom the plant genus *Borrichia* was named.

**Botta,** Dr. Paolo Emilio (1802–1870). Naturalist, archeological explorer and traveller. Travelled and collected for the Museum of Natural History of Paris. One of the very first naturalists to collect in California (1827–1828). He collected the type of the Roadrunner and of Anna's Hummingbird in California. A Pocket Gopher (*Thomomys bottae*) and the Rubber Snake (*Charina bottae*) first described by Blainville in 1835 as *Tortrix bottae*, honor his name.

**Bougainville,** De (1729–1811). French navigator. The gorgeous South American nyctaginaceous genus *Bougainvillia* was named in his honor.

**Boutelou,** Claudio (1774–1842) and his brother, Esteban, were Spanish gardeners. Claudio was Director of the Botanic Garden in Madrid (1776–1839). A genus of grasses, *Bouteloua* was named in their honor by Lagasca.

**Brandegee,** Townsend S. (1843–1925). Pioneer plant explorer of Lower California. Describer of many western plant species. The curcubit genus *Brandegea* commemorates his name.

**Brandt,** Johann Friedrick (1802–1879). German zoologist and at one time director of the Zoological Museum in St. Petersburg. He described a number of birds and mammals from Western America. The name, Brandt's Cormorant is familiar to all bird students.

**Brewer,** Dr. Thomas Mayo (1814–1880). Author of *North American Oology* (1857) and joint author with S. F. Baird, T. M. Brewer, and Robert Ridgway of *A History of North American Birds* (3 vols. devoted to the *Land Birds*). His memory is perpetuated in Brewer's Blackbird (*Quiscalus breweri*) and Brewer's Sparrow (*Spizella breweri*).

**Brewster,** William (1851–1919). One of the founders of the Nuttall Ornithological Club and the American Ornithologist's Union. His name is perpetuated in the name of the California Yellow Warbler (*Dendroica aestiva brewsteri* and a Ganet (*Sula brewsteri*) from the Gulf Coast of Lower California.

**Brickell,** Dr. John. Early botanist of Savannah, Georgia, and correspondent of Muhlenberg; in his honor the plant genus *Brickellia* was named.

**Brodie,** J. J. (1744–1824). Scotch botanist after whom the plant genus *Brodiaea* was named by Smith.

**Buchner,** J. G. (1695–1749). German botanist after whom the plant genus *Buchnera* was named.

**Buffon,** Georges Comte de (1707–1788). French biologist and writer. First to discard concept of special creation. His *Historie Naturelle* (a

sort of general encyclopedia of the sciences) is said to be the first work to present the previously isolated and apparently disconnected facts of natural history in a popular and generally "intelligible form."

**Bullock,** William (1775–?). English traveller and mine owner. While visiting a mine in Mexico he collected a number of birds described by Swainson. Birds named after him include Bullock's Oriole (*Icterus bullocki*) and Bullock's Jay (*Pica bullocki*), a Mexican species.

**Burman,** Johannes (1707–1779). Botanist of Amsterdam after whom the plant genus *Burmannia* was named.

**Burser,** Joachim (1583–1649). German botanist after whom the plant genus *Bursera* was named by Linnaeus.

## C

**Cabanis,** Jean Louis (1816–1906). Eminent German ornithologist. Curator, zoological collections, University of Berlin. The name Cabanis's Woodpecker (*Dryobates villosus hyloscopus*) perpetuates his memory; also the scientific name of the Texas Kingfisher (*Ceryle americana cabanisi*).

**Caesalpinus,** Andreas (1519–1603). Italian botanist after whom the plant genus *Caesalpinia* was named by Carl Linnaeus who highly regarded his writing on plant arrangement.

**Camelli,** George Josef (1661–1706). A Jesuit who is said to have brought the plant now named *Camellia*, from the East.

**Candolle,** Augustin Pyramus de (1778–1841). Swiss botanist and principal founder and early exponent of the natural system of botany. His famous "Prodromus" which he never completed, was finished by his son Alphonse Louis Pierre Pyramus de (1806–1893), also a botanist of note

**Cassin,** John (1813–1869). Curator of birds, Philadelphia Academy of Sciences. He described 193 species of birds. Common names of birds such as Cassin's Auklet, Cassin's Kingbird, Cassin's Purple Finch, and Cassin's Short-eared Owl are familiar to all bird students

**Castilljo,** Juan (1744–1793). Spanish botanist after whom the plant genus *Castilleia* was named by Cervantes.

**Catesby,** Mark (1679?–1749). English naturalist who made rich collections of plants in Virginia and reported on the natural history of Carolina, Florida, and the Bahama Islands. The Bullfrog (*Rana catesbiana*) and the Yellow-billed Tropic Bird (*Phaëthon lepturus catesbyi*) carry his name.

**Clark,** Captain William (1770–1838). Co-leader with Captain Meriweather Lewis of the Lewis and Clark Expedition (1804–06). Clark's Crow, now called Clark's Nutcracker, perpetuates his name in ornithology.

**Clayton,** John (1693?–1773). Among earliest American botanists. A contributor to *Flora*

*Virginica.* The plant genus *Claytonia* perpetuates his name.

**Clinton,** De Witt (1769–1828). Promotor of the Erie Canal. In his honor the plant genus *Clintonia* was named by Rafinesque.

**Cockerell,** Theodore Dru Alison (1866–1948). American naturalist and biologist working in many fields from Mullusca to Insects, Botany and problems of evolution and plant and animal distribution. Described many new species. Among his latest of thousands of journal contributions were those on African bees. He travelled widely, especially after his retirement as long-time head of the Dept. of Zoology, University of Colorado.

**Commelin,** Kaspar (1629–1692). Dutch botanist after whom Linnaeus named the plant genus *Commelina.*

**Condal,** Antonio. Spanish physician. Native of Barcelona who travelled with the scientific expedition sent out by the Spanish government in 1851 under direction of Peter Loefling, the Swedish botanist, to explore its possessions in South America. The Rhamnaceous genus *Condalia,* perpetuates his name. See **Loefling.**

**Cooper,** Dr. James Graham (1830–1902). Army surgeon; author of *Ornithology of California.* The Cooper Ornithological Club was named in his honor. The following birds bear his name: Cooper's Sparrow (*Melospiza melodia cooperi*), Cooper's Tanager (*Piranga rubra cooperi*).

**Cooper,** William (1798?–1864). Father of J. G. Cooper. Active member and one of the founders of the New York Lyceum of Natural History. For him Cooper's Hawk (*Accipiter cooperi*) was named.

**Cope,** Edward D. (1840–1897). American paleontologist; student of the comparative study of living and extinct vertebrates. The herpetological journal *Copeia* was named in his memory. He was one of the early editors of the *American Naturalist* and made known to science many extinct vertebrates.

**Costa,** Louis Marie Pantaleon, Marquis de Beau-Regard (1806–1864). French collector of hummingbirds. In recognition of his interest in these, Bourcier named Costa's Hummingbird (*Calypte costae*).

**Couch,** Lt. Darius Nash (1822–1897). He named and described Scott's Oriole. He probably served with Maj. General Winfield Scott in the Mexican War. Couch's Kingbird (*Tyrannus couchi*), and Couch's Jay (*Cyanocitta ultramarina couchi*) are birds named to perpetuate his ornithological endeavors.

**Coues,** Elliott (1842–1899). Surgeon U. S. Army. Noted American ornithologist; author *Key to North American Birds* (1st edition 1872), *Check-List of North American Birds* (1882), *Birds of the Colorado Valley* (1878), *Birds of the Northwest* (1874), etc. He wrote the definitions of biological and zoological terms for the Century Dictionary (1889–1891). For him was named Coues's Flycatcher (*Myiochanes pertinax pallidiventris*).

**Coues,** Grace Darling. The warbler, *Dendroica graciae,* named in her honor by Prof. Baird, was collected by Dr. Elliott Coues, her brother, near Fort Whipple in Arizona.

**Coulter,** Dr. Thomas (?–1843). Irish botanist who collected in Mexico for many years and in California in 1831–1832. He was the first botanist to penetrate the Colorado Desert. The Coulter Pine (*Pinus coulteri*) which bears the largest and heaviest of cones of all the pines perpetuates his name in botanical science.

**Coville,** Dr. Frederick Vernon (1867–1937). Curator U. S. National Herbarium (1893–1937); botanist Death Valley Expedition, 1890 1891. The genus *Covillea* which includes the Creosote Bush was created to commemorate his name.

**Crantz,** H. J. (1722–1799). Austrian botanist after whom the plant genus *Crantzia* was named.

**Cuvier,** Baron Georges (1769–1832). French comparative anatomist. Founder of the science of Vertebrate Paleontology. He emphasized particularly the relation between structure and function.

# D

**Dale,** Samuel (1659–1739). English physician and botanist. The leguminous genus, *Dalea* was described by the French botanist, Jussieu.

**Dalibard,** T. F. (1703–1779). French botanist after whom Linnaeus named the plant genus *Dalibarda.*

**Dall,** William Healey (1845–1927). American naturalist; paleontologist of the U. S. Geographical Survey (1884–1909); author of many scientific names of mollusks, living and extinct. Dall's sheep (*Ovis dalli*) discovered in 1884, was named in his honor.

**Danthoine,** E. French botanist after whom the plant genus *Danthonia* was named.

**Darwin,** Charles Robert (1809–1882). English naturalist; author of theory of organic evolution. Darwin's *Voyage of H.M.S. Beagle* (1860) is a most interesting account of his voyage of exploration around the world in 1831–36. His most celebrated work was his *Origin of Species by Means of Natural Selection* (1859), a book which revolutionized scientific thought. Darwin spent much of his life at "Down House" in Kent and the place is now almost a shrine for zoologists.

**DeVries,** Hugo (1848–1935). Dutch botanist who developed the Mutation Theory based on experiments with Lamarck's evening primrose.

**Dickey,** Donald Ryder (1887–1932). Ornithologist and patron of science, photographer and writer on birds and mammals. *Branta dickeyi, Peromyscus dickeyi* and *Phaenoptilus dickeyi* are names honoring his activities.

**Diguet,** Leon. Discovered and collected in 1895

on Catalina Island in the Gulf of California, the giant barrel cactus (*Echinocactus digueti*) that bears his name (now listed under the generic name *Ferocactus*).

**Dodoens,** Rembert (1517–1585). Dutch botanist, after whom *Dodonaea viscosa*, the Hop-Bush, which is widely distributed in warmer parts of the world, is named.

**Douglas,** David (1798–1834). Scottish botanist born at Scone, Scotland, and killed at the age of 36 by suspicious natives in the Hawaiian Islands. He made several scientific journeys to America spending some four years (1829–1832), mostly in California, collecting for the Horticultural Society of London. The Douglas Spruce (*Pseudostuga taxifolia*) perpetuates his name in botanical history. Douglas provided the material from which some 300 species of California plants were to be described.

**Duméril,** Prof. Auguste M. C. (1774–1860). French herpetologist, publisher of Erpétologie général (1835–1851) an epoch making work in the history of herpetology.

# E

**Eimer,** Theodore (1843–1898). German zoologist. For him was named the sporozoan genus *Eimeria.*

**Eisen,** Gustav. Made several trips to Lower California as a member of expeditions sent out by the California Academy of Sciences. He published both on the geology and geography of the southern end of the peninsula. "Explorations in the Cape Region of Baja California" appeared in 1894.

**Ellis,** John (1814–1890). Distinguished English naturalist and correspondent of Linnaeus who perpetuated Ellis's name in the generic name *Ellisia*, which includes a number of species of delicate, slender-stemmed annuals of the Water-Leaf Family.

**Emory,** Major William Hemsley (1811–1887). American soldier and later Director of the United States and Mexican Boundary Survey. His well illustrated report makes fascinating reading. Among plants named in his honor by Dr. Engelmann are: *Cereus emoryi, Hyptis emoryi* and *Dalea emoryi*, all southwestern desert species.

**Encel,** Christopher. German authority on oak-galls. Adanson named the genus *Encelia* in his honor.

**Engelmann,** Dr. George (1809–1884). German botanist who came to America to explore the Mississippi Valley. He settled in St. Louis where he set up medical practice. He was the author of many scientific papers (120 of them) in the field of botany, among them treatises on the Cacti, some of them beautifully illustrated. He dealt too with the genus Yucca, Agave, Quercus and the Coniferae. Oaks, cacti, and many other plants bear his name, among them

being *Cereus engelmannii*, and *Quercus engelmannii*.

**Enslen,** A. Austrian botanist who collected in southern United States in the early nineteenth century. Thomas Nuttall perpetuated his memory in the genus *Enslenia*.

**Eschscholtz,** Johann Friedrich (1793–1831). Professor at the University at Dorpat. Acted as naturalist on the ship *Rurik* which sailed around the world (1815–1818) under command of Otto von Kotzebue; a second world trip was made in the ship *Predpriaetae* during 1823–1826. Eschscholtz is remembered for his remarkable animal and plant collections in California. His companion, Adelbert Chamisso, described the California poppy under the name *Eschscholtzia californica*.

**Evermann,** Barton Warren (1853–1932). Ichthyologist, long connected with the California Academy of Sciences. Long-time collaborator with David Starr Jordan and with him author of many papers dealing with the fish fauna of the Pacific.

# F

**Fabricius,** Hieronymus (1537–1619). Italian anatomist and surgeon. Pupil of Fallopius and teacher of the celebrated English physician, William Harvey, who discovered the circulation of the blood.

**Floerke,** H. G. German botanist after whom the plant genus *Floerkea* was named.

**Flower,** Sir William Henry (1831–1899). English biologist. Upon the death of Sir Richard Owen (1884) he became director of the British Museum of Natural History at South Kensington. Comparative anatomist, mammalogist and leading authority on museums. Co-author with Richard Lydekker of *An Introduction to the Study of Mammals Living and Extinct*, and author of *An Introduction to the Osteology of the Mammalia.*

**Forster,** Johann Reinhold (1729–1798). German naturalist who accompanied Captain Cook on one of his voyages to the Pacific. Author of a treatise on the birds of Hudson Bay (1772). Nuttal described the tern called Forster's Tern (*Sterna forsteri*).

**Francke,** Johan (1590–1661). Professor of medicine at Upsala, after whom the plant genus *Frankenia* was named. Frankenia grows in salt marshes of the sea coast and inland deserts.

**Fraser,** John (1750–1811). English botanist after whom the plant genus *Frasera* was named by Thomas Walter.

**Frémont,** Captain John Charles (1813–1890). Early western explorer, naturalist, soldier, and once candidate for President of the United States. He made several western journeys, travelling and collecting plants from the Rocky Mountains to the Pacific. One plant genus (*Fremontia*) and many common and trivial names honor his botanical activities. He issued with Dr. John Torrey an account of the plants he collected and wrote a charming account of his travels under the title *The Exploring Expedition to the Rocky Mountains in 1842 and to Oregon and North California in the years 1843–1844.*

**Froelich,** J. A. 17th Century German botanist after whom the plant genus *Froelichia* was named by K. Moench of the University of Marburg.

# G

**Gabb,** William M. Traveller and explorer in Lower California in 1867. Wrote on his exploration there. Worked on the Geological Survey with Wheeler and did the Geological Survey volume on Palaeontology. *Callisaurus ventralis gabbi* was described by Cope in the Annual Report of the U. S. National Museum, 1898, p. 72.

**Gage,** Sir Thomas (1780?–1820). English botanist after whom the plant genus *Gagea* was named by R. A. Salisbury of the London Horticultural Society.

**Gaillard** de Marentonneau. French patron of botany. The composite genus *Gaillardia* perpetuates his name.

**Galen,** Claudius (130–200? A.D.). Greek physician and medical writer. Galen's writings, of which about 80 have been preserved, were very numerous. His influence and teaching bore absolute sway on medical thought for a thousand years.

**Galinsoga,** M. (1766–1797). Spanish physician and botanist after whom the plant genus *Galinsoga* was named.

**Gambel,** William (1819–1849). Protégé of Thomas Nuttall and early transcontinental traveller, collector and writer on California birds. He died of typhoid fever while trying to cross the Sierra Nevada and was buried on the Feather River. His name is perpetuated in the name of the White-crowned Sparrow (*Zonotrichia gambeli*), Gambel's Quail (*Lophortyx gambeli*), a goose (*Anser gambeli*), and a shrike (*Lanius gambeli*).

**Gapper,** Dr. Anthony (1800?–1883). Collected mammals in Ontario about 1830. The Red-backed Mouse (*Evotomys gapperi* with its many sub-species) which ranges across northern United States from Oregon to the Atlantic Coast, in parts of Europe and Asia, was named in his honor.

**Garry,** N. The plant family *Garryaceae* and the genus *Garrya* which includes the Silk Tassel Bush, was named by David Douglas after this man connected with the Hudson Bay Company and who gave much assistance in his plant collecting expeditions.

**Gaulthier,** M. (1708–1751). Canadian botanist and physician after whom Linnaeus named the mint genus *Gaultheria*, which includes our well known wintergreen.

**Gerard,** John (1545–1612). English herbalist after whom the plant genus *Gerardia* was named by the famous Swedist naturalist Linnaeus.

**Gerber,** Traugott (?–1743). German naturalist after whom the plant genus *Gerbera* was named.

**Gesner,** Konrad von (1516–1565). Celebrated Swiss naturalist and writer of Zurich; author of the first illustrated book on zoology, a very curious publication with quaint illustrations.

**Giard,** Alfred (1846–1908). Biologist in Paris for whom was named the protozoan intestinal parasite of vertebrates, *Giardia.*

**Gilii,** F. L. (1756–1821). Italian botanist; Director of the Vatican Observatory. Some authors say the genus *Gilia* was named after Philip Gil, botanist of Spain who helped Xuarez to write a treatise on exotic plants cultivated in Rome.

**Gill,** Theodore Nicholas (1837–1914). Naturalist, ichthyologist, taxonomist who long worked at the Smithsonian Institution: friend of David Starr Jordan; publisher of numerous papers on mollusks, fishes and mammals.

**Gilman,** M. French (1871–1944). Long active member of the Cooper Ornithological Club and student of California and Arizona birds. He served as first custodian of the Death Valley National Monument (1933–1934) and later as botanist. Among birds, the Saghuaro Screech Owl (*Otus asio gilmani*) commemorates his work. Dr. Frederick Coville created the generic name *Gilmania.* Many rare plants of the Death Valley area honor his name and memory as an assiduous explorer and student of the local flora. See Madroño 1945, pp. 27–29.

**Girard,** Charles (1822–1895). Herpetologist. With Spencer F. Baird he described the lizards in Stansbury's Report entitled: "Exploration and Survey of the Valley of the Great Salt of Utah." With Baird he authored a Catalogue of N. A. Reptiles. A snake, *Coluber taeniatus girardi,* bears his name.

**Gloyd,** Howard Kay (1902–     ). Teacher, specialist on the life-history and embryology of crotaline snakes. Long resident of Kansas and later connected with the Chicago Academy of Science and the department of Zoology of the University of Michigan. *Heterodon gloydi* and *Sonora gloydi* are names perpetuating his memory.

**Godet,** C. H. (1797–1879). Author of *Flora de Jura.* The onagraceus genus, *Godetia,* with many California species commemorates his name.

**Golgi,** Camillo (1844–1926). Italian histologist at Pavia. *Golgi bodies* in the cell, commemorate his name in Cytology.

**Good,** George Brown (1851–1896). Appointed Assistant Secretary of the Smithsonian Institution (1887). One of the chief builders of the science of Oceanography; equally interested in the history of Zoology; Author with Tarleton H. Bean of "Oceanic Ichthyology" (1894).

Commissioner of fisheries 1887–1888. Pupil and friend of Spencer F. Baird.

**Gould,** John (1804–1881). English ornithologist and artist; taxodermist to the London Zoological Society. Author of *Birds of Europe, Birds of Australia, Birds of Great Britain.* These he illustrated with nearly 3000 excellent plates.

**Grant,** Robert E. (1783–1874). Scottish comparative anatomist and student of sponges. The sponge genus *Grantia* was named in his honor.

**Gray,** Asa (1810–1888). Professor of botany, University of Michigan, later professor of natural history at Harvard. Gray's great work in taxonomic botany places him along with Joseph D. Hooker of England, as among the world's greatest botanists. The Gray Herbarium at Harvard is a crowning honor to his work and fame. He laid the foundation of systematic botany in this country. Gray visited California in 1885. The western chenopodiaceous genus *Grayia* was named for the gifted young botanist, then in his 28th year, by Hooker. Famous are Gray's *Manual of Botany,* his *Synoptical Flora of North America,* and *How Plants Grow.*

**Gregg,** Josiah (1806–1850). Dubbed Doctor but never practiced medicine. Santa Fe trader, author and close scientific observer. Preeminently a historian of the Santa Fe trade. Spent some time observing and collecting in Mexico, particularly in Chihuahua. Friend of John M. Bigelow. *Acacia greggii,* the Catsclaw, was named in his honor by Asa Gray. The genus *Greggia* was created by Dr. George Engelmann in 1848 to commemorate Gregg's achievements.

**Grindel,** D. Hieronymus (1776–1836). Russian botanist of Riga after whom the plant genus *Grindelia* was named by K. L. Willdenow, of Berlin.

**Grinnell,** Dr. Joseph (1877–1939). First and long-time editor of *Condor* and Director of the Museum of Vertebrate Zoology at Berkeley, California (1908–1939). Author of numerous special books and papers on California ornithology, herpetology and mammalogy. His valuable work is commemorated in the names of several fossil and recent birds, among them a Shrike (*Lanius ludovinciana grinnelli*) and a Pygmy owl (*Glancidium gnoma grinnelli*), a Kinglet (*Regulus calendula grinnelli*) from Alaska, and an extinct Eagle (*Geranoaetus grinnelli*) from the Rancho La Brea tar pits.

**Gutierrez.** A noble Spanish family after whom the composite genus *Gutierrezia* was named by M. Lagasca, director of the botanical garden at Madrid.

# H

**Haeckel,** Ernst Heinrich (1834–1919). German naturalist and comparative anatomist; author of the biogenetic law or recapitulation theory; the apostle of evolution in Germany.

**Hake,** Baron (1745–1818). Hanovarian patron of science after whom the plant genus *Hakea* was named by Schrader.

**Hall,** Dr. Harvey Monroe (1874–1932). American botanist long connected with the University of California and later with the Carnegie Institution (1919–1932). Special interests: the taxonomy of plants of the genus *Artemisia* and experimental taxonomy, and ecology. Name perpetuated in *Halliophytum hallii,* a shrubby western spurge, curiously carrying a double reference to his scholarly work.

**Hammond,** William Alexander (1828–1900). Surgeon U. S. Army. His name appears in ornithological literature in the bird name, Hammond's Flycatcher (*Empidonax hammondi*).

**Harris,** Edward (1759–1863). Companion of Audubon on his Missouri River trip in 1843. Several birds were named by Audubon to honor their friendship: Harris' Woodpecker (*Dryobates villosus harrisi*) and Harris' Hawk (*Parabuteo unicinctus harrisi*).

**Harvey,** William (1578–1657). English physiologist. Discovered circulation of the blood.

**Havers,** Clopton (1650–1702). English anatomist after whom the Haversian system of canals in bone was named.

**Heermann,** Dr. Adolphus L. (1818?–1865). Oölogist and bird collector. Surgeon and naturalist on one of the Pacific Railroad Surveys (1853–1854). His name is associated with Heermann's Gull (*Larus heermanni*) and a song-sparrow (*Melospiza melodia heermanni*). He is said to have introduced the term "oölogy" into the literature of North American birds.

**Henshaw,** Henry Wetherbee (1850–1930). Chief of the U. S. Biological Survey, 1910–1916, and author of numerous scientific papers on birds. His name occurs in that of the wren-tit (*Chamaea fasciata henshawi*).

**Heucher,** J. Heinrich (1677–1747). German botanist after whom the plant genus *Heuchera* was named. He was director of the botanical garden at Wittemberg.

**Hilare,** Auguste St. (1799–1853). French botanist; traveler in Brazil. In his honor was named *Hilaria,* a genus of western grasses.

**Holboell,** Carl Peter (1795–1856). Lieutenant in the Danish Royal Navy and traveller in Greenland (1822); later Governor of South Greenland. Bird names, *Hierfalco holboelli, Acanthis holboelli* and *Columbus holboelli* were given to perpetuate his memory.

**Holbrook,** Dr. John Edward (1794–1871). Herpetologist; author of "North American Herpetology" (three vols.) and "American Herpetology" (five vols.), Prof. of Anatomy, Medical College of South Carolina (1824–1854). The genus *Holbrookia* was created by Girard in 1851.

**Honckeny,** G. A. (1724–1805). Author of a *Flora of Germany;* after him the plant genus

*Honckenya* was named by the erudite naturalist, Linnaeus.

**Hooke,** Robert (1635–1703). English microscopist. First to report cells in plant tissue. First to use the word "cell" as a name for the unit of tissue structure.

**Hooker,** Sir Joseph Dalton (1817–1911). Son of the famous botanist William J. Hooker, and accounted one of, if not the greatest botanist of his time. His wide travels in New Zealand, Australia, Tasmania, the Himalaya, Sikkim, Syria, Palestine, and North America gave him a wide grasp of knowledge of plants in many parts of the world. With the able Dr. Bentham he published *Genera Plantarum*, a work based on the collections at Kew Gardens of which he was the Director for 20 years. Other works of great interest are his *Flora of British India* and *Himalaya Journals.*

**Hooker,** Sir William Jackson (1785–1865). English botanist and father of the famous Sir J. D. Hooker. Under his directorship the Royal Botanical Gardens at Kew were greatly expanded. Because of his influence many botanists were appointed to accompany British government expeditions to many parts of the world. Author of numerous botanical works including *Botanical Magazine* (38 vols.), *Icones Plantarum* (10 vols.), *Species Filicinum*, etc.

**Hosack,** David (1769–1835). American botanist after whom *Hosackia*, a genus of leguminous plants was named by David Douglas of the Royal Horticultural Society.

**Hotton,** P. (1648–1709). Dutch botanist and professor at Leyden after whom Linnaeus named the plant genus *Hottonia.*

**Howard,** Leland Ossian (1857–1950). Long chief of the U. S. Bureau of Entomology and leader of entomological activities in America. His valuable work in aiding knowledge of flies and mosquitoes in relation to public health and their control won him worldwide renown and gratitude. Many insects bear the specific name *howardi* and one genus, *Howardia*, was created to honor his outstanding leadership in entomology.

**Hubbs,** Carl Leavitt (1894–    ) Ichthyologist, Museum of Zoology, Ann Arbor, Michigan. Now with the Scripp's Institution of Oceanography of the University of California. Specializes in taxonomy, oecology and life history of fishes.

**Hudson,** W. H. (1841–1922). Argentine-born naturalist who left for England when 27 years old and lived there until his death. Books like *Far Away And Long Ago*, describing his boyhood days on the Pampas, *Idle Days in Patagonia, Birds of La Plata,* and *The Naturalist in La Plata* show most ably not only his literary genius but also his superb powers of observation. *The Naturalist in La Plata* was written 25 years after Hudson left the Argen-

tine, yet it has as much accuracy and freshness as though it had been written on the spot.

**Huey,** Laurence Markham (1892–    ). Ornithologist and mammalogist devoting his time to the birds and mammals of southern California and adjacent Baja California, describing therefrom several new species and subspecies. His name will always be associated in the mind of bird students with a Poor-will (*Phalaenoptilus nuttalli hueyi*). He wrote frequently in publications of the San Diego Natural History Museum in which institution he long served.

**Hutton,** William. Field collector of birds in California during the time of the Gold Rush. He also included the Cape Region of Baja California in his travels. John Cassin named a Vireo (*Vireo huttoni*) in his honor.

**Huxley,** Thomas Henry (1825–1895). Brilliant minded English biologist and philosopher, with early training in medicine. Even as Darwin began his active scientific career as he went on the momentous trip on the *H.M.S. Beagle,* so Huxley began his when he was appointed surgeon to *H.M.S. Rattlesnake.* Huxley championed the inductive method of reasoning. Upon publication of the *Origin of Species* (1859) he became one of evolution's greatest proponents. He introduced the laboratory method into the teaching of biology and by his unusually able expository writing and speaking was able to see the biological sciences placed upon an almost wholly new and solid basis. Long before his death he had become the recipient of every honor which English science can bestow.

**Hutchins,** Miss Ellen (1785–1815). Irish cryptogamist after whom Brown named the plant genus *Hutchinsia.*

# I

**Imperato,** Ferrante (1550–1625). Italian naturalist and druggist of Naples, after whom the plant genus *Imperata* was named.

**Isnard,** Antonine Tristan Danty d' (1665?–1743). French botanist of the Jardin des Plantes, after whom the plant genus *Isnardia* was named.

# J

**Jaeger,** Dr. Edmund Carroll (1887–    ). Naturalist, instructor in the biological sciences, writer. Discoverer of hibernation in birds. Special interest, American deserts. At least 10 southwestern U. S. plants, mollusks and arthropods bear in their specific or subspecific appellations reference to his work.

**James,** Dr. Edwin (1797–1861). Botanist of Long's Expedition to the Rocky Mountains in 1820. The plant genus *Jamesia* was named after him by Torrey and Gray.

**Jeffrey,** John. Scotch gardener who collected both plants and seeds in northern California and Oregon (1852–53) for subscribers to the

*Oregon Botanical Expedition* of Edinburgh. A western yellow pine (*Pinus ponderosa jeffreyi*) discovered in California in 1850 bears his name.

**Jepson,** Willis Linn (1867–1946). Professor of botany, University of California; known for his efforts in conservation (a founder of Save-the-Redwoods League) and his taxonomic work on California plants. He described at least 150 native species, a number greater than described by any of the great systematists of his time. He is the author of numerous works, among them: *A Manual of the Flowering Plants of California, A Flora of California,* and *The Silva of California.* The genus *Jepsonia* described by J. K. Small perpetuates his name in California botanical history. Jepson founded the California Botanical Society and launched the official organ of the Society, *Madroño.*

**Jordan,** David Starr (1851–1931). Ichthyologist. First and longtime President of Leland Stanford Jr. University. Interested particularly in the taxonomy and geographical distribution of fishes of the Pacific area. One of the famous pupils of Agassiz at Pennikese Island.

**Jussieu,** Bernard de (1699–1776), and Antoine de Jussieu (1686–1758). French botanists. The plant genus *Jussieua* was named after the brothers. In his *Genera Plantarum* Antoine Laurent de Jussieu (1748–1836) delineated the basis of all natural classification of plants.

# K

**Kaeding,** Henry Barroilhet (1877–1913). California ornithologist and mining engineer, particularly interested in the insular birds off the California and Lower California Pacific Coast. Went with A. W. Anthony to the South Pacific Islands. Prepared the first 10 year index to *Condor.* The White-rumped Petrel carries the name *Oceanodroma kaedingi.*

**Kallstroem.** Probably an obscure botanist after whom the plant genus *Kallstroemia* was named.

**Kalm,** Peter (1715–1799). Pupil of Linnaeus and later professor at Abo in Finland. His travels included the northern United States and Canada. A genus of hardy ornamental evergreen shrubs (*Kalmia*) of wide N. A. distribution was dedicated to him by his former teacher. Kalm published *En resa til Norra Amerika* (Journey to North America) (1753–1761).

**Kellogg,** Dr. Albert (1812–1887). San Francisco physician and botanist. First resident botanist in California. One of the original seven founders of the California Academy of Sciences in 1854. Author "Forest Trees of California" in the Report of the State Mining Bureau, 1882. Dr. Torrey, in recognition of his services to botanical science early dedicated to him the monotypical Rubiaceous genus *Kelloggia. Quercus kelloggii* was described by Dr. Newberry

as a tribute to Kellogg's interest in the genus *Quercus.*

**Kennicott,** Robert (1835–1866). Enthusiastic, friendly and capable naturalist; friend of Spencer F. Baird; valuable collector for the Smithsonian Institution. Travelled in Alaska and the Hudson Bay country. One of the founders of the Chicago Museum of Natural History. Member of the unique Megatherium Club.

**Kirtland,** Jared P. (1793–1877). Well-known Ohio physician and naturalist to whom Charles Pease presented the warbler (*Dendroica kirtlandii*) first discovered as a migrant and taken at Cleveland, Ohio. Dr. Kirtland was founder of the Cleveland Medical College and always much interested in the Natural History of Ohio; especially was he interested in the fishes.

**Klauber,** Dr. Lawrence Monroe (1883–    ). Electrical engineer whose hobby was reptiles, particularly snakes. Author of "Rattlesnakes: Their Habits, Life Histories and Influence on Mankind" (two vols.) Member Zoological Soc. of San Diego. Several reptile names honor his work: *Sonora klauberi, Eusatina klauberi, Hypsiglena klauberi.*

**Knaut,** Christian (1654–1716). German physician of Halle after whom the plant genus *Knautia* was named.

**Kniphof,** Johann Hieronymus (1704–1763). German professor of Medicine at Erfurt after whom the plant genus *Kniphofia* was named.

**Kobres,** Paul Von (Cobres) (1747–1823). Augsburg collector of a rich cabinet of natural history. The plant genus *Kobresia* was named in his honor.

**Koch,** W. D. J. (1771–1849). Director of the Botanical Garden at Erlangen; after him the plant genus *Kochia* was named. He was author of *A Synopsis of the Flora of Germany and Helvetia.*

**Koeberlin,** C. L. Amateur German botanist after whom the plant genus *Koeberlinia* was named.

**Koelle,** J. L. C. German botanist after whom the plant genus *Koellia* was named.

**Koenig,** Charles. Early worker of the British Museum. The plant name *Koeniga* was revived by Robert Brown in commemoration of his endeavors; *Koeniga maritima* is the common and elegant sweet alyssum of gardens.

**Kosteletzky,** V. F. Bohemian botanist after whom the plant genus *Kosteletzkya* was named.

**Krig,** David. Early collector of Maryland plants in whose honor the plant genus *Krigia* was named.

# L

**Lacepede,** Bernard Germain (1756–1825). French naturalist, friend of and co-worker with Buffon. *Menobranchus lacepedii,* a Mud Puppy, has reference to his work.

**Lamark,** Jean Baptiste de (1744–1829). French anatomist and student of classification. Proposed first general theory of evolution.

**Lambert,** Aylmer Bourke (1761–1842). English botanist and vice-president of the Linnean Society. Author of a work on *Pinus*. David Douglas, the Scottish botanist, perpetuates his name in botanical literature by giving to the Sugar Pine the name *Pinus Lambertiana.*

**Lambl,** Wilhelm Dusan (1824–1893). Bohemian physician whose name is perpetuated in *Giardia lamblia* a protozoan causing mechanical interference with absorption, particularly of fats.

**Lapham,** Dr. Increase Allen (1811–1875). Wisconsin botanist, climatologist and geologist after whom the plant genus *Laphamia* was named.

**Lavater,** Johann Kasper (1741–1801). Swiss physician and naturalist after whom the shrub genus *Lavatera* was named by Linnaeus in 1757.

**Lawrence,** George Newbold (1806–1895). A goldfinch (*Astragalinus lawrencei*) was named after him by Cassin. He was particularly known for his interest in aquatic birds. He described the Pacific Loon, Western Grebe, California Gull and Black Brandt.

**Leach,** J. (1704–1764). Swedish botanist after whom the plant genus *Lechea* was named.

**LeConte,** Dr. John Lawrence (1825–1883). Eminent specialist in Coleoptera and cousin of Joseph LeConte, the geologist. Among his numerous interests were birds, and a number were named in his honor, among them a sweet-singing, desert thrasher (*Toxostoma lecontei*).

**Leers,** Johann Daniel (1727–1774). German druggist and botanist in whose honor the plant genus *Leersia* was named.

**Leishman,** Sir William B. (1865–1926). British army surgeon. Name perpetuated in the trypanosome genus *Leishmania. Leishmania donovani* is the causitive agent in dumdum fever.

**Lemaire,** Charles (1801–1871). French cactologist and horticulturalist whose name is commemorated by Britton and Rose in the cactus genus *Lemaireocereus.*

**Lespedes,** V. M. de. Spanish governor of East Florida; after him the plant genus *Lespedeza* was named.

**Lesquereux,** Leo (1806–1889). American bryologist after whom the plant genus *Lesquerella* was named.

**Lewis,** Captain Merriweather (1774–1809). Leader with William Clark in the famous Lewis and Clark Expedition across the continent in 1804–1806. His name is perpetuated in ornithology by the bird name Lewis Woodpecker (*Asyndesmus lewisii*).

**Leydig,** Franz von (1821–1908). German anatomist. Name commemorated in terms: Leydig's duct, Leydig's cells, cylinders of Leydig.

**Linder,** Johann (1678–1723). Swedish botanist after whom the plant genus *Lindera* was named.

**Lindheimer,** Ferdinand. Early botanical explorer of Texas; resident of New Braunfels. Gray and Engelmann created the composite genus, *Lindheimera. Opuntia lindheimeri* was named by George Englemann. A *Veronica* and a *Castillija* both carry the specific name *lindheimeri.*

**Lindley,** John (1799–1865). Prof. of botany at University of London (1829–1860); also noted as a horticulturalist. Editor of *Botanical Register* (1826), *Journal of the Horticultural Society* (1846–1855) and *Gardener's Chronicle* (1841–1865). Lindley was admired for the clearness of his style and his excellent illustrations. The rosaceus genus *Lindleya* is named in his honor.

**Linnaeus,** Karl von (1707–1778). Swedish naturalist and author of the Binomial System of Nomenclature. Of him the Swedish people say, "God made the plants and animals, Linnaeus named them."

**Lippi,** Augustin (1678–1705). French physician and traveler in Abyssinia and Egypt; after him the plant genus *Lippia* was named.

**Lister,** Martin (1638–1712). English physician and naturalist after whom the plant genus *Listera* was named. Peculiarly he regarded fossil mollusks as "inorganic imitations produced in the rocks."

**l'Obel,** Matthias de (1538–1616). Flemish botanist after whom the well known plant genus *Lobelia* was named.

**Loefling,** Peter (1729–1756). Swedish botanist and traveller, friend and pupil of Linnaeus. He accompanied a Spanish exploratory expedition in 1751 to Venezuela where after extensive travels he died of a fever. The plant genus *Loeflingia* was created by Linnaeus. See **Condal.**

**Loiseleur-Deslonchamps** (1774–1849). French botanist in whose honor the plant genera *Loiseleuria* and *Deschampsia* were named.

**Lonicer** (Latinized form of Lonitzer), Adam (1528–1586). German physician and botanist of Frankfort a/M. In his honor the plant genus *Lonicera* was named. It includes the many species of Honeysuckle and Woodbine.

**Ludwig,** C. G. (1709–1773). German botanist of Leipsic and contemporary of Linnaeus. After him the plant genus *Ludwigia* was named.

**Lydekker,** Richard (1849–1915). English naturalist and geologist. Author of the *Royal Natural History* (1896), *Phases of Animal Life* (1892) and *The Great and Small Game of India, Burma, and Tibet, The Sheep and its Cousins, The Horse and its Relatives,* etc. See Flower, Sir William Henry.

# M

**MacDougal,** Dr. Daniel Trembley (1865–1958). Of the Desert Laboratory of the Carnegie

Institution at Tucson, Arizona. Specialist on the vegetation of the American and African deserts, water balance of succulents, etc. One of the ocotillos of Mexico, *Fouquieria macdougallii*, perpetuates the memory of this active student of arid-land plants.

**MacGillivray,** William (1796–1852). Scotch naturalist; author of a *History of British Birds* and coadjutor of John James Audubon in the preparation of his well-known and valuable treatise on American birds. In his honor *Sylvia macgillivrayi* (MacGillivray's Warbler) was named by Audubon. This scientific name was later replaced by *Oporornis tolmiei* by J. K. Townsend who discovered the bird and who dedicated it to his surgeon friend, W. T. Tolmie of Fort Vancouver.

**Maclure,** William (1763–1840). American geologist whose name is perpetuated in the name of the Osage Orange (*Maclura aurantiaca*), a hedge plant armed with stout spines and bearing bright golden yellow fruit. Maclure made the first geological map of the United States, a publication antedating William Smith's geological map of England by six years.

**Malpighi,** Marcello (1628–1694). Italian, founder of microanatomy for both plants and animals. Developed "preformation theory." The *Malpighian tubules* of the kidney are structures familiar to all students of anatomy.

**Marsh,** Othniel Charles (1831–1899). Distinguished American paleontologist, professor at Yale Univ. (1866–1899). He studied and described many of the extinct vertebrates of the United States, particularly those from the Rocky Mountain region. Marsh was long in charge of the division of vertebrate paleontology in the U. S. Geological Survey. He was instrumental in having founded the Peabody Museum at Yale.

**Martyn,** John (1699–1768). English botanist, once professor at Cambridge. The plant genus *Martynia* comprises about ten species of annual and perennial herbs of the warmer parts of North America.

**Matthews,** Dr. W. Member of the U. S. Army stationed in Owens Valley, California, in 1875. One of the loveliest and unique of the small gilias (*Gilia mathewsii*) bears his name.

**Matthioli,** Pierandrea A. (1500–1577). Italian physician in whose honor the plant genus *Matthiola* was named.

**McCall,** George Archibald (1802–1868). General, U. S. Army and much interested in wildlife. While stationed in Texas, he made extensive zoological collections which were deposited with the Smithsonian Institution. The Flat-tailed Horned Lizard was named *Phrynosoma macallii* by Cope = *P. m'calli* Hallowell.

**Mearns,** Dr. Edward Alexander (1856–1916). Long-time army surgeon; enthusiastic, active and efficient student of natural history. Especially was he interested in birds and mammals but neither botany nor ethnology were neglected on his wide travels in southwestern United States, in the Philippines and East Africa. A genus of Philippine birds (*Mearnsia*) was named for him as well as several birds of the southwestern U. S., among them a dove (*Melopelia asiatica mearnsi*) and a gilded flicker (*Colaptes chrysoides mearnsi*) of the Sahuaro belt along the Colorado River. One of his great contributions was his report on the *Mammals of the Mexican Boundary of the U. S.* (1907).

**Mendel,** Gregor (1822–1884). Initiated modern methods of the study of genetics. Discovered the fundamental laws of heredity through his studies with garden peas. *Mendel's Law* and the *Mendelian ratio* are terms known to all students of heredity.

**Mentzel,** Christian (1622–1701). German botanical author and philologist after whom the plant genus *Mentzelia* was named.

**Menzies,** Archibald (1754–1842). Surgeon and naturalist to Vancouver's Pacific Coast expedition. Name commemorated in the dainty western plant called Baby Blue Eyes (*Nemophila menziesii*) and the plant genus *Menziesia.*

**Merriam,** C. Hart (1855–1942). Zoologist, botanist, ethnologist. Interests included mammals and birds of North America, geographical distribution of animals and plants, Indian basketry, mythology, distribution and linguistics of Pacific coast Indian tribes. Chief of U. S. Biological Survey, 1885–1910.

**Merriam,** Dr. John Campbell (1869–1943). Paleontologist of University of California and early director of excavations of the great fossil tar pits at Rancho La Brea. The large raptorial fossil bird, *Teratornis merriami*, found in the tar pits, was named by Dr. Loye Holmes Miller in recognition of his work of exploration.

**Merrill,** James Cushing (1853–1902). Army surgeon; amateur ornithologist of distinction, adding much to our knowledge of western birds. At least three species of birds commemorate his name, among them a Horned Lark (*Octocoris alpestris merrilli*) and a song sparrow *Melospiza melodia merrilli.*

**Mertens,** F. C. (1764–1831). German botanist after whom the plant genus *Mertensia* was named. Plants of this genus are commonly called Lungworts.

**Mikan,** J. C. (1743–1814). Bohemian botanist of Prague after whom the plant genus *Mikania* was named by Willdenow.

**Molina,** Juan Ignazio (1740–1829). Chilean naturalist and writer upon Chilean plants; after him the grass genus *Molinia* was named.

**Monardes,** N. (1512–1588). Spanish botanist after whom the plant genus *Monarda* was named. It includes the wild bergamot, horsemint and Oswego Tea.

**Monti,** Giuseppe (1682–1760). Italian botanist after whom the plant genus *Montia* was named.

**Morgan,** Thomas Hunt (1866–1945). American zoologist; won Nobel Prize in medicine in 1933. Best known for his experimental studies on *Drosophila* and his contributions to the mechanism of heredity.

**Muhlenberg,** G. H. E. (1753–1815). American botanist in whose honor the plant genus *Muhlenbergia* and the journal *Muhlenbergia* were named.

**Müller,** Johannes (1801–1858). Distinguished German comparative physiologist; founder of scientific medicine in Germany. His name is associated with the *Müllerian duct.*

## N

**Nelson,** Dr. Edward William (1855–1934). From 1916–1927, Chief of the U. S. Biological Survey. Travelled extensively and did field work in Alaska, Arizona, California and Mexico including Baja California. Author, *Lower California and Its Natural Resources* (1921); *Larger and Smaller Mammals of North America* (1916, 1918). The names: Nelson's Antelope Squirrel (*Citellus nelsoni*), Arizona Hooded Oriole (*Icterus nelsoni*) and Nelson's Sparrow (*Passerherbulus nelsoni*) remind us of his contributions to a knowledge of our fauna.

**Newberry,** Dr. J. S. (1822–1892). Geologist, palaeontologist; naturalist on the Williamson Survey. He also collected under Lt. Ives along the Colorado River in 1857–58 as far north as the Grand Canyon. *Horsfordia newberryi,* a malvaceous plant of the western deserts, was named after him.

**Nicander** (lit. the man who conquers) (about 150 B.C.). Greek physician and botanist after whom the plant genus *Nicandra* was named.

**Nicot,** Jean (1530–1600). Introduced tobacco into France. The plant genus *Nicotiana* perpetuates his name.

**Nolin,** P. C. French botanist and agricultural writer. *Nolina,* named in his honor by Michaux, is a genus of liliaceous shrubs of the southwestern United States and Mexico.

**Nuttall,** Thomas (1786–1859). English-American ornithologist and botanist, once curator of the Harvard botanical gardens. Two of his literary efforts resulted in *A Manual of the Ornithology of the United States and Canada* (1832–1834) and *The North American Sylva* (1842–1849). His name appears in that of a woodpecker (*Dryobates nuttalli*), Nuttall's Poorwill (*Phalaenoptilus nuttalli*) and a Magpie (*Pica nuttalli*). Nuttall was the naturalist "Old Curious" of Dana's homeward voyage, commemorated in *Two Years Before the Mast.*

## O

**Oberholtzer,** Dr. Harry Church (1870–    ). Active systemist in the field of ornithology. The generic name for the Green-tailed Towhee (*Overholtseria chlorura*) and a Flycatcher (*Empidonax oberholtseri*) recall his ornithological efforts. Curator of ornithology, Cleveland Museum of Natural History, (1941–47): ornithologist biological survey, U. S. Dept. of Agriculture (1895–1914).

**Olney,** Stephen (1812–1878). Rhode Island woolen manufacturer and botanist. His special study was the genus *Carex.* The Desert Ironwood (*Olneya tesota*) calls his name to attention.

**Orcutt,** Charles R. (1864–1929). San Diego botanist and naturalist, editor of the curious journal, *The West-American Scientist,* at the time the only medium that existed in western U. S. for the publication of natural history notes and short articles. He collected widely in southwestern U. S. and in Baja California. Orcutt's Aster (*Aster orcuttii*) and Orcutt's Ceanothus (*Ceanothus orcuttii*) are among the several plant names which perpetuate his memory.

**Owen,** Richard (1804–1892). English research anatomist and paleontologist. First to describe Venus Flower Basket (*Euplectela*); discovered *Trichina spiralis* and the Pearly Nautilus. Author of *Comparative Anatomy and Physiology of Vertebrates* (3 vols.).

## P

**Palmer,** Dr. Edward (1831–1911). Ardent collector of both birds and plants of the southwestern U. S. and Guadalupe Island in Lower California. Palmer's Thrasher (*Toxostoma curvirostre palmeri*) keeps fresh in mind his work in ornithology and nearly two hundred plant names keep alive his important endeavors in botanical lines. Dr. Palmer is said to have discovered more than eleven hundred new species of flowering plants.

**Parkman,** Dr. George (1791–1849). Boston physician and friend of Audubon whom he assisted in getting ready for his trip to Labrador. Audubon showed his appreciation by naming the Western House Wren, sometimes called Parkman's House Wren, *Troglodytes parkmani.* Parkman gave to Harvard University the grounds on which the Harvard Medical School is now located.

**Parry,** Dr. Charles Christopher (1823–1890). American botanist who many times visited the southwestern mountains and deserts and who is commemorated in more than a dozen California plant names alone; among them are *Nolina parryi* and *Lilium parryi.* Many of his interesting observations appeared in the pages of the San Francisco Bulletin. He was once connected with the Mexican Boundary Survey.

**Pavlov,** Ivan Petrovich (1849–1936). Russian; one of the greatest physiologists of modern times and renowned for his work showing the relationship of psychic stimuli to digestion, and for his discovery of conditioned reflexes.

**Pereskia** (Latinization of Peiresc), N. C. F.

(1588–1637), of Aix in Provence. Author of many historical and scientific writings. In 1703, Plumier, Franciscan father, named the cactus genus, *Pereskia*, of the tribe *Opuntieae*, in his honor.

**Perez,** Lorenzo. 16th century medical botanist and writer of Toledo, Spain; after him the plant genus *Perezia* was named.

**Pilsbry,** Dr. Henry Augustus (1862– ). Curator Department of Mollusks, Philadelphia Academy of Sciences; internationally known and preeminent authority on land snails but equally at home in any group of mollusks. Specialist in Cirripedia. Some of his best work appears in the *Manual of Conchology*.

**Planer,** J. J. (1743–1789). German botanist after whom the plant genus *Planera* was named by Gmelin.

**Pliny,** Caius, Pliny the Elder (23–79 A.D.) writer of a *Natural History* (*Historia Mundi*) which Cuvier pronounced "one of the most precious monuments which has come down to us from ancient times." Buffon spoke of it as "a compilation from all that had been written before his time."

**Pluche,** N. A. (1688–1761). French naturalist after whom the plant genus *Pluchea* was named by Cassini.

**Pontedera,** Giulio (1688–1757). Professor of botany at Padua in whose honor the plant genus *Pontederia* was named.

**Pringle,** Cyrus G. (1838–1911). Botanist of Vermont who collected widely in the northern and southwestern United States and Mexico where he spent 26 years. His Mexican work is especially noteworthy. The Pringle Herbarium of more than 150,000 specimens is at the University of Vermont. Pringle was first interested in plant breeding, especially of the potato; in fact, he is considered the pioneer plant breeder of America.

**Prjevalsky,** Nikolai Mikhailovitch (1839–1888). Russian explorer and well-known among zoologists in connection with the name "Prjevalsky's horse." He gained wide fame "as one of the most intrepid, indomitable and successful travellers who ever attempted to penetrate to the 'back-bone' of the Eurasian continent." He made four great expeditions to Central Asia, "bringing back to St. Petersburg 30,000 specimens of natural history objects of which 5,000 were specimens of birds representing 430 species."

**Puccinelli,** Prof. Benedetto. Italian botanist. The grass genus *Puccinellia* perpetuates his memory.

# R

**Rafinesque,** Constantine Samuel (1784–1842). Constantinople-born, eccentric, wanderer-naturalist and self-styled linguist who, after a brief sojourn in America moved to Sicily. He finally settled in the United States and actively collected many objects of natural history from fishes to toads and plants. Many he described as new species. So great was his zeal for naming new things that he claimed to have discovered and gave names to twelve new species of lightning and thunder on the headwaters of the Ohio River! Rafinesque was given to inventing many nonsense generic names of peculiar sound and spelling.

**Ranvier,** Louis Antoine (1835–1922). French histologist; in 1878 he described the interruptions in the myelin sheath of peripheral nerves now known as *nodes of Ranvier*.

**Rathbun,** Dr. Mary Jane (1860–1943). Carcinologist for many years connected with the U. S. National Museum. Her name appears frequently in the literature dealing with crabs, both those of North America and South America. Many of the species she described, and her name thus frequently appears just after the specific or trivial name. The genus *Rathbunia*, given by Nobili in 1896 honors her name and work.

**Rathke,** Martin Heinrich (1793–1860). German anatomist, physiologist and pathologist. He made an important contribution to embryology by his description in 1839 of *Rathke's pocket or pouch*, a diverticulum from the embryonic mouth cavity from which the anterior pituitary body is developed. He published a textbook of embryology in 1861.

**Ray,** John (Wray) (1628–1705). English naturalist. Looked upon as the founder of the modern or "natural" system of plant classification. Ray introduced the idea of species among plants and animals and "dispelled the fabulous monsters and prodigies" of Gesner and other encyclopaedists of medieval times. He introduced many new botanical terms, among them *Monocotyledons*, *Dicotyledons*, and *Leguminosae*.

**Redi,** Francesco (1626–1698). Italian zoologist. His name is perpetuated in *redia*, the second larval stage of certain trematode worms.

**Ricci,** P. Francisco. Italian nobleman; patron of Micheli, who named the liverwort genus *Riccia* in his honor.

**Ridgway,** Robert (1850–1929). Curator of birds, U. S. National Museum (1876–1929); field naturalist and author of 540 publications in the field of ornithology, botany, etc., among them *Birds of North and Middle America* (in collaboration with Drs. Baird and Brewer) and *North American Land Birds*. He was zoologist on the U. S. Geological Exploration of the 40th Parallel (1867–1869) and one of the twenty-three founders of the American Ornithologists Union (A.O.U.). Ridgway's name will be perpetuated in scientific literature in the generic name *Ridgwayi* and 23 specific and 9 sub-specific bird names. Ridgway's Color Key is known to all students of natural history.

**Riley,** Charles Valentine (1843–1895). Foremost of nineteenth century American economic entomologists and long connected with the U. S. Dept. of Agriculture. He published a great number of scientific papers (more than 2400) and was the recipient of many scientific awards, medals, and honors. An unusual number of insects bear the specific name *rileyana*, *rileyella*, or *rileyi*. Riley founded the entomological collections of the U. S. National Museum. He was one of the organizers of the Entomological Society of Washington (1884) and founder of the Association of Economic Entomologists (1889).

**Robin,** Jean and Robin, Vespasien. Father and son. Royal gardeners at Paris who introduced in 1635, the locust tree (*Robinia*) into Europe.

**Roemer,** J. J. (1763–1819). Swiss naturalist after whom the plant genus *Roemeria* was named by De Condole.

**Rose,** Dr. Joseph Nelson (1862–1928). Botanist, Carnegie Institution. Specialist in South American and Mexican plants and botanical features of North American deserts. Author with N. L. Britton: The Cactaceae (four vols.)

**Rothrock,** Joseph Trimble (1839–1922). Student of Asa Gray; graduate in medicine (Uni. of Pennsylvania); botanist and surgeon to the Surveys West of the 100th Meridian under Lieut. G. M. Wheeler; Professor of Botany, Univ. of Pennsylvania (1877–1891). Quite a number of plants carry his name. First there is, from Lower California, a genus *Rothrockia*. Then comes *Artemisia Rothrockii, Halenia Rothrockii, Nama Rothrockii, Stachys Rothrockii, Penstemon Rothrockii,* and *Townsendia Rockrockii.*

**Rudbeck,** Olaf (1660–1740). Father and son, teachers of botany at Upsala after whom the plant genus *Rudbeckia* was named.

**Ruelle,** Jean (1479?–1537). French botanist after whom the plant genus *Ruellia* was named by Linnaeus.

**Rupp,** H. B. (1688–1719). Author of *Flora of Jena.* The plant genus *Ruppia* was named after him by Linnaeus.

**Ruthvan,** Alexander (1883–    ). Onetime director of the University of Michigan Museum, herpetologist, student of variations and relationships in reptiles. *Holbrookia ruthveni, Masticophis ruthveni,* and *Pituophis ruthveni* are names which were given in recognition of his valued work.

# S

**Sabine,** Joseph (1770–1837). Secretary of London Horticultural Society at the time of David Douglas. Douglas named the Sabine or Digger Pine (*Pinus Sabiniana*) for him. The leguminous genus, *Sabinea* also honors his name.

**Sagre,** Cuban historian. In 1837 Duméril and Bibron described *Anolis sagrei* in their Erpétologie général. This is a Cuban lizard now found also in Florida. Sagre's History of Cuba, written in part by Cocteau and after

Cocteau's death, finished by Bibron, appeared both in Spanish and French editions.

**Saint-Hilaire,** Geoffroy (1772–1844). French zoologist and contemporary of Lamarck. Taught theory of sudden transmutation from one species to another in contrast to Lamarck's belief in slow change. One of the "Founders of Evolution."

**Sanford,** Dr. Leonard Cutler (1868–?). Collector, patron of ornithology; organizer of the Whitney-Sanford South Pacific Expedition. An Elf Owl name, *Micropallas whitneyi sanfordi* bears reference to this man of energy.

**Sanseviero,** Prince of (1710–1771). Neapolitan scholar whose proper name was Raimond de Sansgrio. The plant genus *Sanseviera* perpetuates his name.

**Saracha,** Isidore (1733–1803). Spanish Benedictine whose name appears in the generic plant name *Chamaesaracha.*

**Saussure,** Theodore de (1767–1845), and his father, H. B. de Saussure (1740–1799). Swiss writers on botanical science after whom the plant genus *Saussurea* was named.

**Say,** Thomas (1787–1834). Father of American entomology. An unusual number of our important economic as well as other insects were named by him; among them the Colorado Potato Beetle, Hessian Fly, and Chinch Bug. He also named a number of important birds and one, the Say Phoebe named by Bonaparte in his honor (*Sayornis saya*) bears his name twice, something quite unusual. Say was a member of the New Harmony group in Indiana, an association of persons of communistic beliefs and practice.

**Scheuchzer** (1672–1733), (1684–1738). Swiss brothers. The plant genus *Scheuchzeria* was named after these naturalists.

**Schleiden,** Matthias Jacob (1804–1881). German botanist, co-worker of Theodor Schwann in formation of the Cell Theory. In 1838 he published the thesis that cells with their nuclei were the units of structure in plants.

**Schmidt,** Karl Patterson (1880–1957). Curator of Reptiles, Field Museum of Natural History, Chicago, 1922–1957; Marshall Field Expedition, Central America 1923; Brazilian Expedition, 1923. Died from the bite of a venemous snake.

**Schott,** Arthur. One of the naturalists of the Mexican Boundary and other government surveys. *Peucephyllum schottii,* named by Asa Gray, and *Gilia schottii,* named by Soreno Watson, are commemorative names known to all western botanists.

**Senebier,** Jean (1742–1809). Swiss naturalist and writer on plant physiology; after him the plant genus *Senebiera* was named. He was the first investigator to give a connected account of the entire process of plant nutrition in chemical terms.

**Sennett,** George Burritt (1840–1900). Ornithologist, publishing much on Texas Birds.

Name associated with White-tailed Hawk (*Buteo albicaudatus sennetti*).

**Sesler,** Leonardo. Italian botanist after whom the plant genus *Sesleria* was named by Scopoli.

**Sherard,** William (1659–1728). English botanist of Oxford and patron of Dillenius; after him the plant genus *Sherardia* was named.

**Sibbald,** Robert (1641–1722). Professor of Medicine at Edinburgh. Author of *Scotia Illustrata.* Both *Sibbaldia*, a plant genus, and *Sibbaldia*, genus of the blue whale are named after him.

**Sibthorp,** John (1758–1796). English botanist traveler in whose honor the plant genus *Sibthorpia* was named by Linnaeus. He was one of the founders of the Linnean Society.

**Simmonds,** T. W. ( ? –1805). Botanist and explorer who accompanied Lord Seaforth to the West Indies. *Simmondsia* is a genus of shrubs named by Thomas Nuttall in his honor.

**Slevin,** Thomas Edwards (1871–1902). Ardent bird student and long a member of the California Academy of Sciences and of the Cooper Ornithological Club. A Hermit Thrush name (*Hylocichla guttata sleveni*) of n. w. California and Oregon, honors him.

**Sowerby,** George Brettingham (1812–1884). English conchologist and author of *Manual of Conchology* (1839).

**Standley,** Paul Carpenter (1884– ). Student of Mexican and Central American floras. Botanist U. S. National Museum 1909–1928, Chicago Natural History Museum 1928–1950. Co-author with E. O. Wooten of "A Flora of New Mexico," 1915.

**Stanley,** Lord Edward. 18th century English ornithologist. Thomas Nuttal dedicated the plant genus, *Stanleya*, to him.

**Stansbury,** Captain Howard (1806–1863). Explorer. Leader of U. S. Government Exploration of Great Salt Lake. The western lizard name, *Uta stansburiana* honors his contribution to herpetology. His charmingly written *Exploration and Survey of the Valley of the Great Salt Lake* is a classic among books on travel.

**Stejneger,** Leonhard (1851–1943). Herpetologist and ornithologist, dealing particularly with taxonomy. Author of *Herpetology of Japan.* Stejneger's Whiptail Lizard (*Cnemidophorus tigris stejnegeri*) as well as Stejneger's Bluebellied Lizard (*Sceloporus elongatus*) are scientific and popular appellations referring to his name.

**Steller,** George Wilhelm (1709–1746). German zoologist and arctic traveler. He died at Tiumen in Siberia. In his honor were named Steller's Jay, of which several subspecies are recognized, Steller's Eider (*Polysticta stelleri*), Steller's Sea Eagle (*Thallasoäetus pelagicus*), Steller's Sea Lion (*Eumetopias stelleri*), Steller's Sea Cow (*Hydrodamalis stelleri*). There is an avian genus, *Stelleria* (*Stelleria dispar*, Steller's eider).

**Stephens,** Frank (1849–1937). Pioneer, versatile naturalist of the Southwest; collector of birds and mammals; author "California Mammals" and numerous scientific papers published in Condor, Copeia, Buk and Transactions of the San Diego Natural History Museum. His name is associated with three southwestern birds, *Antrostromus vociferus macromystax* (Stephens Poorwill), *Vireo huttoni stephensi*, and *Passerella iliaca stephensi.*

**Stillingfleet,** Benjamin (1702–1771). English botanist in whose honor *Stillingia*, a genus of euphorbiaceus shrubs, was named.

**Swammerdam,** Jan (1637–1680). Dutch physician and microscopist. Famous for his detailed studies of the anatomy of mayflies and the honeybee, also for his description of red corpuscles in 1658. It was he who devised the method of studying the circulatory system by means of injections.

# T

**Tiedemann,** Frederick (1781–1861). Anatomist of Heidelberg. Name perpetuated in names *Tiedemannia*, a plant genus, and Tiedemann's glands=Bartholin's glands, and Tiedemann's body, a structure in the sea-star.

**Tillands,** Elias (1640–1693). Professor of Abo, Finland. *Tillandsia* is the generic name of a large group of chiefly epiphytic plants of the pineapple family, often confined to trees and shrubs of the tropics and sub-tropics of America.

**Tilli,** Michael Angelo (1655–1740). Italian botanist after whom the plant genus *Tillaea* was named.

**Tofield,** Thomas (1730–1779). English botanist in whose honor the plant genus *Tofieldia* was named by W. Hudson with whom he corresponded.

**Tolmie,** W. T. See MacGillivray, William.

**Torrey,** John (1796–1873). New York botanist; teacher of Asa Gray. Describer of many plants collected by John C. Frémont in the western deserts. Sereno Watson named the Nevadan salt bush (*Atriplex torreyi*) as a tribute to his botanical endeavors. Dr. C. C. Parry named in his honor the storm-beaten beach pine (*Pinus torreyana*) of the Pacific coast.

**Tradescant,** John (?–1638). Gardener to Charles I of England. The plant genus *Tradescantia* perpetuates his name.

**Traill,** Dr. Thomas Steward (1781–1862). Scotchman, founder of the Royal Institution of Liverpool and editor of the 8th edition of The Encyclopaedia Brittanica. Audubon, wishing to show appreciation of Traill's kindness to him while in Liverpool, named a Flycatcher (*Empidonax trailli*) in his honor.

**Trelease,** William (1857–1945). One-time director of Missouri Botanical Gardens. Specialist in the genus *Agave, Yucca, Quercus,* and *Phoradendron.* Numerous plant species have been named for him, among them the large

Mexican cactus *Lemaireocereus treleasei. Mt. Trelease*, 12,503 feet altitude, in the Clear Creek area of Colorado was named in his honor.

# V

**Van Denburgh**, John (1872–1927). Graduate of Stanford University, 1904. Curator of Reptiles, California Academy of Sciences. *Trimorphodon vandenburghi*, the California Lyre Snake was named to honor him by Klauber.

**Vasey**, George (1822–1893). American authority on grasses. His name is very familiar to students of botany. A Sage (*Salvia vaseyi*) of southern California, a grass (*Vaseychloa multinevosa*) of Texas, Vasey Grass (*Paspalum urvillei*) and many other plants carry his name in one way or other.

**Vaux**, William Sansom (1811–1882). A Philadelphian and member of the Philadelphia Academy of Natural Sciences. J. R. Towsend, his friend, named the smallest of the N. A. Swifts (*Chaetura vauxi*) after him.

**Veatch**, Dr. John A. First botanical explorer of Cedros Island (1859). His collections were sent to the California Academy of Sciences. Some of the more conspicuous species were described by Albert Kellogg. The generic name *Veatchia* was long applied to the Elephant Tree (now *Pachycormus*) of the Viscaino Desert.

**Venegas**, Miguel (Michael). Jesuit missionary. Wrote on the nature of Lower California. *Venegasia*, a plant genus was described by De Condolle.

**Vernon**, William. English botanist who collected plants in Maryland near the end of the 17th century. For him the plant genus *Vernonia* was named.

**Vesalius**, Andreas (1514–1564). Belgian anatomist and physician. Author of a treatise *On the Structure of the Human Body*, a work of importance because its data were based on direct observations. Vesalius was accounted as most eminent anatomist of the sixteenth century.

**Vigors**, Nicholas Aylward (1785–1840). Zoologist who first used the ending *-idae* for family names of birds. Vigor's Wren of western California is a familiar bird name to many ornithologists. A Pine Warbler name (*Dendroica vigorsi*) honors this describer of the generic name *Colaptes*.

**Viguier**, Alexander. French bookseller and botanist of Montpellier. He wrote a work on Poppies in 1814. *Viguera* is a genus of west-American composites described by Humbolt, Bonpland and Knuth (H.B.K.).

**Vizcaino**, Sebastian (1550?–1615?). Spanish navigator who explored the western Mexican and California coasts. The *Viscaino Desert* of Baja California was named in recognition of his explorations in the peninsula's mid-region. *Viscainoa* is a shrub common on the deserts of Baja California.

# W

**Wagler**, Dr. Johann Georg (1800–1832). Professor of Zoology, University of Munich. Author of "Systema Avium," friend of A. Boucard. *Ortalida wagleri* is the Rufous-bellied Chacalaca of Mexico. He described a number of American reptiles.

**Wahlenberg**, Goran (1780–1851). Swedish botanist of Upsala. Author of *Flora Lapponica*. The genus *Wahlenbergia* includes about 80 species of woody-stemmed herbs of the Southern Hemisphere, especially South Africa. A few inhabit the tropics of America.

**Walcott**, Charles Doolittle (1850–1927). Fourth Secretary of the Smithsonian Institution, geologist and palaeontologist. A number of fossils bear the specific name *walcottiana*.

**Wallace**, Alfred Russell (1823–1913). British naturalist best known for his contributions to the "idea of the survival of the fittest." Author of *Geographical Distribution of Animals* (1876), *Island Life* (1880) and the interesting narrative, *The Malay Archipelago*. Concerning his essay, *Contributions to the Theory of Natural Selection*, it may be said that next to Darwin's *Origin of Species*, perhaps no other single work did more "to promote a clear understanding of natural selection and confidence in its truth."

**Watson**, Sereno (1826–1892). Assistant, and later, successor to Asa Gray as director of the Harvard herbarium. Like Gray he was author of many plant species. Botanist of the King's expedition (a geological exploration of the 40th Parallel under Clarence King) and author of two volumes on the botany of California (1876–1880). *Tricardia watsoni* is an interesting and colorful little plant of the pinyon and juniper forests of the western United States. The New Latin adjective, *"serenanus,"* was coined to honor his name.

**Weismann**, August (1834–1914). German biologist, author of the doctrine of the noninheritance of acquired characters.

**Westwood**, Obadiah (1805–1893). English entomologist; professor of zoology at Oxford. His work: *An Introduction to the Modern Classification of Insects* (1839–1840) is a landmark in entomological history, "connecting the older writers with those of today."

**Wheeler**, William Morton (1865–1937). Often called the "ant man" because of his exceptional knowledge of these insects. In 1910 appeared his now well-known book: *Ants*. Dr. Wheeler was also a philosopher and brilliant linguist. Professor of entomology at Harvard University. Specialist in the embryology, morphology and anatomy of insects; he was also deeply interested in the social life of insects. Many insects, especially ants and flies, from many parts of the world were named in his honor.

**White**, Gilbert (1720–1793). English curate of Selbourne; "first naturalist who clothed his observations in literary style." His book, *The Natural History and Antiquities of Selbourne* (1789) is a classic revered by all men who work in the natural science field. His *Naturalists Calendar* edited by Aikin, was published two years after his death.

**Whitney,** Josiah Wright (1819–1896). Geologist, Director of the Geological Survey of California (1860–1874). Mt. Whitney recorded as highest peak in the Sierra Nevada and the Elf Owl, *Micropallas whitneyi* (described by Dr. J. G. Cooper) were dedicated to this eminent man of science.

**Willdenow,** K. L. (1765–1812). German botanist of Berlin; one of the most eminent and best informed systematists of his time. His name or its abbreviation (Willd.) follows many plant names as describer.

**Wilson,** Alexander (1766–1813). Called the "Father of American Ornithology." He came to the United States from Scotland and later made many foot and other journeys through the American countryside; a nine volume work, *American Ornithology*, appeared as a result of his observations. The Long-eared Owl (*Asio wilsonianus*), Wilson's Plover (*Pagolla wilsonia*) and the generic name of the Pileolated Warbler (*Wilsonia*) help to perpetuate his name in Ornithology.

**Wistar,** Caspar (1761–1818). Anatomist at University of Pennsylvania. Author of the first American textbook of Anatomy. His name is commemorated in the *Wistar Institute of Anatomy and Physiology* in Philadelphia, and in the plant genus, *Wistaria.*

**Wolff,** J. E. (1778–1806). German physician and botanist for whom the plant genus *Wolffia* was named.

**Wolff,** Kaspar F. (1733–1794). German embryologist and anatomist. For him was named the *Wolffian body* and *duct.*

**Woodhouse,** Samuel Washington (1821–1904). Surgeon-naturalist of Creek and Cherokee boundary survey. First naturalist to visit northern Arizona. In 1851, he served under Sitgreaves to investigate the Zuni and the Little Colorado Rivers. Woodhouse's Jay (*Aphelocoma californica woodhousii*), the Rocky Mountain Toad (*Bufo woodhousii*), and a water snake (*Tropidonotus woodhousii*) are named in his honor.

**Woods,** Joseph (1776–1864). English botanist. *Woodsia* is a genus of ferns widely distributed from North America to Japan and Arctic Europe.

**Wright,** Charles (1811–1885). Botanist and collector of birds. Wright's Flycatcher (*Empidonax wrightii*) and the botanical generic name *Carlowrightii* are names which honor him.

**Wucherer,** Otto (1820–1873). German physician in Brazil. *Wuchereria* is a genus of roundworms; *Wuchereria bancrofti* causes elephantiasis.

**Wyeth,** Nathaniel J. (1802–1856). Fellow traveller of Thomas Nuttall. Discovered in the mountains of North America, plants described by Thomas Nuttall under the generic name *Wyethia.*

# X

**Xantus,** John=Xantus de Vesey (1825–1894). Energetic and early collector of birds in California and Lower California. Hungarian who while a young man enlisted in U. S. Army and, while acting as hospital steward at Old Fort Tejon, described a number of new birds. Became custodian of Museum at Budapest. Included among birds named in his honor are: Xantus' Hummingbird (*Basilinia xantusi*) and Xantus' Jay (*Aphelocoma californica hypoleuca*). Perhaps the lizard genus, *Xantusia*, most often recalls to zoologists his connection with herpetology.

# Y

**Yarrow,** Dr. Henry Crecy (1840–?). One-time curator at the Smithsonian Institution. Surgeon and naturalist to the Expedition for the Exploration of the Territory West of the One Hundredth Meridian under Lieut. George M. Wheeler of the U. S. Engineers. *Sceloporus yarrovii* was named for him.

# Z

**Zenaida.** Daughter of Joseph Bonaparte, and wife of Charles Lucian Bonaparte (1803–1853). Her name is preserved in the generic designation *Zenaidura*, under which the Mourning Dove was described.